Vaccine Design

Innovative Approaches and Novel Strategies

Edited by

Rino Rappuoli

and

Fabio Bagnoli

Novartis Vaccines & Diagnostics
Siena, Italy

Caister Academic Press

British Library Cataloguing-in-Publication Data
A catalogue record for this book is available from the British Library

ISBN: 978-1-904455-74-5

Cover image adapted from Figure 16.3

Printed and bound in Great Britain

Contents

Contributors

Jeannette Adu-Bobie
Novartis Vaccines & Diagnostics
Siena
Italy

jeannette.adu-bobie@novartis.com

Beatrice Aricò
Novartis Vaccines & Diagnostics
Siena
Italy

beatrice.arico@novartis.com

Ruth Arnon
Department of Immunology
Weizmann Institute of Science
Rehovot
Israel

ruth.arnon@weizmann.ac.il

Fabio Bagnoli
Project Leader – Vaccine Research
Novartis Vaccines & Diagnostics
Siena
Italy

fabio.bagnoli@novartis.com

Michèle A. Barocchi
Molecular Epidemiology Unit
Novartis Vaccines & Diagnostics
Siena
Italy

michele.barocchi@novartis.com

David R. Bundle
Alberta Ingenuity Centre for Carbohydrate Science
Department of Chemistry
University of Alberta
Edmonton, AB
Canada

dave.bundle@ualberta.ca

Alice G. Cheng
Department of Microbiology
The University of Chicago
Cummings Life Science Center 607
Chicago, IL
USA

agc@uchicago.edu

Roberta Cozzi
Novartis Vaccines & Diagnostics
Siena
Italy

roberta.cozzi@novartis.com

Claudio Donati
Systems Biology Unit
Novartis Vaccines & Diagnostics
Siena
Italy

claudio.donati@novartis.com

Ilaria Ferlenghi
Systems Biology Unit
Novartis Vaccines & Diagnostics
Siena
Italy

ilaria.ferlenghi@novartis.com

Carl E. Frasch
Frasch Biologics Consulting
Martinsburg, WV
USA

crfrasch1@juno.com

Marzia M. Giuliani
Novartis Vaccines & Diagnostics
Siena
Italy

marzia.giuliani@novartis.com

Joanna B. Goldberg
Department of Microbiology
University of Virginia Health System
Charlottesville, VA
USA

jbg2b@virginia.edu

Bart L. Haagmans
Department of Virology
Erasmus MC
GE Rotterdam
The Netherlands

b.haagmans@erasmusmc.nl

Pål Johansen
Department of Dermatology
University Hospital of Zurich
Zürich
Switzerland

pal.johansen@usz.ch

Jerry M. Keith
Eunice Kennedy Shriver National Institute of Child
    Health and Human Development (NICHD)
National Institutes of Health
Bethesda, MD
USA

keithjer@mail.nih.gov

Thomas M. Kündig
Department of Dermatology
University Hospital of Zurich
Zürich
Switzerland

thomas.kuendig@usz.ch

Domenico Maione
Novartis Vaccines & Diagnostics
Siena
Italy

domenico.maione@novartis.com

John R. Middleton
Department of Veterinary Medicine and Surgery
University of Missouri
Columbia, MO
USA

middletonjr@missouri.edu

Dominique Missiakas
Department of Microbiology
The University of Chicago
Cummings Life Science Center 607
Chicago, IL
USA

dmissiak@bsd.uchicago.edu

Moon H. Nahm
Departments of Pathology and Microbiology
University of Alabama at Birmingham
Birmingham, AL
USA

nahm@uab.edu

Nathalie Norais
Structural Mass Spectrometry and Proteomics Unit
Novartis Vaccines & Diagnostics
Siena
Italy

nathalie.norais@novartis.com

Derek T. O'Hagan
Global Head Vaccine Delivery and Formulation
    Research
Novartis Vaccines & Diagnostics
Cambridge, MA
USA

derek.ohagan@novartis.com

James C. Paton
Research Centre for Infectious Diseases
School of Molecular and Biomedical Science
University of Adelaide
Adelaide, SA
Australia

james.paton@adelaide.edu.au

Bali Pulendran
Emory Vaccine Center, and Yerkes National Primate
  Research Center
Atlanta, GA
USA

bpulend@emory.edu

Rino Rappuoli
Global Head – Vaccine Research
Novartis Vaccines & Diagnostics
Siena
Italy

rino.rappuoli@novartis.com

Rajesh Ravindran
Emory Vaccine Center, and Yerkes National Primate
  Research Center
Atlanta, GA
USA

rnair3@emory.edu

Allan Saul
Novartis Vaccines Institute for Global Health
Siena
Italy

allan.saul@novartis.com

Silvana Savino
Protein Purification and Characterization Unit
Novartis Vaccines & Diagnostics
Siena
Italy

silvano.savino@novartis.com

Jennifer M. Scarff
Department of Microbiology
University of Virginia Health System
Charlottesville, VA
USA

Maria Scarselli
Systems Biology Unit
Novartis Vaccines & Diagnostics
Siena
Italy

maria.scarselli@novartis.com

Olaf Schneewind
Department of Microbiology
The University of Chicago
Cummings Life Science Center 607
Chicago, IL
USA

oschnee@bsd.uchicago.edu

Gabriela Senti
Clinical Trials Center
Center for Clinical Research
University Hospital of Zurich
Zürich
Switzerland

gabriela.senti@usz.ch

Davide Serruto
Novartis Vaccines & Diagnostics
Siena
Italy

davide.serruto@novartis.com

David A.G. Skibinski
Head Vaccine Formulation Science
Novartis Vaccines & Diagnostics
Siena
Italy

d.skibinski@novartis.com

John L. Telford
Novartis Vaccines & Diagnostics
Siena
Italy

john.telford@novartis.com

Other Books of Interest

Caister Academic Press   www.caister.com

Introduction

Rino Rappuoli and Fabio Bagnoli

Vaccination, together with the wider availability of potable water, has had the most profound positive effect on the quality of public health of any measure: during the past century, these products essentially eliminated most infectious diseases causing mortality in infants and children. Vaccines against diphtheria, tetanus, polio, measles, mumps, rubella, pneumococcus, hepatitis B and meningitis (*Haemophilus influenzae* and serogroup C meningococcus) have reduced the incidence and mortality of these diseases by > 97–99% (Rappuoli *et al.*, 2002). Nevertheless, perception of vaccines in the public opinion is not completely positive. Many people are still sceptical about the real need of vaccines. This behaviour has been particularly evident during the influenza A (H1N1) pandemic in 2009. This is probably due to the following major reasons: we tend not realize the importance of a cure until we are ill, we are selfish and we do not understand that vaccination is important for the community (herd immunity), we think that a vaccine should only prevents deaths, therefore if a pathogen does not always kill, we believe that a vaccine is not necessary. This book aims to provide a comprehensive illustration of all the reasons why vaccines are needed today more than ever, and of how new and safer vaccines can be developed.

Infectious diseases, remain the leading cause of death worldwide, despite antimicrobial therapies. Many infectious diseases for which a cure is not available or not efficacious are expected to emerge in the coming decades and the development of novel vaccines against these diseases appears urgent (Jones *et al.*, 2008; Rappuoli, 2004). The major vaccines developed to date (e.g. polio, smallpox, pertussis and tetanus) were developed using conventional vaccinology strategies based on the Pasteur's principles: 'isolate, inactivate and inject' the causative microorganism (Rappuoli, 2004; Robbins *et al.*, 2005; and Chapter 1 of this book). Although very successful, in several instances these approaches were not able to deliver vaccines against certain pathogens and on other occasions the vaccines obtained with these classical approaches are no longer adequate due to safety concerns and low efficacy.

Conventional vaccinology has been successful in developing vaccines against pathogens that do not change the vaccine-targeted antigens over time and for which protection is antibody-mediated (Rappuoli, 2007). Classical vaccines such as diphtheria and tetanus, for instance, target toxin antigens that have not had antigenic drift during the past century. To a certain extent, conventional vaccinology was successful in targeting particular pathogens that have many variants of target-vaccine antigens and change them over time. Examples of such vaccines are the ones based on polysaccharide antigens of pneumococcus and meningococcus. Conventional vaccinology has also been relatively successful in developing vaccines against those pathogens that change their surface antigens every year, such as influenza. In this case, to face the antigenic variability of the influenza viruses, the formulation of the vaccine is altered every year and the population is vaccinated annually with this.

However, there are many pathogens for which conventional vaccinology has failed. The reason for these failures are probably mainly based on high antigen variability and protection mediated by mechanisms not clearly understood. Therefore, we have decided to focus this book on novel technologies as well as innovative research to develop new and safer vaccines. There are two major areas that are revolutionizing the way vaccine are developed: (1) the use of new methods to discover novel vaccine candidates and (2) the development of new technologies to express, deliver and formulate antigens.

Antigen discovery today is mainly based on predictive methods such as the so called reverse vaccinology (Rappuoli, 2000), and the antigenome technology (Meinke *et al.*, 2005). The former is based on a 'multiple approach' which comprises *in silico* antigen prediction, functional genomics and comparative genomics, the latter is based on the screening, use of human sera, and use of expression libraries. The availability of genome sequences of any major human pathogen pushed the boundaries in vaccine research. Genome sequences offer unprecedented access to all the possible antigens of an organism and furthermore enable the possibility of rational selection of targets depending on the desired objective. Before genomics, the conventional approach to vaccine development had been based on time-consuming dissection of the pathogen using biochemical, immunological and microbiological methods. The study of pathogen genomes by both computational and experimental approaches has significantly advanced understanding of the physiology and pathogenicity of many microbes and provided insights into the mechanisms of genome evolution as well as microbial population structure. Genome mining and comparative genomics is allowing us to increase the number of candidate vaccine antigens by several orders of magnitude and making possible the selection of antigens that are conserved or specific to a given pathogenic organisms. The ability to globally view the expression of potential antigens on a genome-wide scale with innovative functional genomics technologies (e.g. proteomics an transcriptome analysis) is also significantly changing the field of antigen discovery and vaccine design.

New technologies to express antigens include those based on recombinant DNA, live bacterial vectors, viral vectors, virus-like particles (VLPs), and DNA vaccines. Some of these technologies are already being used in currently available vaccines, while others await approval for use in humans. In particular, recombinant DNA technology was successfully applied to the production of a vaccine against hepatitis B virus (HBV). Indeed, although a vaccine based on purified HBV surface antigen (HBSAg), from the plasma of infected patients, was available since the 1970s, the expression in yeast of the HBSAg vaccine antigen provided the solution for a relatively simple, quick, and inexpensive process to produce a safer vaccine. Another major use of recombinant DNA technology has been the site-specific inactivation of toxins as safe and efficacious alternative to chemical inactivation. If a strong humoral as well as cellular immune response is required, the best way is probably to use either live bacterial vectors, viral vectors or VLPs. There is increased interest in the use of live attenuated bacterial vaccines (LBV) as carriers for the presentation of heterologous antigens for the engineering of live recombinant mucosal vaccines. LBVs allow vaccination through mucosal surfaces and specific targeting of professional antigen-presenting cells located at the inductive sites of the immune system (Daudel *et al.*, 2007). Bacterial species that are being investigated as vector vaccines include attenuated strains of *Salmonella enterica* serovar Typhi and serovar Typhimurium, *Shigella*, *Vibrio cholerae*, *Listeria monocytogenes*, *Mycobacterium bovis* (BCG), and *Yersinia enterocolica*. Other bacterial vectors have included non-pathogenic strains derived from the normal flora such as *Streptococcus gordonii*, *Lactobacillus casei*, and *Lactobacillus lactis* (Mercenier *et al.*, 2000). A broad spectrum of replicating and non-replicating viral vectors are available. A variety of attenuated viruses have been employed as vectors including vaccinia and other pox viruses, adenovirus and single-stranded RNA virus replicon vectors such as alphaviruses, coronaviruses, picornaviruses, flaviviruses, influenza viruses, rhabdoviruses, and paramyxoviruses. VLPs are structures resembling a virus but empty of nucleic acids, which are derived from self-assembling subunits of virus structural antigens. VLP vaccines combine many of the advantages of whole-virus vaccines (induction of strong immune responses) and recombinant subunit vaccines (relative simplicity in manufacturing a safe vaccine). Commercialized VLP-based vaccines have been successful in protecting humans from hepatitis B virus (HBV) and human papillomavirus (HPV) infection.

The discovery in the early 1990s of DNA immunization, radically changed accepted views of the nature of a vaccine. Instead of delivering an antigen *per se*, genetic material that encodes a specific antigen is delivered into mammalian cells, directing expression of the antigen by the host cell itself, essentially working from the inside out. The high expectations associated with DNA vaccination, as a result of promising data obtained in animals, were somewhat tempered by disappointing early results when DNA was tested as a vaccine in humans. A recognized limitation of DNA vaccines is their limited capacity to induce good virus-neutralizing antibody responses. However, new strategies to improve or amplify the immunogenicity of DNA vaccines are being developed.

A rather different strategy to the ones above mentioned, is based on chemical conjugation of polysaccharide antigens to protein carriers. This approach has been used for the prevention of disease caused by encapsulated bacteria such as *Haemophilus influenzae, Neisseria meningitidis* and *Streptococcus pneumoniae.* Conjugation of polysaccharides to protein carriers is essential to engage a T-cell-based immune response providing long-term immunological memory. To date, the polysaccharide used in large-scale vaccine production has been purified from the pathogen itself, grown in large quantities – an approach that is costly and difficult to control. The large-scale production of a conjugate vaccine containing synthetic polysaccharides has been recently achieved (Verez-Bencomo *et al.*, 2004).

New approaches and technologies to deliver and formulate vaccines include the use of alternative immunization routes and new adjuvants. Conventionally, immunization in humans is done by syringe through the subcutaneous or intramuscular route. However, examples of alternative sites of immunization such as the nasal or oral cavities, the dermis and the epidermis have been proven to work and protect against different diseases. At the moment new sites of immunization, such as the intralymphatic, and new vaccine delivery devices,

such as micro-needle patches and needle free injection, are being investigated.

There are several reasons for use of adjuvants, especially in new generation vaccines based on purified recombinant proteins, which although safe, can be poorly immunogenic. The major reasons are the following: to elicit sufficiently high antibody responses to enable protection, antigen dose sparing and reduction in the number of immunizations. Furthermore, adjuvants can increase the breadth of response to cover pathogen diversity, and overcome limited immune responses produced in some populations such as the elderly, young children and the immunocompromised. Finally new adjuvants promise to induce different types of immune response such as more effective T cell responses, which are required for some pathogens for which we do not yet have effective vaccines. The most widely used vaccine adjuvants are insoluble aluminium salts (generically called alum), and until recently these remained the only adjuvants present in licensed vaccines for human use in North America. Despite the success of alum, more potent adjuvants are necessary to enable vaccine development against a significant number of challenging pathogens for which we do not yet have effective vaccines. A significant success story is the vaccine adjuvant MF59. This is a safe and potent oil in water emulsion (o/w) which has been licensed for more than 10 years in a significant number of countries (>20) for use in a seasonal influenza vaccine. Another novel adjuvant that has gained licensure in recent years is AS04.

The book has been divided into two parts. The first part discusses the technologies, and approaches used to identify, generate and test vaccine candidates. Particular attention is given to the new strategies to identify protective antigens, improve efficacy with new adjuvants and alternative immunization routes, increase safety, and find and establish correlates of protection. The second part focuses on ongoing research aimed at developing new vaccines to replace or complement currently available products or for diseases against which prophylactic strategies are missing. Furthermore, it has recently been observed that vaccination strategies against certain infections have been generally ignored. This has happened with diseases that represent a major problem in developing countries only, or those whose importance has only recently become apparent or for bacteria which have developed antibiotic resistance in recent years. The book also includes some examples of strategies for the development of veterinary vaccines. Although most of the book focuses on studies with application in humans, we believe that it is important to give to the reader a look at the same issues from different angles.

During the last decade we have started to think of vaccines in a new way. Several vaccines in development are not envisaged for the general population but for restricted target populations (e.g. nosocomial infections). Furthermore, children are not any longer the primary target of several new vaccines (e.g. *Streptococcus agalactiae* and nosocomial infections). Therefore, the requirements of the vaccines can be very different to the past. If a vaccine is given to hospitalized immuno-compromised patients or to the elderly, it is likely that the use of novel improved adjuvants will be of critical importance. Novel vaccines do not only prevent deaths directly caused by pathogens but that can also prevent malignancies associated with them (e.g. papillomavirus, hepatitis B virus and *Helicobacter pylori*). Furthermore, prevention of infections can increase patient fitness by decreasing the number of infectious episodes and associated sequelae. Indeed, reduction of harmful inflammatory processes happening during infections has been postulated to be a prerequisite of prolonged life expectancy (Crimmins and Finch, 2006). Finally, vaccines can reduce the use of antibiotics and antivirals decreasing the emergence of resistances.

References

Crimmins, E.M., and Finch, C.E. (2006). Infection, inflammation, height, and longevity. Proc. Natl. Acad. Sci. U.S.A. *103*, 498–503.

Daudel, D., Weidinger, G., and Spreng, S. (2007). Use of attenuated bacteria as delivery vectors for DNA vaccines. Expert Rev. Vaccines *6*, 97–110.

Jones, K.E., Patel, N.G., Levy, M.A., Storeygard, A., Balk, D., Gittleman, J.L., and Daszak, P. (2008). Global trends in emerging infectious diseases. Nature *451*, 990–993.

Meinke, A., Henics, T., Hanner, M., Minh, D.B., and Nagy, E. (2005). Antigenome technology: a novel approach for the selection of bacterial vaccine candidate antigens. Vaccine *23*, 2035–2041.

Mercenier, A., Muller-Alouf, H., and Grangette, C. (2000). Lactic acid bacteria as live vaccines. Curr. Issues Mol. Biol. *2*, 17–25.

Rappuoli, R. (2000). Reverse vaccinology. Curr. Opin. Microbiol. 3, 445–450.

Rappuoli, R. (2004). From Pasteur to genomics: progress and challenges in infectious diseases. Nat. Med. *10*, 1177–1185.

Rappuoli, R. (2007). Bridging the knowledge gaps in vaccine design. Nat. Biotechnol. *25*, 1361–1366.

Rappuoli, R., Miller, H.I., and Falkow, S. (2002). Medicine. The intangible value of vaccination. Science *297*, 937–939.

Robbins, J.B., Schneerson, R., Trollfors, B., Sato, H., Sato, Y., Rappuoli, R., and Keith, J.M. (2005). The diphtheria and pertussis components of diphtheria-tetanus toxoids-pertussis vaccine should be genetically inactivated mutant toxins. J. Infect. Dis. *191*, 81–88.

Verez-Bencomo, V., Fernandez-Santana, V., Hardy, E., Toledo, M.E., Rodriguez, M.C., Heynngnezz, L., Rodriguez, A., Baly, A., Herrera, L., Izquierdo, M., et al. (2004). A synthetic conjugate polysaccharide vaccine against *Haemophilus influenzae* type b. Science *305*, 522–525.

Overview of Vaccine Strategies

Ruth Arnon

Abstract

This chapter describes the different strategies applied for vaccination against microbial diseases. These include vaccines against bacterial, viral and parasitic infections which led to tremendous improvement in public health. The use of both live attenuated or killed whole organisms and their subunits is discussed, as well as more novel approaches, such as DNA vaccines, recombinant vaccines and epitope-based, or peptide vaccines. The advantages and disadvantages of each approach are presented, eluding to various considerations, such as efficacy, safety and cost of production. The application of passive vaccination, including the use of pooled IgG (IVIG) is also described in brief. As indicated, these combined strategies led to a long list of vaccines that are presently approved and licensed in the USA, Europe and many other countries, as summarized in detail in Table 1.1 which refers also to the paediatric combination vaccines diphtheria–pertussis–tetanus (DPT) and measles, mumps, rubella (MMR) that are used worldwide, and led to drastic reduction in the incidence of infectious diseases.

Introduction

There is no doubt that the development of vaccines has been one of the most important achievements of immunology and medicine to date and their most significant contribution to human health. Less than two hundred years ago life expectancy was around 40 years of age, a condition that still persists in some undeveloped countries. From about the middle of the nineteenth century, great improvements in living standards were successfully made in industrialized countries, initially by the provision of adequate sanitation facilities and safe drinking water. But this was not enough to prevent epidemics and sometimes pandemics of infectious diseases. The plague, or 'black death', killed almost a third of the population of Europe in the fourteenth century; the 'Spanish flu' in 1918–19 killed over 30 million people, more than all those who died in World War I during 1914–1918. Smallpox was a terrible killer for centuries, responsible for many millions of victims. Coping with these issues became progressively possible through the development of vaccines against some of the major diseases, which led to eradication of smallpox and polio, among others. Today there are tens of vaccines made by traditional techniques, that are based on the disease-inducing agent, used for vaccination of children and adults, as well as newly developed vaccines that are approved for human use. In parallel to vaccine development, life expectancy almost doubled and rose to its present average of late 1970s or early 1980s, in the developed countries. In addition to the classical vaccines, novel strategies for vaccine development hold promise for more efficient and safer vaccines in the future. An overview of these classical and new vaccine strategies is given in this chapter.

Traditional vaccines development

Many of the existing vaccines are intended for children and babies, but most vaccines were and still are used mainly as a means of preventing outbreaks of infectious diseases. Initially, vaccines were developed on a trial-and-error basis, some of

Table 1.1 Approved licensed vaccines for human use

Target disease	Causative agent	Type of vaccine	Comments
Anthrax	*Bacillus anthracis*	**Subunit** cell-free extract of the bacteria, adsorbed on alum.	The only anthrax vaccine for human use
Cervical cancer	Human papillomavirus (HPV)	**Recombinant** DNA vaccine	Quadrivalent vaccine of several viral strains
Cholera	*Vibrio cholerae*	1. **Whole organism** inactivated	The B subunit of cholera toxin serves also as the basis for a vaccine against toxic *E. coli*
		2. **Subunit** – B subunit of cholera toxin	
Diphtheria	*Corynebacterium diphtheriae*	**Subunit toxoid**, alum adsorbed	Used mainly in combination with tetanus toxoid and Acellular pertussis (DPT)[a]
Hib–Induced disease, pneumonia, meningitis	*Haemophilus influenzae* type b	**Conjugate** vaccine, the Hib PRP polysaccharide conjugated to a protein	Two proteins are used in the licensed conjugate vaccines – meningococcal outer membrane protein (OMP) or tetanus toxoid (TT), produced by different manufacturers
Hepatitis type A	Hepatitis A virus	1. **Whole virus** inactivated	Formalin-inactivated virus of several strains
		2. **Whole virus** live attenuated	Prepared by serial passage in cell culture
Hepatitis type B	Hepatitis B virus	1. **Subunit** surface antigen (HBsAg) blood-derived	Approved but used only in Third World countries
		2. **Recombinant** yeast-derived HBsAg	The first recombinant vaccine used also in combination with inactivated hepatitis A vaccine
Influenza[b]	Influenza virus	1. **Whole virus** inactivated trivalent type A and type B vaccine	Seasonal vaccine – the three strains H1N1, H3N2 and B are determined each year by WHO
		2. **Subunit** trivalent vaccine	Seasonal vaccine, trivalent as above, comprising a complex of HA and NA
		3. **Whole virus** H5N1 monovalent inactivated	Prepared only in the USA (for national stockpile)
		4. **Whole virus** H1N1 2009 monovalent	Prepared against pandemic flu of 2009, by many vaccine producers, with or without adjuvant
		5. **Live attenuated** nasal vaccine FluMist	Trivalent, for intranasal use
		6. Subunit proteosome (OMP) – Conjugate FluInsure	Trivalent, for intranasal use

Disease	Organism	Vaccine type	Notes
Japanese encephalitis	Japanese encephalitis virus (JEV)	1. **Whole virus** inactivated vaccine, alum adsorbed 2. **Live attenuated**	Vaccine produced in Japan, and licensed worldwide including the USA. This vaccine is produced and was tested only in China and is not licensed in the USA or Europe.
Measles	Measles virus	**Live attenuated** strains Schwarz or E2–19	Used alone or in combination with: mumps virus vaccine, with mumps and rubella (MMR), or with mumps, rubella and varicella live virus vaccines
Meningitis	*Neisseria meningitidis*	1. **Subunit** capsular polysaccharide (CP) 2. **Conjugate** vaccine – CP with DT 3. **Subunit** outer membrane protein (OMP)	The vaccine consists of different combination of purified high-molecular weight CP from serogroups A, C, Y and W-135 combined The above subunit CP vaccine conjugated to diphtheria toxoid The vaccine containing group B/C OMP produced and licensed only in Cuba
Mumps	Mumps virus	**Live attenuated** strains Jeryl Lynn or Rubini	This vaccine is rarely used as such, but only in combination with measles and rubella (MMR)
Plague	*Yersinia pestis*	**Whole bacteria** inactivated	Very rarely used since disease in humans was practically eradicated
Pneumonia	*Streptococcus pneumoniae*	1. **Subunit M** protein-based 2. **Conjugate** vaccine – cell wall polysaccharides conjugated to a protein	Polyvalent vaccine. The repeat sequence of the pneumococcal M protein of several serotypes Polysaccharides of several strains 7-valent conjugated to either diphtheria (CRM 197 protein) or tetanus toxoid
Polio	Poliovirus	1. **Whole virus** inactivated (salk vaccine) 2. **Live attenuated** virus (Sabin vaccine types 1, 2, & 3)	Known as oral polio vaccine (OPV) and very widely used world-wide
Rabies	Rabies virus	**Live attenuated** virus	Used only post exposure to the virus
Rotavirus infections (mainly diarrhoea)	Rotaviruses	**Live attenuated** oral	Vaccine is safe in infants for protection against diarrhoea and other infections that could be fatal
Rubella	Rubella virus	**Live attenuated** virus (strain RA27/3)	Vaccine is used either as such, or (mainly) in combination with measles and mumps vaccines (MMR).
Smallpox	Vaccinia virus	1. **Live attenuated** dried calf lymph 2. **Live attenuated** virus	As smallpox was declared by WHO an eradicated disease, the routine vaccination with vaccinia vaccine was discontinued worldwide

Table 1.1 continued

Target disease	Causative agent	Type of vaccine	Comments
Tetanus	*Clostridium tetanii*	**Subunit**-toxoid, alum adsorbed	The tetanus toxoid is used as such, mainly as a booster vaccine after injury, but is routinely used as a part of a combined paediatric vaccine DPT (diphtheria, pertussis, tetanus)
Tuberculosis	*Mycobacterium tuberculosis*	**Whole bacteria** live BCG (bacillus Calmette–Güerin)	Three main strains of the BCG are available and in use: grown under different conditions they vary in their characteristics. BCG is used also as an adjuvant and immunopotentiator
Typhus	*Salmonella typhi, Salmonella typhimurium*	1. **Whole cell** inactivated phenol-preserved	New generation of genetically attenuated S. typhi vaccines is currently under way, as candidate live oral vaccines
		2. **Live attenuated** oral Ty21a strain	
		3. **Subunit** *Vi* capsular polysaccharide	
Varicella/zoster	Varicella zoster virus (VZV)	**Live attenuated** virus (Oka strain)	The Oka strain was isolated in Japan from human embryonic lung (HeL) cells by numerous passages
Whooping cough	*Bordetella pertussis*	**Subunit** acellular pertussis vaccine	The acellular pertussis vaccine is composed of 5 components of the bacteria, including pertussis toxin. This vaccine is routinely used in combination with diphtheria and tetanus toxoids (DPT)
Yellow fever	Yellow fever virus	**Live attenuated** whole virus strain 17D	Attenuation is achieved by repeated passages

[a]Several vaccine companies produce a paediatric vaccine that includes DPT combined with one or more of the following: recombinant hepatitis B vaccine, inactivated poliovirus vaccine, conjugated *Haemophilus influenzae*.

[b]Many of the vaccines are produced by more than one manufacturer – in particular Influenza vaccine, of which both the seasonal trivalent vaccines and the pandemic H1N1 vaccine are produced by many manufacturers under different trade names.

them even before it was known that they are based on viruses that cause the disease (e.g. vaccinia by Jenner, or rabies vaccine by Pasteur). The test for efficacy of these vaccines was by their ability to afford protection against the disease in question, resulting in low levels of morbidity and mortality after exposure to the pathogen.

Only at later stages was it realized that the diseases are caused by microorganisms, viruses or bacteria, and that the vaccines as prepared are based on the intact disease-inducing agent – whether killed (inactivated) or attenuated, thereby rendered unable of inducing the disease but capable of affording protection. Still later, when it was realized that vaccination leads to an immune response and when methods for detection and quantitation of specific antibodies became available, the vaccine's efficacy could sometimes be assessed by measuring the extent of seroconversion (increase in antibody level) that occurred following vaccination. Only recently has the role of cell-mediated immunity responses in the control of clearance of an infection been appreciated. As a result of these cumulative techniques it is now possible to evaluate in a reliable manner the efficacy of vaccines.

Classical vaccines

The classical vaccines can be roughly broken down into two main groupings – whole organism vaccines and subunit vaccines. This applies to both bacterial and viral vaccines. However, whereas most bacterial vaccines that are in use today are based on bacterial components as well as subunits, in the case of viral vaccines, most of those available to date are based on the whole organism, and can be grouped into two main types, live attenuated vaccines and killed or inactivated vaccines. Only in relatively few instances subunit viral vaccines are included among the classical vaccines.

Viral vaccines, whole organism, live attenuated

The most effective viral vaccine is usually the live infectious agent that has been somehow modified (attenuated) to make it non-pathogenic. Originally this was done by physicochemical means, such as heating, but more recently genetically

manipulated (crippled) microorganisms have been used, which lack the ability to live for long periods or to cause significant disease. These vaccine agents bear the same broad spectrum of antigens as the pathogen, but because they are alive and can replicate in the host, they persist for longer periods than killed vaccine agents. For this reason they stimulate the most effective and long lasting immunity to the natural infection. They generate both antibody (including secretory IgA if the vaccine agent is inhaled or ingested) and cell-mediated responses to the agent. The downside of live vaccines is that, on rare occasions, they revert back to the pathogenic form. Furthermore, in a patient with an unrecognized immune deficiency, even the 'attenuated' vaccine may cause serious disease. As a result these vaccines cause significant disease, even death, in a small but predictable number of vaccinated people every year (usually fewer than 2 per million). While the incidence of such serious morbidity or mortality is extremely low, in a large country the number of cases can be significant.

Different approaches have been used for attenuation for different viruses: One approach is to use as a human vaccine a similar virus that is pathogenic in another hosts but is not pathogenic in humans. This is actually the basis for the vaccine used by Jenner against smallpox, that consisted of material obtained from cowpox. Another example, although not a human vaccine, is herpes virus from turkeys that can be used to control Marek's disease in chicken.

A second approach is to use as a vaccine a naturally occurring attenuated strain of a virulent virus. One example for that approach is the use of type 2 poliovirus. However, there are additional examples such as several strains of rotavirus, which occur naturally and have been found to be non-virulent in human babies.

The third approach is based on purposely attenuation of virulent viruses and preparation of mutants by prolonged passage of the wild-type virus in another host and/or tissue culture. This principle was the basis of the original rabies vaccine by Pasteur. It may take a long time, even years, to achieve the final product. One example for a vaccine prepared using this approach is the Sabin polio vaccine: this oral polio vaccine was prepared from type 1 strain and type 3 strain of the virus, that had been attenuated by passage in non-human primates and in tissue culture. In the first strain there were many (over 50) base substitutions, and hence this vaccine rarely reverted to virulence (Almond, 1987). In the second, there were less than 10 point mutations, but, interestingly, even one back-mutation was capable of revertance to a neurovirulent form (Evans et al., 1985).

A fourth approach for attenuation is by applying harsh conditions on a virulent virus. An example for that is the cold adaptation of influenza virus which is the basis for a live influenza vaccine (Beyer et al., 2002).

Another approach is to administer a pathogenic virus via a non-pathogenic route of delivery. For example, a virus that infects via the respiratory track, such as adenovirus, can lead to protection when given orally.

Finally, genetic attenuation can be achieved either by cell passage of the virus, or by gene deletion, or using reverse genetics, thereby importing mutations at sites that are associated with virulence.

The advantage of attenuated live vaccines is that since the virus can replicate, the resulting immunity is long lasting. Several such vaccines that have been in practice for decades, such as smallpox, have led after single inoculation to life-long protection with only minor side-effects. Consequently, as already mentioned, global vaccination policy with this vaccine resulted in the eradication of smallpox.

Killed whole virus vaccines

Killed vaccines have the advantage of being much safer than live attenuated vaccines. They cannot revert back to the pathogenic form if they are killed properly. But, though safer than attenuated vaccines, killed vaccines are less effective. Killed 'whole cell' (whole organism) vaccines generally retain the full spectrum of antigens of the pathogen, although some of these antigens are damaged by the killing process. In addition, because the agent is not alive and cannot replicate, it is quickly cleared from the body and does not activate the immune response as well as live organism. Killed vaccines are not as effective as live attenuated vaccines, but they may still give better protection than an isolated component of the pathogen, since they present a wider spectrum of the antigens and activate a greater inflammatory response, which induces more effective antigen presentation by dendritic cells. However, because of this, such vaccines have been reported to cause more side-effects such as local pain, fever, allergic and hypersensitivity reactions.

In spite of the above potential drawbacks, inactivated viral vaccines are in wide use. They include the Salk polio vaccine, influenza vaccines, and Japanese Encephalitis vaccine. In some cases, due to their lower efficacy, more than a single administration may be needed to achieve lasting immunity. Hence, in the 'competition' between live attenuated and killed viral vaccines, different considerations may dictate the preference of one or the other type. For example, in the case of polio, the Sabin live-attenuated oral vaccine is more widely used worldwide due to its lower cost as well as its capability to generate both secretory and systemic antibodies, as well as a more comprehensive T-cell response than the Salk killed vaccine. In the case of influenza vaccines, the killed whole virus is easier to prepare and less time consuming, as well as safer, and hence is more widely used than the live attenuated vaccine. An example for this advantage was noted during the H5N1 avian flu epidemic in 2005 (Beigel et al., 2009), as well as the recent H1N1 epidemic in 2009 when the need to prepare a vaccine in a short time, worldwide use dictated the resort to an inactivated vaccine.

Bacterial, whole-organism vaccines

In general, the availability of antibiotics, which can lead to the elimination of bacterial infections and subsequent therapy, reduced the importance and urgency of developing bacterial vaccines. Yet, bacterial vaccines have been and still are widely used. Since the last quarter of the 19th century, bacteria were revealed as the causative agents of many important and serious diseases, such as the plague, cholera, typhus, diphtheria and tuberculosis. The ability to obtain cultures of these bacteria paved the way to the development of bacterial vaccines.

As in the case of viral vaccines, the first bacterial vaccines to be developed were live attenuated. It started with Pasteur, who developed the veterinarian vaccine against *Pasteurella pestis*, just by using old cultures of the bacteria, and against anthrax by using heated (42–43°C) cultures of the bacillus for vaccination of cows, sheep and goats. In the latter case, he proved the efficacy of vaccination in controlled experiments (Pasteur, 1882). This was followed by the development of other bacterial vaccines for human use, which resulted in several successes, leading to vaccines that are in use to date. The first to be developed was the bacille Calmette–Guérin (BCG) vaccine against tuberculosis. These two scientists obtained the attenuated strain of the bacteria by multiple (over 200) repeated subculturing. The resultant vaccine has been in use for human vaccination (mainly infants) since 1929, with proven efficacy (Great Britain Medical Research Council, 1956) and is still widely used in many developing countries. Another attenuated bacterial vaccine that has proved efficacious in protecting humans against disease is the vaccine against *Salmonella typhi* (Ty21a), which is still in use to date (Levine *et al.*, 2007).

In addition to the live attenuated bacterial vaccines, there are several inactivated whole organism vaccines that are in use for vaccination of humans. These include *Bordetella pertussis*, *Vibrio cholerae* and *Salmonella typhi*. Of those, the only one that is still widely in use is the *Bordetella pertussis* vaccine. However, this vaccine has been associated with significant adverse effects, which led to considerable efforts to produce an acellular vaccine, that has been recently approved by the regulatory agencies. In the case of cholera as well, a killed (heat inactivated) whole organism vaccine is in use in countries inflicted with cholera outbreaks, such as Bangladesh, India, Indonesia and the Philippines. Randomized trials conducted in the 1960s and 1970s in these countries documented that the killed vaccine can confer significant short-term protection in older children and adults. But, more recent developments are towards an improved vaccine that is based on both the heat-inactivated *Vibrio cholera*, of several strains, and the purified B subunit of the cholera toxin. This vaccine is both safer and more effective (Arakawa *et al.*, 1998).

Subunit vaccines

Subunit vaccines refer to vaccine agents that comprise a component of the pathogen rather than the entire pathogen. In general, they are very safe because the immunizing agent is not a whole organism and cannot cause disease and hence they usually lead to fewer unpleasant side-effects. Most, but not all, subunit vaccines consist of a potentially immunodominant protein isolated from the pathogen. They have the advantage of being well-defined and characterized as to both their beneficial properties and any adverse biological activity. They also allow for the targeting of the immune response to specific immunodominant surface antigens, which will lead to effective immune clearance of the infectious agent upon challenge. But they have two serious limitations: Because they are isolated proteins they do not activate a response as broad or as robust as the whole organism vaccines do. Activation of cell-mediated immunity is particularly difficult to achieve. The development of the immune response to these proteins can be enhanced by the use of 'adjuvants' which initiate inflammation and cause prolonged antigen release. In humans, the most commonly used adjuvant in the classical vaccines and the only approved one until recently, is alum, which is composed of aluminium hydroxide crystals. It is usually used to initiate a mild local inflammatory response that will enhance antigen uptake and presentation by the antigen-presenting cells (APCs).

Bacterial subunit vaccines

There are two main types of bacterial subunit vaccines. One area where isolated protein vaccines remain essential is the immunization against microbial toxins. There are several bacteria, for example the tetanus and diphtheria organisms, that are non-invasive, but they secrete powerful toxins that cause major pathology. Such toxins can be manipulated by chemical means, to yield an inactivated form, called a toxoid. Immunization with the toxoids generates a response directed at the pathogenic product of the infectious agent. The aim of the vaccination is therefore to neutralize the toxin by generating antibodies which bind to it and then clears it from the system. It is not needed to immunize against the infectious agent itself because it is only the toxin that causes disease. Immunization with tetanus toxoid and diphtheria toxoid is routine and protects against the potentially fatal effects of respective toxins during infection with *Clostridium tetani* or *Corynebacterium diphtheriae*. The vaccine against pertussis (whooping cough) comprises an acellular product, including the toxin of *Bordetella pertussis* for generating a protective response, while avoiding the serious side-effects of the pertussis whole-organism vaccine that had been used for over 40 years. In view of the neurotoxicity of the pertussis toxin (Pittman, 1986) special efforts were required in order to eliminate completely the neurotoxic effects, that finally led to a safe product. In such cases, genetic manipulation is sometimes required to yield the appropriate mutations. These three toxoids, administered together constitute the DPT (diphtheria–pertussis–tetanus) paediatric vaccine that is routinely used in almost every country worldwide.

Another type of subunit bacterial vaccines is based not on the bacterial proteins, but rather on capsular polysaccharides of encapsulated bacteria. One example is *Streptococcus pneumoniae*, of which over 90 different serotypes are known. A 23-valent vaccine that was licensed in the early 1980s is protective against 85–90 serotypes that infect adults and is used to vaccinate elderly people against pneumonia (Vila-Corcoles *et al.*, 2009). A second example is the Vi antigen of *Salmonella typhi*, which is the capsular material covering the surface of the bacteria. This vaccine conferred highly significant protection that lasted for at least 17 months (Klugman *et al.*, 1987). Actually, basic principles suggest that immunization with polysaccharides will give, at best, an IgM response. This is because a robust IgG response cannot happen without class switch and class switch depends on T-cell activation, which cannot occur unless antigenic peptide is presented to T cells. However, the polysaccharide vaccines do work in humans, although only after the age of 2½–3 years. Two factors may account for this: The first is that the polysaccharides probably bind

to self proteins and serve as 'haptens' and the resultant chimera can be taken up by dendritic cells for activation. The second, and more likely, is that an MHC-like molecule called CD1 is capable of presenting non-peptide antigens to some T cells. It may be that these polysaccharide agents are presented in this manner.

Activation of effective IgG immune responses to the polysaccharides vaccines is of special importance in children, since IgG antibody to polysaccharides is the last to develop during immune maturation. Yet, polysaccharide encapsulated bacteria are major causes of serious and even fatal infections in the first two years of life. One example for such infection is meningococcal disease caused by *Neisseria meningitidis*. The currently licensed vaccines consist of different combinations of the purified high molecular weight capsular polysaccharides. Such vaccines provide protective immunity against the common A and C subgroups, for several years, but only in children from the age of 18 months (Sanborn, 1987). Another example is *Haemophilus influenzae* type b (Hib) which is highly infective in children. This strain is distinguished from other, non-pathogenic, strains by its capsular polysaccharide that has protective properties, namely it induces neutralizing antibodies (Schreiber *et al.*, 1986). This polysaccharide, called PRP, is the basis for the Hib vaccine. However, similarly to other cases (e.g. meningitis), children less than 18 months of age respond poorly to saccharides, as they fail to stimulate adequate cellular immunity.

To address this challenge and increase the efficacy of the capsular polysaccharides, a technology was developed, by which they are conjugated to carrier proteins, thus leading to a new type of vaccines denoted conjugate vaccines. In the case of Hib, either tetanus toxoid or the outer-membrane protein of Hib have been employed, resulting in very successful vaccines for the use of infants as well (Ward *et al.*, 1994). This approach is now applied for the development of other polysaccharide vaccines. These new vaccines have dramatically decreased the frequency of the infections mentioned above, including pneumonia and meningitis, and are now available and recommended for children. This is a major advance because it decreases the need for antibiotic use in an era of increasing problems of evolution of antibiotic-resistant bacteria.

Pure polysaccharide vaccines are still in use for adults (e.g. Pneumovax-23 that contains 23 different polysaccharides from *S. pneumoniae*), but even in adults the conjugate vaccines induce a longer-lived response and better IgG memory than the pure polysaccharide vaccines

Viral subunit and split vaccines

In the case of viruses, especially those viruses where the surface antigens are implicated in eliciting the protective immune response, it is possible to use a 'split' vaccine, namely a preparation where the structure of the virus has been disrupted and it contains a mixture of the various viral components. Alternatively, a 'subunit' vaccine may be employed that consists of a small number of the viral proteins. The most prominent subunit vaccine in use to date is for influenza, where it comprises the isolated two surface antigens of the virus, the haemagglutinin (HA) and the neuraminidase (NA). Each year, for the seasonal flu vaccine these two proteins are isolated from the three strains of the virus as selected by the WHO, and their mixture constitutes the trivalent vaccine, with or without adjuvant. Such vaccines have been approved for human use and proved highly immunogenic and well tolerated in children, young adults and among the elderly (Miehler and Metcalfe, 2002).

Another type of subunit viral vaccine is the hepatitis B vaccine. Since the hepatitis B virus cannot be grown in culture, the first vaccine was composed of the surface antigen, HBsAg, of this virus, an antigen that is present and can be isolated from the blood of infected people. The vaccine's efficacy was tested in chimpanzees and it proved effective. The resultant product became widely used, but initially it was used only in developed countries due to its high cost. At a later stage a recombinant vaccine, in which the viral protein was produced in DNA-transfected yeast cells, became available (Valenzuela *et al.*, 1982). It constitutes the first genetically engineering vaccine produced commercially at affordable cost and used worldwide. This laid the foundation for development of the whole approach of recombinant vaccines, as described in the next section.

The general approach of developing a subunit viral vaccine based on an isolated viral protein with protective potential, is now under investigation for other viral vaccines. For example, an influenza vaccine based on the M2 matrix protein, or its ectodomain, is being developed as a potential universal influenza A vaccine (Schotsaert *et al.*, 2009).

Recombinant vaccines

Similarly to other areas of biology, in vaccinology as well, the application of recombinant DNA approaches has opened up a whole new area of possibilities. Recombinant DNA can be applied both in the identification and towards the isolation of antigens. Thus, by being able to clone and express individually some or all the antigens of an organism, several major problems associated with the traditional approaches to vaccine development are deviated. Firstly, problems related to obtaining enough material can be overcome. Using the traditional technologies, it was often difficult to obtain sufficient quantities of particular antigens in pure form even to allow the testing of this antigen and evaluation of its efficacy. The use of recombinant material has facilitated such testing and has resulted in the screening of large numbers of potential vaccine candidates. Second, the recombinant technologies facilitated the expression of a particular antigen in a desirable vector in order to achieve a more efficient presentation and immunogenicity. Third, these technologies allowed the large scale preparation of the desired antigen for practical vaccine development. Furthermore, recombinant DNA techniques have made the study of pathogenic organisms much safer, since the actual products are single genes and their translated proteins and hence the contamination with the disease-causing organisms is eliminated.

Recombinant vaccines with live bacterial vector

Recombinant DNA techniques with or without classical genetic approaches have also enabled the construction of live vaccines based on avirulent and/or attenuated strains of organism. This applies to both bacteria and viruses. In the case of bacteria, DNA techniques led to the deletion of genes that are associated with virulence, resulting in strains that are safe and efficacious and can serve as vaccines, as was shown for both *Vibrio cholerae* (Levine and Kaper 1993) and *Salmonella typhi* (the Ty21 strain), that have been approved and are in use for vaccination (Germanier and Furer, 1975). Such genetically attenuated agents can also be employed in order to express an external antigen of a desired specificity. This approach can be exemplified by either bacteria or viruses.

In regard to bacteria, both the attenuated *Vibrio cholerae* and *Salmonella typhi* mentioned above have been employed for this purpose by carrying genes encoding heterologous antigens of other microorganisms. For example, a recombinant *V. cholerae* expressing the B subunit of Shiga-like toxin, was prepared in quantities sufficient for large-scale purification (Acheson *et al.*, 1996) and has been utilized as a vaccine. Attenuated *Salmonella* served as a live vector for expressing a variety of foreign antigens including bacterial, viral and even protozoan antigens. Several attenuated *Salmonella* strains served for this purpose; the most extensively used were the Aro mutants, which were employed for the expression of numerous bacterial antigen. The use of *Salmonella* is straightforward due to its similarity to *E. coli*, in which most of the recombinant DNA methodology has been developed. Hence, many of the conventional expression vectors used in *E. coli* can perform well in *Salmonella* and it is relatively easy to transfer recombinant DNA constructs into *Salmonella* vaccine strains. As a result of the efforts in this direction many potential vaccines, for oral immunization against a variety of antigens including bacterial toxins, bacterial enzymes and other protective antigens have been prepared and evaluated as veterinary and/or human vaccines. Furthermore, one of the major advances has been the regulated expression of several 'guest' antigens in the attenuated *Salmonella*, enabling it to protect animals against several diseases after a single oral dose. This paves the way for development of new multivalent bacterial vaccines for human use, using recombinant DNA technology (Shatfield and Dougan, 1997).

In addition to the recombinant vaccines expressing bacterial antigens, a number of viral genes coding for protective antigens have been expressed in avirulent *Salmonella* species, ranging from respiratory syncytial virus (RSV) glycoprotein to HIV gp120 and other HIV proteins, hepatitis B virus proteins, rotavirus capsid antigen and influenza nucleoprotein. In several cases potentially virus-neutralizing antibodies and CD4+ T-cell response were demonstrated after oral immunization with the recombinant *Salmonella*. The only case where such response was accompanied by reduced pulmonary infection is the recombinant vaccine expressing the influenza nucleoprotein, thanks to the availability of an experimental model for influenza in small animals, where *in vivo* protection could be evaluated. Notwithstanding these achievements, however, it remains a challenge to express correctly folded viral surface molecules, that are more relevant protective antigens in prokaryotes for use as practical vaccines.

Last but not least is the interesting approach of using recombinant DNA technologies for expression of protozoan antigens in a live vector. This is of particular importance considering the urgent need for an effective vaccine for diseases such as malaria or other parasitic diseases. In that realm, the successful preparation of recombinant attenuated *Salmonella* expressing proteins of several protozoa and the evaluation of their immunological properties is of interest. The success of this strategy requires that putative protective antigens of the protozoa are known, that the foreign protozoal genes be stabilized in the bacterial live vector and that the initiation and level of expression of the foreign gene is appropriately controlled by the choice of promoter. All these prerequisites have been attended to by using various tactics and sophisticated genetic manipulations. These led to several recombinant *Salmonella* vaccines that elicited relevant immune responses against various protozoa, including *Entamoeba histolytica*, *Leishmania* and, most importantly, several species of *Plasmodium*. The last are based on expressing the circumsporozoite protein (CSP) of either *P. berghei*, *P. yoelii* and even *P. falciparium* in different attenuated mutants of *S. typhi*. The most relevant one is the *P. falciparum* as it is the most infectious in humans. Indeed, using a suitable attenuated *S. typhi* vector vaccine strain CVD 908, as well as an adequate promoter, the foreign gene of *P.*

falciparum CSP was integrated into the chromosome of the vector. The resultant product was well tolerated by volunteers and induced a meaningful anti *P. falciparum* response in a significant proportion of them (Gonzales *et al.*, 1994). The expectation is that similar constructs with improved expression should enhance immunogenicity and may lead to an effective anti-malaria vaccine in the near future.

Recombinant vaccines based on live viral vector

Attenuated viruses can also serve as vectors for recombinant vaccines expressing foreign antigens. The most investigated viral vector is vaccinia virus. As already mentioned, this virus, originally employed by Jenner, was the first vaccine to be used and it eventually led to the eradication of smallpox. As a result, the general need for vaccination against smallpox was eliminated and was discontinued by declaration of the World Health Organization. However, the ability to produce recombinant vaccinia viruses that express genes of other microorganisms led to the possibility of new medical uses and extensive research in this direction towards vaccine development (Moss, 1991). Vaccinia virus can infect most mammalian as well as avian species, making it a useful vector for studying the immune response to proteins of both human and veterinary pathogens. Following the construction of vaccinia virus expression vectors for the expression of foreign genes, it was necessary to develop the technology to isolate the recombinant product and to discriminate it from the parental virus. Excellent results led to a whole series of recombinant vaccinia viruses, that can serve as prophylactic vaccines against a variety of infectious diseases, including hepatitis B, influenza, rabies and measles, respiratory syncytial and herpes simplex viruses. In addition, vaccinia virus vectors expressing tumour antigens have been used for the immunoprophylaxis and immunotherapy of solid tumours, haematogenic and metastatic malignancies (Tsang *et al.*, 1995). Furthermore, recently vaccinia-derived vaccines for HIV infectious (Shiu *et al.*, 2009) and avian influenza (Poon *et al.*, 2009) were prepared and are presently in clinical trials. However, up till now there is no licensed vaccine that is based on this approach.

Synthetic recombinant vaccines

The recombinant DNA vaccines described hitherto are based on the expression of foreign genes in suitable live vectors, leading to the expression of a whole protein that should induce protective immunity. Another approach is to express defined epitopes in an appropriate vector. The main benefit of an epitope-based vaccine is the ability to immunize with a minimal structure, consisting of a well-defined antigen. However, most peptides are of low immunogenicity and hence cannot serve as such as appropriate vaccines. Their immunogenicity can be augmented by the use of macromolecular carriers to which the desired epitope is either complexed or covalently attached. An alternative approach is to express these epitopes by recombinant DNA technology. In that case, a particular protein, usually bacterial envelope protein, rather than an entire organism, can serve as an adequate vehicle: One such vector that has been employed is the flagellin of attenuated *Salmonella* strains. Sites within the protein were selected on the basis of being surface-located, allowing insertion of the foreign DNA without the disruption of the protein structure. In this way a cholera toxin epitope was expressed in *Salmonella* flagellin, leading to effective immune responses (Stocker and Newton, 1994). The same approach was employed for expressing epitopes of influenza virus for the preparation of a synthetic recombinant influenza vaccine. The advantages of flagellin in this context are twofold: (1) its unique properties to form polymers that constitute the flagella, when each flagellum acts as a multivalent antigen, and (2) its capacity to interact with the Toll-like receptor TLR5, thus enhancing the immune response to any foreign epitope it expresses (Takeda *et al.*, 2003). Flagellin is also known for its capacity to induce T-cell responses and hence the potential for the induction of efficient immune response to foreign epitopes expressed in flagellin. Indeed, synthetic recombinant vaccines were developed against various pathogens including influenza. One such vaccine, that expresses several conserved epitopes of the virus, was shown to induce a protective effect against multiple strains of influenza (Levi and Arnon, 1996), even in humanized mice. Hence, the potential of this approach for the development of a universal influenza vaccine for human use.

DNA vaccines

The concept of DNA vaccines is based on the observation in the early 1990s that plasmid DNA could directly transfect animal cells *in vivo* and hence a simple non-replicating plasmid DNA encoding a foreign protein be introduced directly into a living animal for the purpose of immunization, that will induce immune response to the encoded protein. Within one year after the first demonstration of this principle (Tang *et al.*, 1992), several other reports were published, with direct demonstration that immunization with such 'naked' DNA could lead to protection against different pathogens, including influenza virus, HIV, bovine herpes virus and hepatitis B virus. In the case of influenza for example, a plasmid encoding nucleoprotein (NP) gene led to cytotoxic T lymphocytes (CTL) response and to protection extending to different strains of the virus (Ulmer *et al.*, 1993).

Since the first few publications, this approach to immunization has generated sustained interest because of its speed, simplicity, ability to elicit immune responses against native protein antigens with complex structures, and the ability to elicit both humoral and cellular immune responses, without the need for live vectors or complex biochemical techniques. A wide range of methods have been used to deliver plasmids including needle injection, fluid jet injection, injection followed by electroporation, bombardment with gold particles coated with DNA, using 'gene-gun' and topical administration to various mucosal sites including the gut, respiratory tract, skin and eye. The design of the plasmids themselves has been less diverse. The most commonly used plasmids employ a minimal backbone containing a selectable marker, an origin of replication active in *Escherichia coli*, a strong viral promoter such as CMVintA, and a transcriptional chain terminator or polyadenylation signal sequence. The major exception to the trend towards small plasmids

is the addition of alphavirus replicon elements, which were introduced more recently.

It is of interest, however, that both the form of the DNA-expressed antigen (whether it is cell associated or secreted) and the method of the DNA inoculation or delivery, determine the nature of the immune response and, in the case of T-helper cells involvement, whether the response will be primarily type 1 or type 2. Mechanistically, gene-gun-delivered DNA initiates responses by transfected or antigen-bearing epidermal Langerhans cells that move in the lymph from the bombarded skin to the draining lymph nodes. On the other hand, following intramuscular injections, the functional DNA appears to move as free DNA through blood to the spleen, where professional antigen-presenting cells initiate the responses (Robinson and Torres, 1997). As for the potential of a DNA vaccine plasmid to induce an effective immune response, although its sequence and composition may play some role in the immunogenicity, its efficacy is directly related to the level of expression of the encoded protein in eukaryotic cells. In the case of viral antigens, alteration of codon usage to remove regulatory sequences and deletion of functional sequence elements have proven to be important for maximizing the immunogenicity of DNA vaccines. For bacterial, protozoal, and fungal antigens, compactly folded, secreted single polypeptides that express well in eukaryotic cells may be more likely to be successful than the integral membrane proteins or large heteromultimers.

Clinical trials and studies to evaluate the safety and immunogenicity of DNA vaccination in humans began within two years of the first published reports on the protective immune responses against infectious diseases in animals, and many studies are still ongoing. Special emphasis is of course on unmet needs, such as a vaccine for the treatment of melanoma (Nabel *et al.*, 1996), or a broad-spectrum influenza vaccine, a plasmid encoding the malarial circumsporozoite protein of *P. falciparum* (Weiss *et al.*, 2000), and of course an anti-HIV vaccine. Concerning the latter, many efforts were devoted to construct an efficient vaccine, based on the envelope antigens gp120, gp140 and gp160/*rev*, as well as *nef*, but the results to date show little progress. Several factors may account for these disappointing results, including the difficulty to induce protective antibodies. Plasmid DNA vaccines induce mainly T-cell response that does not prevent the early infection. More recently this hurdle has been overcome by using a DNA/MVA vaccination, comprising combined DNA priming and MVA boosting, which led to efficient protection of macaques against SHIV (Ellenberger *et al.*, 2006). These promising results may indicate good prospects for success of an HIV/AIDS vaccine for human use in the not too far future. A successful DNA vaccine that has already been approved and licensed for human use is the human papillomavirus (HPV) vaccine that protects against cervical cancer (Ohlschlager *et al.*, 2009).

In all the clinical trials reported so far, the results indicate that the DNA vaccines are generally safe and well tolerated at doses of up to 2.5 mg of plasmid over three or more injections when given either intramuscularly or intradermally. In the individuals participating in these clinical trials, just as in the animal and non-human primate studies, the DNA vaccination was shown to induce both antibody and cellular responses. However, the response observed in humans is insufficient and the DNA vaccines need to be made much more potent to be considered as a realistic candidate for preventative immunization of humans.

Such attempts to enhance the immunogenicity of DNA vaccine was recently reported for a therapeutic cervical cancer DNA-based vaccine that contains all putative T-cell (HPV) epitopes (denoted HPV-16 E7SH). Co-application of DNA encoded cytokines (IL-2, IL-12, GM-CSF or IFN-γ) as well as the chemokine MIP1-a led to significantly enhanced anti-tumour response (Ohlschlager *et al.*, 2009).

An important consideration to be taken into account concerning DNA vaccines is that most of the clinical trials conducted hitherto were phase II trials and hence, more extended Phase III trials are needed to demonstrate both efficacy and safety. In particular, a safety issue such as the potential for induction of autoimmune diseases and/or integration into the host genome, must be examined carefully. But, all in all, the formal acceptance of this novel technology as a new modality for human vaccines has been remarkable. If potency can be improved and safety established in a satisfactory manner, plasmid or 'naked' DNA vaccines offer many advantages owing to the speed and simplicity of their preparation, as well as the breadth of immune response they promote, that may be useful for the immunization of humans against both infectious diseases and cancer.

Synthetic peptide vaccines

An additional approach is the use of synthetic peptides which constitute epitopes that induce protection. As already mentioned above, the main benefit of immunization with an epitope-based vaccine is the ability to immunize with a minimal structure, consisting of a well-defined antigen that can be thoroughly characterized with respect to its antigenicity and immunogenicity. Comprising the adequate epitopes, such vaccine will stimulate an effective specific immune response, while avoiding potential undesirable effects. For example, antigenic regions that may activate suppressor mechanisms, or a response against self-antigens, can be eliminated from the vaccine preparation, thus providing a safer vaccine. However, the use of small molecules such as peptides may encounter a problem of low immunogenicity, as compared to the multi-epitopes protein antigens, or the entire pathogen, that is used for immunization in the conventional vaccines. Hence, when applying the strategy of peptide-based vaccines, an appropriate macromolecular carrier should be used.

Before attempting to develop synthetic peptide vaccines it is necessary to recognize the immunological properties of the antigens in the pathogen involved and the type of immune response that would lead to a protective effect, whether humoral (involving B cell epitopes) or cellular (involving T-helper or CTL epitopes). Various approaches have been employed for this purpose, including antigen fragmentation and analysis of the resultant fragments, crystallographic analysis for prediction, systematic synthesis of all hepta or decapeptides according to the protein sequence for evaluation of their reactivity and finally computational analysis of either hydrophilicity or flexibility, for the prediction of T-cell epitopes (MHC-binding motifs).

Using these various approaches, potential peptide vaccines against many pathogens have been prepared and evaluated, a few examples of which are described below.

Antibacterial peptide vaccines

Bacterial proteins, including toxins (that serve as the basis for the toxoid vaccines) are natural candidates for the synthetic approach. Indeed, synthetic peptides that correspond to protective fragments of bacterial proteins have shown efficacy in protection. One such example is the M protein of *Streptococcus pyogenes*, which is a type-specific antigen present on the surface of virulent species and is largely responsible for both the virulence and the protective immunity of these bacteria. In the early 1980s it has been demonstrated that a portion of the M24 protein can elicit opsonic and protective antibodies, and consequently the corresponding synthetic peptides demonstrated efficient protective response against several types of group A streptococci without any tissue cross-reactive antibodies (Dale and Beachy, 1982). Furthermore, hybrid peptides led to protection against more than one strain even in the absence of a carrier.

As for the bacterial toxins, several toxins lent themselves to this approach. In the case of diphtheria toxin, a synthetic 14-residue epitope of the native toxin led to a neutralizing immune response against the toxin. When conjugated with the synthetic adjuvant muramyl-dipeptide (MDP) it elicited protective anti-toxin immunity when administered in aqueous physiological solution. For cholera toxin and Shiga toxin, it is the B subunit that is capable of inducing neutralizing antibodies. Two synthetic peptides of cholera toxin, conjugated to tetanus toxoid, led to antibodies that neutralized the biological activity of the toxin (Jacob *et al.*, 1983), as well as that of the heat labile toxin of *E. coli*, due to the homology between the two toxins. Peptides of the B subunit of Shiga toxin, the toxin of *Shigella dysenteriae*, also led to a protective effect. Thus, several peptides were prepared, corresponding to either the amino terminal or the carboxy terminal region of the B subunit, which coincide with the peak of hydrophilicity and surface area. All these peptides, conjugated to an appropriate carrier, induced antibodies which neutralized the toxin's biological activity and hence led to protective immunity (Harari *et al.*, 1998).

Antiviral peptide vaccines

An antiviral response induced by a peptide was first reported as early as 1963, against tobacco mosaic virus. This was followed by demonstration of bacteriophage (MS-2) inactivation by antibodies raised against a synthetic peptide corresponding to a region of its coat protein. These results led to the investigation of the potential of peptide-based vaccines against viral diseases in humans. One example is hepatitis B virus, for which the HBsAg is a protective protein. Synthetic peptides corresponding to almost the entire sequence of the HBsAg have been synthesized, and two of them proved effective in leading neutralizing antibodies and a protective effect in chimpanzees (Gerin *et al.*, 1983). Hepatitis A virus (HAV) was also subjected to the study of the peptide vaccine approach. A HAV motif selected from a combinatorial peptide library, by a well-characterized neutralizing HAV monoclonal antibody, was shown to elicit in mice antibodies that reacted with the virus, and competed with human convalescent antibodies to HAV.

Influenza is another virus for which peptides corresponding to major antigenic regions have been evaluated as potential vaccine components. In regard to the choice of epitopes, one of the major antigens of influenza virus is the surface glycoprotein haemagglutinin (HA). It plays a key role in the initiation of viral infection by binding to sialic acid-containing receptor on host cells and thus mediates the subsequent viral entry and membrane fusion. Oligopeptides corresponding to several regions of the HA have been synthesized and conjugated to tetanus toxoid for immunization. While the immunogenicity of some of these peptides is uncertain, at least one peptide, with the sequence 91–108 was shown to induce antibodies capable of inhibiting the haemagglutination capacity of the virus as well as neutralizing the virus, namely inhibiting its plaque formation in tissue culture. Furthermore, immunization with a tetanus conjugate of this peptide protected mice against challenge infection (Muller *et al.*, 1982). Later it was demonstrated that influenza-specific

MHC I-restricted CTL can be primed *in vivo* by using synthetic peptides derived from the nucleoprotein NP, namely NP 147–158 covalently linked to a lipid (Deres *et al.*, 1989). When in addition to the single B-cell epitope HA91-108, two T-cell epitopes were included in the vaccine, namely the T-helper epitope NP55-69 and a CTL epitope NP147-158, a much more efficient immune response was achieved. This indicates the contribution of both humoral and cellular arms of the immune response to the overall protective effect. Since the B-cell epitope HA91-108 is conserved in H3 influenza strains, and both T-cell epitopes are derived from the nucleoprotein that is an inner viral protein known to be conserved, it is not surprising that the combination of these three epitopes is protective against many influenza viral strains (Levi and Arnon, 1996). Hence, an epitope-based vaccine, containing multiple copies of the relevant epitopes, is presently under development as a universal influenza vaccine, in phase I/II clinical trials.

Of paramount importance is the development of a peptide vaccine against HIV. Several epitopes of HIV proteins could be candidates for this purpose. One example that has been evaluated is an octameric V3 loop peptide of gp120 attached to a heptalysyl core (V3-MAP). In a phase I clinical trial three doses of such a vaccine elicited neutralizing antibodies, but only in 5 out of 15 individuals. Another peptide candidate is the HGP30 epitope of the p17, conjugated to KLH. A phase I clinical trial with this vaccine-induced T-cell response, including CTL, in most of the vaccinees, as well as antibody response both to the peptide and the KLH. The CTL response was CD8+ and SCID mice reconstituted with these cells showed protection against virus challenge (Sarin *et al.*, 1995). These results point to a protective immunity induced by this peptide vaccine. Another strategy is to combine synthetic envelope peptides consisting of helper T-cell sites and B-cell neutralizing epitopes and a *gag* CTL in a multivalent mixture. Such hybrid peptides generated in mice and monkeys high titres of neutralizing antibodies to the homologous HIV-1 strain and also MHC I-specific CTLs. A major drawback of this approach, however, is that multiple T-helper and CTL epitopes would be needed to recognize the broad spectrum of HLA molecules in humans. It may therefore be concluded that although the peptide vaccine approach is promising, the complexity of the HIV-I virus and the immunological parameters that are involved in both the disease itself and the protection against it, does not allow its realization at present.

Antiprotozoan peptide vaccine

The major parasite-inflicted disease for which a peptide vaccine has been developed is malaria. For that purpose, an important target of *Plasmodium*, is the surface antigen circumsporozoite protein (CSP). As mentioned above, this antigen was the basis for a recombinant DNA vaccine. However, peptides of this protein were also evaluated for their protective effect and suitability to serve as a vaccine.

In human malaria, the immunodominant B-cell epitope of the *P. falciparum* CSP is contained within three repeats of the NANP? Sequence. Hence, the first-generation peptide vaccine for malaria comprised $(NANP)_3$ conjugated to tetanus toxoid (TT). Clinical trials with this vaccine showed that most individuals developed antibodies, but no cell-mediated immunity. The second generation synthetic peptide vaccine contained, in addition to this $(NANP)_3$ B-cell epitope, a T-cell epitope of the CSP as well. For optimization, the two types of peptides were built into MAP (multiple antigen peptide), expressing several copies of each peptide, after optimization of the ratio of the T- and B-epitopes. The optimal construct $(T1B)_4$ MAP was tested in volunteers, in phase I and II trials and was shown to elicit high levels of parasite-specific antibodies in individuals expressing different MHC II molecules, with an indication for predictable high responder genotype. The antibody titres were comparable to multiple exposure to *P. falciparum*-infected mosquitoes (Nardin *et al.*, 2000). These results may facilitate the rational design of epitope-based peptide vaccine for malaria.

Another synthetic peptide malaria vaccine which seems quite promising is the SPf66 vaccine against asexual erythrocytic stages of *P. falciparum*. SPf66 is a chimeric molecule consisting of four synthetic epitopes – three erythrocyte stage epitopes derived from amino acid sequences from proteins isolated from the parasite, and one, the PNANP sequence, derived from the repeat domain of the CSP. The final product, namely a polymer, alum-adsorbed, delayed the appearance of malaria symptoms and even elicited a protective response, as was demonstrated first in *Aotus* monkeys and later in humans (Patarroyo *et al.*, 1988). Several human trials were performed with SPf66, to evaluate its safety and efficacy. As for safety, the vaccine has been well tolerated, with minimal side-effects, mostly at the site of injection, and no indications of autoimmunity. A series of trials in almost 10,000 people corroborated that 95% of the vaccinees did not produce any major reaction. Efficacy was demonstrated in several clinical trials and placebo-controlled field trials, conducted in South America and Africa in the 1990s including many thousands of individuals, demonstrating high antibody titre production with levels of protection ranging from 34% to 77% after three doses of the vaccine. More recent reports (Bermudex *et al.*, 2007) demonstrated the advantages of this approach – low cost, scale-up production, reproducibility, stability and safety, which could result in an effective vaccine for the fight against malaria. Although not approved by the FDA, this is the only efficient antimalarial vaccine to date.

Peptide vaccines against autoimmune diseases

Most vaccines for infectious diseases are prophylactic. In case of autoimmune diseases, a therapeutic approach is required, to overcome the immunological process leading to the pathology. In several autoimmune diseases the pathology has been traced to the effect of Th1 cells, which secrete cytokines as TNFα and IFN-γ that are involved in the autoimmune damage to tissues and organs. The role of a therapeutic vaccine is to counteract their effect.

One example is multiple sclerosis, where Th1 response to the myelin proteins leads to the neurological damage in the brain and spinal cord (Liblau *et al.*, 1995). A synthetic polypeptide, glatiramer acetate, was demonstrated to alleviate the symptoms of the disease in both experimental animals and patients (Teitelbaum *et al.*, 1997) by the induction of Th2 cells that secrete beneficial cytokines and neurotrophins and ameliorate the autoimmune damage (Aharoni *et al.*, 1997). It is approved and used worldwide for the treatment of MS patients. Another example is myasthenia gravis, a disease caused by an immune response, mainly specific antibodies, to the acetylcholine receptor (AChR). A synthetic peptide based on two T-cell epitopes of the a subunit of the AChR and denoted a dual analogue, was efficient in interfering with specific autoimmune responses and in alleviating experimental myasthenia gravis (Paas-Rosner *et al.*, 2000). A third example is diabetes type I, caused by the destruction of the insulin-producing pancreatic islet cells. After demonstrating the involvement of a heat-shock protein hsp60 in the induction of the autoimmune damage, an innovative immune-based therapeutic approach was developed using a peptide DiaPep277, which is effective in the treatment of type I diabetic (Raz *et al.*, 2007) and is presently in phase III clinical trials.

It can therefore be surmised that although no peptide-based vaccine has been approved as yet, the bulk of information gathered hitherto illustrates the potential of synthetic peptide vaccines. Such an approach will offer several advantages, including the possibility of synthesizing multivalent broad-spectrum vaccines containing several epitopes, the elimination of undesirable side-effects due to irrelevant epitopes and the possibility of producing vaccines with built-in adjuvanticity. Furthermore, this approach is not limited to infectious diseases and could have a wide-range applicability, such as therapeutic vaccines against autoimmune diseases.

Passive immunization

Passive immunization refers to the transfer of antibody from one individual to another. This procedure does not confer long-term protection, but it is useful for situations of imminent danger when there is no time to generate antibody by the patient and in cases of immune deficiency where antibodies cannot be made. The isotype transferred is IgG because it is the most lasting (with a half-life of 28 days) and it confers the best protection.

Passive immunization is used in the case of tetanus infection. If active protection against tetanus has not been induced owing to lack of immunization, or has waned because it has been many years since the last booster immunization, then immediate protection in the form of anti-tetanus IgG antibody will be required. Passive immunization with anti-rabies IgG is used after an animal bite (such as dog, raccoon and bat) where rabies is prevalent. In both of these examples, since the duration of the passively transferred IgG in the host is short, active immunization should also commence immediately to provide continuing protection by the vaccinee's own antibody. Similarly, antivenin antibodies are used for spider bites or snake bites. Often these antibodies are produced in other species, such as horses and the IgG from their blood is purified for the purpose of vaccination. Several such anti-venins are approved by the regulatory authorities.

Patients who cannot make adequate responses and thus have humoral immune deficiency usually require monthly administration of pooled IgG (also called 'gammaglobulin' or 'immune globulin') to protect them against many types of infections. This is usually given intravenously (IVIG). In other cases, passive protection against a specific infection needs to be given when someone is at high risk of severe disease (for example, chicken pox in an immunosuppressed cancer patient who never had chicken pox or a baby during birth to a hepatitis B-carrying mother). Disease-specific IgG preparations are administered intramuscularly in such settings as passive vaccines.

Immune globulin (pooled IgG used for IVIG) is made from the blood plasma of thousands of human donors and thus confers protection against infections experienced by the population, as long as IgG is sufficient for protection. It is a relatively safe product using current production methods and has also been employed in recent years to overcome Nile-Virus fever infections. Hence, although not considered as vaccines in the conventional sense, this procedure for immune protection should not be overlooked. It has been in use since the beginning of the twentieth century, and is still in use to date.

Concluding remarks

As described in this chapter, vaccines have been in use as a prophylactic strategy for over 200 years. From the initial trial and error basis, more and more accurate and elaborate methodologies were applied towards vaccine production, including the latest genetic manipulation and sophisticated molecular biology approaches. These efforts led to a long list of vaccines that are at present licensed for use in the USA, Europe and many other countries. As a result, the incidence of bacterial and viral infections in the developed world was reduced by orders of magnitude and concomitantly the mortality of babies and young children, which led to almost doubling of the life expectancy. Table 1.1 lists the licensed vaccines that have been approved to date, in alphabetical order of the disease against which they protect. The vaccines that are in very wide use, such as the paediatric vaccine DPT, MMR, the oral polio vaccine (OPV) and influenza vaccine, are produced with slight variations by many manufacturers, under different trade names. But they are all available and have proved their efficacy, leading to a tremendous improvement in public health. There is, however, a whole list of other diseases, such as AIDS caused by HIV, or diseases caused by other viruses such as Dengue virus or cytomegalovirus, against which no vaccines are available as yet. Similarly, there is no efficient vaccine approved against any of the parasitic diseases, of which hundreds of millions of people, mainly in developing countries, suffer and die each year. There is a tremendous effort to develop new vaccines based on new technologies, which are presently at a very advanced stage. These include DNA vaccines and peptide vaccines, as well as other approaches dealt within this book.

References

Acheson, D.W.K., Levine, M.M., and Keusch, J.T. (1996). Protective immunity to Shiga-like toxin I following oral immunization with Shiga-like toxin B-subunit-producing *Vibrio cholerae* CVD 103-HgR. Infect. Immun. *64*, 355–357.

Aharoni, R., Teitelbaum, D., Sela, M., and Arnon, R. (1997). Copolymer 1 induces a T cells of the T helper type 2 that cross react with myelin basic protein and suppress experimental autoimmune encephalomyelitis. Proc. Natl. Acad. Sci. U.S.A. *94*, 10821–10826.

Almond, J.W. (1987). The attenuation of poliovirus neu-rovirulence. Annu. Rev. Microbiol. *41*, 153–180.

Arakawa, T., Chong, D.K., and Langridge, W.H. (1998). Efficacy of a food plant-based oral cholera toxin B subunit vaccine. New Biotechnol. *16*, 292–297.

Beigel, J.H., Voell, J., Huang, C.M., Burbelo, P.D., and Lane, H.C. (2009). Safety and immunogenicity of multiple and higher dose of an inactivated influenza A/H5N1. J. Infect. Dis. *200*, 501–509.

Bermudex, A., Reyes, C., Guzman, F., Vanegas, M., Rosas, J., Amador, R., Rodriguez, R., Patarroyo, M.A., and Pattarroyo, M.E. (2007). Synthetic vaccine update: Applying lessons learned from recent SPf66 malarial vaccine physicochemical structure and immunological characterization. Vaccine *25*, 4487–4501.

Beyer, W., Palache, A., De Jong, J., and Osterkaus, A. (2002). Cold-adapted live influenza vaccine versus inactivated vaccine: Systematic vaccine reactions, local and systemic antibody response and vaccine efficacy. A meta-analysis. Vaccine *20*, 1340–1353.

Dale, J.B., and Beachy, E.H. (1982). Protective antigenic determinant of streptococcal M protein shared with sarcolemmal membrane protein of human heart. J. Exp. Med. *156*, 1165–1176.

Deres, K., Schild, H., Wilsmulker, K.H., Jung, J., and Rammensee, H.G. (1989). In vivo priming of virus-specific cytotoxic lymphocytes with synthetic lipopeptide vaccine. Nature *342*, 561–564.

Ellenberger, D., Otten, R.A., Li, B., Aidoo, M., Rodriguez, I.V., Sariol, C.A., Martinez, M., Monsour, M., Wyatt, L., Hudgens, M.G., Kraiselburd, E., Moss, B., Robinson, H.L., Folks, T., and Butera, S. (2006). HIV DNA/ MVA vaccination reduces the per exposure probability of infection during repeated mucosal SHIV challenges. Virology *352*, 216–225.

Evans, D.M.A., Dunn, G., Minor, P.D., Schild, G.C., Cann, A.J., Stanway, G., and Almond, J.W. (1985). Increased neurovirulence associated with a single nucleotide change in a non coding region of the Sabin type 3 polio vaccine genome. Nature *314*, 548–550.

Gerin, J.L., Alexander, H., Shih, J.W., Purcell, R.H., Dapolito, T., Engle, R., Green, N., Sutcliffe, J.G., Shinnick, T.M., and Lerner, R.A. (1983). Chemically synthesized peptides of hepatitis B surface antigen duplicate the d/y specificities and induce subtype-specific antibodies in chimpanzees. Proc. Natl. Acad. Sci. U.S.A. *80*, 2365–2369.

Germanier, R., and Furer, E. (1975). Isolation and characterization of Gal E mutant Ty21a of *Salmonella Typhus*. A candidate strain for live oral typhoid vaccine. J. Infect. Dis. *141*, 553–558.

Gonzales, C., Hone, D., Noriega, F., Tacket, C.O., Davis, J.R., Losonsky, G., Nataro, J.P., Hoffman, S., Malik, A., Nardin, E., Marcelo, B., Sztein, D., Heppner, G., Fouts, T.R., Isibasi, A., and Levine, M.M. (1994). *Salmonella typhii* vaccine strain CVD 908 expressing the circumsporozoite protein of *P. falciparum*: strain construction and safety and immunogenicity in humans. J. Infect. Dis. *169*, 927–931.

Great Britain Medical Research Council. (1956). BCG and whole bacillus vaccines in the prevention of tuberculosis in adolescence and early life. Br. Med. J. *1*, 413.

Harari, I., Donohue-Rolfe, A., Keusch, G., and Arnon, R. (1998). Synthetic peptides of shiga toxin B subunit induce antibodies which neutralize its biological activity. Infect. Immun. *56*, 1618–1624.

Jacob, C.O., Sela, M., and Arnon, R. (1983). Antibodies against synthetic peptides of the B subunit of cholera toxin: cross-reaction and neutralization of the toxin. Proc. Natl. Acad. Sci. U.S.A. *80*, 7611–7615.

Klugman, K.P., Gilbertson, I.T., and Koornof, H.J.N. (1987). Protective activity of Vi capsular polysaccharide vaccine against typhoid fever. Lancet. *2*, 1165–1169.

Liblau, R.S., Singer, S.M., and McDevitt, H.O. (1995). Th1 and Th2 CD4+T cells in the pathogenesis of organ-specific autoimmune diseases. Immunol. Today *16*, 34–38.

Levi, R., and Arnon, R. (1996). Synthetic recombinant influenza vaccine reduces efficient long-term immunity and cross-strain protection. Vaccine *14*, 85–92.

Levine, M.M., Ferreccio, C., Black, R.E., Lagos, R., San Martin, O., and Blackwelder, W.C. (2007). Ty 21a live oral typhoid vaccine and prevention of paratyphoid fever caused by *Salmonella enterica* serovar Paratyphic Clin. Infect. Dis. *45* (Suppl. 1), S24–8.

Levine, M.M., and Kaper, J.B. (1993). Live oral vaccines against cholera: an update. Vaccine *11*, 207–212.

Miehler, R., and Metcalfe, I. (2002). Inflexal a trivalent virosome subunit influenza vaccine: production. Vaccine *20*, B17–23.

Moss, B. (1991). Vaccinia virus: a tool for research and vaccine development. Science *252*, 1662–1667.

Muller, G.M., Shapira, M., and Arnon, R. (1982). Anti-influenza response achieved by immunization with a synthetic conjugate. Proc. Natl. Acad. Sci. U.S.A. *79*, 569–573.

Nabel, G.A., Gordon, D., Bishop, D.K., Nickoloff, B.J., Young, Z.Y., Aruga, A., Cameron, M.J., Nabel, E.G., and Chang, A.E. (1996). Immune response to human melanoma after transfer of an allogeneic class I major histocompatability complex gene with DNA–liposome complexes. Proc. Natl. Acad. Sci. U.S.A. *93*, 15388–15393.

Nardin, E.H., Oliveira, G.A., Calvo-Calle, J.M., Castro, Z.R., Nussenzweig, R.S., Schmeckpeper, B., Hall, B.F., Diggs, C., Bodison, S., and Edelman, R. (2000). Synthetic malaria peptide vaccine elicits high level of antibodies in vaccines of defined HLA genotypes. J. Infect. Dis. *182*, 1486–1496.

Ohlschlager, P., Quetting, M., Alvarez, G., Durst, M., Gissmann, L., and Kaufmann, A.M. (2009). Enhancement of immunogenicity of a therapeutic cervical cancer DNA-based vaccine by co-application of sequence-optimized genetic adjuvants. Int. J. Cancer *125*, 189–198.

Paas-Rosner, M., Dayan, M., Paas, Y., Changeux, J.-P., Wirguin, J., Sela, M., and Mozes, E. (2000). Oral administration of a dual analog of two myasthenogenic T cell epitopes down regulates experimental autoimmune myasthenia gravis in mice. Proc. Natl. Acad. Sci. U.S.A. 97, 2168–2173.

Pasteur, L. (1882). Une stalistique au sujit de la vaccination preventive contre de charbon, portant sur quatre vingt cingt cing mille animaux. CR Acad. Sci. Paris 95, 1250.

Patarroyo, M.E., Amador, Clavijo, R. et al. (1988). A synthetic vaccine protects humans against challenge with asexual blood stage of Plasmodium falciparum malaria. Nature 332, 158–161.

Pittman, M. (1986). Neurotoxicity of Bordetella pertussis. Neurotoxicity 7, 53–57.

Poon, L.L., Leung, Y.H., Nicholls, J.M., Perera, P.Y., Lichy, J.H., Yamamoto, M., Waldman, T.A., Peiris, J.S., and Perera, L.P. (2009). Vaccinia virus-based multivalent H5N1 avian influenza adjuvanted with IL-15 confer sterile cross-clade protection in mice. J. Immunol. 182, 3063–3071.

Raz, I., Avron, A., Tamir, M., Metzger, M., Symer, L., Eldor, R., Cohen, R.J., and Elias, D. (2007). Treatment of new-onset type I diabetes with peptide DiaPep277 is safe and associated with preserved beta-cell function: extension of a randomized double blind, phase II trial. Diabetes Metals Res. Rev. 23, 292–298.

Robinson, H.L., and Torres, C.A. (1997). DNA vaccines. Semin. Immunol. 9, 271–283.

Sanborn, W. (1987). Development of meningococcal vaccines. In Evolution of Meningococcal Disease, Vol. 2. Vedros, N.A., ed. (Boca Raton, FL, CRC Press), p. 121.

Sarin, P.S., Mora, C.A., Naylor, P.H., Markham, R., Schwartz, D., Kahn, J., Heseltine, P., Gazzard, B., Youle, M., and Rios, A. (1995). HIV-1, p17 synthetic peptide vaccine HGP-30: Induction of immune response in human subjects and preliminary evidence of protection against HIV challenge in SCID mice. Cell. Mol. Biol. 41, 401–407.

Schotsaert, M., De Filette, M., Fiers, W., and Saelens, X. (2009). Universal M2 ectodomain-based influenza A vaccines: Preclinical and clinical developments. Expert Rev. Vaccines 8, 499–508.

Schreiber, J.R., Barrus, V., Cates, K.L., and Siber, G.R. (1986). Functional characterization of human IgG, IgM and IgA antibodies directed to the capsule of Haemophillus influenzae type D. H. J. Infect. Dis. 153, 8–16.

Shatfield, S.N., and Dougan, G. (1997). Attenuated Salmonella as a live vector for expressing bacterial antigens. In New Generation Vaccines, Levine, M.M, Woodrow, G.C., Kaper, J.B., and Capon, G.S., ed. (New York, Marcel Decker, Inc), pp 331–341.

Shiu, C., Cunningham, C.K., Greenough, T., Muresan, P., Sanchez-Meino, V., Carey, V., Jackson, J.B., Ziemniak, C., Fox, L., Belzer, M., Ray, S.C., Luzuriaga, K., and Persaud, D. (2009). Identification of ongoing human immunodeficiency virus type 1 (HIV-1) replication in residual viremia during recombinant HIV-1 poxvirus immunizations in patients with clinically undetectable viral loads on durable suppressive highly active antiretroviral therapy. J. Virol. 83, 9731–9742.

Stocker, B.A.D., and Newton, S.M. (1994). Immune responses to epitopes inserted in Salmonella flagellin. Int. Rev. Immunol. 11, 167–178.

Takeda, K., Kaisho, T., and Akira, S. (2003). Toll-like receptor. Ann. Rev. Immunol. 21, 335–376.

Tang, D., Devit, M., and Johnston, S.A. (1992). Genetic immunization is a simple method for eliciting an immune response. Nature 356, 152–154.

Teitelbaum, D., Sela, M and Arnon. R. (1997). Copolymer 1 from the laboratory to FDA. Israel J. Med. Sci. 33, 280–284.

Tsang, K.Y., Zaremba, S., Nieroda, C.A., Zhu, M.Z., Hamilton, M., and Schlom, J. (1995). Generation of human cytotoxic T cells specific for human carcinoembryonic antigen epitopes from patients immunized with vaccinia-CEA vaccine. J. Natl. Cancer Inst. 87, 982–990.

Ulmer, J.B., Donnelly, J.J., Parker, S.E., Rhodes, G.H., Feigner, P.L., Dwarki, V.J., Gromkowski, R.R., Deck, R.R., DeWitt, C.M., Friedman, A., et al. (1993). Heterologous protection against influenza by injection of DNA encoding a viral protein. Science 259, 1745–1749.

Valenzuela, P., Medina, A., and Rutter, W.J. (1982). Synthesis and assembly of hepatitis B surface antigen particles in yeast. Nature 298, 347–350.

Vila-Corcoles A., Salsench, E., Rodriguez-Blanco, T., Utcho-Gondar, O., de Diego, C., Valdiviesco, A., Hoospital, I., Gomez-Bartomeu, F., and Raga, X. (2009). Clinical effectiveness of 23-valent pneumococcal polysaccharide vaccine against pneumonia in middle-aged and older adults: a matched case-controlled study. Vaccine 27, 1504–1510.

Ward. J., Lieberman, J.M., and Cochi, S.L. (1994). Haemophilus influenzae vaccines. In Vaccines, Plotkin, S.A., Mortimer, E.A., eds (Philadelphia, Saunders), pp. 337–386.

Weiss, R., Leitner, W.W., Scheiblhofer, S., Chan, D., Bernhaupt, A., Mostbock, S., Thalhamer, J., and Lyon, J.A. (2000). Genetic vaccination against malaria infection by intradermal and epidermal injections of a plasmid containing the gene of the Plasmodium berghei circumsporozoite protein. Infect. Immun. 68, 5914–5919.

Designing Vaccines in the Era of Genomics

2

Fabio Bagnoli, Nathalie Norais, Ilaria Ferlenghi, Maria Scarselli, Claudio Donati, Silvana Savino, Michèle A. Barocchi and Rino Rappuoli

Abstract

Genome sequencing has become routine, and modern vaccine design is taking advantage of the accumulating genomic information. Reverse vaccinology, an approach in our institute, is built on genome-based antigen discovery and has largely replaced classical vaccinology methods based on growing and dissecting the microorganism. The main advantage of the approach is the fast prediction of vaccine candidates. Most of the antigens will be surface exposed proteins, since these antigens are most likely accessible to antibodies. This approach can be applied to non-cultivable microorganisms, something difficult or impossible to do with conventional approaches. When the first reverse vaccinology project was started, in the year 2000, antigen identification was mainly based on bioinformatic analysis of one genome. Since then, the technique has shown its full potential, with the first genome-derived vaccine now in clinical trials and several vaccines in preclinical studies. In the meantime the approach has been improved with the support of proteomics, functional genomics and comparative genomics. Herein, we provide a description of the complete process: from antigen prediction to high-throughput purification, screening and selection of the vaccine composition. Furthermore, future applications of structural biology to vaccinology are discussed.

Introduction

Mass vaccination campaigns have dramatically decreased many infectious diseases globally. To date all the major vaccines (e.g. polio, small pox, pertussis and tetanus) were developed using classical vaccinology strategies (Rappuoli, 2004). Indeed, until a few years ago, vaccines were developed by attenuating or killing the infectious agents or isolating immunogenic subunits (see Chapter 1). Although very successful, in several instances these approaches were not able to deliver vaccines against certain pathogens and on other occasions the vaccines obtained with these classical approaches are no longer adequate due to safety concerns and low efficacy. Killed and attenuated vaccines, based on the whole organisms, may contain components that share homology with human antigens and thus have the potential to induce autoimmune reactions. Furthermore, several factors may have reactogenic activity and induce undesirable inflammatory response. The attenuated vaccines could also revert to the virulent status and chemicals used for inactivating pathogens could be present as traces in the final composition. In addition, classical methods used to identify protective subunits are slow and inefficient. The identification is done by fractionating microbial total extracts or culture supernatants using biochemical procedures followed by testing for a reaction with sera from convalescent patients. These components are then tested in appropriate *in vitro* or animal protection models. This strategy for vaccine candidate selection has proved to be laborious and time consuming, and relies on the assumption, which is not always true, that protective antigens are the ones that elicit immune responses during infection.

In the year 2000, a revolutionary approach to identify and develop new vaccines was published

(Pizza *et al.*, 2000). The strategy was called reverse vaccinology (RV) because instead of using experimental observations to identify vaccine candidates, antigen selection was based on a bioinformatic analysis of the pathogens' genome. The first proof of concept project was on the serogroup B *Neisseria meningitidis* (MenB), which led to the development of a multivalent recombinant vaccine that is now in phase III clinical trials (see Chapter 9). The MenB project demonstrated the strength, the speed and all the advantages of the RV methodology over classical vaccine approaches (Table 2.1). The feasibility of the approach relies on the availability of a pathogen's genome. The first complete genome sequence of a free-living organism, *Haemophilus influenzae*, was determined in 1995. Today, most human pathogens have multiple genomes deposited in the public databases. Therefore, the initial and basic information needed for a genome-derived vaccine project is available for most pathogens of interest. Furthermore, the availability of different genomes allows comparative genomics studies to be performed. Comparison of different strains of the same species has shown that significant variability exists between strains, suggesting that a single isolate is not sufficient to describe the genetic complexity of a species (Medini *et al.*, 2005; Tettelin *et al.*, 2005). Therefore, *in silico* antigen prediction is now performed on multiple genomes of the same species. The bioinformatic analysis, however, cannot predict the expression profile of the antigens.

During the past 10 years since the inception of RV, we have implemented several additional techniques to improve the strategy used to

Table 2.1 Comparison of conventional and reverse vaccinology approaches

	Conventional vaccinology	Reverse vaccinology
Essential features	Cultivable microorganisms only	Cultivable and non-cultivable microorganisms
	In vitro expressed antigens	All potential candidates are quickly identified
	Antigens immunogenic during disease	HT antigens expression and screening important
	Animal models essential	Human immunogenic profile of all candidates
		Animal models essential
Advantages	Polysaccharides may be used as antigens	Fast access to virtually all potential candidate
	Lipopolysaccharide-based vaccines are possible	Non-cultivable microorganisms can be approached
		Antigen abundance can be determined
		Antigens transiently expressed during disease can be identified
		Antigens not expressed *in vitro* can be identified
Disadvantages	Long time required for antigen identification	Non-protein antigens are not selected
	Antigen selection is based on too few criteria	
	Safety issues	
	Applicable to cultivable pathogens only	
	Antigens expressed *in vitro* only	

predict vaccine candidates. Importantly, similarly to bioinformatics these methods can be applied without the need of antibodies against the antigen, and can therefore proceed in parallel with the *in silico* analysis. In particular, we established a mass spectrometry-based analyses that allow the identification of proteins expressed on the surface of the pathogen. While, of course, with bioinformatics alone we cannot ascertain if a protein is actually expressed and/or accessible on the surface of the pathogen. In addition, the mass spectrometry-based method can identify secreted factors that are often difficult to predict with the *in silico* screening. Another technique used to identify antigens actually expressed by a pathogen is based on DNA microarray analysis (functional genomics). The strength of this method, once the DNA chip has been generated, is that many samples can be analysed in a relatively short time. Therefore, several growth conditions and host–pathogen interaction studies can be performed in order to identify antigenic candidates most likely expressed during infection. However, this method is not helpful in understanding protein subcellular localization or protein expression levels as can be done with proteomics-based approaches. Importantly, by combining all the prediction methods (bioinformatics, comparative genomics, proteomics and DNA microarray analysis) we can reduce the number of antigens that need to be screened in animal models.

Once the vaccine candidates have been predicted, the second part of the process starts in which the proteins are expressed, utilizing high-throughput technologies, and then screened by several methods. The major assays typically used in the screening process comprise molecular epidemiological analyses, immunoreactivity with human sera and, of course, immunogenicity and protection studies using *in vitro* assays or animal models. From one pathogen to the other, the relevance of the different analyses changes. For examples, for MenB there is no reliable animal model available, therefore most of the screening has been based on an *in vitro* bactericidal activity assay. The antigens that perform the best in these assays are then screened for their epidemiological relevance and their expression profile. These tests are usually done on a number of relevant strains and they aim to determine an antigen's presence and conservation as well as its expression profile. Indeed, the expression level of one antigen can vary considerably between different isolates. Another important aspect of the screening process is the determination of cytotoxicity associated with the candidates. Before going too far with the vaccine development process, antigens are detoxified by site-directed mutagenesis.

At this stage, the number of antigens should have been reduced to a number that is manageable for functional characterization. These studies are based on different techniques that depend on the kind of antigens that have been selected, ranging from biochemical studies to electron or confocal microscopy. However, they typically require the generation of knockout mutants to show the role of the candidates.

At the end of this second phase, from one to five antigens are usually selected for the vaccine composition. The smaller the number of candidates, the simpler the vaccine will be to development and manufacturing. However, in most cases more than one antigen is included in the final vaccine formulation. In fact, during the development of the acellular pertussis vaccines it was shown that protection increased when more antigens were included in the vaccine (Robbins *et al.*, 2005). In addition, during the development of anti-retroviral therapy it was shown that escape mutants against one drug were very common, while escape mutants against three drugs are very rare (Lisziewicz *et al.*, 2000). Before moving from research to development, the candidates need to be 'validated' for the use in humans. This last phase of the process includes proper formulation, confirmatory experiments with tag-less proteins (indeed, the screening is typically done using tagged proteins) and toxicity studies in animals. Concerning formulation studies, most of the preclinical research is still done using alum since it is the most common adjuvant used in humans. However, today novel and more potent adjuvants are becoming available and therefore, an alternative adjuvant is usually also tested in preclinical studies. An adjuvant which has already been licensed in Europe and has been shown to have very good safety and adjuvanticity profiles is the oil in water emulsion MF59 (Ott *et al.*, 1995; Tritto *et al.*, 2009). The importance of testing different adjuvants also depends on the disease. For example, formulation studies depend mainly on the defined target population. Indeed, children, the elderly and immunocompromised can have different requirements from those of healthy adult populations.

During the screening process, immunizations are performed with a fixed amount of the antigen that can vary between 5 and 20 µg per dose. However, a dose range study is done on the final vaccine composition to determine the most appropriate dose for each component. Furthermore, proteins generated with the high-throughput approach contain amino acid sequence tags (usually His tag). For regulatory reasons, the tag must be removed. Hence, the final composition is also tested using tag-less versions of the proteins. In addition, before moving the composition to development, the composition is tested for toxicity studies in animals. Most of the work performed so far using RV has focused on bacterial pathogens. However, the same principles are applicable to any kind of microorganism, including fungi, parasites and viruses. Programmes on fungi and parasites are being evaluated; however, they are still in the early stages of investigation (Kanoi and Egwang, 2007). On the other hand, for most of the viruses, the major problem in developing effective vaccines is not antigen identification but high variability of the surface proteins. Nevertheless, a genome-based approach is instrumental also for viruses. Indeed, the virus responsible for hepatitis C (HCV) was cultivated *in vitro* for the first time only in 2005 (Lindenbach *et al.*, 2005). Therefore, any conventional approach to vaccine development was not applicable before this publication. However, in 1990, the availability of the genome sequence of HCV finally enabled the identification of the envelope proteins E1 and E2, which were then shown to protect chimpanzees from infection with the homologous HCV virus. For other viruses such as HIV, rhinovirus, Dengue, and influenza the development of effective universal vaccines has thus far been impaired by high antigenic diversity (Rappuoli, 2007). In this case, bioinformatics can help to identify putative T-cell epitopes from all known sequences. For Dengue virus it has been found that a limited number of epitopes represents the entire antigenic repertoire recognized by T cells. Given the high variability of many viruses (e.g. HIV and influenza virus), it will be necessary to construct chimeras composed of different epitopes. The same is true for B-cell epitopes. It was recently proposed that antibodies recognizing conserved epitopes could provide immunity to HIV or to diverse influenza subtypes and future pandemic viruses (Ekiert *et al.*, 2009; Zhou *et al.*, 2007). Structural determination of antigen–antibody complexes by X-ray crystallography or electron cryomicroscopy has shown that epitopes recognized by antibodies are formed by large patches on most antigens (Dormitzer *et al.*, 2008). The next step is the development of principles to perform structure-based design of vaccine antigens.

Vaccinology is rapidly evolving by exploiting some of the most advanced and diverse scientific concepts, knowledge and technologies, borrowed from bioinformatics, mathematics, immunology, molecular biology, cell biology, microbiology, systems biology and biochemistry. We expect that in the near future the identification of protective candidates will become even more straightforward. One of the major expectations for the future is the realization of methods to predict the protective domains of antigens. This will most likely depend on structural biology, at least for those pathogens against which protection appears to be mainly antibody-mediated. It is known that antibodies most commonly recognize conformational epitopes. Therefore, *in silico* three-dimensional reconstruction to identify conformational domains conserved in protective antigens might represent, after RV, another revolution in vaccine discovery.

The chapter is divided in three parts. The first two correspond to the flow of the RV process. As illustrated in Fig. 2.1 and discussed above, the process has a prediction phase (part 1), mainly based on bioinformatics, comparative genomics, proteomics and functional genomics, followed by the screening and generation of the antigens (part 2). The last part illustrates future directions of RV.

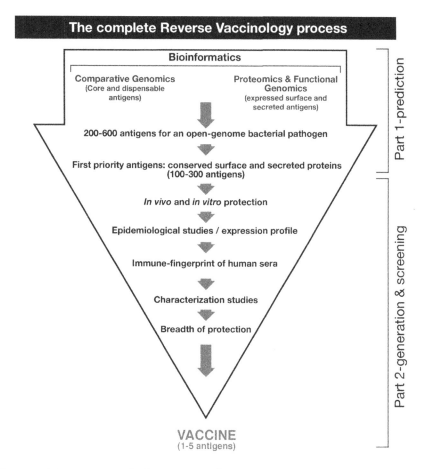

Figure 2.1 The complete reverse vaccinology process. Reverse vaccinology starts with antigen prediction based on *in silico* analysis as well as comparative genomics and functional genomics. The availability of genomes and bioinformatic tools is essential to the *in silico* prediction as well as comparative and functional genomics and proteomics. For an open-pangenome, typically 100–300 antigens are chosen for the second part of the process. During this phase, the antigens are purified and screened for protection efficacy, epidemiological relevance, expression profile, immunoreactivity against human sera. They are also characterized for their function and toxicity. Ideally, a vaccine should comprise from one to five different antigens.

Antigen prediction

This part of the chapter describes the process and the techniques used to predict vaccine candidates.

The basic principles of *in silico* antigen prediction

In the late 1990s, when RV was initially developed, the prediction of putative vaccine candidates was based solely on *in silico* analysis of the genetic information from one strain. Additional selection criteria have been implemented since then, however, *in silico* analysis remains the central step in an RV project (Fig. 2.1).

The first step of the analysis (usually referred to as annotation) consists of the prediction and localization of genes within the genome. This is accomplished by prediction algorithms, which scan the sequence in search of regions that are likely to encode proteins. In prokaryotic systems, the identification of potential coding region or open reading frames (ORFs) relies on a few basic rules. In the simplest formulation, ORFs are identified as segments of the same reading frame contained between one of the three standard start codons (ATG, TTG, GTG) and one of the three standard stop codons (TAA, TAG, TGA).

It is generally accepted that there is approximately one gene every 1000 DNA base pairs; this implies that significantly long start-to-stop segments are likely to encode for proteins.

The genome annotation procedure can be automated to different extents. Automated methods for prokaryotic gene finding like GLIMMER (Delcher *et al.*, 1999), ORPHEUS (Frishman *et al.*,1998a,b) and GeneMark (Besemer *et al.*, 2001) have been used in genome sequencing projects (Fitz-Gibbon *et al.*, 2002; Cerdeno-Tarraga *et al.*, 2003; Wei *et al.*, 2003; McLeod *et al.*, 2004). GLIMMER and GeneMark use interpolated Markov models, while ORPHEUS is mainly based on codon usage and ribosome binding site statistics derived from annotated genes (Box 2.1).

An exhaustive summary of software tools and websites that can be used to obtain bacterial genome annotations has recently been published (Stothard and Wishart, 2006).

The annotation procedure allows the translation of the bacterial genome sequence into a systematic list of all the proteins that a bacterium has the potential to express at any time of its life cycle. Each of these amino acid sequences can then be compared with the content of public databases of proteins or DNA sequences in an attempt to identify related sequences. In the case of obvious sequence similarity, it is reasonable to assign a predicted function to the query. A limiting aspect of this approach is represented by sequences having no or remote homologues filed in databases. ORFs encoding proteins having 20% or less amino acid identity to any sequence found in databases are generally considered to have no reliable homologues. These could represent novel uncharacterized proteins, non-coding regions misidentified as genes. Although homology searches typically identify ORFs that more likely encode functional proteins, the experimental authentication by proteomic techniques is a more powerful approach to distinguish genes from random non-coding ORFs.

The ensemble of hypothetical proteins can be processed with dedicated software to deduce their possible cellular localization. One of the basic assumptions for the search of candidates is that a good antigen is located at the bacterial surface where it is available for antibody recognition.

Several prediction algorithms have been developed to predict the subcellular localization of proteins exclusively on the basis of amino acid sequence and composition (Table 2.2). The basic assumption is that the N-terminal of a protein contains an indication of its cellular destination. The presence of a leader sequence provides evidence that a protein will be exported to extracytoplasmic compartments. Additional signatures are exploited by the major part of these expert systems, for example the presence of a cleavage site immediately after the leader peptide, which generally implies that the protein is released to the extracellular environment of Gram-positive bacteria or into the periplasmic space of Gram negatives. Similarly, proteins that contain a LXXC (where X is any amino acid) motif positioned at the end of the leader peptide are likely to be lipoproteins. The anchoring of a protein to a Gram-positive cell wall requires a specific carboxy-terminal sorting sequence identified by an LPXTG motif followed by approximately 20 hydrophobic amino acids and a charged tail.

In Gram-negative bacteria, additional molecular machineries assist the outer membrane crossing of extracellular proteins. There are at least six distinct extracellular protein secretion systems, reported as type I through VI (T1SS–T6SS), that can deliver proteins through the multilayered bacterial cell membrane and sometimes directly into the target host cell (Economou *et al.*, 2006). This increases the variety and complexity of export signals used by the different secretion mechanisms, sometimes making the identification of outer membrane and secreted proteins a challenging task. Several computational methods have been proposed to predict extracellular proteins in Gram-negative microorganisms (Gardy and Brinkman, 2006) and Psort_B is the most widely used tool for predicting protein localization in Gram-negative bacteria. It uses biological knowledge to elaborate 'if-then' rules, combining information on amino acid composition, similarity to proteins of known localization, presence of signal peptides, transmembrane helices and motifs indicative of specific localization. More recently, two other predictive methods CELLO (Yu *et al.*, 2004) and Proteome Analyst (Lu *et al.*, 2004) have been proposed for Gram-negative

Box 2.1 A summary of the available computational resources for protein annotation

Algorithms for gene identification

ORFinder	Web tool that finds potential ORFs in an RNA sequence	www.nebi.nlm.gov/gorf/gorf.html
Glimmer	Gene locator and interpolated Markov modeller uses interpolated Markov models (IMMs) to identify DNA coding regions	http://www.cbcb.umd.edu/software/glimmer
Geneprediction	A comprehensive web collection of gene finders	http://www.geneprediction.org/software.html

Databases

Nebi Entrez	Ewb portal to the GenBank inventory of protein and nucleic acid sequences	www.ncbi.nlm.nih.gove/Entrez
UniProt	A curated protein sequence database	http://www.expasy.ch/sprot/sprot-top.html
EMBL-Bank	Nucleotide sequence database curated by the European Molecular Biology Laboratory	http://www.ebi.ac.uk/embl/
PROSITE	Database of protein domains, families and functional sites	http://www.expasy.ch/prosite/
PFAM	A large collection of protein families	http://pfam.sanger.ac.uk/
COG	The database of clusters of orthologous groups of proteins	http://www.ncbi.nlm.nih.gov/COG

Search engines

BLAST	The Basic Local Alignment Search Tool available at NCBI	www.ncbi.nlm.nih.gov/BLAST/
FASTA	Acronym of FAST-ALL, reflecting the fact that it can be used for a fast protein comparison or a fast nucleotide comparison	http://www.ebi.ac.uk/Tools/fasta33index.html

Table 2.2 Survey of the algorithms widely used to predict the subcellular localization of bacterial proteins

Method	URL	Reterence
PsortB	http://www.psort.org/psortb/index.html	Gardy (2005)
SignalP	http://www.cbs.dtu.dk/services/SignalP/	Emanuelsson (2007)
Cello	http://cello.life.nctu.edu.tw/	Yu (2004)
PSLpred	http://www.imtech.res.in/raghava/pslpred/	Bahsin (2005)
Proteome Analyst	http://www.cs.ualberta.ca/~bioinfo/PA/Sub/	Lu (2004)
LOCtree	http://cubic.bioc.columbia.edu/services/loctree/	Nair (2008)
AdaBoost	http://chemdata.shu.edu.cn/subcell/	Niu (2008)

bacteria, both providing comparable performances in terms of accuracy and recall with respect to Psort_B (Gardy and Brinkman, 2006).

Despite the recent progresses, *in silico* identification of secretion system components and their effectors still mainly relies on detection of sequence (Tseng *et al.*, 2009) and structural (Tampakaki *et al.*, 2004) similarities. Some caution is necessary when performing such predictions, as sequence similarities can be very weak and not necessarily imply any functional analogy.

In conclusion, availability of genome sequences allows the selection, using bioinformatic tools, of a list of potential antigens without the need of cultivating the microorganism (Fig. 2.2). This is a huge advantage over conventional vaccinology approaches for two major reasons. First of all, *in silico* analysis is very fast and cheap. Secondly, proteins not expressed in *in vitro* conditions are also identified. On the other hand, this analysis provides only a prediction of protein localization and it does not indicate if a protein is expressed and in which conditions. Therefore, the bioinformatic approach has been complemented with other techniques. In particular, with mass spectrometry-based approaches to aid vaccine candidate prediction as discussed later in this chapter. The first RV project was based on a single genome. Indeed, at that time there was only one genome available for MenB. However, in 2009 for most human pathogens there are typically more than five genomes available. Therefore, *in silico* analysis can easily take advantage of comparative genomics.

Comparative genomics and the pangenome concept

Today, as the number of sequenced microbial genomes exceeds 930, it is clear that microbial diversity has been vastly underestimated and a single genome does not exhaust the genomic diversity of any bacterial species. In many cases, there is extensive genomic plasticity; for example, completion of the genome sequence of *E. coli* O157:H7 revealed that this strain possesses >1300 strain-specific genes when compared to *E. coli* K12, that encode proteins that are involved in virulence and metabolic capabilities (Hayashi *et al.*, 2001). Other reports have also revealed an extensive amount of genomic diversity among strains of a single species (Perna *et al.*, 2001).

These early findings were formalized with the definition of the bacterial pangenome as the sum of the genes present in each individual of a species. This concept was originally introduced during the study of the genome variability of eight isolates of *Streptococcus agalactiae* (group B streptococcus, GBS), when it was found that each new sequence contained an average of 30 genes that were not present in any of the previously sequenced genomes. Thus, by mathematical extrapolation, it was predicted that the size of the pangenome of this species continues to grow as the number of sequenced strains increases. Not every bacterial species has same level of complexity as GBS. For instance, the pangenome of *Bacillus anthracis* can be adequately described by just four genome sequences. Hence, scientists refer to certain species as having 'closed' while to others, such as GBS, as having 'open' pangenomes. In species with an open pangenome, an unlimited number of new genes are found by sequencing new genomes, while in the closed pangenomes there are a limited number of strain-specific genes. This difference in the nature of the pangenome reflects several factors, including: the different lifestyles of the two organisms (exposure of GBS to diverse environments versus the occupation of a more isolated biological niche by *B. anthracis*); the ability of the species to acquire and stably incorporate foreign DNA, an advantage in niche adaptation from the acquisition of laterally transferred DNA; and the recent evolutionary history of the species.

The pangenome can be divided into three elements: a core genome that is shared by all strains; a set of dispensable genes that are shared by some but not all isolates; and a set of strain-specific genes that are unique to one isolate. The core genome of GBS encodes the basic functions of the microorganism and represents approximately 80% of the genome of each single isolate. Conversely, the dispensable and strain-specific genes, which are largely composed of hypothetical, phage-related and transposon-related genes (Tettelin *et al.*, 2006), contribute to its genetic diversity. The concept of the pangenome and comparative genomics has practical applications in vaccine research. In fact, while obviously the

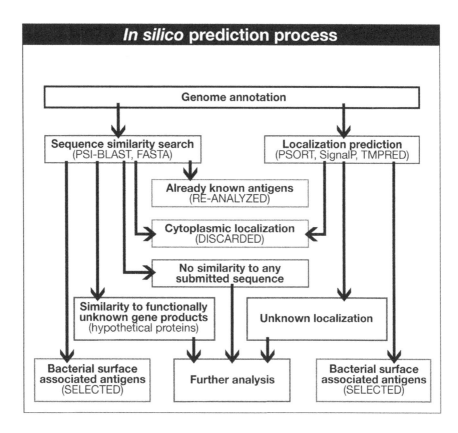

Figure 2.2 Strategy for the identification of vaccine candidates *in silico*. The selection starts from the annotation of all the open reading frames (ORFs) contained in the bacterial genome, from which a subset of potential surface-associated proteins is derived.

ideal vaccine candidate is a conserved protein encoded by a gene present in every isolate of the species, in the case of GBS it was shown that the design of a universal protein-based vaccine against GBS was only possible using dispensable genes (Maione *et al.*, 2005; Ochman and Moran, 2001). Of note, capsular specificity genes and other pathogenicity traits are often identified in the accessory genome. Moving forward, bacterial taxonomy and epidemiology must consider the whole genome sequences instead of considering just a few genetic loci, as is the case with methods based on ribosomal RNA sequences, capsular typing, or multilocus sequence typing (MLST). Indeed, comparison of the whole genome sequences of GBS strains has shown that the genomic diversity does not perfectly correlate with serotypes or MLST sequence types. Employing whole genome sequences will mean that epidemiology will have a more reliable and

systematic correlation between strains and disease, and will allow a standardization of the classification of clinical isolates. These observations are instrumental for the selection of a broad-strain protective vaccine.

Comparative genomics is also important for the identification of pathogenic factors and, of course, these factors could represent good vaccine candidates. The level of distinction and the function played by carrier versus virulent strains of streptococci and neisseriae, for example, has been a matter of discussion for a long time but still lacks an answer. There is, as yet, no clear and strict correlation between the presence of virulence factors and disease caused by these organisms, and epidemiological data are vague. Indeed, it is likely that in species that only rarely cause disease, there are multiple virulence factors and not only one or a few that are uniquely associated with infection. Therefore, comparative genomics can be used to identify the 'pathogenicity signature' associated with the most virulent strains. It is expected that most virulence factors will be found in the accessory genome, at the least the ones that determine increased pathogenicity. However, comparative genomics will not identify expression variability that can also contribute to the different virulence of the strains. Hence, functional studies are still critical to shed light on the relevance of specific virulence factors.

Another potential application of comparative genomics could be the study of certain species of bacterial symbionts, such as mollicutes, rickettsiae and chlamydiae. These species, instead of acquiring genes during evolution, have actually lost a large proportion of genetic information (Ochman and Moran, 2001). In particular, several biosynthetic pathway genes have lost because intracellular bacteria have a relatively constant environment with access to much of what they require for survival, such as amino acids. By applying the concept of pangenomics to these species, we would obtain a 'micro-genome' representative of the set of genes necessary to live in the intracellular niche. Comparing this microgenome with the pangenome of free-living species would likely simplify the identification of genes that are necessary for the microorganism to survive in more variable and difficult environments.

Comparative genomics has become incredibly easy to do in comparison to just a half decade ago given the large number of genomes of different strains available for the most important human pathogens. Therefore, for a new RV study starting today, the level of conservation of selected antigens within a species can be determined from the very beginning.

Proteomics

Ideally, good vaccine candidates, besides being exposed on the bacterial surface should also be abundant, conserved among a large panel of strains, and expressed during infection. Although bioinformatic tools can predict the cellular location of bacterial proteins using homology searches and identification of sorting domains, they might not always provide a perfect indication of protein subcellular localization. Protein abundance, post translational modifications and

surface protein–protein interactions are relevant information for vaccine design that cannot be assessed by bioinformatics. In the last decade, development of mass spectrometers and publication of bacterial genomes have allowed the emergence of the proteomic field. Application of proteomic approaches to bacterial vaccine development has been established and often consists of the identification of the surface proteins, the "surfaceome" (Cullen *et al.*, 2005) or 'surfome' (Doro *et al.*, 2009), of the bacteria. In the first attempts, the methodology was based on protein separation by bi-dimensional electrophoresis (2-DE), and identification of the separated proteins by mass spectrometry (MS), matrix-assisted laser desorption/ionization time-of-flight mass spectrometry (MALDI-TOF MS). Unfortunately, many membrane-associated proteins are hydrophobic and, making them more challenging to study than soluble proteins. Gel-free methods have been developed to overcome many of the problems associated with 2-DE and to provide a comprehensive analysis of bacterial membrane proteins (Wu *et al.*, 2003; Wu and Yates, 2003). Gel-free methods usually utilize two-dimensional liquid chromatography (2-DLC), which frequently involves strong cation exchange in the first dimension followed by reverse-phase chromatography in the second dimension, and tandem mass spectrometry (MS/MS) to separate and identify peptides that have been generated from protease digestion of complex protein mixtures. The 2-DLC-MS/MS approach performs better than 2-DE analysis in processing proteins with the following characteristics: highly hydrophobic or highly basic, poorly expressed, high molecular weight, and/or extreme isoelectric points (Cordwell, 2006).

One of the major challenges in using proteomics for the prediction of vaccine candidates remains the difficulty in isolating the cellular compartment of interest, free of contamination from the other cellular fractions. Because of the different cellular organization of Gram-positive and Gram-negative bacteria, different protocols have been developed to purify either the Gram-positive cell wall or Gram-negative outer membrane fractions. In Gram-positive bacteria, cell wall-associated proteins are usually released after digestion with lysozyme, mutanolyzyn and/ or lysostaphin in the presence of an osmotic protective agent. One of the main limitations of this approach is the high level of contamination with cytoplasmic proteins that makes the assignment of cell wall-associated proteins uncertain. The resistance of Gram-positive bacteria to lysis has permitted the development of an attractive approach for the identification of surface exposed proteins. The procedure is based on the use of a proteolytic enzyme to 'shave' the surface-exposed proteins of live bacteria in conditions that avoid bacterial lysis. The released peptides are then identified by MS/MS (Fig. 2.3). When applied to M1 SF370, a virulent strain of *Streptococcus pyogenes* (group A streptococcus, GAS), 72 surface-exposed proteins were identified. Ninety-five per cent of these proteins were previously predicted to be cell wall anchored proteins, lipoproteins, transmembrane or secreted proteins (Rodriguez-Ortega *et al.*, 2006). Furthermore, this method also allows a semi-quantitative evaluation of protein surface expression. Indeed, the number of peptides identified for a given protein correlates with its expression level as determined by flow cytometry (Rodriguez-Ortega *et al.*, 2006). Moreover, seven of the eleven reported GAS protective antigens were found in the surface proteome. In addition to the reported protective antigens, this method allowed the identification of a new protective antigen, the Spy0416 protein (Rodriguez-Ortega *et al.*, 2006).

In order to demonstrate that the proteomics-based approach represents a reliable and generally applicable strategy for the identification of vaccine components in Gram-positive bacteria, the same protocol has been applied to the hypervirulent GBS strain COH1, that is strongly associated to invasive infections in neonates (Jones *et al.*, 2003; Luan *et al.*, 2005) and therefore of particular interest for vaccine research. Forty-three major proteins belonging to the families of cell-wall proteins, lipoproteins and membrane proteins have been identified. The identified proteins comprise all of the protective antigens described to date in the literature. Moreover, one particular protein, SAN_1485, appeared particularly attractive since it was identified with 25 peptide matches, indicating that it is highly expressed and exposed on the

surface. All peptides fall in the N-terminal part of the molecule. The antigen domain covered by the peptides released from the surface digestion, was expressed in *E. coli* and the polypeptide was used for protection studies using the maternal immunization-neonatal pup challenge mouse model. Over 80% of the offspring generated by immunized adult female mice survived a LD_{90} challenge dose with the two hypervirulent strains COH1 and M781 belonging to the same genetic lineage RDP III-3/ST-17 (Seifert *et al.*, 2006). Therefore, as in the case of GAS, the surface proteome analysis led to a rapid detection of a new, promising protective antigen against GBS infection.

It should also be emphasized that this methodology not only allows the selection of potential protective antigens, but it also identifies the surface-exposed domains of the antigen. This information is important for vaccine design since it can contribute to the ability to recover proteins that cannot be expressed in a full length recombinant form and it may also enable non-protective immunodominant epitopes to be removed. Furthermore, it is also useful to design chimeric proteins composed of surface exposed domains. This strategy allows several domains of different antigens to be fused without the disadvantage of constructing very high molecular weight proteins.

The surfome approach has also been successfully applied to other Gram-positive bacteria by several groups providing promising results for vaccine candidate selection (Rodriguez-Ortega *et al.*, 2006, 2008; Severin *et al.*, 2007; Tjalsma *et al.*, 2008).

Unfortunately, it has been recently reported that the approach failed when applied to the Gram-negative uropathogenic *E. coli* (Walters and Mobley, 2009). The thinner cell walls of Gram-negative *E. coli* may make them more susceptible to lysis during digestion of surface-exposed proteins. Although the authors reported that lysis did not appear to be significant based on determination of CFU pre- and post-digestion, the amount of cytoplasmic proteins released into the sample from even a small number of bacteria lysing may quickly overwhelm the amount of peptides released from surface exposed domains of outer membrane proteins. Bacterial lysis was

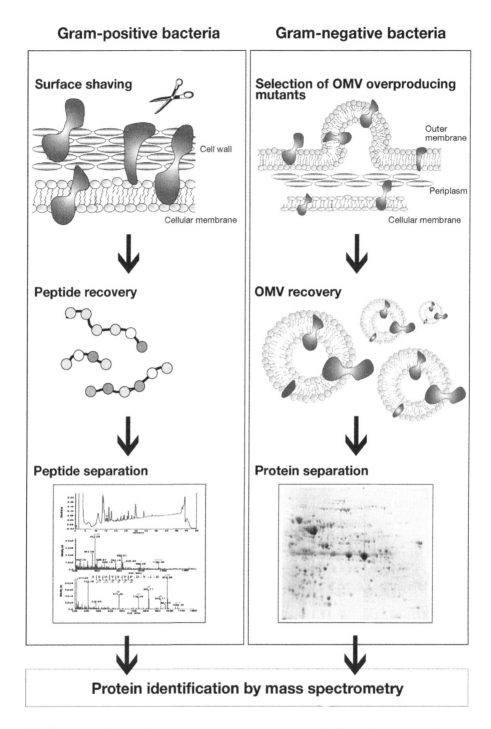

Figure 2.3 Representation of the proteomics strategy used to identify surface-exposed proteins. The methodology developed for Gram-positive bacteria consists of the use of proteases to 'shave' surface-exposed proteins (left panel). Peptides released by the digestion are recovered and then separated and analysed by LC-MS/MS. The methodology developed for Gram-negative bacteria is based on the use of OMV hyperproductive mutants (right panel). The OMVs are recovered, OMV proteins are separated by 2DE and protein spots are identified by MS. An alternative is the digestion of the OMV proteins followed by the separation and analysis of the generated peptides by LC-MS/MS. A colour version of this figure is located in the plate section at the back of the book.

also observed when the methodology was applied to *Neisseria meningitidis*, extraintestinal pathogenic *E. coli* (ExPEC) and *Chlamydia trachomatis* (Norais, personal communication).

A classical way to purify Gram-negative outer membrane proteins consists of using membrane detergents or carbonate extraction (Molloy *et al.*, 2000; Nouwens *et al.*, 2000; Phadke *et al.*, 2001). Although, MenB outer membrane vesicles (OMV) prepared from a deoxycholate extraction of the bacteria have been assessed in efficacy trials to fight MenB infections in Cuba (Sierra *et al.*, 1991), Brazil (de Moraes *et al.*, 1992), New Zealand (Thornton *et al.*, 2006) and Norway (Bjune *et al.*, 1991) with promising results (Holst *et al.*, 2009), this material is not a material of choice for the prediction of vaccine candidates. In fact, proteomic analyses of these vaccines have been reported by different research groups (Ferrari *et al.*, 2006; Vaughan *et al.*, 2006; Vipond *et al.*, 2006). All reported a high level of cytoplasmic and inner membrane proteins. Moreover the ratio of outer membrane proteins present in such vesicles was relatively low compared to native OMVs (Ferrari *et al.*, 2006). A wide variety of Gram-negative bacteria constitutively secrete OMVs during growth (Beveridge, 1999; Devoe and Gilchrist, 1973; Fiocca *et al.*, 1999; Hoekstra *et al.*, 1976; Kadurugamuwa and Beveridge, 1995). Although, the function of OMVs is not well clarified (Mashburn-Warren *et al.*, 2008), they represent a material of choice to identify surface antigens since they are mainly composed of outer membrane proteins and periplasmic proteins encapsulated during vesicles formation but not of cytoplasmic and inner membrane proteins (Berlanda Scorza *et al.*, 2008; Ferrari *et al.*, 2006). The amount of OMVs released in the liquid culture is usually quite minute, and this has prevented their detailed biochemical characterization. In an attempt to increase the amount of OMV production in *N. meningitidis*, a panel of mutants was screened and it was found that inactivation of the *gna33* gene encoding a lytic transglycosylase, an enzyme involved in the integrity of the outer membrane, resulted in massive production of OMVs (Adu-Bobie *et al.*, 2004; Ferrari *et al.*, 2006). The proteomic analysis of the vesicles allowed the identification of 60 proteins,

most of which were classified as outer membrane or periplasmic proteins. In this context, it is worth pointing out that four of the five antigens selected by RV, which are currently in the clinic to test their efficacy as human vaccine (Giuliani *et al.*, 2006), were identified in these vesicles. This result highlights the usefulness of OMV protein characterization as an effective and rapid approach to vaccine candidate discovery (Fig. 2.3). Production of OMVs by non-pathogenic *E. coli* strains mutated in proteins that interact with the murein layer and that form complexes crosslinking the inner and outer membranes have been described a few decades ago. In particular, OMV producing mutants were generated by inactivating genes coding for the Braun's lipoprotein and the proteins belonging to the Tol-Pal system (Bernadac *et al.*, 1998). Recently, McBroom *et al.* (2006) reported that the production of OMV is increased from 5- to 200-fold compared with wild-type levels after the deletion of different families of proteins. However, little proteomic data have been reported from these OMVs (Lee *et al.*, 2008). In order to identify outer membrane proteins for vaccine candidates against ExPEC, a Δ*tolR* mutant of the pathogenic IHE3034 strain has been created, which spontaneously releases a large quantity of OMVs (Berlanda Scorza *et al.*, 2008). The proteomic analysis of this strain led to the identification of one hundred proteins, most of which are localized to the outer membrane and periplasmic compartments. Based on the genome sequences available in the current public databases, seven of the identified proteins appear to be specific for pathogenic *E. coli* and enteric bacteria and therefore are potential targets for vaccine and drug development (Berlanda Scorza *et al.*, 2008).

Nevertheless, in parallel with the concept of pan-genomics and comparative genomics recently described, we are also developing a pan-proteomics-based approach. Antigens suitable for effective vaccines should not only be highly conserved, but should also be equally accessible on the bacterial surface of a large panel of pathogenic strains. Examples of highly conserved antigens presenting variable surface accessibility have been reported (Arenas *et al.*, 2008; Doro *et al.*, 2009). Expression variability and antigen accessibility can depend on the strain but also on the growth conditions. In fact, protein composition of bacteria is constantly adjusted to facilitate survival and growth in ever-changing environments. This point highlights the other challenging aspect of the use of the proteomics for vaccine selection, which is the definition of the bacterial growth conditions that mimic the natural niche of the pathogen in the host.

The pan-proteome should allow the definition of the core and dispensable pool of surface accessible proteins by analysing different strains and the same strain grown in different conditions. This approach, therefore, will allow to determine the set of genes expressed from every strain in every growth conditions as well as the proteins that are more variably expressed.

The development of mass spectrometers that allow the fragmentation of a large number of peptides in association with the improvements in automation, the management of large size databases, and the development of quantitative proteomics, have made the concept of the pan-proteome a reality, and very soon results from the first studies should be available.

New aspects related to the use of proteomics for vaccine design are currently emerging. They are related to the identification of antigen post-translational modifications and the definition of antigen conformation including their involvement in surface-exposed protein complexes.

Traditionally, post-translational modifications have been considered the exclusive property of eukaryotes, but it is now evident that they also play an important role in microorganisms. For example in neisserial species, since the first O-glycosylated pilin protein was reported in 1995, 11 other O-glycoproteins have been reported (Vik *et al.*, 2009). In addition to glycosylation, phosphorylation (Forest *et al.*, 1999), and adduction of phosphoethanolamine, phosphocholine (Hegge *et al.*, 2004; Weiser *et al.*, 1998) and α-glycerophosphate. (Stimson *et al.*, 1996) have been reported on the same family of proteins. Numerous other pathogens have been shown to have post-translationally modified proteins. N-linked glycosylation was reported in bacteria for the first time with the characterization of an N-linked glycan, attached via the eukaryotic Asn-X-Ser/Thr consensus sequence, on multiple proteins in *Campylobacter jejuni* (Young *et al.*, 2002).

Unfortunately, few reports evidence how post-translational modifications could influence antigen immunogenicity and host recognition. The Apa deglycosylated antigen was less active than native molecules in eliciting protective immune response against BCG in animal models (Romain *et al.*, 1999). Evidence suggesting that *Pseudomonas* glycosylated pili provide O-antigen-specific protection via the mucosal and systemic routes of immunity has recently been reported (Horzempa *et al.*, 2008). Many early approaches that were developed to identify peptides and proteins using amino acid sequence information from MS/MS and have been extended to identify modified peptides and proteins. However, many of these methods take into account only a few types of known modifications, ignoring all the others, and the development of new software exploring unknown modifications from MS/MS spectra should be implemented (Na and Paek, 2009).

The growing evidences of the importance of conformational epitopes in the immune response implicate that in order to develop the next generation of therapeutic and vaccine targets, it will be essential to characterize surface protein in their native conformation, including their participation to complexes and evaluate their relevance for the immune response. Epitopes rising from protein–protein interactions could not be induced by immunization with single or non-correlated recombinant proteins. The works provided by the group of CM Ferreirós, on the characterization of MenB PorA and PorB interactions show how the characterization of intact complexes is a crucial step in the search of potential vaccine candidates (Sanchez *et al.*, 2009). To our knowledge, a global characterization of bacterial surface exposed protein complexes has never been reported. Cross-linking of proteins is a powerful method to investigate protein conformation and interactions. A cross-link between two peptides is indicative of spatial proximity of the two linked amino acids at the time of cross-linking. To exploit this information, both peptides that are involved in a cross-link need to be identified. Although this methodology is usually used for single proteins or small, purified protein complexes, recent technical improvements have made the analysis of surface exposed proteins complexes feasible. These improvements include isotopically coded cross-linkers coupled to the improvement in mass spectrometry technology (in particular high mass precision spectrometers) and the creation of new software able to sort and identify cross-linked peptides from very large complex samples and large sequence database (Rinner *et al.*, 2008).

Generation and screening of selected antigens

This part of the chapter describes the process and the techniques used to clone and purify selected antigens in an efficient and fast manner. Indeed, for the RV process, high-throughput generation of the antigens is an essential requisite. Furthermore, this session illustrates how the purified candidates can be tested for immunogenicity/protection and for the epidemiological relevance.

High-throughput antigen generation

As discussed in the first part of this chapter, the antigen selection process is very quick and can lead to the identification of all the best predicted candidates in a matter of three months (Fig. 2.4). However, the generation of the antigens for a RV project normally requires a significant effort.

A solution to this problem is a reiterative approach in which all the selected antigens are processed through a high-speed first round. The antigens that are unexpressed, unsuccessfully purified, have an insufficient yield, or contain many impurities are processed through a second round. Finally, in a third round the most 'difficult' proteins (e.g. insoluble) are treated case by case. By doing so, the antigen generation process does not slow down the screening process since the proteins quickly obtained in the first round (typically 50% success rate) are usually enough to keep the downstream steps in the RV process running. Therefore, in parallel with the screening of the first round proteins, the second and third rounds of expression can produce the rest of the candidates (see Fig. 2.4).

The first step of the antigen generation process is the cloning of the selected genes.

Several strategies are now available to speedup and simplify the cloning. First of all, new cloning methods have been described that reduce the number of steps required to insert a PCR product into a plasmid. Compared with the traditional methods based on PCR/digestion/ligation/transformation, new methods comprise a smaller number of steps. For example site-specific recombination cloning techniques, which eliminate the use of restriction endonucleases and ligase, offer several advantages in the context of high throughput (HT) procedures (Betton, 2004). Another method suitable for high throughput and robotized procedures is the polymerase incomplete primer extension (PIPE) (Klock *et al.*, 2008). In this method the gene of interest is amplified using primers containing 5′ overhanging ends. Primers with complementary tails are used to amplify the cloning vector. In the last step the two unpurified PCR products are mixed and directly transformed into bacteria to create clones without any extra manipulations. The reactions can be done in 96 well plates; therefore, the process is easily amenable to robotization. Compared with traditional methods, estimated to produce about 30 genes in 3 days per operator, the PIPE method is much more efficacious with about 96 genes in 2 days per operator.

Another approach that is rapidly emerging is based on gene synthesis. In this case the gene of interest is not cloned from the DNA of the microorganism, but is chemically synthesized. The major drawback of the method is still the relevant cost per base pair. However, it can be useful and even cheaper in certain cases. Indeed, if the sequence of a gene needs to be modified by inserting different codons, or fused with other genes or if fragments of the sequence need to be deleted or added, PCR-based methods can be very laborious and inefficacious. Therefore, gene synthesis is presently considered a complementary technique to high-throughput cloning programmes. In addition to the cloning methods, it is now possible to use automatic cloning machines that eliminate much of the repetitive work needed for cloning.

In the first round of the HT-antigen generation process, the proteins are typically all cloned in the same vector with a His-tag. The hexahistidine tag is the most common, which binds to immobilized

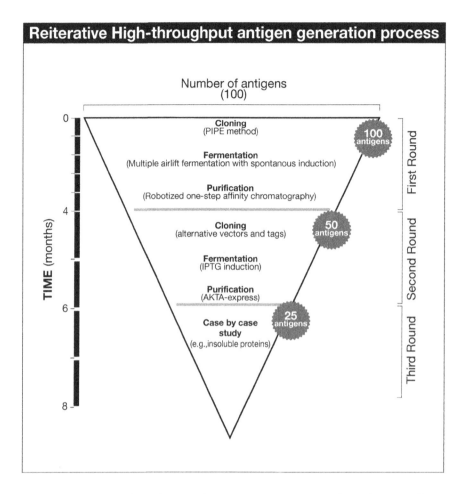

Figure 2.4 The reiterative high-throughput antigen generation process. In order to make a reverse vaccinology project feasible, the availability of techniques and tools to purify selected proteins in a fast and reliable manner is instrumental. Several methods are available in order to speed up cloning, fermentation and purification. However, the rate of success is typically in the range of 50%. Therefore, after a high-through put first round of purification, we usually run a second and a third round in which we generate the rest of the antigens by changing production parameters such as cloning vectors, culture-growth conditions, and purification methods.

metal affinity chromatography (IMAC) beads. However, in the second round, non-expressed clones can be re-cloned in different vectors containing alternative tags such as GST, or even fused with other vaccine candidates.

Once the clones have been obtained and checked by sequencing and for expression, bacteria carrying the plasmid are grown to get a biomass sufficient for protein purification. The HT-expression process is based on multiple airlift fermentation with spontaneous induction. Bacterial cultures are grown in 50 ml conical tubes under agitation for 30 hours. High optical density

is reached (OD = 3) and the wet biomass obtained ranges from 0.5 to 0.75 g. The final purified protein amount is between 1 and 20 mg starting from that biomass. Proteins that are not expressed or poorly expressed can be processed in the second round using IPTG induction and larger bacterial culture volumes. Proteins that are prioritized during the screening process, can be re-prepared using much larger volumes (e.g. 500–1000 ml) to purify a larger amount of protein.

Hopefully in the near future HT-expression will be based on cell-free or *in vitro* protein synthesis. Indeed, these approaches are less laborious, faster and more efficient than cell-based methods. Cell-free techniques are simple and amenable to manipulations and modifications to influence protein folding, disulfide bond formation, incorporation of unnatural amino acids, protein stability (by incorporating protease inhibitors in the system) and even the expression of toxic proteins. Furthermore, this technology can be easily automated. Protocols for expressing recombinant proteins with high yield in a standard 96-well plate format are being set up (Sitaraman and Chatterjee, 2009).

The last step of the HT-antigen generation process is protein purification. High-throughput purification can be performed using automatized one-step affinity chromatography. Robotized machines can start from the clarified bacteria lysate to allow purification of many samples simultaneously. In addition these machines perform automatic determination of protein concentration and sample preparation for electrophoretic analysis. If a higher purity level is needed, an HT-multi step protein purification process can be used. In our institute we use the ÄKTA x-press chromatography system. Different protocols can be set up using this machine. An affinity chromatography (AC), followed by a desalting (DS) and then an ion exchange (IEX) step is the method of choice for many proteins. However, several different programmed protocols can be used such as: AC/DS, AC/gel filtration (GF), AC/DS/IEX/DS, AC/DS/IEX/GF. For several reasons including production and regulatory issues, proteins included in the final vaccine product are not meant to include the His-tag. Therefore, between the sequence encoding for the hexahistidine tag and the protein, a site recognized by specific proteases is introduced for the cleavage. An example is an approach based on the tobacco etch virus (TEV) protease. It recognizes a seven amino acid consensus sequence, Glu-X-X-Tyr-X-Gln/Ser, where X can be various amino acyl residues. Cleavage occurs between the conserved Gln and Ser residues. Methods for overproducing a soluble form of the TEV protease in *Escherichia coli* and purifying it to be used to remove affinity tags have been recently set up (Tropea *et al.*, 2009). In addition a site mutation (S219V) in the protease reduced its rate of autolysis by approximately

100-fold and increased catalytic efficiency compared to the wild-type protease. For His-tag removal, the protocol includes the following steps: AC/DS/TEV digestion/AC/manual elution of tag-less protein. The target protein is first loaded on the column and after the first two steps (AC/DS) the enzyme reaction is performed in batch on the elution fractions. After the overnight digestion, the reaction product is loaded on a AC (IMAC) column, the tag-less product is collected with the elution buffer, while the TEV protease tailed with the His-tag and the His-tag cleaved by the target protein will stick to the IMAC column.

Once the proteins have been purified they are usually checked for identity, purity, concentration and lipopolysaccharide (LPS) quantification. LPS content determination and, if necessary its removal, is an important step before using the proteins in protection studies since LPS is a natural adjuvant and it could give non-specific protection.

Protection efficacy in animal models

Before a vaccine can go through clinical trials it needs to be tested for protection in *in vitro* assays or in animal models. If an established correlate of protection is available, such as in the case of MenB, the use of the *in vitro* assays is preferred over animal models. However, only in a few cases there is an established correlate of protection (see Part 2d and Chapter 9). Therefore, animal models still represent the major tool to show vaccine efficacy at the preclinical level. In a typical RV project, over one hundred proteins are screened in at least, one model. Given the large number of experiments needed, proper facilities and an adequate number of expert operators are necessary. Animals used in vaccine research comprise a vast range of species, however only a few are routinely used and the most common are mice, rats, guinea pigs and rabbits. Depending on the pathogen studied, a different model is used. For example, for pathogens causing pneumonia, the route of infection should enable the microorganism to reach the lungs (e.g. intranasally or intrapulmonary). In some cases, for the same pathogen different routes of infection can be used. For example, for *Staphylococcus aureus* (see Chapter 11), which can cause different diseases including bacteraemia and pneumonia, mice can be challenged either intravenously or intranasally. Another important aspect that can differ between the models is the readout. Most often, protection efficacy of a vaccine is based on the capability of increasing the survival rate of animals after challenge with the pathogen of interest. However, other measures can be used, including bacterial load in the infected organs and histopathology. The immunological response of the animal is also an important readout. Indeed, protection often correlates with high antibody titres, even though higher titres do not necessarily indicate a more protective antigen. The evaluation of antibody titres is also important to investigate the phenomena of immunological interference when a combination of antigens is used, and to analyse the contribution of adjuvants. In the RV process, antigens are usually first tested individually, and often combinations of the best protective antigens are also tested in a second phase. Combinations usually perform better than single antigens, but for reasons that are not well understood some antigens can negatively interfere when combined. We can measure the titres obtained with immunization of a single antigen or in combination with other antigens. Ideally, the antibody titres of an antigen should not vary in combinatorial immunization and an additive protection effect should be obtained. However, this is not always the case and some times a paradoxical reduction in protection efficacy is observed with antigen combinations compared to single antigens. Indeed, we have some times obtained a worse protection with combinations containing antigens performing very well individually than with combinations of poor antigens. An explanation for this phenomenon is still lacking. It is however likely that some proteins interact with each other masking protective epitopes, or some antigens could have a sort of immunodominance and reduce the immunity towards the others. Given the lack of strong and reliable criteria to design combinations on the basis of results obtained with single antigens, an alternative strategy that we have recently used in our animal experiments was based on a multifactorial selection. We did not strictly focus on the best protection data obtained with single antigens, but we took into account every aspect

of the vaccine candidates. In particular, the combinations were designed to be broadly protective against circulating strains, to target important functions and be effective against different clinical outcomes. An approach to satisfy all these criteria is to have antigens with complementary properties. For example, if one antigen protects against strains A and B, we may want to add an antigen to cover strains C and D. The same could be said for disease coverage. If a pathogen is associated with different clinical outcomes, different virulence factors may be involved. Another aspect is the protection mechanism. While for surface proteins protection is generally based on antibody-mediated killing of the pathogen, protection associated with secreted factors is less clear. In the case of secreted toxins, protection may be due, at least in part, to the inhibition of the toxic activity mediated by toxin-specific antibodies. It may therefore be a good strategy to have both surface and secreted proteins in a vaccine composition in order to mediate direct killing of the pathogen as well as inhibit toxic mechanisms.

Antigens demonstrated to be efficacious in a reference animal model are often further tested using different challenge strains (a process called breadth of protection) and with different adjuvant formulations. For the breadth of protection studies, a number of representative isolates is chosen on the basis of the epidemiology of the studied pathogen. The number of different strains needed to be used can also largely vary depending on the antigens used in the vaccine and the genetic as well as expression profile diversity of the pathogen. For example, if the antigens included in the vaccine are known to be conserved and consistently expressed from different strains, it may be sufficient to test a smaller number of strains.

Finally, formulation studies are done to determine the best adjuvant and the antigen doses. Until very recently, the only adjuvant approved for use in vaccines in the USA has been alum, which is a component of many marketed vaccines in the USA. However, additional adjuvants, such as MF59, have been included in vaccines licensed for use in Europe. MF59 has shown very good efficacy and safety profiles in influenza vaccination campaigns and is therefore considered a very important alternative to alum (see Chapter 7).

In addition to MF59 there are several other new adjuvants that have shown good results in pre-clinical studies. Therefore, at the preclinical level is important to perform formulation studies to determine which adjuvant performs better.

Animal studies are considered to be very important to evaluate vaccine efficacy. However, animal models never perfectly correlate with response in humans and the expression of the target antigens and virulence of a microorganism may differ from natural infections. Therefore, given these intrinsic limitations, probably the best approach is to complement the animal experiments with as much as possible different screenings. Among the most important screening to complement the *in vivo* models, are the epidemiology studies and the immunoreactivity of human sera against the selected antigens.

Molecular epidemiology studies on selected antigens

Epidemiological studies are very important to evaluate the strain coverage of a vaccine. In the development of a universal vaccine, capable of inducing protection against virtually all circulating strains, accurate characterization of the selected vaccine candidates is highly recommended. The analysis should be performed on a strain collection representing globally diverse geographic regions that take into account the target population of the vaccine.

In the era of genomics and comparative genomics, the availability of multiple genome sequences in the public databases greatly facilitates the identification of conserved antigens. However, the number of genomes available is usually not sufficient to cover the diversity of a species and a large epidemiological study on a restricted number of molecular antigens is often needed. The number of strains to be analysed depends mainly on the diversity of the species in terms of gene conservation and expression and the size of the target population. There are at least three types of bacterial populations: clonal species, such as *S. aureus*, where mutation is the only active evolutionary force; panmictic (i.e. fully sexual) species, such as *Neisseria gonorrhoeae*, whose population structure is mainly determined by recombination; and epidemic species, such as *N. meningitidis*, composed of highly recombining and unrelated genotypes, as well as single, highly adaptive genotypes that expand clonally giving rise to hypervirulent clusters. Therefore, the number of strains needed to cover the diversity of a clonal species is smaller than that needed to cover panmictic and epidemic species. The number of strains needed to perform a statistically significant analysis depends on the level of clonality of a species. For clonal species the number can be relatively small, while for highly variable bacteria the number can be larger (hundreds).

Molecular epidemiological studies rely on several bacterial typing methods. There are serological methods that are used to determine the capsular polysaccharide type. This method, is considered antiquated because it is restricted to one phenotype and does not provide information about the genetic background of a bacterium. However, capsule type is still used as a typing criterion, often together with one of the following molecular methods: ribosomal RNA-based typing, pulsed-field gel electrophoresis (PFGE), restriction-fragment length polymorphism (RFLP), multilocus sequence typing (MLST), amplified fragment length polymorphism (AFLP). PFGE is highly discriminating, but is labour-intensive and has poor inter-laboratory reproducibility and interpretability (van Leeuwen *et al.*, 2002). MLST, despite its high reproducibility and standardization, is relatively expensive due to the large number of target genes that are sequenced. However, MLST has become prominent in the last decade. MLST is based on the sequencing of internal fragments of seven housekeeping genes, which greatly enhances the typing resolution compared to the other techniques. The analysis defines the sequence type (ST) of the isolate, which may readily be compared to others through the MLST database (http://www.mlst.net). The database contains sequence information from thousands of isolates reported by laboratories worldwide. Although MLST studies provide an accurate representation of the micro-evolution of a bacterial population, recent studies have demonstrated that genes exposed to the selective pressure of the host immune system (such as surface proteins) do not segregate with the genes used for MLST (Russell *et al.*, 2008).

For example, comparison of the whole genome sequences of GBS strains has shown that the genomic diversity does not correlate exactly with serotypes or MLST sequence-types. As a consequence, from the vaccine perspective, MLST analysis based on housekeeping genes is often insufficient to predict variability of surface antigens. Considering that a RV-derived vaccine commonly comprises from one to five antigens, sequencing of the corresponding genes in a panel of representative strains is usually feasible.

Recently, a new approach for typing has emerged that is based on single nucleotide polymorphisms (SNPs). SNPs have recently been used to differentiate *Bacillus anthracis* clinical samples that were collected from a disease outbreak, and also to resolve the population structure and propose a typing scheme for *Mycobacterium tuberculosis*. More recently, the complex evolutionary history of *Salmonella typhi* was reconstructed by analysing 88 biallelic polymorphisms, including 82 SNPs (Read *et al.*, 2002 Gutacker *et al.*, 2002; Roumagnac *et al.*, 2006). In the future, due to the advancement of sequencing technologies, a whole-genome based typing method will soon be possible. Indeed, in this case typing is done by sequencing the entire genome and discriminating one strain from the other by a single nucleotide difference. The recent application of genome-wide single nucleotide polymorphism analysis to *M. tuberculosis* provides a significant contribution in this direction (Stavrum *et al.*, 2009). This method allows precise and very accurate bacterial grouping. This type of analysis can be particularly valid for clonal species characterized by reduced horizontal gene transfer events. Indeed, in such species point mutations are a major source of genetic variability. SNPs can be associated with specific phenotypes, such as pathogenicity.

Another critical aspect that should be considered to generate a widely protective vaccine, is the antigenic expression profile of the antigens used. Even when gene conservation is high, variability can occur in terms of expression. There may be several reasons for such discrepancy between the sequence conservation of an antigen and its expression. First, the genomic organization can vary from strain to strain thus affecting gene expression. Secondly, activity of transcriptional regulators can also vary, and this is not predictable from the sequence of the selected antigens. Expression profiling is commonly done by western blotting or flow cytometry (FACS), however, mass spectrometry can also be used for this purpose. For large screenings, FACS analysis is usually the best approach.

Last but not least, molecular epidemiology studies could help to monitor the effects of the vaccine introduction, to confirm its efficacy, the occurrence of vaccine escape variants or the appearance of novel pathogenic variants as a consequence of the selective pressure imposed. Brueggemann and colleagues recently demonstrated that after the introduction of the heptavalent glycoconjugate vaccine against *Streptococcus pneumoniae*, a serotype not included in the vaccine, had dramatically increased due to capsular switching, in which a genotype previously associated with a vaccine type 4 now expressed a non-vaccine serotype 19A capsule. The serotype 19A is now considered an important medical problem in the USA (Brueggemann *et al.*, 2007). Molecular epidemiological studies are helping to define a new type of vaccine for *S. pneumoniae*. This vaccine will be based on conserved protein antigens that are equally represented in this streptococcal species. The availability of this information will be critical in order to drive vaccine research in new directions, and may be the only effective solution.

Immune response towards selected antigens

The major weapon against extracellular pathogens is humoral immunity. Therefore, determining the human immunological fingerprint against a given pathogen can be very useful for vaccine development. There are several techniques that can be used to evaluate the immune response against a pathogen's antigens. These include common assays such as ELISA and Dot Blot as well as more advanced techniques such as protein microarray chips. The choice will reflect the number of antigens and sera that are needed to be screened. Protein chips are of course used only when a large number of antigens are studied (typically over 100 proteins). An important advantage of protein chips is the small amount of sera needed to perform the experiment. In order to identify immunogenic antigens, the availability of relevant sera is of crucial importance. The most important sera are those of convalescent patients, particularly from individuals that recovered from the disease without specific medical treatments against the pathogen of interest. The antigens recognized by these sera should represent the 'protective fingerprint' of the pathogen. It is of course important to test several sera from patients with the same disease and also from patients with different diseases, if different clinical outcomes are associated with the same pathogenic agent. Variability in the immune response can exist between patients with the same disease, and between patients with different clinical outcomes. Therefore, antigens most frequently recognized could represent good vaccine candidates. In addition, if the pathogen of interest can be carried asymptomatically, it is important to compare the immunogenic profile of the convalescent sera with that of healthy carriers. Antigens expressed during carriage, such as colonization factors, may not induce a protective response in the same manner as the ones specifically associated with disease.

In addition to human sera, screenings could also be done using animal sera, either naturally infected or from animals used in vaccine studies. The first case can be useful for those pathogens that infect both humans and animals, such as *S. aureus, S. agalactiae* and *C. difficile*. Antigens recognized by both humans and animals could be used to develop human and veterinary vaccines in parallel.

Immunogenicity, however, does not necessarily correlate with protection. In order to determine if antibodies are really able to mediate protection, functional assays must be employed. There are two major kinds of *in vitro* assays commonly used to show if sera are functional: bactericidal or opsonophagocytosis assays (see Chapter 4) In the first case, antibodies are able to direct complement mediated killing of bacteria. This happens for example with MenB (see Chapter 9). With other bacteria, such as *S. pneumoniae* or *S. aureus*, opsonic cells and complement are necessary for efficient bacterial killing. In the case of MenB, bactericidal activity of antibodies has been shown to correlate with protection in humans. Therefore, in this case the

bactericidal assay has been established as surrogate of protection against MenB. In the case of *S. pneumoniae*, opsonophagocytosis has been shown to correlate with protection mediated by anti-capsular polysaccharide antibodies. Indeed, the current polysaccharide-based vaccines induce opsonic killing. However, the assay has not yet been established as a correlate of protection for protein antigens. The same is true for *S. aureus* since there are no vaccines on the market yet and it has not been shown that antibodies can protect humans against infection. However, even in those cases in which correlates of protection in humans have not been established, functional assays are still important at the preclinical level for several reasons (Plotkin, 2008). First of all, assays set up preclinically could eventually be used to prove correlation with functional antibodies in humans. Secondly, the assays can provide important information to support protective data obtained with animal models. Opsonization assays have been shown to better predict vaccine efficacy than antibody titres. Thus, functional antibody assays could become surrogates of clinical studies for several vaccines. This would reduce the need of large clinical trials, facilitating development and lowering costs of vaccines.

Characterization studies on selected antigens

Most of the antigens discovered by RV are novel and there is no information available in the literature about them. This has the disadvantage that a lot of work needs to be done in order to understand their function and role in pathogenesis and immunity. At the same time, this has the great advantage that in many instances the characterization of the antigens discovered by RV may bring about important observations. The first was the discovery that GBS, GAS and the pneumococcus harbour pili. These structures, that are essential for the pathogenesis of these bacteria, had escaped the studies of microbiology for a century. Following their initial description, the 3D structure and molecular characterization is now already available. A second case is the discovery that a major antigen of MenB binds the human complement regulatory protein Factor H, while it does not bind Factor H of mouse or rat origin

(Rinaudo, 2009). This finding explains how the meningococcus survives specifically only in the human blood by avoiding the complement attack and why the bacterium is not able to grow in the blood of animals such as mice and rats. These observations also explain the failure of those who tried for decades to develop small animal models for the meningococcus.

Characterization studies of genome-derived vaccines can be classified into functional studies and toxicity studies. To determine an antigen's function there are many different experiments that can be done. The first studies will be driven by the annotation of the selected antigens. For example, if a protein is annotated as an adhesin, host–pathogen interaction studies based on fluorescent and confocal microscopy can be done to confirm its role in adhesion to host cells. If a protein is expected to have an enzymatic activity, biochemical studies to determine substrate binding and catalysis will be done. To prove the specificity of the observations made with the above experiments, knockout isogenic mutants are required. Comparative virulence studies can then be done infecting an appropriate animal model with the wild types and mutant strains. If an antigen is shown to be indispensable for bacterial survival or virulence, the vaccine qualities of the antigen improve.

The other kind of studies done to characterize an antigen aim to determine the cytotoxicity of the candidate. In these experiments, cell lines are exposed to the purified proteins used in the vaccine studies and cell death, apoptosis and cell binding is measured. Another measure to determine the potential toxicity of the proteins is based on the cytokine release from cell lines exposed to the proteins. The cytokines that can be measured in epithelial cells, for example, are IL-6 and IL-8 because these are considered good markers for a proinflammatory response.

As mentioned above, several proteins discovered by RV have been proven to be important for the pathogenesis of the microorganism. Electron microscopy studies demonstrated that pili protrude out of the bacterial surface (Fig. 2.5). Biochemical characterization of pili showed that they are formed by covalently bound pilin

subunits (Kang *et al.*, 2007; Telford *et al.*, 2006). The streptococcal pili have been shown to be very efficacious in *in vivo* protection experiments and to have an important role in virulence (Maione *et al.*, 2005; Barocchi *et al.*, 2006; Gianfaldoni *et al.*, 2007). Furthermore, they have been shown to be involved in adhesion to host cells and to aid biofilm formation (Bagnoli *et al.*, 2008; Manetti *et al.*, 2007; Nelson *et al.*, 2007). Often, among the selected candidates there are adhesins important during colonization and infection. This is the case of the meningococcal protein NadA, shown to be a very protective antigen against MenB and to have a crucial role in adhering to endothelial cells (see Chapter 9). In addition to adhesins, we have identified several proteins with different functions during our vaccine studies. A pullulanase was recently identified in GBS. The protein was shown to be active *in vitro*, being able to degrade α-glucan polysaccharides, such as pullulan, glycogen and starch. Studies performed on whole bacteria indicated that the presence of α-glucan polysaccharides in culture medium up-regulated the expression of the enzyme on bacterial surface as confirmed by FACS analysis and confocal imaging (Santi *et al.*, 2008).

Confocal microscopy is also very suitable for studies for intracellular bacteria or viruses. For example, in the case of hepatitis C virus (HCV), by confocal microscopy we have shown that the envelope proteins E1 and E2 triggered the formation of supramolecular structures co-localizing with the core and the non-structural proteins NS3 and NS5A. Furthermore, the HCV proteins appear to be concentrated in an endoplasmic reticulum-derived subcompartment. Importantly, by labelling *de novo*-synthesized HCV RNA, we showed that these structures constitute a site of viral RNA synthesis, suggesting that these complexes represent not only sites of HCV replication but also potential places of viral pre-budding (Brazzoli *et al.*, 2007).

Another important aspect of functional characterization is to determine if an antigen is dispensable. A gene may be indispensable or become so in particular conditions, such as during infection. Promising vaccine candidates identified by RV have been shown to have important roles for

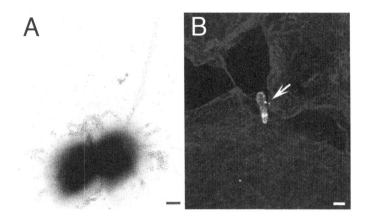

Figure 2.5 Immunoelectron-microscopy and confocal microscopy of streptococcal pili. (A) Pili expressed on the surface of *Streptococcus pneumoniae* strain TIGR4. Whole bacterial cells are incubated with polyclonal antibodies conjugated to 5-nm gold particles. The image shows the pilus backbone stained with gold-labelled antibodies raised against the main *Streptococcus pneumoniae* pilus component (RrgB). The scale bar represents 100 nm. (B) A549 cells were co-cultivated with pneumococcal strain for 1 h and imaged by confocal microscopy. Immunofluorescence shows the pilus during the epithelial cell infection experiment. 3D reconstruction of the confocal images has shown that the pilus extends from the bacterial surface contacting the host cells (arrow). Bacteria were visualized with Alexa Fluor 568 conjugated secondaries (red), A549 cells with phalloidin conjugated to Alexa Fluor 647 secondaries (blue) and the pilus with Alexa Fluor 488 conjugated secondaries (green). Scale bars, 10 μm. A colour version of this figure is located in the plate section at the back of the book.

bacterial viability. For example, staphylococcal proteins encoded in the Isd operon (IsdA and IsdB) have been shown to be part of an iron-uptake system (Skaar *et al.*, 2004; Stranger-Jones *et al.*, 2006). These two proteins contain LPXTG motifs and are surface anchored by a sortase. They have been suggested to mediate iron uptake from haemoglobin. Interestingly, the expression of the Isd proteins is induced by iron deplete conditions reflecting the importance of these antigens during infection. Indeed, there is no free iron in organs, therefore the expression of the Isd locus allows the pathogen to recover iron from haemoglobin. Protection obtained with these proteins is probably not only due to a good immunogenic profile, but also because these factors are indispensable for survival. Indeed, knockout mutants of the genes *isdA* and *isdB* are severely impaired in forming abscesses in mice (Cheng *et al.*, 2009). Therefore, functional studies are also very important for vaccine design. However, these experiments are done only close to the end of the RV process, and they can only contribute to the selection among a small number of candidates.

The future of reverse vaccinology

This section describes the advancements expected to happen in the next decade in the RV field. In particular, we discuss novel strategies to predict protective antigens as well as to identify protective epitopes by structural biology, particularly to develop vaccines against extremely variable pathogens.

In silico prediction of protective epitopes

Antigen selection approaches described in this chapter have been shown to be reliable for the identification of proteins immunogenic and protective in animal models, in *in vitro* assays and in the clinical trials of the MenB vaccine (see Chapter 9). Therefore, the assumption that surface and secreted proteins represent the best candidates has been validated. However, only a small portion of the candidates predicted *in silico* or using functional genomics approaches are usually confirmed to be good vaccine antigens. To make the RV approach even faster and more efficient we are trying to identify 'signatures of protection'. Ideally, we should be able to develop an algorithm able to search for protective epitopes. However, the nature of protective epitopes is not fully understood yet and it seems to depend on several factors. In order to shed light on these factors we first have to look at the antigens known to be protective. A comparison of these antigens may reveal consensus motifs at the basis of protection. A search of all the known protective antigens could also reveal if certain kind of proteins are more likely to be good vaccine candidates. In particular, it is not clear yet if pathogenic factors are more protective than other antigens such as colonization factors. An example of such an effort is 'Violin' (Vaccine Investigation and Online Information Network; http://www.violinet. org/). This is a web site aimed to store, annotate, compare, and analyse vaccine information for different pathogens. It contains research data for commercial vaccines and vaccines in clinical trials or in early stages of research. The site integrates vaccine literature and research data including studies using humans and laboratory animals. Interestingly, it also provides bioinformatic tools to predict vaccine candidates. The algorithm is based on RV criteria.

Comparison of vaccine data obtained with different pathogens may give some additional indication to drive candidate prediction.

Structural biology

Structural biology stems from molecular biology, biochemistry and biophysics pertaining to the molecular structure of biological macromolecules. The goal of structural biology, in a broad definition, can be taken as understanding the relationship between structure and function – from the atomic structure of the molecule, its assembly into complex arrays, and on to the microstructure regime. Quantitative understanding of protein folding and biogenesis of their interactions with other molecules and how alterations in their structures affect their functions is required to provide a framework for developing new therapeutic concepts and will be the pivotal challenge in molecular biology in the years ahead. This subject is of great interest because macromolecules carry out most of the functions of cells, and because it

is only by folding into specific three-dimensional shapes that they are able to perform these functions. Central processes such as cell signalling and cell–cell interactions in both healthy and diseased tissue can be investigated. Proteins deregulated in diseases can thus be used as targets for drug design. Efficient experimental structure determination of soluble and in particular membrane proteins, and the biophysical study of the protein folding process is a central component and provides the basis for the future development of basic research and practical applications such as drug design and delivery. Structural biology assists the definition of gene function since structural information can be used to predict the preferred substrate of a protein, and thereby greatly improve annotation of the corresponding gene. Furthermore, structural studies enable the effects of amino acid substitutions in enzymes to be better understood with respect to enzyme function, thereby providing insights into natural variation in genes. Structural biology can also significantly aid the rational design of future vaccines. The application of structural biology principles in vaccine research may result in the creation of a powerful new approach, called structural vaccinology. This approach can use the atomic resolution of the structures of potential antigens to enable the rational design of target epitopes to use as vaccine candidates.

Although most of the work discussed in structural biology is now at the atomic level, as we build the database of molecular structures it becomes more important to understand how they act together to carry out larger scale functions. Optical microscopy has served to define the larger scale structures, but it has always been limited by the diffraction limit of light and often suffers from poor contrast. Fluorescence microscopy solves much of the contrast problem, although it introduces a requirement for labelling. Despite their great importance in modern cell biology, conventional light microscopy and confocal fluorescence scanning light microscopy cannot be used for the study of individual proteins. These techniques cannot resolve distances smaller than 100–200 nm, which is about 1000 times the length of a typical chemical bond. During the past few years, a number of exciting developments in fundamental fluorescence microscopy techniques have led to improvements in resolution, getting below the 'diffraction limit' by an order of magnitude. At this resolution, features well below 100 Å can be seen, which almost allows structures to be solved at the atomic level. The connections between atomic-scale and micro-scale structures are critical to understanding many of the large-scale dynamic features that are the essence of function at the cellular level. The evolution of the methodology is fascinating, and the application of multi-photon and interference microscopy is now under evaluation. The methods that structural biologists use to determine protein structures generally involve dimension measurements on vast numbers of identical molecules at the same time. The methods that have proven successful for reaching beyond the wavelength limits of visible light and into a resolution range in which individual molecules and their internal structure can be visualized include X-ray crystallography, NMR, ultra fast laser spectroscopy, electron microscopy, electron cryomicroscopy (cryo-EM), circular dichroism and mass spectrometry. Most often researchers use these methods to study the static 'native states' of macromolecules. But variations to these methods are also used to watch nascent or denatured molecules assuming or reassuming their native states. A third approach that structural biologists use to understand structure is bioinformatics to look for patterns among the diverse sequences that give rise to particular shapes. Researchers often can deduce aspects of the structure of integral membrane proteins based on the membrane topology predicted by hydrophobicity analysis. In the past few years it has become possible for highly accurate physical molecular models to be developed that complement the *in silico* study of biological structures.

The contribution of structural biology to vaccine design

Antibodies bind to their corresponding antigens at discrete sites known as antigenic determinants or epitopes. Proteins represent the most abundant and various repertoires of potential antigens, and for this reason the knowledge of the key features responsible for crucial events such as the ability to be targeted by antibodies and to induce an effective protective immunity still represents a challenging issue. The importance of protein folding for antibody recognition is now widely recognized since the pioneer studies of Atassi (1975, 1978). However, what actually constitutes the antigenic determinant of a protein is still at the centre of a debate in which two opposite views have been proposed. One suggests that a globular protein is a continuum of potential antigenic sites, all capable of eliciting an immune response. The opposite theory asserts that protein surfaces contain a limited number of exclusive sites that can be immunogenic. A compromise between these views suggests that all sites on the molecule can be antigenic, although some have a greater potential to stimulate the immune system. Then, the question is raised about what are the protein traits that define a strong epitope.

Many bio-physical parameters have been proposed to correlate with immunogenicity, such as hydrophilicity (Hopp and Woods, 1981), backbone flexibility (Novotny *et al.*, 1986; Westhof *et al.*, 1984), accessible surface area (Novotny *et al.*, 1986) and protrusion from the protein surface (Thornton *et al.*, 1986).

Recent advances in structure determination technologies in the 1990s led to a remarkable increase in the number of structures of protein complexes, including antibody–antigen complexes, that have been solved and submitted to the Protein Data Bank database (www.pdb.org). Comprehensive analyses of these structures has become possible, allowing an evaluation of the biophysical properties of epitopes that are no longer limited to the level of primary structure. Although the conclusions from these type of analyses could be still biased towards proteins for which the three-dimensional structure is available, the size of the data set makes it possible to perform the analyses in a statistically more robust manner in order to reliably determine whether some features significantly distinguish epitopes from the remaining protein surface. In addition, the availability of native antigen structure makes it now possible to examine possible changes that proteins experience upon antibody binding.

An example of such large scale analysis has been performed by Rubistein and colleagues on a total of 53 antigen–antibody complexes

(Dormitzer *et al.*, 2008). Their main conclusion was that protein epitopes consist of defined molecular areas that can be distinguished by the remaining antigen surface on the basis of amino acid preferences, secondary structure composition, geometrical shape and evolutionary conservation. Specifically, known protein epitopes consist on average of 20 residues, located on a variable number of linear segments (generally five or more) that occupy a continuous patch of 600–1000 Å. This outlines the conformational nature of epitopes and stresses the relevance of the three dimensional structure of a protein in determining the immunogenicity of a given portion of its surface. Moreover, epitope surfaces are more accessible than protein portions not involved in antibody binding, and as such are commonly enriched in loops rather than helices and beta strands. These unorganized elements confer flexibility, and this correlates with the observation that epitopes undergo compression upon antibody binding. Another important distinctive feature is the major evolutionary variability observed in epitopes with respect to the 'silent' regions of the protein surface. Also this property can be partially explained by the enrichment of loops, which are generally more tolerant to amino acid replacements.

The problem of intrinsic variability is one of the major challenges in the development of new generation vaccines. While protein-based vaccines are already available against some pathogens whose surfaces antigens have not undergone significant antigenic change over many decades (e.g. diphtheria, polio, hepatitis B), a more challenging goal is now represented by micro-organisms (e.g. HCV, HIV, Dengue virus) that adopt hypervariability of their surface proteins as a major mechanism of escape from the humoral immune system. One important goal of structural vaccinology is to use structural analysis to selectively present to the immune system antigenic determinants that are as stable and conserved as possible. An essential step in this approach is the identification of the parts of the antigen that must be retained to preserve the immunogenic power of the native antigen. Several techniques can be used to this end (Rubinstein *et al.*, 2008), including nuclear magnetic resonance and X-ray crystallography which are methods that can provide fine epitope mapping, by defining the protein zones bound to antibodies (NMR) or directly solving the structure of the antigen–antibody complex at the atomic level (X-ray).

An example of a structural investigation aimed to clarify the essential epitopes involved in protective immunity is provided by the series of studies carried out on the factor H binding protein (fHbp) of MenB. fHbp is a component of a recently proposed recombinant vaccine against meningococcal infection and its name reflects the ability to bind human factor H, a negative regulator of the complement cascade (Giuliani *et al.*, 2006; Madico *et al.*, 2006). Following its identification by RV (Pizza *et al.*, 2000), the first serological studies revealed that the C-terminal portion of the molecule was responsible for the protective immune response. Mutagenesis experiments coupled to peptide scanning revealed the conformational character of the protective epitopes contained in the C-terminus domain of this protein (Giuliani *et al.*, 2005). The resolution by NMR of this protein portion allowed the localization of critical residues identified by mutagenesis as crucial for antibody recognition, and their clustering within a defined zone of the protein suggested the existence of a well-defined protective epitope (Cantini *et al.*, 2009; Giuliani *et al.*, 2005). This hypothesis was confirmed by further NMR studies in which the fHbp residues targeted by a protective monoclonal antibody were identified, leading to the delineation of a surface area involved in the formation of the immune complex (Cantini *et al.*, 2009).

More generally, identification of protective epitopes coupled to the knowledge of the amino acid variability observed within the microorganism population is particularly useful in the case of variable antigens, which could be engineered to enable them to cross-protect against microbial strains containing mutations in the neutralizing epitope(s). These 'artificial' molecules could represent possible candidates with increased coverage.

Another example of successful application of structural biology to vaccinology was the engineering of the lipoprotein OspA of *Borrelia burgdorferi* reported by Koide and colleagues (Koide *et al.*, 2005). The native protein was engineered to drastically reduce its overall size, while at the same time retaining the structure and stability of the critical neutralization epitope, with the aim to focus the host immune response away from non-neutralizing protein sites.

On the other hand, the outer-layer VP7 protein story is a perfect example of how basic structural research can help in the development of an effective treatment for a viral infection. Rotaviruses are multilayered, non-enveloped particles where the VP7-like trimers form the outermost virion layer and participate both in a membrane-displacing assembly step and in a membrane-disrupting entry step. Rotavirus infection has been identified as the principal cause of severe dehydrating diarrhoea in infants. Already introduced live-attenuated vaccines present an efficacy and practicality that have not yet been established specially in poor countries in which most infant deaths occur. VP7 is the target (together with VP4) of neutralizing and protective antibodies, and the structure and immunogenicity of this protein underlies ongoing efforts to produce next generation subunits vaccines. Epitopes of a number of neutralizing monoclonal antibodies have been determined, but the lack of a three-dimensional structure has precluded systematic study of neutralization mechanisms. The crystal structure at 3.4 Å resolution of the VP7 bound with the Fab fragment of a neutralizing antibody (4F8) was recently reported (Aoki et al., 2009). The three-dimensional structure sheds light on the regions (across the outer surface of the VP7 inter-subunit contact, which contains two Ca^{2+} sites) of this antigen that may act as epitopes. The monovalent Fab is sufficient to neutralize infectivity. The authors propose that neutralizing antibodies against VP7 act by stabilizing the trimer, thereby inhibiting the uncoating trigger for VP4 rearrangement. Furthermore, they identify a disulfide-linked trimer as a potential subunit immunogen.

Another important contribution of basic research to antiviral treatments is the generation of inhibitors targeting the fusion step of enveloped viruses. Enveloped viruses, like HIV-1, Semliki forest virus, Dengue virus, and influenza virus, encode specialized fusion proteins which promote the merger of viral and cell membranes,

permitting the release of the viral cores in cytosol. Understanding the molecular details of this process is essential for antiviral strategies. Structural studies reveal an impressive diversity of viral fusion proteins in their native state. In spite of this diversity, the post-fusion structures of these proteins share a common trimeric hairpin motif in which the amino- and carboxy-terminal hydrophobic domains are positioned at the same end of a rod-shaped molecule. The converging hairpin motif implies that different viral proteins promote membrane merger via a universal 'cast-and-fold' mechanism (Melikyan, 2008). According to this model, fusion proteins first anchor themselves to the target membrane through their hydrophobic segments and then fold back, bringing the viral and cellular membranes together and forcing their merger. One principal example is the identification of key conformational states of the HIV-1 envelope glycoprotein en route to fusion allowed the generation of specific inhibitors targeting distinct steps of HIV-1 entry.

Conclusions

In contrast to classical vaccinology approaches, which traditionally require a time-consuming series of experiments to identify immunogenic antigens, RV predicts all the putative candidates, which are then tested in protection studies. The first RV project was mainly based on in silico prediction of candidates using one genome of MenB (see Chapter 9). Ten years after this seminal work, the process with additional analyses, tools and techniques was implemented. Candidate prediction, the first phase of the process, is now mainly based on bioinformatics as well as comparative genomics, proteomics and functional genomics. The main criteria used for the candidate prediction are the following: surface exposed or secreted (determined by the presence of sorting domains and by proteomics), and conserved proteins (determined by comparative genomics). The importance of proteomics and functional genomics to aid in silico candidate prediction is based on several considerations. First of all, similar to the bioinformatic analysis, they can be done prior to antigen selection. Indeed, to identify vaccine candidates, they only need a bacterial culture and the genomic sequence, for the mass spectrometry-based approach, and a DNA chip and bacterial cDNA, for the microarray-based approach. Second, they complement bioinformatics in the identification of proteins actually expressed and surface exposed or secreted. Meanwhile, comparative genomics has become easy to do, given that for most of the human pathogens multiple genomes are publicly available. This provides very important information to identify conserved antigens as well as to have some indication of antigens that are more exposed to the immune system on the basis of variability profiles. Given the wealth of the public genomic databases, for a given RV vaccine project starting now, we can have the list of predicted vaccine candidates in a matter of one or two months. Predicted antigens are then generated by a high throughput process, which involves the latest techniques for cloning, fermentation and purification. Purified proteins are then screened with *in vitro* and *in vivo* assays for immunogenicity and protection. Later, a smaller number of selected antigens are screened for epidemiological relevance and immunoreactivity to human sera. Finally, a handful of antigens are characterized in order to provide information on their function and possible toxicity.

Besides technical improvements, we have learned how to make the process more efficient by using a few strategies. In particular we use a reiterative process to generate the antigens. By doing so we get approximately 50% of the predicted candidates in the first round of purification. This means that in four months we will have roughly 50 antigens to be tested in protection assays. In the mean time, a second and third round of purification will generate the more 'difficult' proteins. Furthermore, a good vaccine often requires a combination of antigens. This is due to several reasons. First, there is often a synergistic protective effect associated with combinations. Second, the inclusion of more antigens is more likely to provide protection against different strains. This can be due to sequence or expression variability of the antigens. Sequence variability is simple to determine, while expression can be much more difficult to determine. Indeed, *in vitro* culture conditions may not properly resemble *in vivo* conditions. Therefore, the larger the number of antigens in the combination, the greater the

possibility to get a universal vaccine. Finally, we generally select antigens with different functions. In particular, within a combination we may have antigens important for bacterial survival and others for infection. A vaccine targeting survival factors would theoretically inhibit bacterial growth while immunity against pathogenic factors would block disease. Efficacy of a vaccine composition, however, has to be shown by protection experiments, either *in vitro* or using animal models. Protection is considered the most important criterion to choose a vaccine candidate. However, candidate selection should take into account all the criteria important for a human vaccine. Therefore, combinations should not necessarily be composed of the most protective antigens, rather of antigens that together complement each other in terms of the characteristics important for the vaccine. In particular, combinations should be broadly protective against circulating strains, to target important functions and ideally be effective against a wide range of clinical outcomes associated with a pathogen. Furthermore, all of the antigens should have good safety profiles: no toxic activity and no homologies with human antigens. If promising candidates do not fulfil these requirements, they could be mutagenized to remove the problem. Based on this assumption, our strategy is to start assessing protection of the single antigens. Then, protective proteins that appear more interesting from all the other vaccinology criteria are used to design combinations and then tested for protection. Combinations able to induce good protection will therefore represent developable vaccines.

Despite the progress made in the last decade, we cannot yet say that we have reached the point of a true rational design of vaccines. Indeed, it still depends on the experimental studies and empirical observations. Furthermore, efforts are still failing in the development of efficacious vaccines against several viruses. Viral antigen prediction is simple because of the small number of genes present in their genomes and lack of expression variability. However, the antigen sequence of many viruses is extremely variable and protective epitopes are difficult to identify. The solution to these issues appears to be based on structural biology studies. This is mainly owing to the following reasons. Often, the first line of defence against a pathogen is mediated by antibodies. Structural biology is showing that antibodies usually recognize conformational epitopes that cannot be predicted by sequence analysis. Several recent studies have shown that protective epitopes can be found by structural studies of antigen–antibody complexes. We expect that these studies will provide the information necessary to develop vaccines against variable viruses as well as the first information to build prediction methods to identify protective epitopes.

Acknowledgements

We are grateful to Kate Seib and Lisa Danzig for critical reading, to Giorgio Corsi for artwork, to Luigi Fiaschi for permission to use Fig. 2.2, and to Catherine Mallia for manuscript editing.

References

Adu-Bobie, J., Lupetti, P., Brunelli, B., Granoff, D., Norais, N., Ferrari, G., Grandi, G., Rappuoli, R., and Pizza, M. (2004). GNA33 of *Neisseria meningitidis* is a lipoprotein required for cell separation, membrane architecture, and virulence. Infect. Immun. 72, 1914–1919.

Aoki, S.T., Settembre, E.C., Trask, S.D., Greenberg, H.B., Harrison, S.C., and Dormitzer, P.R. (2009). Structure of rotavirus outer-layer protein VP7 bound with a neutralizing Fab. Science 324, 1444–1447.

Arenas, J., Abel, A., Sanchez, S., Marzoa, J., Berron, S., van der Ley, P., Criado, M.T., and Ferreiros, C.M. (2008). A cross-reactive neisserial antigen encoded by the NMB0035 locus shows high sequence conservation but variable surface accessibility. J. Med. Microbiol.57, 80–87.

Atassi, M.Z. (1975). Antigenic structure of myoglobin: the complete immunochemical anatomy of a protein and conclusions relating to antigenic structures of proteins. Immunochemistry 12, 423–438.

Atassi, M.Z. (1978). Precise determination of the entire antigenic structure of lysozyme: molecular features of protein antigenic structures and potential of 'surface-simulation' synthesis–a powerful new concept for protein binding sites. Immunochemistry 15, 909–936.

Bagnoli, F., Moschioni, M., Donati, C., Dimitrovska, V., Ferlenghi, I., Facciotti, C., Muzzi, A., Giusti, F., Emolo, C., Sinisi, A., et al. (2008). A second pilus type in *Streptococcus pneumoniae* is prevalent in emerging serotypes and mediates adhesion to host cells. J. Bacteriol. 190, 5480–5492.

Barocchi, M.A., Ries, J., Zogaj, X., Hemsley, C., Albiger, B., Kanth, A., Dahlberg, S., Fernebro, J., Moschioni, M., Masignani, V., et al. (2006). A pneumococcal pilus influences virulence and host inflammatory responses. Proc. Natl. Acad. Sci. U.S.A. 103, 2857–2862.

Besemer, J., Lomsadze, A., and Borodovsky, M. (2001). GeneMarkS: a self-training method for prediction of gene starts in microbial genomes. Implications for finding sequence motifs in regulatory regions. Nucleic Acids Res. 29, 2607–2618.

Berlanda Scorza, F., Doro, F., Rodriguez-Ortega, M.J., Stella, M., Liberatori, S., Taddei, A.R., Serino, L., Gomes Moriel, D., Nesta, B., Fontana, M.R., et al. (2008). Proteomics characterization of outer membrane vesicles from the extraintestinal pathogenic *Escherichia coli* DeltatolR IHE3034 mutant. Mol. Cell Proteomics 7, 473–485.

Bernadac, A., Gavioli, M., Lazzaroni, J.C., Raina, S., and Lloubes, R. (1998). *Escherichia coli* tol-pal mutants form outer membrane vesicles. J. Bacteriol. 180, 4872–4878.

Betton, J.M. (2004). High throughput cloning and expression strategies for protein production. Biochimie 86, 601–605.

Beveridge, T.J. (1999). Structures of gram-negative cell walls and their derived membrane vesicles. J. Bacteriol. 181, 4725–4733.

Bhasin, M., Garg, A., and Raghava, G.P.S. (2005). PSLpred: prediction of subcellular localization of bacterial proteins. Bioinformatics 21, 2522–2524.

Bjune, G., Hoiby, E.A., Gronnesby, J.K., Arnesen, O., Fredriksen, J.H., Halstensen, A., Holten, E., Lindbak, A.K., Nokleby, H., Rosenqvist, E., et al. (1991). Effect of outer membrane vesicle vaccine against group B meningococcal disease in Norway. Lancet 338, 1093–1096.

Brazzoli, M., Crotta, S., Bianchi, A., Bagnoli, F., Monaghan, P., Wileman, T., Abrignani, S., and Merola, M. (2007). Intracellular accumulation of hepatitis C virus proteins in a human hepatoma cell line. J. Hepatol. 46, 53–59.

Brueggemann, A.B., Pai, R., Crook, D.W., and Beall, B. (2007). Vaccine escape recombinants emerge after pneumococcal vaccination in the United States. PLoS Pathog. 3, e168.

Cantini, F., Veggi, D., Dragonetti, S., Savino, S., Scarselli, M., Romagnoli, G., Pizza, M., Banci, L., and Rappuoli, R. (2009). Solution structure of the factor H-binding protein, a survival factor and protective antigen of *Neisseria meningitidis*. J. Biol. Chem. 284, 9022–9026.

Cerdeno-Tarraga, A.M., Efstratiou, A., Dover, L.G., Holden, M.T., Pallen, M., Bentley, S.D., Besra, G.S., Churcher, C., James, K.D., De Zoysa, A., et al. (2003). The complete genome sequence and analysis of *Corynebacterium diphtheriae* NCTC13129. Nucleic Acids Res. 31, 6516–6523.

Cheng, A.G., Kim, H.K., Burts, M.L., Krausz, T., Schneewind, O., and Missiakas, D.M. (2010). Genetic requirements for *Staphylococcus aureus* abscess formation and persistence in host tissues. FASEB J. (In press).

Cordwell, S.J. (2006). Technologies for bacterial surface proteomics. Curr. Opin. Microbiol. 9, 320–329.

Cullen, P.A., Xu, X., Matsunaga, J., Sanchez, Y., Ko, A.I., Haake, D.A., and Adler, B. (2005). Surfaceome of *Leptospira* spp. Infect. Immun. 73, 4853–4863.

Delcher, A.L., Kasif, S., Fleischmann, R.D., Peterson, J., White, O., and Salzberg, S.L. (1999). Alignment of whole genomes. Nucleic Acids Res. *27*, 2369–2376.

de Moraes, J.C., Perkins, B.A., Camargo, M.C., Hidalgo, N.T., Barbosa, H.A., Sacchi, C.T., Landgraf, I.M., Gattas, V.L., Vasconcelos Hde, G., *et al.* (1992). Protective efficacy of a serogroup B meningococcal vaccine in Sao Paulo, Brazil. Lancet *340*, 1074–1078.

Devoe, I.W., and Gilchrist, J.E. (1973). Release of endotoxin in the form of cell wall blebs during *in vitro* growth of *Neisseria meningitidis*. J. Exp. Med. *138*, 1156–1167.

Dormitzer, P.R., Ulmer, J.B., and Rappuoli, R. (2008). Structure-based antigen design: a strategy for next generation vaccines. Trends Biotechnol. *26*, 659–667.

Doro, F., Liberatori, S., Rodriguez-Ortega, M.J., Rinaudo, C.D., Rosini, R., Mora, M., Scarselli, M., Altindis, E., D'Aurizio, R., Stella, M., *et al.* (2009). Surfome analysis as a fast track to vaccine discovery: identification of a novel protective antigen for group B streptococcus hypervirulent strain COH1. Mol. Cell Proteomics *8*, 1728–1737.

Economou, A., Christie, P.J., Fernandez, R.C., Palmer, T., Plano, G.V., and Pugsley, A.P. (2006). Secretion by numbers: Protein traffic in prokaryotes. Mol. Microbiol. *62*, 308–319.

Ekiert, D.C., Bhabha, G., Elsliger, M.A., Friesen, R.H., Jongeneelen, M., Throsby, M., Goudsmit, J., and Wilson, I.A. (2009). Antibody recognition of a highly conserved influenza virus epitope. Science *324*, 246–251.

Emanuelsson, O., Brunak, S., von Heijne, Henrik, G., and Nielsen, H. (2007). Locating proteins in the cell using TargetP, SignalP, and related tools. Nature Protocols *2*, 953–971.

Fitz-Gibbon, S.T., Ladner, H., Kim, U.J., Stetter, K.O., Simon, M.I., and Miller, J.H. (2002). Genome sequence of the hyperthermophilic crenarchaeon Pyrobaculum aerophilum. Proc. Natl. Acad. Sci. U.S.A. *99*, 984–989.

Ferrari, G., Garaguso, I., Adu-Bobie, J., Doro, F., Taddei, A.R., Biolchi, A., Brunelli, B., Giuliani, M.M., Pizza, M., Norais, N., *et al.* (2006). Outer membrane vesicles from group B *Neisseria meningitidis* delta gna33 mutant: proteomic and immunological comparison with detergent-derived outer membrane vesicles. Proteomics *6*, 1856–1866.

Fiocca, R., Necchi, V., Sommi, P., Ricci, V., Telford, J., Cover, T.L., and Solcia, E. (1999). Release of *Helicobacter pylori* vacuolating cytotoxin by both a specific secretion pathway and budding of outer membrane vesicles. Uptake of released toxin and vesicles by gastric epithelium. J. Pathol. *188*, 220–226.

Forest, K.T., Dunham, S.A., Koomey, M., and Tainer, J.A. (1999). Crystallographic structure reveals phosphorylated pilin from Neisseria: phosphoserine sites modify type IV pilus surface chemistry and fibre morphology. Mol. Microbiol. *31*, 743–752.

Frishman, D., Heumann, K., Lesk, A., and Mewes, H.W. (1998a). Comprehensive, comprehensible,

distributed and intelligent databases: current status. Bioinformatics *14*, 551–561.

Frishman, D., Mironov, A., Mewes, H.W., and Gelfand, M. (1998b). Combining diverse evidence for gene recognition in completely sequenced bacterial genomes. Nucleic Acids Res. *26*, 2941–2947.

Gardy, J.L., and Brinkman, F.S.L. (2006). Methods for predicting bacterial protein subcellular localization. *4*, 741–751.

Gardy, J.L., Laird, M.R., Chen, F., Rey, S., Walsh, C.J., Ester, M., and Brinkman, F.S.L. (2005). PSORTb v.2.0: Expanded prediction of bacterial protein subcellular localization and insights gained from comparative proteome analysis. Bioinformatics *21*, 617–623.

Gianfaldoni, C., Censini, S., Hilleringmann, M., Moschioni, M., Facciotti, C., Pansegrau, W., Masignani, V., Covacci, A., Rappuoli, R., Barocchi, M.A., *et al.* (2007). *Streptococcus pneumoniae* pilus subunits protect mice against lethal challenge. Infect. Immun. *75*, 1059–1062.

Giuliani, M.M., Adu-Bobie, J., Comanducci, M., Arico, B., Savino, S., Santini, L., Brunelli, B., Bambini, S., Biolchi, A., Capecchi, B., *et al.* (2006). A universal vaccine for serogroup B meningococcus. Proc. Natl. Acad. Sci. U.S.A. *103*, 10834–10839.

Giuliani, M.M., Santini, L., Brunelli, B., Biolchi, A., Arico, B., Di Marcello, F., Cartocci, E., Comanducci, M., Masignani, V., Lozzi, L., *et al.* (2005). The region comprising amino acids 100 to 255 of *Neisseria meningitidis* lipoprotein GNA 1870 elicits bactericidal antibodies. Infect. Immun. *73*, 1151–1160.

Gutacker, M.M., Smoot, J.C., Migliaccio, C.A., Ricklefs, S.M., Hua, S., Cousins, D.V., Graviss, E.A., Shashkina, E., Kreiswirth, B.N., and Musser, J.M. (2002). Genome-wide analysis of synonymous single nucleotide polymorphisms in *Mycobacterium tuberculosis* complex organisms: resolution of genetic relationships among closely related microbial strains. Genetics *162*, 1533–1543.

Hayashi, T., Makino, K., Ohnishi, M., Kurokawa, K., Ishii, K., Yokoyama, K., Han, C.G., Ohtsubo, E., Nakayama, K., Murata, T., *et al.* (2001). Complete genome sequence of enterohemorrhagic *Escherichia coli* O157:H7 and genomic comparison with a laboratory strain K-12. DNA Res. *8*, 11–22.

Hegge, F.T., Hitchen, P.G., Aas, F.E., Kristiansen, H., Lovold, C., Egge-Jacobsen, W., Panico, M., Leong, W.Y., Bull, V., Virji, M., *et al.* (2004). Unique modifications with phosphocholine and phosphoethanolamine define alternate antigenic forms of *Neisseria gonorrhoeae* type IV pili. Proc. Natl. Acad. Sci. U.S.A. *101*, 10798–10803.

Hoekstra, D., van der Laan, J.W., de Leij, L., and Witholt, B. (1976). Release of outer membrane fragments from normally growing *Escherichia coli*. Biochim. Biophys. Acta *455*, 889–899.

Holst, J., Martin, D., Arnold, R., Huergo, C.C., Oster, P., O'Hallahan, J., and Rosenqvist, E. (2009). Properties and clinical performance of vaccines containing outer membrane vesicles from *Neisseria meningitidis*. Vaccine *27* (Suppl. 2), B3–12.

Hopp, T.P., and Woods, K.R. (1981). Prediction of protein antigenic determinants from amino acid sequences. Proc. Natl. Acad. Sci. U.S.A. *78*, 3824–3828.

Horzempa, J., Held, T.K., Cross, A.S., Furst, D., Qutyan, M., Neely, A.N., and Castric, P. (2008). Immunization with a *Pseudomonas aeruginosa* 1244 pilin provides O-antigen-specific protection. Clin. Vaccine Immunol. *15*, 590–597.

Jones, A.L., Needham, R.H., Clancy, A., Knoll, K.M., and Rubens, C.E. (2003). Penicillin-binding proteins in Streptococcus agalactiae: a novel mechanism for evasion of immune clearance. Mol. Microbiol. *47*, 247–256.

Kadurugamuwa, J.L., and Beveridge, T.J. (1995). Virulence factors are released from *Pseudomonas aeruginosa* in association with membrane vesicles during normal growth and exposure to gentamicin: a novel mechanism of enzyme secretion. J. Bacteriol. *177*, 3998–4008.

Kang, H.J., Coulibaly, F., Clow, F., Proft, T., and Baker, E.N. (2007). Stabilizing isopeptide bonds revealed in gram-positive bacterial pilus structure. Science *318*, 1625–1628.

Kanoi, B.N., and Egwang, T.G. (2007). New concepts in vaccine development in malaria. Curr. Opin. Infect. Dis. *20*, 311–316.

Klock, H.E., Koesema, E.J., Knuth, M.W., and Lesley, S.A. (2008). Combining the polymerase incomplete primer extension method for cloning and mutagenesis with microscreening to accelerate structural genomics efforts. Proteins *71*, 982–994.

Koide, S., Yang, X., Huang, X., Dunn, J.J., and Luft, B.J. (2005). Structure-based design of a second-generation Lyme disease vaccine based on a C-terminal fragment of *Borrelia burgdorferi* OspA. J. Mol. Biol. *350*, 290–299.

Lee, E.Y., Choi, D.S., Kim, K.P., and Gho, Y.S. (2008). Proteomics in gram-negative bacterial outer membrane vesicles. Mass Spectrom. Rev. *27*, 535–555.

Lindenbach, B.D., Evans, M.J., Syder, A.J., Wolk, B., Tellinghuisen, T.L., Liu, C.C., Maruyama, T., Hynes, R.O., Burton, D.R., McKeating, J.A., et al. (2005). Complete replication of hepatitis C virus in cell culture. Science *309*, 623–626.

Lisziewicz, J., Zeng, G., Gratas, C., Weinstein, J.N., and Lori, F. (2000). Combination gene therapy: synergistic inhibition of human immunodeficiency virus Tat and Rev functions by a single RNA molecule. Hum. Gene Ther. *11*, 807–815.

Luan, S.L., Granlund, M., Sellin, M., Lagergard, T., Spratt, B.G., and Norgren, M. (2005). Multilocus sequence typing of Swedish invasive group B streptococcus isolates indicates a neonatally associated genetic lineage and capsule switching. J. Clin. Microbiol. *43*, 3727–3733.

Lu, Z.S.D., Greiner, R., Lu, P., Wishart, D.S., Poulin, B., Anvik, J., Macdonell, C., and Eisner, R. (2004). Predicting subcellular localization of proteins using machine-learned classifiers. Bioinformatics *20*, 547–556.

Madico, G., Welsch, J.A., Lewis, L.A., McNaughton, A., Perlman, D.H., Costello, C.E., Ngampasutadol, J., Vogel, U., Granoff, D.M., and Ram, S. (2006). The meningococcal vaccine candidate GNA1870 binds the complement regulatory protein factor H and enhances serum resistance. J. Immunol. *177*, 501–510.

Maione, D., Margarit, I., Rinaudo, C.D., Masignani, V., Mora, M., Scarselli, M., Tettelin, H., Brettoni, C., Iacobini, E.T., Rosini, R., et al. (2005). Identification of a universal Group B streptococcus vaccine by multiple genome screen. Science *309*, 148–150.

Manetti, A.G., Zingaretti, C., Falugi, F., Capo, S., Bombaci, M., Bagnoli, F., Gambellini, G., Bensi, G., Mora, M., Edwards, A.M., et al. (2007). *Streptococcus pyogenes* pili promote pharyngeal cell adhesion and biofilm formation. Mol. Microbiol. *64*, 968–983.

Mashburn-Warren, L., McLean, R.J., and Whiteley, M. (2008). Gram-negative outer membrane vesicles: beyond the cell surface. Geobiology *6*, 214–219.

McBroom, A.J., Johnson, A.P., Vemulapalli, S., and Kuehn, M.J. (2006). Outer membrane vesicle production by *Escherichia coli* is independent of membrane instability. J. Bacteriol. *188*, 5385–5392.

Medini, D., Donati, C., Tettelin, H., Masignani, V., and Rappuoli, R. (2005). The microbial pan-genome. Curr. Opin. Genet. Dev. *15*, 589–594.

Melikyan, G.B. (2008). Common principles and intermediates of viral protein–mediated fusion: the HIV-1 paradigm. Retrovirology *5*, 111.

Molloy, M.P., Herbert, B.R., Slade, M.B., Rabilloud, T., Nouwens, A.S., Williams, K.L., and Gooley, A.A. (2000). Proteomic analysis of the *Escherichia coli* outer membrane. Eur. J. Biochem. *267*, 2871–2881.

McLeod, M.P., Qin, X., Karpathy, S.E., Gioia, J., Highlander, S.K., Fox, G.E., McNeill, T.Z., Jiang, H., Muzny, D., Jacob, L.S., et al. (2004). Complete genome sequence of Rickettsia typhi and comparison with sequences of other rickettsiae. J. Bacteriol. *186*, 5842–5855.

Na, S. and Paek, E. (2009). Prediction of novel modifications by unrestrictive search of tandem mass spectra. J. Proteome Res. *8*, 4418–4427.

Nair, R. and Rost, B. (2008). Predicting protein subcellular localization using intelligent systems. Methods Mol. Biol. *484*, 435–463.

Nelson, A.L., Ries, J., Bagnoli, F., Dahlberg, S., Falker, S., Rounioja, S., Tschop, J., Morfeldt, E., Ferlenghi, I., Hilleringmann, M., et al. (2007). RrgA is a pilus-associated adhesin in *Streptococcus pneumoniae*. Mol. Microbiol. *66*, 329–340.

Niu, B., Yu-huan, J., Kai-Yan, F., Wen-Cong, L., Yu-Dong, C., and Guo-Zheng, L. (2008). Using AdaBoost for the prediction of subcellular location of prokaryotic and eukaryotic proteins. Mol. Divers. *12*, 41–45.

Nouwens, A.S., Cordwell, S.J., Larsen, M.R., Molloy, M.P., Gillings, M., Willcox, M.D., and Walsh, B.J. (2000). Complementing genomics with proteomics: the membrane subproteome of *Pseudomonas aeruginosa* PAO1. Electrophoresis *21*, 3797–3809.

Novotny, J., Handschumacher, M., Haber, E., Bruccoleri, R.E., Carlson, W.B., Fanning, D.W., Smith, J.A., and

Rose, G.D. (1986). Antigenic determinants in proteins coincide with surface regions accessible to large probes (antibody domains). Proc. Natl. Acad. Sci. U.S.A. *83*, 226–230.

Ochman, H., and Moran, N.A. (2001). Genes lost and genes found: evolution of bacterial pathogenesis and symbiosis. Science *292*, 1096–1099.

Ott, G., Barchfeld, G.L., Chernoff, D., Radhakrishnan, R., van Hoogevest, P., and Van Nest, G. (1995). MF59. Design and evaluation of a safe and potent adjuvant for human vaccines. Pharm. Biotechnol. *6*, 277–296.

Perkins, D.N., Pappin, D.J., Creasy, D.M., and Cottrell, J.S. (1999). Probability-based protein identification by searching sequence databases using mass spectrometry data. Electrophoresis *20*, 3551–3567.

Perna, N.T., Plunkett, G., 3rd, Burland, V., Mau, B., Glasner, J.D., Rose, D.J., Mayhew, G.F., Evans, P.S., Gregor, J., Kirkpatrick, H.A., et al. (2001). Genome sequence of enterohaemorrhagic *Escherichia coli* O157:H7. Nature *409*, 529–533.

Phadke, N.D., Molloy, M.P., Steinhoff, S.A., Ulintz, P.J., Andrews, P.C., and Maddock, J.R. (2001). Analysis of the outer membrane proteome of *Caulobacter crescentus* by two-dimensional electrophoresis and mass spectrometry. Proteomics *1*, 705–720.

Pizza, M., Scarlato, V., Masignani, V., Giuliani, M.M., Arico, B., Comanducci, M., Jennings, G.T., Baldi, L., Bartolini, E., Capecchi, B., et al. (2000). Identification of vaccine candidates against serogroup B meningococcus by whole-genome sequencing. Science *287*, 1816–1820.

Plotkin, S.A. (2008). Vaccines: correlates of vaccine-induced immunity. Clin. Infect. Dis. *47*, 401–409.

Rappuoli, R. (2004). From Pasteur to genomics: progress and challenges in infectious diseases. Nat. Med. *10*, 1177–1185.

Rappuoli, R. (2007). Bridging the knowledge gaps in vaccine design. Nat. Biotechnol. *25*, 1361–1366.

Read, T.D., Salzberg, S.L., Pop, M., Shumway, M., Umayam, L., Jiang, L., Holtzapple, E., Busch, J.D., Smith, K.L., Schupp, J.M., et al. (2002). Comparative genome sequencing for discovery of novel polymorphisms in *Bacillus anthracis*. Science *296*, 2028–2033.

Rinaudo, C.D., Telford, J.L., Rappuoli, R., and Seib, K.L. (2009). Vaccinology in the genome era. J. Clin. Invest. *119*, 2515–2525.

Rinner, O., Seebacher, J., Walzthoeni, T., Mueller, L.N., Beck, M., Schmidt, A., Mueller, M., and Aebersold, R. (2008). Identification of cross-linked peptides from large sequence databases. Nat. Methods *5*, 315–318.

Robbins, J.B., Schneerson, R., Trollfors, B., Sato, H., Sato, Y., Rappuoli, R., and Keith, J.M. (2005). The diphtheria and pertussis components of diphtheria-tetanus toxoids-pertussis vaccine should be genetically inactivated mutant toxins. J. Infect. Dis. *191*, 81–88.

Rodriguez-Ortega, M.J., Luque, I., Tarradas, C., and Barcena, J.A. (2008). Overcoming function annotation errors in the Gram-positive pathogen *Streptococcus suis* by a proteomics-driven approach. BMC Genomics *9*, 588.

Rodriguez-Ortega, M.J., Norais, N., Bensi, G., Liberatori, S., Capo, S., Mora, M., Scarselli, M., Doro, F., Ferrari, G., Garaguso, I., et al. (2006). Characterization and identification of vaccine candidate proteins through analysis of the group A *Streptococcus* surface proteome. Nat. Biotechnol. *24*, 191–197.

Romain, F., Horn, C., Pescher, P., Namane, A., Riviere, M., Puzo, G., Barzu, O., and Marchal, G. (1999). Deglycosylation of the 45/47-kilodalton antigen complex of *Mycobacterium tuberculosis* decreases its capacity to elicit *in vivo* or *in vitro* cellular immune responses. Infect. Immun. *67*, 5567–5572.

Roumagnac, P., Weill, F.X., Dolecek, C., Baker, S., Brisse, S., Chinh, N.T., Le, T.A., Acosta, C.J., Farrar, J., Dougan, G., et al. (2006). Evolutionary history of *Salmonella typhi*. Science *314*, 1301–1304.

Rubinstein, N.D., Mayrose, I., Halperin, D., Yekutieli, D., Gershoni, J.M., and Pupko, T. (2008). Computational characterization of B-cell epitopes. Mol. Immunol. *45*, 3477–3489.

Russell, J.E., Urwin, R., Gray, S.J., Fox, A.J., Feavers, I.M., and Maiden, M.C. (2008). Molecular epidemiology of meningococcal disease in England and Wales 1975–1995, before the introduction of serogroup C conjugate vaccines. Microbiology *154*, 1170–1177.

Sanchez, S., Abel, A., Marzoa, J., Gorringe, A., Criado, T., and Ferreiros, C.M. (2009). Characterisation and immune responses to meningococcal recombinant porin complexes incorporated into liposomes. Vaccine *27*, 5338–5343.

Santi, I., Pezzicoli, A., Bosello, M., Berti, F., Mariani, M., Telford, J.L., Grandi, G., and Soriani, M. (2008). Functional characterization of a newly identified group B *Streptococcus pullanase* eliciting antibodies able to prevent alpha-glucans degradation. PLoS One *3*, e3787.

Seib, K.L., Serruto, D., Oriente, F., Delany, I., Adu-Bobie, J., Veggi, D., Arico, B., Rappuoli, R., and Pizza, M. (2009). Factor H-binding protein is important for meningococcal survival in human whole blood and serum and in the presence of the antimicrobial peptide LL-37. Infect. Immun. *77*, 292–299.

Seifert, K.N., Adderson, E.E., Whiting, A.A., Bohnsack, J.F., Crowley, P.J., and Brady, L.J. (2006). A unique serine-rich repeat protein (Srr-2) and novel surface antigen (epsilon) associated with a virulent lineage of serotype III *Streptococcus agalactiae*. Microbiology *152*, 1029–1040.

Severin, A., Nickbarg, E., Wooters, J., Quazi, S.A., Matsuka, Y.V., Murphy, E., Moutsatsos, I.K., Zagursky, R.J., and Olmsted, S.B. (2007). Proteomic analysis and identification of *Streptococcus pyogenes* surface-associated proteins. J. Bacteriol. *189*, 1514–1522.

Sierra, G.V., Campa, H.C., Varcacel, N.M., Garcia, I.L., Izquierdo, P.L., Sotolongo, P.F., Casanueva, G.V., Rico, C.O., Rodriguez, C.R., and Terry, M.H. (1991). Vaccine against group B *Neisseria meningitidis*: protection trial and mass vaccination results in Cuba. NIPH Ann *14*, 195–207; discussion 208–110.

Sitaraman, K., and Chatterjee, D.K. (2009). High-throughput protein expression using cell-free system. Methods Mol. Biol. *498*, 229–244.

Skaar, E.P., Humayun, M., Bae, T., DeBord, K.L., and Schneewind, O. (2004). Iron-source preference of *Staphylococcus aureus* infections. Science *305*, 1626–1628.

Stavrum, R., Myneedu, V.P., Arora, V.K., Ahmed, N., and Grewal, H.M. (2009). In-depth molecular characterization of Mycobacterium tuberculosis from New Delhi–predominance of drug resistant isolates of the 'modern' (TbD1) type. PLoS One *4*, e4540.

Stimson, E., Virji, M., Barker, S., Panico, M., Blench, I., Saunders, J., Payne, G., Moxon, E.R., Dell, A., and Morris, H.R. (1996). Discovery of a novel protein modification: alpha-glycerophosphate is a substituent of meningococcal pilin. Biochem. J. *316*, 29–33.

Stranger-Jones, Y.K., Bae, T., and Schneewind, O. (2006). Vaccine assembly from surface proteins of *Staphylococcus aureus*. Proc. Natl. Acad. Sci. U.S.A. *103*, 16942–16947.

Tampakaki, A., Fadouloglou, V.E., Gazi, A.D., Panopoulos, N.J., and Kokkinidis, M. (2004). Conserved features of type III secretion. Cell Microbiol. *6*, 805–806.

Telford, J.L., Barocchi, M.A., Margarit, I., Rappuoli, R., and Grandi, G. (2006). Pili in gram-positive pathogens. Nat. Rev. Microbiol. *4*, 509–519.

Tettelin, H., Medini, D., Donati, C., and Masignani, V. (2006). Towards a universal group B Streptococcus vaccine using multistrain genome analysis. Expert Rev. Vaccines *5*, 687–694.

Tettelin, H., Masignani, V., Cieslewicz, M.J., Donati, C., Medini, D., Ward, N.L., Angiuoli, S.V., Crabtree, J., Jones, A.L., Durkin, A.S., et al. (2005). Genome analysis of multiple pathogenic isolates of *Streptococcus agalactiae*: implications for the microbial 'pan-genome'. Proc. Natl. Acad. Sci. U.S.A. *102*, 13950–13955.

Thornton, J.M., Edwards, M.S., Taylor, W.R., and Barlow, D.J. (1986). Location of 'continuous' antigenic determinants in the protruding regions of proteins. EMBO J. *5*, 409–413.

Thornton, V., Lennon, D., Rasanathan, K., O'Hallahan, J., Oster, P., Stewart, J., Tilman, S., Aaberge, I., Feiring, B., Nokleby, H., et al. (2006). Safety and immunogenicity of New Zealand strain meningococcal serogroup B OMV vaccine in healthy adults: beginning of epidemic control. Vaccine *24*, 1395–1400.

Tjalsma, H., Lambooy, L., Hermans, P.W., and Swinkels, D.W. (2008). Shedding & shaving: disclosure of proteomic expressions on a bacterial face. Proteomics *8*, 1415–1428.

Tritto, E., Mosca, F., and De Gregorio, E. (2009). Mechanism of action of licensed vaccine adjuvants. Vaccine *27*, 3331–3334.

Tropea, J.E., Cherry, S., and Waugh, D.S. (2009). Expression and purification of soluble His(6)-tagged TEV protease. Methods Mol. Biol. *498*, 297–307.

Tseng, T., Tyler, B.M., and Setubal, J.C. (2009). Protein secretion systems in bacterial-host associations, and their description in the Gene Ontology. BMC Microbiol. *9*.

van Leeuwen, W.B., Snoeijers, S., van der Werken-Libregts, C., Tuip, A., van der Zee, A., Egberink, D.,

de Proost, M., Bik, E., Lunter, B., Kluytmans, J., et al. (2002). Intercenter reproducibility of binary typing for *Staphylococcus aureus*. J. Microbiol. Methods *51*, 19–28.

Vaughan, T.E., Skipp, P.J., O'Connor, C.D., Hudson, M.J., Vipond, R., Elmore, M.J., and Gorringe, A.R. (2006). Proteomic analysis of *Neisseria lactamica* and *Neisseria meningitidis* outer membrane vesicle vaccine antigens. Vaccine *24*, 5277–5293.

Vik, A., Aas, F.E., Anonsen, J.H., Bilsborough, S., Schneider, A., Egge-Jacobsen, W., and Koomey, M. (2009). Broad spectrum O-linked protein glycosylation in the human pathogen *Neisseria gonorrhoeae*. Proc. Natl. Acad. Sci. U.S.A. *106*, 4447–4452.

Vipond, C., Suker, J., Jones, C., Tang, C., Feavers, I.M., and Wheeler, J.X. (2006). Proteomic analysis of a meningococcal outer membrane vesicle vaccine prepared from the group B strain NZ98/254. Proteomics *6*, 3400–3413.

Walters, M.S., and Mobley, H.L. (2009). Identification of uropathogenic *Escherichia coli* surface proteins by shotgun proteomics. J. Microbiol. Methods *78*, 131–135.

Wei, J., Goldberg, M.B., Burland, V., Venkatesan, M.M., Deng, W., Fournier, G., Mayhew, G.F., Plunkett, G., 3rd, Rose, D.J., Darling, A., et al. (2003). Complete genome sequence and comparative genomics of Shigella flexneri serotype 2a strain 2457T. Infect. Immun. *71*, 2775–2786.

Weiser, J.N., Goldberg, J.B., Pan, N., Wilson, L., and Virji, M. (1998). The phosphorylcholine epitope undergoes phase variation on a 43-kilodalton protein in *Pseudomonas aeruginosa* and on pili of *Neisseria meningitidis* and *Neisseria gonorrhoeae*. Infect. Immun. *66*, 4263–4267.

Westhof, E., Altschuh, D., Moras, D., Bloomer, A.C., Mondragon, A., Klug, A., and Van Regenmortel, M.H. (1984). Correlation between segmental mobility and the location of antigenic determinants in proteins. Nature *311*, 123–126.

Wu, C.C., MacCoss, M.J., Howell, K.E., and Yates, J.R., 3rd (2003). A method for the comprehensive proteomic analysis of membrane proteins. Nat. Biotechnol. *21*, 532–538.

Wu, C.C., and Yates, J.R., 3rd (2003). The application of mass spectrometry to membrane proteomics. Nat Biotechnol. *21*, 262–267.

Young, N.M., Brisson, J.R., Kelly, J., Watson, D.C., Tessier, L., Lanthier, P.H., Jarrell, H.C., Cadotte, N., St Michael, F., Aberg, E., et al. (2002). Structure of the N-linked glycan present on multiple glycoproteins in the Gram-negative bacterium, *Campylobacter jejuni*. J. Biol. Chem. *277*, 42530–42539.

Yu, C., Lin, C.J., and Hwang, J.K. (2004). Predicting subcellular localization of proteins for Gram-negative bacteria by support vector machines based on n-peptide compositions. Protein Sci. *13*, 1402–1406.

Zhou, T., Xu, L., Dey, B., Hessell, A.J., Van Ryk, D., Xiang, S.H., Yang, X., Zhang, M.Y., Zwick, M.B., Arthos, J., et al. (2007). Structural definition of a conserved neutralization epitope on HIV-1 gp120. Nature *445*, 732–737.

New Analytical Approaches for Measuring Protective Capacity of Antibodies

3

Moon H. Nahm and Carl E. Frasch

Abstract

Antibodies to the pneumococcal polysaccharide capsule protect the host by opsonizing pneumococci for host phagocytes, while antibodies to the meningococcal polysaccharide capsule protect by directly killing meningococci in the presence of complement. *In vitro* measurement of serum bactericidal antibody (SBA) against the meningococcus has been used for a long time as a measure of protective immunity. Technical developments of pneumococcal opsonophagocytosis assays (OPA) in the past decade permit measurements of opsonic capacity of sera from persons immunized with pneumococcal vaccines. Experience with OPAs shows that opsonic capacities of antisera are better than their antibody levels in predicting vaccine efficacy. Thus, measurements of opsonic capacity could be a surrogate of clinical studies of pneumococcal vaccines. By being the surrogate for clinical studies, the assays for protective function of antibodies would reduce the need for large clinical trials and facilitate vaccine developments and improvements.

Introduction

The goal of vaccination is to alter the adaptive immune system to obtain clinical benefits. While direct demonstration of the clinical benefits following the introduction of a vaccine is the best way to establish effectiveness of the vaccine, clinical effects may not be easy to demonstrate. Clinical trials of a vaccine are often costly and account for a major fraction of its development cost. Clinical trials may not be feasible since the effect of a vaccine may require a very long time of follow up (e.g. cancer development). In some cases, completely randomized clinical trials may be ethically impossible. For instance, three pneumococcal conjugate vaccines have been licensed and ethical considerations require one of the approved vaccines to be given to the control group when a new pneumococcal conjugate vaccine is clinically tested.

An alternative to clinical trials is to assess the alteration in the adaptive immune system that provides clinical benefits. The vaccine-induced alteration may include enumeration of new memory T or B cells as well as antibodies specific to the vaccine. Technological tools for enumerating memory T and B cells are being developed but measuring memory B cells may not correlate with protection against rapidly progressing bacterial infections (e.g. *Hemophilus* vaccine) (Kelly *et al.*, 2005). In contrast, protection by many currently available bacterial vaccines appears to be antibody mediated and the amounts of antibodies are often correlated with protection (Kayhty *et al.*, 1983). In some cases, however, the antibody levels cannot be correlated with protection (Lee *et al.*, 2009) perhaps because antibodies bind an epitope too weakly or bind a rarely found epitope. Thus, it is highly desirable in vaccine developments to understand the mechanism of protective immunity and to have *in vitro* tests that directly measure the protective immunity.

To distinguish different measures of immune responses, two terms are often used. A 'correlate of protection' is any immunological measurement, such as antibody concentration, that is strongly associated with a true clinical outcome (protection) in many clinical situations. A 'surrogate of protection' is an *in vitro* measure that simulates the way the immune response provides protection *in vivo*. Antibodies to pneumococcal capsular polysaccharide (PS) provide protection by opsonophagocytic killing *in vivo* as described below. For the pneumococcus, the concentration of type specific antibody may be a correlate of protection but *in vitro* measurement of opsonophagocytic killing by antibodies is a surrogate for the *in vivo* protection. For meningococci, *in vitro* measurement of bactericidal ability of antibodies is a surrogate. The surrogate will be statistically correlated with protection and can take the place of a true clinical endpoint. Therefore, assays for surrogates (assays of antibody function) are very useful in vaccine evaluations.

Importance of meningococcal bactericidal assay in developing meningococcal vaccines

In classic studies Goldschneider and his colleagues showed an inverse relationship between the incidence of meningococcal disease and the age-specific prevalence of serum bactericidal antibody (SBA) to serogroups A, B, and C (Goldschneider et al., 1969). They found that the frequency of meningococcal disease was highest among infants four to 18 months of age, an age group with the fewest number of individuals with detectable SBA. They followed their pioneering studies in development of natural immunity and produced a serogroup C PS vaccine (Gotschlich et al., 1969a,b). This vaccine was used to immunize incoming Army recruits and had an efficacy of about 90%. Importantly, similar serogroup C PS vaccines induced increases in bactericidal antibodies in over 90% of adults (Weibel et al., 1976). These studies thus provided strong evidence for the role of bactericidal antibody as the protective mechanism, the surrogate for protective immunity to the meningococcus. This conclusion is also supported by clinical observations of frequent meningococcal infections among patients with defective membrane attack complex of the complement cascade.

There are now highly effective PS–protein conjugate vaccines against several invasive bacterial diseases of children including those caused by *Haemophilus influenzae* type b, pneumococci and meningococci. Polysaccharides are found on most bacteria causing invasive disease in young children. The reason is that, while PSs are immunogenic and induce protective immunity in older children and adults, they are poorly immunogenic in infants. For example, those pneumococcal serotypes that are the least immunogenic in young children are among the principal serotypes responsible for invasive disease in infants (Robbins et al., 1983). For a PS to induce sufficient antibodies to be protective in an infant, the immunogen must elicit the help of T cells. A conjugate vaccine is generally composed of an intact bacterial PS, size-reduced PS, or oligosaccharides that are covalently attached to a T cell epitope containing polypeptide, generally a protein (Finn, 2004). The immune response to a conjugate resembles closely that to a protein. The result is that T cell help is elicited and young infants make strong IgG responses to a conjugated PS, and generate immunological memory.

The meningococcal group C conjugate vaccine was first introduced into routine use in young children in the UK in 1999 on the basis of immunogenicity and safety data without direct efficacy evaluation (Miller et al., 2001), based upon the just described pioneering work of Goldschneider and his colleagues. Induction of SBA in a high percentage of vaccinees when evaluated with rabbit complement was accepted as indicative of protective efficacy. However, an assay using rabbit serum complement gives higher titres relative to using human complement (Borrow et al., 2001; Santos et al., 2001). This difference is thought to be due to the presence of human complement factor H binding protein on most meningococcal disease isolates, which renders meningococci relatively resistant to killing by human complement, but not by rabbit complement (Granoff et al., 2009). The agreed upon primary immunogenicity endpoint for licensure in the UK was the percent of vaccinees achieving rabbit SBA titres of 8 or greater (Borrow et al., 2001). This was based upon the sentinel observations of Goldschneider et al. (Goldschneider et al., 1969) who showed that protection was correlated with presence of detectable SBA.

Importantly, scientists in the UK were able to validate the use of SBA titres using rabbit complement as a reliable predictor of protection against serogroup C meningococcal disease by comparing vaccine induced bactericidal seroconversion rates with effectiveness found through post-licensure surveillance (Andrews *et al.*, 2003; Balmer *et al.*, 2002). Effectiveness during the first year after immunization in toddlers was closely predicted by the proportion achieving SBA titre higher than 1:8 at 1 month post vaccination. These efforts again established SBA as the surrogate assay and titre of 1:8 as the correlate of protection for group C. A four-valent meningococcal conjugate vaccine against serogroups A, C, Y and W135 is now licensed in the United States, based upon ability of the conjugate to induce bactericidal titres and seroconversion rates to all four serogroups comparable to that achieved by the PS vaccine.

The standardized serum bactericidal assay for meningococci is based upon the assay first described by Maslanka *et al.* (1997) and in detail in 2001 (Borrow, 2001), excepting that human complement should be used to obtain a truer assessment of presence of protective antibody. Briefly, the assay is initiated the evening before by inoculation of the target meningococcal strain to a solid growth medium. Early in the next day, the meningococci grown on the solid growth medium are used to inoculate either another plate or broth for a 3- to 4-h incubation to obtain meningococci in early log phase growth. For the assay, the test sera are serially diluted from 1:4 in flat bottom 96-well tissue culture plates. The target bacteria are then diluted to contain about 100 cfu/25 µl in Dulbecco's phosphate buffered saline containing calcium and magnesium salts, plus 1% glucose and 0.1% gelatin. It is important that full viability of the bacterial inoculum be maintained for 30 to 60 min. Twenty-five microlitres of the target bacteria is added to individual wells containing 25 µl of diluted serum (or control wells without test sera) along with 20 µl of human complement serum. The 96-well plates are covered and then incubated at 37°C for 60 min (optimal time may vary by meningococcal serogroup). Viable counts are determined either by adding semisolid agar directly to the wells (manual counting), or by plating 10 µl aliquots using the 'tilt plate' method

to obtain isolated colonies on one plane for automated counting. The endpoint titre is the highest dilution with > 50% reduction in bacterial count over the zero time control.

Importance of opsonophagocytosis in resistance against pneumococcal infections

While complement-mediated direct killing of meningococci is the primary protective mechanism of meningococcal antibodies, many observations suggest that *in vivo* defence against pneumococcal infections depends on opsonophagocytosis involving phagocytes, early complement components, and antibodies. The need for neutrophils in defence against pneumococci is clearly shown in neutropenic patients receiving chemotherapy (Johansson *et al.*, 1992) or in animals made neutropenic (Chen *et al.*, 2001; Zuluaga *et al.*, 2006). Also, the need for antibodies to pneumococcal capsular PS is shown by patients with Wiskott–Aldrich syndrome (Blaese *et al.*, 1968) or Bruton's agammaglobulinaemia (Lederman and Winkelstein, 1985). These patients are unable to make antibodies to pneumococcal capsular PS and are thus susceptible to pneumococcal infections but become resistant to pneumococcal infections after gamma globulin treatments. In addition, the need for C3 and CR3 has been clearly established in defence against pneumococcal infections (Prince *et al.*, 2001; Ren *et al.*, 2004; Winkelstein, 1981). Studies of C1q deficiencies showed that the classical pathway is the main complement activation pathway required for the host protection, and the lectin pathway may only slightly influence C3 deposition in some cases (Hostetter, 2004). Complement-mediated bacteriolysis is probably not important in pneumococcal infections since deficiencies in the lytic cascade involving components C5 through C9 are associated with an increased risk of meningococcal but not pneumococcal infections (Ross and Densen, 1984).

The *in vivo* importance of opsonophagocytosis is further supported by the fact that several pneumococcal virulence molecules enhance opsonization avoidance. Pathogenic pneumococci have carbohydrate capsules which do not fix complement and thus act as a shield that covers the inner structures attracting host opsonins. The thickness of the capsule can be associated with resistance to opsonization (Kim *et al.*, 1999) and pathogenicity (MacLeod and Kraus, 1950). Many pathogenic pneumococci also express molecules like PspA and PspC (CbpA) that are associated with deactivating complement (Hostetter, 2004; Quin *et al.*, 2005; Ren *et al.*, 2004). Considered together, all of these observations indicate that the antibody's ability to enhance opsonophagocytosis should be a good measure of pneumococcal vaccine-induced immunity.

To study the cellular processes involved in opsonophagocytosis of pneumococci *in vivo*, many investigators have developed an *in vitro* assay for opsonophagocytosis (Stuart and Ezekowitz, 2005). The *in vitro* assay showed that the antibodies immobilized on the bacteria can bind to one of several receptors for the Fc portions of IgG antibodies (FcgRs) on the host phagocytes, activate them, induce pseudopodia formation, and initiate the phagocytic process (Caron and Hall, 1998). CD64 (FcγRI) is the high-affinity receptor for antibodies of the IgG1, IgG3, and IgG4 subclasses (van de Winkel and Anderson, 1991). CD32 (FcγRII) is the low-affinity receptor for IgG1, IgG2, and IgG3 antibodies (van de Winkel and Anderson, 1991). CD32 has two alleles: allele H131 has histidine at residue 131; and allele R131 has arginine at the same location. R131 allele has a low affinity for IgG2 antibodies (van de Winkel and Anderson, 1991). HL-60 cell line that is used as phagocytes in the opsonophagocytic assay (OPA) is homozygous for the R131 allele (Fleck *et al.*, 2005). FcR-mediated opsonization may have some *in vivo* relevance (Sanders *et al.*, 1994), but the deletion of all FcRs has no effect on the ability of anticapsular antibodies to passively protect against pneumococcal challenge (Sael *et al.*, 2003).

The immobilized antibodies further opsonize bacteria by activating complement proteins and coating the bacteria with C3b, iC3b, and C3d, which represent C3 at three different stages of degradation. Of these, iC3b strongly binds to CR3 (CD11b/CD18) and is the most powerful opsonin. Upon recognizing iC3b, phagocytes take up bacteria without inducing pseudopodia

formation and pro-inflammatory mediator production (Caron and Hall, 1998). Affinity of CR3 to iC3b increases when the phagocytes are activated (Brown and Gresham, 2003) and complement-mediated phagocytosis occurs most effectively with activated phagocytes. C3b and C3d also bind, respectively, to CD35 (CR1) on granulocytes and CR2 on B cells, but their role in phagocytosis seems to be minor. After phagocytes ingest the bacteria, the resulting phagosomes containing the bacteria are fused with granules containing enzymes in granulocytes or with lysosomes in macrophages. This fusion is followed by a prominent burst of oxidation, activation of phagocytic enzymes, and killing of the bacteria. Although some bacteria are killed slowly within phagocytes (Gresham *et al.*, 2000), the entire process of recognition, ingestion, and killing of bacteria is generally very rapid (15–20 min) (Hampton and Winterbourn, 1999). Consequently, *in vitro* OPA for pneumococci requires relatively short incubation periods.

Development of analytical technologies for *in vitro* opsonophagocytosis assay

The *in vitro* OPA was widely used to investigate the biological process required for opsonophagocytosis. However, despite its attractiveness as a surrogate assay for measuring pneumococcal vaccine-induced immunity, till recently, it was generally accepted that the *in vitro* OPA was not sufficiently robust, efficient, reproducible, and standardized enough to be capable of producing precise results on multiple serotypes for many samples with only small amounts of sera. Thus, the initial licensure of Prevnar was based on a clinical trial and serological data produced with ELISA (Rennels *et al.*, 1998; Shinefield and Black, 2000). The *in vitro* OPA was used only as the supportive measure of vaccine-induced protective immunity. Classic efficacy trials with new pneumococcal vaccines with additional serotype coverage have become impossible to perform due to logistic and ethical reasons when three conjugate vaccines are licensed and clinically used. Also, experience showed that ELISA is not sufficient. For instance, Prevnar induces high levels of anti-19A antibody (Lee *et al.*, 2009) but Prevnar did not provide protection against 19A. These circumstances have provided strong impetus to develop and improve an *in vitro* OPA. For the last 10 years, investigations of analytical technologies have been under way to develop a robust and well standardized *in vitro* OPA. As a result of those efforts, OPA methods have significantly changed and a historical approach was used to describe the changes.

Opsonophagocytic killing process can be replicated *in vitro*

In the classical *in vitro* OPA system, target bacteria are exposed to antibodies, complement, and phagocytes, followed by enumeration of the number of surviving bacteria. However, this classical approach is very tedious to perform, primarily due to the counting of colonies. Consequently, many researchers have developed various alternative OPA methods requiring no colony counting, including a radiolabelled bacteria uptake assay (Nowak-Wegrzyn *et al.*, 2000; Vakevainen *et al.*, 2001), a fluorescent bacteria uptake assay (Jansen *et al.*, 1998; Martinez *et al.*, 1999; Vakevainen *et al.*, 2001), chemiluminescence (Bortolussi *et al.*, 1981), and an oxidative burst generation assay (Shah *et al.*, 2006). While these assays have been used with varying degrees of success, all the methods were sufficiently cumbersome and none of them gained a wide acceptance.

A killing type OPA was developed as a reference assay

Many investigators have been improving the classical killing type OPA over a long time period. Several investigators began to use 96-well microtitre plates instead of individual test tubes (Gray, 1990). Use of these plates not only reduced the volume of the required reagents but also facilitated testing of a large number of samples. Some investigators began to use frozen pneumococci as target bacteria for infections (Aaberge *et al.*, 1995; Gray, 1990). This has not only greatly simplified the preparation of target bacteria but also standardized bacterial counts and greatly reduced assay variability. Dr. Porter Anderson in Rochester NY introduced the use of differentiated HL-60 cells as phagocytes (Romero-Steiner *et al.*, 1997) and this innovation permitted standardization of

phagocytes, which is one of the most significant variables in the OPA.

Recognizing the importance of these innovations in developing an OPA suitable for vaccine studies, investigators at the Centres for Disease Control and Prevention (CDC) incorporated all these innovations and developed a 'standardized' OPA (Romero-Steiner *et al.*, 1997). These investigators also introduced use of micro-colonies of pneumococci as assay readouts. The counting of micro-colonies was important in handling a large number of samples, because it dramatically reduced biological wastes by allowing one to place reaction mixtures from many reaction wells in a single Petri dish. These investigators found that, for their OPA method, human complement can be replaced with baby rabbit complement, a commercially available material. In addition, the CDC made their target bacteria widely available and published a detailed protocol enabling others to duplicate their OPA procedure. Since no other pneumococcal antibody OPA has been so extensively evaluated and standardized at the time, this OPA method was widely adopted by various investigators for clinical studies and has served as the 'gold standard' reference method until now (Romero-Steiner *et al.*, 2006).

In addition, the CDC OPA served as the basis for assays developed by two pneumococcal vaccine manufacturers: Wyeth and GSK (Henckaerts *et al.*, 2006; Hu *et al.*, 2005). While these two widely used OPAs are both single-serotype killing assays using HL-60 cells and use rabbit complement, they differ in several aspects from the CDC OPA. For instance, the assay was modified at Wyeth to add complement early in the opsonization reaction when the reaction volume is only half of its final level. This modification results in the transient exposure of target bacteria to a 2-fold higher complement concentration. In the GSK assay, opsonization and phagocytosis phases were merged. Also, both assays use different target bacterial strains than does the CDC OPA. The significance of these differences has not yet been determined.

While development of the single-serotype killing OPA was an important milestone, the assay had several limitations. First, many investigators in Europe had difficulty using HL-60 cells as phagocytes. Further investigation later showed that HL-60 cell lines provided by European cell banks were not useful as phagocytes whereas HL-60 cells from the USA are (Fleck *et al.*, 2005). This example further illustrates the need to standardize the assay. Second, this OPA did not employ a standard serum to normalize the results. Consequently, when the CDC investigators led the first attempt to standardize OPA among five different laboratories, inter-laboratory variation in OPA results was still significant: 75% of the titres were within one dilution of the median titre and 88% of the titres were within two dilutions of the median titre (Romero-Steiner *et al.*, 2003a). Lastly, the most serious limitation of the method was that micro-colonies were visualized under the dissection microscope and colony counting was too tedious to be used routinely.

Several groups of investigators overcame the difficulties of colony counting. One approach was to use chromogenic (Lin *et al.*, 2001) or fluorogenic dyes (Bieging *et al.*, 2005), which produce optical signals proportional to the number of bacteria in the reaction wells. While this approach worked, it was cumbersome and direct counting of bacterial colonies was preferable. Ultimately, several groups independently found ways to rapidly count micro-colonies. Two groups succeeded in adapting counters developed for Spot ELISA to count OPA's microcolonies (Henckaerts *et al.*, 2006; Liu *et al.*, 2004). Another group found that a tetrazolium dye (TTC) added to agar plates will be taken up by the pneumococci and stain the colonies red (Kim *et al.*, 2003; Romero-Steiner *et al.*, 2003b) thus making the microcolonies discernible to conventional bacterial colony counters (Kim *et al.*, 2003). In fact, the TTC-stained microcolonies are so distinct that digital camera images of the TTC-stained microcolonies can be e-mailed for colony counting (Putman *et al.*, 2005). Public domain software for colony counting has been designed to work with images obtained with widely available inexpensive digital cameras and is available from US NIST (nice@nist.gov). The software will make automated colony counting readily available. In summary, by 2009, colony counting had become as fast and easy to read as the optical densities of ELISA plates.

Development of a flow cytometer-based uptake OPA

Because colony counting was so cumbersome, several groups adapted flow cytometers to determine the number of fluorescent bacteria that had been ingested by phagocytes (Jansen *et al.*, 1998; Martinez *et al.*, 1999). This uptake assay obviated the need to count colonies and could be adapted to handle a large number of samples. Also, the uptake assay could use either killed bacteria or antigen-coated latex particles and could analyse serum samples containing antibiotics (Martinez *et al.*, 2006). Studies found that the uptake assay results correlate with those of the classical killing assay, although the uptake assays were not as sensitive as the killing assay (Vakevainen *et al.*, 2001). Both the killing and uptake assays measure the process of opsonophagocytosis, however the uptake assays use non-viable targets and therefore do not measure the proportion of killed bacteria. Furthermore, the speed of counting ingested bacteria with a flow cytometer is significantly slower than is automated micro-colony counting.

Development of multiplex OPA methodologies

Currently available pneumococcal conjugate vaccines contain capsular PSs from 7 to 13 different serotypes, and conjugate vaccine evaluations require performing OPAs on a large number of serotypes for each serum sample. There is a need for assays that can perform many assays with a small volume of serum since one cannot obtain much serum from young children. To meet this need, two different multiplexed OPAs have been developed.

One approach is to measure phagocytosis using a multiplexed flow cytometry assay. In this uptake assay, different bacteria are tagged with different fluorochromes and a mixture of the different bacteria is allowed to be ingested by phagocytes in the presence of antibody and complement. The number of each type of target bacteria in the phagocytes can then be determined by multi-colour flow cytometry (Martinez *et al.*, 2006). In some cases, fluorescent target bacteria are replaced with fluorescent latex particles coated with capsular PS (Martinez *et al.*, 2006). One multiplex assay using four different types of fluorescent latex particles has been developed for seven serotypes (4, 6B, 9V, 14, 18C, 19F, and 23F). The results correlated with those obtained with the reference killing assay, with r values ranging 0.68 and 0.92 (Martinez *et al.*, 2006). Compared with automated colony counting, this assay is relatively slow in enumerating ingested target particles (Burton and Nahm, 2006; Martinez *et al.*, 2006). Also, one must assume that PS-coated latex particles faithfully mimic the phagocytosis and PS epitope expression of natural bacteria. However, this assay has the advantages of not requiring bacterial culture and of being able to be performed with samples containing antibiotics.

Another approach is based on the killing type OPA using a mixture of antibiotic-resistant pneumococci as target bacteria (Bogaert *et al.*, 2004; Burton and Nahm, 2006; Kim *et al.*, 2003; Nahm *et al.*, 2000). For this assay, different target bacteria are prepared to be resistant to different antibiotics. A mixture of target bacteria is then combined with phagocytes, antibodies, and complement. The number of surviving bacteria in each serotype is determined by plating the reaction mixture on agar plates containing antibiotics for selection of the organisms that are resistant. This multiplexed OPA (MOPA) method is very similar to the classical killing assay, its biological relevance is easy to accept, and it requires no special equipment. Consequently, many laboratories can readily adopt the assay method. Indeed, this approach is robust enough that one can perform 2-fold (Nahm *et al.*, 2000) to 7-fold (Bogaert *et al.*, 2004) multiplexed assays. Also, this approach can be applied to an OPA using a dye whose fluorescence is proportional to the number of bacteria (Bieging *et al.*, 2005). The multiplex killing method was favoured in the 2005 OPA meeting held in Atlanta (Romero-Steiner *et al.*, 2006).

Recently, a well-characterized 4-fold multiplexed, killing type OPA was described for 13 serotypes (1, 3, 4, 5, 6A, 6B, 7F, 9V, 14, 18C, 19A, 19F, and 23F) (Burton and Nahm, 2006). The assay procedure is very similar to the CDC OPA (Romero-Steiner *et al.*, 1997), but its assay conditions were further optimized and the assay performance was validated. Consequently, the multiplex assay results are highly correlated with the single-serotype OPA results (r>0.97) and

are relatively precise. The 4-fold multiplexed OPA reduces the required amount of serum significantly enough to be useful for studying infants. Also, the detailed assay protocol and critical reagents are readily available. This 4-fold multiplex OPA may be useful as a reference assay in the future.

Currently available pneumococcal MOPA is efficient and robust enough to be useful in analysing a large number of samples. In my (MHN) laboratory, one person can analyse about 40 serum samples for 4 serotypes in one assay run and can analyse a large number of samples. GSK also has independently developed an OPA protocol that can handle a large number of samples. Thus, even though pneumococcal OPA is a bioassay, it can be used in a large scale.

At a WHO workshop in Geneva in 2007 a multilab OPA study was organized to compare OPAs as then performed in five different labs all using a set of 24 blinded samples. The results were presented in 2008 at a WHO/Health Canada meeting in Ottawa (Feavers *et al.*, 2009). The study showed that, despite the lack of any standardization, all different OPA systems are satisfactory and produced comparable (acceptable) results. Additional work should be done to standardize the assay by additional international collaborations.

Experience with OPA and ELISA in pneumococcal vaccine trials

Experience in the last few years with pneumococcal conjugate vaccines has produced several occasions to compare usefulness of OPA and ELISA. A recent study used both ELISA and OPA methods to compare two pneumococcal conjugate vaccines in a large number of children (about 1600) (Vesikari *et al.*, 2009). These examples have established that OPA can be used in clinical studies. Several specific examples where OPA has been particularly useful are discussed below.

Cross-reactive serotypes (serotype 19A)

Pneumococcal vaccines are formulated to provide protection against cross-reactive serotypes. For instance, PCV7 contains 6B and 19F PSs, and is expected to provide cross-protection against serotypes 6A and 19A. One of the most surprising findings with PCV7 has been the absence of cross-protection against serotype 19A. ELISA-based serological studies showed that PCV7 from Wyeth (Lee *et al.*, 2009) and its closely related experimental 5-valent conjugate vaccine (Yu *et al.*, 1999) elicits a robust antibody response against both 19F and 19A serotypes. Thus, it was widely believed that PCV7 from Wyeth would provide a good protection against both serotypes. However, soon after the introduction of Prevnar, the prevalence of serotype 19A unexpectedly increased although the prevalence of 19F decreased as expected (Beall, 2007; Moore *et al.*, 2008; Pai *et al.*, 2005; Park *et al.*, 2008).

Although the increase may have been partially due to the fact that 19A is often antibiotic resistant (Choi *et al.*, 2008), OPA results predict low cross-protection. In 1999, it was observed that a pentavalent conjugate vaccine available from Wyeth elicited poor opsonic titres to serotype 19A even though it elicited good antibody response to serotype 19A when the response was measured with ELISA (Yu *et al.*, 1999). The pentavalent vaccine is substantially similar to Prevnar except for two serotypes found only in Prevnar. In a more recent study, Lee *et al.* directly compared antibody responses and opsonic indices induced in young children with Prevnar. They found that Prevnar elicits strong antibody responses to 19A as measured by ELISA but protective immunity measured with OPA is strikingly less in 19A serotype (Lee *et al.*, 2009). Similar findings were reported by two other groups (Fernsten *et al.*, 2008) (Vesikari *et al.*, 2009). These examples clearly show that OPA predicts immune protection better than ELISA when cross-reactive antibodies are assessed.

Elderly adults

While ELISA and OPA results are correlated well for young children ($r = 0.5$–0.9) (Ekstrom *et al.*, 2007; Puumalainen *et al.*, 2003), most studies with adults show low correlations between the two assays. For instance, ELISA and OPA results showed r values ranging from 0.18 to 0.6 for serum samples from adults with renal transplant (Kumar *et al.*, 2003) and from 0.3 to 0.6 for

samples from elderly adults (> 65 years old), who were vaccinated with PPV23 (Romero-Steiner *et al.*, 1999). Limited information does support the contention that OPA may be more reflective of immune protection than ELISA. In elderly adults PPSV23 is not as effective as it is in young adults (Shapiro *et al.*, 1991). Yet, ELISA-based studies showed that old adults can produce as much antibody as young adults (Romero-Steiner *et al.*, 1999). However, elderly adults produced antibodies that are less opsonic than those produced by young adults (Romero-Steiner *et al.*, 1999; Schenkein *et al.*, 2008). Also, a recent study found that PCV7-CRM elicits antibodies with superior functional capacity than PPSV23 in old adults (Jackson *et al.*, 2007). Opsonization depends on antibody avidity as well as on antibody concentration (Romero-Steiner *et al.*, 1999; Sun *et al.*, 2001; Usinger and Lucas, 1999). These discrepancies between ELISA and OPA results may be due to the low avidity of the antibodies made by old adults (Romero-Steiner *et al.*, 1999). These findings support the usefulness of OPA in assessing pneumococcal vaccines among adults.

HIV patients

HIV infected persons are prone to pneumococcal infections and pneumococcal vaccines can be beneficial to this population. Clinical trials found that PCVs are less effective in HIV-infected children than in normal children (Klugman *et al.*, 2003). Yet, PCVs elicit equivalent levels of pneumococcal antibodies in both healthy and HIV-infected children as determined by ELISA (King *et al.*, 1997; Madhi *et al.*, 2005). However, HIV-infected children have lower OPA titres than HIV-uninfected children in response to PCV immunization (Madhi *et al.*, 2005). Thus, OPA may be better than ELISA in reflecting clinical trial results.

Future directions

Over the last several years, we have seen dramatic changes in assays for antibody function. Increasing evidence now supports the usefulness of MOPA in pneumococcal vaccine evaluations. Thus, there are several urgent requirements for pneumococcal MOPA. Also, there are increasing interests to expand the pneumococcal MOPA experience to other pathogens and other assay types. These are described below.

Future directions of pneumococcal MOPA

Development of a simple and reliable assay alone is not sufficient for a clinically useful assay. As the assays are used at multiple locations by multiple users, one should recognize that assays are altered for various reasons – local conditions, improving efficiency. Thus, assay conditions were investigated in detail to identify critical parameters. The critical parameters of MOPA were investigated by a reference laboratory and made available to the public. Also, a preliminary study showed that MOPA and other opsonization assay formats produce results that are largely comparable (Feavers *et al.*, 2009). Further standardization in MOPA would be achievable with the use of a reference serum, which is being developed under the leadership of FDA and NIH. Wider clinical uses of an assay require the assay to be correlated with clinical outcomes. It is likely that many vaccine trials incorporate MOPA to assess their vaccine response and these studies will yield the minimal level of opsonization index required for protection against different diseases in various populations.

Another application of MOPA in the future is to study antibodies to various pneumococcal surface exposed protein antigens that are being investigated as a potential candidate for pneumococcal vaccines. For instance, PspA has been extensively studied as vaccine candidate (Briles *et al.*, 2000). Since PspA is thought to provide protection by reducing complement deposition on bacteria (Ren *et al.*, 2004), antibodies to PspA should enhance the opsonophagocytosis and opsonophagocytosis assay may be appropriate. With these antigens, it may be important to recognize that the increased opsonization is weak and one may have to use special target bacteria to observe small degree of opsonization *in vitro*. Multiplexed OPA may be relevant because these proteins are serologically heterogeneous and protection is likely serotype dependent.

However, there are other candidate pneumococcal proteins such as pneumolysin, pili or PsaA, which are involved in bacterial adhesion

or a metal ion transporter (Briles *et al.*, 2000). Antibodies to these protein antigens may provide protection by ways other than opsonizing bacteria. For instance, antibodies to pneumolysin may provide protection by neutralizing the toxin and other antibodies may be involved in reducing inflammation or activating T cells, or interfering with nutrient uptake or adhesion. Thus, application of the *in vivo* functional assays requires our understanding of the pathogenic role that each antigen provides.

Application of the MOPA technology to other target bacteria and other assay formats

While there is much work still needed to be done to develop pneumococcal antibody MOPA into a standardized assay useful in pneumococcal vaccine evaluations, the example has shown that MOPA is a robust technological platform useful for other target bacteria. The other target bacteria may be staphylococci or group A or group B streptococci. MOPA may be particularly important for the group B streptococcus as clinical disease endpoint efficacy will not be practical. As MOPA technology is applied to additional target bacteria, one must be aware that each bacterial species behave differently. For instance, staphylococci tend to avoid lysosome fusion and may require longer time for killing.

Another interesting approach may be to engineer bacteria that are each dependent on a specific nutrient. This would make the multiplexed assay operational but this approach would be especially useful when the target bacteria are virulent. Additional improvements may be possible. For instance, phagocytes may be cryopreserved in a central facility and can be distributed to users. This will eliminate the need to perform tissue culture in the vaccine evaluation laboratory and make the assay much more widely available.

In addition to opsonophagocytosis, the assay technology can be used for complement mediated bactericidal assays. For instance, functional capacity of *Hemophilus* antibodies can be measured by the ability of antibodies to fix complement and kill *Hemophilus* bacteria (Nahm *et al.*, 1995; Tomlinson *et al.*, 1989).

MOPA may not be the only way to measure antibody function

Because of strong desire to develop assays that measure functional aspects of antibodies, several alternative approaches have been developed. An alternative approach may be to measure the level of antibodies binding to a specific epitope that is associated with protective antibodies. This is useful when the protective capacities of antibodies cannot be measured easily. For instance, there are monoclonal antibodies that have been shown to neutralize HPV. Since it is not easy to perform *in vitro* neutralization assays for antibodies to HPV, one may assume that any antibody binding to this protective epitope should be protective. Using this logic, investigators at Merck developed an inhibition type assay for antibodies recognizing the specific epitope (Palker *et al.*, 2001). By using a multiplex inhibition assay, investigators at Merck adapted this assay to measure antibodies to multiple epitopes of HPV (Opalka *et al.*, 2003). The assay was used to successfully develop their HPV vaccine. Nevertheless, a simple HPV ELISA correlates well with the functional antibody levels (Dessy *et al.*, 2008). Thus, the usefulness of this new approach requires additional experience.

Acknowledgements

The work was supported by NIH NO1-AI-30021 (MHN), which supports the NIH bacterial respiratory pathogen reference laboratory.

References

Aaberge, I.S., Eng, J., Lermark, G., and Lovik, M. (1995). Virulence of *Streptococcus pneumoniae* in mice: a standardized method for preparation and frozen storage of the experimental bacterial inoculum. Microb. Pathog. *18*, 141–152.

Andrews, N., Borrow, R., and Miller, E. (2003). Validation of serological correlate of protection for meningococcal C conjugate vaccine by using efficacy estimates from postlicensure surveillance in England. Clin. Diagn. Lab. Immunol. *10*, 780–786.

Balmer, P., Borrow, R., and Miller, E. (2002). Impact of meningococcal C conjugate vaccine in the UK J. Med. Microbiol. *51*, 717–722.

Beall, B. (2007). Vaccination with the pneumococcal 7–valent conjugate: a successful experiment but the species is adapting. Expert Rev. Vaccines 6, 297–300.

Bieging, K., Rajam, G., Holder, P., Udoff, R., Carlone, G.M., and Romero-Steiner, S. (2005). A fluorescent multivalent opsonophagocytic assay for the measurement of functional antibodies to *Streptococcus*

pneumoniae. Clin. Diagn. Lab. Immunol. *12*, 1238–1242.

Blaese, R.M., Strober, W., Brown, R.S., and Waldmann, T.A. (1968). The Wiskott–Aldrich syndrome. A disorder with a possible defect in antigen processing or recognition. Lancet *1*, 1056–1061.

Bogaert, D., Sluijter, M., De Groot, R., and Hermans, P.W. (2004). Multiplex opsonophagocytosis assay (MOPA): a useful tool for the monitoring of the 7-valent pneumococcal conjugate vaccine. Vaccine 22, 4014–4020.

Borrow, R., Andrews, N., Goldblatt, D., and Miller, E. (2001). Serological basis for use of meningococcal serogroup C conjugate vaccines in the United Kingdom: reevaluation of correlates of protection. Infect. Immun. 69, 1568–1573.

Borrow, R., and Carlone, G.M. (2001). Serogroup B and C Serum Bactericidal Assays. (Totowa, NJ, Humana Press).

Bortolussi, R., Marrie, T.J., Cunningham, J., and Schiffman, G. (1981). Serum antibody and opsonic responses after immunization with pneumococcal vaccine in kidney transplant recipients and controls. Infect. Immun. 34, 20–25.

Briles, D.E., Paton, J.C., Swiatlo, E., and Nahm, M.H. (2000). Pneumococcal vaccines. In Gram-positive pathogens, Fischetti, V.A., Novick, R.P., Ferretti, J.J., Portnoy, D.A., and Rood, J.I., eds (Washington DC, ASM Press), pp. 244–250.

Brown, E.J., and Gresham, H.D. (2003). Phagocytosis. In Fundamental Immunology, Paul, E., ed. (Philadelphia, Lippincott Williams & Wilkins), pp. 1105–1126.

Burton, R.L., and Nahm, M.H. (2006). Development and validation of a fourfold multiplexed opsonization assay (MOPA4) for pneumococcal antibodies. Clin. Vaccine Immunol. *13*, 1004–1009.

Caron, E., and Hall, A. (1998). Identification of two distinct mechanisms of phagocytosis controlled by different Rho GTPases. Science 282, 1717–1721.

Chen, S.C., Mehrad, B., Deng, J.C., Vassileva, G., Manfra, D.J., Cook, D.N., Wiekowski, M.T., Zlotnik, A., Standiford, T.J., and Lira, S.A. (2001). Impaired pulmonary host defense in mice lacking expression of the CXC chemokine lungkine. J. Immunol. 166, 3362–3368.

Choi, E.H., Kim, S.H., Eun, B.W., Kim, S.J., Kim, N.H., Lee, J., and Lee, H.J. (2008). *Streptococcus pneumoniae* serotype 19A in Children, South Korea. Emerg. Infect. Dis. *14*, 275–281.

Dessy, F.J., Giannini, S.L., Bougelet, C.A., Kemp, T.J., David, M.P., Poncelet, S.M., Pinto, L.A., and Wettendorff, M.A. (2008). Correlation between direct ELISA, single epitope-based inhibition ELISA and pseudovirion-based neutralization assay for measuring anti-HPV-16 and anti-HPV-18 antibody response after vaccination with the AS04-adjuvanted HPV-16/18 cervical cancer vaccine. Hum. Vaccin. *4*, 425–434.

Ekstrom, N., Vakevainen, M., Verho, J., Kilpi, T., and Kayhty, H. (2007). Functional antibodies elicited by two heptavalent pneumococcal conjugate vaccines in the Finnish Otitis Media Vaccine Trial. Infect. Immun. 75, 1794–1800.

Feavers, I., Knezevic, I., Powell, M., and Griffiths, E. (2009). Challenges in the evaluation and licensing of new pneumococcal vaccines, 7–8 July 2008, Ottawa, Canada. Vaccine 27, 3681–3688.

Fernsten, P., Hu, B., and Yu, X. (2008). Specificity of the opsonic response to serotype 19A and 19F conjugate vaccines. Paper presented at: 6th International Symposium on Pneumococcal and Pneumococcal Diseases (Reykjavik, Iceland).

Finn, A. (2004). Bacterial polysaccharide–protein conjugate vaccines. Br. Med. Bull. 70, 1–14.

Fleck, R.A., Romero-Steiner, S., and Nahm, M.H. (2005). Use of HL-60 cell line to measure opsonic capacity of pneumococcal antibodies. A review. Clin. Diagn. Lab. Immunol. 12, 19–27.

Goldschneider, I., Gotschlich, E.C., and Artenstein, M.S. (1969). Human immunity to the meningococcus. I. The role of humoral antibodies. J. Exp. Med. 129, 1307–1326.

Gotschlich, E.C., Goldschneider, I., and Artenstein, M.S. (1969a). Human immunity to the meningococcus. IV. Immunogenicity of group A and group C meningococcal polysaccharides in human volunteers. J. Exp. Med. 129, 1367–1384.

Gotschlich, E.C., Liu, T.Y., and Artenstein, M.S. (1969b). Human immunity to the meningococcus. III. Preparation and immunochemical properties of the group A, group B, and group C meningococcal polysaccharides. J. Exp. Med. 129, 1349–1365.

Granoff, D.M., Welsch, J.A., and Ram, S. (2009). Binding of complement factor H (fH) to Neisseria meningitidis is specific for human fH and inhibits complement activation by rat and rabbit sera. Infect. Immun. 77, 764–769.

Gray, B.M. (1990). Opsonophagocidal activity in sera from infants and children immunized with Haemophilus influenzae type b conjugate vaccine (meningococcal protein conjugate). Pediatrics 85, 694–697.

Gresham, H.D., Lowrance, J.H., Caver, T.E., Wilson, B.S., Cheung, A.L., and Lindberg, F.P. (2000). Survival of Staphylococcus aureus inside neutrophils contributes to infection. J. Immunol. 164, 3713–3722.

Hampton, M.B., and Winterbourn, C.C. (1999). Methods for quantifying phagocytosis and bacterial killing by human neutrophils. J. Immunol. Methods 232, 15–22.

Henckaerts, I., Goldblatt, D., Ashton, L., and Poolman, J. (2006). Critical differences between pneumococcal polysaccharide enzyme-linked immunosorbent assays with and without 22F inhibition at low antibody concentrations in pediatric sera. Clin. Vaccine Immunol. 13, 356–360.

Hostetter, M.K. (2004). Interactions of Streptococcus pneumoniae with the proteins of the complement pathways. In The Pneumococcus, Tuomanen, E.I., ed. (Washington, D.C., ASM Press), pp. 201–210.

Hu, B.T., Yu, X., Jones, T.R., Kirch, C., Harris, S., Hildreth, S.W., Madore, D.V., and Quataert, S.A. (2005). Approach to validating an opsonophagocytic assay for Streptococcus pneumoniae. Clin. Diagn. Lab. Immunol. 12, 287–295.

Jackson, L.A., Neuzil, K.M., Nahm, M.H., Whitney, C.G., Yu, O., Nelson, J.C., Starkovich, P.T., Dunstan, M., Carste, B., Shay, D.K., et al. (2007). Immunogenicity of varying dosages of 7-valent pneumococcal polysaccharide–protein conjugate vaccine in seniors previously vaccinated with 23-valent pneumococcal polysaccharide vaccine. Vaccine 25, 4029–4037.

Jansen, W.T., Gootjes, J., Zelle, M., Madore, D.V., Verhoef, J., Snippe, H., and Verheul, A.F. (1998). Use of highly encapsulated Streptococcus pneumoniae strains in a flow-cytometric assay for assessment of the phagocytic capacity of serotype-specific antibodies. Clin. Diagn. Lab. Immunol. 5, 703–710.

Johansson, P.J., Sternby, E., and Ursing, B. (1992). Septicemia in granulocytopenic patients: a shift in bacterial etiology. Scand. J. Infect. Dis. 24, 357–360.

Kayhty, H., Peltola, H., Karanko, V., and Makela, P.H. (1983). The protective level of serum antibodies to the capsular polysaccharide of Haemophilus influenzae type b. J. Infect. Dis. 147, 1100.

Kelly, D.F., Pollard, A.J., and Moxon, E.R. (2005). Immunological memory: the role of B cells in long-term protection against invasive bacterial pathogens. JAMA 294, 3019–3023.

Kim, J.O., Romero-Steiner, S., Sorensen, U.B., Blom, J., Carvalho, M., Barnard, S., Carlone, G., and Weiser, J.N. (1999). Relationship between cell surface carbohydrates and intrastrain variation on opsonophagocytosis of Streptococcus pneumoniae. Infect. Immun. 67, 2327–2333.

Kim, K.H., Yu, J., and Nahm, M.H. (2003). Efficiency of a pneumococcal opsonophagocytic killing assay improved by multiplexing and by coloring colonies. Clin. Diagn. Lab. Immunol. 10, 616–621.

King, J.C., Jr., Vink, P.E., Farley, J.J., Smilie, M., Parks, M., and Lichenstein, R. (1997). Safety and immunogenicity of three doses of a five-valent pneumococcal conjugate vaccine in children younger than two years with and without human immunodeficiency virus infection. Pediatrics 99, 575–580.

Klugman, K.P., Madhi, S.A., Huebner, R.E., Kohberger, R., Mbelle, N., and Pierce, N. (2003). A trial of a 9-valent pneumococcal conjugate vaccine in children with and those without HIV infection. N. Engl. J. Med. 349, 1341–1348.

Kumar, D., Rotstein, C., Miyata, G., Arlen, D., and Humar, A. (2003). Randomized, double-blind, controlled trial of pneumococcal vaccination in renal transplant recipients. J. Infect. Dis. 187, 1639–1645.

Lederman, H.M., and Winkelstein, J.A. (1985). X-linked agammaglobulinemia: an analysis of 96 patients. Medicine 64, 145–156.

Lee, H., Nahm, M.H., Burton, R., and Kim, K.H. (2009). Immune response in infants to the heptavalent pneumococcal conjugate vaccine against vaccine-related serotypes 6A and 19A. Clin. Vaccine Immunol. 16, 376–381.

Lin, J.S., Park, M.K., and Nahm, M.H. (2001). A chromogenic assay measuring opsonophagocytic killing capacities of anti-pneumococcal antisera. Clin. Diagn. Lab. Immunol. 8, 528–533.

Liu, X., Wang, S., Sendi, L., and Caulfield, M.J. (2004). High-throughout imaging of bacterial colonies grown on filter plates with application to serum bactericidal assays. J. Immunol. Methods *292*, 187–193.

MacLeod, C.M., and Kraus, M.R. (1950). Relation of virulence of pneumococcal strains for mice to the quantity of capsular polysaccharide formed *in vitro*. J. Exp. Med. *92*, 1–9.

Madhi, S.A., Kuwanda, L., Cutland, C., Holm, A., Kayhty, H., and Klugman, K.P. (2005). Quantitative and qualitative antibody response to pneumococcal conjugate vaccine among African human immunodeficiency virus-infected and uninfected children. Pediatr. Infect. Dis. J. *24*, 410–416.

Martinez, J.E., Clutterbuck, E.A., Li, H., Romero-Steiner, S., and Carlone, G.M. (2006). Evaluation of multiplex flow cytometric opsonophagocytic assays for determination of functional anticapsular antibodies to *Streptococcus pneumoniae*. Clin. Vaccine Immunol. *13*, 459–466.

Martinez, J.E., Romero-Steiner, S., Pilishvili, T., Barnard, S., Schinsky, J., Goldblatt, D., and Carlone, G.M. (1999). A flow cytometric opsonophagocytic assay for measurement of functional antibodies elicited after vaccination with the 23-valent pneumococcal polysaccharide vaccine. Clin. Diagn. Lab. Immunol. *6*, 581–586.

Maslanka, S.E., Gheesling, L.L., Libutti, D.E., Donaldson, K.B., Harakeh, H.S., Dykes, J.K., Arhin, F.F., Devi, S.J., Frasch, C.E., Huang, J.C., *et al.* (1997). Standardization and a multilaboratory comparison of *Neisseria meningitidis* serogroup A and C serum bactericidal assays. The Multilaboratory Study Group. Clin. Diagn. Lab. Immunol. *4*, 156–167.

Miller, E., Salisbury, D., and Ramsay, M. (2001). Planning, registration, and implementation of an immunisation campaign against meningococcal serogroup C disease in the UK: a success story. Vaccine *20* (Suppl. 1), S58–67.

Moore, M.R., Gertz, R.E., Jr., Woodbury, R.L., Barkocy-Gallagher, G.A., Schaffner, W., Lexau, C., Gershman, K., Reingold, A., Farley, M., Harrison, L.H., *et al.* (2008). Population snapshot of emergent *Streptococcus pneumoniae* serotype 19A in the United States, 2005. J. Infect. Dis. *197*, 1016–1027.

Nahm, M.H., Briles, D.E., and Yu, X. (2000). Development of a multi-specificity opsonophagocytic killing assay. Vaccine *18*, 2768–2771.

Nahm, M.H., Kim, K.H., Anderson, P., Hetherington, S.V., and Park, M.K. (1995). Functional capacities of clonal antibodies to *Haemophilus influenzae* type b polysaccharide. Infect. Immun. *63*, 2989–2994.

Nowak-Wegrzyn, A., Winkelstein, J.A., Swift, A.J., and Lederman, H.M. (2000). Serum opsonic activity in infants with sickle-cell disease immunized with pneumococcal polysaccharide protein conjugate vaccine. The Pneumococcal Conjugate Vaccine Study Group. Clin. Diagn. Lab. Immunol. *7*, 788–793.

Opalka, D., Lachman, C.E., MacMullen, S.A., Jansen, K.U., Smith, J.F., Chirmule, N., and Esser, M.T. (2003). Simultaneous quantitation of antibodies to neutralizing epitopes on virus-like particles for human papillomavirus types 6, 11, *16*, and 18 by a multiplexed luminex assay. Clin. Diagn. Lab. Immunol. *10*, 108–115.

Pai, R., Moore, M.R., Pilishvili, T., Gertz, R.E., Whitney, C.G., and Beall, B. (2005). Postvaccine genetic structure of *Streptococcus pneumoniae* serotype 19A from children in the United States. J. Infect. Dis. *192*, 1988–1995.

Palker, T.J., Monteiro, J.M., Martin, M.M., Kakareka, C., Smith, J.F., Cook, J.C., Joyce, J.G., and Jansen, K.U. (2001). Antibody, cytokine and cytotoxic T lymphocyte responses in chimpanzees immunized with human papillomavirus virus-like particles. Vaccine *19*, 3733–3743.

Park, S.Y., Moore, M.R., Bruden, D.L., Hyde, T.B., Reasonover, A.L., Harker-Jones, M., Rudolph, K.M., Hurlburt, D.A., Parks, D.J., Parkinson, A.J., *et al.* (2008). Impact of conjugate vaccine on transmission of antimicrobial-resistant *Streptococcus pneumoniae* among Alaskan children. Pediatr. Infect. Dis. J. *27*, 335–340.

Prince, J.E., Brayton, C.F., Fossett, M.C., Durand, J.A., Kaplan, S.L., Smith, C.W., and Ballantyne, C.M. (2001). The differential roles of LFA-1 and Mac-1 in host defense against systemic infection with *Streptococcus pneumoniae*. J. Immunol. *166*, 7362–7369.

Putman, M., Burton, R., and Nahm, M.H. (2005). Simplified method to automatically count bacterial colony forming unit. J. Immunol. Methods *302*, 99–102.

Puumalainen, T., Ekstrom, N., Zeta-Capeding, R., Ollgren, J., Jousimies, K., Lucero, M., Nohynek, H., and Kayhty, H. (2003). Functional antibodies elicited by an 11-valent diphtheria-tetanus toxoid-conjugated pneumococcal vaccine. J. Infect. Dis. *187*, 1704–1718.

Quin, L.R., Carmicle, S., Dave, S., Pangburn, M.K., Evenhuis, J.P., and McDaniel, L.S. (2005). *In vivo* binding of complement regulator factor H by *Streptococcus pneumoniae*. J. Infect. Dis. *192*, 1996–2003.

Ren, B., McCrory, M.A., Pass, C., Bullard, D.C., Ballantyne, C.M., Xu, Y., Briles, D.E., and Szalai, A.J. (2004). The virulence function of *Streptococcus pneumoniae* surface protein A involves inhibition of complement activation and impairment of complement receptor-mediated protection. J. Immunol. *173*, 7506–7512.

Rennels, M.B., Edwards, K.M., Keyserling, H.L., Reisinger, K.S., Hogerman, D.A., Madore, D.V., Chang, I., Paradiso, P.R., Malinoski, F.J., and Kimura, A. (1998). Safety and immunogenicity of heptavalent pneumococcal vaccine conjugated to CRM197 in United States infants. Pediatrics *101*, 604–611.

Robbins, J.B., Austrian, R., Lee, C.J., Rastogi, S.C., Schiffman, G., Henrichsen, J., Makela, P.H., Broome, C.V., Facklam, R.R., Tiesjema, R.H., and Parke, J.C., Jr. (1983). Considerations for formulating the second-generation pneumococcal capsular polysaccharide vaccine with emphasis on the cross-reactive types within groups. J. Infect. Dis. *148*, 1136–1159.

Romero-Steiner, S., Frasch, C., Concepcion, N., Goldblatt, D., Kayhty, H., Vakevainen, M., Laferriere, C., Wauters, D., Nahm, M.H., Schinsky, M.F., *et al.* (2003a).

Multilaboratory evaluation of a viability assay for measurement of opsonophagocytic antibodies specific to the capsular polysaccharides of *Streptococcus pneumoniae*. Clin. Diagn. Lab. Immunol. *10*, 1019–1024.

Romero-Steiner, S., Frasch, C.E., Carlone, G., Fleck, R.A., Goldblatt, D., and Nahm, M.H. (2006). Use of opsonophagocytosis for the serological evaluation of pneumococcal vaccines. Clin. Vaccine Immunol. *13*, 165–169.

Romero-Steiner, S., Libutti, D., Pais, L.B., Dykes, J., Anderson, P., Whitin, J.C., Keyserling, H.L., and Carlone, G.M. (1997). Standardization of an opsonophagocytic assay for the measurement of functional antibody activity against *Streptococcus pneumoniae* using differentiated HL-60 cells. Clin. Diagn. Lab. Immunol. *4*, 415–422.

Romero-Steiner, S., Musher, D.M., Cetron, M.S., Pais, L.B., Groover, J.E., Fiore, A.F., Plikaytis, B.D., and Carlone, G.M. (1999). Reduction in functional antibody activity against *Streptococcus pneumoniae* in vaccinated elderly individuals highly correlates with decreased IgG antibody avidity. Clin. Infect. Dis. *29*, 281–288.

Romero-Steiner, S., Pilishvili, T., Sampson, J.S., Johnson, S.E., Stinson, A., Carlone, G.M., and Ades, E.W. (2003b). Inhibition of pneumococcal adherence to human nasopharyngeal epithelial cells by anti-PsaA antibodies. Clin. Diagn. Lab. Immunol. *10*, 246–251.

Ross, S.C., and Densen, P. (1984). Complement deficiency states and infection: epidemiology, pathogenesis and consequences of neisserial and other infections in an immune deficiency. Medicine (Baltimore) *63*, 243–273.

Saeland, E., Vidarsson, G., Leusen, J.H., Van Garderen, E., Nahm, M.H., Vile-Weekhout, H., Walraven, V., Stemerding, A.M., Verbeek, J.S., Rijkers, G.T., *et al.* (2003). Central role of complement in passive protection by human IgG1 and IgG2 anti-pneumococcal antibodies in mice. J. Immunol. *170*, 6158–6164.

Sanders, L.A.M., van de Winkel, J.G.J., Rijkers, G.T., Voorhorst-Ogink, M.M., de Haas, M., Capel, P.J.A., and Zegers, B.J.M. (1994). Fcg receptor IIa (CD32) heterogeneity in patients with recurrent bacterial respiratory tract infections. Infect. Dis. *170*, 854–861.

Santos, G.F., Deck, R.R., Donnelly, J., Blackwelder, W., and Granoff, D.M. (2001). Importance of complement source in measuring meningococcal bactericidal titers. Clin. Diagn. Lab. Immunol. *8*, 616–623.

Schenkein, J.G., Nahm, M.H., and Dransfield, M.T. (2008). Pneumococcal vaccination for patients with COPD: current practice and future directions. Chest *133*, 767–774.

Shah, C., Hari-Dass, R., and Raynes, J.G. (2006). Serum amyloid A is an innate immune opsonin for Gram-negative bacteria. Blood *108*, 1751–1757.

Shapiro, E.D., Berg, A.T., Austrian, R., Schroeder, D., Parcells, V., Margolis, A., Adair, R.K., and Clemens, J.D. (1991). The protective efficacy of polyvalent pneumococcal polysaccharide vaccine. N. Engl. J. Med. *325*, 1453–1460.

Shinefield, H.R., and Black, S. (2000). Efficacy of pneumococcal conjugate vaccines in large scale field trials. Pediatr. Infect. Dis. J. *19*, 394–397.

Stuart, L.M., and Ezekowitz, R.A. (2005). Phagocytosis: elegant complexity. Immunity *22*, 539–550.

Sun, Y., Hwang, Y., and Nahm, M.H. (2001). Avidity, potency, and cross-reactivity of monoclonal antibodies to pneumococcal capsular polysaccharide serotype 6B. Infect. Immun. *69*, 336–344.

Tomlinson, S., Taylor, P.W., Morgan, B.P., and Luzio, J.P. (1989). Killing of gram-negative bacteria by complement. Fractionation of cell membranes after complement C5b-9 deposition on to the surface of *Salmonella minnesota* Re595. Biochem. J. *263*, 505–511.

Usinger, W.R., and Lucas, A.H. (1999). Avidity as a determinant of the protective efficacy of human antibodies to pneumococcal capsular polysaccharides. Infect. Immun. *67*, 2366–2370.

Vakevainen, M., Jansen, W., Saeland, E., Jonsdottir, I., Snippe, H., Verheul, A., and Kayhty, H. (2001). Are the opsonophagocytic activities of antibodies in infant sera measured by different pneumococcal phagocytosis assays comparable? Clin. Diagn. Lab. Immunol. *8*, 363–369.

van de Winkel, J.G., and Anderson, C.L. (1991). Biology of human immunoglobulin G Fc receptors. J. Leukoc. Biol. *49*, 511–524.

Vesikari, T., Wysocki, J., Chevallier, B., Karvonen, A., Czajka, H., Arsene, J.P., Lommel, P., Dieussaert, I., and Schuerman, L. (2009). Immunogenicity of the 10-valent pneumococcal non-typeable *Haemophilus influenzae* protein D conjugate vaccine (PHiD-CV) compared to the licensed 7vCRM vaccine. Pediatr. Infect. Dis. J. *28*, S66–76.

Weibel, R.E., Villarejos, V.M., Vella, P.P., Woodhour, A.F., McLean, A.A., and Hilleman, M.R. (1976). Clinical and laboratory investigations of monovalent and combined meningococcal polysaccharide vaccine, groups A and C. Proc. Soc. Exp. Biol. Med. *153*, 436–440.

Winkelstein, J.A. (1981). The role of complement in the host's defense against *Streptococcus pneumoniae*. Rev. Infect. Dis. 3, 289–298.

Yu, X., Gray, B., Chang, S.J., Ward, J.I., Edwards, K.M., and Nahm, M.H. (1999). Immunity to cross-reactive serotypes induced by pneumococcal conjugate vaccines in infants. J. Infect. Dis. *180*, 1569–1576.

Zuluaga, A.F., Salazar, B.E., Rodriguez, C.A., Zapata, A.X., Agudelo, M., and Vesga, O. (2006). Neutropenia induced in outbred mice by a simplified low-dose cyclophosphamide regimen: characterization and applicability to diverse experimental models of infectious diseases. BMC Infect. Dis. *6*, 55.

New Frontiers in the Chemistry of Glycoconjugate Vaccines

4

David R. Bundle

Abstract

Methods for single point attachment of polysaccharides and oligosaccharides to protein carriers and T-cell peptides are briefly reviewed with emphasis on contemporary approaches that involve synthetic oligosaccharides with linker or tether chemistry designed for compatibility with synthetic strategies. The synthesis and evaluation of conjugate vaccines designed to combat infectious bacterial and fungal diseases, as well as promising attempts to design and test therapeutic cancer vaccine are summarized. The prevailing dogma that protective B-cell epitopes should be comprised of 10–20 monosaccharides is confirmed for several experimental vaccines including those directed towards *Shigella flexneri* and *Shigella dysenteriae*. However, several small epitopes composed of 3–5 monosaccharide residues are sufficient to induce antibody against the whole organism and to confer protection.

Introduction

The covalent attachment of polysaccharides to immunogenic carrier proteins such as tetanus toxoid or diphtheria toxoid has provided polysaccharide conjugate vaccines, which have demonstrated efficacy in preventing potentially deadly bacterial diseases and at the same time unequivocally established the commercial viability and profitability of this approach to disease prevention (Giebink *et al.*, 1993; Lindberg, 1999; Whitney *et al.*, 2003; Ada and Isaacs, 2003). This success has spawned renewed interest not only in carbohydrate based prophylactic vaccines against microbial, viral (Scanlan *et al.*, 2005) and parasitic

(Schofield *et al.*, 2002) infections, but also in therapeutic vaccines for specific applications such as the treatment of cancer (Danishefsky and Allen, 2000; Ragupathi *et al.*, 2003; Freire *et al.*, 2006; Livingston and Ragupathi, 2006; Warren *et al.*, 2007) and certain fungal infections (Han *et al.*, 1999; Torosantucci *et al.*, 2005).

Polysaccharides with the exception of zwitterionic polysaccharides (Brubaker *et al.*, 1999) are T-cell independent antigens that are unable to elicit a T-cell assisted immune response (Nahm *et al.*, 1999; Kelly *et al.*, 2006). Although adults respond to pure polysaccharide vaccines with protective IgM and to a limited extent IgG levels, infants do not (Jennings, 1983; Kelly *et al.*, 2006). As opposed to large polysaccharides, oligosaccharides are simply not immunogenic. However, covalent attachment of pneumococcal, meningococcal and *Haemophilus influenzae* capsular polysaccharides to protein (Giebink *et al.*, 1993; Lindberg, 1999; Whitney *et al.*, 2003; Ada and Isaacs, 2003; Prymula *et al.*, 2008) and notably a relatively short synthetic *Haemophilus influenzae* oligosaccharide (Verez-Bencomo *et al.*, 2004) have created vaccines that drastically reduced the incidence of meningitis in infants and young adults. New data on the processing of glycoconjugate vaccines suggests that polysaccharide–protein conjugates are processed in a similar fashion to the zwitterionic polysaccharides in that oxidative degradation reduces the molecular weight of the polysaccharide component of the glycopeptide to around 10 kDa (Duan *et al.*, 2008). This and proteolytic digestion of the protein carrier result in a peptide with attached low molecular weight oligosaccharide that is presented by MHC-II to the T-cell receptor (TCR), an event which may also involve T-cell recognition of the saccharide component (Kasper *et al.*, unpublished results; Lai and Schreiber, 2009). Precedence for this type of glycoconjugate presentation and saccharide recognition by TCR can be found for glycolipid antigens presented by the CD1 complex (Moody *et al.*, 1997, 1999). Precisely how glycoconjugates are bound by MHC-II and how or if the protruding saccharide engages parts of the MHC-II and TCR is far from clear, but may be decisive in determining the immune response.

How glycoconjugates are processed by the immune system is essential knowledge when designing optimal conjugate vaccines. Exactly which part of a carbohydrate antigen is able to provide functionally protective antibody is a vital piece of information that will determine through which residue it should be linked to protein. For example, for the majority of bacterial O-antigens protective antibodies frequently recognize terminal non-reducing epitopes (Fernandez and Sverremark, 1994) (Fig. 4.1).

Attaching such antigens to proteins via their reducing end groups is therefore logical. By comparison, conjugate vaccines, prepared from capsular antigens such as the meningococcal group C polysaccharide, are made by attachment through the non-reducing end of the polymer (Fig. 4.2), presumably because protective

antibodies can recognize internal repeating epitopes of this capsular antigen. There are now several examples of protective antigens identified and validated by monoclonal antibodies. These provide a basis to 'reverse engineer' a vaccine that is capable of inducing such protective antibodies in patients. Often the monoclonal antibody is generated by a whole cell vaccine and then its precise specificity is delineated by inhibition with chemically defined oligosaccharides. Examples include *Candia albicans* (Nitz *et al.*, 2002), HIV/ AIDs (Scanlan *et al.*, 2005) and mycobacterial arabinogalactans and lipoarabinomannans (Murase *et al.*, 2009). Several specific examples will be described.

Increasingly a key requirement for generating a conjugate vaccine is a well defined chemistry that allows the carbohydrate epitope to be activated and covalently conjugated to reactive groups on the protein. This can be achieved in several ways: direct conjugation of the carbohydrate to protein, e.g. via the aldehyde group of the reducing terminal sugar of the poly or oligosaccharide and the lysine residue of a protein, or indirectly via a bifunctional tether that is attached to the poly- or oligosaccharide and then subsequently reacted with the protein (William and Beurret, 1989). Less well-defined glycoconjugates result when conjugation chemistry yields polysaccharides attached to protein via multiple conjugation sites. Although not as desirable, the latter types of vaccines have been approved for use but there

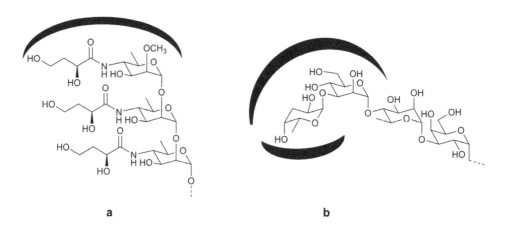

a   b

Figure 4.1 The terminal immunodominant epitopes of (a) *Vibrio cholerae* Ogawa O-antigen (Kenne *et al.*, 1979) and (b) the *Salmonella typhi* O-antigen, O-factor 4.

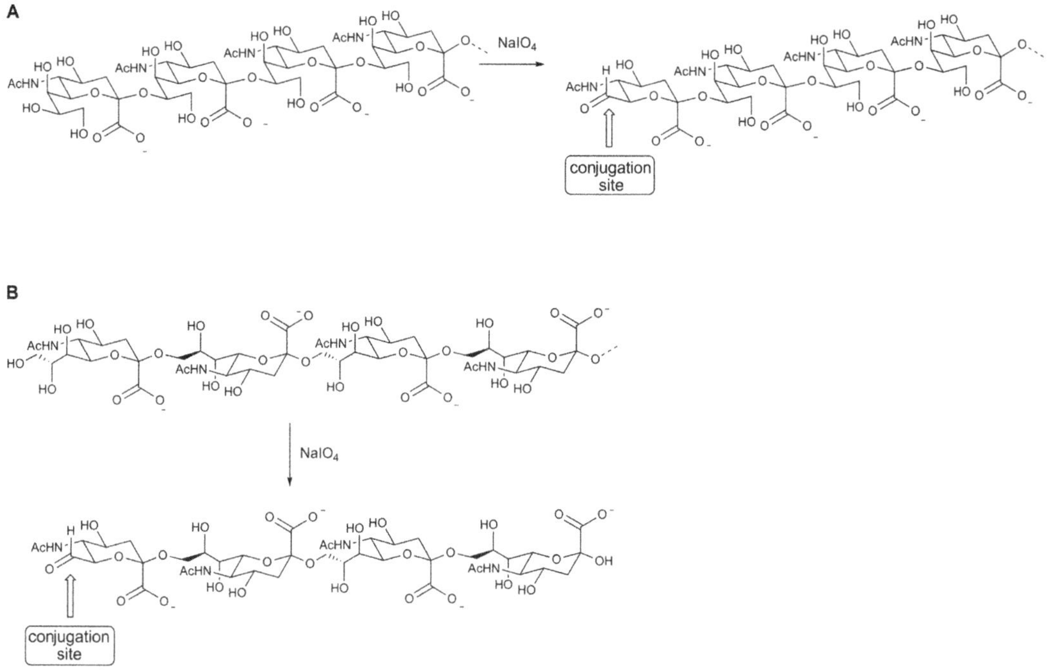

Figure 4.2 (A) Periodate oxidation of the exocyclic chain of the terminal NeuNAc residue of the meningococcal serogroup B polysaccharide. The terminal aldehyde is then used for conjugation to carrier protein by reductive amination. (B) Although the meningococcal serogroup C polysaccharide is in principle labile to periodate cleavage between the vicinal diol at C7 and C8 of the exocyclic chain, the terminal NeuNAc residue is selectively oxidized, thereby providing a similar site of conjugation.

is a growing sentiment and desire for conjugates produced by well defined chemistry and site of attachment to protein. The challenge in meeting this objective depends to a large extent on the carbohydrate antigen (polysaccharide vs. oligosaccharide and the source, synthetic vs. bacterially derived) and the carrier protein or peptide.

The techniques and methods to produce polysaccharide conjugates have been well reviewed (William and Beurret, 1989; Vliegenthart, 2006; Lucas et al., 2005). Several detailed reviews of the synthesis and conjugation of oligosaccharide conjugates have been published (Pozsgay, 2000, 2003, 2008; Vliegenthart, 2006; Hecht et al., 2009). Over the last decade relatively few conceptually different approaches have appeared for conjugating polysaccharides to protein. This review briefly touches on methods that can lead to single point attachment of polysaccharides and oligosaccharides to protein. For the most part, contemporary approaches that involve synthetic oligosaccharides with linker or

tether chemistry, specifically designed for compatibility with synthetic strategies, are highlighted here, with special emphasis on conjugating oligosaccharides to protein carriers and T-cell peptides.

Conjugation chemistry

Bacterial conjugate vaccines in current use are manufactured by two separate fermentation runs to produce, after downstream purification, pure bacterial polysaccharide and toxin such as tetanus toxin or diphtheria toxin. In a third step the polysaccharide and protein are conjugated by chemical methods to produce a covalent bond between the two components. The functional groups used to create this covalent bond may be intrinsic to the polysaccharide (Roy et al., 1984) and protein but to improve the efficiency and speed of conjugation these are more often introduced by selective chemistry conducted on, for example, the polysaccharide (Jennings and Lugowski, 1981). Alternatively a chemical linker such as adipic acid dihydrazide may be used (Bartoloni et al., 1995). Examples of other homobifunctional and heterobifunctional cross-linking reagents that achieve such conjugation have been reviewed (William and Beurret, 1989).

While practical considerations restrict the range of conjugation chemistry available for polysaccharide-protein conjugation, the chemistry available to conjugate synthetically derived oligosaccharides to protein carriers is much broader. This flexibility derives from the fact that functional groups and tethers (linkers) are incorporated during chemical synthesis of the oligosaccharide. To date there is only one example of an approved fully synthetically derived oligosaccharide conjugate vaccine (Verez-Bencomo et al., 2004), although there are intensive research efforts targeting prophylactic and therapeutic vaccines.

Since synthetic oligosaccharides especially as they become larger are very expensive to produce, an optimal conjugation process must be highly efficient. A conjugating reagent should possess sufficiently high reactivity to maximize the yield of glycoconjugate, while readily and reproducibly achieving the desired number of hapten groups attached to carrier protein and avoiding the need for a large excess of valuable saccharide or protein partners. The intermediates formed during the conjugation method should be relatively resistant to the aqueous buffer conditions in which conjugation occurs and the activated saccharide/protein pair should react in an equimolar ratio of oligosaccharide and reactive functionalities on the protein. This reactivity profile should ideally be combined with the property that the linker or new bonds created should be 'immunologically silent'. This refers to an absence of any propensity for the linker or tether component to induce high titres of linker specific antibody. Very few reagents and conjugation chemistries satisfy all criteria and a compromise has to be reached.

Polysaccharide and oligosaccharide conjugates

Conjugating polysaccharides to proteins by homo and even heterobifunctional conjugating reagents can result in conjugates with multipoint attachment and relatively large molecular weight

conjugates of poorly defined size and stoichiometry (William and Beurret, 1989; Wessels et al., 1998; Michon et al., 2006). Reductive amination (Roy et al., 1984) continues to be a frequently used conjugation method since aldehyde groups are intrinsic functional groups at the reducing end of polysaccharides and in certain cases can be readily introduced elsewhere in the polymer under controlled conditions.

Chemistry of this type that facilitates single point attachment is especially attractive and could be achieved for the meningococcal serogroup B and C capsular antigens, which are homopolysaccharides of 2,8- and 2,9-linked sialic acid. It is possible by selective periodate oxidation to cleave the vicinal diols of only the terminal exocyclic C7–C9 chain (Jennings and Lugowski, 1981) (Fig. 4.2). This approach exploits the known higher reactivity of periodate with open chain polyols compared to cyclic polyols and pyranosides (Lambert and Neish, 1950; Hanahan and Olley, 1958; Jennings and Lugowski, 1981). In the examples of the meningococcal group B and C antigens, the resultant polysaccharide bearing an aldehyde group at the terminal non-reducing end is then conjugated to protein via reductive amination using sodium cyanoborohydride (Fig. 4.2). This approach produces an effective meningococcal group C conjugate vaccine (Trotter et al., 2004). In principle if antibodies against the terminal non-reducing end of the antigen are important for protection this method of polysaccharide activation would be unsuitable. Another single point attachment chemistry involves a homobifunctional disuccinimidyl adipate linker introduced at the terminal amino groups, previously generated at the reducing ends of oligosaccharides by reductive amination. This method produces very well defined, characterized and immunogenic meningococcal group A, C, W_{135} and Y conjugates (Bardotti et al., 2008; Bröker et al., 2009; Perrett et al., 2009).

To avoid destruction of essential terminal recognition elements, sodium borohydride reduction of the reducing residue of a polysaccharide can provide a polyol which should be amenable to selective periodate oxidation to generate an aldehyde for reductive amination conjugation (Fig. 4.3). This approach has been used to synthesize

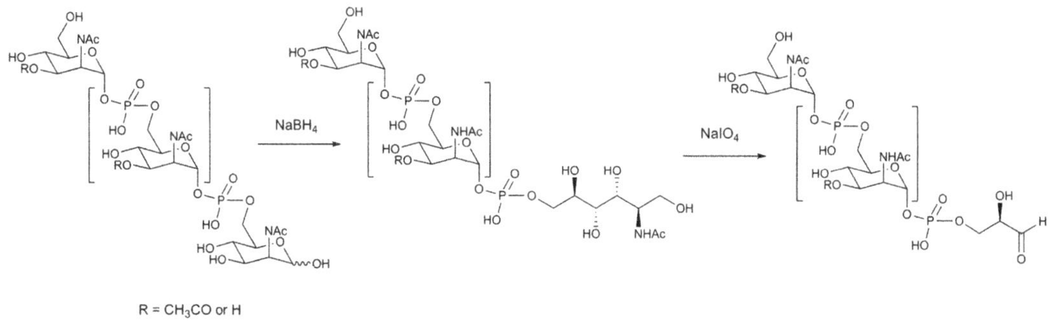

R = CH₃CO or H

Figure 4.3 Sodium borohydride reduction of the terminal reducing residue of the meningococcal serogroup A polysaccharide creates a polyol which is susceptible to oxidation introducing an aldehyde for reductive amination conjugation with carrier protein.

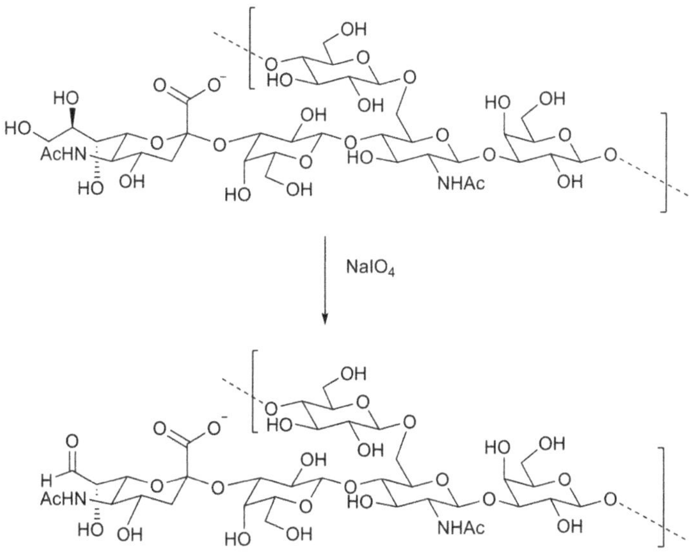

Figure 4.4 Limited periodate oxidation of GBS type II polysaccharide results in approximately 25% of sialic acid residues being oxidized. The presence of multiple aldehyde groups results in conjugates with several cross links.

the meningococcal group A conjugate vaccine (Jennings and Lugowski, 1981). A caveat is that antigens should lack periodate sensitive saccharide residues bearing exo-cyclic carbon chains, such common examples being sialic acids, hexofuranose, KDO or heptoses. Streptoccocal group B conjugate vaccines are an example of antigens that can pose a challenge for defined glycoconjugate synthesis since the complex repeating units of these antigens often contain sialic acid residues as side chains (Wessels *et al.*, 1990). Periodate

oxidation of these leads to derivatized polysaccharides with multiple conjugation sites and potentially complex cross-linked polysaccharide–protein conjugates (Fig. 4.4). The role of the sialic acid residue in modulating recognition of these antigens by protective antibody is complex (Zou *et al.*, 1999; Zou and Jennings, 2001; Johnson *et al.*, 2003; Kadirvelraj *et al.*, 2006) and the success of such experimental vaccines likely derives from the intentional preservation of intact sialic acid side chains through controlled periodate oxidation of only 25% of the sialic acid side chains (Wessels *et al.*, 1990; Paoletti and Kasper, 2003).

The recently discovered *N* and *O* glycosylation machinery used by bacteria to glycosylate proteins appears to hold the potential to generate many glycoconjugate vaccines by fermentation of bacteria that have been engineered to incorporate the genes that code for the biosynthesis of appropriate capsular or O-antigen polysaccharides (Wacker *et al.*, 2002; Feldman *et al.*, 2005; Faridmoayer *et al.*, 2008).

Synthetic oligosaccharide conjugates

The modern era of glycoconjugate chemistry can be traced back to pioneering work of Lemieux's group (Lemieux *et al.*, 1975). He introduced a nine carbon tether based on 8-methoxycarbonyloctanol, derived by borane reduction of the half ester of the nine carbon dicarboxylic acid, azalaic acid. The oligosaccharide is built on this tether and after deprotection the ester group I is converted to an acyl hydrazide II and by reaction with nitrous acid is converted to an acyl azide III,

which is not isolated and reacts rapidly with the ε-amino group of protein lysine residues, when a dimethylformamide solution is added to an aqueous buffered solution of protein (Fig. 4.5A) (Lemieux *et al.*, 1977). The efficiency of the coupling reaction is high. Perceived disadvantages are the need to preserve a relatively reactive ester moiety during oligosaccharide assembly, and the multi-reagent cocktail employed to activate the tether via generation of nitrous acid in DMF with subsequent conversion of the hydrazide to an acyl azide. A shorter *in situ* generation of acyl azide can be achieved using stoichiometric amounts of dinitrogen tetroxide (Pinto and Bundle, 1983). Other adaptations of this approach involve the use of 5-methoxycarbonylpentanol which is readily obtained by transesterification of caprolactone (Pozsgay, 1998a; Ogawa and Kováč, 1997).

Generation of the acyl azide can be sidestepped and instead ethylene diamine is used to convert the ester group I to an amide IV (Kamath and Hindsgaul, 1995). The resultant amino terminated tether can be reacted with a hetero- or

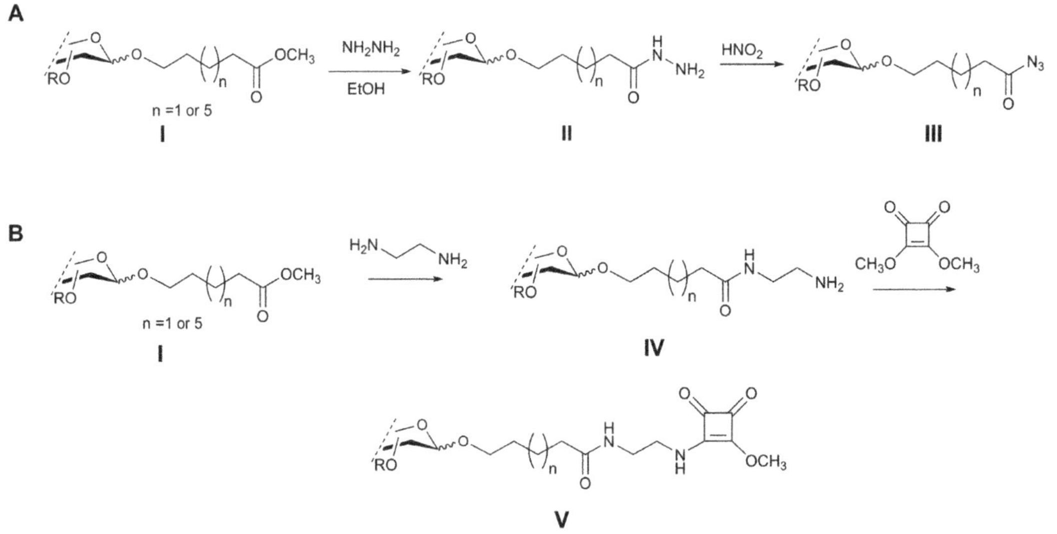

Figure 4.5 The conjugation chemistry of the 'Lemieux' linker, 8-methoxycarbonyloctyl glycosides, $n = 5$. (A) The ester of the tether is converted to an acyl hydrazide which reacts with nitrous acid in anhydrous conditions to give a reactive acyl azide (Lemieux *et al.*, 1975). Pinto and Bundle used a stoichiometric amount of dinitrogen tetroxide to convert the hydrazide to acyl azide (Pinto and Bundle, 1983). (B) Hindsgaul (Kamath and Hindsgaul 1995) introduced the variant method of reacting the ester group with 1,2-diaminoethane to obtain the amino amide derivative (Zhao *et al.*, 1994). This could then be easily conjugated to protein using dimethyl squarate. This homobifunctional conjugating reagent can be used to first prepare the half ester which is then added to a solution of protein. The method of conjugation is one of the most efficient in current use. Several groups have opted to use a 6 carbon version of this tether, $n = 2$.

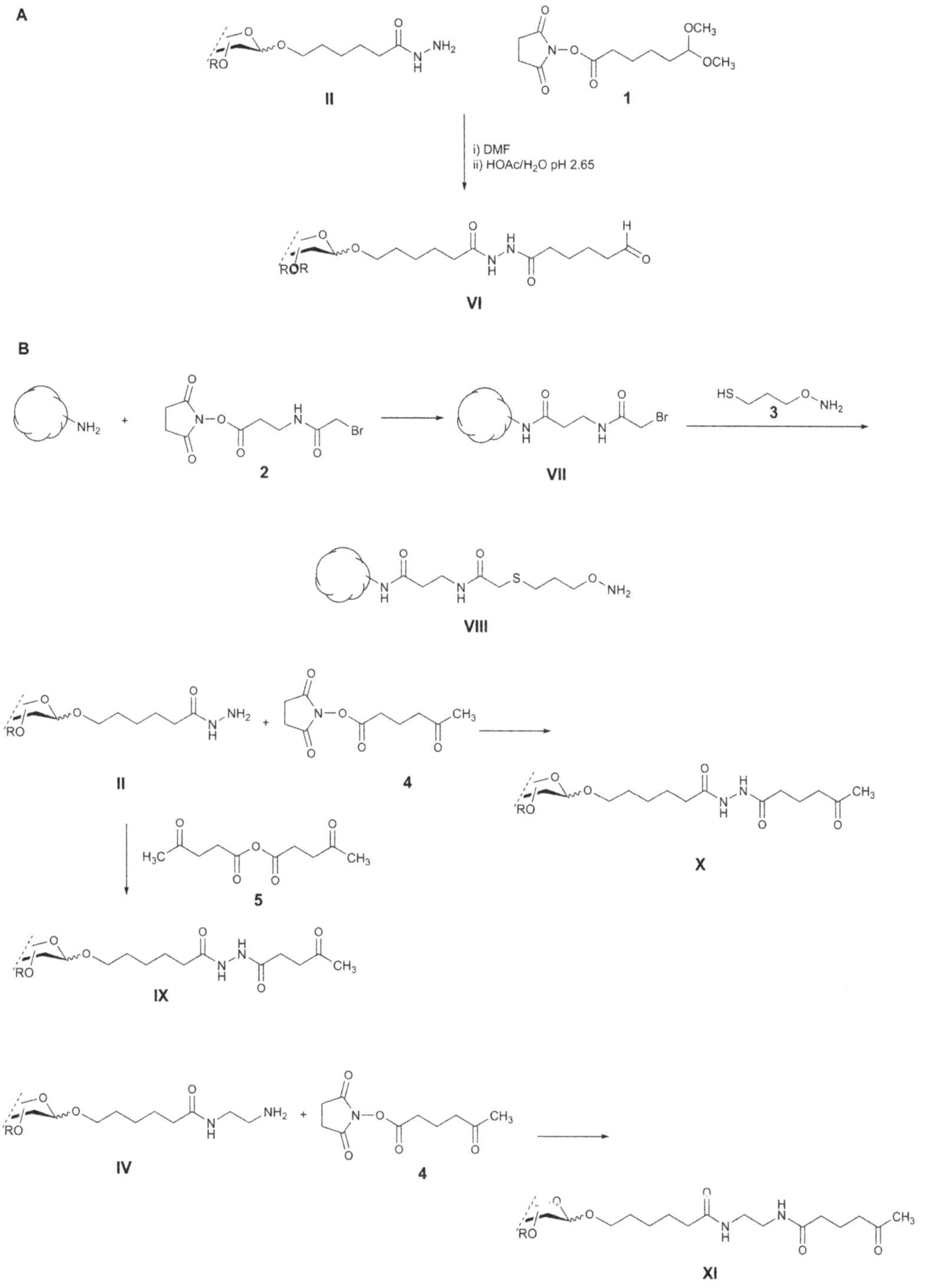

Figure 4.6 (A) Introduction of an aldehyde group by extension of the tether after chemical synthesis of the oligosaccharide. The dimethylacetal is hydrolysed in aqueous acetic acid at pH 2.65 over 6 hours (Pozsgay, 1998a). (B) Oxime conjugation (Kubler-Kielba and Pozsgay, 2005). A bromoacetate group is introduced into the protein followed by reaction with a heterobifunctional reagent bearing aminooxy and thiol moieties. The reactive carbonyl group is introduced to the tether either as described in Fig. 4.5 for the aldehyde or via ω-ketocarboxylic acids.

homobifunctional conjugating reagent to create a glycoconjugate (Fig. 4.5B). A highly efficient reagent and perhaps the most effective homobifunctional reagent is dimethyl or diethyl squarate (Tietze *et al.*, 1991; Kamath and Hindsgaul, 1995). It reacts to form the half ester V which is sufficiently stable to be isolated but is more usually added directly to the carrier protein. Hapten incorporation rates by this approach are amongst the highest seen in glycoconjugate chemistry. Optimization of the conjugation protocol with this reagent has been studied in detail by the group of Kováč (Saksena *et al.*, 2002; Hou *et al.*, 2008). This group has also reported a mass spectrometer based method for the real-time monitoring of the progress of conjugation and a method for the recovery of unreacted oligosaccharide (Chernyak *et al.*, 2001; Saksena *et al.*, 2002).

The functional group embedded in a tether and carried through oligosaccharide assembly may serve as the focal point for chain extension and incorporation of functional groups that would not easily survive the protection deprotection strategies of oligosaccharide synthesis. One example is the introduction of a masked aldehyde group (Fig. 4.6A). The dimethylacetal succinimide ester 1 readily reacts with the hydrazide II to generate a new amide bond and subsequent acid hydrolysis of the dimethylacetal releases an aldehyde VI, which is then used to create a conjugate by efficient reductive amination chemistry.

A further elaboration and apparent improvement of this general concept is based on oxime formation between an *O*-alkyl hydroxylamine and aldehydo or keto groups (Kubler-Kielba and Pozsgay, 2005). Protein carrier is equipped with a spacer that carries an aminooxy group, while the oligosaccharide has an aldehyde or keto group incorporated in the tether (Fig. 4.6B). The commercially available bromoacetylating reagent, succinimidyl 3-(bromoacetamido) propionate 2 introduces a thiol reactive bromoacetamido group to the carrier protein yielding VII. This derivatized protein reacts with the thiol group of 1-aminooxy-3-mercaptopropane 3 to provide the aminoxylated protein VIII. The oligosaccharide with a 5-methoxycarbonylpentanyl tether is converted to either the corresponding hydrazide II or the amide IV. Then an aldehyde or a keto group is

introduced by *N*-acylation with the succinimide ester of 5-ketohexanoic acid 4 or laevulinic anhydride 5 to give either IX, X or XI. Conjugation to the aminoxylated protein VIII by reaction with IX, X or XI proceeded equally well at either pH 5.5 or pH 7.0 and was >90% complete within 6 hours. Importantly an excess of hapten was not necessary for good incorporation levels. For example incorporation of 17 haptens/BSA was achieved whether a tetrasaccharide hapten was used in ~50% molar (over aminooxy groups) or in equimolar amounts. When only 0.55 molar equivalents of the hapten were used the incorporation levels dropped to a still respectable 11 haptens/BSA. The method appears to hold great promise and has been applied to polysaccharide-protein glycoconjugates (Kubler-Kielba *et al.*, 2008).

Allyl glycosides provide a frequently employed example of a flexible anomeric centre protecting group that can also function as an extendable tether (Fig. 4.7A). The photochemical reaction of the double bond of XII with the thiol group of 2-aminoethanol 6 (cysteamine) creates a tether with a terminal amino group XIII, thereby making available a wide range of conjugation chemistry via homo or heterobifunctional coupling reagents (Lee and Lee, 1974; van Seeventer *et al.*, 1997).

The allyl or pentenyl groups of XIV may also be truncated by oxidative cleavage to generate an aldehyde group (Danishefsky and Allen, 2000) XV for reductive amination. Direct reductive amination conjugation of hindered aldehydes to protein can be very slow (Roy *et al.*, 1984). Danishefsky's group has used aldehydes generated in the tether by ozonolysis of pentenyl and allyl glycosides to conjugate to the maleimide group of 7. Reductive amination of XV by 7 proceeds via a Schiff base to yield the extended tether XVI, the maleimide functionality of which reacts with thiolated carrier protein XVII to yield the conjugate XVIII. Thiolates are strong nucleophiles generated under mild conditions and react well in conjugate addition reactions. This type of chemistry has become a method of choice in glycoconjugate chemistry (Fig. 4.7). Frequently seen examples employ a maleimide group which undergoes addition of sulphur. The approach is quite flexible and may involve a thiol terminated

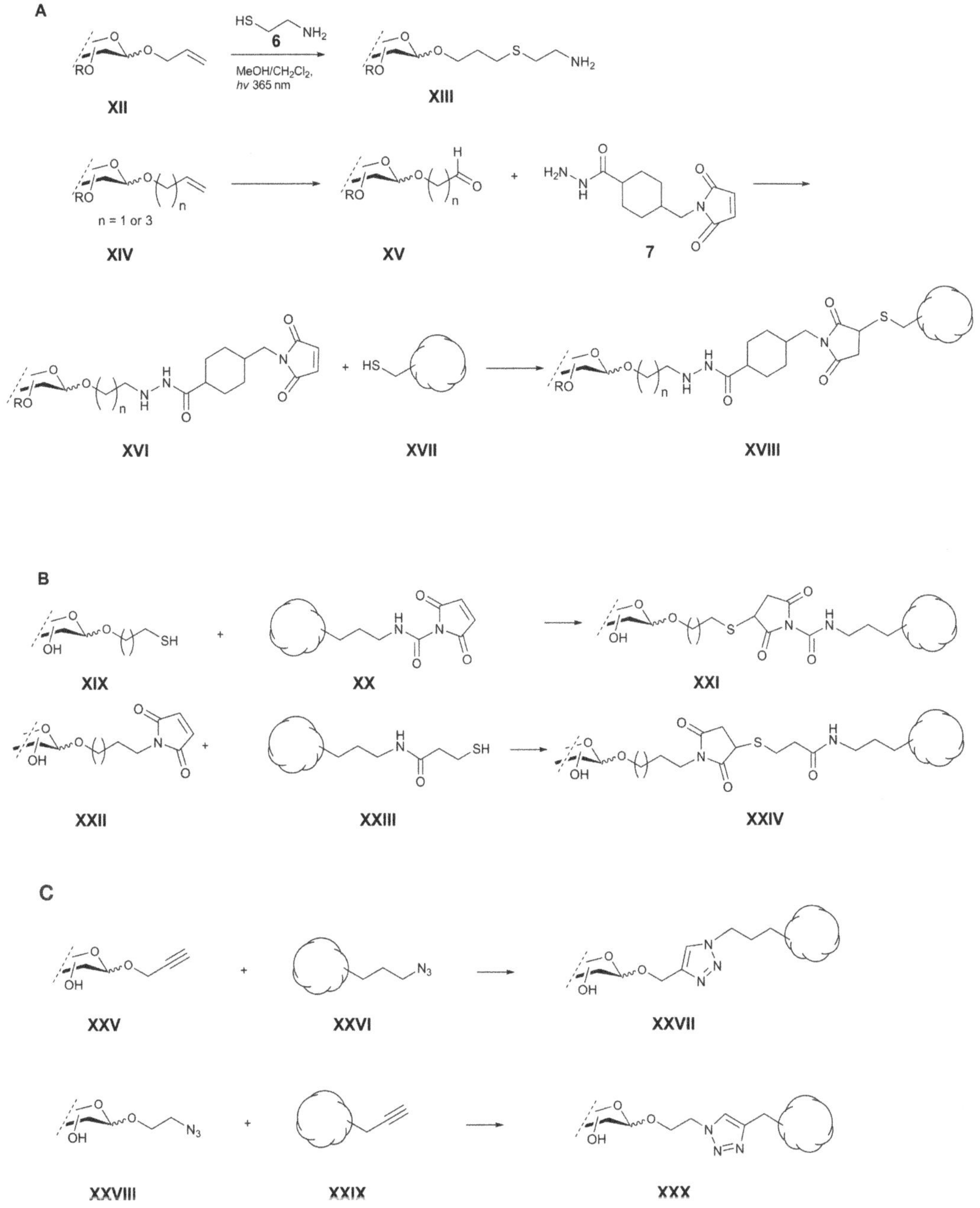

Figure 4.7 (A) Photochemical addition of cysteamine to an allyl glycoside yields an amino terminated tether or an aldehyde group via ozonolysis. Other functional group may also be introduced by addition of thiols containing a second functional group (van Seeventer, *et al.*, 1997). (B) Thiol addition to a maleimide. The thiol may be attached to either the protein or the oligosaccharide. (C) Cycloaddition of alkynes to azides provides a highly efficient conjugation strategy.

tether XIX reacting with maleimide functionality attached to protein XX, or a thiolated protein XXIII with the maleimide functionality attached to the oligosaccharide XXII. In both approaches, disulphide bond formation may become a complicating issue necessitating the exclusion of oxygen and or reducing conditions to ensure the presence of thiol.

The copper(I)-catalysed cycloaddition of azides to alkynes ('Click' chemistry) recently effectively exploited by the groups of Meldal (Tornoe *et al.*, 2002) and Sharpless (Kolb, *et al.*, 2001) offers an attractive and highly efficient method for synthesis of glycoconjugates (Fig. 4.7C). However, the introduction of an aromatic triazole moiety is an obvious potential drawback since tethers incorporating this functionality could be expected to generate significant qualities of linker specific antibody. Nevertheless, there are situations where one could envisage that the triazole group could be buried as for example in the formation of a polysaccharide protein conjugate or a conjugate with highly branched oligosaccharide. Depending on the circumstances, the alkyne or azide moieties may be attached to the oligosaccharide XXV or to the functionalized protein XXVI or the inverse situation for XXVIII and XXIX. The reaction to give XXVII or XXX is promoted by copper I salts and these can degrade proteins (Sereikaite *et al.*, 2006). However, by incorporating bathophenanthroline these deleterious effects can be eliminated or minimized (Gupta *et al.*, 2005). Alternatively, highly strained alkynes based on cyclooctynes can be employed to avoid the use of copper I salts (Agard *et al.*, 2006). However, these bulky groups are even more likely to attract a linker specific response.

Instead of covalently attaching saccharides to protein in order to render them immunogenic other approaches such as attachment to virus particles (Liu *et al.*, 2006; Kaltgrad *et al.*, 2007; Miermont *et al.*, 2008) or to nanoparticles (Ojeda *et al.*, 2007) have received attention.

The various conjugation approaches discussed above as well as others are illustrated by recent examples applied to the synthesis of prophylactic and therapeutic vaccines.

A licensed synthetic oligosaccharide vaccine against *Haemophilus influenzae*

Without doubt the most impressive and resounding success in the field of synthetic glycoconjugate vaccines has been the work of Verez-Bencomo and Roy and their co-workers (Verez-Bencomo *et al.*, 2004), who successfully created a synthetic oligosaccharide for *Haemophilus influenzae* type b and which after conjugation to tetanus toxoid or outer membrane protein (OMP) from *Neisseria meningitidis* afforded a conjugate vaccine that matched the efficacy of conjugate vaccine derived from the native bacterial polysaccharide (Yogev *et al.*, 1990).

The chemical synthesis of this teichoic acid type oligomer had previously been studied by van Boom (Evenberg *et al.*, 1992) and Just (Chan and Just, 1990). In the approach by the Verez-Bencomo and Roy groups the *H*-phosphonate 8 was copolymerized with the corresponding azido terminated diethylene glycol tether derivative 9 in a ratio of 10:1 (Fig. 4.8A). The tethered monomer 9 reacted with excess 8 to give, after oxidation with iodine, the fully protected oligomer 10 containing 6–9 repeating units. Hydrogenolysis removed the benzyl ethers and reduced the azide of the tether to amine 11. The latter group was acylated with the succinimide ester of the maleimido proprionic acid 12 to provide an oligosaccharide tether derivative 13 activated for conjugation with a thiol. Thiol groups were introduced into the carrier protein 15 (either OMP or tetanus toxoid) by reaction with dithiobis(succinimidyl propionate) 14 (Fig. 4.8B). Reduction of the disulphide by dithiothreitol afforded carrier proteins bearing thiolpropionate groups 16. The thiol derivatized carrier proteins reacted with the maleimide derivatized oligomer 13 to give the conjugate vaccine 17.

It is interesting to note that the tetanus toxoid conjugate was poorly immunogenic in mice and Verez-Bencomo tested the conjugate on himself. After a prime and two booster injections a high level of anti-Hib polysaccharide antibodies was observed. By comparison, conjugates with OMP carrier protein gave good immune responses in rats and mice. It was also observed that polysaccharide specific responses were higher in animals vaccinated with conjugates bearing 12 oligosaccharide substituents compared to conjugates bearing only four such haptenic groups. The synthetic Hib oligosaccharide-tetanus toxoid conjugate vaccine has been used extensively and successfully in Cuba (Fernandez-Santana *et al.*, 2004).

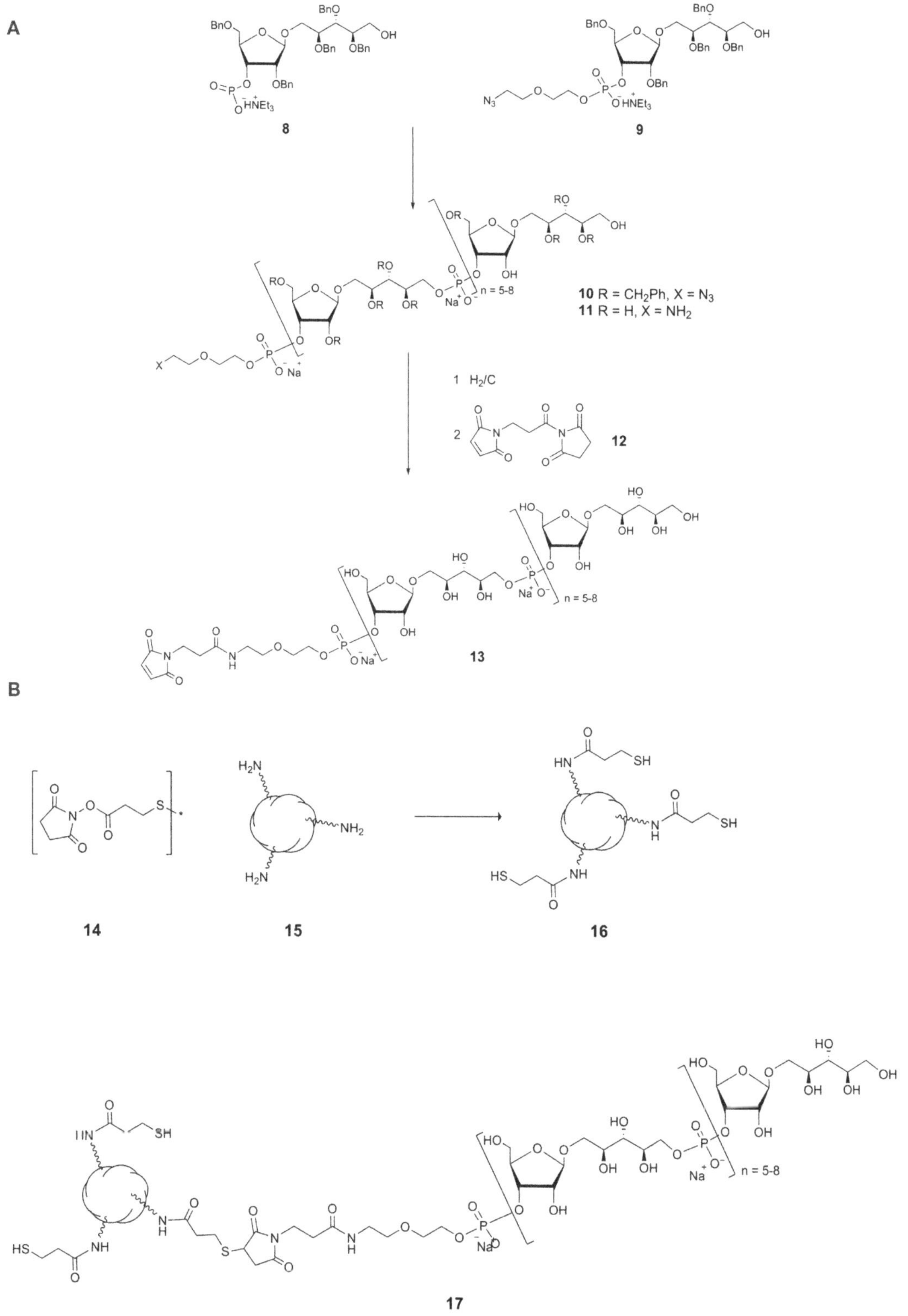

Figure 4.8 (A) Synthesis of the *Haemophilus influenzae* oligosaccharide. (B) Conjugation of the *Haemophilus influenzae* oligosaccharide bearing a maleimide terminated tether to thiolated carrier protein, tetanus toxoid.

O-antigen vaccine synthesis

Lemieux type tethers with a variety of novel conjugation chemistries have been used in several projects targeting bacterial O-antigen vaccines.

Vibrio cholerae vaccines

Detailed and thorough synthetic studies of the O-antigen of *Vibrio cholerae* O:1 serotypes Ogawa and Inaba have yielded important findings that address key issues such as conjugation chemistry (Hou *et al.*, 2008), real time monitoring of the conjugation reaction by mass spectrometry (Saksena *et al.*, 2002), as well as an important crystal structure of serotype Ogawa specific antibody Fab complexed with its oligosaccharide antigen (Villeneuve *et al.*, 2000).

In the structure of its O-antigen *V. cholerae* resembles the O-antigen of *Yersinia enterocolitica* O9 (Caroff *et al.*, 1984a) and that of *Brucella abortus* (Caroff *et al.*, 1984b) with the crucial difference that in the last two antigens the *N*-acylation moiety is a formate while for *V. cholerae* it is (2*S*)-glycerotetronic acid (Kenne *et al.*, 1979).

Kováč's group have synthesized di- through hexasaccharides 16 (Fig. 4.9) of this O-antigen by a variety of glycosylation chemistries (Ogawa *et al.*, 1996: Ogawa and Kováč, 1997; Zhang *et al.*, 1999a,b; Ma *et al.*, 2003; Adamo *et al.*, 2007). These include protecting group manipulation that allowed the 2-*O*-methyl ether to be attached to the terminal non-reducing residue. This is the immunodominant feature of *Vibrio cholerae* O:1 Ogawa serotype (Fig. 4.1a). With collaborators the crystal structure of a murine monoclonal antibody Fab revealed that it is primarily this methyl ether and only the terminal hexosamine to which it is attached that contacts antibody (Villeneuve *et al.*, 2000). Immune responses to various conjugates examining the length of the tether, interval between immunization, nature of the carrier protein all provide evidence supportive of the potential for a synthetic Ogawa specific vaccine construct (Chernyak *et al.*, 2002; Saksena *et al.*, 2005; Rollenhagen *et al.*, 2006; Wade *et al.*, 2006; Saksena *et al.*, 2006). However, further work confirmed earlier indications that the antigenic determinant of the *Vibrio cholerae* Inaba serotype is much larger and includes elements of the LPS core (Meeks *et al.*, 2004).

Synthetic *V. cholerae* O-antigens 18 were synthesized on the 5-methoxycarbonylpentanol tether and conjugation was achieved either via the hydrazide or the aminoamide 19 (Fig. 4.9). Dimethyl squarate reacts with 19 to give the half ester derivative 20 which is then added to a buffered solution of the protein (Hou *et al.* 2008). Conjugations involving haptens prepared from hydrazides were generally slower and less efficient than those with compounds which were made from amino amide derivatives such as 19. The authors demonstrated using surface-enhanced laser-desorption/ionization-time-of-flight mass spectrometry that it was possible to determine the extent of reaction within 20 min from the time of sampling for samples spotted on and adsorbed on a ProteinChip® System (a chemically active surface that permits retention of the protein conjugate). By monitoring the progress of conjugation the precise degree of hapten incorporation can be accurately determined (Saksena *et al.*, 2002) and the degree of hapten loading may be controlled quite accurately. Since this is an important parameter in the magnitude of the immune response this approach could have wide application.

Shigella vaccines

Shigella species cause bacillary diarrhoea and dysentery. Among the *Shigella* species the dominating serotypes are *Shigella dysenteriae* type 1, *Shigella flexneri* types 1a, 2a, 3a and 6 (local differences occur) and *Shigella sonnei*. It has been proposed that an efficacious *Shigella* vaccine most likely has to contain *S. dysenteriae* 1, *S. sonnei* and two or more serotypes of *S. flexneri* (Lindberg, 1998).

Shigella dysenteriae vaccine

The remarkable achievements of Pozsgay in the synthesis of a *Shigella dysenteriae* conjugate vaccine is one of the earliest examples of large oligosaccharide synthesis and essentially represents 'state of the art' in oligosaccharide synthesis (Pozsgay, 1998a,b; Fig. 4.10). The tetrasaccharide repeating unit was constructed from 4 monosaccharide synthons 21–24. The terminal reducing rhamnose 21 was protected as a thiophenyl glycoside. Stepwise addition of a glucosamine

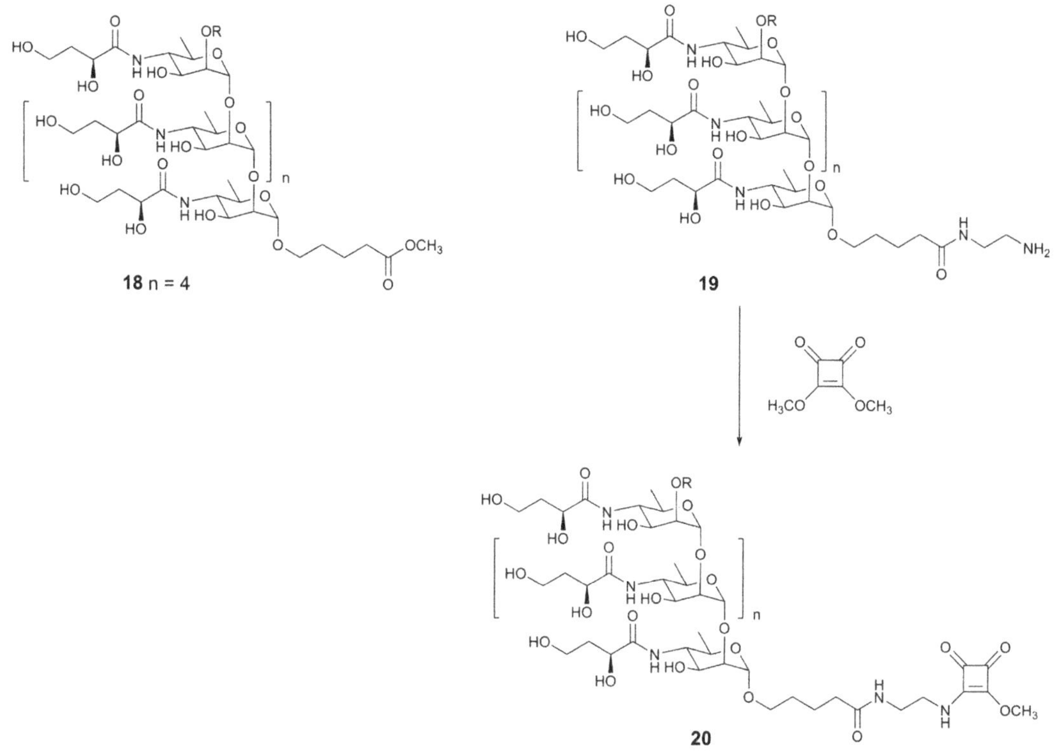

Figure 4.9 Structure of the synthetic *Vibrio cholerae* oligosaccharide epitopes and their conjugation chemistry using dimethyl squarate.

residue as its 2-azido-2-deoxy derivative 22 afforded a disaccharide which required a series of interconversions to provide selective protection of the glucosamine residue prior to glycosylation by the galactose derivative 23 which was first converted to a galactosyl chloride derivative. The fourth residue was added as a rhamnose imidate 24 and the completely protected tetrasaccharide thioglycoside 25 was converted to the more reactive imidate 26 and used to glycosylate 5-methoxycarbonylpentanol 27. A dimeric repeating unit was prepared by selective removal of the monochloroacetate from the terminal rhamnose residue of 28 and the resulting tetrasaccharide 29 was glycosylated by the tetrasaccharide imidate 26 to provide octasaccharide 30. Two iterations of this protection-glycosylation sequence afforded octasaccharide 31, which reacted with 26 to give dodecasaccharide 32 and then hexadecasaccharide 33. After removal of the protecting groups to give 34 hydrazinolysis afforded the hydrazide 35, and this tether was further extended by

reaction with the heterobifunctional spacer 1. The resulting hexadecasaccharide dimethylacetal was hydrolysed to the corresponding aldehyde 36 under acidic conditions and this could then be conjugated to human serum albumin by reductive amination. A keto derivative was also investigated in the reductive amination reaction. In addition, a tetrasaccharide was conjugated by a highly efficient reaction employing a significantly longer linker and oxime formation with either aldehyde or ketone (see Fig. 4.6B derivatives X, XI and XII) (Kubler-Kielba and Pozsgay, 2005).

All synthetic saccharide–albumin conjugates with the exception of the tetrasaccharide conjugate evoked statistically significant increases in anti-LPS IgG titre after a third injection. The highest antibody response to the saccharide hapten was obtained when an average of ten copies of the hexadecasaccharide were coupled to HSA. Lower and higher loadings with the dodeca- and hexadecasaccharides resulted in a significantly diminished anti-oligosaccharide response.

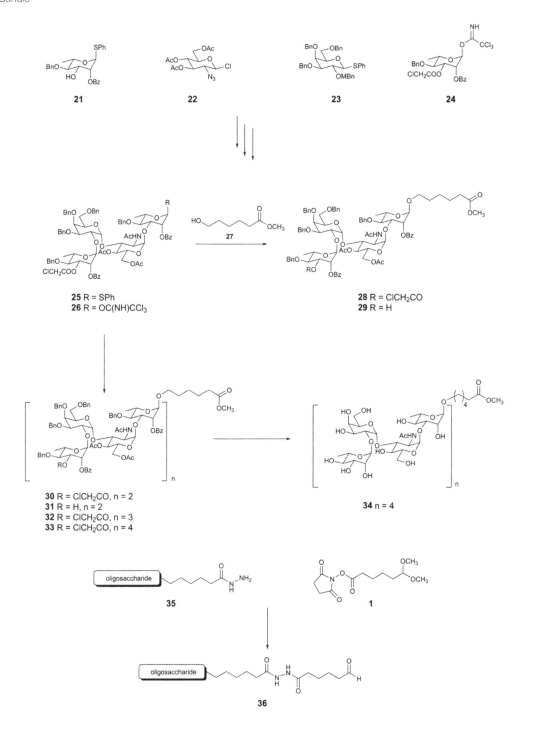

Figure 4.10 Synthetic strategy used to synthesize the hexadecasaccharide (4 repeating units) of the *Shigella dysenteriae* type 1 O-antigen. The deprotected oligosaccharide was conjugated to a heterobifunctional tether bearing a latent aldehyde as its dimethyl acetal.

The general trend of these data later confirmed by other studies such as the synthetic studies on *S. flexneri* 2a cited below show that a complex

oligosaccharide hapten as small as an octasaccharide can efficiently function as an immunogen as part of the conjugate vaccine, provided that it is incorporated in an optimum loading. Similar findings were also reported in the 1970s for phage derived oligosaccharide conjugate vaccines against *Salmonella* (reviewed in Lindberg *et al.*, 1983). (In special cases such as the *V. cholerae* serotype Ogawa and more recently for *Candida albicans* even smaller epitopes may suffice.)

Shigella flexneri vaccines

The O-antigenic polysaccharides of the LPS of *S. flexneri* serotype 2a and *S. sonnei* were covalently linked to exoprotein A of *Pseudomonas aeruginosa* as a carrier and used to vaccinate Israeli solders (Passwell *et al.*, 2001) and more recently children (Ashkenazi *et al.*, 1999). In the latter study serum levels of O-polysaccharide specific IgG antibodies correlated with protection.

A considerable body of work has been reported on synthetic *Shigella flexneri* epitopes. Oligosaccharides from disaccharides up to a heptasaccharide of the *Shigella flexneri* variant Y antigen have been synthesized on the 8-methoxy-carbonyloctanol tether (Pinto *et al.*, 1987; 1989; 1990a; 1990b). Combined NMR and modelling studies established the solution conformation of these antigenic determinants (Bock *et al.*, 1982) and also established the biological repeating unit as the sequence Rha-Rha-Rha-GlcNAc (Carlin *et al.*, 1984). Several crystal structures of this antigen with a murine monoclonal Fab were determined together with a significant body of thermodynamic data for native epitope and its analogues including a peptide mimeotope (Vyas *et al.*, 1993, 2002, 2003).

The variant Y antigen is the simplest of the various *Shigella flexneri* serogroups (Kenne *et al.*, 1977) and extensive work on synthetic analogues of the serogroup 2a antigen investigated the immunogenicity and protection achieved with conjugate vaccines constructed from oligosaccharides composed of 1, 2 and 3 repeating units of the O-antigen (Bélot *et al.*, 2004; 2005; Guerreiro *et al.*, 2004; Phalipon *et al.*, 2006). The synthesis of these glycoconjugate vaccines employed 2-aminoethyl glycosides as the tether, and this group was carried through the oligosaccharide synthesis as a 2-azidoethylglycoside 37 (Fig. 4.11A). This impressive synthetic accomplishment employed tetra- and pentasaccharide trichloroacetimidate

donors 38 and 40. Glycosylation of 37 by the tetrasaccharide imidate 38 afforded a single repeating unit 39 of the *S. flexneri* 2a O-antigen. Two repeating units can be assembled by combination of 40 with 37 to afford a hexasaccharide, which is glycosylated by the tetrasaccharide donor 38 to give the decasaccharide 41. The pentadecasaccharide 42 was synthesized via glycosylation of 37 by 40, and after selective de-*O*-acetylation, this hexasaccharide was glycosylated by 38. Following a similar selective deprotection step, the undecasaccharide was glycosylated by tetrasaccharide donor 38 to yield the pentadecsaccharide 42. Deprotection was achieved by acid hydrolysis of the acetals, transesterification of the acyl groups, followed by hydrogenolysis of the benzyl groups and concomitant conversion of the azidoethyl group to the corresponding aminoethyl glycoside 43. The thioacetyl acetamido linker was introduced to the aminoethyl glycosides via the activated pentafluorophenyl ester to give 44 (Fig. 4.11B).

Chemoselective ligation of the oligosaccharide epitope with PADRE T-cell peptide was achieved through coupling of the tether thiol function of the carbohydrate hapten with a maleimido group introduced at the C terminus of the T-helper peptide 45 to give the chemically defined glycopeptide 46 (Bélot *et al.*, 2004).

The penta-, deca- and pentadeca-saccharides were also conjugated to maleimide-activated tetanus toxoid with an average of 12 saccharide chains per protein (Bélot *et al.*, 2005; Guerreiro *et al.*, 2004). Mice were inoculated with and without an adjuvant, three times at three-week intervals (Phalipon *et al.*, 2006). A fourth injection was given 1 month after the third inoculation. Antibody levels were assayed one week after the third and fourth injections. All of the mice produced anti-LPS titres in response to the pentadecamer conjugate but only 85% did so in the decasaccharide group and only 28% of mice in the pentasaccharide group. The anti-LPS O-antigen titre elicited by the decasaccharide conjugate was significantly lower than that evoked by the pentadecasaccharide conjugate, leading the authors to conclude that the pentadecasaccharide 43 is an appropriate functional mimic of the native PS

Repeating Unit of *Shigella flexneri* serogroup 2a

37

38

39

40

41

42

Figure 4.11A Synthetic strategy used to synthesize the pentadecasaccharide (3 repeating units) of the *Shigella flexneri* serogroup 2a O-antigen.

and that it is a candidate for a synthetic vaccine against *S. flexneri* 2a.

The crystal structures of a monoclonal antibody Fab complexed with the aminoethyl glycosides of the decasaccharide 41 and the pentadecasaccharide 43 were recently solved (Vulliez-Le Norm *et al.*, 2008). The monoclonal antibody F22-4 is a protective IgG that protects against homologous, but not heterologous, challenge in an experimental animal model.

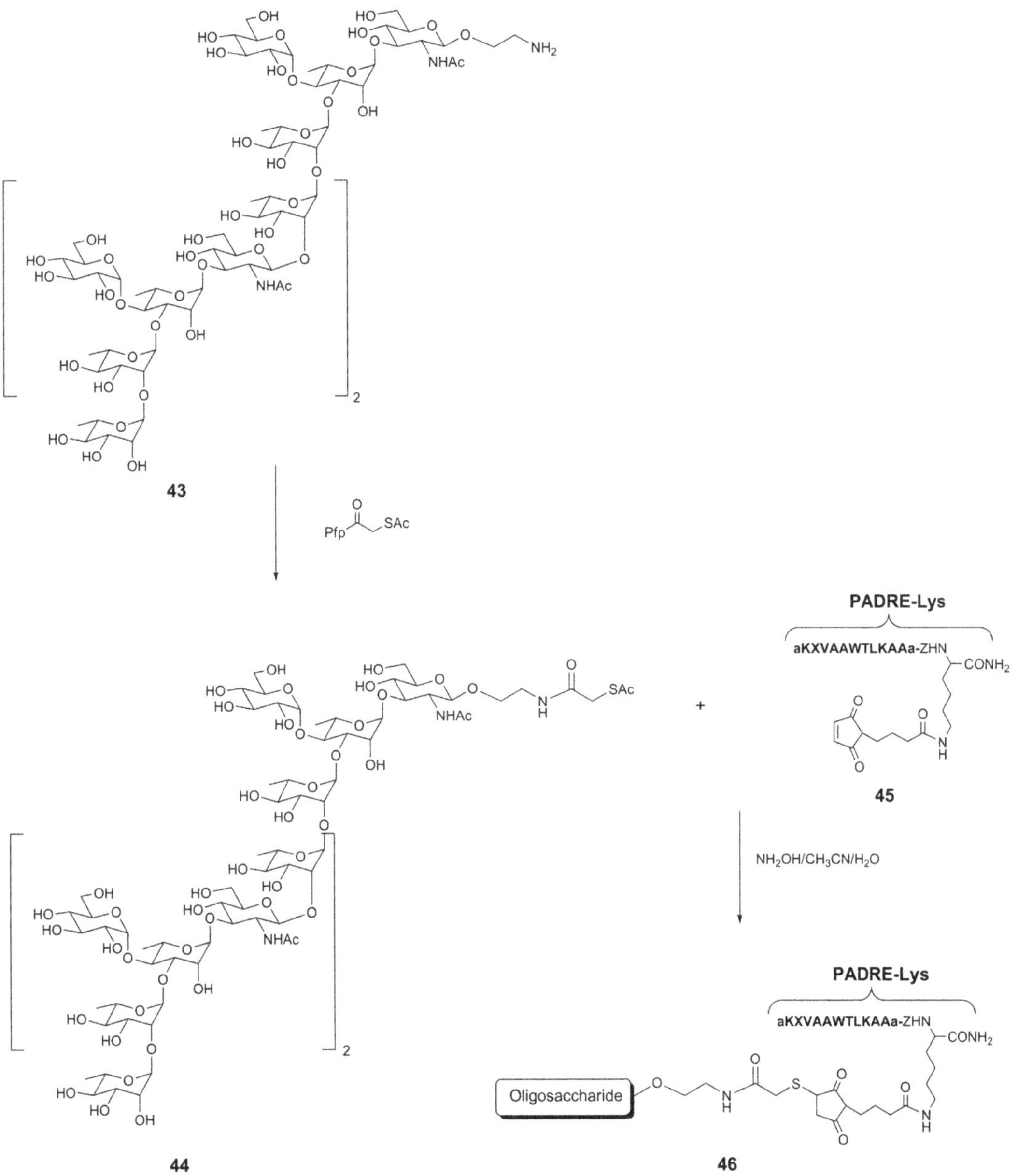

Figure 4.11B Conjugation of *Shigella flexneri* serogroup 2a pentadecasaccharide to PADRE peptide. Z denotes a carboxybenzyl protecting group.

The crystal structures are consistent with the immunochemistry of the antibodies produced by the decasaccharide and pentadecasaccharide vaccines. F22-4 binds to an epitope contained within two consecutive 2a serotype pentasaccharide repeat units. Six sugar residues from a contiguous nine-residue segment make direct contacts with the antibody, including the non-reducing rhamnose and both branching glucosyl residues from the two repeating units. The glucosyl residue, whose position of attachment to the tetrasaccharide backbone of the O-antigen repeating unit defines the serotype 2a O-antigen, is critical for recognition by F22-4. Although the complete decasaccharide is visible in the electron density maps, the last four pentadecasaccharide residues from the reducing end are not, and these residues do not contact the antibody. The antibody V_H germline gene of F22-4 is similar to that of SYA/J6, which binds the *S. flexneri* serogroup Y antigen (Vyas *et al.*, 2002). The bound conformation of the 2a and Y variant antigens share common features.

Streptococcus pneumoniae

Experimental carbohydrate-conjugate vaccines composed of the 13 amino acid universal Pan HLA-DR Epitope (PADRE) and *Streptococcus pneumoniae* capsular polysaccharides from serotypes 6B, 9V and 14 have been prepared by simple carbodiimide-mediated condensation chemistry (Alexander *et al.*, 2004). Three chemically different capsular polysaccharides were coupled to PADRE molecules in a 1:1 molar ratio. The immunogenicity of the PADRE peptide component of the conjugate vaccines was confirmed by the induction of PADRE-specific CD4+ helper T cell (HTL) responses following immunization of C57BL/6 mice. High titre antibody responses specific for polysaccharides of *S. pneumoniae* serotypes 6B, 9V and 14 were induced using Complete Freund's Adjuvant (CFA) and alum formulations. The functional or potential protective value of the polysaccharide-specific antibodies was measured as a function of opsonophagocytic activity for the 6B serotype. High titres of opsonophagocytic activity were measured in sera from mice immunized with formulations containing both adjuvants. These data demonstrate that the PADRE synthetic peptide can induce the HTL responses needed to support the development of antibodies specific for bacterial carbohydrates used in conjugate vaccines (Fig. 4.12).

Streptococcus pneumoniae type 3

As a further example of the utility of relatively small saccharide epitopes in vaccine conjugates, synthetic di- 47, tri- 48, and tetrasaccharide 49 fragments of the capsular polysaccharide of *Streptococcus pneumoniae* type 3 were conjugated to the fragment of diphtheria toxin (CRM197) by the squarate method (Benaissa-Trouw *et al.*, 2001). The average hapten incorporation of oligosaccharide was in the 3 to 12 range. All mice immunized with the tri- and tetrasaccharide conjugates developed type 3 specific IgG antibodies after the second immunization. All of the mice vaccinated with the tri- and tetrasaccharide as well as some mice immunized with the disaccharide conjugate were resistant to intraperitoneal challenge with a lethal dose of *S. pneumoniae* type 3. This provides

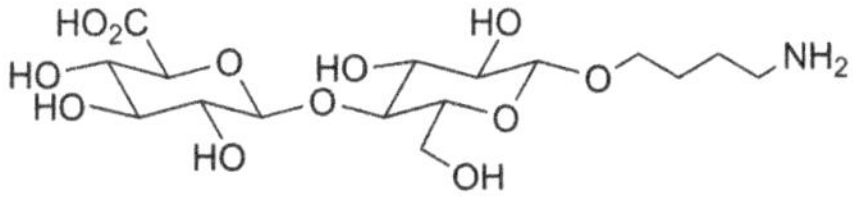

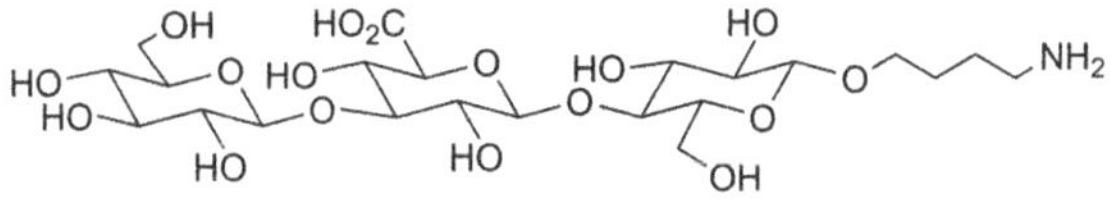

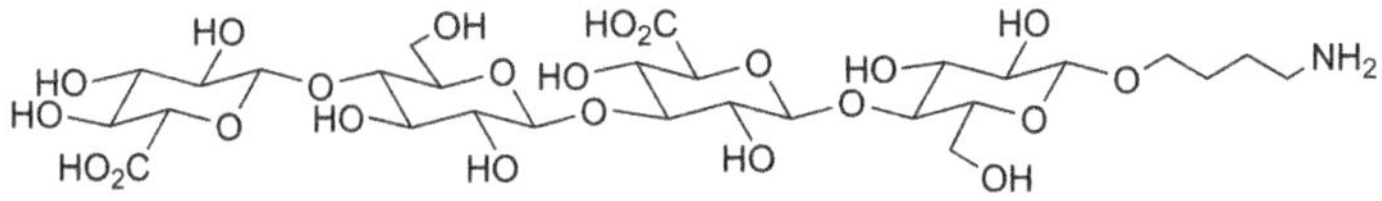

Figure 4.12 Synthetic epitopes of *Streptococcus pneumoniae* type 3.

additional evidence that conjugate vaccines composed of relatively short oligosaccharides can produce protective antibody.

Streptococcus pneumoniae type 14

Several studies of the immune response to *S. pneumoniae* type 14 synthetic oligosaccharides of various lengths and conjugated to CRM197 have been reported (Joosten *et al.*, 2003a,b; Mawas *et al.*, 2002; Safari *et al.*, 2008). Conjugates containing the adipic acid linker were significantly more immunogenic that those prepared with the dialkyl squarate reagent. As with other studies higher saccharide hapten:protein carrier ratios resulted in high levels of antibody directed towards the capsular polysaccharide. In a broad study synthetic overlapping oligosaccharide fragments of *Streptococcus pneumoniae* serotype 14 capsular polysaccharide, {6)-[β-D-GalP-(1→4)-]β-D-GlcPNAc-(1→3)-β-D-GalP-(1→4)-β-D-Glcp-(1→)}ₙ (Fig. 4.13), were conjugated to CRM197 and injected into mice to determine the smallest immunogenic structure. The resulting antibodies were then tested for serotype 14 capsular polysaccharide specificity and for their ability to promote phagocytosis of *S. pneumoniae* type 14 bacteria (Safari *et al.*, 2008). Earlier studies have reported that the oligosaccharide corresponding to one repeating unit of

S. pneumoniae type 14 polysaccharide, i.e. Gal-Glc-(Gal-)GlcNAc, induces a specific antibody response to the polysaccharide antigen (Mawas *et al.*, 2002). In the evaluation of 16 oligosaccharides it was shown that the branched trisaccharide element Glc-(Gal-)GlcNAc is essential in inducing serotype 14 capsular polysaccharide-specific antibodies and the neighbouring galactose unit at the non-reducing end contributes to the immunogenicity of the epitope. Only the oligosaccharide conjugates that produce antibodies recognizing the capsular polysaccharide were capable of promoting the phagocytosis of *S. pneumoniae* type 14. This study suggests that the branched tetrasaccharide Gal-Glc-(Gal-)GlcNAc 50 may be a candidate synthetic oligosaccharide for a conjugate vaccine against infections caused by *S. pneumoniae* type 14 (Safari *et al.*, 2008).

Bacillus anthracis

The recent incidence of bioterrorism involving *B. anthracis* has promoted research on potential conjugate vaccines and carbohydrate based diagnostics. Anthrax is a severe mammalian disease caused by the spore-forming bacterium *B. anthracis* (Mock and Fouet, 2001; Moayeri and Leppla, 2004). There is no vaccine available for general use. However, the outer surface layer of *B. anthracis*, termed exosporium, contains a glycoprotein that expresses multiple copies of a tetrasaccharide

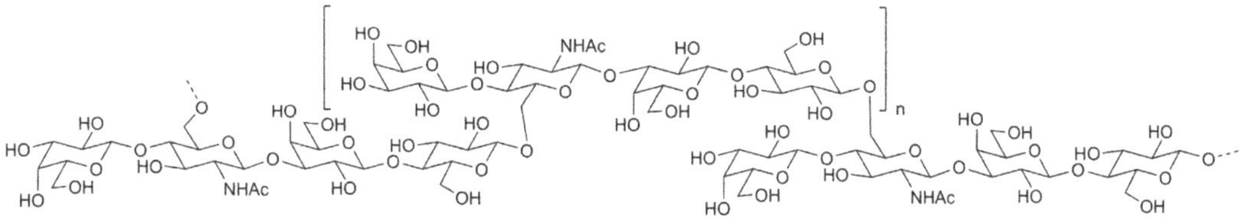

Structure of the *Streptococcus pneumoniae* type 14 polysaccharide

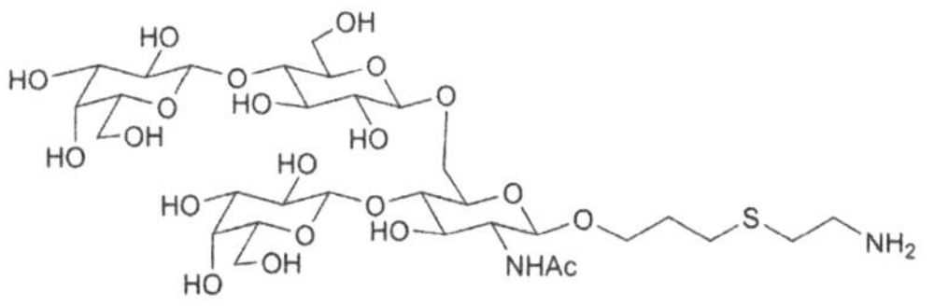

50

Figure 4.13 Synthetic epitopes of *Streptococcus pneumoniae* type 14.

sequence 51 (Fig. 4.14). An unusual monosaccharide, anthrose, is the terminal non-reducing residue and has hydroxy-methyl-butyric acid N-linked to the 4-amino group (Daubenspeck et al., 2004). Tetrasaccharide 51 is unique to *B. anthracis* and since serum antibodies against the glycan 51 can kill *B. anthracis*, it was identified as a potential candidate for a conjugate vaccine.

The first syntheses of 51 were reported with pentenyl 52 (Werz and Seeberger, 2005) and 5-methoxycarbonylpentanol tethers 53 (Adamo et al., 2006). The combination of two disaccharides was used to synthesize the pentenyl glycoside of 52 and ozonolysis of the pentenyl double bond afforded an aldehyde, which could be used for reductive amination conjugation to keyhole limpet haemocyanin (KLH) (Werz and Seeberger, 2005). The synthesis of the 5-methoxycarbonylpentanyl tetrasaccharide glycoside 53 was achieved by stepwise addition of monosaccharides (Adamo et al., 2006). Antibodies raised to the KLH conjugate recognized the spores of *B. anthracis* and may provide a diagnostic test for the bacterium (Tamborrini et al., 2006; Seeberger, 2008).

The group of Boons has synthesized terminal non-reducing trisaccharide epitopes 54–57 with and without the O-methyl ether of the terminal residue and with varied N-acyl groups (Mehta et al., 2006). The epitopes were conjugated to KLH and bound antibodies raised to spores of *B.*

anthracis. Not surprisingly the structural features of the terminal non-reducing residue were critical for antibody recognition. The O-methyl ether was critical (cf. *Vibrio cholerae* Ogawa) and the N-acyl moiety was a key structural requirement.

Although vaccines are a potential outcome of this work, to date there are no reports of challenge experiments (Werz and Seeberger, 2005; Seeberger, 2008).

Mycobacterium tuberculosis

Tuberculosis is a major global health problem aggravated by the emergence of multiple-drug resistant strains (Coker, 2004). There is no vaccine available to prevent this infectious disease in adults. The bacille Calmette–Guerin (BCG) is given to 100 million newborn children each year but its efficacy varies widely (Fine, 1995). Unlike many pathogenic bacteria, *M. tuberculosis* does not synthesize a surface polysaccharide with a regular repeating unit structure such as the O-antigens and capsular polysaccharides of Gram-negative and Gram-positive bacteria. Instead, the cell wall is composed of two glycoconjugates arabinogalactan and lipoarabinomannan. The arabinogalactan is esterified by mycolic acids. These domains are made even more complex by the presence of frequent branches in the saccharide chains as well as by phosphoglycerolipids and glycosylated inositol moieties (Kaur et al., 2009). In addition, whereas pyranose rings dominate the

51 R^1 = H or OH, R^2 = OH or H
52 R^1 = 4-pentenyloxy R^2 = H
53 R^1 = H, R^2 = methoxycarbonylpentanyloxy

54 R^1 = CH$_3$, R^2 =
55 R^1 = H, R^2 =
56 R^1 = CH$_3$, R^2 =
57 R^1 = CH$_3$, R^2 =

Figure 4.14 Synthetic tri- and tetrasaccharide of the *B. anthracis* exosporium glycoprotein.

structures of bacterial O-antigens and capsular polysaccharides, in mycobacteria dominant ring forms are arabinofuranose and galactofuranose.

There is good evidence that antibodies against lipoarabinomannan (LAM) afford passive immunity and protection in mice (Teitelbaum et al., 1998; Hamasur et al., 2004). Oligosaccharides derived from bacterial LAM have been conjugated to mycobacterial proteins and to tetanus toxoid and the conjugates shown to be highly immunogenic in rodents. These conjugate vaccines showed good protective efficacy in mice (C57BL/6) and guinea pigs in terms of prolonged survival and reduced pathology, when administered in a novel adjuvant using a subcutaneous priming-intranasal boost regime (Hamasur et al., 1999, 2003). However, there is as yet no conclusive correlation of immunity to tuberculosis with serum antibody levels against these carbohydrate antigens. There is a considerable body of evidence suggesting that a successful vaccine will have to engage and induce multiple aspects of immunity (Kallenius et al., 2008).

Several oligosaccharides that are derivatized for glycoconjugate synthesis have been reported. Lowary's group has reported extensively on the synthesis of mycobacterial LAM (Gadikota et al., 2003; Joe, et al., 2007; Kim et al., 2007). Fraser Reid's laboratory has synthesized a 28-mer oligosaccharide core of mycobacterial LAM (Fraser-Reid et al., 2006). Seeberger's group have synthesized arabinomannan (Hölemann et al., 2006) and phosphatidylinositol mannosides (PIMs), a key class of antigenic glycolipids found on the cell wall of Mycobacterium tuberculosis (Liu et al., 2006; Boonyarattanakalin et al., 2008). The arabinomannan and each synthetic PIM was equipped with a thiol-linker for immobilization on microarray slides and carrier proteins for biological and immunological studies (Bernardes et al., 2009).

Lowary's group has used the epitope binding motif and structure of LAM fragments complexed with monoclonal antibody CS35 to calibrate and refine the design of potential vaccine candidates. They determined the first crystal structure of a furanose-containing antigen with a M. tuberculosis monoclonal antibody and mapped the specificity of this MAb with a suite

of oligosaccharides (Rademacher et al., 2007; Murase et al., 2009). The crystal structures of CS35 Fab with three oligosaccharides 58–60 were solved. A triangular binding pocket in the CS-35Fab accommodates the Y-shaped, branched hexasaccharide sequence AB(DF)CE 58 which is found at the non-reducing terminus of LAM (Fig. 4.15). This pocket is flanked by a broad groove which would accommodate the non-reducing end-extended structures of native LAM; hence residue A lies next to this groove while residues C and E are in the triangular pocket. Residues D and F extend out of the binding site, and in fact, residue F makes no contact with antibody. Notably, this binding pattern was first observed by solution NMR spectroscopy, which also revealed an alternative binding mode for 60 in which the DF branch could replace the CE backbone residue in the triangular pocket (Rademacher et al., 2007). This binding mode would be precluded by the larger, more extended structure of native LAM. Such observations emphasize the utility of detailed structural and immunochemical studies in glycoconjugate vaccine development.

The crystal structures of LAM fragments bound to CS-35 Fab provide a basis for the rational design of novel vaccines. Residues A and E are most intimately associated with the Fab and are the major determinants of specificity and affinity. As has been observed for other branched antigens, complimentarity for the branch point residue (here residue B) is also important in conjunction with branching as a cause of greater conformational rigidity and hence preorganization for binding. These three structural elements would thus appear to comprise a minimal epitope suitable for the design of novel vaccines. Tetanus toxoid conjugates of several of these oligosaccharides including hexasaccharide 61 are currently being evaluated for their ability to bind patient antisera and in protection in challenge experiments.

Malaria

The power of synthesis is vividly illustrated in the case of a malaria vaccine candidate. Using tiny amounts of isolated mixtures of GPI anchor glycolipids, Schofield proposed that glycans of this type may confer protection from death by

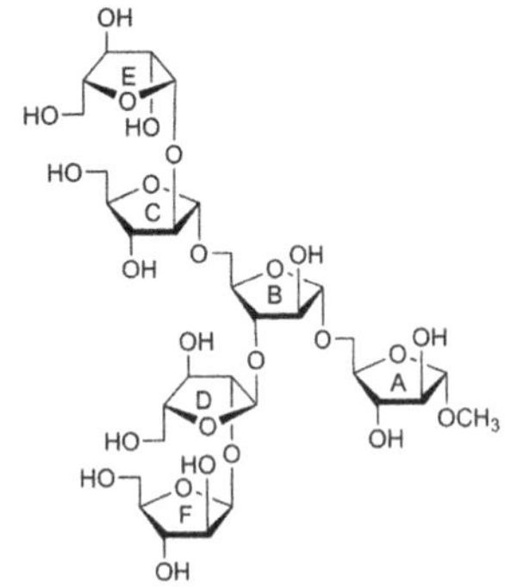

Figure 4.15 *Mycobacterium tuberculosis* lipoarabinomannan fragments used for co-crystallization with the Fab of the monoclonal antibody CS-35.

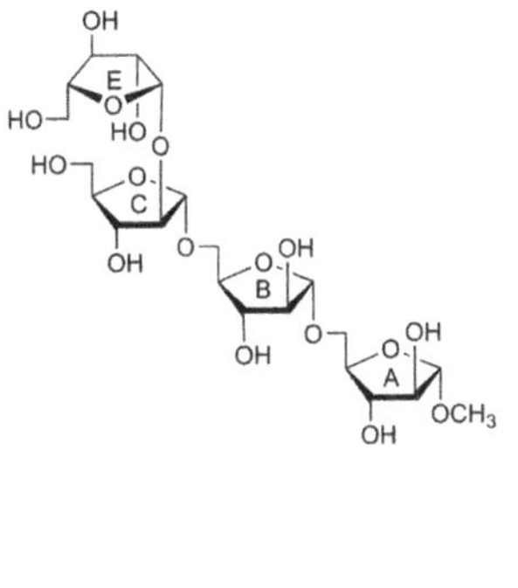

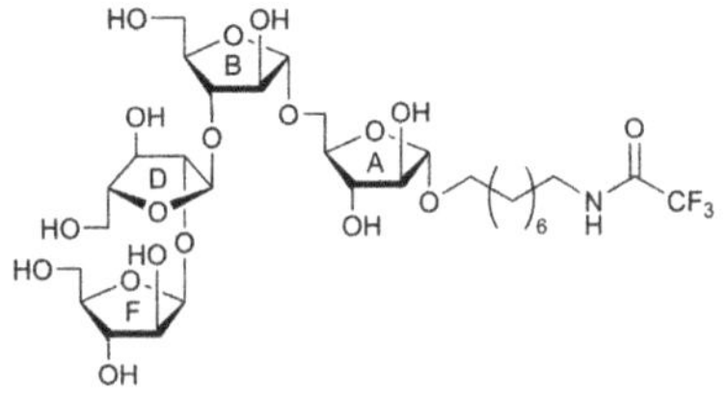

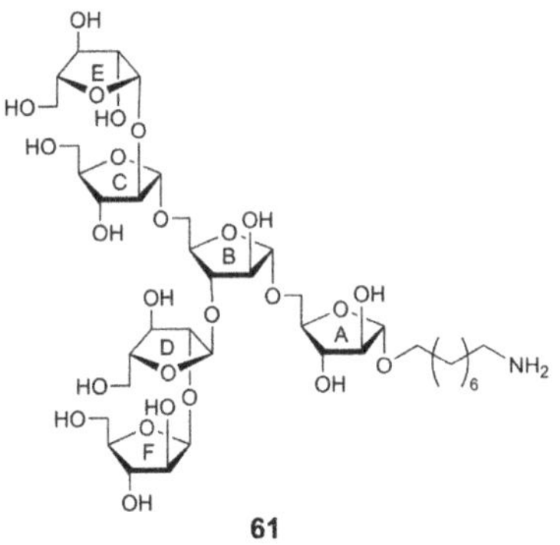

Plasmodium falciparum, the parasite that causes malaria (Schofield and Hackett, 1993; Schofield *et al.*, 1993). Although the exact glycan structure that conferred protection was not known the *P. falciparum* GPI glycan sequence $NH_2\text{-}CH_2\text{-}CH_2\text{-}PO_4\text{-}$($\alpha$Man1→2)6$\alpha$Man1→2$\alpha$Man1→6$\alpha$Man1→4 αGlcNH$_2$1→6*myo*-inositol-1,2-cyclic-phosphate was synthesized, conjugated to a carrier protein, and helped to protect mice from a challenge with cerebral malaria (Schofield *et al.*, 2002). Using the six building blocks 62–67 (Fig. 4.16) and an automated synthesizer Seeberger's group created the fully protected hexasaccharide 68. Selective removal of the TIPS group allowed for specific introduction of an ethanolamine phosphate residue at this site. Following a series of deprotection steps, 69 was obtained and reacted with 2-iminothiolane to introduce a sulfhydryl group at the ethanolamine 70. This provided the conjugation functionality for reaction with maleimide-activated ovalbumin (OVA), or KLH.

An antitoxin effect rather than killing of the parasite provided protection. Precedence for this concept of a Malaria vaccine can be found in both diphtheria and tetanus vaccines which operate on the principle of neutralizing toxin. This proposal was further supported by several observations. Adults in endemic malaria areas are protected from severe disease via a resistance mechanism (Naik *et al.* 2000) but children under the age of two years are victims of cerebral malaria – suggesting that the inability of infants up to this age to form anti-carbohydrate antibodies could explain this observation. High-throughput screening of thousands of human sera on a carbohydrate microarray carrying defined GPI-glycan structures was used to screen sera from populations in different endemic countries as well as naive

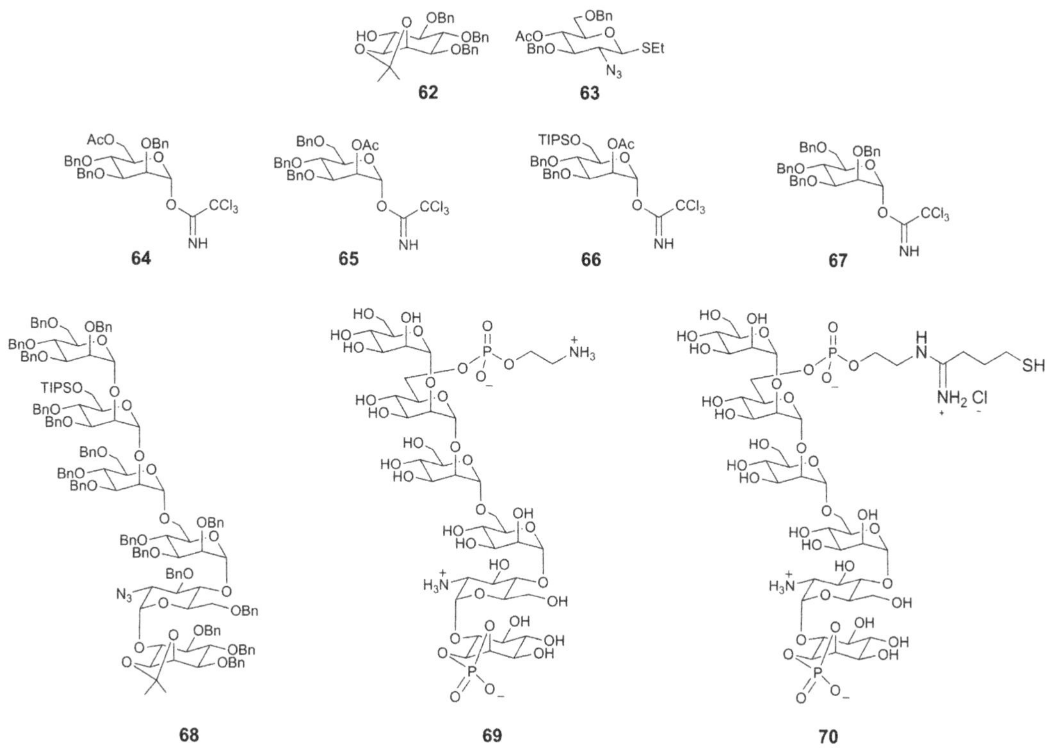

Figure 4.16 The synthetic strategy used to construct the first generation synthetic conjugate vaccine consisting of a hexasaccharide epitope of the *P. falciparum* GPI glycan.

individuals (Seeberger, 2008). This revealed that a specific response against particular GPI-glycans protects adults in endemic areas from severe disease (Kamena *et al.*, 2008). Children under two years of age and naive individuals failed to show antibody titres. Some adults that took part in a challenge trial developed anti-GPI antibodies after infection with *P. falciparum*.

Fungal vaccines: *Candida* and *Aspergillus*

Candida albicans has emerged as the third or fourth leading cause of bloodstream infections in hospitalized patients. The dramatic increase in the rates of infection, the rise of antibiotic-resistant strains and the small differences between therapeutic and toxic dose levels for many antibiotics has renewed interest in specific antifungal vaccines.

Experimental and clinical evidence has demonstrated the promise of acquired immunity as a means of enhancing the host defence mechanisms against candidiasis, and convincing evidence has established the protective potential of relevant antibody (Casadevall, 1995; Casadevall *et al.*, 1998; Cutler *et al.*, 2002; Matthews and Burnie, 1996). Immunoprophylactic approaches are therefore of interest, since a successful vaccine could be used to provide protection for individuals at high risk. Candidates would include patients scheduled to receive abdominal surgery: bone marrow, kidney or heart transplantations, or immunosuppressive cancer therapy, as well as those exposed to long term hospitalization.

The cell wall of *C. albicans* consists mainly of chitin, glucan and mannan. The mannan is highly immunogenic (Cassone, 1989) and has been associated with the adhesion of yeast to different cell types (Edwards *et al.*, 1992; Kanbe and Cutler, 1994). Suzuki *et al.* used NMR and mass spectrometry to elucidate the structure of the mannan of *C. albicans* (Kobayashi *et al.*, 1990; Shibata *et al.*, 1995). The complex N-linked components have extended α1–6 linked mannopyranan backbones containing α1–2 mannopyranan branches. β1–2 mannopyranan oligomers are attached through a phosphodiester bridge (Fig. 4.17). The position of attachment of this phosphodiester has yet to be determined. In addition, β-mannan structures linked directly to the α-mannan have been reported (Shibata *et al.*, 2003). Recent work has also identified the β1–3 glucan of *Candida* as a promising protective antigen (Torosantucci *et al.*, 2005).

Chemical synthesis of the β-mannan represented one of the unconquered challenges in glycoside synthesis. Crich developed an effective solution to the problem and succeeded in synthesizing oligosaccharides from di- up to octasaccharide (Crich *et al.*, 2001). Although none of these molecules were functionalized with a tether the work did provide a crystal structure for a protected tetrasaccharide. This conformation of the protected saccharide was similar to that deduced by NMR and modelling studies for penta and hexasaccharide antigens in Bundle's laboratory (Nitz *et al.*, 2002). This group had synthesized oligosaccharide using a ulosyl bromide intermediate (Nitz and Bundle, 2001) and the di- through hexasaccharides were used to characterize the binding profile of two monoclonal antibodies developed in Cutler's laboratory (Han *et al.*, 1999; Han *et al.*, 2000; Cutler *et al.*, 2002) that could confer passive immunity on mice in *Canida albicans* challenge

experiments (Han *et al.*, 2000). An especially significant observation was the optimal inhibition of both antibodies by tri- and disaccharides while larger homo-oligomers were progressively worse inhibitors (Nitz *et al.*, 2002). This led the group to propose a conjugate vaccine (Fig. 4.17) based on a trisaccharide epitope covalently linked to tetanus toxoid (Nitz and Bundle, 2001; Wu and Bundle, 2005). The antigen prepared on a multigram scale used an allyl glycoside building block 71 which was glycosylated by a glucosyl donor 72 to give 73 (Fig. 4.18). The glucose residue was selectively *O*-deacetylated and an oxidation reduction sequence provided the acceptor 74. A second round of this glycosylation-selective deprotection-oxidation/reduction sequence afforded the target trisaccharide 75. Photochemical addition of 2-aminoethanol and deprotection by removal of the benzyl ethers gave 76. The deprotected trisaccharide 76 was reacted with the *p*nitrophenyl diester of adipic acid 77 to give the activated half-ester 78 which was added to a buffered solution of tetanus toxoid. Two injections of this glycoconjugate antigen raised high-titre sera in rabbits when administered with alum. The vaccine reduced fungal load in some organs in a immunosuppressed rabbit model of disseminated candidiasis (Bundle *et al.*, unpublished results). However, the same conjugate was not sufficiently

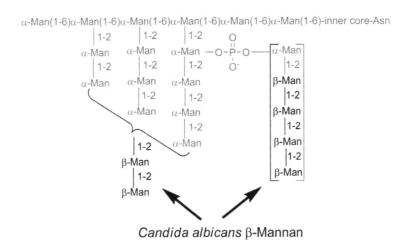

Candida albicans β-Mannan

Figure 4.17 The structure of the *Candida albicans* phosphomannan and the antigenic β1,2-linked mannose side chains.

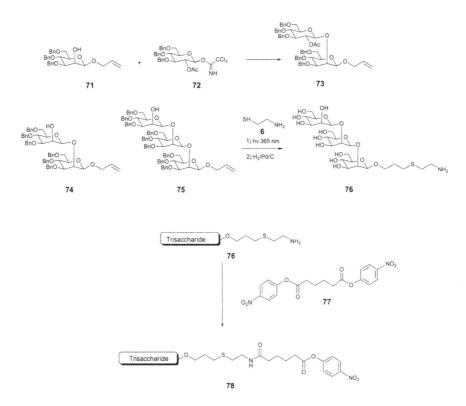

Figure 4.18 Synthetic strategy used to construct the trisaccharide epitope of a proposed *C. albicans* glycoconjugate vaccine. The trisaccharide is activated for conjugation to tetanus toxoid.

immunogenic in mice for challenge experiments to be conducted in this animal model.

In a recent study a novel glycopeptide was synthesized from the β-mannan trisaccharide 76 and a new linker 79 to give the tether trisaccharide 80. A T-cell peptide 81 terminated at its N-terminus by thioproprionic acid reacted with 80 to give the glycopeptide 82. A series of six glycopeptides selected from T-cell peptides present in the cell wall proteins of *Candida albicans* were investigated (Dziadek *et al.*, 2008). Three of the glycopeptide vaccines induced both a humoral and cell mediated response and were effective in conferring protection in a mouse model of candidiasis (Xin *et al.*, 2008). The limiting factor from a practical perspective was the requirement to perform *ex vivo* dendritic cell stimulation (Fig. 4.19).

An Italian research group have developed a very promising conjugate vaccine prepared from the β1–3 glucan, laminarin conjugated to the diphtheria toxoid CRM197, carrier protein. This vaccine has been shown to confer protection against not only *Candida albicans* but also against aspergillosis, a potentially far more life threatening fungal infection (Torosantucci *et al.*, 2005).

Therapeutic cancer vaccines

It has been known for a long time that tumour-associated carbohydrate antigens are over expressed on the cell membrane of many cancers (Hakomori, 1985, 1989). The list is extensive but the most frequently studied antigens include GM_2, GD_2, GD_3, Lewis-Y, STn, TF and Tn antigens. Since it is rarely the case that vaccines alone can induce tumour regression and elimination, the rationale for therapeutic vaccines that target tumour-associated carbohydrate antigens is to induce an immune response to prevent tumour reoccurrence and metastasis following surgery to remove primary tumours. The ganglioside antigens have been targets for the treatment of melanoma, while mucin-related epitopes such as the Thomsen Friedenreich, Tn and sialylTn

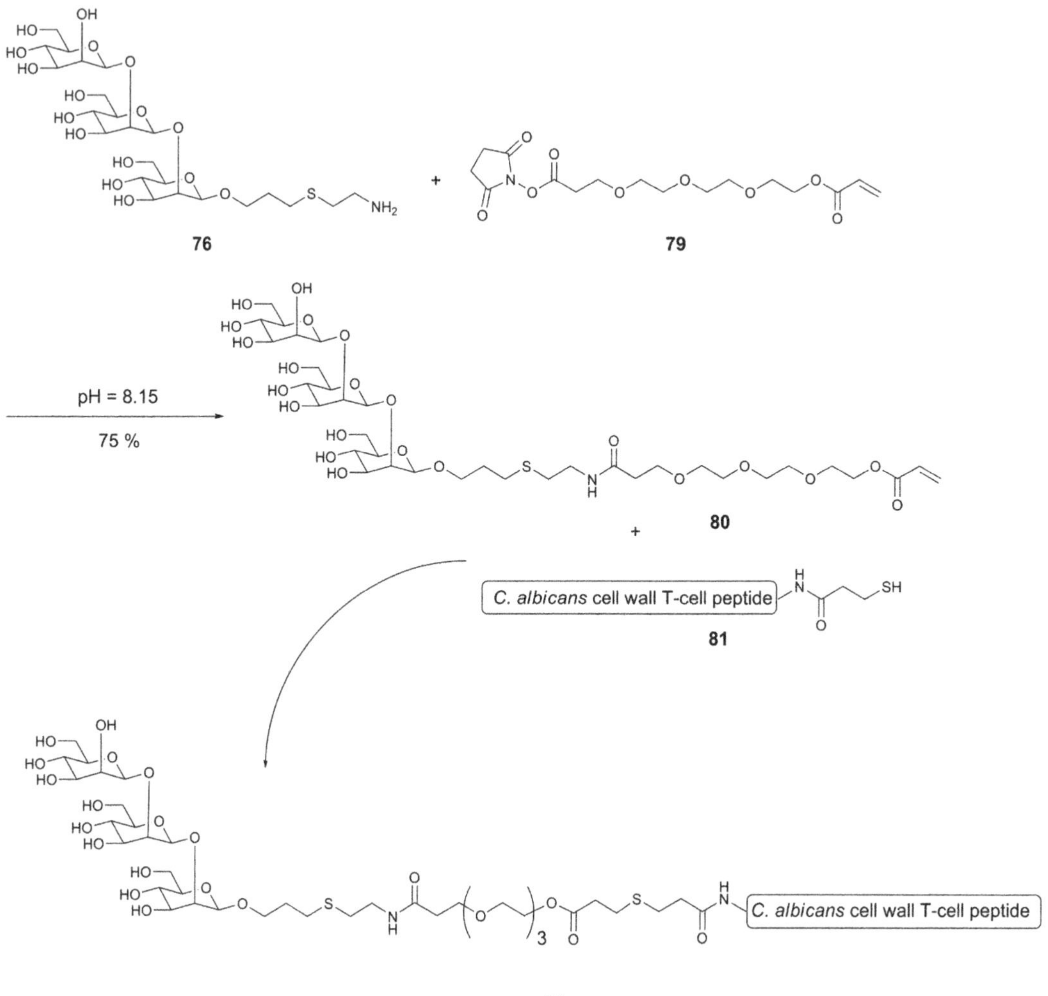

Figure 4.19 The components and novel tether employed to synthesize the trisaccharide epitope of a *C. albicans* glycopeptide vaccine.

antigens have been the target of breast cancer vaccines. Globo-H is a prostate antigen that is receiving significant attention. Since few, if any, cancer specific antigens exist and virtually all cancer-associated antigens can be found on normal cells, intrinsic difficulties in generating an immune response against self antigen necessitate the use of highly immunogenic carrier proteins and strong adjuvants.

Over the last 2–3 decades, immunologists at the memorial Sloan Kettering Institute have pioneered human trials using oligosaccharides isolated from naturally occurring glycosphingolipids and gangliosides conjugated to KLH (Livingston, 1989; Helling *et al.*, 1995; Ragupathi and Livingston, 2002; Ragupathi *et al.*, 2003). More recently the group of Danishefsky has picked up the formidable challenge of producing synthetic vaccines for many of the promising target antigens. Their approach has been comprehensive and is described in two reviews (Danishefsky and Allen, 2000; Warren *et al.*, 2007). The approaches pursued have involved monomeric vaccines (Danishefsky *et al.*, 1995; Deshpande and Danishefsky, 1997), clustered vaccines including both mono (Kuduk *et al.*, 1998; Kagan *et al.*, 2005) and poly-antigen clustered vaccines (Zhu *et al.*, 2009).

A common theme in securing the covalent linkage to protein in monovalent vaccines has involved ozonolysis of an oligosaccharide allyl or pentenyl glycoside followed by reductive amination to a maleimide tether derivative (Fig. 4.7, derivatives XV–XIX). The suitably derivatized glycoside was then conjugated to thiolated KLH as the carrier protein (Danishefsky *et al.*, 1995; Deshpande and Danishefsky, 1997).

In a recent improvement on conjugation chemistry this group reported the synthesis of a pentavalent vaccine incorporating Globo-H, GM$_2$, STn, TF and Tn epitopes using synthetic epitopes displayed on non-natural amino acids (Zhu *et al.*, 2009). Initial coupling of the Tn glycosylamino acid 83 with *tert*-butyl *N*-(3-aminopropyl)carbamate 84 was followed by a series of iterative Fmoc deprotections and glycosylamino acid peptide couplings with the glycosylamino acids 85–88 to yield the protected polyvalent construct (Fig. 4.20). Removal of the Boc group allowed reaction of the terminal amino group with *S*-acetylthioglycolic acid pentafluorophenyl ester 89. Global deprotection gave the thiol terminated pentavalent complex glycopeptide 90. Carrier protein, KLH, was incubated with sulfo-MBS (*m*-maleimidobenzoyl-*N*-hydroxysuccinimide) 91 and after chromatography the maleimide activated-KLH 92 was obtained free of MBS (Fig. 4.21). When freshly prepared maleimide activated-KLH 92 was stirred with freshly prepared glycopeptide 90, the conjugate 93 that resulted had 505 copies of glycopeptide per KLH, a conjugation yield of 50%. This was approximately a two fold improvement over the first generation conjugation method (Keding and Danishefsky, 2004).

In one of the most impressive contributions to the recent literature on cancer vaccines (Buskas *et al.*, 2005; Ingale *et al.* 2007) have designed a single synthetic construct with all the desirable features of a conjugate vaccine – a tumour-associated glycopeptide from human mucin (MUC1) 94, a T-cell helper peptide epitope 95, and the adjuvant, a lipopeptide Toll-like receptor (TLR) ligand 96 (Fig. 4.22). The three-component vaccine construct 97 was assembled by liposome-mediated native chemical ligation (NCL). When conducted in solution, NCL was very slow but when the glycopeptide (B-cell epitope) 94 and T-cell epitope 95 were incorporated into dodecylphosphocholine vesicles, NCL proceeded much more rapidly than in organic solvents. The TLR lipopeptide 96 was then added through a second NCL reaction. The completed construct was incorporated in a liposome. In this formulation the vaccine induced a robust immune response with high titres of IgG antibodies. Elements of this approach have been tried by others but the three components work extremely well when presented as a liposome. It appears the concept may be generally applicable to other antigens, not just carbohydrate-specific antigens. A fully synthetic approach of this type allows one to conceive of a vaccine optimization strategy whereby any of the three components could be varied, much as drug candidates are optimized on the basis of structure–activity relationships. The general strategy was recently described for other tumour-associated antigens (Ingale *et al.*, 2009). The Pam$_3$Cys adjuvant has been used by others (Kudryashov *et al.*, 2001) and it seems likely that the liposomal presentation of the Boons vaccine may also be a critical factor.

The group of Jennings has extended their work on the meningococcal group B capsular antigen to a candidate cancer vaccine. The *N*-propionylated group B meningococcal polysaccharide (NPrG-BMP) mimics a unique protective epitope on the surface of group B meningococci (GBM). It has been hypothesized that the formation of the protective epitope on the surface of GBM is due to the interaction of helical segments of the GBMP with another molecule and that the protective epitope is mimicked by the NPrGBMP. While the native 2,8-linked sialic acid antigen fails to yield protective antibody even when conjugated to protein the *N*-propionylated group B polysaccharide can induce protective antibody. A vaccine consisting of *N*-propionylated polysialic acid conjugated to KLH has been tested in patients with small cell lung cancer.

Polysialic acid expression has been noted in malignancies of neuroectodermal origin, particularly small cell lung cancer (SCLC). Several investigators have found immunostaining for

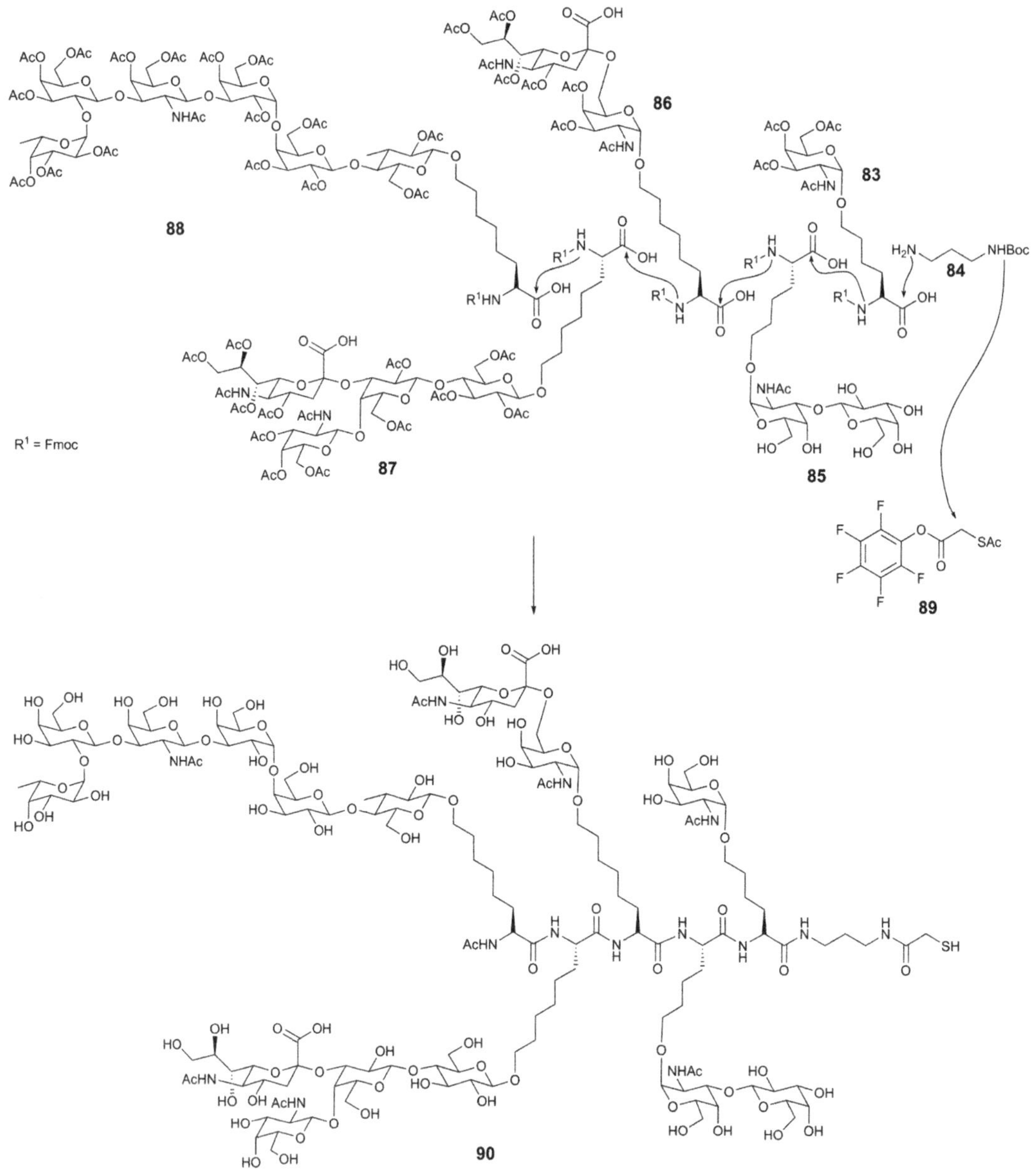

Figure 4.20 The assembly of pentavalent cancer vaccine incorporating Globo-H, GM₂, STn, TF and Tn epitopes.

polysialic acid in nearly every SCLC tumour tested (Komminoth *et al.*, 1991; Lantuejoul *et al.*, 1998; Zhang *et al.*, 1997). In two of these reports, the level of polysialic acid expression was greater in SCLC than in carcinoid tumours, generating the hypothesis that this moiety may contribute to the clinically aggressive nature of SCLC (Komminoth *et al.*, 1991; Lantuejoul *et al.*, 1998). The propensity for early metastases may derive from

the inhibitory effect of polysialic acid on cell adhesion.

Vaccination with NPrGBMP-KLH, but not polysialic acid–KLH conjugate, resulted in a consistent high titre antibody response. The immunogenicity, toxicities, and optimal dose of this vaccine are being assessed.

Jennings' group has also investigated the potential of targeting tumour cells for

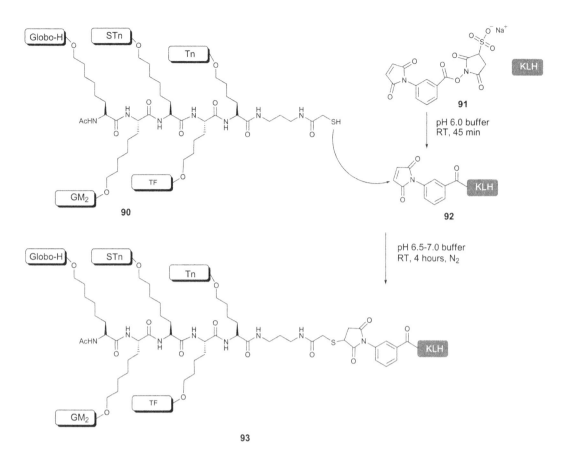

Figure 4.21 The conjugation chemistry used to secure a glycoconjugate with 505 copies of glycopeptide per KLH.

immunotherapy by altering the epitopes on the surface polysialic acid of tumour cells. A precursor (N-propionylmannosamine), when incubated with leukaemic cells, RBL-2H3 and RMA, resulted in substitution of the N-acetyl groups of surface α2–8 polysialic acid with N-propionyl groups. Expression of the altered α2–8 N-propionylpolysialic acid on the surface of tumour cells induced their susceptibility to cell death mediated by antibody to the NPrGBMP antigen. *In vivo*, antibody effectively controlled metastasis of leukaemic cells RMA when mice were administered precursor of the modified sialic acid, N-propionylated-D-mannosamine (Liu *et al.*, 2000).

HIV/AIDS vaccines

The recent structural elucidation of a human monoclonal antibody 2G12 that neutralizes the HIV virus by way of its unique recognition of the virus's N-linked envelope glycoprotein gp120, has spawned considerable interest in the potential of a glycan based vaccine using 2G12 as a template (Scanlan *et al.*, 2005). 2G12 binds to high-mannose-type oligosaccharides on gp120 (Fig. 4.23) (Lee *et al.*, 2004) but it does so in a unique manner. Biochemical, biophysical and crystallographic evidence indicates that the V_H domains of antibody 2G12 exchange in the two Fab regions of the single IgG molecule so as to form an extensive multivalent binding surface composed of the two conventional V_L/V_H combining sites and a homodimeric V_H/V_H interface (Calarese *et al.*, 2003). The interdigitation of Fab domains within an antibody uncovers a previously unappreciated mechanism for high-affinity recognition of carbohydrate. This oligomeric structure of 2G12 can account for the unusually high affinity

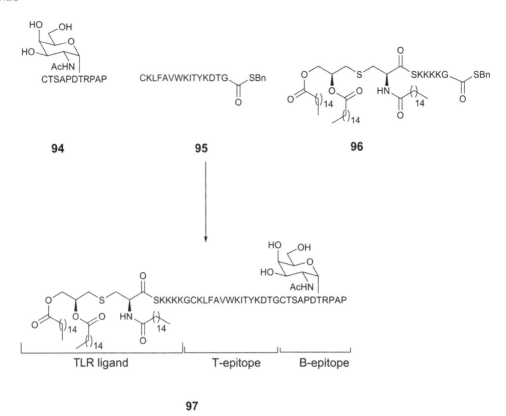

94   **95**   **96**

97

Figure 4.22 Liposome-mediated native chemical ligation was used to assemble this tri-component cancer vaccine targeting the tumour-associated glycopeptide from human mucin (MUC1).

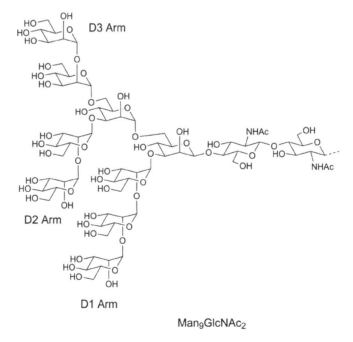

Man₉GlcNAc₂

Figure 4.23 The structure of the high-mannose-type glycan present on the HIV gp120 glycoprotein.

(nanomolar) recognition for a carbohydrate antigen by providing a virtually continuous surface for multivalent recognition with interaction sites that match the geometrical spacing of the carbohydrate cluster on gp120. Given the unique and rare nature of this antibody, it seems highly unlikely that a vaccine based on a simple epitope could reliably and reproducibly induce antibodies with such properties.

Several synthetic mannose antigens have been created as 2G12 epitope mimics for the development of a carbohydrate-based HIV-1 vaccine component (Krauss *et al.*, 2007; Lee *et al.*, 2003; Warren *et al.*, 2007; Geng *et al.*, 2004; Li and Wang, 2004; Ni *et al.*, 2006; Wang *et al.*, 2007; Wang *et al.*, 2004). Multivalent presentations of $Man_9GlcNAc_2$ on various scaffolds and, more recently, the D1 arm tetramannoside on a regioselectively addressable functionalized template have been synthesized and shown to be more effective in binding to 2G12 than monomeric $Man_9GlcNAc_2$ (or $Man_9GlcNAc_2Asn$) (Wang *et al.*, 2007). So far, no glycoconjugate vaccine has yet successfully elicited reasonable titres against the displayed glycans that also cross-react with gp120 (Astronomo *et al.*, 2008; Ni *et al.*, 2006).

Conclusions

Recent accomplishments in the synthesis of highly complex glycans as epitopes for totally synthetic conjugate vaccines are rapidly advancing the prospect of commercially viable, chemically defined vaccines. Reverse engineering of vaccine constructs based on the specificity of well defined protective antibodies provides important insights into the minimum size of protective epitopes, which in certain instances may require quite short oligosaccharides.

References

Ada, G., and Isaacs, D. (2003). Carbohydrate–protein conjugate vaccines. Clin. Microbiol. Infect. 9, 79–85.

Adamo, R., and Kováč, P. (2007). Glycosylation under thermodynamic control. Synthesis of the di and the hexasaccharide fragments of the O-SP of Vibrio cholerae O:1, serotype Ogawa, from fully functionalized building blocks. Eur. J. Org. Chem. 988–1000.

Adamo, R., Saksena, R., and Kováč, P. (2006). Studies towards a conjugate vaccine for anthrax: synthesis of the tetrasaccharide side chain of the Bacillus anthracis exosporium. Helv. Chim. Acta 89, 1074–1089.

Agard, N.J., Baskin, J.M., Prescher, J.A., Lo, A., and Bertozzi, C.R. (2006). A comparative study of bioorthogonal reactions with azides. ACS Chem. Biol. 1, 644–648.

Alexander, J., del Guercio, M.-F., Barbara Frame, B., Maewal, A., Sette, A., Nahmb, M.H., and Newman, M.J. (2004). Development of experimental carbohydrate-conjugate vaccines composed of Streptococcus pneumoniae capsular polysaccharides and the universal helper T-lymphocyte epitope (PADRE®). Vaccine 22, 2362–2367.

Ashkenazi, S., Passwell, J.P., Harlev, E., Miron, D., Dagan, R., Farzan, N., Ramon, R., Majadly, F., Bryla, D.A., Karpas, A.B., Robbins, J.B., Schneerson, R., and the Israel Pediatric Shigella Study Group. (1999). Safety and immunogenicity of Shigella sonnei and Shigella flexneri 2a O-specific polysaccharide conjugates in children. J. Infect. Dis. 179, 1565–1568.

Astronomo, R.D., Lee, H.-K., Scanlan, C.N., Pantophlet, R., Huang, C.-Y., Wilson, I.A., Blixt, O., Dwek, R.A., Wong, C.-H., and Burton, D.R. (2008). A glycoconjugate antigen based on the recognition motif of a broadly neutralizing human immunodeficiency virus antibody, 2G12, is immunogenic but elicits antibodies unable to bind to the self glycans of gp120. J. Virol. 82, 6359–6368.

Bardotti, A., Averani, G., Berti, F., Berti, S., Carinci, V., D'Ascenzi, S., Fabbri, B., Giannini, S., Giannozzi, A., Magagnoli, C., Proietti, D., Norelli, N., Rappuoli, R., Ricci, S., and Costantino, P. (2008). Physicochemical characterisation of glycoconjugate vaccines for prevention of meningococcal diseases. Vaccine 26, 2284–2296.

Bartoloni, A., Norelli, F., Ceccarini, C., Rappuoli, R., and Costantino, P. (1995). Immunogenicity of meningococcal B polysaccharide conjugated to tetanus toxoid or CRM197 via adipic acid dihydrazide. Vaccine 13, 463–470.

Bélot, F.; Wright, K.; Costachel, C.; Phalipon, A.; and Mulard, L.A. (2004). Blockwise approach to fragments of the O-specific polysaccharide of Shigella flexneri serotype 2a: convergent synthesis of a decasaccharide representative of a dimmer of the branched repeating unit. J. Org. Chem. 69, 1060–1074.

Bélot, F., Guerreiro, C., Baleux, F., and Mulard, L.A. (2005). Synthesis of two linear PADRE conjugates bearing a deca- or pentadecasaccharide B epitope as potential synthetic vaccines against Shigella flexneri serotype 2a infection. Chem. Eur. J. 11, 1625–1635.

Benaissa-Trouw, B., Lefeber, D.J., Kamerling, J.P., Vliegenthart, J.F.G., Kraaijeveld, K., and Snippe, H. (2001). Synthetic polysaccharide type 3-related di-, tri-, and tetrasaccharide–CRM197 conjugates induce protection against Streptococcus pneumoniae Type 3 in mice. Infect. Immun. 69, 4698–4701.

Bernardes, G.C., Castagner, B., and Seeberger, P.H. (2009). Combined approaches to the synthesis and study of glycoproteins. ACS Chem. Biol. 4, 703–713.

Bock, K., Josephson, S., and Bundle, D.R. (1982). Lipopolysaccharide solution conformation: Antigen shape inferred from high resolution ^1H and ^{13}C

nuclear magnetic resonance spectroscopy and hard-sphere calculations. J. Chem. Soc. Perkin II, 59–70.

Boonyarattanakalin, S., Liu, X., Michieletti, M., Lepenies, B., and Seeberger, P.H. (2008). Chemical synthesis of all phosphatidylinositol mannoside (PIM) glycans from Mycobacterium tuberculosis. J. Am. Chem. Soc. 130, 16791–16799.

Bröker, M., Dull, P.M., Rappuoli, R., and Costantino, P. (2009). Chemistry of a new investigational quadrivalent meningococcal conjugate vaccine that is immunogenic at all ages. Vaccine 27, 5574–5580.

Brubaker, J.O., Li. Q., Tzianabos, A.O., Kasper, D.L., and Finberg, R.W. (1999). Mitogenic activity of purified capsular polysaccharide A from Bacteroides fragilis: Differential stimulatory effect on mouse and rat lymphocytes in vitro. J. Immunol. 162, 2235–2242.

Bundle, D.R., Nitz, M., Wu, X., and Sadowska, J.M. (2007). A uniquely small, protective carbohydrate epitope may yield a conjugate vaccine for Candida albicans. Amer. Chem. Soc. Symp. Ser. 989, 163–183.

Buskas, T., Ingale, S., and Boons, G.-J. (2005). Towards a fully synthetic carbohydrate-based anticancer vaccine: synthesis and immunological evaluation of a lapidated glycopeptide containing the tumor-associated Tn antigen. Angew Chem. Int. Ed. 44, 5985–5988.

Calarese, D.A., Scanlan, Zwick, C.N., M.B., Deechongkit, S., Mimura, Y., Kunert, R., Zhu, P., Wormald, M.R., Stanfield, R.L., Roux, K.H., Kelly, J.W., Rudd, P.M., Dwek, R.A., Katinger, H., Burton, D.R., and Wilson, I.A. (2003). Antibody domain exchange is an immunological solution to carbohydrate cluster recognition. Science 300, 2065–2071.

Carlin, N.I.A., Lindberg, A.A., Bock, K., and Bundle, D.R. (1984). The Shigella flexneri O-antigenic polysaccharide chain: nature of the biological repeating unit. Eur. J. Biochem. 139, 189–194.

Caroff, M., Bundle, D.R., and Perry, M.B. (1984). Structure of the O-chain of the phenol-phase soluble cellular lipopolysaccharide of Yersinia enterocolitica serotype O:9. Eur. J. Biochem. 139, 195–200.

Caroff, M., Bundle, D.R., Perry, M.B., Cherwonogrodsky, J.W., and Duncan, J.R. (1984). Antigenic S-type lipopolysaccharide of Brucella abortus 1119–3. Infect. Immun., 46, 384–388.

Casadevall, A. (1995). Antibody immunity and invasive fungal infections. Infect. Immun. 63, 4211–4218.

Casadevall, A., Cassone, A., Bistono, F., Cutler, J.E., Magliani, W., Murphy, J.W., Polonelli, L., and Romani, L. (1998). Antibody and/or cell-mediated immunity, protective mechanisms in fungal disease: An ongoing dilemma or an unnecessary dispute? Med. Mycol. 36 (Suppl. 1), 95–105.

Cassone, A. (1989). Cell wall of Candida albicans: Its functions and its impact on the host. Curr. Top. Med. Mycol. 3, 248–314.

Chan, L., and Just, G. (1990). Syntheses of oligomers of the capsular polysaccharide of the Haemophilus influenzae type b bacteria. Tetrahedron 46, 151–162.

Chernyak, A., Karavanov, A., Ogawa, Y., and Kováč, P. (2001). Conjugating oligosaccharides to proteins by squaric acid diester chemistry; rapid monitoring of the progress of conjugation, and recovery of the unused ligand. Carbohydr. Res. 330, 479–486.

Chernyak, A., Kondo, S., Wade, T., Meeks, M., Alzari, P.M., Fournier, J.-M., Taylor, R.K., Kováč, P., and Wade, W.F. (2002). Induction of protective immunity by synthetic Vibrio cholerae hexasaccharide derived from Vibrio cholerae O:1 Ogawa lipopolysaccharide bound to a protein carrier. J. Infect. Dis. 185, 950–962.

Coker, R.J. (2004). Multidrug-resistant tuberculosis: public health challenges. Trop. Med. Int. Health 9, 25–40.

Crich, D., Li, H., Yao, Q., Wink, D.J., Sommer, R.D., and Rheingold, A.L. (2001). Direct synthesis of beta-mannans. A hexameric [–>3)-beta-D-Man-(1](3) subunit of the antigenic polysaccharides from Leptospira biflexa and the octameric (1→2)-linked beta-D-mannan of the Candida albicans phospholipomannan. X-ray crystal structure of a protected tetramer. J. Am. Chem. Soc. 123, 5826–5828.

Cutler, J.E., Granger, B.L., and Han. Y. (2002). Immunoprotection against candidiasis. In Candida and Candidiasis, Calderone, R.A., ed. (ASM Press, Washington, DC).

Daffe, M., and Reyrat, J.M. (2008). The Mycobacterial Cell. (ASM Press, Washington, DC).

Danishefsky, S.J., and Allen, J.R. (2000). From the laboratory to the clinic: A retrospective on fully synthetic carbohydrate-based anticancer vaccines. Angew. Chem. Intl. Ed. 39, 836–863.

Danishefsky, S.J., Behar, V., Randolph, J.T., and Lloyd, K.O. (1995). Application of the glycal assembly method to the concise synthesis of neoglycoconjugates of Le^y and Le^b blood group determinants of H-type I and H-type II oligosaccharides. J. Am. Chem. Soc. 117, 5701–5711.

Daubenspeck, J.M., Zeng, H., Chen, P., Dong, S., Steichen, C.T., Krishna, N.R., Pritchard, D.G., and Turnbough, C.L. (2004). Novel oligosaccharide side chains of the collagen-like region of bcla, the major glycoprotein of the Bacillus anthracis exosporium. J. Biol. Chem. 279, 30945–30953.

Deshpande, P.P., and Danishefsky, S.J. (1997). Total synthesis of the potential anticancer vaccine KH-1 adenocarcinoma antigen. Nature 387, 164–166.

Duan, J., Avci, F.Y., and Kasper, D.L. (2008). Microbial carbohydrate depolymerization by antigen-presenting cells: Deamination prior to presentation by the MHCII pathway. Proc. Natl. Acad. Sci. U.S.A. 105, 5183–5188.

Dziadek, S., Jacques, S., and Bundle, D.R. (2008). A novel linker methodology for the synthesis of tailored conjugate vaccines composed of complex carbohydrate antigens and specific T$_H$-cell peptide epitopes. Chem. Eur. J. 14, 5908–5917.

Edwards, J.E., Mayer, C.L., Riller, S.G., Wadsworth, E., and Calderone. R.A. (1992). Cell extracts of Candida albicans block adherence of the organisms to endothelial cells. Infect. Immun. 60, 3087–3091.

Evenberg, D., Hoogerhout, P., van Boeckel, C.A.A., Rijkers, G.T., Beuvery, E.C., van Boom, J.H., and Poolman, J.T. (1992). Preparation, antigenicity, and immunogenicity of synthetic ribosylribitol phosphate oligomer–protein conjugates and their potential use for vaccination against *Haemophilus influenzae* type b disease. J. Infect. Dis. *165* (Suppl. 1), S152–S155.

Faridmoayer, A., Fentabil, M.A., Haurat, M.F., Yi, W., Woodward, R., Wang, P.G., and Feldman, M.F. (2008). Extreme substrate promiscuity of the *Neisseria* oligosaccharyl transferase involved in protein o-glycosylation. J. Biol. Chem. *283*, 34596–34604.

Feldman, M.F., Wacker, M., Hernandez, M., Hitchen, P.G., Marolda, C.L., Kowarik, M., Morris, H.R., Dell, A., Valvano, M.A., and Aebi., M. (2005). Engineering N-linked protein glycosylation with diverse O antigen lipopolysaccharide structures in *Escherichia coli*. Proc. Natl. Acad. Sci. U.S.A. *102*, 3016–3021.

Fernandez, C., and Sverremark, E. (1994). Immune responses to bacterial polysaccharides: Terminal epitopes are more immunogenic than internal structures. Cell Immunol. *153*, 67–78.

Fernandez-Santana, V., Cardoso, F., Rodriguez, A., Carmenate, T., Peña, L., Valdes, Y., Hardy, E., Mawas, F., Heynngnezz, L., Rodriguez, M.C., *et al.* (2004). Antigenicity and immunogenicity of a synthetic oligosaccharide-protein conjugate vaccine against *Haemophilus influenzae* Type b. Infect. Immun. *72*, 7115–7123.

Fine, P.E. (1995). Variation in protection by BCG: implications of and for heterologous immunity. Lancet *346*, 1339–1345.

Fraser-Reid, B., Lu, J., Jayaprakash, K.N., and Cristóbal López, J. (2006). Synthesis of a 28-mer oligosaccharide core of *Mycobacterial lipoarabinomannan* (LAM) requires only two *n*-pentenyl orthoester progenitors. Tetrahedron: Asym. *17*, 2449–2463.

Freire, T., Bay, S., Vichier-Guerre, S., Lo-Man, R., and Leclerc, C. (2006). Carbohydrate antigens: Synthesis aspects and immunological applications in cancer. Mini-Rev. Med. Chem. *6*, 1357–1373.

Gadikota, R.R., Callam, C.S., Appelmelk, B.J., and Lowary, T.L. (2003). Studies toward the development of anti-tuberculosis vaccines based on mycobacterial lipoarabinomannan. J. Carbohydr. Chem. *22*, 459–480.

Geng, X., Dudkin, V.Y., Mandal, M., and Danishefsky, S.J. (2004). In pursuit of carbohydrate-based HIV vaccines, Part 2: The total synthesis of high-mannose-type gp120 fragments – evaluation of strategies directed to maximal convergence. Angew. Chem. Int. Ed. *43*, 2562–2565.

Giebink, G.S., Koskela, M., Vella, P.P., Harris, M., and Le, C.T. (1993). Pneumococcal capsular polysaccharide-meningococcal outer membrane protein complex conjugate vaccines: immunogenicity and efficacy in experimental pneumococcal otitis media. J. Infect. Dis. *167*, 347–355.

Gupta, S.S., Kuzelka, J., Singh, P., Lewis, W.G., Manchester, M., and, Finn, M.G. (2005). Accelerated bioorthogonal conjugation: a practical method for the ligation of diverse functional molecules to a polyvalent virus scaffold. Bioconjugate Chem. *16*, 1572–1579.

Hakomori, S.-i. (1985). Aberrant glycosylation in cancer cell membranes as focused on glycolipids: overview and perspectives. Cancer Res. *45*, 2405–2414.

Hakomori, S.-i. (1989). Aberrant glycosylation in tumors and tumor-associated carbohydrate antigens. Adv. Cancer Res. *52*, 257–331.

Hamasur, B., Kallenius, G., and Svenson, S.B. (1999). Synthesis and immunologic characterization of *Mycobacterium tuberculosis* lipoarabinomannan specific oligosaccharide–protein conjugates. Vaccine *17*, 2853–2861.

Hamasur, B., Haile, M., Pawlowski, A., Schroder, U., Williams, A., Hatch, G., Hall, G., Mash, P., Kallenius, G., and Svenson, S.B. (2003). *Mycobacterium tuberculosis* arabinomannanprotein conjugates protect against tuberculosis. Vaccine *21*, 4081–4093.

Hamasur, B., Haile, M., Pawlowski, A., Schroder, U., Kallenius, G., and Svenson, S.B. (2004). A mycobacterial lipoarabinomannan specific monoclonal antibody and its F(ab′)₂ fragment prolong survival of mice infected with *Mycobacterium tuberculosis*. Clin. Exp. Immunol. *138*, 30–38.

Han, Y., Ulrich, M.A., and Cutler, J.E. (1999). *Candida albicans* mannan extract–protein conjugates induce a protective immune response against experimental candidiasis. J. Infect. Dis. *179*, 1477–1484.

Han, Y., Riesselman, M., and Cutler, J.E. (2000). Protection against candidiasis by an immunoglobulin G3 (IgG3) monoclonal antibody specific for the same mannotriose as an IgM protective antibody. Infect. Immun. *68*, 1649–1654.

Hanahan, D.J., and Olley, J.N. (1958). Chemical nature of monophosphoinositides. J. Biol. Chem. *231*, 813–828.

Hecht, M.L., Stallforth, P., Silva, D.V., Adibekian, A., and Seeberger, P.H. (2009). Recent advances in carbohydrate-based vaccines. Curr. Opin. Chem. Biol. *3*, 354–359.

Helling, F., Zhang, S., Shang, A., Adluri, S., Calves, M., Koganty, R., Longenecker, B.M., Yao, T.-Y., Oettgen, H.F., and Philip Livingston, O., P.O. (1995). GM₂-KLH Conjugate vaccine: increased immunogenicity in melanoma patients after administration with immunological adjuvant QS-21. Cancer Res. *55*, 2783–2788.

Hölemann, A., Stocker, B.L., and Seeberger. P.H. (2006). Synthesis of a core arabinomannan oligosaccharide of *Mycobacterium tuberculosis*. J. Org. Chem. *71*, 8071–8088.

Hou, S.-J., Saksena, R., and Kováč, P. (2008). Preparation of glycoconjugates by dialkyl squarate chemistry revisited. Carbohydr. Res. *343*, 196–210.

Ingale, S., Wolfert, M.A., Gaekwad, J., Buskas, T., and Boons, G.-J. (2007). Robust immune responses elicited by a fully synthetic three-component vaccine. Nat. Chem. Biol. *3*, 663–667.

Ingale, S., Wolfert, M.A., Buskas, T., and Boons, G.-J. (2009). Increasing the antigenicity of synthetic tumor-associated carbohydrate antigens by targeting toll-like receptors. ChemBioChem. *10*, 455–463.

Jennings, H.J. (1983). Capsular polysaccharides as human vaccines. Adv. Carbohydr. Chem. Biochem. *41*, 155–208.

Jennings, H.J., and Lugowski, C. (1981). Immunochemistry of groups A, B, and C meningococcal polysaccharide-tetanus toxoid conjugates. J. Immunol. *127*, 1011–1018.

Joe, M., Bai, Y., Nacario, R.C., and Lowary, T.L. (2007). Synthesis of the docosanasaccharide arabinan domain of mycobacterial arabinogalactan and a proposed octadecasaccharide biosynthetic precursor J. Am. Chem. Soc. *129*, 9885–9901.

Johnson, M.A., Jaseja, M., Zou, W., Jennings, H.J., Copié, V., Pinto, M.B., and Pincus, S. (2003). NMR studies of carbohydrates and carbohydrate-mimetic peptides recognized by an anti-group B *Streptococcus* antibody. J. Biol. Chem. *278*, 24740–24752.

Joosten, J.A.F., Kamerling, J.P., and Vliegenthart, J.F.G. (2003). Chemo-enzymatic synthesis of a tetra- and octasaccharide fragment of the capsular polysaccharide of *Streptococcus pneumoniae* type 14. Carbohydr. Res. *338*, 2611–2627.

Joosten, J.A.F., Lazet, B.J., Kamerling, J.P., and Vliegenthart, J.F.G. (2003). Chemo-enzymatic synthesis of tetra-, penta-, and hexasaccharide fragments of the capsular polysaccharide of *Streptococcus pneumoniae* type 14. Carbohydr. Res. *338*, 2629–2651.

Kadirvelraj, R., Gonzalez-Outeirino, J., Lachele, Foley, B., B.L., Beckham, M.L., Jennings, H.J., Foote, S., Ford, M.G., and Woods, R.J. (2006). Understanding the bacterial polysaccharide antigenicity of *Streptococcus agalactiae* versus *Streptococcus pneumoniae*. Proc. Natl. Acad. Sci. U.S.A. *103*, 8149–8154.

Kanbe, T., and Cutler, J.E. (1994). Evidence for adhesin activity in the acid-stable moiety of the phosphomannoprotein cell wall complex of *Candida albicans*. Infect. Immun. *62*, 1662–1668.

Kagan, E., Ragupathi, G., Yi, S.S., Reis, C.A., Gildersleeve, J., Kahne, D., Clausen, H., Danishefsky, S.J., and Livingston, P.O. (2005). Comparison of antigen constructs and carrier molecules for augmenting the immunogenicity of the monosaccharide epithelial cancer antigen Tn. Cancer Immunol. Immunother. *54*, 424–430.

Källenius, G., Pawlowski1, A., Hamasur, B., and Svenson, S.B. (2008). Mycobacterial glycoconjugates as vaccine candidates against tuberculosis. Trends Microbiol. *16*, 456–462.

Kaltgrad, E., Sen Gupta, S., Punna, S., Huang, C.Y., Chang, A., Wong, C.-H., Finn, M.G., and Blixt, O. (2007). Anti-carbohydrate antibodies elicited by polyvalent display on a viral scaffold. Chembiochem *8*, 1455–1462.

Kamath, V.P., and Hindsgaul, O. (1995). Use of diethyl squarate for the coupling of oligosaccharide amines to carrier proteins and characterization of the resulting neoglycoproteins by MALDI-TOF mass spectrometry. Glycoconj. J. *13*, 315–319.

Kamena, F., Tamborrini, M., Liu, X., Kwon, Y.-U., Thompson, F., Pluschke, G., and Seeberger, P.H. (2008). Synthetic GPI array to study antitoxic malaria response. Nat. Chem. Biol. *4*, 238–240.

Kaur, D., Guerin, M.E., Škovierová, H., Brennan, P.J., and Jackson, M. (2009). Biogenesis of the cell wall and other glycoconjugates of *Mycobacterium tuberculosis*. Adv. Appl. Microbiol. *69*, 23–78.

Keding, S.J., and Danishefsky, S.J. (2004). Prospects for total synthesis: A vision for a totally synthetic vaccine targeting epithelial tumors. Proc. Natl. Acad. Sci. U.S.A. *101*, 11937–11942.

Kelly, D.F., Snape, M.D., Cutterbuck, E.A., Green, S., Snowden, C., Diggle, L., Yu, L.-M., Borkowski, A., Moxon, E.R., and Pollard, A.J. (2006). CRM197-conjugated serogroup C meningococcal capsular polysaccharide, but not the native polysaccharide, induces persistent antigen-specific memory B cells. Blood *108*, 2642–2647.

Kenne, L., Lindberg, B., and Petersson, K. (1977). Basic structure of the oligosaccharide repeating-unit of the *Shigella flexneri* O-antigens. Carbohydr. Res. *5*, 363–370.

Kenne, L., Lindberg, B., Unger, P., Holme, T., and Holmgren. J. (1979). Structural studies of the *Vibrio cholerae* O-antigen. Carbohydr. Res. *68*, C14–C16.

Kim, H.-S., Ng, E.S.M., Zheng, R.B., Whittal, R., Schriemer, D.C., and Lowary, T.L. (2007). Studies toward the development of anti tuberculosis vaccines based on mycobacterial lipoarabinomannan. ACS Symp. Ser. *989*, 184–197.

Kobayashi, H., Shibata, N., Nakada, M., Chaki, S., Mizugami, K., Ohkubo, Y., and Suzuki, S. (1990). Structural study of cell wall phosphomannan of *Candida albicans* NIH B-792 (serotype B) strain, with special reference to ^{1}H and ^{13}C NMR analyses of acid-labile oligomannosyl residues. Arch. Biochem. Biophys. *278*, 195–204.

Kolb, H.C., Finn, M.G., and Sharpless, K.B. (2001). Click Chemistry: diverse chemical function from a few good reactions. Angew. Chem. Interl. Ed. *40*, 2004–2021.

Komminoth, P., Roth, J., Lackie, P.M., Bitter-Suermann, D., and Heintz, P.U. (1991). Polysialic acid of the neural cell adhesion molecule distinguishes small cell lung carcinoma from carcinoids. Am. J. Pathol. *139*, 297–304.

Krauss, I.J., Joyce, J.G., Finnefrock, A.C., Song, H.C., Dudkin, V.Y., Geng, X., Warren, J.D., Chastain, W.M., Shiver, J.W., and Danishefsky, S.J. (2007). Fully synthetic carbohydrate HIV antigens designed on the logic of the 2G12 antibody. J. Am. Chem. Soc. *129*, 11042–11044.

Krug, L.M., Ragupathi, G., Ng, K.K., Hood, C., Jennings, H.J., Guo, Z., Kris, M.G., Miller, V., Pizzo, B., Tyson, L., *et al.* (2004). Vaccination of small cell lung cancer patients with polysialic acid or N-propionylated polysialic acid conjugated to keyhole limpet hemocyanin. Clin. Cancer Res. *10*, 916–923.

Kubler-Kielba, J., and Pozsgay, V. (2005). A new method for conjugation of carbohydrates to proteins using an aminooxy-thiol heterobifunctional linker. J. Org. Chem. *70*, 6987–6990.

Kubler-Kielba, J., Vinogradovb, E., Ben-Menachema, Pozsgay, G., V., Robbins, J.B., and Schneerson, R. (2008). Saccharide/protein conjugate vaccines for *Bordetella* species: Preparation of saccharide, development of new conjugation procedures, and physico-chemical and immunological characterization of the conjugates. Vaccine *26*, 3587–3593.

Kuduk, S.D., Schwarz, J.B., Chen, X.-T., Glunz, P.W., Sames, D., Ragupathi, G., Livingston, P.O., and Danishefsky, S.J. (1998). Synthetic and immunological studies on clustered modes of mucin-related Tn and TF O-linked antigens: The preparation of a glycopeptide-based vaccine for clinical trials against prostate cancer. J. Am. Chem. Soc. *120*, 12474–12485.

Kudryashov, V., Glunz, P.W., Williams, L.J., Hintermann, S., Danishefsky, S.J., and Lloyd, K.O. (2001). Toward optimized carbohydrate-based anticancer vaccines: Epitope clustering, carrier structure, and adjuvant all influence antibody responses to Lewis y conjugates in mice. Proc. Natl. Acad. Sci. U.S.A. *98*, 3264–3269.

Lai, Z.Z., and Schreiber, J.R. (2009). Antigen processing of glycoconjugate vaccines; the polysaccharide portion of the pneumococcal CRM197 conjugate vaccine co-localizes with MHC II on the antigen processing cell surface. Vaccine *27*, 3137–3144.

Lambert, M., and Neish, A.C. (1950). Rapid method for estimation of glycerol in fermentation solutions. Cancer J. Res. *28*, 83–89.

Lantuejoul, S., Moro, D., Michalides, R.J., Brambilla, C., and Brambilla, E. (1998). Neural cell adhesion molecules (NCAM) and NCAMPSA expression in neuroendocrine lung tumors. Am. J. Surg. Pathol. *22*, 1267–1276.

Lee, H.-K., Scanlan, C.N., Huang, C.-Y., Chang, A.Y., Calarese, D.A., Dwek, R.A., Rudd, P.M., Burton, D.R., Wilson, I.A., and Wong, C.-H. (2004). Reactivity-based one-pot synthesis of oligomannoses: defining antigens recognized by 2G12, a broadly neutralizing anti-HIV-1 antibody. Angew. Chem. Intl. Ed. *43*, 1000–1003.

Lee, R.T., and Lee, Y.C. (1974). Synthesis of 3-(2-aminoethylthio)propyl glycosides. Carbohydr. Res. *37*, 193–201.

Lemieux, R.U., Baker, D.A., and Bundle, D.R. (1977). A methodology for the production of carbohydrate-specific antibody. Cancer J. Biochem. *55*, 507–512.

Lemieux, R.U., Bundle, D.R., and Baker, D.A. (1975). The properties of a synthetic antigen related to the human blood group Lewis-a. J. Am. Chem. Soc. *97*, 4076–4083.

Li, H., and Wang, L.X. (2004). Design and synthesis of a template-assembled oligomannose cluster as an epitope mimic for human HIV-neutralizing antibody 2G12. Org. Biomol. Chem. *2*, 483–488.

Lindberg, A.A., Wollin, R., Bruse, G., Ekwall, E., and Svenson, S.B. (1983). Immunology and immunochemistry of synthetic and semisynthetic *Salmonella* O-antigen-specific glycoconjugates. In Bacterial Lipopolysaccharides. ACS Symp. Ser., pp. 83–118.

Lindberg, A.A. (1998). Vaccination against enteric pathogens: from science to vaccine trials. Curr. Opin. Microbiol. *1*, 116-l 24.

Lindberg, A.A. (1999). Glycoprotein conjugate vaccines. Vaccine *17* (Suppl. 2), S28–S36.

Livingston, P.O. (1989). The basis for ganglioside vaccines in melanoma. In Human Tumor Antigens and Specific Tumor Therapy, Metzgar, R., and Mitchell, M., eds (New York, Alan Liss, R., Inc.), Vol. 99, pp. 287–296.

Lui, T., Guo, Z., Yang, Q., Sad, S., and Jennings, H.J. (2000). Biochemical engineering of surface alpha 2–8 polysialic acid for immunotargeting tumor cells. J. Biol. Chem. *275*, 32832–32836.

Liu, X., Kwon, Y.-U., and Seeberger, P.H. (2005). Convergent synthesis of a fully lipidated glycosylphosphatidylinositol anchor of *Plasmodium falciparum*. J. Am. Chem. Soc. *127*, 5004–5005.

Liu, X., Bridget Stocker, L., B.L., and Seeberger, P.H. (2006). Total synthesis of phosphatidylinositol mannosides of mycobacterium tuberculosis. J. Am. Chem. Soc. *128*, 3638–3648.

Liu X., Siegrist, S., Amacker, M., Zurbriggen, R., Pluschke, G., and Seeberger, P.H. (2006). Enhancement of the immunogenicity of synthetic carbohydrates by conjugation to virosomes: a leishmaniasis vaccine candidate. ACS Chem. Biol. *1*, 161–164.

Lucas, A.H., Apicella, M.A., and Taylor, C.E. (2005). Carbohydrate moieties as vaccine candidates. Clin. Infect. Dis. *41*, 705–712.

Ma, X., Saksena, R., Chernyak, A., and Kovác, P. (2003). Neoglycoconjugates from synthetic tetra and hexasaccharides that mimic the terminus of the O-PS of *Vibrio cholerae* O:1, serotype Inaba, Org. Biomol. Chem. *1*, 775–784.

Matthews, R., and Burnie. J. (1996). Antibodies against *Candida*: Potential therapeutics? Trends Microbiol. *4*, 354–358.

Mawas, F., Niggemann, J., Jones, C., Corbel, M.J., Kamerling, J.P., and Vliegenthart, J.F.G. (2002). Immunogenicity in a mouse model of a conjugate vaccine made with a synthetic single repeating unit of type 14 pneumococcal polysaccharide coupled to CRM197. Infect. Immun. *70*, 5107–5114.

Meeks M.D., Saksena, R., Ma, X., Wade, T.K., Taylor, R.K., Kovác, P., and Wade, W.F. (2004). Synthetic fragments of *Vibrio cholerae* O1 Inaba O-SP bound to a protein carrier are immunogenic in mice but do not induce protective antibodies. Infect. Immun. *72*, 4090–4101.

Mehta, A.S., Saile, E., Zhong, W., Buskas, T., Carlson, R., Kannenberg, E., Reed, Y., Quinn, C.P., and Boons, G.-J. (2006). Synthesis and antigenic analysis of the bcla glycoprotein oligosaccharide from the *Bacillus anthracis* exosporium. Chem. Eur. J. *12*, 9136–9149.

Miermont, A., Barnhill, H., Strable, E., Lu, X., Wall, K.A., Wang, Q., Finn, M.G., and Huang, X. (2008). Cowpea mosaic virus capsid, a promising carrier for the development of carbohydrate based anti-tumor vaccines. Chem. Eur. J. *14*, 4939–4947.

Moayeri, M., and Leppla, S.H. (2004). The roles of anthrax toxin in pathogenesis. Curr. Opin. Microbiol. *7*, 19–24.

Mock, M., and Fouet, A. (2001). Anthrax. Annu. Rev. Microbiol. *55*, 647–671.

Moody, B.D., Reinhold, B.B., Guy, M.R., Beckman, E.M., Frederique, D.E., Furlong, S.T., Ye, S., Reinhold, V.N., Sieling, P.A., Modlin, R.L., *et al.* (1997). Structural requirements for glycolipid antigen recognition by CD1b-restricted T cells. Science 278, 283–286.

Moody, D.B., Besra, G.S., Wilson, I.A., and Porcelli, S.A. (1999). The molecular basis of CD1-mediated presentation of lipid antigens. Immunol. Rev. 172, 285–296.

Michon, F., Uitz, C., Sarkar, A., D'Ambra, A.J., Laude-Sharp, M., Moore, S., and Fusco, P.C. (2006). Group B streptococcal type II and III conjugate vaccines: Physicochemical properties that influence immunogenicity. Clin. Vaccine Immunol. 13, 936–943.

Murase, T., Zheng, R.B., Joe, M., Bai, Y., Marcus, S.L., Lowary, T.L., and Ng, K.K.S. (2009). Structural insights into antibody recognition of mycobacterial polysaccharides. J. Mol. Biol. 392, 381–392.

Nahm, M.H., Apicella, M.A., and Briles, D.E. (1999). Immunity to extracellular bacteria. In Fundamental Immunology, Paul, W.E., ed. (Philadelphia, Lipincott-Williams & Wilkins), pp. 1373–1386.

Naik, R.S., Branch, O.H., Woods, A.S., Vijaykumar, M., Perkins, D.J., Nahlen, B.L., Lal, A.A., Cotter, R.J., Costello, C.E., Ockenhouse, C.F., et al. (2000). Glycosylphosphatidylinositol anchors of Plasmodium falciparum: molecular characterization and naturally elicited antibody response that may provide immunity to malaria pathogenesis. J. Exp. Med. 192, 1563–1576.

Ni, J., Song, H., Wang, Y., Stamatos, N.M., and Wang, L.X. (2006). Toward a carbohydrate-based HIV-1 vaccine: synthesis and immunological studies of oligomannose-containing glycoconjugates. Bioconjug. Chem. 17, 493–500.

Nitz, M., and Bundle, D.R. (2001). Synthesis of di- to hexasaccharide 1, 2-linked β-mannopyranan oligomers, a terminal S-linked tetrasaccharide congener and the corresponding BSA glycoconjugates. J. Org. Chem. 66, 8411–8423.

Nitz, M., Ling, C.C., Otter, A., Cutler, J.E., and Bundle, D.R. (2002). The unique solution structure and immunochemistry of the Candida albicans β-1, 2-mannopyranan cell wall antigen. J. Biol. Chem. 277, 3440–3446.

Ogawa, Y., and Kovác, P. (1997). Synthesis of the dodecasaccharide fragment representing the O-polysaccharide of Vibrio cholerae O:1, serotype Ogawa, bearing an aglycon offering flexibility for chemical linking to proteins. Glycoconj. J. 14, 433–438.

Ogawa, Y., Lei, P.-S., and Kovác, P. (1996). Synthesis of four glycosides of a disaccharide fragment representing the terminus of the O-polysaccharide of Vibrio cholerae O:1, serotype Inaba, bearing aglycons suitable for linking to proteins. Carbohydr. Res. 288, 85–98.

Ojeda, R., de Paz, J.L., Barrientos, A.G., Martín-Lomas, M., and Penadé, S. (2007). Preparation of multifunctional glyconanoparticles as a platform for potential carbohydrate-based anticancer vaccines. Carbohydr. Res. 342, 448–459.

Paoletti, L.C., and Kasper, D.L. (2003). Glycoconjugate vaccines to prevent group B streptococcal infections. Expert Opin. Biol. Ther. 3, 975–984.

Passwell, J.H., Harlev, E., Ashkenazi, S., Chu, C., Miron, D., Ramon, R., Farzan, N., Shiloach, J., Bryla, D.A., Majadly, F., et al. (2001). Safety and immunogenicity of improved Shigella O.-specific polysaccharide–protein conjugate vaccines in adults in Israel. Infect. Immun. 69, 1351–1357.

Perrett, K.P., Snape, M.D., Ford, K.J., John, T.M., Yu, L.M., Langley, J.M., McNeil, S., Dull, P.M., Ceddia, F., Anemona, A., et al. (2009). Immunogenicity and immune memory of a nonadjuvanted quadrivalent meningococcal glycoconjugate vaccine in infants. Pediatr. Infect. Dis. J. 28, 186–193.

Phalipon, A., Costachel, C., Grandjean, C., Thuizat, A., Guerreiro, C., Tanguy, M., Nato, F., Normand, V.V.L., Bélot, F., Wright, K., et al. (2006). Characterization of functional oligosaccharide mimics of the Shigella flexneri serotype 2a O-antigen: implications for the development of a chemically defined glycoconjugate vaccine. J. Immunol. 176, 1686–1694.

Pinto, B.M., and Bundle, D.R. (1983). Preparation of glycoconjugates for use as artificial antigens: A simplified procedure. Carbohydr. Res. 124, 313–318.

Pinto, B.M., Morrisette, D.G., and Bundle, D.R. (1987). Synthesis of oligosaccharides corresponding to biological repeating units of Shigella flexneri Y polysaccharide. Part 1. Overall strategy, synthesis of a key trisaccharide intermediate, and synthesis of a pentasaccharide. J. Chem. Soc. Perkin 1, 9–14.

Pinto, B.M., Reimer, K.B., Morissette, D.G., and Bundle, D.R. (1989). Oligosaccharides corresponding to biological repeating units of Shigella flexneri variant Y polysaccharide. Synthesis and two-dimensional NMR analysis of a hexasaccharide hapten. J. Org. Chem. 54, 2650–2656.

Pinto, B.M., Reimer, K.B., Morissette, D.G., and Bundle, D.R. (1990a). Oligosaccharides corresponding to biological repeating units of Shigella flexneri variant Y polysaccharide: Part III. Synthesis and 2D NMR analysis of a heptasaccharide hapten. J. Chem. Soc. Perkin 1, 293–299.

Pinto, B.M., Reimer, K.B., Morissette, D.G., and Bundle, D.R. (1990b). Synthesis and 2D-n.m.r. analysis of a pentasaccharide glycoside of the biological repeating units of Shigella flexneri variant Y polysaccharide and the preparation of a synthetic antigen. Carbohydr. Res. 196, 156–166.

Pozsgay, V. (1998a). Synthesis of glycoconjugate vaccines against Shigella dysenteriae type 1. J. Org. Chem. 63, 5983–5999.

Pozsgay, V. (1998b). Synthetic Shigella Vaccines: A carbohydrate–protein conjugate with totally synthetic hexadecasaccharide haptens. Angew. Chem. Intl. Ed., 37, 138–142.

Pozsgay, V. (2000). Oligosaccharide–protein conjugates as vaccine candidates against bacteria. Adv. Carbohydr. Chem. Biochem. 56, 153–199.

Pozsgay, V. (2008). Recent developments in synthetic oligosaccharide-based bacterial vaccines. Curr. Top. Med. Chem. 8, 126–140.

Pozsgay, V. (2003). Chemical synthesis of bacterial carbohydrates. In Immunobiology of Carbohydrates, Wong, S.Y.C., and Arsequell, G., eds (Georgetown, TX; New

York, NY, Landes Bioscience/Eurekah.com), pp. 192–273.

Prymula, R., Chlibek, R., Splino, M., Kaliskova, E., Kohl, I., Lommel, P., and Schuerman, L. (2008). Safety of the 11-valent pneumococcal vaccine conjugated to non-typeable *Haemophilus influenzae*-derived protein D in the first 2 years of life and immunogenicity of the co-administered hexavalent diphtheria, tetanus, acellular pertussis, hepatitis B, inactivated polio virus, *Haemophilus influenzae* type b and control hepatitis A vaccines. Vaccine *26*, 4563–4570.

Wang, Q., Chan, T.R., Hilgraf, R., Fokin, V.V., Sharpless, K.B., and Finn, M.G. (2003). Bioconjugation by copper(I)-catalyzed azide-alkyne [3 + 2] cycloaddition. J. Am. Chem. Soc. *125*, 3192–3193.

Rademacher, C., Shoemaker, G.K., Kim, H.-S., Zheng, R.B., Taha, H., Liu, C., Nacario, R.C., Schriemer, D.C., Klassen, J.S., Peters, T., *et al.* (2007). Ligand specificity of CS-35, a monoclonal antibody that recognizes mycobacterial lipoarabinomannan. A model system for oligofuranoside-protein recognition. J. Am. Chem. Soc. *129*, 10489–10502.

Ragupathi, G., and Livingston, P. (2002). The case for polyvalent cancer vaccines that induce antibodies. Expert Rev. Vaccines *1*, 193–206.

Ragupathi, G., Koide, F., Sathyan, N., Kagan, E., Spassova, M., Bornmann, W., Gregor, P., Reis, C.A., Clausen, H., Danishefsky, S.J., *et al.* (2003). A preclinical study comparing approaches for augmenting the immunogenicity of a heptavalent KLH-conjugate vaccine against epithelial cancers. Cancer Immunol. Immunother. *52*, 608–616.

Rollenhagen, J.E., Kalsy, A., Saksena, R., Sheikh, A., Alam, M.M., Qadri, F., Calderwood, S.B., Kovác, P., and Ryan, E.T. (2006). Transcutaneous immunization with a neoglycoconjugate containing a *Vibrio cholerae* hexasaccharide derived from *V. cholerae* O1 Ogawa lipopolysaccharide bound to a protein carrier. Am. J. Trop. Med. Hyg. 75 (Suppl.). 84–85.

Roy, R., Katzenellenbogen, E., and Jennings, H.J. (1984). Improved procedures for the conjugation of oligosaccharides to protein by reductive amination. Can. J. Biochem. Cell Biol. *62*, 270–275.

Safari, D., Dekker, H.A.T., Joosten, J.A.F., Michalik, D., De Souza, A.C., Adamo, R., Lahmann, M., Sundgren, A., Oscarson, S., Kamerling, J.P., *et al.* (2008). Identification of the smallest structure capable of evoking opsonophagocytic antibodies against *Streptococcus pneumoniae* type 14. Infect. Immun. *76*, 4615–4623.

Saksena, R., Chernyak, A., Karavanov, A., and Kováč, P. (2002). Development of a conjugate vaccine for cholera from synthetic carbohydrate antigens: chemistry, monitoring of the conjugation, and recovery of the excess ligand. Glycobiology *12*, 670–671.

Saksena, R., Ma, X.W., Wade, T.K., Kováč, P., and Wade, W.F. (2005). Effect of saccharide length on the immunogenicity of neoglycoconjugates from synthetic fragments of the O-SP of *Vibrio cholerae* O1, serotype Ogawa. Carbohydr. Res. *340*, 2256–2269.

Saksena, R., Ma, X., Wade, T.K., Kováč, P., and Wade, W.F. (2006). Length of the linker and the interval between immunizations influences the efficacy of *Vibrio cholerae* O1, Ogawa hexasaccharide neoglycoconjugates, FEMS *47*, 116–128.

Scanlan, C., Calarese, D., Lee, H.-K., Blixt, O., Wong, C.-H., Wilson, I., Burton, D., Dwek, R., and Rudd, P. (2005). Antibody recognition of a carbohydrate epitope: A template for HIV vaccine design. Adv. Exp. Med. Biol. *564*, 7–8.

Schofield, L., and Hackett, F. (1993). Signal transduction in host cells by a glycosylphosphatidylinositol toxin of malaria parasites. J. Exp. Med. *177*, 145–153.

Schofield, L., Hewitt, M.C., Evans, K., Siomos, M.-A., and Seeberger, P.H. (2002). Synthetic GPI as a candidate antitoxic vaccine in a model of malaria. Nature *418*, 785–789.

Schofield, L., Vivas, L., Hackett, F., Gerold, P., Schwarz, R.T., and Tachado, S. (1993). Neutralizing monoclonal antibodies to glycosylphosphatidylinositol, the dominant TNF-alpha-inducing toxin of *Plasmodium falciparum*: prospects for the immunotherapy of severe malaria. Annu. Trop. Med. Parasitol. *87*, 617–626.

Seeberger, P.H. (2008). Automated carbohydrate synthesis as platform to address fundamental aspects of glycobiology-current status and future challenges. Carbohydr. Res. *343*, 1889–1896.

Sereikaite, J., Jachno, J., Santockyte, R., Chmielevski, P., Bumelis, V.-A., and Dienys, G. (2006). Protein scission by metal ion-ascorbate system. Protein J. *25*, 369–378.

Shibata, N., Ikuta, K., Imai, T., Satoh, Y., Richi, S., Suzuki, A., Kojima, C., Kobayashi, H., Hisamichi, K., and Suzuki, S. (1995). Existence of branched side chains in the cell wall mannan of pathogenic yeast, *Candida albicans.* Structure–antigenicity relationship between the cell wall mannans of *Candida albicans* and *Candida parapsilosis.* J. Biol. Chem. *270*, 1113–1122.

Shibata, N., Kobayashi, H., Okawa, Y., and Suzuki, S. (2003). Existence of novel beta-1, 2 linkage-containing side chain in the mannan of *Candida lusitaniae*, antigenically related to *Candida albicans* serotype A. Eur. J. Biochem. *270*, 2565–2575.

Tamborrini, M., Werz, D.B., Frey, J., Pluschke, G., and Seeberger, P.H. (2006). Anti-carbohydrate antibodies for the detection of anthrax spores. Angew. Chem. Int. Ed. *45*, 6581–6582.

Teitelbaum, Glatman-Freedman, A., Chen, B., Robbins, J.B., Unanue, E., Casadevall, A., and Bloom, B.R. (1998). A mAb recognizing a surface antigen of *Mycobacterium tuberculosis* enhances host survival. Proc. Natl. Acad. Sci. U.S.A. *95*, 15688–15693.

Tietze, L.F., Schröter, C., Gabius, S., Brinck, U., Goerlach-Graw, A., and Gabius, H.-J. (1991). Conjugation of *p*-aminophenyl glycosides with squaric acid diester to a carrier protein and the use of neoglycoprotein in the histochemical detection of lectins. Bioconjugate Chem. *2*, 148–153.

Tornoe, C.W., Christensen, C., and Meldal, M. (2002). Peptidotriazoles on solid phase: [1, 2, 3]-triazoles by regiospecific copper(I)-catalyzed 1, 3-dipolar cycloadditions of terminal alkynes to azides. J. Org. Chem. *67*, 3057–3064.

Torosantucci, A., Bromuro, C., Chiani, P., De Bernardis, F., Berti, F., Galli, C., Norelli, F., Bellucci, C., Polonelli, L., Costantino, P., *et al.* (2005). A novel glyco-conjugate vaccine against fungal pathogens. J. Exp. Med. *202*, 597–606.

Trotter, C.L., Andrews, N.J., Kaczmarski, E.B., Miller, E., and Ramsay, M.E. (2004). Effectiveness of meningococcal serogroup C conjugate vaccine 4 years after introduction. Lancet *364*, 365–367.

van Seeventer, P.B., van Dorst, A.L.M., Siemerink, J.F., Kamerling, J.P., and Vliegenthart, J.F.G. (1997). Thiol addition to protected allyl glycosides: an improved method for the preparation of spacer-arm glycosides. Carbohydr. Res. *300*, 369–373.

Verez-Bencomo, V., Fernández-Santana, V., Hardy, E., Toledo, M.E., Rodriguez, M.C., Heynngnezz, L., Rodriguez, A., Baly, A., Herrera, L., Izquierdo, M., *et al.* (2004). A synthetic conjugate polysaccharide vaccine against *Haemophilus influenzae* type b. Science *305*, 522–525.

Villeneuve, S., Souchon, H., Riottot, M.M., Mazie, J.C., Lei, P.-S., Glaudemans, C.P.J., Kováč, P., Fournier, J.M., and Alzari, P.M. (2000). Crystal structure of an anticarbohydrate antibody directed against *Vibrio cholerae* 01 in complex with antigen: molecular basis for serotype specificity. Proc. Natl. Acad. Sci. U.S.A. *97*, 8433–8438.

Vliegenthart, J.F.G. (2006). Carbohydrate based vaccines. FEBS Lett. *580*, 2945–2950.

Vulliez-Le Normand, B., Saul, F.A., Phalipon, A., Bélot, F., Guerreiro, C., Mulard, L.A., and Bentley, G.A. (2008). Structures of synthetic O-antigen fragments from serotype 2a *Shigella flexneri* in complex with a protective monoclonal antibody. Proc. Natl. Acad. Sci. U.S.A. *105*, 9976–9981.

Vyas, M.N., Vyas, N.K., Meikle, P.J., Sinnott, B., Pinto, B.M., Bundle, D.R., and Quiocho, F.A. (1993). Preliminary crystallographic analysis of a Fab specific for the O-antigen of *Shigella flexneri* cell surface lipopolysaccharide with and without bound saccharides. J. Mol. Biol. *231*, 133–136.

Vyas, N.K., Vyas, M.N., Chervenak, M.C., Bundle, D.R., Pinto, B.M., and Quiocho, F.A. (2003). Structural basis of peptide-carbohydrate mimicry in an antibody-combining site. Proc. Natl. Acad. Sci. U.S.A. *100*, 15023–15028.

Vyas, N.K., Vyas, M.N., Chervenak, M.C., Johnson, M.A., Pinto, B.M., Bundle, D.R., and Quiocho, F.A. (2002). Molecular recognition of oligosaccharide epitopes by a monoclonal Fab specific for *Shigella flexneri* Y lipopolysaccharide: X-ray structures and thermodynamics. Biochemistry *41*, 13575–13586.

Wacker, M., Linton, D., Hitchen, P.G., Nita-Lazar, M., Haslam, S.M., North, S.J., Panico, M., Morris, H.R., Dell, A., Wren, B.W., *et al.* (2002). N-linked glycosylation in *Campylobacter jejuni* and its functional transfer into *Escherichia coli*. Science *298*, 1790–1793.

Wade T.K., Saksena, R., Shiloach, J., Kovácv, P., and Wade, W.F. (2006). Immunogenicity of synthetic saccharide fragments of *Vibrio cholerae* O1 (Ogawa and Inaba) bound to Exotoxin A. FEMS Immunol. Med. Microbiol. *48*, 237–251.

Wang, J., Li, H., Zou, G., and Wang, L.X. (2007). Novel template-assembled oligosaccharide clusters as epitope mimics for HIV-neutralizing antibody 2G12. Design, synthesis, and antibody binding study. Org. Biomol. Chem. *5*, 1529–1540.

Wang, L.X., Ni, J., Singh, S., and Li, H. (2004). Binding of high-mannose-type oligosaccharides and synthetic oligomannose clusters to human antibody 2G12: implications for HIV-1 vaccine design. Chem. Biol. *11*, 127–134.

Warren, J.D., Geng, X., and Danishefsky, S.J. (2007). Synthetic glycopeptide-based vaccines. Top. Curr. Chem. *267*, 109–141.

Werz, D.B., and Seeberger, P.H. (2005). Total synthesis of antigen *Bacillus anthracis* tetrasaccharide – creation of an anthrax vaccine candidate. Angew. Chem. Int. Ed. *44*, 6315–6318.

Wessels, M.R., Paoletti, L.C., Kasper, D.L., DiFabio, J.L., Michon, F., Holme, K., and Jennings, H.J. (1990). Immunogenicity in animals of a polysaccharide–protein conjugate vaccine against type III group B Streptococcus. J. Clin. Invest. *86*, 1428–1433.

Wessels, M.R., Paoletti, L.C., Guttormsen, H.K., Michon, F., D'Ambra, A.J., and Kasper, D.L. (1998). Structural properties of group B streptococcal type III polysaccharide conjugate vaccines that influence immunogenicity and efficacy. Infect. Immun. *66*, 2186–2192.

Whitney, C.G., Farley, M.M., Hadler, J., Harrison, L.H., Bennett, N.M., Lynfield, R., Reingold, A., Cieslak, P.R., Pilishvili, T., Jackson, D., *et al.*, for the Active Bacterial Core Surveillance of the Emerging Infections Program Network. (2003). Decline in invasive pneumococcal disease after the introduction of protein–polysaccharide conjugate vaccine. N. Engl. J. Med. *348*, 1737–1746.

William, E.D., and Beurret, M. (1989). Glycoconjugates of bacterial carbohydrate antigens. Contrib. Microbiol. Immunol. *10*, 48–114.

Wright, K., Guerreiro, C., Laurent, I., Baleux, F., and Mulard, L.A. (2004). Preparation of synthetic glycoconjugates as potential vaccines against *Shigella flexneri* serotype 2a disease. Org. Biomol. Chem. *2*, 1518–1527.

Wu, X., and Bundle, D.R. (2005). Synthesis of glycoconjugate vaccines for *Candida albicans* using novel linker methodology. J. Org. Chem. *70*, 7381–7388.

Xin, H., Dziadek, S., Bundle, D.R., and Cutler, J. (2008). Synthetic glycopeptide vaccines combining β-mannan and peptide epitopes induce protection against candidiasis. Proc. Natl. Acad. Sci. U.S.A. *105*, 13526–13531.

Yogev, R., Arditi, M., Chadwick, E.G., Amer, M.D., and Sroka, P.A. (1990). *Haemophilus influenzae* Type b conjugate vaccine (meningococcal protein conjugate): immunogenicity and safety at various doses. Pediatrics *85*, 690–693.

Zhang, S., Cordon-Cardo, C., Zhang, H., Reuter, V., Adluri, S., Hamilton, W., Lloyd, K., and Livingston, P. (1997). Selection of tumor antigens as targets for immune attack using immunuhistochemistry: I. Focus on gangliosides. Int. J. Cancer *73*, 42–49.

Zhang, J., and Kováč, P. (1999). A highly efficient preparation of neoglycoconjugate vaccines using subcarriers that bear clustered carbohydrate antigens. Bioorg. Med. Chem. Lett. *9*, 487–490.

Zhang, J., and Kováč, P. (1999). Studies on vaccines against cholera. Synthesis of neoglycoconjugates from the hexasaccharide determinant of *Vibrio cholerae* O:*1*, serotype Ogawa by single point attachment or by attachment of the hapten in the form of clusters. Carbohydr. Res. *321*, 157–167.

Zhang, J., Yergey, A., Kowalak, J., and Kováč, P. (1998). Studies towards neoglycoconjugates from the monosaccharide determinant of *Vibrio cholerae* O:*1*, serotype Ogawa using the diethyl squarate reagent. Carbohydr. Res. *313*, 15–20.

Zhao, J.Y., Dovichi, N.J., Hindsgaul, O., Gosselin, S., and Palcic, M.M. (1994). Detection of 100 molecules of product formed in a fucosyltransferase reaction. Glycobiology *4*, 239–242.

Zhu, J., Wan, Q., Lee, D., Yang, G., Spassova, M.K., Ouerfelli, O., Ragupathi, G., Damani, P., Livingston, P.O., and Danishefsky, S.J. (2009). From synthesis to biologics: preclinical data on a chemistry derived anticancer vaccine. J. Am. Chem. Soc. *131*, 9298–9303.

Zou, W., Mackenzie, R., Thérien, L., Hirama, T., Yang, Q., Gidney, M.A., and Jennings, H.J. (1999). Conformational epitope of the type III group B *Streptococcus* capsular polysaccharide. J. Immunol. *163*, 820–825.

Zou, W., and Jennings, H.J. (2001). The conformational epitope of type III group B Streptococcus capsular polysaccharide. Adv. Exp. Med. Biol. *491*, 473–484.

Bacterial Protein Toxins Used in Vaccines

5

Jerry M. Keith

Abstract

At first glance, the idea of using protein toxins as vaccines against bacterial human diseases seems somewhat of a paradox. However, in some diseases, the severe pathological effects manifested by the causative agents are mediated entirely by protein toxins. Thus, it seems reasonable to expect that if antibodies could be induced against the protein toxin, they should be effective at preventing severe disease. Of course, the obvious challenge is to detoxify the protein toxin activity without destroying its ability to induce neutralizing antibodies. From an academic point of view, it is ironic that early vaccines against diphtheria, tetanus, and whooping cough were successful without understanding what made them work. One of the keys to this puzzle was uncovered quite by accident when it was discovered that diphtheria toxin stock preparations stored in large earthenware jars too large to be autoclaved were being detoxified by the residual formalin that leached into the preparations from the formalin-sterilized jars. It took two decades for this discovery to be understood and appreciated to a point at which formalin treatment could be applied to produce toxoid preparations for vaccination. It then took another half a century to develop the scientific tools and knowledge needed to bring forth the new generation of vaccines, which are highly effective and less reactogenic. This chapter traces the scientific history, controversies, and development of diphtheria, tetanus and pertussis vaccines.

Introduction

Vaccination is the most effective defence against infectious diseases. It is believed that the practice of inoculation began in China as early as 200 BC and was prevalent in both India and China during the seventeenth century. The procedure consisted of placing a small amount of serous fluid containing a weakened infectious agent under the skin. This procedure produced an attenuated form of the disease, but most importantly it induced immunity to the native infectious agent. Its more modern-day inception started in 1798 when Edward Jenner (1749–1823) began inoculations (originally referred to as variolation) for smallpox after he observed that milkmaids became resistant to cowpox virus. Since this time, society has come to accept (although not without controversy) vaccination as an inseparable part of public health policy. In fact vaccines have been so successful, that the public expects medical science to produce vaccines for every infectious agent that emerges to attack human and animal populations. This notion has certainly been true for smallpox (1798), rabies (1885), diphtheria (1923), pertussis (1926), tetanus (1927), influenza (1945), polio (1955), measles (1963), mumps (1967), rubella (1969), anthrax (1970), meningococcal polysaccharide (1981), hepatitis B (1982), 23-valent pneumococcal pneumonia (1983), *Haemophilus influenzae* type B (1990), typhoid (1990), acellular pertussis (1991), varicella (1995), hepatitis A (1995), rotavirus (1998), and human papillomavirus (2006) (CDC, 2009; Lombard *et al.*, 2007; Mandell *et al.*, 2010; Warren, 1986). However, effective vaccines against many other infectious diseases such as malaria, tuberculosis, cholera, acquired immune deficiency syndrome, hepatitis C, and norovirus have remained elusive, not withstanding years of dedicated research.

This chapter is devoted to the history and recent advances of three vaccines in which the protective immunogen is derived from bacterial toxins. The concept of bacterial toxin-mediated diseases is clearly relevant to diphtheria and tetanus as well as pertussis, which was eloquently discussed in 1979 and 1984 by Margaret Pittman (1901–1995) at the Food and Drug Administration in Bethesda, Maryland (Pittman, 1979, 1984). In each of these three diseases, the causative agent that is indicative of the disease is a bacterial protein toxin. The toxins themselves are the three essential components of paediatric, adolescent, and adult diphtheria–tetanus–pertussis (DTP) vaccines.

The history and events that mark the discovery of diphtheria and tetanus toxins and their ability to produce protective immunity is remarkable, even by today's standards. As early as the 4th Century BC, Hippocrates, often considered the 'father of clinical medicine', observed diseases with characteristics typical of diphtheria. A detailed description of the clinical presentation of the disease including the classic sore throat with the accompanying pseudo-membrane production formed by local tissue necrosis was provided in 1821 by Pierre Bretonneau (1778–1862). This was followed with the 1884 discovery of the Gram-positive, aerobic bacterium *Clostridium tetani* by Arthur Nicolaier (1862–1942). However, the events that directly led to our current diphtheria and tetanus vaccines began with the work of Japanese scientist, Shibasaburo Kitasato (1853–1931), who isolated the bacterium in 1889, and Emil von Behring (1854–1917) working in the laboratory of Robert Koch (1843–1910) in Berlin with his discovery of the bacteria responsible for diphtheria (*Corynebacterium diphtheriae*), anthrax (*Bacillus anthracis*), and with Alexandre Yersin (1863–1943) bubonic plague (*Yersinia pestis*). Working together, Kitasato and Behring found that blood as well as cell-free serum from animals immunized with diphtheria and tetanus toxins produced *antitoxins*

that neutralized the biological activities of the toxins and produced immunity to the disease. They also demonstrated that serum from non-immunized animals lacked this antitoxin activity, but antitoxin activity could be passively transferred to these animals. The antitoxin remained stable in these animals and conferred immunity. From this work, our understanding of immunology expanded and the modern age of vaccination began (Grundbacher, 1992; Kantha, 1991).

At the time of these remarkable medicine discoveries, the notion that a biological toxin capable of causing major biological damage and even death, could, under certain condition, induce immunity to the disease was paradoxical and seemed to defy logic and intuition. However, serendipity, which often plays a pivotal role in great discoveries, apparently solved this paradox. Historical documents suggest that shortly after research effort began on diphtheria toxin, efficient production methods were developed resulting in the generation of large stocks of toxin. This efficient production necessitated storing the toxin preparations in large earthenware jars. Because the earthenware containers could not be autoclaved, they were treated with formaldehyde as part of the cleaning process before they were reused. When the earthenware jars were use again to store toxins, residual formaldehyde, which leached from the jars, inactivated toxicity without altering the immunogenicity of the toxin molecule (Oakley, 1966). It was Alexander Glenny (1882–1965) working at the Wellcome Physiological Research Laboratories, in England, who observed this toxoid effect in 1904 but apparently did not fully appreciate the significance of his discovery (Glenny and Hopkins, 1923; Glenny *et al.*, 1924; Glenny and Sudmerson, 1921; Glenny and Walpole, 1915). However, Gaston Romon (1886–1963) furthered the understanding of diphtheria at the Institut Pasteur in France by developing a method of determining the potency of the diphtheria antitoxin by the precipitation formed between diphtheria toxin and the antitoxin. The basis of this method is still used to determine potency of diphtheria vaccines. By 1923, Ramon was well on the way to standardizing a diphtheria toxoid vaccine (Ramon, 1923). Glenny again contributed to our current vaccine formulation, when in 1926 he developed the method of absorbing toxoid to aluminium hydroxide (alum), thus increasing its immunogenicity (Baxby, 2005; Dean and Webb, 1928; Glenny et al., 1926; Gronski et al., 1991; Lombard et al., 2007), which today is still the only vaccine adjuvant approved for use in the United States (Mandell et al., 2010).

During this same era in medical science, Bordetella pertussis, a Gram-negative bacillus and the causative agent of whooping cough or pertussis, was isolated in 1906 by Jules Borget (1870–1961) and his brother-in-law Octave Gengou (1875–1957) at the Pasteur Institute in Brussels, where Bordet served as director of the institute (Bordet and Gengou, 1906; Legon, 1998). Pertussis is a highly contagious respiratory disease which attacks both children and adults and is still responsible for an estimated mortality rate of 350,000 deaths per year worldwide. The name pertussis for the disease and toxin is derived from the Latin word per- (intensive) + tussis (cough) i.e. 'intensive or severe cough' which reflects the classic clinical symptoms associated with whooping cough when the respiratory tract is infected with the bacteria (Haubirch, 2003). In 1914, a survey of about 17,000 cases on the lower East Side of Manhattan showed that the highest incidence of pertussis occurred in children 3 to 5 years old, but most of the deaths occurred in children less than one year of age (Luttinger, 1916). Our current immunization schedule for the DTP vaccine of 2, 4, and 6 months is intended to prevent this high death rate in infants.

Two studies published in 1933, one from Denmark and the other from United States, showed that there was a clear prophylactic value to vaccination for whooping cough (Madsen, 1933; Sauer, 1933). During this time, there were those who questioned the usefulness of vaccination with Bordetella pertussis, believing that whooping cough, which had some characteristics of influenza, could be caused by a filterable agent (virus) and not a bacterium. However, this notion was dispelled through carefully planned and executed whooping cough vaccination studies conducted by Pearl Kendrick (1890–1980) and Grace Eldering (1900–1988) at the Michigan Department of Health as they continued to refine and characterize conditions for preparation of a 'whole culture' vaccine consisting of Bordetella pertussis Phase I cells grown on Bordet–Gengou agar plates supplemented with sheep blood. Bacterium were removed from the agar, chemically inactivated, washed once with normal saline, tested for biological properties and suspended for use as the vaccine (Kendrick and Eldering, 1935). Kendrick later developed the mouse intracerebral challenge (IC) model (similar to one used for diphtheria) for use in determining the potency of the vaccine (Kendrick et al., 1947). Margaret Pittman (1902–1995) at the National Institutes of Health, Bureau of Biological Standards in Bethesda, Maryland (now the Food and Drug Administration) devised the United States national standards for vaccine quantitative potency and immune response for the pertussis whole cell vaccine (Pittman, 1956). Clearly the whole cell pertussis vaccine has not been without its problems. Compared to diphtheria and tetanus toxoid vaccines, which were routinely given to adults as booster, pertussis vaccines were not given to adolescences and adults because of side-effects (Keitel, 1999). With the introduction of the new acellular pertussis vaccines which are much less reactogenic, adolescent and adult pertussis vaccines are now licensed by the FDA and are available in the United States.

What drove the development of new generation vaccines?

Clearly by the late 1920s, diphtheria, tetanus, and pertussis vaccines were being used effectively in public health vaccination programmes. One might ask the question, what happened in the 70 intervening years between these early life-saving discoveries and the introduction of the first new generation vaccines? Why did it take so long to bring about new generation vaccines for these important diseases? Was it the philosophy of 'If it is not broken, don't try to fix it.' or was it a lack of scientific knowledge and technical ability? I prefer to believe that it was the latter explanation.

The pioneering achievements against disease, which took place during the 19th and 20th centuries, resulted in a basic understanding of vaccination and immunity principles. Together with the discovery of antibiotics in the 1940s, better health care was established, social networks formed, longevity increased, and populations expanded as the disease burden decreased. However, populations now had unprecedented mobility and control of infectious diseases became a challenge not only from the traditional human diseases, but also through introduction of emerging new pathogens. As a result, the demand for vaccines increased and the challenges of product safety, stability, standardization, and monitoring effectives became important issues. Along with these developments came public perception and expectations. Public confidence and perceptions have enormous impact on vaccination programmes. When vaccines are efficacious and coverage rates remain high, unvaccinated individuals who reside within the immunized population are protected by from disease because the disease agent cannot circulate in the highly immunized community. This simple concept is referred to as 'herd immunity', which is reflective of its veterinarian origin (Fine, 1993). Ironically, because of this effective vaccination concept, the public becomes complacent after a period of time without experiencing disease among their community or general populations. Since they no longer hear reports or experience the disease, their perception becomes, 'Why should I or my children be vaccinated against a disease that I never hear anything about?' As history has shown us, this misperception has serious consequences for public health. It has been documented many times that complacency and negative public perception can seriously undermining the best thought out vaccination policies. In the early years of vaccines, there was public paranoia against the use of vaccines going so far as political ruling bodies outlawing the use of vaccination. With increased literacy and public health education programmes, most governmental bodies now appreciate the necessity and public health benefit vaccination bring to their communities. Although not as prevalent today, public paranoia unfortunately still occasionally prevails with a few highly vocal, ardent sceptics leading entire anti-vaccination movements (just search the term 'vaccines' on the internet) (Omer et al., 2009). In some instances, there has been rational reason to pause and re-evaluate – often

leading to new insight and better vaccines. One such example was an unfortunate incident that occurred in Japan in 1974–1975 when an increase in the number of adverse reactions were reported, including the deaths of two children due to neurological illness after receiving the whole cell pertussis vaccine. Forced by media and public pressure, vaccination for pertussis was temporarily suspended in Japan and the vaccination age was raised from 3 months to 2 years. However, because of the public's lack of confidence and the mistaken perception that vaccination for pertussis was no longer needed because very few pertussis cases were seen in their communities, vaccine acceptance dropped to near 10%. Almost immediately, a dramatic rise in pertussis cases was seen with a 28-fold increase in morbidity rising from 0.4 in 1974 to 11.3 per 100,000 population in 1979 when a pertussis epidemic occurred with more than 13,000 cases and 41 deaths (Gangers et al., 1998; Sato and Sato, 1999; Watanabe and Nagai, 2005). A call went out for the development of a safer whooping cough vaccine. This call for action was answered by Yuji Sato and Hiroko Sato at the National Institute of Health – Japan with the development of an acellular pertussis vaccine consisting of pertussis toxin (PT) and fibril haemagglutinin (FHA). Introduction of this acellular toxoid vaccine in 1981 immediately restored confidence and acceptance rates climbed to >80% by 1982 and brought the rise in pertussis under control (Kino-Sakai and Kimura, 2004; Kuno-Sakai et al., 2004; Sato and Sato, 1999; Sato et al., 1984b). Clearly, the pioneering work by the Sato's resulting in the use of the first acellular pertussis vaccine, which is now collectively referred to as the 'Japanese experience', had important implications for world health.

During this time, several nations had stopped vaccinating for whooping cough or were in legal turmoil with vaccine manufacturers. In fact, with the television broadcast of *Vaccine Roulette* and publication of the book *A Shot in the Dark* in the United States in the early 1980s, some companies simply stopping producing vaccines to avoid becoming entangled in a quagmire of legal issues. However, strong support from paediatric and primary-care organizations along with highly efficacious vaccines and a mandatory vaccine policy kept the coverage rate at 90–95% for children of primary-school age.

Other examples of the difficulties presented by the pertussis whole cell vaccine occurred in both Sweden and England when the public lost confidence in the safety of the vaccine and began questioning the need for vaccination. In Sweden by 1975, parents were refusing to have their children vaccinated after paediatricians and an influential medical leader, Justus Ström, questioned the need to continuing pertussis vaccination programmes because he believed that pertussis was a relatively mild disease and was sceptical of the vaccine's effectiveness, especially when weight against the risk of side-effects (Strom, 1960, 1967). These concerns combined with strong media coverage perpetuated parents fears resulting in a dramatic decrease in coverage rate from 90% in 1974 to only 12% by 1979. That same year the vaccination programme against pertussis, which begun in Sweden in the 1950s, was discontinued. By 1980–83 the annual incidence of pertussis in children 0–4 years climbed to 3370 cases/100,000 population. After this time, Sweden continued to record 10,000 cases of pertussis annually – equivalent to the 100 cases per 100,000 rate seen in developing countries (Gangarosa et al., 1998).

In the United Kingdom, a 1974 report attributing 36 neurological reactions to whole-cell pertussis vaccination drew widespread television and media coverage (Kulenkampff et al., 1974). A few years later Gordon Stewart, a prominent public-health educator in Scotland, deemed that the risks associated with the pertussis vaccine out weighed the marginal protection provided by the vaccine (Stewart, 1977). These events seriously undermine a highly successful pertussis vaccination programme that had begun in the 1950s. Pertussis vaccine coverage immediately decreased from 81% to 31% because of safety concerns of the public related to the vaccine. This non-acceptance of vaccination resulted in pertussis epidemics with incidence rates climbing to over 100 per 100,000 population during the 1975–1983 period. Finally with a national reassessment of vaccine efficacy, public perception was finally reversed and the population began

having their children vaccinated bring national acceptance rate to 93% by the mid-1990s and with it a dramatic decline in pertussis (Gangarosa et al., 1998).

Similar scenarios were experienced in the Russian Federation, Ireland, Italy, Australia and West Germany (Gangarosa et al., 1998). As a result of these worldwide events, entire childhood vaccine programmes were undermined, which then furthered social concerns for public health, especially for health care providers and policy administrators. These events began setting the path for development of new generation pertussis vaccines.

The Sato's dedicated efforts led the way to re-evaluate pertussis vaccines and encourage research efforts for development of safer vaccine products. A recent review of the Japanese acellular pertussis vaccine programme showed that through the uses of this two-component acellular vaccine, pertussis cases in Japan have now declined to an unprecedented low (Sato and Sato, 1999; Watanabe and Nagai, 2005). More recent acellular vaccine trials have now shown that the pertussis toxin component of acellular vaccines is essential for immunity and somewhat ironically, clinical studies have now concluded that the FHA component does not contribute to protection (Cherry et al., 1998; Cherry and Olin, 1999; Robbins et al., 2009; Storsaeter et al., 1998). In fact, a single component pertussis toxoid whooping cough vaccine was licensed for use in the United States and several countries have used pertussis toxoid as a monocomponent acellular whooping cough vaccine (Hviid et al., 2004; Robbins et al., 2009; Taranger et al., 2001b).

In retrospect, this is an intriguing conclusion and brings up the question of why FHA was included in acellular pertussis vaccines? Perhaps because FHA functions as an adhesion factor during colonization, it seemed beneficial to include it in the vaccine. Add this to the fact that it is very difficult to completely separate FHA from pertussis toxin during purification of the two components from the culture supernatant and one could argue, why not leave FHA in the formulation? Perhaps this explains how both pertussis toxin and fimbrial haemagglutinin arose in two-component acellular pertussis vaccines – it was just too difficult to completely separate the two antigens. The extent of this problem became evident during acellular pertussis vaccine trials where 'purified' pertussis antigens were used in the formulation of five different vaccines consisting of one, two, three, and four components (all five formulations contained pertussis toxin). Unexpectedly, in the formulation where pertussis toxin was the single target antigen, antibody response was also detected against both FHA and fimbrial agglutinogens (Fim). In fact, in four out of the five formulations, antibody responses were detected against pertussis antigens not known to be present in the vaccines (Keitel *et al.*, 1999). Similar unexpected antibody responses where seen in a clinical trial involving 12 acellular pertussis vaccines where children were given the acellular vaccine as the fourth dose after receiving three primary doses of the whole cell pertussis vaccine (Pichichero *et al.*, 1997). These results undoubtedly reflect the production difficulty of separating the pertussis toxin from the fimbrial haemagglutinin antigen as well as fimbrial agglutinogens, all of which are found in the culture media during bacterial growth. It is also an interesting fact that the licensed whole cell pertussis vaccine produced by Lederle Laboratories and extensively used in the United States until the introduction of acellular vaccines lacked fimbrial haemagglutinin because of a genetic defect in the *Bordetella pertussis* strain that they used for vaccine production (Edwards *et al.*, 1995; Pichichero *et al.*, 1997).

These events, taken together with relatively recent major advancements in the biological sciences, seem to have emerged at almost the same time, opening doors and emphasizing new public health awareness and at the same time recognizing an opportunity to apply this newly acquired scientific knowledge to an important problem. The result was an expansion and marriage of both basic and applied sciences. These efforts have culminated in development of new generation vaccines and therapeutics. And, so this part of the story begins with the events that brought about breakthrough scientific discoveries and new technologies. The dawning of an exciting new phase of the scientific discipline known as molecular biology was about to begin.

Jump starting the sciences and a stalled industry – new discoveries

In 1968, Gunther Stent (1924–2008), a professor at the University of California at Berkeley and one of the founding pioneers of the scientific discipline known as molecular biology, proclaimed in an essay on its history, that we had learned all there was to know about molecular biology and therefore the field was dead (Stent, 1968). In academic circles, Gunther's prognosis caused a great deal of consternation. Was molecular biology *really* dead? In comparison to what? Was he right or was he 'dead' wrong? The answer is another science paradox – he was both right and wrong. From his prospective, molecular biology had proceeded as far as bacteriophage genetics and the research 'tools' that were available at the time could take it. Although it appeared to have stopped breathing and to be in dire striates, in reality it was not dead – it still had a faint heartbeat – just suspended, waiting for development of a ventilator and a transfusion of new 'whatever'. As Stent mused a few years later, molecular biology wasn't really dead – it was just a remarkable era in time where scientific discovery (which eventually became the basis of molecular biology) was 'ahead of its time' or 'premature' because we did not have enough information to appreciate the implications of the discoveries and thus we could not fill in the steps that would lead to general acceptance and established knowledge (Stent, 1972). Using Stent's perception, we can see that insight and technology were slow to catch up to the remarkable advances that had already been made. As this gap closed, the doorway to new frontiers swung open and the new revolution in molecular biology began. So what were these events that established molecular biology as a discipline that had seemingly stalled by 1968? What brought about its revival in the form of a new molecular biology revolution and how did this lead to the development of new generation vaccines?

Certainly there were many contributors to expansion of molecular biology as a major scientific discipline, which must have begun in 1944 when Avery, MacLeod, and McCarty discovered that DNA was responsible for the transforming principle (Avery *et al.*, 1944) followed 8 years later the Hershey-Chase experiments that elegantly brought attention to DNA and not protein as the ultimate carrier of the code of life (Hershey and Chase, 1952). Then within a year James Watson and Francis Crick (1916–2004) at the University of Cambridge in England showed in 1953 how the four nucleotide building blocks of DNA fit together as long anti-parallel chains in the form of a double helix (Watson and Crick, 1953). So now there was DNA, but where did RNA fit in the picture and the question everyone was asking, where did protein come from anyway? Was there an adaptor (now known as transfer RNA – tRNA) as proposed by Crick between nucleic acid chains and protein? Did protein come first and then DNA or was it the other way around? Finally, in 1955 a gigantic step forward took place – the genetic blueprint for life began to take shape when Heinz Fraenkel-Conrat at the University of California at Berkeley showed that RNA alone carried all the instructions necessary to produce the protein-coated tobacco mosaic virus (TMV) (Fraenkel-Conrat *et al.*, 1957; Fraenkel-Conrat and Williams, 1955). He then went on in 1958 to solve the complete 158 amino acid sequence of the TMV coat protein (Tsugita *et al.*, 1960). So now it was clearly established that there was a link demonstrated between RNA and protein. Next, Oxford University's Sydney Brenner's 1950s 'codon' terminology and messenger RNA (mRNA) concept for the undefined link between RNA and protein was adapted by Marshall Nirenberg at the National Institutes of Health in Bethesda, Maryland, when he broke the genetic code in 1966. By using the four RNA bases (G, A, U, C) Nirenberg devised a three-letter RNA 'codon' scheme for each of the 20 amino acids that make up the universal genetic code used to write the book of life (NLM). A suburb account of this most important and dramatic period of biological science is presented in the late Lily Kay's book *Who Wrote the Book of Life?* (Kay, 2000). Then in 1968 using extremely laborious methods, Fred Sanger and his colleagues at the Medical Research Council Laboratory of Molecular Biology in Cambridge, England showed that a small *E. coli* 5S ribosomal RNA molecule of 124

nucleotides could be sequenced using enzymes that specifically cleaved RNA after G, C, and U nucleotides (Brownlee *et al.*, 1968). However, the concept of sequencing such a large molecule as DNA where there were no equivalent cleavage enzymes was stifling and the enormity of such an undertaking seemed impossible. Some progress was made in 1972 when Walter Fiers and his colleagues at the University of Ghent in Belgium published the sequence of the gene encoding the coat protein of bacteriophage MS2 (Min Jou *et al.*, 1972). Shortly after that, Allan Maxam and Wally Gilbert at Harvard University revolutionized sequencing with a new concept in 1976–1977. Using 5′-end radiolabelled DNA restriction fragments and chemical cleavage to generate a ladder of DNA fragments that could be separated using 'King Kong' polyacrylamide gel electrophoresis, they demonstrated that extensive regions of DNA could be sequenced (Maxam and Gilbert, 1977). The next major sequencing breakthrough, again from Fred Sanger, used oligonucleotide primers to initiate DNA synthesis from specific sites and G, A, T, and C dideoxynucleotides in four separate reactions to terminate replication of the growing DNA chains, which were then separated according to their length (Sanger and Coulson, 1975; Sanger *et al.*, 1977). For their work on these astounding sequence technologies, Wally Gilbert and Fred Sanger shared the 1980 Nobel Prize in Chemistry. These sequencing technologies along with advances in DNA cloning and translation/transcription protein expression system led to a new era in molecular biology. It would no longer take months to sequence 23 nucleotides (Beemon and Keith, 1976). Long DNA stretches covering several genes now could be sequenced in less than a week and site-directed mutagenesis allowed structure–function experiments with amino acid substitutions in biologically synthesized proteins. Taken together, these molecular biology breakthroughs exponentially expanded our knowledge and understanding of biological processes to a level where it became possible to design new protein and even modify animal models to begin to test and confirm ideas that had only been dreams in the past. This was a true revolution in the sciences that took off and has never looked back.

Development of new generation bacterial toxin vaccines

With these scientific and technological break-throughs in molecular biology and increased concern from health care specialists and the general public regarding safety of some of the paediatric vaccines, especially the pertussis component of the combined DTP vaccine, the stage was set for the development of new generation vaccines. This chapter is organized with a discussion of pertussis vaccines first because greater progress has been made on development of highly immunogenic, genetically detoxified acellular pertussis toxin vaccines, compared to those for diphtheria and tetanus. This is likely due to the higher visibility of pertussis control as a public health problem and as discussed above, public response to a whole cell vaccine that was more reactogenic that either diphtheria or tetanus toxoid vaccines. Diphtheria toxin and tetanus toxin are discussed separately although from their discovery and implementation as toxoid vaccines they have taken a somewhat parallel path. At the end of the chapter, a summary of current DTP vaccines with a generalized description for each including the recommended immunization schedule adopted by the United States is included. In addition, Table 5.1 lists the known specific genetic mutations that affect toxicity of the three bacterial toxins.

And finally, suggestions will be proposed as to how we can do better in development of safer and cheaper new generation, genetically detoxified, bacterial toxin vaccines in the future.

As with any review, it is highly likely that work by some authors that have contributed to our knowledge of bacterial protein toxins have not been referred to in this chapter. To those authors please accept my sincere apology and I assure you that any omission was not in any way intentional.

Pertussis toxin

Pertussis toxin is one of several bacterial exotoxins defined as A-B toxins where A represents an enzymatic active subunit that modifies components of the eukaryote host cell and B represents the subunit portion of the toxin that binds to receptors on the host cell and facilitates translocation of the A subunit across the cellular membrane and

into the cell cytoplasm. The pertussis holotoxin is a 105,000 dalton protein complex consisting of five different subunits, S1, S2, S3, S4, and S5, in a ratio of 1:1:1:2:1, respectively. The A portion of the holotoxin is the S1 subunit, which is an ADP-ribosyltransferase. The B portion is comprised of the S2 and S3 subunits, which have a high degree of homology to each other, two copies of the S4 subunit, and a S5 subunit for a total of five polypeptides. My laboratory showed that pertussis toxin is encoded by a single copy of an operon carried on the chromosome of Gram-negative *Bordetella pertussis* (Locht *et al.*, 1986). The gene encoding the S1 subunit is located directly behind the transcriptional promoter and is followed by the genes for S2, S4, S5, and S3 in that order (Locht and Keith, 1986). It is not known how two copies of the S4 subunit are produced for incorporation into the holotoxin when only once copy of the gene is present in the toxin operon. We also showed that the Tn5 transposon insertion responsible for a toxin negative strain of *Bordetella pertussis*, mapped to the S3 subunit and that the toxin subunits are present in the mutant strain, but are not released into the culture growth media. We suggested that a downstream non-structural gene encoding a protein involved in transport of the toxin across the bacterial membrane had been affected by the Tn5 insertion into the toxin operon (Marchitto *et al.*, 1987a). In fact, additional genes essential for toxin liberation from the bacteria immediately follow the S3 subunit gene and are part of the same toxin operon as shown later by Drusilla Burns and her colleagues at the Food and Drug Administration (FDA) in Bethesda, Maryland when they identified a series of genes immediately following the S3 subunit gene that are essential for pertussis toxin liberation (*ptl*-locus) from the bacteria and identified these genes as a part of the same toxin operon, now known as the *ptx-ptl* operon. These liberation genes encode nine different proteins that make up a type IV secretion system known as the Ptl transporter. This transporter is responsible for moving the holotoxin (assembled from the individual toxin subunits as they cross through the inner membrane) across the outer membrane and releasing the completed pertussis holotoxin

Table 5.1 Select mutations in bacterial protein toxins that modify function

Mutant name	Toxin structure	Mutation	Reference
Pertussis toxin analogue			
Y8F	S1 subunit	Tyr8Phe	Burnette *et al.* (1988), Cieplak *et al.* (1990)
Y8D	S1 subunit	Tyr8Asp	Lobet *et al.* (1989)
R9Δ	S1 subunit	Arg9Δ	Lobet *et al.* (1989)
R9A/H/K	S1 subunit	Arg9Ala/His/Lys	Lobet *et al.* (1989)
R9K	S1 subunit	Arg9Lys	Burnette *et al.* (1988), Pizza *et al.* (1989), Cieplak *et al.* (1990)
R9K/H/Δ	S1 subunit	Arg9Lys/His/Δ	Loosmore *et al.* (1990)
D11E	S1 subunit	Asp11Glu	Burnette *et al.* (1988), Cieplak *et al.* (1990)
S12G	S1 subunit	Ser12Gly	Burnette *et al.* (1988), Cieplak *et al.* (1990)
R13K	S1 subunit	Arg13Lys	Burnette *et al.* (1988), Cieplak *et al.* (1990)
R13E/K	S1 subunit	Arg13Glu/Lys	Loosmore *et al.* (1990)
R13L	S1 subunit	Arg13Leu	Pizza *et al.* (1989)
W26T/Δ	S1 subunit	Trp26Thr/Δ	Locht *et al.* (1989)
W26A/C	S1 subunit	Trp26Ala/Cys	Loosmore *et al.* (1990)
H35A	S1 subunit	His35Ala	Loosmore *et al.* (1990)
S40Δ	S1 subunit	Ser40Δ	Locht *et al.* (1990)
C41S/G/E/P/D/N/Δ	S1 subunit	Cys41Ser/Gly/Glu/Pro/Asp/Asn/Δ	Locht *et al.* (1990)
C41A/S	S1 subunit	Cys31Ala/Ser	Loosmore *et al.* (1990)
R58K	S1 subunit	Arg58Lys	Lobet *et al.* (1989)
R58E/H/K	S1 subunit	Arg58Glu/His/Lys	Loosmore *et al.* (1990)
E106D/Δ	S1 subunit	Glu106Asp/Δ	Locht *et al.* (1989)
E106Q	S1 subunit	Glu106Gln	Loosmore *et al.* (1990)
E129D/Δ	S1 subunit	Glu129Asp/Δ	Locht *et al.* (1989)
E129Q/G	S1 subunit	Glu129Gln/Gly	Cockle *et al.* (1988), Loosmore *et al.* (1993)
E129G	S1 subunit	Glu129Gly	Pizza *et al.* (1989)
E129D/C	S1 subunit	Glu129Asp/Cys	Antoine *et al.* (1993)
E129G/Q/D/N/K/H/P/S/Δ	S1 subunit	Glu129Gly/Gln/Asp/Asn/Lys/His/Pro/Ser/Δ	Loosmore *et al.* (1990)
Y130F/S	S1 Subunit	Tyr130Phe/Ser	Loosmore *et al.* (1990)
Y8L; R9E	S1 subunit	Tyr8Leu; Arg9Glu	Burnette *et al.* (1988)
Y8Δ; R9H	S1 subunit	Tyr8Δ; Arg9His	Lobet *et al.* (1989)

Table 5.1 continued

Mutant name	Toxin structure	Mutation	Reference
R9N; S12G	S1 subunit	Arg9Asn; Ser12Gly	Burnette *et al.* (1988), Cieplak *et al.* (1990)
R9K; E129G	S1 subunit	Arg9Lys; Glu129Gly	Loosmore *et al.* (1990)
R9K; R58E; E129G	S1 subunit	Arg9Lys; Arg58Glu; Glu129Gly	Loosmore *et al.* (1990)
R9K; E129A	S1 subunit	Arg9Lys; Glu129Ala	Brown *et al.* (1991)
R9K; E129G	S1 subunit	Arg9Lys; Glu129Gly	Pizza *et al.* (1989)
D11P; P14D	S1 subunit	Asp11Pro; Pro14Asp	Burnette *et al.* (1988)
R13L; E129G	S1 subunit	Arg13Leu; Glu129Gly	Pizza *et al.* (1989)
C41A; E129G	S1 subunit	Cys41Ala; Glu129Gly	Loosmore *et al.* (1990)
R58E; E129G	S1 subunit	Arg58Glu; Glu129Gly	Loosmore *et al.* (1990)
E129G; Y130A/F	S1 subunit	Glu129Gly; Tyr130A/F	Loosmore *et al.* (1990)
Truncated AA#1–6	S1 subunit	(AA#1–6)Δ	Cieplak *et al.* (1988), Cieplak *et al.* (1990)
Truncated AA#1–14	S1 subunit	(AA#1–14)Δ	Cieplak *et al.* (1988), Cieplak *et al.* (1990)
Truncated AA#1–16	S1 subunit	(AA#1–16)Δ	Cieplak *et al.* (1988), Cieplak *et al.* (1990)
Deletion AA9–13	S1 subunit	(AA#9–13)Δ	Loosmore *et al.* (1990)
Y102A, Y103A	S2 subunit	Tyr102Ala, Tyr103Ala	Loosmore *et al.* (1993)
Deletion N105	S2 subunit	Asn105Δ	Lobet *et al.* (1993)
Deletion T91, R92, N93	S2 subunit	(T91, R92, N93)Δ	Loosmore *et al.* (1993)
Y82A	S3 subunit	Y82A	Loosmore *et al.* (1993)
Deletion K102	S3 subunit	Lys105Δ	Lobet *et al.* (1993)
Deletion I91, Y92, K93	S3 subunit	(I91, Y92, K93)Δ	Loosmore *et al.* (1993)
K54A, K57A	S4 subunit	Lys54Ala, Lys57Ala	Loosmore *et al.* (1993)
E129G/Y83A	S1/S3 subunits	E129G/Y83A	Loosmore *et al.* (1993)
E129G/deletion I91,Y92,K93	S1/S3 subunits	E129G/(I91,Y92,K93)Δ	Loosmore *et al.* (1993)
R9K,E129G/K82A	S1/S3 subunits	R9K,E129G/K82A	Loosmore *et al.* (1993)
R9K,E129G/I91,Y92,K93	S1/S3 subunits	R9K,E129G/ (I91,Y92,K93)Δ	Loosmore *et al.* (1993)

Diphtheria toxin analogue

CRM 197	A fragment	Gly52Gln	Uchida *et al.* (1973), Giannini *et al.* (1984)
CRM 228	A fragment	Gly79Asp	Kaczorek *et al.* (1983), Johnson and Nichols (1994a)
E148D	A fragment	Glu148Asp	Carroll and Collier (1984), Carroll (1985)
H21-all 20 amino acids	A fragment	His21-all 20 amino acids	Johnson and Nichols (1994b)

Table 5.1 continued

Mutant name	Toxin structure	Mutation	Reference
Tetanus toxin analogue			
H233C/V	Light chain	His233Cys/Val	Hohne-Zell et al. (1993)
E234A	Light chain	Glu234Ala	Arora et al. (1994), Li et al. (1994)
E234Q	Light chain	Glu234Gln	Hohne-Zell et al. (1993)
H237A	Light chain	His237Ala	Arora et al. (1994)

from the bacteria cell. (Cheung et al., 2004; Kotob et al., 1995; Weiss et al., 1993).

Surprisingly, two other members of the Bordetella family also carry the gene for the holotoxin. Bordetella parapertussis, which infects humans but does not cause severe disease, and Bordetella bronchiseptica, which primarily infects animals, e.g. kennel cough in dogs, both carry the gene (Marchitto et al., 1987b). However, changes in the promoter region of the operon render the gene transcriptionally silent (Arico and Rappuoli, 1987). Recent genome sequencing studies at The Sanger Institute, Wellcome Trust Genome Campus in Hinxton, Cambridge, UK, suggests that Bordetella pertussis evolved from Bordetella bronchiseptica perhaps as little as a few thousand years ago. In addition, their data suggest that recent mutations in the promoter and BvgA activator regions in Bordetella pertussis likely are responsible for increased regulated expression of pertussis toxin (Parkhill et al., 2003). However, it is not clear why similar mutations did not occur leaving the pertussis toxin operon silent in Bordetella parapertussis and Bordetella bronchiseptica. In Bordetella pertussis, the assembled holotoxin is released by the bacteria during growth. Pertussis toxin has several potential targets, but perhaps the best characterized is the membrane-bound adenylate cyclase complex where the ADP-ribosyltransferase activity of pertussis toxin S1 subunit ADP-ribosylates the inhibition regulatory Gi protein of the adenylate cyclase complex. This covalent modification effectively blocks inhibitory signals to the enzyme complex

resulting in uncontrolled increases in the cellular level of cyclic AMP (cAMP). This increase, in what is often called the body's second messenger, starts a cascade of biological reaction detrimental to the host. One example is increased insulin production. Before its identification as a protein toxin, pertussis toxin was named by the biological activity that it elicited. Therefore, one can see in the Bordetella pertussis literature references to 'factors' that reflected what was observed in the bioreactivity of the toxin, i.e. histamine-sensitizing factor, islet-activating protein (IAP – insulin secretion), leucocyte-promoting factor (LPF – lymphocytosis promotion), mouse-protective antigen, late-appearing toxic factor, immunopotentiating factor, as well as pertussigen (Bergman and Munoz, 1968; Clausen et al., 1968; Munoz and Bergman, 1968; Munoz et al., 1981; Sato and Arai, 1972).

As the new molecular biology technology became established, it quickly became apparent that these applications could be used in an attempt to develop genetically detoxified toxins for use in new generation DTP vaccines. Perhaps the most challenging would be pertussis toxin. Unlike diphtheria and tetanus vaccines where it was well established that the chemical treatment of both tetanus toxin and diphtheria toxin produced highly effective toxoids that were stable and relatively easy to produce, whooping cough vaccines were comprised of the whole Bordetella pertussis bacterial cells. Whole cell pertussis vaccines were produced in large bacterial fermentors in which the bacteria were removed from the culture media by centrifugation (which ironically containing the majority of the essential protective antigen – pertussis toxin). The bacterial cells were then treated with formalin and heat and this became what is known as the whooping cough 'whole cell vaccine'. Vaccine lots were tested for potency using the intracranial mouse challenge model. In these tests, mice were immunized with the pertussis whole cell vaccine and then challenged by injecting *Bordetella pertussis* standard strain 18323 directly into the cranial space. If the mice survived bacterial infection through the intracranial challenge, the vaccine lot passed. This assay was the 'gold standard' for whole cell pertussis vaccines. From what we now know about pertussis, it is likely that the assay showed such a remarkable correlation between vaccine potency and protection due to the fine balance of pertussis toxin that remain associated with the whole bacterial cell preparations after they were removed from the culture media by centrifugation. Indeed, in my laboratory we found that injection of small amounts of purified pertussis toxin applied immediately before the intracranial challenge shifted response of the mice to protection against the challenge. In addition to this phenomenon, we and others have demonstrated that under certain conditions it is possible for formalin-treated pertussis toxin to revert to its original biological activity (ADP-ribosylation of an acceptor substrate). Since formalin treatment readily induces covalent linkages with primary amines, the observed reversion of enzymatic toxin activity is likely due to the fact that the catalytically active S1 subunit of the toxin is completely void of lysine residues (Fowler *et al.*, 2003; Locht and Keith, 1986).

With the work of Margaret Pittman at the Food and Drug Administration in Bethesda, Maryland, Alison Weiss in Stan Falkow's laboratory at Stanford, Yuji and Hiroko Sato at the Japanese National Institute of Health in Tokyo, and the John Robbins/Rachel Schneerson laboratory at the National Institutes of Health in Bethesda, Maryland, all pointing the way, it was clear that pertussis toxin, the one essential component of vaccines against *Bordetella pertussis,* should be the first target for this new technology (Pittman, 1979, 1984; Sato *et al.*, 1984b; Weiss *et al.*, 1983).

In late 1983, Ken Sell, the scientific director of the intramural research programme at the National Institute of Allergy and Infectious Diseases (NIAID), established my laboratory at the institute's Rocky Mountain Laboratories (RML), in Hamilton, Montana. Our mission was to establish a research programme dedicated to development of new generation biologics for use against infectious diseases with an initial focus on *Bordetella pertussis*. These efforts were highly successful. Using a sample of pertussis toxin provided by Jack Munoz at RML, the individual subunits were separated by HPLC and the S4 subunit was subjected to N-terminal protein sequencing. Using reverse genetics, an oligonucleotide probe was synthesized encoding the derived S4 subunit protein sequence. This probe was then radiolabelled at the 5'-end and used to identify a cloned 4.5-kb *Eco*RI/*Bam*HI DNA restriction fragment encoding the entire pertussis toxin operon (Locht *et al.*, 1986). Using Maxam–Gilbert and Sanger's dideoxy-sequencing technology, the Keith laboratory (Locht and Keith, 1986) and the Rappuoli laboratory at Biocine-Sclavo R & D Vaccines, Siena, Italy (Nicosia *et al.*, 1986), independently published the complete DNA sequence and gene organization of the pertussis toxin operon in June and July 1986, respectively. From these data and the biological characterization of the toxin that had been previously described (Bergman and Munoz, 1968; Clausen *et al.*, 1968; Munoz and Bergman, 1968; Munoz *et al.*, 1981; Sato and Arai, 1972), the target for genetic inactivation was identified as the enzymatic S1 subunit of the toxin – an ADP-ribosylating enzyme specific for eukaryote host cell components (CHO cell membranes, rhodopsin and the inhibition regulatory Gi protein of the membrane bound adenylyl cyclase enzyme complex, which produces cyclic AMP, the second messenger, etc.).

Because of a striking similarity between an 8-amino acid 'homology box' identified by Camille Locht (Locht and Keith, 1986) in my laboratory in the S1 subunit of pertussis toxin, *Vibrio cholerae* toxin, and *E. coli* heat-labile toxin and the fact that all three were ADP-ribosylating toxins conforming to the classic A-B toxin model using nicotinamide adenine dinucleotide (NADH) as a donor substrate rather than its traditional biological role as a co-enzyme hydrogen donor, our focus became the structure/function relationship of the 8-amino acid homology box. By this time, my laboratory had already shown that an enzymatically active S1 subunit could be produced in *E. coli* using recombinant DNA technology and that the recombinant protein was recognized by a powerful protective monoclonal antibody developed by Hiroko Sato (Locht *et al.*, 1987; Sato *et al.*, 1984a). In an effort to move our research forward more rapidly and avoid reinventing the proverbial wheel (i.e. a high-yield protein expression system), my laboratory established a collaborative effort in 1987 with Amgen, Incorporated in Thousand Oaks, California. Using their proprietary vector designed for high-yield protein expression, Amgen's research goal, led by Neil Burnette, was to use their system to produce recombinant pertussis toxin subunits, which would then be assembled *in vitro* into a holotoxin configuration for use as a pertussis vaccine. During this collaboration, truncation experiments were designed to eliminate hairpin loops at the 5'-end of the mRNA encoding the S1 subunit. It was thought that these hairpin structures might interfere with translation of the message and account for the poor protein expression that we were experiencing. As it turned out, the yield problem was quickly dispatched by a competent and experienced Amgen fermentor crew, but it left us with a series of very well characterized recombinant S1 subunits truncated from the amino terminal. My laboratory decided to assay these truncated S1 subunits for enzymatic activity to determine whether they retained enzymatic activity as we had found earlier with a 48-amino acid truncation at the carboxyl terminal (Locht *et al.*, 1987). From these experiments, it was determined that loss of enzymatic activity occurred only when the amino terminal truncations extended into the 'homology box' and this loss of activity correlated with the loss of binding to the 1B7 protective monoclonal antibody (Cieplak *et al.*, 1988). This work immediately led to site-directed mutagenesis of individual amino acids comprising the homology box and to the discovery of a conservative amino acid change of arginine at position 9 to lysine (R9K) that effectively genetically detoxified pertussis toxin

by reducing the enzymatic toxicity to background level, yet fully retained its immunological properties (Burnette *et al.*, 1988; Lobet *et al.*, 1989).

Through a series of experiments, my laboratory determined that recombinant pertussis holotoxin (subunits S1 through S5) could not be produced using *E. coli* because the host lacked specific genes responsible for assembly and transport of the holotoxin from the bacterial cell. Since *Bordetella pertussis* provided all of the cellular machinery to produce holotoxin and release it into the culture media, the decision was made to remove the portion of pertussis toxin operon encoding the S1 subunit from the *Bordetella pertussis* strain 3779 and replace it with an identical copy except that the arginine residue at position 9 of the mature protein was replaced with lysine (R9K). Using the 'return to pertussis' (RTP) protocols developed by Scott Stibitz in Stan Falkow's laboratory at Stanford (Stibitz *et al.*, 1986), our work generated the first *Bordetella pertussis* strain that produced genetically detoxified pertussis toxin (Brown *et al.*, 1991).

Based on the work published in 1984 by John Collier at Harvard University in Cambridge, Massachusetts, that showed modification of glutamic acid residue number 148 near the carboxyl terminus of diphtheria toxin reduced toxin enzymatic activity (Carroll and Collier, 1984), it seemed likely that since both pertussis toxin and diphtheria toxin have similar NAD ribosylation enzymatic activities (although different acceptor substrates) modification of a glutamic acid residue in the carboxyl terminus of the S1 subunit may have a similar outcome. In fact it did. Michael Klein's laboratory at Connaught Laboratories in Canada produced a pertussis toxin S1 subunit with a glutamic acid 129 to glutamine (E129Q) mutation that reduced enzymatic activity to about 5% of the native toxin (Cockle, 1988, 1989; Loosmore *et al.*, 1990). However, because this mutation still retained some enzymatic activity and the potential problem of spontaneous deamination of glutamine to glutamic acid, my laboratory chose to incorporate alanine at position 129 into our double mutant (R9K; E129A) holotoxin because of alanine's ability to preserve protein structure by preserving the peptide bond angle of the protein chain with a minimal methyl side chain (Brown *et al.*, 1991; Wells, 1991). My group went on to develop a vaccine production strain of bacteria using *Bordetella bronchiseptica*, which has a transcriptional/translational silent copy of the pertussis toxin operon including the genes necessary for assembly and liberation of the toxin from the bacteria into the growth media. In this work, the entire coding region including the promoter region was removed and replaced with the equivalent segment of the pertussis toxin operon from *Bordetella pertussis* with the double mutant (R9K; E129A). In addition, the gene for fimbrial haemagglutinin (FHA), which is also released into the culture medium, was functionally removed to improve purification of genetically detoxified pertussis from the media. These modifications resulted in a new *Bordetella bronchiseptica* strain TY-178 that is easier to work with and produces higher yields of toxin with maximum production in 14 to 15 hours opposed to the 36 to 72 hours required by *Bordetella pertussis* strains (Merkel *et al.*, 2006).

With dedication and perseverance, the vaccine research group led by Rino Rappuoli at Biocine-Sclavo R & D Vaccines in Siena, Italy, brought the first genetically inactivated pertussis vaccine to clinical trials (Nencioni *et al.*, 1991a,b; Peppoloni *et al.*, 1995; Pizza *et al.*, 1988, 1989; Rappuoli, 1996; Rappuoli *et al.*, 1991, 1992a–c, 1994, 1996). The pertussis toxin component was inactivated using the R9K/E129G double mutation in the S1 subunit (Brown *et al.*, 1991; Burnette *et al.*, 1988; Cockle, 1989; Loosmore *et al.*, 1990). The formulation of the purified pertussis toxoid included treatment with a very low concentration of formaldehyde, which presumably helped stabilize the protein. Data from the myriad of clinical trials using this genetically detoxified antigen clearly demonstrated that this product was a superior immunogen (Edwards and Decker, 1996; Edwards *et al.*, 1995; Hewlett, 1996; Keitel, 1999; Keitel *et al.*, 1999; Peppoloni *et al.*, 1995; Pichichero *et al.*, 1997).

With the availability of the pertussis toxin operon sequence (Locht and Keith, 1986; Nicosia *et al.*, 1986) and techniques for genetically manipulate the chromosome of *Bordetella pertussis*, several research groups set out on a quest to produce new generation pertussis vaccines. From those efforts, unique strains were developed, which produce toxin molecules designed to be less reactogenic, yet retain immunogenicity. The list of genetic modifications made to pertussis toxin is extensive, each with interesting characteristics. Many of these genetic mutations are summarized in Table 5.1. However, none have been proven more effective than the R9K and E129G/A double mutant. While the majority of these genetic alterations have targeted the S1 subunit, the B-oligomer has also received some attention. It has been shown that deletion of Asn 105 in the S2 subunit and Lys 105 in the S3 subunit drastically reduce binding to haptoglobin and CHO cells, respectively. Mitogenic levels induced by holotoxin with these mutations were undetectable and the mutant toxin is recognized by various monoclonal antibodies against native toxin (Lobet *et al.*, 1993). Extensive genetic modifications to pertussis toxin were made and the biological and immunogenic activities characterized by Michel Klein's laboratory at Connaught Centre for Biotechnology Research in Ontario, Canada. These mutations include multiple amino acid deletions and substitutions in the S1, S2, S3 and S4 subunits of the toxin (Cockle *et al.*, 1989, 1991; Loosmore *et al.*, 1990, 1991, 1993).

The lack of public confidence in the effectiveness of the whole cell pertussis vaccines and increased safety concerns throughout many of the industrialized nations, made it clear that new research and development efforts had to be implemented to specifically address these issues. Following Japan's example, research began focusing on acellular vaccines capable of inducing high levels of protection with less reactogenic acellular components. One of the first to be developed came for the laboratory of John Robbins and Rachel Schneerson at the National Institute of Child Health and Human Development in Bethesda, Maryland. Their vaccine was a monocomponent pertussis toxoid consisted of purified pertussis toxin inactivated with hydrogen peroxide (Sekura *et al.*, 1988). Clinical trials of the vaccine were carried out in Gothenburg, Sweden, in the 1980s and 1990s, which was an ideal location, since Sweden had stopped vaccination for pertussis in 1979 (Krantz *et al.*, 1990). The vaccine was initially given to infants and results of the trial showed that the monocomponent pertussis toxoid was well tolerated and induced an immune response with an efficacy of 71% (Taranger *et al.*, 1997, 1999; Trollfors *et al.*, 1995, 1998). The distribution of the vaccine was then expanded to include mass vaccination of children in 11 communities participating in a mass vaccination project. Remarkably, these studies showed that as a result of mass vaccination of children with just pertussis toxoid, the incidence of pertussis in the entire population was decreased (Taranger *et al.*, 2001a,b). This monocomponent vaccine technology was licensed by the National Institutes of Health to Amvax, Inc. in Laurel, Maryland, which became North American Vaccines, Beltsville, Maryland, where it was further developed into the combined DTaP and approved for use by the U.S. Food and Drug Administration in 1998.

In an effort to support the development of new generation acellular pertussis vaccines, the National Institutes of Allergy and Infectious Diseases in Bethesda, Maryland, sponsored large clinical trials in the early 1990s where vaccine manufactures were asked to submit acellular pertussis vaccine candidates. From 1991 to 1992, a phase 1–2 clinical trial evaluated 13 acellular pertussis vaccines from various manufacturers. The results indicated that the acellular vaccines were less reactogenic and the majority of these vaccines producing antibody response equal or better that the whole cell vaccine. From these results and other data, vaccines were then selected for large NIAID-sponsored efficacy trials in Sweden and Italy. Addition efficacy trials were concurrently initiated by other sponsors in Erlangen, Mainz, and Munich, Germany; Gothenburg, Sweden; and Senegal. In these clinical trials, the acellular vaccines formulations all containing purified pertussis toxoid with either one, two or four other acellular pertussis antigens. All of the pertussis toxin components in these vaccines were inactivated with either formaldehyde or glutaraldehyde, except for the vaccine produced by Biocine (formally Sclavo R & D Vaccines, Siena, Italy, which later became Chiron Vaccines and is now Novartis Vaccines). In the Biocine-Sclavo vaccine, in which the pertussis toxin was genetically inactivated with R9K and E129G mutation in the S1 subunit. The results

of these large clinical trials clearly demonstrated that acellular pertussis vaccines were safe, highly efficacious, and much less reactogenic. However, one vaccine candidate containing the genetically detoxified pertussis toxin stood out from all the others. Using only 5 µg of toxin compared to 25 to 50 µg of the chemically inactivated toxin, the Biocine vaccine induced higher titers of anti-toxin activity and showed an efficacy of 90% protection against whooping cough. (Edwards and Decker, 1996; Edwards *et al.*, 1995). As a result of various successful clinical trials, the Center for Disease Control and Prevention's Advisory Committee on Immunization Practices (ACIP) recommended that acellular pertussis vaccines replace the whole cell pertussis vaccine in the paediatric vaccine schedule.

Acellular pertussis vaccines were first licensed for use in the United States in 1996 for use in the 4th and 5th booster dose of the paediatric DTaP combined vaccine. It was intended for children who had already received the whole cell pertussis vaccine for the 2, 4, and 6 month dose. Typically the 4th and 5th booster doses are given to children between 15 months and 6 years, with the last dose given just before entering primary school. Then in 1997, after the acellular pertussis vaccines were successfully introduced as booster doses, the Center for Disease Control and Prevention's Advisory Committee on Immunization Practices (ACIP) recommended that acellular pertussis vaccines replace the whole cell pertussis vaccine in the paediatric vaccine schedule. Finally in 2005, the Food and Drug Administration licensed the ACIP recommended the use of acellular pertussis vaccines (all containing at least pertussis toxoid) to boost the immune response in adolescents and adults with the goal of eradicating pertussis by eliminating it from its only known reservoir, the human host (Kretsinger *et al.*, 2006). Early results from these efforts are very encouraging (Cherry, 2009; Ward *et al.*, 2005).

Diphtheria toxin

A chapter involving diphtheria toxin would be incomplete without acknowledgement of the contributions made by the late Alwin Pappenheimer, Jr. (1908–1995) at Harvard University in Cambridge, Massachusetts. Perhaps more than any other modern-day scientist, he provided relentless support and encouragement throughout his career and lead the way to our current understanding of diphtheria disease and the molecular mechanism responsible for its toxicity. Beginning with experiments on cultivation of *Corynebacterium diphtheriae* and purification of diphtheria toxin at the Antitoxin and Vaccine Laboratory of the Massachusetts Department of Public Health in Jamaica Plain, his dedication, guidance and mentoring have shown us that the gene which encodes diphtheria toxin is carried by a beta-phage and that maximum production of the toxin is obtained under limiting iron conditions. Before we had molecular cloning and DNA sequencing, his pioneering research using chemical mutagenesis to generate cross-reactive material (CRM) strains of *Corynebacterium diphtheriae* led the way in understanding toxicity by providing non-toxic or reduced toxic forms of diphtheria toxoid for use in immunology and biological studies. Later research includes experiments that defined our understanding of the molecular mechanism of diphtheria toxicity by demonstrating that the enzymatic activity of the toxin covalently modified elongation factor 2 (EF-2) resulting in inhibition of protein synthesis in eukaryote cells. Clearly, our knowledge and understanding of diphtheria and its toxicity has been built around Pappenheimer's legacy (Boquet and Pappenheimer, 1976; Boquet *et al.*, 1977; Collier and Pappenheimer, 1964; Gill *et al.*, 1973; Lawrence, 1999; Moynihan and Pappenheimer, 1981; Murphy *et al.*, 1974; Pappenheimer, 1937, 1948, 1977, 1979, 1980; Pappenheimer and Gill, 1973; Pappenheimer *et al.*, 1972; Uchida and Pappenheimer, 1974; Uchida *et al.*, 1972, 1973).

Diphtheria toxin is produced as a single chain 58,000 Dalton polypeptide that is released from certain pathogenic strains of Gram-negative *Corynebacterium diphtheriae* that are infected with a phage carrying the toxin gene. Upon entering the host cell, the toxin is cleaved by a trypsin-like protease producing an A and B fragment. Diphtheria toxin's mechanism of action is the ADP-ribosyltransferase activity of the A fragment. This enzymatic activity inhibits host cell protein synthesis by ADP-ribosylating a post-translationally modified histidine residue in elongation factor 2 (eEF-2), which is associated with the host ribosome complex resulting in death of the cell (Honjo *et al.*, 1968). The name is of the disease and toxin derived from the Greek word *diphtheria* meaning 'pair of leather scrolls' – this name reflects the appearance of the upper respiratory tract when it is infected by the bacteria (Haubirch, 2003).

Diphtheria toxin was one of the first bacterial protein toxins identified and is perhaps one of the most studied and best characterized. Its use as the essential immunogen in diphtheria vaccines was facilitated by the discovery over 100 year ago that treatment with formaldehyde could eliminate toxicity yet the detoxified toxin, commonly referred to as diphtheria toxoid, retain its ability to stimulate a protective immune response. The mechanism of formaldehyde detoxification is based on the reactivity of this simple single carbon molecule with primary amine groups on the protein (i.e. side-chain of lysine and the amino-terminal of a polypeptide chain). The reaction forms a methylol intermediate, which condenses with water to form a Schiff-base. The Schiff-base then reacts with the 5-position on a tyrosine ring or to a lesser degree with other amino acids to form stable, covalent methylene bridges. In detoxification protocols for vaccine production, the resulting Schiff-base is stabilized using precise concentrations of either glycine or lysine. In the presence of a reducing agent the Schiff-base forms a methyl group resulting in ε-methyllysine. Even with its use for over 100 years, the exact mechanism of formaldehyde detoxification is not fully understood because of partly unknown chemical reactions that are possible due to the complexity of protein (Fraenkel-Conrat and Olcott, 1948a,b; Metz *et al.*, 2005; Thaysen-Andersen *et al.*, 2007).

The complete sequence of the diphtheria toxin structural gene was determined in 1983 (Greenfield *et al.*, 1983). Based on its structural arrangement and catalytic activity, diphtheria toxin is classified as an A-B toxin. The catalytically active A-domain is an ADP-ribosyltransferase that modifies elongation factor 2 (eEF2). The enzyme activity covalently transfers the ADP-ribose moiety of nicotinamide adenine dinucleotide (NAD+) to a post-translational modified His residue on EF2. This modification effectively inhibits protein synthesis by the host ribosome resulting in death of the cell. The NAD⁺ binding site domain has been identified and involves His21, Tyr54, Tyr65, and Glu148 (Blanke *et al.*, 1994; Papini *et al.*, 1989). The host cellular receptor has been identified as a heparin-binding epidermal growth factor-like protein. This protein interacts with the receptor-binding domain on the toxin, which consists of an amino acid loop structure between residues 511 and 530. Within this receptor-binding loop, it has been shown that Tyr514, Lys516, Val 523, Asn524, Lys526 and Phe530 participate in binding to the cell receptor (Naglich *et al.*, 1992; Shen *et al.*, 1994). The crystal structure of diphtheria toxin complexed with an extracellular portion of the host receptor has been determined (Louie *et al.*, 1997). The sites in diphtheria toxin that are affected by the detoxification process correlate well with the type of chemical modification induced by formaldehyde and the physical structure defined by the primary amino acid sequence and crystal structure (Metz *et al.*, 2005).

Initially it was discovered that under certain conditions, chemical treatment of pathogenic bacteria produced mutant strains with unique characteristics and reduced toxicity. These early experiments using the nucleic acid mutagen nitrosoguanidine produced bacterial toxins with altered toxic function, yet retained cross-reactive activity with polyclonal anti-toxin antibodies (Uchida *et al.*, 1971). These cross-reactive mutants (CRM) provided insight into the structure–function relationship of protein toxins and produced a source of non-toxic proteins to use as vaccine candidates and as alternatives to toxoid (formalin-treated toxins) to use as non-toxic carrier proteins for the development of highly immunogenic conjugate polysaccharide vaccines (Anderson *et al.*, 1985; Schneerson *et al.*, 1980a,b; Uchida *et al.*, 1973).

John Collier and his colleagues at Harvard University in Cambridge, Massachusetts, furthered our early structure-function knowledge of bacterial protein toxins with the discovery of the NAD binding site and the specific amino acid mutations that reduced the biological toxicity of diphtheria toxin and *Pseudomonas aeruginosa* exotoxin A using photochemical affinity labelling with NAD and site-directed mutagenesis. Using these techniques, they were able to demonstrate that a single amino acid mutation of glutamic acid 148 to aspartic acid in diphtheria toxin fragment A and glutamic acid 553 to aspartic acid in exotoxin A dramatically reduced the enzymatic activity and toxicity of these protein toxins (Carroll and Collier, 1984, 1987; Carroll *et al.*, 1985). A number of investigators successfully pursued this approach and as the new molecular biology tools became available the detailed structure of CRM toxins could be defined at the molecular level. For example, this type of analysis showed that single amino acid changes in the A chain of diphtheria toxin effectively detoxified the diphtheria toxin molecule: CRM 197 glycine 52 to glutamine (G52Q) (Giannini *et al.*, 1984). Similarly, comparative sequence analysis of CRM 228 and wild-type toxin showed several amino acid changes in the A and B chains that correlated with nitrosoguanidine mutagenesis treatement (Kaczorek *et al.*, 1983). Later, Virginia Johnson at the Food and Drug Administration (FDA) and Peter Nicholls in my laboratory showed that a single amino acid mutation of glycine 79 to aspartic acid (G79D) in the CRM 228 diphtheria toxin A chain correlated with lose of enzymatic activity (Johnson and Nicholls, 1994b). Johnson and Nicholls then went on to look closely at another diphtheria toxin mutant. In this work they substituted all 20 amino acids in to the His21 site of the enzymatically active diphtheria toxin fragment A and demonstrated that various amino acids could reduce toxin activity, but that a His21Gly mutation reduced toxin activity to essentially background and yet conserved its ability to stimulate a protective immune response. The reduction in enzymatic activity exhibited by the various amino acid substitutions suggest that the His21 residue is important for preserving a steric conformation required for catalytic activity rather that participating in acid-base or electrostatic type of exchange (Johnson and Nicholls, 1994a).

Relatively recently, a rather unanticipated application of a CRM mutant has been found. It has been shown the CRM197 diphtheria toxin mutant strongly inhibits tumour growth in nude mice and promising results in humans (Buzzi *et al.*, 2004; Miyamoto *et al.*, 2004, 2006). Recent studies demonstrate that this tumorigenic activity is potentiated by a weak EF2-ADP-ribosyl activity still present in the CRM197 mutant (Kageyama et al., 2007). The CRM197 mutant and the His21Gly mutant have been widely used as an immunological adjuvant carrier protein for conjugate bacterial polysaccharide vaccines and protein/protein conjugate vaccines (Daum et al., 1997; Kayhty et al., 1989; Schneider et al., 1990). The use of genetically detoxified toxins, such as the diphtheria toxin His21Gly and the pertussis toxin Arg9Lys mutant as vaccine immunogens will be an important step in the development of safe effective vaccines (Robbins et al., 2005, 2007, 2009).

Tetanus toxin

Tetanus toxin is an extremely potent protein neurotoxin produced by the anaerobic Gram-positive bacterium Clostridium tetani. The toxin is structurally and functionally related to botulinum toxins produced by Clostridium botulinum. Clostridium tetani is an abundant, natural inhabitant of the soil usually in the form of spores. One of the toxins it produces is tetanospasmin, which is the causative agent of the potentially fatal disease commonly known as 'lockjaw' or tetanus. The bacterium is typically introduced into the host through puncture wounds or skin abrasions where spores can then germinate due to abundant nutrients and very low oxygen tension; however, unlike diphtheria and pertussis, tetanus is not contagious between individuals. Tetanus is a vaccine preventable disease and responds to post-exposure prophylaxis. Once across the protective layer of skin, the germinating spores produce several toxins. Biologically, the most detrimental is tetanospasmin toxin, which targets nerve cells and disrupts the inhibitory motor nerve endings. This action effectively blocks inhibitory impulses leading to unopposed muscle contraction and spasm. Tetanus dates back to the time of Hippocrates and the name is derived from the Greek word tetavos meaning 'tension', which reflects the clinical symptoms of arched, elongated position with convulsive rigid muscle spasms associated with the disease when the toxin reaches the motor neurons cells (Haubirch, 2003). These symptoms usually occur in 5–10 days, but can appear anywhere from 2 to 50 days after infection and can result in respiratory paralysis and death.

Tetanus toxin was another one of the first bacterial protein toxins identified and has been well characterized although possibly not as well as diphtheria toxin due to its extreme toxicity. The toxin is the essential antigen in tetanus vaccines and is detoxified to tetanus toxoid using formaldehyde in the same manner as described above for diphtheria toxin. The complete sequence of the tetanus toxin structural gene was determined in 1986 (Eisel et al., 1986; Fairweather and Lyness, 1986). Based on its structural arrangement and catalytic activity, tetanus toxin is metalloproteinase. The catalytically active light chain domain is a zinc-dependent endo-metalloproteinase that targets the junction of synaptic nerve endings making them incapable of receiving neurotransmitters signals. This modification causes muscle contractions and repeated spasms which can interrupt respiration and result in death. A zinc binding motif has been identified and it has been shown that mutations within this motif result in a protein that totally lacks neurotoxin activity (Arora et al., 1994). The host cellular receptor has been identified as a ganglioside, which interacts with the receptor-binding domain on the toxin. The amino acid domain on the toxin has now been identified. It has now been shown that mutations within the zinc binding motif inactivate the toxin. The sites in tetanus toxin that are effected by the detoxification process correlate well with the type of physical structure defined by the primary amino acid sequence and crystal structure (Emsley et al., 2000; Fotinou et al., 2001; Umland et al., 1997, 1998).

The toxin is synthesized as a 150,000 dalton polypeptide that is cleaved by clostridial proteases (Helting et al., 1979) into its tetanospasmin active form consisting of a 50,000 dalton amino-terminal L-chain referred to as the 'light chain' and a 100,000 dalton carboxyl-terminal H-chain known as the 'heavy chain' (Montecucco and Schiavo, 1994). The light chain is a zinc endoprotease (Schiavo et al., 1992a,b), which is covalently attached to the heavy chain by a single disulfide bond. The heavy chain undergoes further cleavage into H_N (N for amino-terminal) and H_C (C for carboxyl-terminal) fragments. All of the toxin fragments have unique activities. The H_C fragment is likely responsible for binding to sensitive cells and internalization of the fragment into vesicles (Halpern and Loftus, 1993; Lalli et al., 1999). Crystal structure studies of the Hc fragment suggest that the fragment binds the host cell through interaction with a ganglioside receptor (Fotinou et al., 2001; Herreros et al., 2000a,b). Translocation of the L-chain across the vesicular membrane relies on the H_N fragment (Halpern and Neale, 1995; Lalli et al., 1999; Montecucco, 1986). Neurotoxicity of tetanus toxin is the result of proteolytic cleavage of vesicle-associate membrane protein by the metalloprotease activity of the L-chain. This cleavage blocks neurotransmitter release by preventing formation of the synaptic SNARE complex (Sollner et al., 1993). With the arrival of DNA sequencing and cloning technology, protein expression systems, and site-directed mutagenesis, extensive research studies have allowed us to look at the biological activity of bacterial protein toxins at a molecular level. By examining the similarities between the amino acid sequences of a class of metalloendoproteases and the light chain of tetanus toxin, it was discovered that the tetanus toxin chain contain the sequence 233-HELIH-237, which matches the classic consensus zinc binding motif defined as HEXXH. Naveen Arora and Steve Leppla from my laboratory with colleagues from NICHD and the FDA used this information to produce a chimeric fusion protein consisting of the first 254 amino acids of anthrax toxin lethal factor and the tetanus light chain to study the cytotoxic effects on cells that lack the receptors for tetanus toxin. From this work, they concluded that cytotoxicity was caused by inhibition of a specific intracellular membrane fusion event mediated by a synaptobrevin (by experimental design, the cells they used lacked the tetanus toxin receptor and the actual target was cellubrevin, which is a homologue to synaptobrevin and is a known substrate for tetanus toxin). With this work, they also demonstrated that mutations in zinc binding motif of Glu234 to alanine or His237 to alanine produced a toxin protein molecule with no biological activity (Arora et al., 1994). Other investigators, using site-directed mutagenesis, showed that tetanus toxin light chain activity could be abolished by

changing His233 to either Cys or Val or Glu234 to Gln (Hohne-Zell et al., 1993) and Glu234 to Ala (Li et al., 1994).

Recently, a tetanus vaccine consisting of the tetanus toxin Hc fragment delivered by a transcutaneous immunization strategy outperformed the traditional tetanus toxoid in its ability to induce a tetanus toxin neutralizing antibody response. This approach suggests that it may be possible to deliver recombinant toxin antigens containing protective epitopes as an effective transcutaneous vaccine administered directly on the surface of the skin (Johnston et al., 2009).

Use of toxins as carrier protein for development of protein/polysaccharides conjugate vaccines and protein/protein conjugate vaccines

Beginning in the 1980s it became apparent that polysaccharides from pathogenic bacteria could be made more immunogenic by chemically conjugating them to carrier proteins. This resulted in a vaccine that effectively induces an immune response against the conjugated polysaccharide. This approach was an extremely important breakthrough for paediatric polysaccharide vaccine because unlike adults, children lacked the ability to produce a protective immune response when the polysaccharide moiety was used by itself. The choice of carrier protein for such a purpose is somewhat limited because ideally one does not want to introduce a foreign protein into a vaccine unless it is an essential component of the immune response that is being sought. Therefore, some of the favoured choices for a carrier protein have been diphtheria and tetanus toxoids and a few viral proteins. However, toxoids by the very nature of the chemical process of detoxification are not ideal candidate carrier protein because many potential conjugation sites are blocked by the formaldehyde treatment (especially lysine). The ability to produce genetically detoxified diphtheria and tetanus toxin has provided the ideal carrier protein platform for the development of conjugate polysaccharide and protein/protein conjugate vaccines. Some examples of genetically inactivated bacterial toxin proteins with potential use as conjugate carrier proteins are diphtheria toxin (His21Gly) (Johnson and Nicholls, 1994a) and tetanus toxin light chain with either Glu234Ala or His237Ala mutation, which fall within the zinc-binding motif (Arora *et al.*, 1994).

Numerous CRM strains of diphtheria and tetanus have been developed some of which are suitable for use as vaccine candidates or as carrier proteins for conjugate vaccine. However, these are strains of the pathogenic bacteria and still harbour other potential toxins which must be removed. In addition, characterization of the mutation is sometimes difficult and there is a possibility of reversion of the mutation back to a wild type. For these reasons, recombinant genetically inactivated diphtheria toxin, tetanus toxin, and pertussis toxin have been proposed for the development of the next generation of DTaP vaccines (Mooi, 2010; Robbins *et al.*, 2995, 2007, 2009).

Summation DT and TT

Taken together these studies provide a comprehensive description at diphtheria toxoid and tetanus toxoid which have a long proven history as the essential immunogen in diphtheria and tetanus vaccines. These toxoids are immunogenic and they induce protective immunity. However, producing a toxoid still involves growing and processing large volumes of pathogenic strains of *Corynebacterium diphtheriae* and *Clostridium tetani*, purification of these extremely potent toxins from the growth media, and then chemical detoxification using just the right combination of formaldehyde, glycine or lysine. The chemical reaction typically takes two to four weeks for completion and requires constant monitoring and control of pH and temperature. Even with standardized protocols and precision monitoring, it is difficult to predict the exact outcome; therefore, before the vaccine can be released for use, each production lot of vaccine undergoes extensive vaccine potency test requiring large numbers of animals to insure that the toxicity has been neutralized and that the toxoid can still stimulate a protective immune response.

Formaldehyde detoxification of diphtheria toxin and tetanus toxin is still an active area of research. Concerns related to lot consistency and the use of animals in testing safety and effectiveness have driven this investigation to understand and define an exact description of a vaccine that really has not had much change in production in the nearly 100 years since its inception. There are still unknown chemical reactions produced by this procedure which causes concern when producing a vaccine product.

The molecular biology revolution has now provided the ability to produce recombinant diphtheria toxin and tetanus in unprecedented amounts using *E. coli*-based protein expression systems and industrial fermentors. This technology has the advantage that these protein toxins can be genetically modified so that they are enzymatically inert, yet retains a nearly native, highly immunogenic protein structure capable of inducing a protective immune response using a minimal amount of toxin protein. These scientific advances can provide reliable source of safe and extremely well defined protein immunogen for use in new generation vaccines.

In this chapter, I have discussed a broad swath of medical science discoveries related to three bacterial protein toxins responsible for countless deaths and enormous human suffering. All were known well over a hundred years ago – before there was any concept of DNA, RNA, amino acids, or proteins – they were mysterious factors that could not be fully explained, but were to be avoided. Fortunately, through the dedicated efforts of the forebears of our scientific lineage and our mentors, scientific knowledge has evolved creating disciplines with overlapping borders that merge the intellectual resources of physics, chemistry, biochemistry, and molecular biology to produce stunning breakthroughs over the last 100 years. Building on these breakthrough, we have begun to unravel those 'mysterious factors'. By applying our new knowledge towards understanding and defining the structure and molecular interactions and mechanisms responsible for toxicity, it is now possible to design new, precisely defined innovative biological products for medical intervention, which can be in the form of vaccines or therapeutic biologics that can target, invade, and modify or kill selected cells which are detrimental to living organisms.

Current formulation of diphtheria, tetanus and acellular pertussis vaccines

In the United States, the three vaccines are given as a single dose DTaP combination vaccine in a paediatric immunization series of five single-dose injections at 2, 4, 6 and 15–16 months, and a preschool booster at 4–6 years of age. The DTaP combination vaccine is also available as an adolescent and adult booster vaccine and is given as a single dose containing a reduced amount of antigens. Tetanus and diphtheria toxoid potency is determined by measuring the amount of neutralizing antitoxin in previously immunized guinea pigs. The potency of the acellular pertussis vaccine antigens pertussis toxin, filamentous haemagglutinin, and pertactin is determined by enzyme-linked immunosorbent assay (ELISA) on sera from previously immunized mice.

The DTaP vaccine is a turbid, white suspension in water and is given as a single intramuscular injection. Each manufacturer prepares their vaccines using proprietary methods in accordance with protocols that were established at the time the vaccine was licensed for use by the FDA. The following is a general description of the vaccine manufacturing process.

Diphtheria vaccine

Diphtheria toxoid (D): 2.5 Lf (0.5 ml dose)

Diphtheria vaccine is prepared from *Corynebacterium diphtheriae* grown in either modified Mueller's, Fenton, or Linggoud growth medium sometimes containing bovine extract (depending on manufacturer). When bovine extracts are used, they are obtained from countries which have been determined to be free or not at risk of bovine spongiform encephalopathy (BSE) by the United States Department of Agriculture (USDA). The diphtheria toxin, which is released from the bacterium during growth, is purified from the growth medium by ammonium sulfate fractionation and detoxified with formaldehyde. Diphtheria toxoid is then absorbed onto aluminium phosphate (alum) and formulated with water to the appropriate dose. A serum diphtheria antitoxin level of > 0.1 IU/ml measured by neutralization assay is regarded as protective.

Tetanus vaccine

Tetanus toxoid (T): 5 Lf (0.5 ml dose)

Tetanus vaccine is prepared from *Clostridium tetani* grown in either modified Mueller–Miller or modified Latham medium with casamino acids without beef heart infusion growth medium. The tetanus toxin, which is released from the bacterium during growth, is purified from the growth medium by ammonium sulfate fractionation and detoxified with formaldehyde. Tetanus toxoid is then absorbed onto aluminium phosphate (alum) and formulated with water to the appropriate dose. A serum tetanus antitoxin level of ≥0.1 IU/ml measured by neutralization assay is regarded as protective.

Acellular pertussis vaccines

Pertussis vaccines are comprised either of one, two, three or five pertussis antigens, depending on the manufacture. However, all pertussis vaccines contain pertussis toxin.

- pertussis toxin (detoxified) (PT): 8–25 μg (0.5 ml);
- filamentous haemagglutinin (FHA): 5–8 μg;
- pertactin (PRN): 2.5–3 μg;
- fimbrial agglutinogens types 2 and 3 (FIM): 5 μg.

Pertussis vaccines are prepared from *Bordetella pertussis* grown in Stainer–Scholte medium modified with casamino acid and dimethyl-beta-cyclodextrin. The pertussis toxin, filamentous haemagglutinin, and pertactin antigens are isolated separately from the culture medium supernatant. There have been several chemical methods used to produce pertussis toxoid for use in whooping cough vaccines. The most common is either formaldehyde or glutaraldehyde, both of which react similarly with their aldehyde groups combining with primary amine groups. However, glutaraldehyde, with two aldehydes separated by a flexible three-carbon methylene bridge, can more readily crosslink proteins. Tetranitromethane has been used by the Massachusetts Public Health Biologic Laboratories in Boston, Massachusetts, to chemically inactivated pertussis toxin in their acellular pertussis vaccine. Treatment of proteins with tetranitromethane results in modification of SH groups with the formation of cysteic acid and tyrosyl nitration. Treatment of toxin with hydrogen peroxide results in modification of the R-groups of amino acids – specifically formation of cysteic acid from cystine and cysteine and sulfones from methionine. In addition, there is a reduction of number of tyrosine and tryptophan residues (Means and Feeney, 1971). This treatment is effective at inactivating the toxin, but it also affects the immunogenicity of the protein antigen. Certainly all of the chemical treated pertussis toxins are as a whole much less immunogenic that the genetically modified R9K/E129G pertussis toxin used in acellular pertussis vaccine formulations (i.e. 5 µg for the genetically inactivated toxin; 25 µg for the formaldehyde and glutaraldehyde treated toxin, 40 µg for the hydrogen peroxide treated toxin, and 50 µg for the tetranitromethane treated toxin (Edwards et al., 1995; Hewlett, 1996). Based on overwhelming data, we have recommended that the chemically detoxified pertussis toxin component of acellular pertussis vaccines be replaced by the genetically detoxified pertussis toxin (Brown et al., 1991; Burnette et al., 1988; Mooi, 2010; Peppoloni et al., 1995; Robbins et al., 2005, 2007, 2009). Fimbriae are extracted and co-purified from the bacterial cells. The vaccine components are purified by various methods including sequential filtration, salt-precipitation, chromatography and ultrafiltration. Filamentous haemagglutinin is treated with formaldehyde. All of the individual antigens are absorbed onto aluminium phosphate (alum) and formulated with water to the appropriate dose.

What lies ahead and recommendations for the future – we can do better

With the advances in molecular biology and cloning, we can produce genetically detoxified diphtheria toxin, tetanus toxin, and pertussis toxin. These genetically modified protein toxins are highly immunogenic and lack toxicity making them outstanding immunogens for use in new generation vaccines for diphtheria, tetanus, pertussis, and as carrier proteins for both polysaccharide/carrier conjugate and protein/carrier conjugate vaccines. With less toxicity and engineering out troublesome contaminating cellular proteins they are easy to purify and can be handled with less difficulty than their native counterparts. With these ideal characteristics, they can be scaled up to meet downstream industrial vaccine produced levels making them ideal for vaccine manufacturing. We have written several articles encouraging a dialogue between FDA regulators, vaccine manufacturers and research scientist in which we have documented and argued the benefits of using genetically inactivated mutant toxins as the primary immunogenic components for acellular diphtheria, tetanus, and pertussis toxin vaccines (Mooi, 2010; Rappuoli, 1996; Robbins et al., 2005, 2007, 2009). The future of new generation vaccines for old, but too familiar diseases is now. We have the know how and technology, the science backs these discussions and development, so now it is up to those who can make it happen without negative pressure from the bottom line on a corporate balance sheet. This is too important and populations, especially children, should have the protection afforded by these truly superior vaccines.

References

Anderson, P., Pichichero, M.E., and Insel, R.A. (1985). Immunogens consisting of oligosaccharides from the capsule of Haemophilus influenzae type b coupled to diphtheria toxoid or the toxin protein CRM197. J. Clin. Invest. 76, 52–59.

Antoine, R., Tallett, A., van Heyningen, S., and Locht, C. (1993). Evidence for a catalytic role of glutamic acid 129 in the NAD-glycohydrolase activity of the pertussis toxin S1 subunit. J. Biol. Chem. 268, 24149–24155.

Arico, B., and Rappuoli, R. (1987). Bordetella parapertussis and Bordetella bronchiseptica contain transcriptionally silent pertussis toxin genes. J. Bacteriol. 169, 2847–2853.

Arora, N., Williamson, L.C., Leppla, S.H., and Halpern, J.L. (1994). Cytotoxic effects of a chimeric protein consisting of tetanus toxin light chain and anthrax toxin lethal factor in non-neuronal cells. J. Biol. Chem. 269, 26165–26171.

Avery, O.T., Macleod, C.M., and McCarty, M. (1944). Studies on the chemical nature of the substance inducing transformation of pneumococcal types: induction of transformation by a desoxyribonucleic acid fraction isolated from pneumococcus type Iii. J. Exp. Med. *79*, 137–158.

Baxby, D. (2005). The discovery of diphtheria toxoid and the primary and secondary immune response. Epidemiol. Infect. *133*, S21–S22.

Beemon, K.L., and Keith, J.M. (1976). Structure of Rous Sarcoma Virus RNA: 1) Localization of N6-methyladenosine; 2) the sequence of 23 nucleotides following the 5' capped terminus m7GpppGmp. In Animal Virology, Baltimore, D., Huang, A.S., and Fox, C.F., eds (New York, Academic Press), pp. 97–105.

Bergman, R.K., and Munoz, J. (1968). Action of the histamine sensitizing factor from *Bordetella pertussis* on inbred and random bred strains of mice. Int. Arch. Allerg. Appl. Immunol. *34*, 331–338.

Blanke, S.R., Huang, K., and Collier, R.J. (1994). Active-site mutations of diphtheria toxin: role of tyrosine-65 in NAD binding and ADP-ribosylation. Biochemistry *33*, 15494–15500.

Boquet, P., and Pappenheimer, A.M., Jr. (1976). Interaction of diphtheria toxin with mammalian cell membranes. J. Biol. Chem. *251*, 5770–5778.

Boquet, P., Silverman, M.S., and Pappenheimer, A.M., Jr. (1977). Interaction of diphtheria toxin with mammalian cell membranes. Prog. Clin. Biol. Res. *17*, 501–509.

Bordet, J., and Gengou, O. (1906). Le Microbe De La Coqueluche. Ann. Inst. Pasteur *20*, 731–741.

Brown, D.R., Keith, J.M., Sato, H., and Sato, Y. (1991). Construction and characterization of genetically inactivated pertussis toxin. Dev. Biol. Standard. *73*, 63–73.

Brownlee, G.G., Sanger, F., and Barrell, B.G. (1968). The sequence of 5s ribosomal ribonucleic acid. J. Mol. Biol. *34*, 379–412.

Burnette, W.N., Cieplak, W., Mar, V.L., Kaljot, K.T., Sato, H., and Keith, J.M. (1988). Pertussis toxin S1 mutant with reduced enzyme activity and a conserved protective epitope. Science *242*, 72–74.

Buzzi, S., Rubboli, D., Buzzi, G., Buzzi, A.M., Morisi, C., and Pironi, F. (2004). CRM197 (nontoxic diphtheria toxin): effects on advanced cancer patients. Cancer Immunol. Immunother. *53*, 1041–1048.

Carroll, S.F., and Collier, R.J. (1984). NAD binding site of diphtheria toxin: identification of a residue within the nicotinamide subsite by photochemical modification with NAD. Proc. Natl. Acad. Sci. U.S.A. *81*, 3307–3311.

Carroll, S.F., and Collier, R.J. (1987). Active site of *Pseudomonas aeruginosa* exotoxin A. Glutamic acid 553 is photolabeled by NAD and shows functional homology with glutamic acid 148 of diphtheria toxin. J. Biol. Chem. *262*, 8707–8711.

Carroll, S.F., McCloskey, J.A., Crain, P.F., Oppenheimer, N.J., Marschner, T.M., and Collier, R.J. (1985). Photoaffinity labeling of diphtheria toxin fragment A with NAD: structure of the photoproduct at position 148. Proc. Natl. Acad. Sci. U.S.A. *82*, 7237–7241.

C.D.C. (2009). (Center for Disease Control and Prevention, United States Government, HHS).

Centers for Disease Control and Prevention (CDC) (1997). Pertussis vaccination: use of acellular pertussis vaccines among infants and young children. Recommendations of the Advisory Committee on Immunization Practices (ACIP). MMWR Recomm. Rep. *46*, 1–25.

Cherry, J.D. (2009). How can we eradicate pertussis. Adv. Exp. Med. Biol. *634*, 41–51.

Cherry, J.D., Gornbein, J., Heininger, U., and Stehr, K. (1998). A search for serologic correlates of immunity to *Bordetella pertussis* cough illnesses. Vaccine *16*, 1901–1906.

Cherry, J.D., and Olin, P. (1999). The science and fiction of pertussis vaccines. Pediatrics *104*, 1381–1383.

Cheung, A.M., Farizo, K.M., and Burns, D.L. (2004). Analysis of relative levels of production of pertussis toxin subunits and Ptl proteins in *Bordetella pertussis*. Infect. Immun. *72*, 2057–2066.

Cieplak, W., Burnette, W.N., Mar, V.L., Kaljot, K.T., Morris, C.F., Chen, K.K., Sato, H., and Keith, J.M. (1988). Identification of a region in the S1 subunit of pertussis toxin that is required for enzymatic activity and that contributes to the formation of a neutralizing antigenic determinant. Proc. Natl. Acad. Sci. U.S.A. *85*, 4667–4671.

Cieplak, W., Jr., Locht, C., Mar, V.L., Burnette, W.N., and Keith, J.M. (1990). Photolabelling of mutant forms of the S1 subunit of pertussis toxin with NAD+. Biochem. J. *268*, 547–551.

Clausen, C., Munoz, J., and Bergman, R.K. (1968). Lymphocytosis and histamine sensitization of mice by fractions from *Bordetella pertussis*. J. Bacteriol. *96*, 1484–1487.

Cockle, S., Loosmore, S., Radika, K., Zealey, G., Boux, H., Phillips, K., and Klein, M. (1989). Detoxification of pertussis toxin by site-directed mutagenesis. Adv. Exp. Med. Biol. *251*, 209–214.

Cockle, S., Zealey, G., Loosmore, S., Yacoob, R., Fahim, R., Yang, Y.P., Jackson, G., Boux, H., Boux, L., and Klein, M. (1991). Development of non-toxigenic vaccine strains of *Bordetella pertussis* by gene replacement. Adv. Exp. Med. Biol. *303*, 221–225.

Cockle, S.A. (1988). Pertussis Toxin NAD Photocrosslinking. Paper presented at: Miami Biotechnology Winter Symposium (Miami, Florida).

Cockle, S.A. (1989). Identification of an active-site residue in subunit S1 of pertussis toxin by photocrosslinking to NAD. FEBS Lett. *249*, 329–332.

Collier, R.J., and Pappenheimer, A.M., Jr. (1964). Studies on the mode of action of diphtheria toxin. I. Phosphorylated intermediates in normal and intoxicated Hela cells. J. Exp. Med. *120*, 1007–1018.

Daum, R.S., Hogerman, D., Rennels, M.B., Bewley, K., Malinoski, F., Rothstein, E., Reisinger, K., Block, S., Keyserling, H., and Steinhoff, M. (1997). Infant immunization J. Infect. Dis. *176*, 445–455.

Dean, H.R., and Webb, R.A. (1928). The Determination of the rate of antibody (precipitin) production in rabbit's blood by method of 'optimal proportions.' J. Path. Bact. *31*, 89–99.

Edwards, K.M., and Decker, M.D. (1996). Acellular pertussis vaccines for infants. N. Engl. J. Med. *334*, 391–392.

Edwards, K.M., Meade, B.D., Decker, M.D., Reed, G.F., Rennels, M.B., Steinhoff, M.C., Anderson, E.L., Englund, J.A., Pichichero, M.E., and Deloria, M.A. (1995). Comparison of 13 acellular pertussis vaccines: overview and serologic response. Pediatrics 96, 548–557.

Eisel, U., Jarausch, W., Goretzki, K., Henschen, A., Engels, J., Weller, U., Hudel, M., Habermann, E., and Niemann, H. (1986). Tetanus toxin: primary structure, expression in E. coli, and homology with botulinum toxins. EMBO J. 5, 2495–2502.

Emsley, P., Fotinou, C., Black, I., Fairweather, N.F., Charles, I.G., Watts, C., Hewitt, E., and Isaacs, N.W. (2000). The structures of the H(C) fragment of tetanus toxin with carbohydrate subunit complexes provide insight into ganglioside binding. J. Biol. Chem. 275, 8889–8894.

Fairweather, N.F., and Lyness, V.A. (1986). The complete nucleotide sequence of tetanus toxin. Nucleic Acids Res. 14, 7809–7812.

Fine, P.E. (1993). Herd immunity: history, theory, practice. Epidemiol. Rev. 15, 265–302.

Fotinou, C., Emsley, P., Black, I., Ando, H., Ishida, H., Kiso, M., Sinha, K.A., Fairweather, N.F., and Isaacs, N.W. (2001). The crystal structure of tetanus toxin Hc fragment complexed with a synthetic GT1b analogue suggests cross-linking between ganglioside receptors and the toxin. J. Biol. Chem. 276, 32274–32281.

Fowler, S., Xing, D.K., Bolgiano, B., Yuen, C.T., and Corbel, M.J. (2003). Modifications of the catalytic and binding subunits of pertussis toxin by formaldehyde: effects on toxicity and immunogenicity. Vaccine 21, 2329–2337.

Fraenkel-Conrat, H., and Olcott, H.S. (1948a). The reaction of formaldehyde with proteins; cross-linking between amino and primary amide or guanidyl groups. J. Am. Chem. Soc. 70, 2673–2684.

Fraenkel-Conrat, H., and Olcott, H.S. (1948b). Reaction of formaldehyde with proteins; cross-linking of amino groups with phenol, imidazole, or indole groups. J. Biol. Chem. 174, 827–843.

Fraenkel-Conrat, H., Singer, B., and Williams, R.C. (1957). Infectivity of viral nucleic acid. Biochim. Biophys. Acta 25, 87–96.

Fraenkel-Conrat, H., and Williams, R.C. (1955). Reconstitution of active tobacco mosaic virus from its inactive protein and nucleic acid components. Proc. Natl. Acad. Sci. U.S.A. 41, 690–698.

Gangarosa, E.J., Galazka, A.M., Wolfe, C.R., Phillips, L.M., Gangarosa, R.E., Miller, E., and Chen, R.T. (1998). Impact of anti-vaccine movements on pertussis control: the untold story. Lancet 351, 356–361.

Giannini, G., Rappuoli, R., and Ratti, G. (1984). The amino-acid sequence of two non-toxic mutants of diphtheria toxin: CRM45 and CRM197. Nucleic Acids Res. 12, 4063–4069.

Gill, D.M., Pappenheimer, A.M., Jr., and Uchida, T. (1973). Diphtheria toxin, protein synthesis, and the cell. Fed. Proc. 32, 1508–1515.

Glenny, A.T., and Hopkins, B.E. (1923). Diphtheria toxoid as an immunizing agent. Br. J. Exp. Pathol. 4, 283–288.

Glenny, A.T., Hopkins, B.E., and Pope, C.G. (1924). Further notes on modification of diphtheria toxin by formaldehyde. J. Path. Bact. 27, 262–270.

Glenny, A.T., Pope, C.G., Waddington, U., and Wallace, H. (1926). Immunological notes, XIII; the antigenic value of toxoid precipitated by potassium alum. J. Path. Bact. 29, 38–39.

Glenny, A.T., and Sudmerson, H.J. (1921). Notes on the production of immunity to diphtheria toxin. J. Hyg. 20, 176–220.

Glenny, A.T., and Walpole, G.S. (1915). Detection and concentration of antigens by ultrafiltration, pressure dialysis, etc. with special reference to diphtheria and tetanus toxins. Biochem. J. 9, 291, 298–308.

Greenfield, L., Bjorn, M.J., Horn, G., Fong, D., Buck, G.A., Collier, R.J., and Kaplan, D.A. (1983). Nucleotide sequence of the structural gene for diphtheria toxin carried by corynebacteriophage beta. Proc. Natl. Acad. Sci. U.S.A. 80, 6853–6857.

Gronski, P., Seiler, F.R., and Schwick, H.G. (1991). Discovery of antitoxins and development of antibody preparations for clinical uses from 1890 to 1990. Mol. Immunol. 28, 1321–1332.

Grundbacher, F.J. (1992). Behring's discovery of diphtheria and tetanus antitoxins. Immunol. Today 13, 188–190.

Halpern, J.L., and Loftus, A. (1993). Characterization of the receptor-binding domain of tetanus toxin. J. Biol. Chem. 268, 11188–11192.

Halpern, J.L., and Neale, E.A. (1995). Neurospecific binding, internalization, and retrograde axonal transport. Curr. Top. Microbiol. Immunol. 195, 221–241.

Haubirch, W.S. (2003). Medical Meanings: A Glossary of Word Origins, 2nd edn (American College of Physicians).

Helting, T.B., Parschat, S., and Engelhardt, H. (1979). Structure of tetanus toxin. Demonstration and separation of a specific enzyme converting intracellular tetanus toxin to the extracellular form. J. Biol. Chem. 254, 10728–10733.

Herreros, J., Lalli, G., Montecucco, C., and Schiavo, G. (2000a). Tetanus toxin fragment C binds to a protein present in neuronal cell lines and motoneurons. J. Neurochem. 74, 1941–1950.

Herreros, J., Lalli, G., and Schiavo, G. (2000b). C-terminal half of tetanus toxin fragment C is sufficient for neuronal binding and interaction with a putative protein receptor. Biochem. J. 347, 199–204.

Hershey, A.D., and Chase, M. (1952). Independent functions of viral protein and nucleic acid in growth of bacteriophage. J. Gen. Physiol. 36, 39–56.

Hewlett, E.L. (1996). Acellular pertussis trial. Pediatrics 98, 800.

Hohne-Zell, B., Stecher, B., and Gratzl, M. (1993). Functional characterization of the catalytic site of the tetanus toxin light chain using permeabilized adrenal chromaffin cells. FEBS Lett. 336, 175–180.

Honjo, T., Nishizuka, Y., and Hayaishi, O. (1968). Diphtheria toxin-dependent adenosine diphosphate ribosylation of aminoacyl transferase II and inhibition of protein synthesis. J. Biol. Chem. *243*, 3553–3555.

Hviid, A., Stellfeld, M., Andersen, P.H., Wohlfahrt, J., and Melbye, M. (2004). Impact of routine vaccination with a pertussis toxoid vaccine in Denmark. Vaccine *22*, 3530–3534.

Johnson, V.G., and Nicholls, P.J. (1994a). Histidine 21 does not play a major role in diphtheria toxin catalysis. J. Biol. Chem. *269*, 4349–4354.

Johnson, V.G., and Nicholls, P.J. (1994b). Identification of a single amino acid substitution in the diphtheria toxin A chain of CRM 228 responsible for the loss of enzymatic activity. J. Bacteriol. *176*, 4766–4769.

Johnston, L., Mawas, F., Tierney, R., Qazi, O., Fairweather, N., and Sesardic, D. (2009). Transcutaneous delivery of tetanus toxin Hc fragment induces superior tetanus toxin neutralizing antibody response compared to tetanus toxoid. Human Vaccines *5*, 230–236.

Kaczorek, M., Delpeyroux, F., Chenciner, N., Streeck, R.E., Murphy, J.R., Boquet, P., and Tiollais, P. (1983). Nucleotide sequence and expression of the diphtheria tox228 gene in *Escherichia coli*. Science *221*, 855–858.

Kageyama, T., Ohishi, M., Miyamoto, S., Mizushima, H., Iwamoto, R., and Mekada, E. (2007). Diphtheria toxin mutant CRM197 possesses weak EF2-ADP-ribosyl activity that potentiates its anti-tumorigenic activity. J. Biochem. *142*, 95–104.

Kantha, S.S. (1991). A centennial review; the 1890 tetanus antitoxin paper of von Behring and Kitasato and the related developments. Keio J. Med. *40*, 35–39.

Kay, L.E. (2000). Who Wrote the Book of Life? A History of the Genetic Code (Stanford, Stanford University Press).

Kayhty, H., Peltola, H., Eskola, J., Ronnberg, P.R., Kela, E., Karanko, V., and Makela, P.H. (1989). Immunogenicity of *Haemophilus influenzae* oligosaccharide-protein and polysaccharide–protein conjugate vaccination of children at 4, 6, and 14 months of age. Pediatrics *84*, 995–999.

Keitel, W.A. (1999). Cellular and acellular pertussis vaccines in adults. Clin. Infect. Dis. *28* (Suppl. 2), S118–123.

Keitel, W.A., Muenz, L.R., Decker, M.D., Englund, J.A., Mink, C.M., Blumberg, D.A., and Edwards, K.M. (1999). A randomized clinical trial of acellular pertussis vaccines in healthy adults: dose–response comparisons of 5 vaccines and implications for booster immunization. J. Infect. Dis. *180*, 397–403.

Kendrick, P., and Eldering, G. (1935). Significance of bacteriological methods in the diagnosis and control of whooping cough. Am. J. Publ. Hlth Nation's Hlth *25*, 147–155.

Kendrick, P.L., Eldering, G., Dixon, M.K., and Misner, J. (1947). Mouse protection tests in the study of pertussis vaccine: a comparative series using the intracerebral route for challenge. Am. J. Publ. Hlth Nation's Hlth *37*, 803–810.

Kotob, S.I., Hausman, S.Z., and Burns, D.L. (1995). Localization of the promoter for the ptl genes of *Bordetella pertussis*, which encode proteins essential for secretion of pertussis toxin. Infect. Immun. *63*, 3227–3230.

Krantz, I., Sekura, R., Trollfors, B., Taranger, J., Zackrisson, G., Lagergard, T., Schneerson, R., and Robbins, J. (1990). Immunogenicity and safety of a pertussis vaccine composed of pertussis toxin inactivated by hydrogen peroxide, in 18- to 23-month-old children. J. Pediatr. *116*, 539–543.

Kretsinger, K., Broder, K.R., Cortese, M.M., Joyce, M.P., Ortega-Sanchez, I., Lee, G.M., Tiwari, T., Cohn, A.C., Slade, B.A., Iskander, J.K., *et al.* (2006). Preventing tetanus, diphtheria, and pertussis among adults: use of tetanus toxoid, reduced diphtheria toxoid and acellular pertussis vaccine recommendations of the Advisory Committee on Immunization Practices (ACIP) and recommendation of ACIP, supported by the Healthcare Infection Control Practices Advisory Committee (HICPAC), for use of Tdap among health-care personnel. MMWR Recomm. Rep. *55*, 1–37.

Kulenkampff, M., Schwartzman, J.S., and Wilson, J. (1974). Neurological complications of pertussis inoculation. Arch. Dis. Child. *49*, 46–49.

Kuno-Sakai, H., and Kimura, M. (2004). Safety and efficacy of acellular pertussis vaccine in Japan, evaluated by 23 years of its use for routine immunization. Pediatr. Int. *46*, 650–655.

Kuno-Sakai, H., Kimura, M., and Watanabe, H. (2004). Verification of components of acellular pertussis vaccines that have been distributed solely, been in routine use for the last two decades and contributed greatly to control of pertussis in Japan. Biologicals *32*, 29–35.

Lalli, G., Herreros, J., Osborne, S.L., Montecucco, C., Rossetto, O., and Schiavo, G. (1999). Functional characterisation of tetanus and botulinum neurotoxins binding domains. J. Cell Sci. *112*, 2715–2724.

Lawrence, H.S. (1999). Alwin Max Pappenheimer, Jr. Biographical memoirs. Nat. Acad. Sci. *77*, 1–18.

Legon, L. (1998). Jules Bordet: pioneer researcher in immunology and pertussis (1870–1961). Semin. Pediatr. Infect. Dis. *9*, 163–167.

Li, Y., Foran, P., Fairweather, N.F., de Paiva, A., Weller, U., Dougan, G., and Dolly, J.O. (1994). A single mutation in the recombinant light chain of tetanus toxin abolishes its proteolytic activity and removes the toxicity seen after reconstitution with native heavy chain. Biochemistry *33*, 7014–7020.

Lobet, Y., Cieplak, W., Jr., Smith, S.G., and Keith, J.M. (1989). Effects of mutations on enzyme activity and immunoreactivity of the S1 subunit of pertussis toxin. Infect. Immun. *57*, 3660–3662.

Lobet, Y., Feron, C., Dequesne, G., Simoen, E., Hauser, P., and Locht, C. (1993). Site-specific alterations in the B oligomer that affect receptor-binding activities and mitogenicity of pertussis toxin. J. Exp. Med. *177*, 79–87.

Locht, C., Barstad, P.A., Coligan, J.E., Mayer, L., Munoz, J.J., Smith, S.G., and Keith, J.M. (1986). Molecular cloning of pertussis toxin genes. Nucleic Acids Res. *14*, 3251–3261.

Locht, C., Capiau, C., and Feron, C. (1989). Identification of amino acid residues essential for the enzymatic activities of pertussis toxin. Proc. Natl. Acad. Sci. U.S.A. *86*, 3075–3079.

Locht, C., Cieplak, W., Marchitto, K.S., Sato, H., and Keith, J.M. (1987). Activities of complete and truncated forms of pertussis toxin subunits S1 and S2 synthesized by *Escherichia coli*. Infect. Immun. *55*, 2546–2553.

Locht, C., and Keith, J.M. (1986). Pertussis toxin gene: nucleotide sequence and genetic organization. Science *232*, 1258–1264.

Locht, C., Lobet, Y., Feron, C., Cieplak, W., and Keith, J.M. (1990). The role of cysteine 41 in the enzymatic activities of the pertussis toxin S1 subunit as investigated by site-directed mutagenesis. J. Biol. Chem. *265*, 4552–4559.

Lombard, M., Pastoret, P.P., and Moulin, A.M. (2007). A brief history of vaccines and vaccination. Rev. Sci. Tech. (International Office of Epizootics) *26*, 29–48.

Loosmore, S., Cockle, S., Zealey, G., Boux, H., Phillips, K., Fahim, R., and Klein, M. (1991). Detoxification of pertussis toxin by site-directed mutagenesis: a review of Connaught strategy to develop a recombinant pertussis vaccine. Mol. Immunol. *28*, 235–238.

Loosmore, S., Zealey, G., Cockle, S., Boux, H., Chong, P., Yacoob, R., and Klein, M. (1993). Characterization of pertussis toxin analogs containing mutations in B-oligomer subunits. Infect. Immun. *61*, 2316–2324.

Loosmore, S.M., Zealey, G.R., Boux, H.A., Cockle, S.A., Radika, K., Fahim, R.E., Zobrist, G.J., Yacoob, R.K., Chong, P.C., Yao, F.L., *et al.* (1990). Engineering of genetically detoxified pertussis toxin analogs for development of a recombinant whooping cough vaccine. Infect. Immun. *58*, 3653–3662.

Louie, G.V., Yang, W., Bowman, M.E., and Choe, S. (1997). Crystal structure of the complex of diphtheria toxin with an extracellular fragment of its receptor. Mol. Cell *1*, 67–78.

Luttinger, P. (1916). The epidemiology of pertussis. Am J. Dis. Child. *12*, 296–307.

Madsen, T. (1933). Vaccination against whooping cough. J. Am. Med. Assoc. *101*, 187–188.

Mandell, G.L., Bennett, J.E., and Dolin, R. (2010). Mandell, Douglas, and Bennett's Principles and Practice of Infectious Diseases, 7 edn (Philadelphia, Elsevier).

Marchitto, K.S., Munoz, J.J., and Keith, J.M. (1987a). Detection of subunits of pertussis toxin in Tn5-induced *Bordetella* mutants deficient in toxin biological activity. Infect. Immun. *55*, 1309–1313.

Marchitto, K.S., Smith, S.G., Locht, C., and Keith, J.M. (1987b). Nucleotide sequence homology to pertussis toxin gene in *Bordetella bronchiseptica* and *Bordetella parapertussis*. Infect. Immun. *55*, 497–501.

Maxam, A.M., and Gilbert, W. (1977). A new method for sequencing DNA. Proc. Natl. Acad. Sci. U.S.A. *74*, 560–564.

Means, G.E., and Feeney, R.E. (1971). Chemical Modification of Proteins (San Francisco, Holelen-Day).

Merkel, T.J., Keith, J.M., and Yang, x. (2006). A High Yield Pertussis Vaccine Production Strain and Method for Making Same, Office, U.S.P., ed. (United States of America, National Institutes of Health). US Patent 7, *101*, 558.

Metz, B., Kersten, G.F.A., Jong, A.D., Meiring, H., Hove, J.T., Hennink, W.E., Crommelin, D.J.A., and Jiskott, W. (2005). Identification of formaldehyde-induced modifications in proteins: reactions with diphtheria toxin. In Structural Characterisation of Diphtheria Toxoid (Utrecht, Universiteit Utrecht), pp. 140–154.

Min Jou, W., Haegeman, G., Ysebaert, M., and Fiers, W. (1972). Nucleotide sequence of the gene coding for the bacteriophage MS2 coat protein. Nature *237*, 82–88.

Miyamoto, S., Hirata, M., Yamazaki, A., Kageyama, T., Hasuwa, H., Mizushima, H., Tanaka, Y., Yagi, H., Sonoda, K., Kai, M., *et al.* (2004). Heparin-binding EGF-like growth factor is a promising target for ovarian cancer therapy. Cancer Res. *64*, 5720–5727.

Miyamoto, S., Yagi, H., Yotsumoto, F., Kawarabayashi, T., and Mekada, E. (2006). Heparin-binding epidermal growth factor-like growth factor as a novel targeting molecule for cancer therapy. Cancer Sci. *97*, 341–347.

Montecucco, C. (1986). How do tetanus and botulinum toxins bind to neuronal membranes? Trends Biochem. Sci. *11*, 314–317.

Montecucco, C., and Schiavo, G. (1994). Mechanism of action of tetanus and botulinum neurotoxins. Mol. Microbiol. *13*, 1–8.

Mooi, F.R. (2010). *Bordetella pertussis* and vaccination: The persistence of a genetically monomorphic pathogen. Infect. Genet. Evol. *10*, 36–49.

Moynihan, M.R., and Pappenheimer, A.M., Jr. (1981). Kinetics of adenosinediphosphoribosylation of elongation factor 2 in cells exposed to diphtheria toxin. Infect. Immun. *32*, 575–582.

Munoz, J., and Bergman, R.K. (1968). Histamine-sensitizing factors from microbial agents, with special reference to *Bordetella pertussis*. Bacteriol. Rev. *32*, 103–126.

Munoz, J.J., Arai, H., Bergman, R.K., and Sadowski, P.L. (1981). Biological activities of crystalline pertussigen from *Bordetella pertussis*. Infect. Immun. *33*, 820–826.

Murphy, J.R., Pappenheimer, A.M., Jr., and de Borms, S.T. (1974). Synthesis of diphtheria tox-gene products in *Escherichia coli* extracts. Proc. Natl. Acad. Sci. U.S.A. *71*, 11–15.

Naglich, J.G., Metherall, J.E., Russell, D.W., and Eidels, L. (1992). Expression cloning of a diphtheria toxin receptor: identity with a heparin-binding EGF-like growth factor precursor. Cell *69*, 1051–1061.

Nencioni, L., Pizza, M., Volpini, G., Podda, A., and Rappuoli, R. (1991a). Genetic approaches to a vaccine for pertussis. Adv. Exp. Med. Biol. *303*, 119–127.

Nencioni, L., Volpini, G., Peppoloni, S., Bugnoli, M., De Magistris, T., Marsili, I., and Rappuoli, R. (1991b). Properties of pertussis toxin mutant PT-9K/129G after formaldehyde treatment. Infect. Immun. *59*, 625–630.

Nicosia, A., Perugini, M., Franzini, C., Casagli, M.C., Borri, M.G., Antoni, G., Almoni, M., Neri, P., Ratti, G., and Rappuoli, R. (1986). Cloning and sequencing of the pertussis toxin genes: operon structure and gene duplication. Proc. Natl. Acad. Sci. U.S.A. *83*, 4631–4635.

NLM. Profiles in Science (National Library of Medicine, National Institutes of Health).

Oakley, C.L. (1966). Alexander Thomas Glenny. Biographical Memoirs of the Royal Society *12*, 162–180.

Omer, S.B., Salmon, D.A., Orenstein, W.A., deHart, M.P., and Halsey, N. (2009). Vaccine refusal, mandatory immunization, and the risks of vaccine-preventable diseases. N. Engl. J. Med. *360*, 1981–1988.

Papini, E., Schiavo, G., Sandona, D., Rappuoli, R., and Montecucco, C. (1989). Histidine 21 is at the NAD+ binding site of diphtheria toxin. J. Biol. Chem. *264*, 12385–12388.

Pappenheimer, A.M., Jr. (1937). Diphtheria toxin. J. Biol. Chem. *120*, 543–553.

Pappenheimer, A.M., Jr. (1948). The iron enzymes of *C. diphtheriae* and their possible relation to diphtheria toxin. Bull. N.Y. Acad. Med. *24*, 331.

Pappenheimer, A.M., Jr. (1977). Diphtheria toxin. Annu. Rev. Biochem. *46*, 69–94.

Pappenheimer, A.M., Jr. (1979). Transport of diphtheria toxin A fragment across the plasma membrane. Progr. Clin. Biol. Res. *31*, 669–674.

Pappenheimer, A.M., Jr. (1980). Diphtheria: studies on the biology of an infectious disease. Harvey Lectures *76*, 45–73.

Pappenheimer, A.M., Jr., and Gill, D.M. (1973). Diphtheria. Science *182*, 353–358.

Pappenheimer, A.M., Jr., Uchida, T., and Harper, A.A. (1972). An immunological study of the diphtheria toxin molecule. Immunochemistry *9*, 891–906.

Parkhill, J., Sebaihia, M., Preston, A., Murphy, L.D., Thomson, N., Harris, D.E., Holden, M.T., Churcher, C.M., Bentley, S.D., Mungall, K.L., *et al.* (2003). Comparative analysis of the genome sequences of *Bordetella pertussis*, *Bordetella parapertussis* and *Bordetella bronchiseptica*. Nat. Genet. *35*, 32–40.

Peppoloni, S., Pizza, M., De Magistris, M.T., Bartoloni, A., and Rappuoli, R. (1995). Acellular pertussis vaccine composed of genetically inactivated pertussis toxin. Physiol. Chem. Phys. Med. NMR *27*, 355–361.

Pichichero, M.E., Deloria, M.A., Rennels, M.B., Anderson, E.L., Edwards, K.M., Decker, M.D., Englund, J.A., Steinhoff, M.C., Deforest, A., and Meade, B.D. (1997). A safety and immunogenicity comparison of 12 acellular pertussis vaccines and one whole-cell pertussis vaccine given as a fourth dose in 15- to 20-month-old children. Pediatrics *100*, 772–788.

Pittman, M. (1956). Pertussis and pertussis vaccine control. J. Washington Acad. Sci. *46*, 234–243.

Pittman, M. (1979). Pertussis toxin: the cause of the harmful effects and prolonged immunity of whooping cough. A hypothesis. Rev. Infect. Dis. *1*, 401–412.

Pittman, M. (1984). The concept of pertussis as a toxin-mediated disease. Pediatr. Infect. Dis. *3*, 467–486.

Pizza, M., Bartoloni, A., Prugnola, A., Silvestri, S., and Rappuoli, R. (1988). Subunit S1 of pertussis toxin: mapping of the regions essential for ADP-ribosyl-

transferase activity. Proc. Natl. Acad. Sci. U.S.A. *85*, 7521–7525.

Pizza, M., Covacci, A., Bartoloni, A., Perugini, M., Nencioni, L., De Magistris, M.T., Villa, L., Nucci, D., Manetti, R., Bugnoli, M., *et al.* (1989). Mutants of pertussis toxin suitable for vaccine development. Science *246*, 497–500.

Ramon, G. (1923). Sur le pouvoir et sur les proprieties immunisantes d'une toxine diphtherique rendue anatoxique (anatoxine). Comp. Rends. Acad. Sci. Paris *177*, 1338–1340.

Rappuoli, R. (1996). Acellular pertussis vaccines: a turning point in infant and adolescent vaccination. Infect. Agents Dis. *5*, 21–28.

Rappuoli, R., Pizza, M., Covacci, A., Bartoloni, A., Nencioni, L., Podda, A., and De Magistris, M.T. (1992a). Recombinant acellular pertussis vaccine–from the laboratory to the clinic: improving the quality of the immune response. FEMS Microbiol. Immunol. *5*, 161–170.

Rappuoli, R., Pizza, M., De Magistris, M.T., Podda, A., Bugnoli, M., Manetti, R., and Nencioni, L. (1992b). Development and clinical testing of an acellular pertussis vaccine containing genetically detoxified pertussis toxin. Immunobiology *184*, 230–239.

Rappuoli, R., Pizza, M., Douce, G., and Dougan, G. (1996). New vaccines against bacterial toxins. Adv. Exp. Med. Biol. *397*, 55–60.

Rappuoli, R., Pizza, M., Podda, A., De Magistris, M.T., Ceccarini, C., and Nencioni, L. (1994). Development of the new acellular recombinant pertussis vaccine. Arch. Inst. Pasteur Tunis *71*, 557–563.

Rappuoli, R., Pizza, M., Podda, A., De Magistris, M.T., and Nencioni, L. (1991). Towards third-generation whooping cough vaccines. Trends Biotechnol. *9*, 232–238.

Rappuoli, R., Podda, A., Pizza, M., Covacci, A., Bartoloni, A., de Magistris, M.T., and Nencioni, L. (1992c). Progress towards the development of new vaccines against whooping cough. Vaccine *10*, 1027–1032.

Robbins, J.B., Schneerson, R., Keith, J.M., Miller, M.A., Kubler-Kielb, J., and Trollfors, B. (2009). Pertussis vaccine: a critique. Pediatric Infect. Dis. J. *28*, 237–241.

Robbins, J.B., Schneerson, R., Keith, J.M., Shiloach, J., Miller, M., and Trollors, B. (2007). The rise in pertussis cases urges replacement of chemically-inactivated with genetically-inactivated toxoid for DTP. Vaccine *25*, 2811–2816.

Robbins, J.B., Schneerson, R., Trollfors, B., Sato, H., Sato, Y., Rappuoli, R., and Keith, J.M. (2005). The diphtheria and pertussis components of diphtheria-tetanus toxoids-pertussis vaccine should be genetically inactivated mutant toxins. J. Infect. Dis. *191*, 81–88.

Sanger, F., and Coulson, A.R. (1975). A rapid method for determining sequences in DNA by primed synthesis with DNA polymerase. J. Mol. Biol. *94*, 441–448.

Sanger, F., Nicklen, S., and Coulson, A.R. (1977). DNA sequencing with chain-terminating inhibitors. Proc. Natl. Acad. Sci. U.S.A. *74*, 5463–5467.

Sato, H., Ito, A., Chiba, J., and Sato, Y. (1984a). Monoclonal antibody against pertussis toxin: effect on toxin activity and pertussis infections. Infect. Immun. *46*, 422–428.

Sato, H., and Sato, Y. (1999). Experience with diphtheria toxoid-tetanus toxoid-acellular pertussis vaccine in Japan. Clin. Infect. Dis. *28* Suppl. 2, S124–130.

Sato, Y., and Arai, H. (1972). Leucocytosis-promoting factor of *Bordetella pertussis*. I. Purification and characterization. Infect. Immun. *6*, 899–904.

Sato, Y., Kimura, M., and Fukumi, H. (1984b). Development of a pertussis component vaccine in Japan. Lancet *1*, 122–126.

Sauer, L. (1933). Whooping cough: a study in immunization. J. Am. Med. Assoc. *100*, 239–241.

Schiavo, G., Benfenati, F., Poulain, B., Rossetto, O., Polverino de Laureto, P., DasGupta, B.R., and Montecucco, C. (1992a). Tetanus and botulinum-B neurotoxins block neurotransmitter release by proteolytic cleavage of synaptobrevin. Nature *359*, 832–835.

Schiavo, G., Poulain, B., Rossetto, O., Benfenati, F., Tauc, L., and Montecucco, C. (1992b). Tetanus toxin is a zinc protein and its inhibition of neurotransmitter release and protease activity depend on zinc. EMBO J. *11*, 3577–3583.

Schneerson, R., Barrera, O., Sutton, A., and Robbins, J.B. (1980a). Preparation, characterization, and immunogenicity of *Haemophilus influenzae* type b polysaccharide–protein conjugates. J. Exp. Med. *152*, 361–376.

Schneerson, R., Robbins, J.B., Barrera, O., Sutton, A., Habig, W.B., Hardegree, M.C., and Chaimovich, J. (1980b). *Haemophilus influenzae* type B polysaccharide–protein conjugates: model for a new generation of capsular polysaccharide vaccines. Progr. Clin. Biol. Res. *47*, 77–94.

Schneider, L.C., Insel, R.A., Howie, G., Madore, D.V., and Geha, R.S. (1990). Response to a *Haemophilus influenzae* type b diphtheria CRM197 conjugate vaccine in children with a defect of antibody production to *Haemophilus influenzae* type b polysaccharide. J. Allergy Clin. Immunol. *85*, 948–953.

Sekura, R.D., Zhang, Y.L., Roberson, R., Acton, B., Trollfors, B., Tolson, N., Shiloach, J., Bryla, D., Muir-Nash, J., Koeller, D., *et al.* (1988). Clinical, metabolic, and antibody responses of adult volunteers to an investigational vaccine composed of pertussis toxin inactivated by hydrogen peroxide. J. Pediatr. *113*, 806–813.

Shen, W.H., Choe, S., Eisenberg, D., and Collier, R.J. (1994). Participation of lysine 516 and phenylalanine 530 of diphtheria toxin in receptor recognition. J. Biol. Chem. *269*, 29077–29084.

Sollner, T., Whiteheart, S.W., Brunner, M., Erdjument-Bromage, H., Geromanos, S., Tempst, P., and Rothman, J.E. (1993). SNAP receptors implicated in vesicle targeting and fusion. Nature *362*, 318–324.

Stent, G.S. (1968). That was the molecular biology that was. Science *160*, 390–395.

Stent, G.S. (1972). Prematurity and uniqueness in scientific discovery. Sci. Am. *227*, 84–93.

Stewart, G.T. (1977). Vaccination against whooping-cough. Efficacy versus risks. Lancet *309*, 234–237.

Stibitz, S., Black, W., and Falkow, S. (1986). The construction of a cloning vector designed for gene replacement in *Bordetella pertussis*. Gene *50*, 133–140.

Storsaeter, J., Hallander, H.O., Gustafsson, L., and Olin, P. (1998). Levels of anti-pertussis antibodies related to protection after household exposure to *Bordetella pertussis*. Vaccine *16*, 1907–1916.

Strom, J. (1960). Is universal vaccination against pertussis always justified? Br. Med. J. *2*, 1184–1186.

Strom, J. (1967). Further experience of reactions, especially of a cerebral nature, in conjunction with triple vaccination: a study based on vaccinations in Sweden 1959–65. Br. Med. J. *4*, 320–323.

Taranger, J., Trollfors, B., Bergfors, E., Knutsson, N., Lagergard, T., Schneerson, R., and Robbins, J.B. (2001a). Immunologic and epidemiologic experience of vaccination with a monocomponent pertussis toxoid vaccine. Pediatrics *108*, E115.

Taranger, J., Trollfors, B., Bergfors, E., Knutsson, N., Sundh, V., Lagergard, T., Lind-Brandberg, L., Zackrisson, G., White, J., Cicirello, H., *et al.* (2001b). Mass vaccination of children with pertussis toxoid–decreased incidence in both vaccinated and nonvaccinated persons. Clin. Infect. Dis. 33, 1004–1010.

Taranger, J., Trollfors, B., Lagergard, T., and Robbins, J.B. (1997). Protection against pertussis with a monocomponent pertussis toxoid vaccine. Int. J. Infect. Dis. *1*, 148–151.

Taranger, J., Trollfors, B., Lagergard, T., and Robbins, J.B. (1999). Protection against pertussis with a monocomponent pertussis toxoid vaccine. Biologicals 27, 89.

Thaysen-Andersen, M., Jorgensen, S.B., Wilhelmsen, E.S., Petersen, J.W., and Hojrup, P. (2007). Investigation of the detoxification mechanism of formaldehyde-treated tetanus toxin. Vaccine 25, 2213–2227.

Trollfors, B., Taranger, J., Lagergard, T., Lind, L., Sundh, V., Zackrisson, G., Lowe, C.U., Blackwelder, W., and Robbins, J.B. (1995). A placebo-controlled trial of a pertussis-toxoid vaccine. N. Engl. J. Med. 333, 1045–1050.

Trollfors, B., Taranger, J., Lagergard, T., Sundh, V., Bryla, D.A., Schneerson, R., and Robbins, J.B. (1998). Immunization of children with pertussis toxoid decreases spread of pertussis within the family. Pediatr. Infect. Dis. J. 17, 196–199.

Tsugita, A., Gish, D.T., Young, J., Fraenkel-Conrat, H., Knight, C.A., and Stanley, W.M. (1960). The complete amino acid sequence of the protein of tobacco mosaic virus. Proc. Natl. Acad. Sci. U.S.A. 46, 1463–1469.

Uchida, T., Gill, D.M., and Pappenheimer, A.M., Jr. (1971). Mutation in the structural gene for diphtheria toxin carried by temperate phage. Nature: New Biol. 233, 8–11.

Uchida, T., and Pappenheimer, A.M., Jr. (1974). Diphtheria toxin and mutant proteins. Japan J. Med. Sci. Biol. 27, 93–95.

Uchida, T., Pappenheimer, A.M., Jr., and Greany, R. (1973). Diphtheria toxin and related proteins. I. Isolation and properties of mutant proteins serologically related to diphtheria toxin. J. Biol. Chem. 248, 3838–3844.

Uchida, T., Pappenheimer, A.M., Jr., and Harper, A.A. (1972). Reconstitution of diphtheria toxin from two nontoxic cross-reacting mutant proteins. Science *175*, 901–903.

Umland, T.C., Wingert, L., Swaminathan, S., Schmidt, J.J., and Sax, M. (1998). Crystallization and preliminary X-ray analysis of tetanus neurotoxin C fragment. Acta Crystallogr. *54*, 273–275.

Umland, T.C., Wingert, L.M., Swaminathan, S., Furey, W.F., Schmidt, J.J., and Sax, M. (1997). Structure of the receptor binding fragment HC of tetanus neurotoxin. Nat. Struct. Biol. *4*, 788–792.

Ward, J.I., Cherry, J.D., Chang, S.J., Partridge, S., Lee, H., Treanor, J., Greenberg, D.P., Keitel, W., Barenkamp, S., Bernstein, D.I., *et al.* (2005). Efficacy of an acellular pertussis vaccine among adolescents and adults. N. Engl. J. Med. *353*, 1555–1563.

Warren, K.S. (1986). New scientific opportunities and old obstacles in vaccine development. Proc. Natl. Acad. Sci. U.S.A. *83*, 9275–9277.

Watanabe, M., and Nagai, M. (2005). Acellular pertussis vaccines in Japan: past, present and future. Expert Rev. Vaccines *4*, 173–184.

Watson, J.D., and Crick, F.H. (1953). Molecular structure of nucleic acids; a structure for deoxyribose nucleic acid. Nature *171*, 737–738.

Weiss, A.A., Hewlett, E.L., Myers, G.A., and Falkow, S. (1983). Tn5-induced mutations affecting virulence factors of *Bordetella pertussis*. Infect. Immun. *42*, 33–41.

Weiss, A.A., Johnson, F.D., and Burns, D.L. (1993). Molecular characterization of an operon required for pertussis toxin secretion. Proc. Natl. Acad. Sci. U.S.A. *90*, 2970–2974.

Wells, J.A. (1991). Systematic mutational analyses of protein–protein interfaces. Methods Enzymol. *202*, 390–411.

Adjuvants

David A.G. Skibinski and Derek T. O'Hagan

Abstract

The development of new effective vaccines, especially those consisting of highly purified antigens, will increasingly require the inclusion of an adjuvant. With over half a century of experience, aluminium containing adjuvants (alum) will continue to be widely used and until very recently remained the only vaccine adjuvant approved for human use in the US. In recent years a number of studies have started to reveal a more detailed understanding of alum's mechanism of action. Here we review these recent advances as well as discussing considerations for optimal formulation of the adjuvant. We will also address the need for more potent adjuvants than alum, with particular emphasis on the discovery and development of MF59, an emulsion based vaccine adjuvant which as been licensed for more than 10 years in more than 20 countries, for use in an influenza vaccine focused on elderly subjects (Fluad®).

Introduction

The term vaccine adjuvant, was first used by Gaston Ramon to describe components which when added to a specific antigen produce a stronger immune response (Ramon, 1924). The most widely used vaccine adjuvants are insoluble aluminium salts (generically called alum), and until recently these remained the only adjuvants present in licensed vaccines for human use in North America (Fig. 6.1). Alum has contributed significantly to many currently available vaccines, which could not otherwise be developed. Over half a century of experience has shown that alum is effective, safe and will continue to be used in

current and new vaccines for some time. In the first part of this chapter we will review the history of this adjuvant, recent advances in understanding its mechanism of action and approaches to ensure optimal use.

Despite the success of alum more potent adjuvants are necessary to enable vaccine development against a significant number of challenging pathogens for which we do not yet have effective vaccines (Rappuoli, 2007). However, the inclusion of a new adjuvant in a vaccine product is a demanding regulatory hurdle and success in this area has been limited. A significant success story is the vaccine adjuvant MF59. A safe and potent oil in water emulsion (o/w) which has been licensed for more than 10 years in a significant number of countries (>20) for use in a seasonal influenza vaccine (O'Hagan and De Gregorio, 2009). MF59 is a new adjuvant, at least relative to alum, and its discovery and development emphasizes a number of key lessons. The second half of this chapter will focus on MF59 and what we have learnt so far with this adjuvant. We will also touch on AS04, another novel adjuvant that has gained licensure in recent years.

Why and when we need adjuvants?

The need for a vaccine adjuvant is a key question that needs to be addressed by any vaccine development programme. The inclusion of an adjuvant must be justified via demonstration that it gains clear benefit over the unadjuvanted alternative. Of course simplistically adjuvants are incorporated to enhance the immune response to the vaccine, traditionally in the form of increased antibody

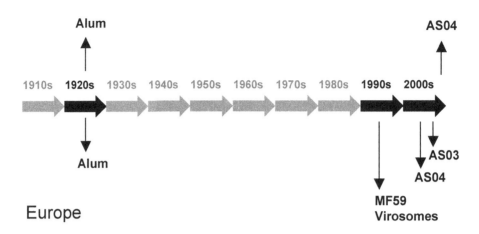

Figure 6.1 Timeline: approval of adjuvants for inclusion in licensed vaccines for human use in the US and Europe. Alum was licensed in the 1920s, was the first adjuvant to be licensed and until the recent approval of Ceravix (containing AS04) remained the only adjuvant approved for human use in the US and Europe. MF59 has been included in a licensed flu vaccine Fluad since 1997. More recently, AS04, a combination of monophosphoryl lipid A with alum was included in vaccines licensed for HBV and HPV (Fendrix in 2005 and Cervarix in 2007, respectively). In 2008 oil in water emulsion AS03 was approved for a pandemic flu vaccine, Prepandrix. Virosomes have been licensed as a component of Inflexal V since 1997 but it is unclear whether they significantly enhance immune responses and can therefore truly be classified as adjuvants.

titres. Therefore the main reason for inclusion is to elicit sufficiently high antibody responses to enable protection. However, adjuvant inclusion may result in more pragmatic effects such as antigen dose sparing, reduction in the number of immunizations necessary, overcoming antigen competition in combination vaccines or stabilizing antigens. Furthermore, adjuvants can increase the breadth of response to cover pathogen diversity, and overcome limited immune responses in some populations such as the elderly, young children and the immunocompromised. Adjuvants will become increasingly important in new generation vaccines which will likely comprise purified recombinant proteins, which although safe are poorly immunogenic. Finally new adjuvants promise to induce different types of immune response such as more effective T cell responses, which are required for some pathogens for which we do not yet have effective vaccines (Rappuoli, 2007). Taken together, the advantages an adjuvant has the potential to offer a vaccine, are summarized in Table 6.1.

Universally approved adjuvants: insoluble aluminium salts

Alum was the first adjuvant to be included in licensed vaccines and until recently remained the only adjuvant approved for human use in the US. This has changed with the very recent approval of Cervarix, a vaccine for HPV containing the novel adjuvant AS04, a combination of monophosphoryl lipid A with alum. In Europe a handful of other adjuvants have recently gained approval for human use (Fig. 6.1). MF59 had been included in a licensed flu vaccine Fluad since 1997. As well as its inclusion in Cervarix, AS04 is also included in a vaccine licensed for HBV in Europe (Fendrix). Last year oil in water emulsion AS03 was approved for a pandemic flu vaccine, Prepandrix.

The newly approved adjuvants are the rare success stories in a long list of unsuccessful attempts to develop vaccines with experimental adjuvants. Adjuvant development is a slow and difficult process and regulatory hurdles significant. Inclusion of an experimental adjuvant in a vaccine

Table 6.1 Advantages an adjuvant can afford a vaccine

Elicit sufficient immune response to afford protection

Extend the duration of the immune response

Lower antigen dose requirements

Reduce number of immunizations

Overcome antigen competition in combination vaccines

Increase antigen stability

Increase breadth of response

Overcome limited immune response in some populations

under development will slow the path of licensure for that vaccine and a clear and consistent benefit over established adjuvants (alum) will need to be demonstrated. Clearly alum, when sufficiently potent for efficacy, will continue to be included in new vaccines as the regulatory hurdles are lower and path to licensure quicker than any alternatives.

Alum is a generic name, describing preformed hydrated gels of aluminium hydroxide or aluminium phosphate which for more than half a century have been formulated with protein antigens to increase their immunogenicity in humans. They are known for their ability to induce robust antibody titres through the induction of a T helper cell type 2 (Th2) immune response (Grun and Maurer, 1989). Billions of doses have been administered over this time and generally these adjuvants are considered safe when used according to current immunization schedules (Edelman, 1980; World Health Organization, 1976).

The origins of alum

Glenny and coworkers in 1926 were the first to demonstrate the adjuvant activity of insoluble aluminium salts. They found that precipitation of diphtheria toxoid with potassium alum lead to significantly increased immune responses against the toxoid (Glenny *et al.*, 1926). Vaccines prepared by this approach were named alum precipitated vaccines and examples of use in practical vaccination can be found in older literature; reviewed in (Lindblad, 2004). However, these vaccines are highly heterogeneous depending on the buffer anions present at the time of precipitation and

have since been superseded by alum adsorbed vaccines. These consist of preformed aluminium hydroxide hydrated gels that are able to adsorb soluble antigens from aqueous solution (Lindblad, 2004).

Aluminium phosphate was introduced as an adjuvant when in 1946 Erikson and co-workers co-precipitated the adjuvant with diphtheria toxoid (Ericsson, 1946). Shortly after this, Holt demonstrated increased adjuvanticity of diphtheria toxoid adsorbed to preformed aluminium phosphate (Holt, 1947).

A third type of aluminium-containing adjuvant in clinical use is Merck's proprietary amorphous aluminium hydroxyphosphate sulfate (AAHS), which is included in Merck's human papillomavirus vaccine Gardasil (Bryan, 2007).

The mechanism of action of alum

Adjuvanticity of alum is characterized by a Th2-cell response and elevated titres of Th2-cell-associated antibody isotypes IgG1 and IgE (Grun and Maurer, 1989). An overview of the current understanding of how alum potentiates this process is presented in Fig. 6.2. It has long been proposed that alum functions by formation of a depot with antigens adsorbed to alum retained at the injection site. Also it has been proposed that antigens adsorbed onto alum show increased uptake by antigen-presenting cells (APC). Beyond functioning as a delivery system, alum has been shown to directly activate the immune system with a number of recent studies reporting that alum induces activation of NLRP3 inflammasome either directly or indirectly.

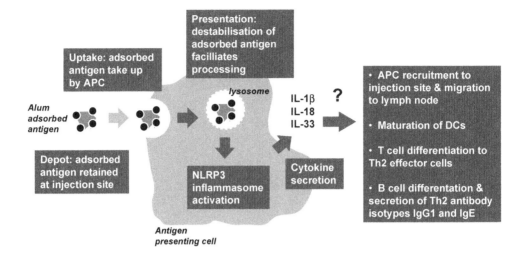

Figure 6.2 Overview of the currently proposed mechanisms of action of alum. Antigens adsorbed to alum are retained at the injection site (depot effect) and show increased uptake by antigen-presenting cells (APC). Adsorption has also been shown to destabilize antigen structure facilitating processing and presentation by APCs. Uptake leads to rupture of the lysosome and release of the antigen into the cytosol. Alum induces activation of NLRP3 inflammasome either directly or indirectly (through the production of uric acid and the formation of monosodium urate (MSU) crystals). NLRP3 activation leads to the secretion of pro-inflammatory cytokines IL-1β, IL-18 and IL-33. The contribution of all these events to alum adjuvanticity and the induction of a Th2 response are not yet fully understood.

Adsorption to alum delivers the antigen to the immune system

As mentioned previously original experiments by Glenny (Glenny *et al.*, 1931) with diphtheria toxin found that precipitation of diphtheria toxoid with potassium alum led to significantly increased protection against challenge in guinea pigs than a soluble solution of the toxoid. Analysis of the vaccine found that none of the antigen was present in the supernatant and led to the belief that adjuvant function was through the formulation of an antigen 'depot' at the site of injection. Glenny's depot theory led to the belief that antigen adsorption to alum was necessary for adjuvant function, a concept that was widely accepted for more than half a century. This is not without basis as slow release of the adsorbed antigen from the adjuvant would give time for newly recruited antigen-presenting cells to arrive (Hem and HogenEsch, 2007). However, evidence exists in the literature both supporting (Glenny *et al.*, 1931; Harrison, 1935) and opposing (Holt, 1950; White *et al.*, 1955) this hypothesis and definitive proof remains elusive (Marrack *et al.*, 2009).

Antigen adsorption to alum is also important if alum adjuvants are to act as a 'delivery system', promoting uptake into antigen-presenting cells. Studies *in vitro* have found that the uptake by dendritic cells of antigen adsorbed to alum was more efficient than the uptake of unadsorbed antigen (Mannhalter *et al.*, 1985; Morefield *et al.*, 2005b). More recently (Jones *et al.*, 2005; Kool *et al.*, 2008b), demonstrated *in vivo*, the increased uptake of adsorbed antigen by dendritic cells and B cells compared to soluble antigen alone. Adsorption has also been shown to alter antigen tertiary structure and a hypothesis has been put forward that this could facilitate antigen processing and presentation by antigen-presenting cells (Jones *et al.*, 2005).

However, a search of the literature, provides numerous examples where adsorption is not necessary for immunopotentiation (Romero Méndez *et al.*, 2007) and clearly other mechanisms of action must exist to explain the adjuvanticity of alum. In this study the authors propose that unabsorbed antigen may become trapped in the porous aluminium phosphate aggregates thus forming a depot in the absence of adsorption (Romero Méndez *et al.*, 2007). Another study by the same authors found that the degree of antigen adsorption in interstitial fluid correlated more closely with antibody titres than the degree of adsorption in the vaccine itself (Chang *et al.*, 2001). Recent studies have also found a negative impact of very tight adsorption binding on immunogenicity. If antigens are too strongly adsorbed there is a negative effect on antibody responses as the tight binding effects antigen processing and presentation by dendritic cells (Egan *et al.*, 2009; Hansen *et al.*, 2007).

The importance of adsorption is likely to be antigen specific and its requirement for alum adjuvanticity dependent on the nature of the antigen. For example the Hib conjugate vaccine consisting of capsular polysaccharide conjugated to a non-toxic mutant of diphtheria toxin does not require adsorption for aluminium phosphate to enhance anti-polysaccharide titres (Kanra *et al.*, 2003). Adsorption is likely to be important for new-generation vaccines comprising purified recombinant proteins and as a general rule is still considered important in formulation prior to immunization (Gupta *et al.*, 1995). Not least, regulatory requirements outline that antigen must be adsorbed to aluminium containing adjuvants (Romero Méndez *et al.*, 2007).

Beyond delivery, alum is able to activate the immune system

As well as delivery of antigen to antigen-presenting cells, alum has also been found to have immunostimulatory properties.

A number of studies have found that alum promotes the recruitment of immune cells, including antigen-presenting cells, to the site of injection (Goto and Akama, 1982; Goto *et al.*, 1997; Kool *et al.*, 2008b). Kool *et al.* (2008b) demonstrated the local production of chemoattractants like CCL2 and CXCL1 and the recruitment of neutrophils, eosinophils, monocytes and dendritic cells following intraperitoneal injection. Expression profiling of mouse muscle after alum injection identified gene expression patterns compatible with these observations such as the up regulation of chemokines (Mosca *et al.*, 2008). Another recent study found that alum induces the

secretion from human immune cells of chemokines involved in cell recruitment and acts mainly on macrophages and monocytes (Seubert *et al.*, 2008).

Evidence exists that alum also has a role in the maturation of antigen-presenting cells. *In vitro* this maturation has been characterized by the aluminium hydroxide dependent expression of co-stimulatory molecule CD86 in human PBMCs (Rimaniol *et al.*, 2004; Ulanova *et al.*, 2001; Wilcock *et al.*, 2004) and the aluminium hydroxide and aluminium phosphate-dependent expression of CD86 and CD80 in mouse dendritic cells (Sokolovska *et al.*, 2007). Kool *et al.* (2008b) demonstrated maturation of APCs *in vivo*, with the finding that monocytes recruited following intra-peritoneal injection with alum, trafficked antigen to the draining lymph node and differentiated into monocytic DC precursor cells. These recruited cells were responsible for priming naïve CD4+ T cells in the lymph node. Seubert *et al.* (2008) found that in monocytes alum induced phenotypic changes consistent with differentiation towards dendritic cells.

A key feature of the Th2 response is the production of IL-4, and Th2 cells can be characterized by the secretion of IL-4 (Hem and HogenEsch, 2007). Induction of IL-4 by alum has been found to play a key role in the induction of the Th2 response in large part through the inhibition of Th1-cell-associated responses (Brewer *et al.*, 1996, 1999). Another study found alum induced secretion of IL-4 by Gr 1+ myeloid cells increased B cell proliferation in the spleen (Jordan *et al.*, 2004). This Gr1+ cell population has since been identified as eosinophils, thus indicating that these cells have an accessory function in adjuvanticity (Wang and Weller, 2008).

Molecular target for alum immunostimulatory activity

The immunostimulatory properties described above are alum-dependent observations, and the molecular mechanism by which alum initiates these events is still unknown. However, a number of recent studies have identified a molecular target for alum immunostimulatory activity.

These studies have implicated alum in the activation of the NLR family, pyrin domain containing 3 (NLRP3) inflammasome (Eisenbarth *et al.*, 2008; Franchi and Nunez, 2008; Hornung *et al.*, 2008; Kool *et al.*, 2008a). NLRP3 is an intracellular protein that can sense various pathogens, pathogen products and crystals such as monosodium urate (Ogura *et al.*, 2006). Together with apoptosis-associated speck-like protein (ASC) and caspase-1, NLRP3 forms a protein complex called the inflammasome. NLRP3 and ASC have been found to be required for the activation of caspase 1 and subsequently its target cytokines IL-1β, IL-18 and IL-33 (Eisenbarth *et al.*, 2008; Franchi and Nunez, 2008; Hornung *et al.*, 2008; Kool *et al.*, 2008a) (Fig. 6.2).

It is thought that alum activates the NLRP3 inflammasome following engulfment by phagocytic cells. This engulfment leads to the rupture of the lysosome containing alum and antigen release into the cytosol. As well as releasing antigen, lysosomal enzymes are released such as cathepsin B, an enzyme which as been found to be partially dependent for IL-1β induction (Hornung *et al.*, 2008; Sharp *et al.*, 2009). An alternative pathway for activation has been suggested where aluminium salt cytotoxicity leads to the release of uric acid and formation of monosodium urate (MSU) crystals, which in turn may be phagocytosed and activate the NLRP3 inflammasome via the same mechanism (Marrack *et al.*, 2009).

Toll-like receptors (TLRs) are pathogen recognition receptors expressed on antigen-presenting cells, including macrophages and dendritic cells, involved in the activation of T cells. Immunopotentiation by alum is a TLR-independent mechanism as genetic inactivation of two key adaptors in the TLR signalling pathway, MyD88 and TRIF, did not affect adjuvant function (Gavin *et al.*, 2006; Schnare *et al.*, 2001). TLR agonists, like LPS, can activate the transcription of pro- IL-1β through the NF-κB pathway, but unlike alum, do not target the inflammasome and therefore do not induce efficient IL-1β secretion *in vitro*. Conversely alum cannot activate IL-1β transcription. This indicates the requirement of both alum and TLR agonists for the secretion of mature IL-1β (De Gregorio *et al.*, 2008). This may explain why the NLRP3 and caspase-1-dependent *in vitro* production of IL-1β in response to alum required macrophages to be pretreated with LPS

(Eisenbarth *et al.*, 2008; Franchi and Nunez, 2008; Li *et al.*, 2008).

Some controversy surrounds the link between the NLRP3 inflammasome and alum induced antibody production as some studies have found that NLRP3 is required for alum induced antibody production (Eisenbarth *et al.*, 2008; Li *et al.*, 2008) whilst others not (Franchi and Nunez, 2008; Marrack *et al.*, 2009). Separate studies looking at the role of IL-1β (Nakae *et al.*, 2003; Schmitz *et al.*, 2003) or IL-18 (Reddy, 2004) in the induction of adaptive immune responses in response to alum found that these cytokines were not required. It is possible that they act redundantly or that another cytokine or target of caspase 1 is involved. To conclude, at present it appears that NLRP3 is sometimes but not always required for alum induced antibody production and the connection between the NLRP3 inflammasome and the induction of a Th2 response is unknown. The contrasting results could be a result of different immunization schedules, mice backgrounds or alum adjuvant forms. The models used across these studies were not standardized and other variables such as vaccine formulation and antigen adsorption to alum could have affected the outcome.

The composition of aluminium-containing adjuvants

Aluminium hydroxide and aluminium phosphate gels are the most widely used aluminium based adjuvants and as in this review, often referred to simply as 'alum'. Strictly speaking these are not an example of an alum and even the names aluminium hydroxide and aluminium phosphate do not correctly describe the structures of the adjuvants.

Alum only applies to aluminium potassium sulphate as used by Glenny and co-workers in 1926 (Lindblad, 2004). When used in an alum-precipitated vaccine, the resulting precipitate is amorphous aluminium hydroxysulfate as some sulfate anions are substituted for hydroxyls (Shirodkar *et al.*, 1990).

Aluminium hydroxide is correctly termed crystalline aluminium oxyhydroxide, whilst aluminium phosphate is amorphous aluminium hydroxyphosphate (Hem and HogenEsch, 2007). Aluminium phosphate is aluminium hydroxide whereby some hydroxyls have been substituted for phosphate (Hem and HogenEsch, 2007). This forms a non-stoichiometric compound where the degree of phosphate substitution depends on excipients and method of preparation (Hem and HogenEsch, 2007).

Manufacturing of aluminium adjuvant

Aluminium hydroxide and aluminium phosphate are prepared by exposing soluble aluminium salts to alkaline conditions (Lindblad, 2004). The results of this process are strongly influenced by a number of factors including the presence of anions, temperature, concentration and rate of reagent addition (Lindblad, 2004). Aluminium phosphate, for example, is produced when soluble aluminium salts are exposed to alkaline conditions in the presence of phosphate ions (Lindblad, 2004). Preformed gels of aluminium hydroxide and aluminium phosphate adjuvant are commercially available.

Aluminium adjuvants are sterilized by autoclaving. This process has been shown to have a negative impact on adsorptive capacity of the adjuvant and measures to produce more uniform heating, e.g. rotation of containers, and the avoidance of repeated autoclaving, is recommended (Hem and HogenEsch, 2007).

Formulation of aluminium-containing adjuvants

As discussed above, adsorption to alum is thought to be important to maximize antigen exposure to the immune system and aiding uptake into APCs. Furthermore, we have found that adsorption can stabilize antigens and overcome stability issues observed for the soluble antigens alone (unpublished observations). Physical adsorption of the antigen onto the surface of alum is therefore still considered important for function of these adjuvants and considered as a general rule for correct formulation prior to immunization.

Optimizing adsorption

The most widespread approach for the preparation of alum adsorbed vaccines is the adsorption of antigens onto the surface of preformed aluminium hydroxide or aluminium phosphate gels (Bomford, 1989; Gupta et al., 1995). A number of attractive mechanisms contribute to this adsorption including electrostatic attraction, ligand exchange, hydrophobic interactions, hydrogen bonding and van der Waals forces (Peek et al., 2007). Extensive studies by Stanley Hem at Purdue have identified electrostatic attraction and ligand exchange as the predominant forces in adsorption, and common factors effecting adsorption are outlined in Table 6.2. Complete adsorption can be achieved through the optimal choice of pH, buffer, ionic strength of the medium and type of aluminium salt.

Electrostatic interactions occur when the antigen and the adjuvant have opposite charges (Hem and HogenEsch, 2007). Aluminium hydroxide adjuvant has an isoelectric point (IEP) of 11 and is positively charged at neutral pH, whilst aluminium phosphate has an IEP of 5 and is negatively charged. Therefore, as a general rule antigens with low IEP bind to aluminium hydroxide adjuvant at neutral pH and those with high IEPs bind to aluminium phosphate (Hem and HogenEsch, 2007). Best adsorption is often when pH of the formulation is near the IEP of the antigen (Matheis et al., 2001), as interactions between the antigen molecules are minimized. Increasing ionic strength of the vaccine formulation reduces the contributions of the surface charges and therefore reduces electrostatic attraction (Hem and HogenEsch, 2007). Furthermore, the IEP of aluminium hydroxide adjuvant decreases in the presence phosphate anions and other ions (Egan et al., 2009; Hem and HogenEsch, 2007) through substitution of hydroxyl groups. This surface modification of alum has consequences also for the pH microenvironment at the surface of the adjuvant particle (Hem and HogenEsch, 2007), which will differ from that of the bulk formulation. Recently (Jezek et al., 2009) demonstrated that the presence of phosphate in an alum based hepatitis B vaccine formulation increased the thermal stability of the HBsAg, by lowering the microenvironment pH to a value optimal for stability of the adsorbed antigen.

Ligand exchange is the strongest adsorption force and can drive adsorption even when electrostatic repulsion occurs between antigen and adjuvant (Morefield et al., 2005a). It occurs when hydroxyl groups on the surface of aluminium

Table 6.2 Common formulation parameters influencing antigen adsorption to alum

Formulation parameter	Influence on adsorption mechanism	
	Electrostatic attraction	Ligand exchange
pH	Positive or negative (depending on IEP of antigen)	Less influence
Phosphate anions (also sulphate and fluoride)	Positive or negative (will lower IEP of aluminium hydroxide) Negative (increase ionic strength)	Negative (occupies hydroxyl groups on adjuvant)
NaCl	Negative (increase ionic strength)	Positive (increase ionic strength)

hydroxide and aluminium phosphate are replaced by more strongly binding phosphate groups contained on the antigen (Hem and HogenEsch, 2007). Both aluminium hydroxide and aluminium phosphate possess free hydroxyls and can therefore undergo ligand exchange with phosphorylated antigens, however aluminium hydroxide possesses more surface hydroxyls. Similarly, phosphate treatment of aluminium adjuvants reduces the capacity and strength of adsorption via this mechanism by diminishing the number of ligand exchange sites (Iyer *et al.*, 2003). Increasing ionic strength of the formulation increases adsorption via ligand exchange due to shielding of antigen charge, allowing antigen of same charge to pack more closely together on the adjuvant surface (Hem, unpublished communication). Substitution by ligand exchange is not limited to phosphate ions as other highly charged ions such as fluoride, sulphate, carbonate or citrate can substitute the free hydroxyls on the surface of aluminium containing adjuvants (Matheis *et al.*, 2001). Chloride and nitrate ions are adsorbed only weakly so formulation components such as NaCl have limited impact on this mechanism (Matheis *et al.*, 2001).

Characterizing adsorption

In order to achieve consistency in the formulation of alum adjuvanted vaccines it is important to characterize these formulations for chemical environment (e.g. pH and osmolality) and adsorption. The degree of adsorption is most commonly measured by centrifuging the adsorbed vaccine and assaying the supernatant for unabsorbed antigens by BCA assay, SDS-PAGE or other assays for quantifying proteins. To establish the identity

of the adsorbed antigen, desorption from the adjuvant can be achieved by adding phosphate, salt or altering the pH to a point that does not support adsorption (Gupta and Rost, 2000). Dissolution of aluminium hydroxide and aluminium phosphate can also be performed to release adsorbed antigen by incubation at 37°C with 5–10% sodium citrate (World Health Organization, 1977). A novel method involving 48 h incubation with 4 M GnHCl and 0.85% H_3PO_4 (final pH of ~1) was found to have no deleterious effect on BSA and multiple group A *Streptococcus* vaccine protein controls, as judged by multiple physiochemical, structural and functional tests (Hutcheon *et al.*, 2006).

Efforts have been made to characterize the effect of adsorption on protein antigen structures. Dong *et al.* (2006) found that the secondary structure of six model antigens was unaffected by adsorption to aluminium hydroxide adjuvant. However, another study found changes in the tertiary structure and thermal stability of model antigens following the adsorption to aluminium adjuvants (Jones *et al.*, 2005).

Direct *in situ* quantification of antigen whilst still adsorbed to alum has been described for ELISA using aluminium-adsorbed antigens (Katz, 1987; Thiele *et al.*, 1990). More recently near-infrared spectroscopy was used to measure the concentration of bovine serum adsorbed onto aluminium hydroxide. This method is non-destructive and is able to directly measure the adsorbed protein concentration in suspension over the linear range of 0.1–1.75 mg/ml (Lai *et al.*, 2008). Another recent study developed a direct alum formulation immunoassay (DAFIA) which was able to quantify alum-bound malaria antigen AMA1-C1 over a linear detection range of 0.16 μg/ml to 10 μg/ml (Zhu *et al.*, 2009a). Furthermore, the same laboratory has also developed a generic method for direct quantification of protein in adsorbed alum formulations (Zhu *et al.*, 2009b).

Mixing and storage

Optimal adsorption is usually ensured by slow stirring of the formulation for a few hours to overnight (Lindblad, 1995). A detailed study of antigen adsorption to aluminium hydroxide adjuvant found that alpha-casein, bovine serum albumin (BSA), myoglobin and recombinant protective antigen (rPA) were all adsorbed with high efficiency (>99%) after 1 min mixing with a magnetic stir bar (Morefield *et al.*, 2004). Of these antigens, those that adsorbed by ligand exchange or strong electrostatic attraction required minimal mixing (1 min is sufficient) for uniform distribution on the adjuvant surface. However, myoglobin which adsorbs by weak electrostatic attraction required longer mixing for uniform distribution (60 min).

As touched on above, vaccines containing aluminium adjuvants should be stored at 2–8°C (Gupta and Rost, 2000). Special care should be taken not to freeze these as this causes irreversible coagulation of the adjuvant and loss of potency (Rowe, 2006; World Health Organization, 2006). The phenomenon of maturation, a process that takes up to several months, has been described for diphtheria and tetanus vaccines whereby adsorption is stabilized by time (Matheis *et al.*, 2001). The effect is attributed to increased contact between the antigens and the adjuvant, either through conformational or orientational changes, and is associated with an increase in vaccine potency (Brash and Horbett, 1995; Matheis *et al.*, 2001).

Clinical experience with alum

Today alum adjuvanted vaccines are used widely in licensed vaccines worldwide and until recently they were the only adjuvants present in licensed vaccines in North America.

Table 6.3 summaries the alum-based vaccines currently licensed in the US. Licensed parentally administered vaccines containing alum include;

diphtheria toxoid, tetanus toxoid and acellular pertussis based vaccine combinations, hepatitis B, pneumococcal conjugate vaccine, hepatitis A and human papillomavirus vaccine (Kenney and Edelman, 2003). There are only a few vaccine trials which have examined the effectiveness of alum by comparing populations administered with a given vaccine with or without the adjuvant, and those trials that have been performed were done over 50 years ago with soluble toxoids (Baylor *et al.*, 2002). Results varied significantly in these early studies but consensus indicated that alum adsorbed diphtheria and tetanus toxoids were significantly more effective in primary immunization than plain toxoids (Baylor *et al.*, 2002). Whereas for booster immunizations there was little difference between the alum adjuvanted and un-adjuvanted vaccines.

Upper limits exist for the amount of aluminium that can be administered as an adjuvant to humans. This limit is 1.25 mg of aluminium per dose in Europe (European Pharmacopoeia, 1997), and 1.25 mg in the USA if justifiable by safety and efficacy data (Code of Federal Regulations, 2003).

Combination of alum with immunopotentiators

A limitation of alum is the clear Th2 profile of these adjuvants, which is unlikely to enable protection against diseases for which Th1 immunity and MHC class I restricted CTL are essential for protection. Many studies have explored the possibility of combining alum with other immunopotentiators to create novel adjuvants that are safe and induce enhanced and more effective immune responses (Hem and HogenEsch, 2007)). For example combination of TLR9 agonist CpG oligodeoxynucleotides with alum induce higher antibody titres than either individual adjuvant component alone (De Gregorio *et al.*, 2009). Furthermore, this immunostimulatory effect is dependent on the association of CpG with alum (Aebig *et al.*, 2007; Mullen *et al.*, 2007). A phase I trial with CpG 7909 formulated with recombinant *Plasmodium falciparum* protein AMA1-C1 adsorbed to aluminium hydroxide adjuvant elicited increased antibody titres capable of increased *in vitro* growth inhibition levels (Mullen *et al.*,

Table 6.3 Aluminium content of available vaccines licensed for human use in the US[a]

Vaccine	Trade Name	Manufacturer	Al per dose (mg)	Chemical form of Al	No. of doses in series	Total Al for series (mg)	Comments
Childhood vaccines							
Diphtheria and tetanus toxoids adsorbed	No trade name	Sanofi Pasteur, Inc.	≤0.17	Alum	5	0.85	
Diphtheria and tetanus toxoids and acellular pertussis vaccine adsorbed	Tripedia	Sanofi Pasteur Inc.	≤0.17	Alum	5	0.85	
	Infanrix	GlaxoSmithKline Biologicals	≤0.625	Hydroxide	5	3.1	
	DAPTACEL	Sanofi Pasteur, Ltd	0.33	Phosphate	5	1.65	
Diphtheria and tetanus toxoids and acellular pertussis vaccine adsorbed, hepatitis B (recombinant) and inactivated poliovirus vaccine combined	Pediarix	GlaxoSmithKline Biologicals	≤0.85	Hydroxide and phosphate	3	2.55	
Diphtheria and tetanus toxoids and acellular pertussis adsorbed and inactivated poliovirus vaccine	KINRIX	GlaxoSmithKline Biologicals	≤0.6	Hydroxide	1	0.6	
Diphtheria and tetanus toxoids and acellular pertussis adsorbed, inactivated poliovirus and *Haemophilus* b conjugate (tetanus toxoid conjugate) vaccine	Pentacel	Sanofi Pasteur Ltd	0.33	Phosphate	4	1.32	
Haemophilus b conjugate vaccine (meningococcal protein conjugate) and hepatitis B vaccine (recombinant)	Comvax	Merck & Co., Inc.	0.225	AAHS	3	0.68	
Hepatitis B vaccine (recombinant)	Recombivax HB	Merck & Co., Inc.	0.25	AAHS	3	0.75	
	Engerix-B	GlaxoSmithKline Biologicals	0.25	Hydroxide	3	0.75	
Pneumococcal 7-valent conjugate vaccine (diphtheria CRM197 protein)	Prevnar	Wyeth Pharmaceuticals Inc.	0.125	Phosphate	4	0.5	Older infants and children get fewer doses

Adult

Vaccine	Trade name	Manufacturer		Adjuvant			Notes
Anthrax vaccine adsorbed	Biothrax	Emergent BioDefense Operations Lansing Inc.	0.6	Hydroxide	5	3	Additional booster every year
Hepatitis A vaccine, inactivated	Havrix	GlaxoSmithKline Biologicals	0.5	Hydroxide	2	1	Adults
	Havrix	GlaxoSmithKline Biologicals	0.25	Hydroxide	2	0.5	Children and adolescents
	VAQTA	Merck & Co, Inc	0.45	AAHS	2	0.9	Adults
	VAQTA	Merck & Co, Inc	0.225	AAHS	2	0.45	Children and adolescents
Hepatitis A inactivated and hepatitis B (recombinant) vaccine	Twinrix	GlaxoSmithKline Biologicals	0.45	Hydroxide and phosphate	3	1.35	Additional booster may be given
Hepatitis B vaccine (recombinant)	Recombivax HB	Merck & Co, Inc	0.5	AAHS	3	1.5	Only two doses for adolescents
	Engerix-B	GlaxoSmithKline Biologicals	0.5	Hydroxide	3	1.5	
Human papillomavirus (types 6, 11, 16, 18) recombinant vaccine	Gardasil	Merck and Co, Inc.	0.225	Hydroxide	3	0.68	Women 9–27 years of age
Human papillomavirus bivalent (types 16, 18) vaccine, recombinant	Cervarix	GlaxoSmithKline Biologicals	0.5	Hydroxide[b]	3	1.5	Females 10–25 years of age
Japanese encephalitis virus vaccine, inactivated, adsorbed	IXIARO	Intercell Biomedical	0.25	Hydroxide	2	0.5	
Tetanus and diphtheria toxoids adsorbed for adult use	No trade name	MassBiologics	≤0.45	Phosphate	3	1.35	Additional booster every 10 years
	DECAVAC	Sanofi Pasteur, Inc.	≤0.28	Alum	3	0.84	Additional booster every 10 years

Table 6.3 continued

Vaccine	Trade Name	Manufacturer	Al per dose (mg)	Chemical form of Al	No. of doses in series	Total Al for series (mg)	Comments
Tetanus toxoid adsorbed	No trade name	Sanofi Pasteur, Inc.	0.25	Alum	3	0.75	Additional booster every 10 years
Tetanus toxoid, reduced diphtheria toxoid and acellular pertussis vaccine, adsorbed	Adacel	Sanofi Pasteur, Ltd	0.33	Phosphate	1	0.33	To be used as a booster at ages 11 to 64
	Boostrix	GlaxoSmithKline Biologicals	≤0.39	Hydroxide	1	0.39	To be used as a booster at ages 10 to 64

aInformation taken from 'Complete List of Vaccines Licensed for Immunization and Distribution in the US' and supporting documents at www.fda.com.
bCervarix also contains the TLR4 agonist monophosphoryl lipid A (MPL).

2008). However, also reported in this trial was the fact that local and systemic adverse events were significantly more likely to be of higher severity with the addition of CPG 7909 (Mullen *et al.*, 2008).

Clearly most successful example in this category is AS04, which consists of a TLR4 agonist monophosphoryl lipid A (MPL) adsorbed to aluminium hydroxide. MPL is derived from the cell wall LPS of Gram-negative *Salmonella* Minnesota R595 and is detoxified by mild hydrolytic treatment and purification. Activation of TLR4 by MPL is more selective than LPS resulting in reduced toxicity (Mata-Haro *et al.*, 2007). Developed by GSK, AS04 has been licensed in a significant number of countries, as a component of a vaccine against human papillomavirus (Harper *et al.*, 2006) and in an improved vaccine against hepatitis B virus (Bol *et al.*, 2004).

Given the extensive experience we have with alum they are likely to continue to be an essential component of vaccine formulations. Future research and exploratory studies with the adjuvant is likely to increase our understanding and lead to optimal use and improved performance of alum.

The development of new generation adjuvants: lessons learned from MF59

Emulsions as adjuvants

Emulsions are defined as liquid dispersions of two immiscible phases, usually an oil and water, either of which may comprise the dispersed phase or the continuous phase to provide water in oil, or oil in water emulsions respectively. Emulsions are generally unstable and need to be stabilized by surfactants, which lower interfacial tension, and prevent coalescence of the dispersed droplets. Stable emulsions can be prepared through the use of surfactants, which orientate at the interface between the two phases and reduce inter facial tensions, since surfactants comprise both hydrophobic and hydrophilic components. Although charged surfactants are excellent stabilizers, nonionic surfactants are widely used in pharmaceutical emulsions due to their lower toxicity and their lower sensitivity to the destabilizing effects of

formulation additives. Surfactants can be defined by their ratio of hydrophilic to hydrophobic components (hydrophile to lipophile balance – HLB), which gives information on their relative affinity for water and oil phases. At the high end of the scale, surfactants are predominantly hydrophilic and can be used to stabilize oil in water (o/w) emulsions. In contrast, oil-soluble surfactants are at the lower end of the scale and are used mainly to stabilize water in oil (w/o) emulsions. Polysorbates (Tweens) are commonly used surfactants with HLB values in the 9–16 range, while sorbitan esters (Spans) have a HLB in the range of 2–9. Extensive pharmaceutical experience has shown that a mixture of surfactants offers maximum emulsion stability, probably due to the formation of more rigid films at the interface. The physicochemical characteristics of emulsions, including droplet size, viscosity etc, are controlled by a variety of factors, including the choice of surfactants, the ratio of continuous to dispersed phases and the method of preparation. For an emulsion to be used for administration as an injection, stability and viscosity are important parameters, as too is sterility of course. In general, stability is enhanced by having smaller sized droplets, while viscosity is decreased by having a lower volume of the dispersed phase. Sterility can be achieved more easily if droplet size is small enough to allow sterilization via filtration.

Emulsions have a long history of use as adjuvants in both human and animal vaccines. Over 70 years ago, Freund demonstrated the adjuvant effect of mineral (paraffin) oil combined with mycobacterial cells, and this adjuvant came to be known as Freund's complete adjuvant (FCA) (Freund *et al.*, 1937). The w/o emulsion, without bacterial cells (Freund's incomplete adjuvant – FIA) was subsequently used in veterinary vaccines (Hilleman, 1966) and even in humans. Although w/o emulsions containing mineral oils like FIA have been used as vaccine adjuvants in humans, including influenza vaccines (Jansen *et al.*, 2005), they are generally considered as too reactogenic for human use in prophylactic vaccines (Edelman, 1980). Nevertheless, long-term follow-up has established that there are no significant long-term adverse effects following FIA use in humans, although local reactogenicity was very common (Page, 1993). More recently, w/o emulsions with high oil content, based on mineral and non-mineral oils have been evaluated as vaccine adjuvants for malaria and HIV vaccines (Aucouturier *et al.*, 2002). Clinical trials have demonstrated that these newer generation w/o emulsions induce potent immune responses, but also induce a significant number of local reactions, which can occasionally be severe (Audran *et al.*, 2005). Owing to the reactogenicity of w/o emulsions, o/w approaches were evaluated as alternatives, and were initially promoted as delivery systems for immune potentiators (Allison and Byars, 1986).

The development of MF59 o/w adjuvant for flu vaccines

In the 1980s a number of groups worked on the development of novel adjuvant formulations with the potential to be more potent and effective adjuvants than insoluble aluminium salts, which were the only adjuvants included in licensed human vaccines at that time (Vogel and Pruett, 1995). Amongst these, Syntex developed an o/w emulsion adjuvant (Syntex adjuvant formulation – SAF) using the biodegradable oil, squalane, to deliver a synthetic immune potentiator, called N-acetyl-muramyl-L-threonyl-D-isoglutamine (threonyl-MDP) (Allison and Byars, 1986). The closely related immune potentiator, N-acetyl-L-alanyl-D-isoglutamine (MDP), had been originally identified in 1974 as the minimal structure isolated from the peptidoglycan of mycobacterial cell walls, which had adjuvant activity (Ellouz *et al.*, 1974). However, MDP was pyrogenic and induced uveitis in rabbits (Waters *et al.*, 1986), making it unacceptable as an adjuvant for human vaccines. Therefore, various synthetic derivatives of MDP were produced, in an effort to identify an adjuvant molecule with an acceptable safety profile; threonyl-MDP was one of these synthetic compounds. More recently, it has been shown that MDP actually activates immune cells through interaction with the nucleotide-binding domain (NOD), which acts as an intracellular recognition system for bacterial components (Fritz *et al.*, 2006). In addition to threonyl-MDP, SAF also contained a pluronic polymer surfactant (L121), which was included to help bind antigens to the surface of the emulsion droplets. Unfortunately, clinical evaluations of SAF as an adjuvant for a HIV vaccine showed it to have an unacceptable profile of reactogenicity (Kenney and Edelman, 2004). As an alternative to SAF, Chiron Vaccines used squalene, a similar biodegradable oil, to develop an o/w emulsion as a delivery system for an alternative synthetic MDP derivative, muramyl-tripeptide phosphatidylethanolamine (MTP-PE). MTP-PE was lipidated to allow it to be more easily incorporated into lipid like formulations and to reduce toxicity (Wintsch *et al.*, 1991). Unfortunately, clinical testing also showed that emulsions of MTP-PE displayed an unacceptable level of reactogenicity, which made them unsuitable for routine clinical use (Keitel *et al.*, 1993). Although the emulsion formulation of MTP-PE enhanced antibody responses against influenza vaccine in humans, the level of adverse effects observed made this adjuvant unsuitable for widespread clinical use (Keitel *et al.*, 1993). Nevertheless, additional clinical studies undertaken at the same time highlighted that the squalene based emulsion alone (MF59), without any added immune potentiator, was well tolerated and had comparable immunogenicity to the formulation containing the MTP-PE (Kahn *et al.*, 1994; Keefer *et al.*, 1996). These observations resulted in the further development of the MF59 o/w emulsion vehicle alone as an injectable adjuvant.

In preclinical studies with influenza vaccine, it was confirmed that the immune potentiator, MTP-PE, was not required for MF59 to be an effective adjuvant (Ott *et al.*, 1995c). A key early study highlighted the ability of MF59 adjuvant to enhance protective immunity to flu virus challenge (Cataldo and Van Nest, 1997). The use of MF59 adjuvant allowed a dose reduction of flu vaccine (50- to 200-fold lower doses) and improved protection against challenge for more than six months after vaccination (Cataldo and Van Nest, 1997). MF59 induced enhanced antibody titres in comparison to flu vaccine alone, even at very low antigen dose (Fig. 6.3). Moreover, the addition of MF59 to flu vaccine offered improved survival against challenge with influenza virus in mice (Fig. 6.4) and also reduced viral titres in the lungs of challenged mice (Fig. 6.5). The enhanced protection afforded by the

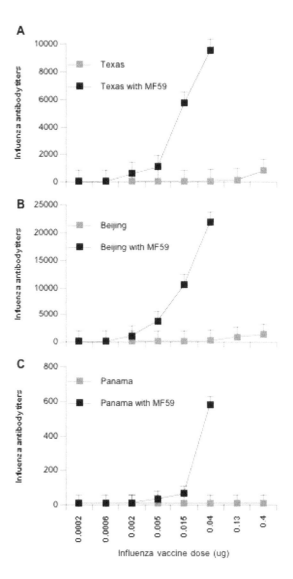

Figure 6.3 Serum antibody titres in mice at immunized with decreasing dose of influenza vaccine with and without MF59 adjuvant. Antibody titres to Texas (A), Beijing (B) or Panama (C) vaccine strains.

inclusion of MF59 in the vaccine was long lived and allowed a significant dose reduction in the amount of antigen needed to induce protection (Figs. 6.4 and 6.5). In follow-up studies, it was shown that MF59 was able to enhance the immune response to flu vaccines in both young and old animals (Higgins *et al.*, 1996). Old mice (18 months old in these studies) typically have poor responses to flu vaccines, as do elderly humans, but the inclusion of MF59 in the vaccine restored the response of the old mice back up to the level of response achieved in young mice (Fig.

6.6). Moreover, MF59 was also shown to induce a potent T-cell response to the flu vaccine, both in young and old mice (Fig. 6.7). Pushing the mouse model further, MF59 was also shown to be an effective adjuvant in old mice, which had previously been infected with influenza (Fig. 6.8), a situation more similar to that found in humans, who are often re-infected annually with circulating flu strains (Higgins *et al.*, 1996). These pre-clinical studies highlighted the huge potential of MF59 to be used as an adjuvant for an improved flu vaccine, potentially allowing antigen dose

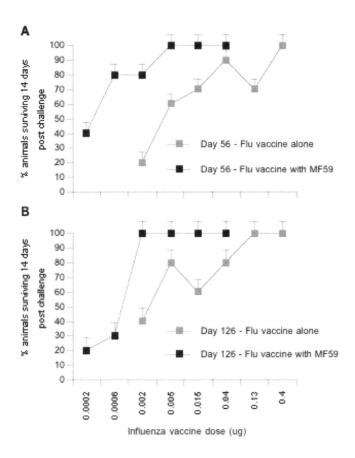

Figure 6.4 Survival after influenza virus challenge in mice immunized with decreasing doses of influenza vaccine with and without MF59. Mice challenged at 56 days (A) and 126 days (B) post first immunization.

reduction, while enhancing protective antibody and T-cell responses, for extended time periods. The ability of MF59 adjuvant to offer a significant reduction in the protective dose for flu vaccines subsequently became very important in the pandemic flu vaccine setting.

The small droplet size of MF59 adjuvant emulsion, generated through the use of a micro-fluidizer in the preparation process, was crucial to the potency of the adjuvant, but also enhanced emulsion stability and allowed the formulation to be sterile filtered for clinical use.

The mechanism of action of MF59 adjuvant

Early studies designed to determine the mechanism of action of MF59 focused on the possibility of the creation of a 'depot' effect for

co-administered antigen, since there had been suggestions that emulsions may retain antigen at the injection site. However, early work showed that an antigen depot was not established at the injection site and that the emulsion was cleared rapidly (Ott *et al.*, 1995b). The lack of an antigen depot with MF59 was confirmed in later studies (Dupuis *et al.*, 2001), which also established that MF59 and antigen were cleared independently. Subsequently, it was thought that perhaps the emulsion acted as a 'delivery system' and was responsible for promoting the uptake of antigen into antigen-presenting cells (APC). This theory was linked to earlier observations with SAF, which contained a pluronic surfactant that was thought to be capable of binding antigen to the emulsion droplets to promote antigen uptake (Allison and Byars, 1986). However, studies with

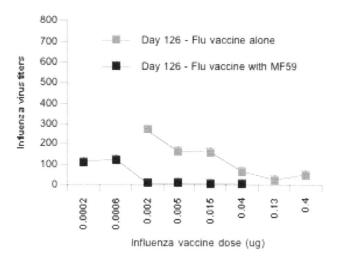

Figure 6.5 Influenza virus titres post challenge in the lungs of mice immunized with increasing doses of influenza vaccine with and without MF59 (virus titre in naïve challenged mice was 346±104).

recombinant antigens showed that MF59 was an effective adjuvant, despite no evidence of binding of the antigens to the oil droplets. Moreover, an adjuvant effect was still observed if MF59 was injected up to 24 hours before the antigen and up to 1 hour after, confirming that direct association was not required for an adjuvant effect (Ott *et al.*, 1995a). Nevertheless, administration of MF59 24 hours after the antigen resulted in a much reduced adjuvant effect, suggesting that the emulsion was activating immune cells, which were then able to better process and present the co-administered antigen. A direct effect of MF59 on cytokine levels *in vivo* was also observed in separate studies, suggesting that the delivery method alone was too simplistic an explanation (Valensi *et al.*, 1994). In order to better understand the mechanism of action of MF59, we have recently studied the early steps of the immune response on human cells *in vitro* and in mouse muscle *in vivo*. We have shown that there are at least two human target cells for MF59, monocytes and granulocytes, and that MF59 has a range of effects, including; increased antigen uptake, the release of chemoattractants and cell differentiation (Seubert, 2008). The observation of increased antigen uptake is in line with previous findings in mice (Dupuis *et al.*, 1998). The most readily induced chemoattractant was the chemokine, CCL2, which is involved

in cell recruitment. Previous work had shown a reduction of MF59-induced cell recruitment into the muscle in CCR2 deficient mice (Dupuis *et al.*, 2001), which is consistent with our observations on human cells. Moreover, ongoing experiments on gene expression profiles at the injection site are also consistent with the key role of chemokines (Mosca *et al.*, 2008). In addition, CCL2 was also found in serum after injection of MF59 into mouse muscle, providing further consistency between *in vitro* and *in vivo* observations (Seubert *et al.*, 2008). MF59 also induces phenotypic changes on human monocytes that are consistent with a maturation process towards immature dendritic cells (Seubert *et al.*, 2008). So far there appears to be an impressive consistency between data obtained *in vitro* with human cells, and the *in vivo* data from mouse. These observations suggest that MF59 induces a local pro-inflammatory environment within the muscle, which promotes the induction of potent immune responses to co-administered vaccines.

Hence, we conclude that during vaccination, adjuvants like MF59 augment the immune response at a range of intervention points: Through induction of chemokines, they increase recruitment of immune cells to the injection site, they augment Ag uptake by monocytes at the injection site, and they enhance differentiation

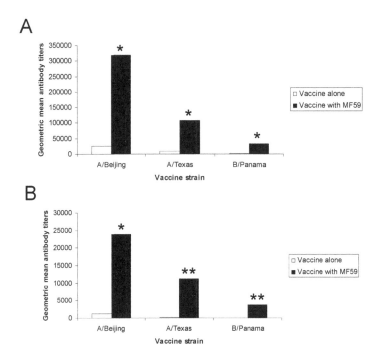

Figure 6.6 Geometric mean antibody titres in young (A; 8 weeks) or old (B; 18 months) mice after immunizations with influenza vaccine with and without MF59 in naive mice. *P-value, ANOVA (Fisher's PSLD) <0.0001 vs. non-adjuvanted vaccine. **P-value, ANOVA (Fisher's PSLD) <0.01 vs. non-adjuvanted vaccine.

of monocytes into DCs which represent the gold-standard cell type for priming naïve T cells. A particularly important feature of MF59 is that it strongly induces the homing receptor CCR7 on maturing DCs, thus facilitating their migration into draining lymph nodes where they can trigger off the adaptive immune response specific to the vaccine antigen. Nevertheless, further studies are necessary to better define the precise mechanism of action of MF59 and these studies are ongoing.

The composition of MF59

MF59 is a low oil content o/w emulsion. The oil used for MF59 is squalene, which is a naturally occurring substance found in plants and in the livers of a range of species, including humans. Squalene is an intermediate in the human steroid hormone biosynthetic pathway and is a direct synthetic precursor to cholesterol. Therefore, squalene is biodegradable and biocompatible, since it is naturally occurring. Eighty per cent of shark liver oil is squalene and shark liver provides the natural source of the squalene, which is used

to prepare MF59. MF59 also contains two non-ionic surfactants Tween 80 and Span 85, which are designed to optimally stabilize the emulsion droplets. Fig. 6.9 shows the composition of MF59 adjuvant. Citrate buffer is also used in MF59 to stabilize pH. Although single vial formulations can be developed with vaccine antigens dispersed directly in MF59, MF59 can also be added to antigens immediately prior to their administration. Although a less favourable option, combination prior to administration may be necessary to ensure optimal antigen stability for some antigens, but not for flu.

Manufacturing of MF59

Details of the manufacturing process for MF59 at the 50 litre scale have previously been described (Ott, 2000). The process involves dispersing Span 85 in the squalene phase and Tween 80 in the aqueous phase, before high speed mixing to form a coarse emulsion. The coarse emulsion is then passed repeatedly through a microfluidizer to produce an emulsion of uniform small droplet size

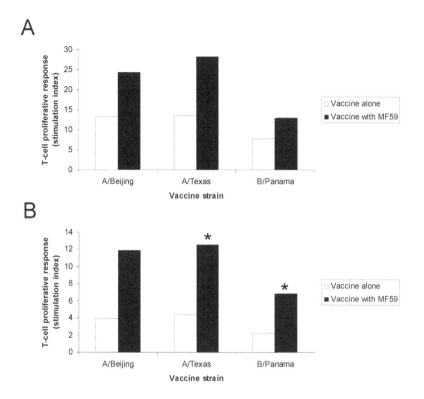

Figure 6.7 T-cell proliferative responses from splenocytes of mice immunized with influenza vaccine with and without MF59. Values are shown as geometric mean stimulation indices in young (A; 8 weeks) or old (B; 18 months) mice. * P value, ANOVA (Fisher's PSLD) <0.05 vs. non-adjuvanted vaccine.

(165 nm), which can be sterile filtered and filled into vials. Methods have also been published previously to allow the preparation of MF59 at small scale, for use in research studies (Traquina *et al.*, 1996). MF59 is extensively characterized by various physicochemical criteria after preparation.

Preclinical experience with MF59

Preclinical experience with MF59 is extensive and has been reviewed on several occasions previously (O'Hagan and Podda, 2008; Ohagan, 2007; Ott, 2000; Podda and Del Giudice, 2003; Podda *et al.*, 2005). MF59 has been shown to be a potent adjuvant in a diverse range of species, in combination with a broad range of vaccine antigens, to include recombinant protein antigens, isolated viral membrane antigens, bacterial toxoids, protein polysaccharide conjugates, peptides and virus like particles. MF59 is particularly effective for inducing high levels of antibodies, including functional titres (neutralizing, bactericidal and opsonophagocytic titres) and is generally more potent than alum.

In a recent study, we directly compared MF59 and alum for several different vaccines and confirmed that MF59 was more potent, although alum performed well for bacterial toxoid antigens, particularly diphtheria toxoid (Singh *et al.*, 2006). MF59 has also shown enhanced potency over alum when directly compared in non-human primates with protein polysaccharide conjugate vaccines (Granoff *et al.*, 1997) and with a recombinant viral antigen (Traquina *et al.*, 1996).

In pre-clinical studies, MF59 is the most potent adjuvant for flu vaccines, in comparison to various alternatives (Fig. 6.10). In a recent study, we compared a number of adjuvants for flu vaccine in mice, and showed that MF59 significantly outperforms alternatives, including alum, for both antibody and T cell responses (Wack *et al.*, 2008). Moreover, we have recently shown that MF59 offers enhanced protection against challenge with

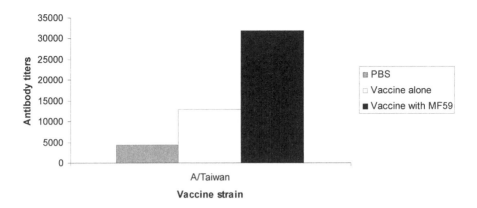

Figure 6.8 Geometric mean antibody titres to A/Taiwan virus following immunization with influenza vaccine with and without MF59 in mice pre infected with antibody titres in old mice (18 months). *P*-value, ANOVA (Fisher's PSLD) <0.05 for difference between HA alone and MF59+HA.

- **Appearance:** milky white oil in water (o/w emulsion)

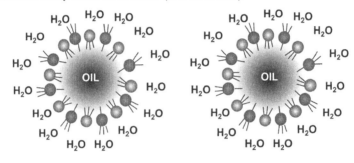

- **Composition:** 0.5% Polysorbate 80 water-soluble surfactant, 0.5% Sorbitan Triolate oil-soluble surfactant, 4.3% Squalene oil, Water for injection, 10 nM Na-citrate buffer.

- **Density:** 0.9963 g/ml

- **Size:** 160nm

- **Viscosity:** close to water, easy to inject

Figure 6.9 The composition of MF59 emulsion adjuvant.

pandemic flu strains in mice (unpublished observations, Kanta Subbarao), which is consistent with our earlier work on inter pandemic strains (Cataldo and Van Nest, 1997). A recent study in the ferret model found that MF59 enhanced cross clade antibody responses and protection after challenge with H5N1 virus (Forrest *et al.*, 2009). In addition to immunogenicity studies, extensive

pre-clinical toxicology studies have been undertaken with MF59, in combination with a range of different antigens in a number of species. In these studies, it has been shown that MF59 is neither mutagenic, nor teratogenic, and did not induce sensitization in an established guinea pig model to assess contact hypersensitivity.

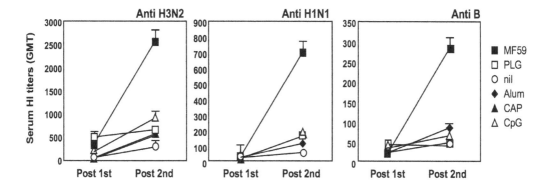

Figure 6.10 Serum haemagglutination titres against the three stains of influenza virus included in vaccines (H3N2, H1N1 and B) for flu cell culture vaccine in combination with various adjuvants. The adjuvants evaluated included MF59, Aluminium salt (alum), calcium phosphate (CAP), poly-lactide co-glycolide microparticles (PLG) CpG oligonucleotide (CpG) and vaccine alone without adjuvant (nil). MF59 was the most potent adjuvant for all three strains.

Clinical experience with MF59 adjuvant

Clinical testing of MF59 adjuvant has resulted in the registration in more than 20 countries of an effective and well-tolerated influenza vaccine for use in the elderly population. In addition clinical trials have demonstrated that MF59 can be safely administered with a range of antigens, to diverse age groups, including the paediatric population.

Fluad® in elderly subjects

Fluad®, which was initially licensed as an inter-pandemic influenza vaccine in Italy in 1997 and is now licensed in more than 20 countries, contains MF59, which was the first novel adjuvant accepted for human use after the registration of insoluble alum salts in the first part of the twentieth century. The registration of Fluad® was based on the results of a large clinical development plan which showed, in more the 20,000 subjects, that the MF59 adjuvanted vaccine was well tolerated with only a low incidence of local mild reactions and more immunogenic than conventional non-adjuvanted influenza vaccines (Podda, 2001). Most of the early clinical trials with Fluad® were performed in elderly subjects, in which conventional influenza vaccines are cost effective, but do not provide optimal protection (Strassburg *et al.*, 1986). In this population, the adjuvanted vaccine consistently induced enhanced geometric mean titres, seroconversion rates and seroprotection rates. An increased response against heterovariant influenza strains, different from the strains included in the vaccine, was also consistently observed, which is particularly beneficial whenever the vaccine antigens do not match perfectly those of the circulating viruses (Camilloni *et al.*, 2009; De Donato *et al.*, 1999; Del Giudice *et al.*, 2006b; Minutello *et al.*, 1999; Podda, 2001). Most encouragingly, clinical effectiveness data indicate that the increased immunogenicity of MF59 also translates into improved protection. Puig-Barbera *et al.* (2004) described the improved effectiveness of MF59 in preventing emergency admissions for pneumonia, while Iob *et al.* (2005) reported superior clinical protection against influenza like illness. The data showing improved clinical effectiveness of MF59 adjuvanted influenza vaccine was recently reviewed (Puig Barbera and Gonzalez Vidal, 2007), and showed significantly reduced influenza-like disease, and reduced hospitalization rates for pneumonia, cardiovascular and cerebrovascular disease in the elderly.

With more than 50 million doses distributed, the safety of Fluad® is supported by extensive pharmacovigilance data which show that vaccination with Fluad® is associated with a very low frequency of adverse reactions and the safety profile is broadly similar to that for non-adjuvanted flu vaccines (D'Agosto *et al.*, 2006; Schultze *et al.*, 2008). A large-scale analysis of safety data from 64 clinical trials involving MF59 adjuvanted seasonal and pandemic influenza vaccines supports the good safety profile of MF59 and suggests a clinical benefit over non-adjuvanted influenza vaccines (Pellegrini *et al.*, 2009). Moreover, it has been shown that the rate of Guillain–Barré syndrome after Fluad® immunization is within the normal range of rates found in the US after immunization with conventional non-adjuvanted influenza vaccines (Schultze *et al.*, 2008). In addition, a recent study highlighted that immunization with MF59 adjuvant neither raises the levels of pre-existing antibodies, nor induces new antibody responses against squalene, although antibodies to squalene are already naturally occurring in many subjects (Del Giudice *et al.*, 2006a).

Fluad® in young children

More recently, the potency of an MF59 adjuvanted subunit influenza vaccine in young children (6–36 months of age) was evaluated and directly compared with a licensed split influenza vaccine product. The MF59 adjuvanted vaccine (Fluad®) was significantly more immunogenic than the comparator for all three strains included in the vaccine. Moreover, the MF59 adjuvanted vaccine met all three EMEA CHMP criteria to gain vaccine approval for healthy adults (seroconversion, seroprotection and Mean geometric increase in response), while the comparator vaccine did not (unpublished data). In addition, the Fluad® vaccine offered significantly enhanced protection against heterovariant strains not included in the vaccine, in this vulnerable population and antibody titres persisted at higher levels than the non-adjuvanted comparator for at least 6 months post vaccination (Vesikari *et al.*, 2009). Again, Fluad® was well tolerated in this young population and had a similar reactogenicity profile to the split vaccine product (Vesikari *et al.*, 2009).

The data on the safety of MF59 in toddlers, infants and newborns are very encouraging and suggest that the adjuvant will have an acceptable safety profile in young children with similar reactogenicity levels observed for the split vaccine product (McFarlet al., 2001; Mitchell *et al.*, 2002). Nevertheless, despite these encouraging data, the safety of MF59 adjuvant needs to be carefully evaluated in this young population, prior to consideration for product approval.

Pandemic flu vaccines containing MF59 (Focetria® and Aflunov®)

The response to potential pandemic flu vaccines is much improved by the addition of MF59 adjuvant (Atmar *et al.*, 2006; Galli *et al.*, 2009b; Nicholson *et al.*, 2001; Stephenson *et al.*, 2003) (Fig. 6.11). Studies have established that, since humans are naïve to potential pandemic influenza strains, they respond poorly to conventional vaccines and large doses are needed in these vaccines to induce acceptable levels of antibodies (Treanor *et al.*, 2006). This is particularly important to increase the pandemic vaccine production and allows for significant dose sparing. Antigen sparing has been demonstrated by MF59-adjuvanted H5N1 vaccine Aflunov®, with two 7.5-µg doses inducing higher microneutralization assay titres than two 15-µg doses of the non-adjuvanted vaccine (Galli *et al.*, 2009b). Furthermore, MF59-adjuvanted H5N1 vaccine elicited significantly greater immune response than higher doses of alum-adjuvanted vaccine (Bernstein *et al.*, 2008). Similar antigen sparing of MF59 over a non-adjuvanted comparator has been demonstrated for an MF59-adjuvanted H9N2 vaccine (Atmar *et al.*, 2006). The dose-sparing aspect of MF59 is particularly attractive for pandemic flu vaccines, given the current limited capacity worldwide for influenza vaccine production, which is not sufficient to deal with a pandemic, given the high dose requirements for unadjuvanted influenza vaccines (Daems *et al.*, 2005). As already shown for the interpandemic vaccine, broader cross-neutralization against heterovariant pandemic strains was an additional benefit of an MF59 adjuvanted vaccine (Stephenson *et al.*, 2005, 2008). Some of these data were used to gain approval for a pandemic influenza vaccine containing the MF59 adjuvant (Focetria®) in May 2007 in all 27 member states of the European Union and to submit for approval as a pre-pandemic vaccine (Aflunov®). The use of a pre-pandemic flu vaccine could significantly reduce the potential catastrophic public health consequences of a pandemic, since large subsets of the population might be already protected or at

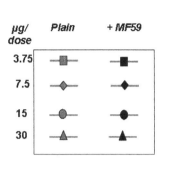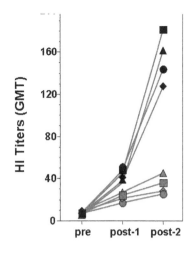

Figure 6.11 Serum haemagglutination inhibition titres against H9N2 influenza at varying vaccine doses (3.75–30 μg) with and without MF59 adjuvant after one and two vaccine doses (post-1 and post-2) in healthy adults (18–34 years of age). All dose levels of vaccine were significantly more potent with MF59 than without (plain). Moreover, one dose of vaccine with MF59 adjuvant was comparable to two doses of the plain vaccine.

least primed against a pandemic virus at the initiation of the pandemic threat. Pre-pandemic vaccination strategies to induce immune responses to novel influenza antigens are supported by clinical data with adults primed with an MF59-adjuvanted H5N3 vaccine (Galli *et al.*, 2009a). Boosting these individuals years later with a mismatched H1N5 MF59-adjuvanted vaccine generated significantly higher titres of cross-reactive neutralizing antibodies than the non-adjuvanted vaccine (Galli *et al.*, 2009a). Approval of Focetria® was on the basis that it would be manufactured to contain an influenza strain declared by the World Health Organization (WHO) as a pandemic strain. On 12 June 2009, the WHO declared a pandemic, with the pandemic strain type A/H1N1 influenza, first isolated in April 2009 from patients in Mexico. Studies found that a single dose of pandemic vaccine adjuvanted with MF59 was able to elicit functional antibody titres equivalent to those associated with protection of humans from seasonal influenza (Dormitzer *et al.*, 2009). Without MF59, two doses of the vaccine were required for a robust antibody response (Dormitzer *et al.*, 2009). On 25 September 2009, pandemic vaccine Focetria® containing MF59 received approval for persons of 6 months of age and older in the European Union. This was followed in November

2009 by approval in Germany and Switzerland for Celtura®, an MF59-adjuvanted cell culture-based influenza A (H1N1) pandemic vaccine.

Beyond influenza: clinical experience with MF59
Several trials have been performed also with other MF59-adjuvanted investigational vaccines, including HIV, HSV, CMV, HBV and HCV. These studies have provided additional evidence of the safety, tolerability and adjuvanticity of MF59 (Corey *et al.*, 1999; Heineman *et al.*, 1999; Langenberg *et al.*, 1995). Seronegative toddlers immunized with the CMV gB vaccine showed antibody titres that were higher than those found in adults naturally infected with CMV. Phase 2 evaluation in women within 1 year after childbirth found that gB vaccine adjuvanted with MF59 had the potential to decrease incident cases of maternal and congenital CMV infection with a reported vaccine efficacy of 50% (Pass *et al.*, 2009). Additionally, an MF59 adjuvanted HIV vaccine was evaluated in newborns, born to HIV positive mothers (Borkowsky *et al.*, 2000; Cunningham *et al.*, 2001; McFarlet al., 2001). The vaccine was very well tolerated and, despite the presence of maternal antibodies, induced an antibody response in 87% of the immunized infants

(Cunningham *et al.*, 2001; McFarlet al., 2001). Moreover, the MF59 vaccine was significantly more potent than alum for the induction of cell mediated immune responses (proliferative T cell responses) against homologous and heterologous strains of HIV (Borkowsky *et al.*, 2000).

Combination of MF59 with immunopotentiators

MF59 is particularly effective for enhancing antibody and T cell proliferative responses (Ott, 2000; Ott *et al.*, 1995a). However, it is not a potent adjuvant for the induction of Th1 cellular immune responses in pre clinical models involving naïve mice (Ott, 2000; Ott *et al.*, 1995a). Potent Th1 responses may be required to provide protective immunity against some viruses and additional intracellular pathogens. Th1 immune potentiators, including CpG oligonucleotides (Klinman, 2004) have been successfully added to MF59 to improve its potency and to alter the kind of response induced (O'Hagan *et al.*, 2002). Although the formulation of MF59 can be modified to promote the association of CpG with the oil droplets (O'Hagan *et al.*, 2002), more recent studies suggest that this may not be necessary, and simple addition of CpG to MF59 may be sufficient in some situations (Wack *et al.*, 2008). More recently co-delivery of E6020 with MF59 was found to shift antibody titres towards a Th1 based immune response (Baudner *et al.*, 2009). Furthermore, evaluation of this formulation with meningococcus vaccine antigens showed enhanced serum IgG and bactericidal titres for both Men B antigens and for Men B combined with Men ACYW conjugates (Singh *et al.*, unpublished). Delivery of E6020 within MF59 emulsion oil droplets led to a more potent response than simply adding the TLR agonist to MF59 (Singh *et al.*, unpublished). Careful consideration is required in relation to which immune potentiators to add to MF59 emulsion and how best to formulate them. Our early experience in the clinic showed that MTP-PE added to MF59 gave an unacceptable level of reactogenicity (Keefer *et al.*, 1996; Keitel *et al.*, 1993). Unfortunately, the animal models available at the time were not able to predict the poor tolerability of MTP-PE in humans.

In addition to immune potentiators, alternative delivery systems, including microparticles can also be added to MF59 to enhance its potency (O'Hagan *et al.*, 2000). However, the level of enhancement achieved would need to be highly significant and probably enabling for vaccine efficacy, to justify development of such a complex formulation.

The use of MF59 in prime/ boost settings

As an alternative to the inclusion of immune potentiators in MF59 to promote a Th1 response, MF59 can also be used as a booster vaccine with proteins once a Th1 response has already been established by immunization with DNA (Cherpelis *et al.*, 2001). Recently, this strategy has been shown to be highly promising for the development of a vaccine against HIV, since all arms of the immune response, including CTL responses, T helper responses and neutralizing antibodies were induced by this combination approach (Otten *et al.*, 2003, 2004, 2005, 2006). A similar approach of DNA prime and protein boost with MF59 has also shown significant promise in non-human primates as a vaccine strategy against HCV (O'Hagan *et al.*, 2004). Alternatively, MF59 and protein antigens can also be used to boost Th1 responses primed by immunization with attenuated viral vectors. The concept of an attenuated viral vector prime followed by MF59 boost has been established in the clinic using canarypox vectors, as a strategy for both HIV (AIDS, 2001) and CMV (Bernstein *et al.*, 2002).

Future perspectives on the use of MF59

MF59 is the first commercially available oil-in-water vaccine adjuvant and has a heritage of over 12 years' clinical experience and over 45 million commercial doses distributed in Fluad®. In addition to improving immune responses, MF59 is well tolerated, with a safety profile comparable to non-adjuvanted vaccines (Pellegrini *et al.*, 2009). Pre-clinical data have firmly established that MF59 is a more potent adjuvant than alum for a wide range of vaccines, including recombinant proteins and protein polysaccharide conjugates (Singh *et al.*, 2006). The potential for MF59 to be used in recombinant vaccines has been high-lighted in a study on a new generation serogroup B meningococcus vaccine candidate (Giuliani *et al.*, 2006). Previously, MF59 had been shown to enhance immune responses to protein polysac-charide conjugate vaccines, including one against *Neisseria meningitidis* serogroup C (Granoff *et al.*, 1997). Hence, MF59 has broad potential to be used as a vaccine adjuvant for a range of vaccines, in diverse age groups, including infants. Although MF59 has been mainly used in elderly adults to date, recent data have highlighted the significant potential of MF59 to be used in an improved influenza vaccine for young children. Following recent approval of Focetria® for use in persons of 6 months of age and older we confidently expect that the adjuvant will gain approval for wider use in the forthcoming years.

References

Aebig, J.A., Mullen, G.E.D., Dobrescu, G., Rausch, K., Lambert, L., Ajose-Popoola, O., Long, C.A., Saul, A., and Miles, A.P. (2007). Formulation of vaccines containing CpG oligonucleotides and alum. J. Immunol. Methods *323*, 139–146.

A.I.D.S. (2001). Vaccine Evaluation Group 022 Protocol Team: Cellular and humoral immune responses to a canarypox vaccine containing human immunodeficiency virus type 1 Env, Gag, and Pro in combination with rgp120. J. Infect. Dis. *183*, 563–570.

Allison, A.C., and Byars, N.E. (1986). An adjuvant formulation that selectively elicits the formation of antibodies of protective isotypes and of cell-mediated immunity. J. Immunol. Methods *95*, 157–168.

Atmar, R.L., Keitel, W.A., Patel, S.M., Katz, J.M., She, D., El Sahly, H., Pompey, J., Cate, T.R., and Couch, R.B. (2006). Safety and immunogenicity of nonadjuvanted and MF59-adjuvanted influenza A/H9N2 vaccine preparations. Clin. Infect. Dis. *43*, 1135–1142.

Aucouturier, J., Dupuis, L., Deville, S., Ascarateil, S., and Ganne, V. (2002). Montanide ISA 720 and 51: a new generation of water in oil emulsions as adjuvants for human vaccines. Expert Rev. Vaccines *1*, 111–118.

Audran, R., Cachat, M., Lurati, F., Soe, S., Leroy, O., Corradin, G., Druilhe, P., and Spertini, F. (2005). Phase I malaria vaccine trial with a long synthetic peptide derived from the merozoite surface protein 3 antigen. Infect. Immun. *73*, 8017–8026.

Baudner, B.C., Ronconi, V., Casini, D., Tortoli, M., Kazzaz, J., Singh, M., Hawkins, L.D., Wack, A., and O'Hagan, D.T. (2009). MF59 emulsion is an effective delivery system for a synthetic TLR4 agonist (E6020). Pharm. Res. *26*, 1477–1485.

Baylor, N.W., Egan, W., and Richman, P. (2002). Aluminum salts in vaccines – US perspective. Vaccine *20*, S18–23.

Bernstein, D.I., Edwards, K.M., Dekker, C.L., Belshe, R., Talbot, H.K.B., Graham, I.L., Noah, D.L., He, F., and Hill, H. (2008). Effects of adjuvants on the safety and immunogenicity of an avian influenza H5N1 vaccine in adults. J. Infect. Dis. *197*, 667–675.

Bernstein, D.I., Schleiss, M.R., Berencsi, K., Gonczol, E., Dickey, M., Khoury, P., Cadoz, M., Meric, C., Zahradnik, J., Duliege, A.M., *et al.* (2002). Effect of previous or simultaneous immunization with canary-pox expressing cytomegalovirus (CMV) glycoprotein B (gB) on response to subunit gB vaccine plus MF59 in healthy CMV-seronegative adults. J. Infect. Dis. *185*, 686–690. Epub 6 Feb 2002.

Boland, G., Beran, J., Lievens, M., Sasadeusz, J., Dentico, P., Nothdurft, H., Zuckerman, J.N., Genton, B., Steffen, R., and Loutan, L. (2004). Safety and immunogenicity profile of an experimental hepatitis B vaccine adjuvanted with AS04. Vaccine *23*, 316–320.

Bomford, R. (1989). Aluminium salts: perspectives in their use as adjuvants. NATO ASI series Series A: life sciences *179*, 35–41.

Borkowsky, W., Wara, D., Fenton, T., McNamara, J., Kang, M., Mofenson, L., McFarland, E., Cunningham, C., Duliege, A.M., Francis, D., *et al.* (2000). Lymphoproliferative responses to recombinant HIV-1 envelope antigens in neonates and infants receiving gp120 vaccines. AIDS Clinical Trial Group 230 Collaborators. J. Infect. Dis. *181*, 890–896.

Brash, J.L., and Horbett, T.A. (1995). Proteins at interfaces: an overview. In Proceedings of the ACS Symposium Series on Proteins at Interfaces II, Horbett, T.A, Brash, J.L., eds, vol. 602, pp. 1–23.

Brewer, J.M., Conacher, M., Hunter, C.A., Mohrs, M., Brombacher, F., and Alexander, J. (1999). Aluminium hydroxide adjuvant initiates strong antigen-specific Th2 responses in the absence of IL-4- or IL-13-mediated signaling. J. Immunol. *163*, 6448–6454.

Brewer, J.M., Conacher, M., Satoskar, A., Bluethmann, H., and Alexander, J. (1996). In interleukin-4-deficient mice, alum not only generates T helper 1 responses equivalent to Freund's complete adjuvant, but continues to induce T helper 2 cytokine production. J. Immunol. *26*, 2062–2066.

Bryan, J.T. (2007). Developing an HPV vaccine to prevent cervical cancer and genital warts. Vaccine *25*, 3001–3006.

Camilloni, B., Neri, M., Lepri, E., and Iorio, A.M. (2009). Cross-reactive antibodies in middle-aged and elderly volunteers after MF59-adjuvanted subunit trivalent influenza vaccine against B viruses of the B/Victoria or B/Yamagata lineages. Vaccine *27*, 4099–4103.

Cataldo, D.M., and Van Nest, G. (1997). The adjuvant MF59 increases the immunogenicity and protective efficacy of subunit influenza vaccine in mice. Vaccine *15*, 1710–1715.

Chang, M., Shi, Y., Nail, S.L., HogenEsch, H., Adams, S.B., White, J.L., and Hem, S.L. (2001). Degree of antigen adsorption in the vaccine or interstitial fluid and its effect on the antibody response in rabbits. Vaccine *19*, 2884–2889.

Cherpelis, S., Srivastava, I., Gettie, A., Jin, X., Ho, D.D., Barnett, S.W., and Stamatatos, L. (2001). DNA vaccination with the human immunodeficiency virus type 1 SF162DeltaV2 envelope elicits immune responses that offer partial protection from simian/human immunodeficiency virus infection to CD8(+) T-cell-depleted rhesus macaques. J. Virol. 75, 1547–1550.

Code of Federal Regulations (2003). Code 21.

Corey, L., Langenberg, A.G., Ashley, R., Sekulovich, R.E., Izu, A.E., Douglas, J.M., Jr., Handsfield, H.H., Warren, T., Marr, L., Tyring, S., et al. (1999). Recombinant glycoprotein vaccine for the prevention of genital HSV-2 infection: two randomized controlled trials. Chiron HSV Vaccine Study Group [see comments]. JAMA 282, 331–340.

Cunningham, C.K., Wara, D.W., Kang, M., Fenton, T., Hawkins, E., McNamara, J., Mofenson, L., Duliege, A.M., Francis, D., and McFarland, E.J. (2001). Safety of 2 recombinant human immunodeficiency virus type 1 (HIV-1) envelope vaccines in neonates born to HIV-1-infected women. Clin. Infect. Dis. 32, 801–807.

D'Agosto, V., Berardi, S., Burroni, D., and Hennig, R. (2006). Tolerability and safety of an MF59-adjuvanted subunit influenza vaccine (FLUAD®). IVW 2006 – The Second International Conference on Influenza Vaccines for the World.

Daems, R., Del Giudice, G., and Rappuoli, R. (2005). Anticipating crisis: towards a pandemic flu vaccination strategy through alignment of public health and industrial policy. Vaccine 23, 5732–5742.

De Donato, S., Granoff, D., Minutello, M., Lecchi, G., Faccini, M., Agnello, M., Senatore, F., Verweij, P., Fritzell, B., Podda, A., et al. (1999). Safety and immunogenicity of MF59-adjuvanted influenza vaccine in the elderly; ISCOMs: an adjuvant with multiple functions. Vaccine 17, 3094–3101.

De Gregorio, E., D'Oro, U., and Wack, A. (2009). Immunology of TLR-independent vaccine adjuvants. Curr. Opin. Immunol. 21, 339–345.

De Gregorio, E., Tritto, E., and Rappuoli, R. (2008). Alum adjuvanticity: unraveling a century old mystery. Eur. J. Immunol. 38, 2068–2071.

Del Giudice, G., Fragapane, E., Bugarini, R., Hora, M., Henriksson, T., Palla, E., O'Hagan, D., Donnelly, J., Rappuoli, R., and Podda, A. (2006a). Vaccines with the MF59 adjuvant do not stimulate antibody responses against squalene. Clin. Vaccine Immunol. 13, 1010–1013.

Del Giudice, G., Hilbert, A.K., Bugarini, R., Minutello, A., Popova, O., Toneatto, D., Schoendorf, I., Borkowski, A., Rappuoli, R., and Podda, A. (2006b). An MF59-adjuvanted inactivated influenza vaccine containing A/Panama/1999 (H3N2) induced broader serological protection against heterovariant influenza virus strain A/Fujian/2002 than a subunit and a split influenza vaccine. Vaccine 24, 3063–3065.

Dong, A., Jones, L.S., Kerwin, B.A., Krishnan, S., and Carpenter, J.F. (2006). Secondary structures of proteins adsorbed onto aluminum hydroxide: infrared spectroscopic analysis of proteins from low solution concentrations. Anal. Biochem. 351, 282–289.

Dormitzer, P.R., Rappuoli, R., Casini, D., O'Hagan, D., Runham, C., Montomoli, E., Baudner, B.C., Donnelly, J.J., Lapini, G., and Wack, A. (2009). Adjuvant is necessary for a robust immune response to a single dose of H1N1 pandemic flu vaccine in mice. PLoS Curr. RRN1025.

Dupuis, M., Denis-Mize, K., LaBarbara, A., Peters, W., Charo, I.F., McDonald, D.M., and Ott, G. (2001). Immunization with the adjuvant MF59 induces macrophage trafficking and apoptosis. Eur. J. Immunol. 31, 2910–2918.

Dupuis, M., Murphy, T.J., Higgins, D., Ugozzoli, M., van Nest, G., Ott, G., and McDonald, D.M. (1998). Dendritic cells internalize vaccine adjuvant after intramuscular injection. Cell. Immunol. 186, 18–27.

Edelman, R. (1980). Vaccine adjuvants. Rev. Infect. Dis. 2, 370–383.

Egan, P.M., Belfast, M.T., Giménez, J.A., Sitrin, R.D., and Mancinelli, R.J. (2009). Relationship between tightness of binding and immunogenicity in an aluminum-containing adjuvant-adsorbed hepatitis B vaccine. Vaccine 27, 3175–3180.

Eisenbarth, S.C., Colegio, O.R., O'Connor, W., Sutterwala, F.S., and Flavell, R.A. (2008). Crucial role for the Nalp3 inflammasome in the immunostimulatory properties of aluminium adjuvants. Nature 453, 1122–1126.

Ellouz, F., Adam, A., Ciorbaru, R., and Lederer, E. (1974). Minimal structural requirements for adjuvant activity of bacterial peptidoglycan derivatives. Biochem. Biophys. Res. Commun. 59, 1317–1325.

Ericsson, H. (1946). Purification and adsorption of diphtheria toxoid. Nature 158, 350–351.

European Pharmacopoeia (1997). Vaccines for human use, 3rd edn.

Forrest, H.L., Khalenkov, A.M., Govorkova, E.A., Kim, J.K., Del Giudice, G., and Webster, R.G. (2009). Single-and multiple-clade influenza A H5N1 vaccines induce cross protection in ferrets. Vaccine 27, 4187–4195.

Franchi, L., and Nunez, G. (2008). The Nlrp3 inflammasome is critical for aluminium hydroxide-mediated IL-1β secretion but dispensable for adjuvant activity. Eur. J. Immunol. 38, 2085–2089.

Freund, J., Casals, J., and Hosmer, E.P. (1937). Sensitization and antibody formation after injection of tubercle bacilli and paraffin oil. Proc. Soc. Exp. Biol. Med. 37, 509–513.

Fritz, J.H., Ferrero, R.L., Philpott, D.J., and Girardin, S.E. (2006). Nod-like proteins in immunity, inflammation and disease. Nat. Immunol. 7, 1250–1257.

Galli, G., Hancock, K., Hoschler, K., DeVos, J., Praus, M., Bardelli, M., Malzone, C., Castellino, F., Gentile, C., and McNally, T. (2009a). Fast rise of broadly cross-reactive antibodies after boosting long-lived human memory B cells primed by an MF59 adjuvanted prepandemic vaccine. Proc. Natl. Acad. Sci. U.S.A. 106, 7962.

Galli, G., Medini, D., Borgogni, E., Zedda, L., Bardelli, M., Malzone, C., Nuti, S., Tavarini, S., Sammicheli, C., and Hilbert, A.K. (2009b). Adjuvanted H5N1 vaccine induces early CD4+ T cell response that predicts long-term persistence of protective antibody levels. Proc. Natl. Acad. Sci. U.S.A. *106*, 3877.

Gavin, A.L., Hoebe, K., Duong, B., Ota, T., Martin, C., Beutler, B., and Nemazee, D. (2006). Adjuvant-enhanced antibody responses in the absence of toll-like receptor signaling. Science's STKE *314*, 1936.

Giuliani, M.M., Adu-Bobie, J., Comanducci, M., Arico, B., Savino, S., Santini, L., Brunelli, B., Bambini, S., Biolchi, A., Capecchi, B., *et al.* (2006). A universal vaccine for serogroup B meningococcus. Proc. Natl. Acad. Sci. U.S.A. *103*, 10834–10839.

Glenny, A.T., Buttle, A.H., and Stevens, M.F. (1931). Rate of disappearance of diphtheria toxoid injected into rabbits and guinea-pigs: Toxoid precipitated with alum. J. Pathol. Bacteriol. *34*, 267–275.

Glenny, A.T., Pope, C.G., Waddington, H., and Wallace, U. (1926). The antigenic value of toxoid precipitated by potassium alum. J. Pathol. Bacteriol. *29*, 31–40.

Goto, N., and Akama, K. (1982). Histopathological studies of reactions in mice injected with aluminum-adsorbed tetanus toxoid. Microbiol. Immunol. *26*, 1121.

Goto, N., Kato, H., Maeyama, J., Shibano, M., Saito, T., Yamaguchi, J., and Yoshihara, S. (1997). Local tissue irritating effects and adjuvant activities of calcium phosphate and aluminium hydroxide with different physical properties. Vaccine *15*, 1364–1371.

Granoff, D.M., McHugh, Y.E., Raff, H.V., Mokatrin, A.S., and Van Nest, G.A. (1997). MF59 adjuvant enhances antibody responses of infant baboons immunized with *Haemophilus influenzae* type b and *Neisseria meningitidis* group C oligosaccharide–CRM197 conjugate vaccine. Infect. Immun. *65*, 1710.

Grun, J.L., and Maurer, P.H. (1989). Different T helper cell subsets elicited in mice utilizing two different adjuvant vehicles: the role of endogenous interleukin 1 in proliferative responses. Cell Immunol. *121*, 134.

Gupta, R.K., and Rost, B.E. (2000). Aluminum compounds as vaccine adjuvants. In Vaccine Adjuvants: Preparation Methods and Research Protocols, D.T. O'Hagan, ed. (Humana Press).

Gupta, R.K., Rost, B.E., Relyveld, E., and Siber, G.R. (1995). Adjuvant properties of aluminum and calcium compounds. In Vaccine Design: The Subunit and Adjuvant Approach, Powell, M.F., and Newman, M.J., eds (New York, Plenum Press), pp. 229–248.

Hansen, B., Sokolovska, A., HogenEsch, H., and Hem, S.L. (2007). Relationship between the strength of antigen adsorption to an aluminum-containing adjuvant and the immune response. Vaccine *25*, 6618–6624.

Harper, D.M., Franco, E.L., Wheeler, C.M., Moscicki, A.B., Romanowski, B., Roteli-Martins, C.M., Jenkins, D., Schuind, A., Costa Clemens, S.A., and Dubin, G. (2006). Sustained efficacy up to 4.5 years of a bivalent L1 virus-like particle vaccine against human papillomavirus types 16 and 18: follow-up from a randomised control trial. Lancet *367*, 1247–1255.

Harrison, W.T. (1935). Some observations on the use of alum-precipitated diphtheria toxoid. Am. J. Publ. Health *25*, 298–300.

Heineman, T.C., Clements-Mann, M.L., Poland, G.A., Jacobson, R.M., Izu, A.E., Sakamoto, D., Eiden, J., Van Nest, G.A., and Hsu, H.H. (1999). A randomized, controlled study in adults of the immunogenicity of a novel hepatitis B vaccine containing MF59 adjuvant. Vaccine *17*, 2769–2778.

Hem, S.L., and HogenEsch, H. (2007). Relationship between physical and chemical properties of aluminum-containing adjuvants and immunopotentiation. Expert Rev. Vaccines *6*, 685–698.

Higgins, D.A., Carlson, J.R., and Van Nest, G. (1996). MF59 adjuvant enhances the immunogenicity of influenza vaccine in both young and old mice. Vaccine *14*, 478–484.

Hilleman, M.R. (1966). Critical appraisal of emulsified oil adjuvants applied to viral vaccines. Prog. Med. Virol. *8*, 131–182.

Holt, L.B. (1947). Purified precipitated diphtheria toxoid of constant composition. Lancet *8*, 282–285.

Holt, L.B. (1950). Developments in diphtheria prophylaxis. (London, Heinemann), p. 99.

Hornung, V., Bauernfeind, F., Halle, A., Samstad, E.O., Kono, H., Rock, K.L., Fitzgerald, K.A., and Latz, E. (2008). Silica crystals and aluminum salts activate the NALP3 inflammasome through phagosomal destabilization. Nat. Immunol. *9*, 847.

Hutcheon, C.J., Becker, J.O., Russell, B.A., Bariola, P.A., Peterson, G.J., and Stroop, S.D. (2006). Physiochemical and functional characterization of antigen proteins eluted from aluminum hydroxide adjuvant. Vaccine *24*, 7214–7225.

Iob, A., Brianti, G., Zamparo, E., and Gallo, T. (2005). Evidence of increased clinical protection of an MF59-adjuvant influenza vaccine compared to a non-adjuvant vaccine among elderly residents of long-term care facilities in Italy. Epidemiol. Infect. *133*, 687–693.

Iyer, S., HogenEsch, H., and Hem, S.L. (2003). Effect of the degree of phosphate substitution in aluminum hydroxide adjuvant on the adsorption of phosphorylated proteins. Pharm. Dev. Technol. *8*, 81–86.

Jansen, T., Hofmans, M.P., Theelen, M.J., and Schijns, V.E. (2005). Structure–activity relations of water-in-oil vaccine formulations and induced antigen-specific antibody responses. Vaccine *23*, 1053–1060.

Jezek, J., Chen, D., Watson, L., Crawford, J., Perkins, S., Tyagi, A., and Braun, L.T.J. (2009). A heat-stable hepatitis B vaccine formulation. Hum. Vaccines *5*, 1–7.

Jones, L.T.S., Peek, L.J., Power, J., Markham, A., Yazzie, B., and Middaugh, C.R. (2005). Effects of adsorption to aluminum salt adjuvants on the structure and stability of model protein antigens. J. Biol. Chem. *280*, 13406–13414.

Jordan, M.B., Mills, D.M., Kappler, J., Marrack, P., and Cambier, J.C. (2004). Promotion of B cell immune responses via an alum-induced myeloid cell population. Science *304*, 1808–1810.

Kahn, J.O., Sinangil, F., Baenziger, J., Murcar, N., Wynne, D., Coleman, R.L., Steimer, K.S., Dekker, C.L., and Chernoff, D. (1994). Clinical and immunologic responses to human immunodeficiency virus (HIV) type 1SF2 gp120 subunit vaccine combined with

MF59 adjuvant with or without muramyl tripeptide dipalmitoyl phosphatidylethanolamine in non-HIV-infected human volunteers. J. Infect. Dis. *170*, 1288–1291.

Kanra, G., Viviani, S., Yurdakök, K., Ozmert, E., Anemona, A., Yalcin, S., Demiralp, O., Bilgili, N., Kara, A., and Cengiz, A.B.L., *et al.* (2003). Effect of aluminum adjuvants on safety and immunogenicity of *Haemophilus influenzae* type b–CRM197 conjugate vaccine. Pediatr. Int. *45*, 314.

Katz, J. (1987). Desorption of porcine parvovirus from aluminum hydroxide adjuvant with subsequent viral immunoassay or hemagglutination assay. Vet. Res. Commun. *11*, 83–92.

Keefer, M.C., Graham, B.S., McElrath, M.J., Matthews, T.J., Stablein, D.M., Corey, L., Wright, P.F., Lawrence, D., Fast, P.E., Weinhold, K., *et al.* (1996). Safety and immunogenicity of Env 2–3, a human immunodeficiency virus type 1 candidate vaccine, in combination with a novel adjuvant, MTP-PE/MF59. NIAID AIDS Vaccine Evaluation Group. AIDS Res. Hum. Retroviruses *12*, 683–693.

Keitel, W., Couch, R., Bond, N., Adair, S., Van Nest, G., and Dekker, C. (1993). Pilot evaluation of influenza virus vaccine (IVV) combined with adjuvant. Vaccine *11*, 909–913.

Kenney, R.T., and Edelman, R. (2003). Survey of human-use adjuvants. Expert Rev. Vaccines *2*, 167–188.

Kenney, R.T., and Edelman, R. (2004). New Generation Vaccines, 3rd edn (New York, Marcel Dekker, Inc.).

Klinman, D.M. (2004). Use of CpG oligodeoxynucleotides as immunoprotective agents. Expert Opin. Biol. Ther. *4*, 937–946.

Kool, M., Petrilli, V., De Smedt, T., Rolaz, A., Hammad, H., van Nimwegen, M., Bergen, I.M., Castillo, R., Lambrecht, B.N., and Tschopp, J. (2008a). Cutting edge: alum adjuvant stimulates inflammatory dendritic cells through activation of the NALP3 inflammasome. J. Immunol. *181*, 3755.

Kool, M., Soullie, T., van Nimwegen, M., Willart, M.A.M., Muskens, F., Jung, S., Hoogsteden, H.C., Hammad, H., and Lambrecht, B.N. (2008b). Alum adjuvant boosts adaptive immunity by inducing uric acid and activating inflammatory dendritic cells. Science's STKE *205*, 869.

Lai, X., Zheng, Y., Jacobsen, S., Larsen, J.N., Ipsen, H., Løwenstein, H., and Søndergaard, I.B. (2008). Determination of adsorbed protein concentration in aluminum hydroxide suspensions by near-infrared transmittance spectroscopy. Appl. Spectrosc. *62*, 784–790.

Langenberg, A.G., Burke, R.L., Adair, S.F., Sekulovich, R., Tigges, M., Dekker, C.L., and Corey, L. (1995). A recombinant glycoprotein vaccine for herpes simplex virus type 2: safety and immunogenicity [corrected] [published erratum appears in Ann. Intern. Med. 1995, 123, 395]. Ann. Intern. Med. *122*, 889–898.

Li, H., Willingham, S.B., Ting, J.P.Y., and Re, F. (2008). Cutting edge: inflammasome activation by alum and alum's adjuvant effect are mediated by NLRP3. J. Immunol. *181*, 17.

Lindblad, E.B. (1995). Aluminium adjuvants. In The Theory and Practical Application of Adjuvants, Stewart-Tull, D.E.S., ed. (New York, John Wiley & Sons), pp. 21–35.

Lindblad, E.B. (2004). Aluminium compounds for use in vaccines. Immunol. Cell Biol. *82*, 497–505.

Mannhalter, J.W., Neychev, H.O., Zlabinger, G.J., Ahmad, R., and Eibl, M.M. (1985). Modulation of the human immune response by the non-toxic and non-pyrogenic adjuvant aluminium hydroxide: effect on antigen uptake and antigen presentation. Clin. Exp. Immunol. *61*, 143–151.

Marrack, P., McKee, A.S., and Munks, M.W. (2009). Towards an understanding of the adjuvant action of aluminium. Nat. Rev. Immunol. *9*, 287–293.

Mata-Haro, V., Cekic, C., Martin, M., Chilton, P.M., Casella, C.R., and Mitchell, T.C. (2007). The vaccine adjuvant monophosphoryl lipid A as a TRIF-biased agonist of TLR4. Science *316*, 1628.

Matheis, W., Zott, A., and Schwanig, M. (2001). The role of the adsorption process for production and control combined adsorbed vaccines. Vaccine *20*, 67–73.

McFarland, E.J., Borkowsky, W., Fenton, T., Wara, D., McNamara, J., Samson, P., Kang, M., Mofenson, L., Cunningham, C., Duliege, A.M., *et al.* (2001). Human immunodeficiency virus type 1 (HIV-1) gp120-specific antibodies in neonates receiving an HIV-1 recombinant gp120 vaccine. J. Infect. Dis. *184*, 1331–1335. Epub 13 Oct 2001.

Minutello, M., Senatore, F., Cecchinelli, G., Bianchi, M., Andreani, T., Podda, A., and Crovari, P. (1999). Safety and immunogenicity of an inactivated subunit influenza virus vaccine combined with MF59 adjuvant emulsion in elderly subjects, immunized for three consecutive influenza seasons. Vaccine *17*, 99–104.

Mitchell, D.K., Holmes, S.J., Burke, R.L., Duliege, A.M., and Adler, S.P. (2002). Immunogenicity of a recombinant human cytomegalovirus gB vaccine in seronegative toddlers. Pediatr. Infect. Dis. J. *21*, 133–138.

Morefield, G.L., HogenEsch, H., Robinson, J.P., and Hem, S.L. (2004). Distribution of adsorbed antigen in mono-valent and combination vaccines. Vaccine *22*, 1973–1984.

Morefield, G.L., Jiang, D., Romero-Mendez, I.Z., Geahlen, R.L., Hogenesch, H., and Hem, S.L. (2005a). Effect of phosphorylation of ovalbumin on adsorption by aluminum-containing adjuvants and elution upon exposure to interstitial fluid. Vaccine *23*, 1502–1506.

Morefield, G.L., Sokolovska, A., Jiang, D., HogenEsch, H., Robinson, J.P., and Hem, S.L. (2005b). Role of aluminum-containing adjuvants in antigen internalization by dendritic cells *in vitro*. Vaccine *23*, 1588–1595.

Mosca, F., Tritto, E., Muzzi, A., Monaci, E., Bagnoli, F., Iavarone, C., O'Hagan, D., Rappuoli, R., and De Gregorio, E. (2008). Molecular and cellular signatures of human vaccine adjuvants. Proc. Natl. Acad. Sci. U.S.A. *105*, 10501.

Mullen, G.E.D., Aebig, J.A., Dobrescu, G., Rausch, K., Lambert, L., Long, C.A., Miles, A.P., and Saul, A. (2007). Enhanced antibody production in mice to the malaria antigen AMA1 by CPG 7909 requires physical association of CpG and antigen. Vaccine *25*, 5343–5347.

Mullen, G.E.D., Ellis, R.D., Miura, K., Malkin, E., Nolan, C., Hay, M., Fay, M.P., Saul, A., Zhu, D., and Rausch, K. (2008). Phase 1 trial of AMA1-C1/Alhydrogel plus CPG 7909: An asexual blood-stage vaccine for Plasmodium falciparum malaria. PLoS One 3, 2940.

Nakae, S., Komiyama, Y., Yokoyama, H., Nambu, A., Umeda, M., Iwase, M., Homma, I., Sudo, K., Horai, R., and Asano, M. (2003). IL-1 is required for allergen-specific Th2 cell activation and the development of airway hypersensitivity response. Int. Immunol. *15*, 483–490.

Nicholson, K.G., Colegate, A.E., Podda, A., Stephenson, I., Wood, J., Ypma, E., and Zambon, M.C. (2001). Safety and antigenicity of non-adjuvanted and MF59-adjuvanted influenza A/Duck/Singapore/97 (H5N3) vaccine: a randomised trial of two potential vaccines against H5N1 influenza. Lancet *357*, 1937–1943.

Ogura, Y., Sutterwala, F.S., and Flavell, R.A. (2006). The inflammasome: first line of the immune response to cell stress. Cell *126*, 659–662.

O'Hagan, D.T., and Podda, A. (2008). MF59: A safe and potent oil in water emulsion adjuvant for influenza vaccines, which induces enhanced protection against virus challenge. Influenza Vaccines for the Future, 221.

O'Hagan, D.T., Singh, M., Dong, C., Ugozzoli, M., Berger, K., Glazer, E., Selby, M., Wininger, M., Ng, P., Crawford, K., et al. (2004). Cationic microparticles are a potent delivery system for a HCV DNA vaccine. Vaccine *23*, 672–680.

O'Hagan, D.T., Singh, M., Kazzaz, J., Ugozzoli, M., Briones, M., Donnelly, J., and Ott, G. (2002). Synergistic adjuvant activity of immunostimulatory DNA and oil/water emulsions for immunization with HIV p55 gag antigen. Vaccine *20*, 3389–3398.

O'Hagan, D.T., and De Gregorio, E. (2009). The path to a successful vaccine adjuvant–'The long and winding road'. Drug Discov. Today *14*, 541–551.

O'Hagan, D.T., Ugozzoli, M., Barackman, J., Singh, M., Kazzaz, J., Higgins, K., Vancott, T.C., and Ott, G. (2000). Microparticles in MF59, a potent adjuvant combination for a recombinant protein vaccine against HIV-1. Vaccine *18*, 1793–1801.

O'Hagan, D.T. (2007). MF59 is a safe and potent vaccine adjuvant that enhances protection against influenza virus infection. Expert Rev. Vaccines *6*, 699–710.

Ott, G. (2000). Vaccine adjuvants: preparation methods and research protocols. In Vaccine Adjuvants: Preparation Methods and Research Protocols, D. O'Hagan, ed. (Totowa, NJ, Humana Press).

Ott, G., Barchfeld, G.L., Chernoff, D., Radhakrishnan, R., Van Hoogevest, P., and Van Nest, G. (1995a). MF59. Design and evaluation of a safe and potent adjuvant for human vaccines. Pharm. Biotechnol. *6*, 277.

Ott, G., Barchfeld, G.L., Chernoff, D., Radhakrishnan, R., van Hoogevest, P., and Van Nest, G. (1995b). MF59: Design and evaluation of a safe and potent adjuvant for human vaccines. In Vaccine Design: The Subunit and Adjuvant Approach, Powell, M.F., and Newman, M.J., eds (New York, Plenum Press), pp. 277–296.

Ott, G., Barchfeld, G.L., and Nest, G.V. (1995c). Enhancement of humoral response against human influenza vaccine with the simple submicron oil/water emulsion adjuvant MF59. Vaccine *13*, 1557–1562.

Otten, G., Schaefer, M., Doe, B., Liu, H., Srivastava, I., zur Megede, J., O'Hagan, D., Donnelly, J., Widera, G., Rabussay, D., et al. (2004). Enhancement of DNA vaccine potency in rhesus macaques by electroporation. Vaccine *22*, 2489–2493.

Otten, G.R., Schaefer, M., Doe, B., Liu, H., Megede, J.Z., Donnelly, J., Rabussay, D., Barnett, S., and Ulmer, J.B. (2006). Potent immunogenicity of an HIV-1 gag–pol fusion DNA vaccine delivered by in vivo electroporation. Vaccine *24*, 4503–4509.

Otten, G.R., Schaefer, M., Doe, B., Liu, H., Srivastava, I., Megede, J., Kazzaz, J., Lian, Y., Singh, M., Ugozzoli, M., et al. (2005). Enhanced potency of plasmid DNA microparticle human immunodeficiency virus vaccines in rhesus macaques by using a priming-boosting regimen with recombinant proteins. J. Virol. *79*, 8189–8200.

Otten, G.R., Schaefer, M., Greer, C., Calderon-Cacia, M., Coit, D., Kazzaz, J., Medina-Selby, A., Selby, M., Singh, M., Ugozzoli, M., et al. (2003). Induction of broad and potent anti-HIV immune responses in rhesus macaques by priming with a DNA vaccine and boosting with protein-adsorbed PLG microparticles. J. Virol. *77*, 6087–6092.

Page, W. (1993). Long-term followup of Army recruits immunized with Freund's incomplete adjuvanted vaccine. Vaccine Res. *2*, 141–149.

Pass, R.F., Zhang, C., Evans, A., Simpson, T., Andrews, W., Huang, M.L., Corey, L., Hill, J., Davis, E., and Flanigan, C. (2009). Vaccine prevention of maternal cytomegalovirus infection. N. Engl. J. Med. *360*, 1191.

Peek, L.J., Martin, T.T., Elk Nation, C., Pegram, S.A., and Middaugh, C.R. (2007). Effects of stabilizers on the destabilization of proteins upon adsorption to aluminum salt adjuvants. J. Pharm. Sci. *96*, 547–557.

Pellegrini, M., Nicolay, U., Lindert, K., Groth, N., and Della Cioppa, G. (2009). MF59-adjuvanted versus non-adjuvanted influenza vaccines: Integrated analysis from a large safety database. Vaccine *27*, 6959–6965.

Podda, A. (2001). The adjuvanted influenza vaccines with novel adjuvants: experience with the MF59-adjuvanted vaccine. Vaccine *19*, 2673–2680.

Podda, A., and Del Giudice, G. (2003). MF59-adjuvanted vaccines: increased immunogenicity with an optimal safety profile. Expert Rev. Vaccines *2*, 197–203.

Podda, A., Del Giudice, G., and T., O.H.D. (2005). Chapter 9. MF59: A safe and potent adjuvant for human use. In Immunopotentiators in Modern Vaccines, Schijns, V., and T. O'Hagan D., eds (Elsevier Press), p. 149.

Puig-Barbera, J., Diez-Domingo, J., Perez Hoyos, S., Belenguer Varea, A., and Gonzalez Vidal, D. (2004). Effectiveness of the MF59-adjuvanted influenza vaccine in preventing emergency admissions for pneumonia in the elderly over 64 years of age. Vaccine *23*, 283–289.

Puig Barbera, J., and Gonzalez Vidal, D. (2007). MF59™-adjuvanted subunit influenza vaccine: an improved interpandemic influenza vaccine for vulnerable populations. Expert Rev. Vaccines 6, 659–665.

Ramon, G. (1924). Sur la toxine et surranatoxine diphtheriques. Ann. Inst. Pasteur 38, 1.

Rappuoli, R. (2007). Bridging the knowledge gaps in vaccine design. Nat. Biotechnol. 25, 1361–1366.

Reddy, P. (2004). Interleukin-18: recent advances. Curr. Opin. Hematol. 11, 405.

Rimaniol, A.C., Gras, G., Verdier, F., Capel, F., Grigoriev, V.B., Porcheray, F., Sauzeat, E., Fournier, J.G., Clayette, P., Siegrist, C.A., et al. (2004). Aluminum hydroxide adjuvant induces macrophage differentiation towards a specialized antigen-presenting cell type. Vaccine 22, 3127–3135.

Romero Méndez, I.Z., Shi, Y., HogenEsch, H., and Hem, S.L. (2007). Potentiation of the immune response to non-adsorbed antigens by aluminum-containing adjuvants. Vaccine 25, 825–833.

Rowe, R.C. (2006). Handbook of Pharmaceutical Excipients, 5th edn (London, Pharmaceutical Press).

Schmitz, N., Kurrer, M., and Kopf, M. (2003). The IL-1 receptor 1 is critical for Th2 cell type airway immune responses in a mild but not in a more severe asthma model. Eur. J. Immunol. 33, 991–1000.

Schnare, M., Barton, G.M., Holt, A.C., Takeda, K., Akira, S., and Medzhitov, R. (2001). Toll-like receptors control activation of adaptive immune responses. Nat. Immunol. 2, 947–950.

Schultze, V., D'Agosto, V., Wack, A., VandenBossche, G., Novicki, D., Zorn, J., and Hennig, R. (2008). Safety of MF59 adjuvant. Vaccine 26, 3209.

Seubert, A., Monaci, E., Pizza, M., O'Hagan D., T., and Wack, A. (2008). The adjuvants MF59 and alum induce monocyte chemoattractants and enhance monocyte differentiation towards dendritic cells. J. Immunol. 180, 5402.

Sharp, F.A., Ruane, D., Claass, B., Creagh, E., Harris, J., Malyala, P., Singh, M., O'Hagan, D.T., Pétrilli, V., and Tschopp, J. (2009). Uptake of particulate vaccine adjuvants by dendritic cells activates the NALP3 inflammasome. Proc. Natl. Acad. Sci. U.S.A. 106, 870.

Shirodkar, S., Hutchinson, R.L., Perry, D.L., White, J.L., and Hem, S.L. (1990). Aluminum compounds used as adjuvants in vaccines. Pharm. Res. 7, 1282–1288.

Singh, M., Ugozzoli, M., Kazzaz, J., Chesko, J., Soenawan, E., Mannucci, D., Titta, F., Contorni, M., Volpini, G., Del Guidice, G., et al. (2006). A preliminary evaluation of alternative adjuvants to alum using a range of established and new generation vaccine antigens. Vaccine 24, 1680–1686.

Sokolovska, A., Hem, S.L., and HogenEsch, H. (2007). Activation of dendritic cells and induction of CD4+ T cell differentiation by aluminum-containing adjuvants. Vaccine 25, 4575–4585.

Stephenson, I., Nicholson, K.G., Bugarini, R., Podda, A., Wood, J., Zambon, M., and Katz, J. (2005). Cross reactivity to highly pathogenic avian influenza H5N1 viruses following vaccination with non-adjuvanted and MF-59-adjuvanted influenza A/Duck/Singapore/97 (H5N3) vaccine: a potential priming strategy. J. Infect. Dis. 191.

Stephenson, I., Nicholson, K.G., Colegate, A., Podda, A., Wood, J., Ypma, E., and Zambon, M. (2003). Boosting immunity to influenza H5N1 with MF59-adjuvanted H5N3 A/Duck/Singapore/97 vaccine in a primed human population. Vaccine 21, 1687–1693.

Stephenson, I., Nicholson, K.G., Hoschler, K., Zambon, M.C., Hancock, K., DeVos, J., Katz, J.M., Praus, M., and Banzhoff, A. (2008). Antigenically distinct MF59-adjuvanted vaccine to boost immunity to H5N1. N. Engl. J. Med. 359, 1631.

Strassburg, M.A., Greenland, S., Sorvillo, F.J., Lieb, L.E., and Habel, L.A. (1986). Influenza in the elderly: report of an outbreak and a review of vaccine effectiveness reports. Vaccine 4, 38–44.

Thiele, G.M., Rogers, J., Collins, M., Smith, D., McDonald, T.L., and Yasuda, N. (1990). An enzyme-linked immunosorbent assay for the detection of antitetanus toxoid antibody using aluminum-absorbed coating antigen. J. Clin. Lab. Anal. 4, 126–129.

Traquina, P., Morandi, M., Contorni, M., and Van Nest, G. (1996). MF59 adjuvant enhances the antibody response to recombinant hepatitis B surface antigen vaccine in primates. J. Infect. Dis. 174, 1168–1175.

Treanor, J.J., Campbell, J.D., Zangwill, K.M., Rowe, T., and Wolff, M. (2006). Safety and immunogenicity of an inactivated subvirion influenza A (H5N1) vaccine. N. Engl. J. Med. 354, 1343–1351.

Ulanova, M., Tarkowski, A., Hahn-Zoric, M., Hanson, L.A., and Moingeon, P. (2001). The common vaccine adjuvant aluminum hydroxide up-regulates accessory properties of human monocytes via an interleukin-4-dependent mechanism. Infect. Immun. 69, 1151–1159.

Valensi, J.P., Carlson, J.R., and Van Nest, G.A. (1994). Systemic cytokine profiles in BALB/c mice immunized with trivalent influenza vaccine containing MF59 oil emulsion and other advanced adjuvants. J. Immunol. 153, 4029–4039.

Vesikari, T., Pellegrini, M., Karvonen, A., Groth, N., Borkowski, A., O'Hagan, D.T., and Podda, A. (2009). Enhanced immunogenicity of seasonal influenza vaccines in young children using MF59 adjuvant. Pediatr. Infect. Dis. J. 28, 563.

Vogel, F.R., and Pruett, M.F. (1995). A compendium of vaccine adjuvants and excipients. In Vaccine Design: The subunit and adjuvant approach, Powell, M.F., and Newman, M.J., eds (New York, Plenum Press), pp. 141–228.

Wack, A., Baudner, B.C., Hilbert, A.K., Scheffczik, H., Ugozzoli, M., Singh, M., Kazzaz, J., Del Giudice, G., Rappuoli, R., and O'Hagan D., T. (2008). Combination adjuvants for the induction of potent, long-lasting antibody and T cell responses to influenza vaccine. Vaccine 26, 552–561.

Wang, H.B., and Weller, P.F. (2008). Pivotal Advance: Eosinophils mediate early alum adjuvant-elicited B cell priming and IgM production. J. Leukoc. Biol. 83, 817.

Waters, R.V., Terrell, T.G., and Jones, G.H. (1986). Uveitis induction in the rabbit by muramyl dipeptides. Infect. Immun. 51, 816–825.

White, R.G., Coons, A.H., and Connolly, J.M. (1955). Studies on antibody production. III. The alum granuloma. J. Exp. Med. *102*, 73–82.

Wilcock, L.K., Francis, J.N., and Durham, S.R. (2004). Aluminium hydroxide down-regulates T helper 2 responses by allergen-stimulated human peripheral blood mononuclear cells. Clin. Exp. Allergy *34*, 1373–1378.

Wintsch, J., Chaignat, C.L., Braun, D.G., Jeannet, M., Stalder, H., Abrignani, S., Montagna, D., Clavijo, F., Moret, P., Dayer, J.M., *et al.* (1991). Safety and immunogenicity of a genetically engineered human immunodeficiency virus vaccine. J. Infect. Dis. *163*, 219–225.

World Health Organization (1976). Immunological Adjuvants. In Tech Rep Series no 595 (World Health Organization, Geneva), pp. 3–40.

World Health Organization (1977). Manual for the Production and Control of Vaccines – Tetanus Toxoid, BLG/UNDP/77.2 Rev.1.

World Health Organization (2006). Temperature Sensitivity of Vaccines (Geneva).

Zhu, D., Huang, S., Gebregeorgis, E., McClellan, H., Dai, W., Miller, L., and Saul, A. (2009a). Development of a direct alhydrogel formulation immunoassay (DAFIA). J. Immunol. Methods *344*, 73–78.

Zhu, D., Saul, A., Huang, S., Martin, L.B., Miller, L.H., and Rausch, K.M. (2009b). Use of o-phthalaldehyde assay to determine protein contents of Alhydrogel-based vaccines. Vaccine *27*, 6054–6059.

Mucosal Vaccines

Rajesh Ravindran and Bali Pulendran

7

Abstract

The term 'mucosal vaccination' has traditionally been used to describe strategies in which a vaccine is administered via the mucosal route. Unlike parenteral vaccination, mucosal vaccines do not require the use of needles, thus enabling vaccine compliance and reducing logistical challenges and the risks of acquiring blood-borne infections. However, despite the great success of mucosal vaccines such as the polio vaccine, several formidable challenges hinder the effective elicitation of immunity against pathogens that invade mucosal sites. First, in humans the mucosal surfaces of the gut, lung, oral cavity and reproductive tracts are estimated to cover an area of 400 square metres, and thus represent the largest portal of entry for pathogens. Second, the acidic environments of many mucosal sites, and the delineation of mucosal sites by the epithelial barrier, pose challenges to the effective delivery of vaccines. Third, the mucosal immune system is faced with a somewhat schizophrenic challenge of having to launch robust immunity against mucosal pathogens, whilst restraining immune reactivity to commensals and food antigens. Fourth, the induction of the appropriate type of immune response is critical for effective protection against different pathogens. Fifth, the accurate quantitation of mucosal T and B cell responses pose unique challenges. Despite these challenges, recent advances in our understanding of the innate immunity and its regulation of adaptive immunity at mucosal sites, are beginning to offer new insights into strategies that result in immune protection at mucosal surfaces. In particular, several recent studies demonstrate that parenteral vaccination with the appropriate adjuvants can induce migration of antigen-specific T and B lymphocytes to mucosal sites. These advances promise to accelerate the development and testing of new mucosal vaccines against many diseases including HIV/AIDS. In this chapter, we will review the present mucosal immunization strategies and look at opportunities for exploiting newer developments for devising effective oral vaccines.

Most infectious agents that infect humans do so via mucosal sites, principally the digestive, respiratory and genitourinary tracts. Immune defences at mucosal surfaces therefore constitute a very vital part of the overall protective responses against these invading pathogens. Vaccines that are administered via the oral routes most proficiently induce the mucosal immune responses. In contrast, parenterally administered vaccines are generally poor inducers of mucosal immunity and are therefore less efficient against infections originating at mucosal surfaces (Lamm, 1997; Levine, 2000). However, only a few mucosal vaccines have been approved for human use (Table 7.1) (Levine, 2000). However, progress in research aimed at understanding the molecular and cellular mechanisms of the mucosal system is presently accelerating, allowing us to design innovative strategies for the development of mucosal vaccines.

The mucosal immune system

Mucosa-associated lymphoid tissue (MALT) is scattered along mucosal linings, measuring roughly about $400\,m^2$ (Santacroce et al., 1997). The lymphoid tissues associated with mucosal surfaces may be further subdivided into

Table 7.1 Mucosal vaccines licensed for human use

Disease	Vaccine	Type of vaccine	Trademarks	Route	Status
Polio	Inactivated polio vaccine (IPV)	Inactivated	IPOL (Sanofi Pasteur), POLIOVAX (Sanofi Pasteur)	Injectable (arm, leg)	Less immunogenic at mucosal surfaces than OPV
	OPV vaccine	Live attenuated	Orimune (Lederle Labs/ Wyeth)	Oral	The presently used vaccine. 95% effectiveness leading to life long immunity.
Cholera	Killed Vibrio cholerae O1 strains+rCTB	Killed Vibrio cholerae	Dukoral (SBL Vaccines/ Crucell)	Oral	56–95% protection, Duration of protection is about 1–3 years
	Vibrio cholerae O1 and O139 without CTB	Killed Vibrio cholerae	ORC-Vax (Vabiotech), International Vaccine Institute, Shanta Biotech (India)	Oral	14–79% efficacy. Duration of protection is 3–5 years The vaccine is licensed only in Vietnam and India
	Vibrio cholerae O1 classical Inaba strain 569B (CVD 103. HgR expressing CTB)	Live attenuated genetically modified	Orochol/Mutachol (Berna, Swiss Serum and Vaccine Institute, Crucell)	Oral	60–90% protection for approximately 3 months. Berna has discontinued production
Rotavirus	Monovalent human rotavirus	Live attenuated	RotaRix (GlaxoSmithKline)	Oral	Licensed in US, Europe, Mexico, Dominican Republic, Brazil, Panama, Venezuela and Kuwait. 85% protection against severe infection. No association with intussusception
	Pentavalent	Live attenuated	RotaTeq (Merck)	Oral	Provides protection against 5 serotypes G1,G2, G3, G4 and P1. 98% protection against severe disease. No association with intussusception
	Lanzhou lamb vaccine	Live attenuated	Lanzhou Institute for Biological Products (China)	Oral	Licensed for use only in China. Efficacy unknown
Influenza	Influenza strains	Live attenuated	FluMist/Live Attenuated Influenza Vaccine (MedImmune)	Nasal	Approved for healthy people between 5 and 49 years
Typhoid fever	Ty21a	Live attenuated	Vivotif Berna, Swiss Serum and Vaccine Institute	Oral	50–80% effective
	Vi polysaccharide vaccine	Subunit	Typherix (Glaxo) Typhim Vi (Sanofi Pasteur)	Injectable	70% efficacy lasting for 3 years

inductive sites where antigens are encountered and responses initiated, and effector sites where local immune responses occur. In the digestive system, the intestinal epithelium, lamina propria (LP), Peyer's patches (PP), appendix and the mesenteric lymph nodes (MLN) make up the gut-associated lymphoid tissue (GALT). The GALT represents the largest mass of immuno-competent cells within the human body. The inductive sites in the GALT are organized into specialized aggregations of lymphoid follicles called Peyer's Patches, while effector sites are more diffusely dispersed. PPs contain specialized cells for uptake of antigens known as Microfold or M cells, which are located in the dome region of the epithelial layer known as follicle-associated epithelium (FAE). These specialized cells facilitate engulfment of microorganisms and

transcellular passage of luminal antigens (Fig. 7.1). After the uptake of antigens through M cells, the antigens are processed by professional APCs present at the subepithelial dome region (SED) and presented to T cells present in inter follicular region (IFR) (Fig. 7.1). Myeloid DCs within the SED are capable of migrating within the PPs, acquiring the antigen and then moving to the IFR area and presenting them to the T cells directly or indirectly via the lymphoid DCs (Fig. 7.1).

Beneath the SED of the PPs is a network of immune cells comprising of distinct B cell follicles surrounded by T cell enriched regions. The germinal centres (GCs), which are rich in B cells, are considered to be a major site for active μ to α class switch recombination (CSR) involving the enzyme activation-induced cytidine deaminase (AID). In addition, antigen presentation also

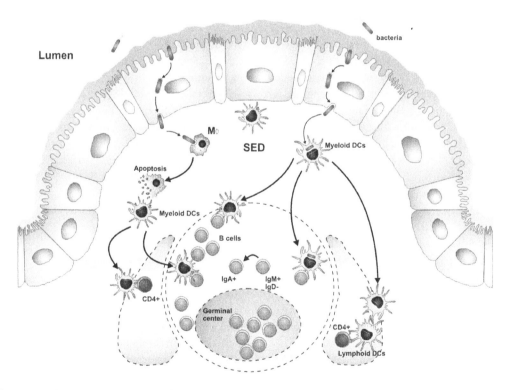

Figure 7.1 Proposed mechanism of immune induction in peyers patches. In Peyer's patches, bacteria or their fragments are internalized by M cells through FAE. The myeloid DCs acquire bacterial antigens either directly or released by apoptotic MQ and then migrate to the IFR to effectively present antigens to the T cells. Bacterial presentation to immune cells in PPs, leads to enhanced cytokine production, as well the switch from IgM to IgA B cells. The IgA+ cells then migrate to MLN and then via thoracic duct to circulation, arriving LP. PPs, Peyer's patches; MLN, mesentric lymph node; LP, lamina propria; DC, dendritic cells; MØ, macrophages; APC, antigen-presenting cells; FAE, follicle-associated epithelial cells; SED, subepithelial dome; IFR, interfollicular region.

occurs in the T cell regions adjoining the B cell follicles thus facilitating the transformation of naïve CD4 T cells into either Th1 or Th2 type cells (Fig. 7.1). GALT also has a diffuse component comprising of large number of lymphocytes in the mucosa lining the gut. Several reviews contain detailed descriptions of the anatomic structures and cell subtypes observed in the GALT (McGhee *et al.*, 2007; Mason *et al.*, 2008).

In the respiratory tract, the nasal associated lymphoid tissue (NALT) in the nasopharynx is the principal inductive site and often the target tissue for nasal vaccination (Kiyono and Fukuyama, 2004). So far NALT has been characterized in rodents as an aggregation of organized lymphoid tissue found on both sides of the nasopharyngeal duct which is considered analogous to the Waldeyer's ring in humans (Perry and Whyte, 1998). They are important inductive sites for priming of immunocompetent cells to induce antigen-specific mucosal immune responses in the upper respiratory tract and other mucosal effector tissues in rodents (Shimoda *et al.*, 2001; Zuercher *et al.*, 2002). In experimental models, nasal immunization induces antigen-specific IgA responses in nasal passages and at other distal mucosal sites indicating the importance of NALT in host immune defence (Imaoka *et al.*, 1998; Kurono *et al.*, 1999; Hou *et al.*, 2002). In humans, NALT-like structures of lymphocyte exist in nasal mucosa of infants (Debertin *et al.*, 2003). Therefore, NALT tissues might be the likely immune targets for inhalative vaccination strategies in young children and play a vital role in mucosal host defence (Imaoka *et al.*, 1998; Kurono *et al.*, 1999; Hou *et al.*, 2002; Hussell and Humphreys, 2002; Debertin *et al.*, 2003). The cellular composition of NALT is similar to that of the GALT, although the relative numbers and subtypes of B and T cells differ (Heritage *et al.*, 1997).

Bronchus-associated lymphoid tissue (BALT), consists of discrete lymphoid aggregates located mostly at bifurcations of the bronchus. The BALT is present during fetal stages and rapidly develops following birth, especially in the presence of antigens (Luhrmann *et al.*, 2002; Bienenstock and McDermott, 2005). BALT has an organized component following airways with B and T cell-rich areas. These aggregations are clustered in follicle-like structures (Tirouvanziam *et al.*, 2002) and are composed of B cells bordered by a parafollicular area composed primarily of T cells. The organization of the BALT is similar to that of PP as both provide for the entry of mucosal pathogens through special M cells, and are also involved in local immune protection (Sminia *et al.*, 1989). There is a diffuse component in BALT similar to the LP in gut containing multiple lymphocyte populations. Further, there is a constant circulation of lymphocytes that enter the BALT from blood via high endothelial venules (HEVs) and leave via lymphatics.

In adults, BALT is absent in healthy individuals, but its development is induced under various disease conditions. In mice, BALT is only occasionally present (Moyron-Quiroz *et al.*, 2004) but repetitive inhalations with heat-killed bacteria induced its development (Toyoshima *et al.*, 2000). This demonstrates that BALT is inducible by infection or inflammation in rodents very much similar to that in humans. Thus, it seems conceivable that induced BALT facilitates primary immune responses to respiratory infections.

Several other subdivisions of MALT have been used in literature, but very little is known about them. These are TALT (Eustachian tube-associated lymphoid tissue), SALT/DALT (salivary gland- or duct-associated lymphoid tissue), CALT (conjunctiva-associated lymphoid tissue), LDALT (lachrymal-drainage-associated lymphoid tissue) and LALT (larynx-associated lymphoid tissue) (Brandtzaeg *et al.*, 2008).

Innate immunity at mucosal surfaces

The innate immune system provides the first line of protection against ingested and inhaled infectious agents. The cells of the innate system act both as physical and immunological barriers, playing a critical role in maintaining the structural integrity of the intestine and also in protecting the host against a vast number of potential pathogens.

Innate sensing of pathogens, commensals or vaccines

The innate system primarily senses via the pattern recognition receptors (PRRs), an essential component for recognizing the microbial motifs

(Takeda *et al.*, 2003). The PRRs recognize evolutionarily conserved molecular patterns present in microorganisms, collectively known as the pathogen-associated molecular patterns (PAMPs) (Takeda *et al.*, 2003). The interaction between the PRRs and the PAMPs result in induction and expression of genes required to eliminate the pathogens and to initiate adaptive immune responses (Takeda *et al.*, 2003). The Toll-like receptors (TLRs), the cytoplasmic nucleotide oligomerization domain (NOD) signalling, mannose and other scavenger receptors all serve as the PRRs in the innate immune defence system. Additionally, the epithelia also produce various antimicrobial peptides or defensins, which represent a crucial element of the host innate mucosal immune defence. Any dysfunction of the mucosal innate system may impair the host's ability to mount an immune defence leading to an invasion of pathogens resulting in chronic inflammation, thus illustrating the pivotal role of innate immunity in maintaining immune protection. Additionally, the innate immune machinery at mucosal sites serves as a link to the cells of the adaptive immune system.

TLRs

TLRs play a vital role in distinct recognition of pathogen receptors and in subsequent initiation of innate immune responses at mucosal surfaces. TLRs are widely expressed in a variety of mucosal cell types, including epithelial cells (Cario and Podolsky, 2000; Cario *et al.*, 2000; Abreu *et al.*, 2001; Gewirtz *et al.*, 2001a; Haller *et al.*, 2002; Hornef *et al.*, 2002), macrophages (Smith *et al.*, 2001; Hausmann *et al.*, 2002) and DCs (Cerovic *et al.*, 2009). Some TLRs (1, 2, 4, 5, 6, 10 and 11) are found on the cell surface, while other TLRs (3, 7, 8 and 9) are known to reside within the endosomal compartments (Takeda and Akira, 2007). TLR signaling is linked to MyD88-dependent and TRIF-dependent signaling pathways that regulate the activation of different transcription factors, such as NFκB. The PAMPs recognized by TLRs are not unique to pathogens, but rather are shared with other benign bacteria, which inhabit the mucosal surfaces in large numbers. Hence, it's intriguing that innate immune responses in the mucosa are regulated by the requirement to

identify pathogens specifically and yet remain inert to harmless bacteria.

The expression and activation of TLRs in mucosal surfaces are highly dynamic and tightly regulated in order to exclusively respond to microbial PAMPs. In humans, primary intestinal epithelial cells have been shown to express most of the TLRs (1–9), but despite the presence of these receptors, TLR signalling in the normal gut appears to be muted (Melmed *et al.*, 2003; Otte *et al.*, 2004; Abreu *et al.*, 2005). The expression of TLRs 4 and 2 is dampened in isolated primary intestinal epithelial cells, whereas TLR3 and TLR5 are expressed at higher levels (Cario and Podolsky, 2000). These levels of expression may change with disease within the colonic mucosa, as there are reports of increased TLR2 and TLR4 during intestinal inflammation (Cario and Podolsky, 2000).

Similarly, human LP macrophages that express mRNA for TLR2 and TLR4, lack the expression of CD14, a GPI-linked glycoprotein that acts as a receptor for complexes of LPS (Smith *et al.*, 2001). The absence of CD14 on LP macrophages and the absence of a LPS mediated inflammatory response offer a likely explanation for the refractory nature of TLR induced signalling in normal gut despite its close proximity to bacterial derived LPS. Likewise, murine LP DCs that express mRNA for most TLRs, express low TLR4 mRNA by RT-PCR in the steady state and are non-responsive to LPS stimulation *in vitro* (Uematsu *et al.*, 2008). Although down regulation of TLRs in intestinal immune cells seems to be a interesting mechanism to distinguish commensals from microbes, recent evidence does suggest that murine intestinal DCs and epithelial cells do express high levels of TLR5 that recognizes flagellin (Bambou *et al.*, 2004; Begue *et al.*, 2006; Uematsu *et al.*, 2008). Activation of LP DCs via TLR5 leads to a subsequent generation of IgA+ B cells and antigen-specific IL-17 and IFN-γ producing CD4+ T cells demonstrating a critical role for TLR5 in gut immunity. In addition, flagellin the only known PAMP for TLR5 derived from pathogenic *Salmonella* species but not commensal bacteria is able to stimulate secretion of the chemokines such as CCL20 by IEC, leading to the recruitment of immature DCs (Sierro *et al.*,

2001). Thus, signalling by TLR5 in response to pathogen-derived flagellin results in the recruitment of inflammatory cells and the initiation of an adaptive immune response to pathogens.

For most currently licensed mucosal vaccines, the contribution of TLRs has not been studied. Vaccines containing killed or attenuated pathogens most likely contain multiple TLR ligands, but there is no evidence to date that the immunogenicity of these vaccines is influenced by TLR engagement. In contrast, there is emerging data that parenterally administered live vaccines do stimulate TLRs and that the TLR triggering may in part mediate their immunogenicity (Querec et al., 2006; Querec and Pulendran, 2007). Future studies will determine if TLR signalling has a significant role in protective immune responses to established mucosal vaccines, including the highly successful oral polio vaccine.

PAMPs recognition by cytosolic PRRs

TLR signalling occurs via interactions with adaptor proteins including MyD88 and toll-receptor associated activator of interferon (TRIF). In addition to TLR signalling, PRRs such as Nod1, Nod2 and Ipaf are involved in cytosolic detection of microbes. NOD1, also known as caspase recruitment domains (CARD4) senses diaminopimelic acid- containing peptide glycan, mainly found in Gram-negative bacteria (Chamaillard et al., 2003) while NOD2 known as CARD15 is senses the muramyl dipeptide (MDP) an immunogen present in both Gram-positive and Gram-negative peptide glycan (Inohara et al., 2003). The NOD signalling acts via the inflammasome and involves activation of Rip-like interactive clarp kinase (RICK) (Park et al., 2007).

Numerous intracellular pathogens such as *Shigella flexneri* (Girardin et al., 2001) and enteroinvasive pathogens such as enteropathogenic *Escherichia coli* (EPEC) (Kim et al., 2004), *Chlamydophila pneumoniae* (Opitz et al., 2005) have been shown to be sensed via the NOD1 signalling. NOD2 has also been implicated in intracellular bacterial sensing. Mice deficient in NOD2 were defective in innate mucosal immune defence against oral infection with *Listeria monocytogenes* indicating a vital role for NOD2 signalling in host defence (Kobayashi et al., 2005). These studies

suggest a potential important link between Nod proteins in the innate immune response to bacterial infection.

Recently, flagellin was shown to have a TLR5-independent proinflammatory activity that depends on two related intracellular pattern recognition receptors, Naip5 (Lightfield et al., 2008) and Ipaf (Franchi et al., 2006). In contrast to TLR5 mediated signalling, NLR proteins do not act through MyD88, and while NF-κB activation does occur after stimulation, the major effect of these receptors appears to be an effect on activation of caspases (Franchi et al., 2006). The principal mechanisms of Naip5 or Ipaf-dependent flagellin signalling remain largely unknown, but their contribution to the detection of flagellin is important to initiate innate immunity in the gut.

These intracellular detectors of PAMPs are well adapted to recognize the presence of pathogenic bacteria as compared to commensal organisms. Targeting such intracellular pathways may be vital in designing an effective mucosal vaccine. In this respect, it is to be noted that most bacteria infecting via the mucosal routes possess various TLR ligands including peptide glycan, which can be sensed via Nod1 and/or Nod2 after degradation by host enzymes. Therefore, the host innate immune responses to invading bacteria could potentially possess dual signalling through both extracellular TLR molecules and intracellular Nod molecules leading to the synergistic activation of host immune cells.

Epithelial cells and Paneth cells

Non-haematopoietic cells actively contribute to the vibrant immune responses at mucosal sites by producing various protective factors and also through their interactions with the cells of the adaptive immune system (Ayabe et al., 2000; Cario et al., 2002). For example, the mucosal epithelial cells and Paneth cells both produce a variety of antimicrobial peptides and bacteriolytic enzymes that protect mucosal surfaces against microbes (Porter et al., 2002). These antimicrobial peptides include various bactericidal peptides/proteins such as defensins, cathelicidins, cryptidin-related sequence (CRS) peptides, bactericidal/permeability-increasing protein (BPI), chemokines as well as antimicrobial enzymes, lysozyme and group IIA phospholipase A2 (Ganz, 2003). The cationic antimicrobial molecules (CAMs) have broad spectrum antimicrobial activity and act via damaging the integrity of the bacterial membranes by pore formation (Zasloff, 2002; Ganz, 2003).

Defensins are a class of peptide products made by either neutrophils (α-defensins) or epithelia (β-defensins) having broad spectrum antimicrobial activity (Ganz, 2003). The antimicrobial activity of defensins is similar to that of other cationic antimicrobial peptides and is thought to involve disruption of the membrane integrity of the microbial targets (Ganz, 2003). In addition to defensins, Paneth cells also secrete other antimicrobial proteins and peptides such as lysozyme, Reg3γ (Ismail and Hooper, 2005), and secretory phospholipase A$_2$ (Keshav, 2006). Unlike the α-defensins, which are expressed in human Paneth cells prenatally (Wehkamp and Stang, 2006), the expression of Reg3γ sharply responds to the bacterial presence in the intestinal lumen (Ismail and Hooper, 2005). Additionally defensins also have chemoattractant properties for cells expressing the chemokine CCR6 such as dendritic cells (DCs), suggesting that these defensins could serve as a bridge between the innate immunity at the intestinal mucosa and subsequent adaptive immune responses (Yang *et al.*, 1999).

The epithelial cells are the immediate interface between the pathogens and the immune system, since they function as sensors that can bind to microbial components through TLRs (Fig. 7.2) (Ganz, 2003). Activation of TLRs on epithelial cells triggers an array of responses, including production and release of cytokines/chemokines and anti-microbial peptides (Modlin and Cheng, 2004; Sansonetti, 2004; Eckmann and Kagnoff, 2005). The expression of TLRs by the epithelial layer is tightly regulated to distinguish the invading pathogens from commensals. For example, the induction of human β-defensin 2 (HBD-2) by pathogenic and not commensal bacteria appeared to be involved in discriminating pathogens from commensals (Chung and Dale, 2004). The HBD-2 induction by epithelial cells leads to subsequent NF-κB mediated signalling and gene transcription, which typically are associated with cytokine production. The DCs and macrophages inhabiting the underlying mucosal tissues respond to these signals and promote adaptive immune responses. Hence, the ability of the epithelial cells to modulate specific immune functions in the airway and gastrointestinal tract offers the prospect of novel vaccination strategies to target multiple diseases.

Antigen presentation and intestinal immunity

The layer of epithelial cells joined by tight junctions produces innate defences in the form of mucins and antimicrobial proteins. In most instances, these secretions can contain and eliminate the local infections. Nevertheless, foreign antigens and microorganisms frequently breach the epithelial barrier and the underlying mucosal tissues are sites of intense immunological activity. The deeper mucosal tissues are rich in specialized antigen-presenting cells (APCs) capable of sensing these pathogens and mounting appropriate local responses. In the absence of pathogens, mucosal APCs are known to get into a state of unresponsiveness ignoring the antigen or inducing a regulatory response. Upon recognition of microbes that breach the mucosal barrier, mucosal APCs are capable of recognizing and mounting a robust protective response (Fig. 7.2). In this process, pathogens are phagocytosed and presented to T cells, initiating long-lasting adaptive immunity.

Mucosal dendritic cells

DCs are unique APCs capable of initiating an adaptive immune response by efficiently activating naïve T cells (Pulendran, 2005). In mucosal tissues they are strategically situated at sites of pathogen entry and are among the first cells to recognize incoming pathogens through a set of PRRs (Iwasaki, 2007; Pulendran *et al.*, 2008; Manicassamy and Pulendran, 2009). Upon direct activation of PRRs, DCs mature and migrate to T cell rich regions of the organized lymphoid tissues in both PP and MLN where they activate and stimulate naïve T cells (Fig. 7.1) (Iwasaki, 2007). In transit, DCs undergo a maturation programme that results in the up regulation of surface major histocompatibility complex (MHC) molecules

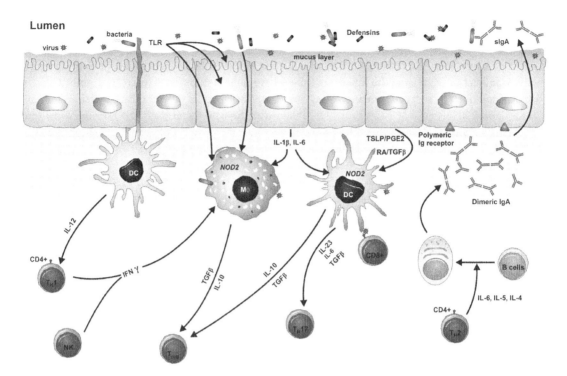

Figure 7.2 Model for intestinal bacteria-mediated differentiation of adaptive immune cells in the LP. A single layer of epithelial cells separates the intestinal LP from the pathogens present in the lumen of the gut. The preliminary protection is mediated by a thick layer of mucus, bactericidal defensins, and large amounts of antigen-specific secretory IgA present on the surfaces. Intestinal microflora interacts with immune cells in the LP through variety of mechanisms. Following innate immune induction via the TLR induction, epithelial cells produce factors such as TSLP and PGE2 that trigger the maturation of DCs and induce the polarization of inflammatory Th2 cells. After the interaction with the epithelial cells, bacteria or their fragments are internalized. The first cells that interact with them are the antigen-presenting cells such as macrophages and dendritic cells. Bacterial TLR ligands activate specific subsets of APCs, thereby inducing the differentiation of IgA+ plasma cells and Th17 cells. CX3CR1+ dendritic cells extend their dendrites into the lumen, sense the luminal bacteria, and produce cytokines (IL-12, IL-6 and TGF-β) promoting differentiation of Th1/Th17 or Tregs cells. A subset of intestinal DCs (CD103+ DCs) or CD11b+ macrophages produce retinoic acid and other factors to induce differentiation of Tregs. Cytokines such as IL-6, IL-5 and IL-4 made by activated CD4+ T cells would favour the clonal expansion of IgA B lymphocytes, resulting in their conversion to plasmatic cells in the LP of the gut. TLR, Toll-like receptor; APC, antigen-presenting cell; LP, lamina propria; DCs, dendritic cells; TSLP, thymic stromal lymphopoietin; PGE2, prostaglandin E2.

(Pulendran, 2005; Steinman and Banchereau, 2007). Once in the local draining lymph node, DCs present antigens derived from the pathogens to naive T cells and induce differentiation of these cells by secreting appropriate cytokines.

DCs are abundant throughout the MALT and lymphoid organs and there is constant recirculation of the antigen loaded DCs from the mucosal wall to the local draining lymph nodes in steady state conditions (Iwasaki, 2007). In the GALT, DCs are found in LP beneath the conventional villus epithelium, as well as in the organized lymphoid tissues of PP and MLN (Iwasaki and Kelsall, 1999). Several mucosal DC subsets have been described, but their contributions to immune function at mucosal sites and tolerance induction to commensals have begun to unravel only recently.

The mouse LP DCs have been reported to consist of two major subsets of DCs: CD11c^bright CD8α^dull CD11b+ DCs, which stimulate Th17 responses (Denning *et al.*, 2007), and CD11c^bright CD8α+ CD11b− whose function is unclear at present (Chirdo *et al.*, 2005). In addition, a small proportion (2–5%) of plasmacytoid DCs (pDCs) have been described in the LP

(Chirdo *et al.*, 2005). Lamina propria DCs can also be classified, based on the expression of the chemokine receptor CX3CR1 (the receptor of CX3CL1, fractalkine) (Niess *et al.*, 2005) and into DCs that express the α-integrin CD103 (Annacker *et al.*, 2005; Johansson and Kelsall, 2005; Johansson-Lindbom *et al.*, 2005; Coombes *et al.*, 2007; Sun *et al.*, 2007). The connection between CX3CR1, CD103 and the 'conventionally defined' subsets is at present murky, but our recent work suggests that CD11c+CD8α−CD11b+ DCs in the LP are CX3CR1+ and a major fraction of this subset also expresses CD103 (Denning *et al.*, 2007). In addition, a proportion of the CD11c+CD8α+CD11b− DCs in the LP are CX3CR1− and CD103bright (Denning *et al.*, 2007). The functional properties of various subsets of DCs in the LP are only now beginning to be appreciated. CX3CR1+ CD11b+ DCs directly access the intestinal lumen by extending transepithelial dendrites (Niess *et al.*, 2005). Furthermore, both the CD103+ and CD103− DC subsets in the LP seemed capable of converting naïve T cells to T regulatory cells via a mechanism dependent on the conversion of vitamin A to retinoic acid, which stimulates T regulatory cells (Sun *et al.*, 2007). Whereas the ability to convert vitamin A to retinoic acid seems to be constitutively present on intestinal antigen-presenting cells (Benson *et al.*, 2007; Denning *et al.*, 2007; Sun *et al.*, 2007), it has recently been demonstrated that such a mechanism can be induced in splenic DCs by TLR2 dependent stimuli (Manicassamy *et al.*, 2009). In the MLN, in addition to the CD11cbrightCD8α+CD11b− and CD11cbrightCD8α−CD11b+ two additional DC subsets can be defined using the differential expression of CD4 and DEC-205 (Johansson and Kelsall, 2005). However, the functional significance of these remains unknown.

The PP are organized mucosal inductive sites vital in mucosal immune responses to antigens, pathogens, and vaccines in the intestine. The antigens and vaccines transported across the microfold cells encounter multiple DC populations (CD8α−CD11b+B220-DCs (myeloid) and CD8α−CD11b−B220-DCs) immediately under the FAE in the 'subepithelial dome' (SED) region (Iwasaki and Kelsall, 2001; Johansson and

Kelsall, 2005). The FAE also expresses high levels of the chemokines CCL9, CXCL16 and CCL20 following infection thus facilitating the migration of CCR6+ DCs towards the FAE (Iwasaki and Kelsall, 2000; Zhao *et al.*, 2003b; Salazar-Gonzalez *et al.*, 2006). The CCR6+ DCs could potentially facilitate microbial recognition and the subsequent local immune responses (Salazar-Gonzalez *et al.*, 2006). Further, the inter-follicular region (IFR) contains CD11b−CD8α− double negative (DN DCs) and CD11b−CD8α+ DCs (Lymphoid) (Iwasaki and Kelsall, 2001). These DC subpopulations secrete a distinct pattern of cytokines upon exposure to T cell and microbial stimuli. The myeloid DCs produce high levels of IL-10 upon soluble CD40 ligand, or microbial (e.g. *Staphylococcus aureus*) stimulation. In contrast, both the lymphoid and the DN DCs, produce IL-12p70 following microbial stimulation (Iwasaki and Kelsall, 2001). The myeloid DCs due to their ability to make IL-10 are competent in priming naive T cells to secrete high levels of IL-4 and IL-10, while lymphoid and DN DCs prime a Th1 response (Iwasaki and Kelsall, 2001). This suggests that DC subsets in the PP are functionally different in the setting of either T cell activation or in their ability to respond to a microbial signal.

The mucosal system also comprises DCs resident in the genitourinary (GU) tracts. In steady state, the main DC populations in the genital mucosa consist of those that reside within the stratified squamous epithelium (Langerhans cells) and those that reside in the LP of the GU tracts (submucosa DCs) (Iijima *et al.*, 2008). In a murine model of herpes simplex virus (HSV)-2, the blood-derived Gr-1(hi) monocytes precursors were found to repopulate into vaginal epithelial DCs (Iijima *et al.*, 2007). All the blood derived DCs expressed both classical DC marker 'CD11c' and MHC class II but consisted of CD11b+F4/80hi, CD11b+F4/80int, and CD11b−F4/80− populations (Iijima *et al.*, 2007). It is unknown whether these distinct populations possess any specific immune functions at the vaginal mucosa. During HSV-2 infection, distinct CD11b+ submucosal DCs but not Langerhans or CD8α+ DCs migrate rapidly to the LP bordering the infected vaginal epithelial cells, phagocytose viral antigens and migrate to local lymph nodes to induce protective Th1 CD4+ T cell responses (Zhao *et al.*, 2003a). While Th1-inducing immune response depends on Myd88 expression by DCs the maturation and migratory capacity of these cells is independent of Myd88 signalling (Sato and Iwasaki, 2004). In addition to the submucosal DCs, intravaginal infection with HSV-2 recruits pDCs to the vaginal tissues that are capable of producing large amounts of IFN-α, capable of inducing innate defence (Lund *et al.*, 2006). Future studies should be aimed at understanding the function of both the resident and blood-recruited populations in the local GU tract following infection and intravaginal vaccination.

Studies have shown that an extensive network of DCs reside within the mucosa of the nasal tract and the large conducting airways, the alveolar lumen and septum (Lambrecht *et al.*, 1998; GeurtsvanKessel and Lambrecht, 2008). There exists an extensive phenotypic and functional heterogeneity in DCs within the lungs and its adjoining compartments. In murine lungs, DCs can be broadly subdivided into conventional (cDCs) and plasmacytoid DCs (pDCs) (GeurtsvanKessel and Lambrecht, 2008). The cDCs can be further subdivided into CD11b+ or CD11b- DCs (GeurtsvanKessel and Lambrecht, 2008). In addition intraepithelial DCs resembling Langerhan cells exist in the trachea and large conducting airways (GeurtsvanKessel and Lambrecht, 2008). The local lung-draining lymph node contains three distinct DC populations, namely CD11b+F4/80+, CD11b-F4/80+ and CD11b-F4/80- (GeurtsvanKessel and Lambrecht, 2008).

Compared to the DCs that reside in the periphery, lung DCs reside in largely immature state specialized for antigen uptake and recognition. These DCs are capable of activating naive T cells and differentiate them into predominantly tolerizing T regulatory cells (Lambrecht *et al.*, 2001). In steady state, resting lung myeloid DCs (CD11b+) induce a Th2 response to inhaled harmless antigen (Lambrecht *et al.*, 2000). This response is dramatically changed following inflammatory or infectious condition. Depending on the stimulus, lung DCs can induce Th1, Th2, Th17 or cytotoxic CD8 T cells (Lambrecht *et al.*, 2001). The concept that lung dendritic cells could

be specifically targeted by intranasal vaccination is attractive.

The functional insights provided by characterization of mucosal DCs will undoubtedly advance strategies for mucosal vaccine design. Experimental approaches to target specific subsets based on their ability to activate a protective immune response will be important in future vaccine strategies. Further, understanding the *in vivo* mechanism for the immunostimulatory activity of cholera toxin and other mucosal adjuvants on DCs and its effect in modulating DCs would be crucial in the design of a future mucosal vaccine.

In addition to understanding the location of DCs, it is important to understand the nature of the DCs, which determine how DCs can be activated to allow mucosal priming. The local microenvironment plays a vital role in shaping DC function. Thymic stromal lymphopoietin (TSLP) was identified as one of the conditioning factors released by epithelial cells. Interestingly, TSLP conditioned DCs have been shown to support IgA class switching of B cells in the LP by their ability to make both BAFF (B cell-activating factor) and APRIL (a proliferation inducing ligand) (Castigli *et al.*, 2004; Xu *et al.*, 2007). Additionally both mouse and human epithelial cells also release retinoic acid (RA) and TGF-β that can drive the development of a regulatory or tolerogenic phenotype on DCs (Iliev *et al.*, 2009). Altogether these findings indicate that factors available in the local environment are vital in conditioning the phenotypic changes in DCs.

Furthermore, intestinal DCs are subject to tissue-specific conditioning in the GALT that endow them the capacity to establish T cell gut tropism (Johansson-Lindbom *et al.*, 2003). DCs play a vital role in the generation of gut-homing T cell populations, by providing GALT-specific signals, leading to the generation of CCR9+ $\alpha_4\beta_7^+$ T cells (Johansson-Lindbom *et al.*, 2003). Importantly, in addition to this modulation of the homing properties of lymphocytes to mucosal sites, gut DC-derived RA was identified as an important mediator of IgA secretion in B cells (Mora *et al.*, 2006). The ability of mucosal DCs in efficient generation of gut tropic signals can in turn orchestrate the downstream adaptive immune responses that are designed for defence against live pathogens.

Macrophages

Macrophages, the major population of tissue-resident mononuclear phagocytes, play key roles in bacterial recognition and elimination at mucosal sites. They are abundant in the mucosal tracts of both humans and mice. In general, intestinal macrophages express low levels of MHC II in steady state and are refractory to any external stimuli (Smythies *et al.*, 2005; Kelsall, 2008). Macrophages are capable of upregulating the level of MHC II on their surface and thereby gaining enhanced phagocytic capability, critical in elimination of pathogens within the mucosal tissues (Kelsall, 2008). In general, intestinal macrophages have poor expression of most innate response receptors, leading to suppressed NF-κB signalling, possibly due to the presence of suppressive cytokines, such as IL-10 or TGF-β (Kelsall, 2008). One reason for the refractory nature of mouse intestinal macrophages could possibly be its ability to constitutively produce IL-10 (Denning T *et al.*, 2007). Additionally, these macrophages have the ability to convert naïve CD4 T cells to regulatory T cells *in vitro*, due to their ability to make both IL-10 and RA (Denning *et al.*, 2007; Manicassamy and Pulendran, 2009). In summary, intestinal macrophages in the steady state lack an ability to induce proinflammatory mediators but have retained their capacity to phagocytose and kill invading microbes. Macrophages also play an important role in maintaining the intestinal homeostasis by negatively regulating excess immune responses by their ability to induce T_{regs}.

Adaptive immune protection at mucosal surfaces

Immune defence against mucosal pathogens involves multiple strategies that operate on both sides of the epithelial barrier. The deep tissues beneath the epithelial barrier have a unique and complex immune system composed of a variety of adaptive immune cell populations, including IgA-producing plasma cells, γδ T cells, CD4+ T cells and cytotoxic CD8+ T cells. The present section will review our current knowledge about the nature of the adaptive immune responses at mucosal sites.

Mucosal B cell responses

The hallmark of the mucosal immune system is its ability to produce and secrete dimeric or multimeric IgA that unlike other antibodies are resistant to degradation in the protease rich mucosal microenvironment (Macpherson *et al.*, 2008). In humans, more IgA is produced than any other immunoglobulin isotypes combined, and high amounts of IgA antibodies are present in the secretions that are associated with mucosal surfaces (Mora and von Andrian, 2008). High-affinity IgA acts to exclude microorganisms and toxins from entering the body, whereas low-affinity IgA is thought to inhibit the binding of commensal bacteria to epithelial cells. Furthermore, the latter may be important in the maintenance of intestinal microbiota (Cerutti, 2008). The importance of secretory IgA (sIgA) is evidenced by the fact that mice deficient in the secretory component (polymeric Ig receptor), exhibit a higher susceptibility to primary *Salmonella* infections (Wijburg *et al.*, 2006), although they have normal recall responses (Uren *et al.*, 2005).

Intestinal IgA+ B cells are generated from IgM+ B cells by *in situ* class switching in two separate gut microenvironments: organized follicular structures (PP, MLN) (Fig. 7.1) and in the LP (Fig. 7.2) (Fagarasan *et al.*, 2001). The LP microenvironment aids both class switching and differentiation of IgM+ B cells to IgA plasma cells (Fagarasan *et al.*, 2001). In the LP, TLR-dependent recombination-inducing factors, APRIL and its homologue, BAFF made by epithelial cells, DCs, and B cells, are thought to drive T-cell-independent IgA B-cell differentiation (Castigli *et al.*, 2004). Naive B cells are also induced to undergo class switch recombination and somatic hypermutation in organized lymphoid tissues such as PP (Fig. 7.1) and MLN of the intestine in addition to the processes in LP. Such switching requires specific signals in the form of costimulating molecules, cytokines and the presence of activated CD4+ T cells. In organized tissues, availability of cytokines such as TGF-β, IL-6, and IL-10, drive the expression of AID, an enzyme essential in isotype switching

(Cerutti, 2008). AID expression in these organized lymphoid tissues is thought to be responsible for T-cell-dependent IgA B-cell differentiation (Cerutti, 2008). IgA B cells or IgA plasmablasts then migrate from the organized lymphoid tissues to the LP of the gut (Cerutti, 2008). The key receptor–ligand pair involved in traffic of IgA B cells is $\alpha_4\beta_7$ integrin expressed by lymphocytes and MadCAM-1 adhesion molecules expressed by LP high endothelial venules (HEV) (Bowman *et al.*, 2002). Mucosal exposure to antigens seems to favour expression of $\alpha_4\beta_7$ integrins on mucosal B cells, and intranasal immunization has been shown to induce expression of L-selectin as well as $\alpha_4\beta_7$ integrins. However, systemic immunization is generally restricted to the expression of L-selectin (Quiding-Jarbrink *et al.*, 1997). The preferential homing of IgA B cells to gut is also explained by their high expression of CCR9 and the subsequent up regulation of its ligand CCL25/TECK in the small intestine (Mora *et al.*, 2006; Mora and von Andrian, 2008). Migration of B cells to LP is critical for conferring protection against intestinal pathogens (Kuklin *et al.*, 2001). The antigen-sensitized B cells undergo terminal differentiation in the mucosal LP to IgA-producing plasma cells. Such differentiation involves interaction with a variety of cytokines and T-cell subsets (Fig. 7.2). The resulting sIgA molecules are designed to participate in immune exclusion and other immunological functions at the mucosal surface. There is currently growing interest in the possibility of developing oral vaccines that activate sIgA against infections in mucosal tracts.

Mucosal T-cell responses

Although sIgA has been shown to be an important effector molecule to protect mucosal surfaces, mucosal protection by T cells is also likely to play a pivotal role. The mucosa has a large reservoir of T cells that play a vital role in maintaining intestinal homeostasis and defence against intestinal pathogens. These T cells are localized within the various organized lymphoid organs or diffused throughout the intestine (Cheroutre and Lambolez, 2008). The intestine harbours unique subsets of T cells which differ phenotypically and functionally from conventional T cells seen

elsewhere in the body. In the gut, local defence depends in part on a subclass of mucosa-homing T lymphocytes called intraepithelial lymphocytes (IELs) that are nestled among gut epithelial cells protecting mucosal surfaces from infectious pathogens (Cheroutre and Lambolez, 2008). Many of these cells display unique phenotypic markers that reflect intermediate stages of T-cell differentiation (Cheroutre and Lambolez, 2008). The heterogeneous mucosal T-cell populations differ in both phenotype and function, hence can be subdivided based on T-cell receptor (TCR) and co-receptor expression (van Wijk and Cheroutre, 2009). The first group 'type a' includes CD4+ and CD8αβ+, αβ– IELs that primarily recognize antigens presented by the classical major histocompatability complex (MHC) class I and class II molecules and are primed within the systemic circulation. The second group 'type b' IELs includes CD8αβ +, αβ– and γδ– IELs that respond to antigens not restricted by classical MHC molecules (van Wijk and Cheroutre, 2009). There are probably multiple functions associated with the IELs, but presently very little is known about them. Several studies indicate that IELs play a pivotal role in immune regulation and maintaining the normal gut homeostasis (van Wijk and Cheroutre, 2009). Additionally, IELs have also been implicated in protective responses to certain pathogens (van Wijk and Cheroutre, 2009).

The gastro intestinal tract is abundant in IL-17-producing-cells, mostly represented mostly by CD4+TCRαβ and CD8ααTCRγδ T cells in the LP (Ivanov *et al.*, 2008). Similar to the intestinal tract, IL-17-producing cells are also present in high numbers at steady-state conditions within the airways, mostly been produced by invariant NK T cells, TCRγδ T cells, and non-T cells (Ivanov *et al.*, 2008). Mucosal APCs, especially CD11b+CD11c+ or CD103–CD11c+ DCs, are able to induce the production of Th17 cells in the LP (Coombes *et al.*, 2007; Denning *et al.*, 2007). Despite being involved in various autoimmune disorders, not all Th17 responses are detrimental to the host. IL-17-producing cells can play an important role in host defence against bacterial and fungal pathogens (Wu *et al.*, 2007). In case of gastrointestinal pathogens such as *H. pylori* (Luzza *et al.*, 2000), *S. typhimurium* (Raffatellu *et al.,* 2008) and *B. fragilis* (Chung *et al.,* 2003), IL-17 is known to play a protective role. Furthermore, Th17 cells have been implicated in respiratory infections caused by *Klebsiella pneumoniae, Mycoplasma pulmonis, Pseudomonas aeruginosa* and *Mycobacterium tuberculosis* (Huang *et al.,* 2004; Acosta-Rodriguez *et al.,* 2007; Zelante *et al.,* 2007). Recent evidence has shown that Th17-mediated protective responses are vital in vaccine-induced immunity (Higgins *et al.,* 2006; Khader *et al.,* 2007). In summary there is growing evidence that Th17 cells contributes to the overall immunity to mucosal pathogens.

Regulatory T cells

The immune system has evolved a variety of mechanisms to achieve and maintain tolerance against self-antigens and the plethora of environmental antigens present in mucosal surfaces. There is mounting evidence that T regulatory cells (T_{regs}) play an indispensable role in maintaining immunological tolerance and suppressing excessive immune responses in the host (Honda and Takeda, 2009). T_{regs} arise and mature in the thymus but there is growing evidence that they can develop extrathymically under certain conditions, particularly in the intestinal mucosa (Honda and Takeda, 2009). Induction of regulatory cells in the intestinal LP after mucosal delivery of antigens has been reported in several animal studies (Honda and Takeda, 2009). In LP, several subsets of APCs have been implicated in T_{regs} formation at steady state conditions (Coombes *et al.,* 2007; Denning *et al.,* 2007; Sun *et al.,* 2007). These APCs express retinal dehydrogenase (RALDH) and have the ability to produce RA, which mediates the expression of the gut homing receptors on T cells in addition to enhancing T_{reg} differentiation in the gut by cooperating with TGF-β. A better understanding of the mechanisms by which these regulatory T cells are activated in the intestinal mucosa may provide clues as how to use them as novel vaccine strategies against autoimmune diseases existing in the mucosal surfaces.

Mucosal vaccination

Mucosal vaccination has traditionally relied on approaches involving the mucosal delivery of antigens with adjuvants. Mucosal exposure to infectious agents and other foreign antigens often culminates in the development of mucosal immune responses. Therefore, the use of mucosal immunization strategies with specific vaccine antigens continues to remain the most attractive possibility for immunization against a vast majority of infections. Despite its important role, there are only a handful of mucosal vaccines that exist (Ogra *et al.,* 2001; Table 7.1). However, recent advances demonstrate that parenteral vaccination can also elicit antigen-specific T and B cells that migrate to mucosal sites (Masopust *et al.,* 2006), therefore there is much emphasis on understanding the innate parameters that determine homing of lymphocytes to mucosal sites. Therefore, in the current discussion, we consider 'mucosal vaccination' to be any strategy that stimulates T and B cell immunity at mucosal sites, either by mucosal delivery of a vaccine, or via induction of antigen-specific T and B lymphocytes.

Advantages of mucosal vaccination

Most vaccines available today are delivered by injection, with the associated problems of low vaccine compliance, logistics and safety. In recent years, there have been significant developments in needle-free vaccine delivery, among which, mucosal delivery is of particular interest (Lavelle and O'Hagan, 2006). The administration of mucosal vaccines does not require the use of needles, thus facilitating better compliance, reducing logistical challenges and potential safety issues from blood transmissible infections (Aziz *et al.,* 2007). Additionally, reduced adverse effects and the potential for frequent boosting may also represent further advantages over injectable vaccines. Finally, oral vaccines can be delivered outside a clinical setting without the need for skilled health personnel.

The characteristics of immune responses induced through the mucosal route confer several additional advantages over traditional systemic vaccination. Mucosal vaccines have the ability to induce specific mucosal immunity in addition to systemic immunity. Oral vaccination generates a large amount of sIgA, which plays a major role in mucosal defence (Lavelle and O'Hagan, 2006). In addition to a strong IgA response in mucosal surfaces, oral vaccination also induces systemic

IgG responses that represent an additional line of protection against the invading pathogens (Lavelle and O'Hagan, 2006). The concurrent stimulation of both mucosal and systemic responses is important for protection against opportunistic pathogens such as HIV that can infect the host through both systemic and mucosal route. Furthermore, mucosal immunization can stimulate cell-mediated responses including helper CD4+ T cells and cytotoxic T lymphocytes (Lavelle and O'Hagan, 2006). Thus, mucosal vaccines have the potential to activate all the different arms of the immune system and could be potentially exploited for combating pathogens acquired through non-mucosal routes.

Limitations of mucosal vaccination

Despite the many attractive features of oral vaccination, the real challenge is in overriding the harsh conditions prevailing in the gastrointestinal tracts such as acidic pH, proteolytic enzyme activity and the deterrent effects of bile salts. The instability and poor absorption of antigens in gastrointestinal tract are major obstacles in the development of oral vaccines (Jepson et al., 2004; George and Abraham, 2006). Nevertheless, several polymeric delivery systems are currently been designed to shield the antigen in the gut and to target the antigen efficiently to the effector sites in GALT. Further, administration of antigens to human gastrointestinal or respiratory tracts is known to induce a profound reduction in cell-mediated immunity as manifested by diminished DTH (Delayed type hypersensitivity), T cell proliferation, and secretion of cytokines with suppressor activity (Husby et al., 1994; Waldo et al., 1994; Matsui et al., 1996; Kraus et al., 2004). This phenomenon known as oral tolerance can occur via multiple mechanisms such as active suppression, anergy, or apoptosis of antigen-reactive cells (Mowat et al., 2004). The general belief is that mucosal vaccination strategies that lack the ability to induce costimulatory molecules on GALT APCs are programmed for tolerance (Bendelac, 1995; Kagnoff, 1996). Hence, current efforts to overcome obstacles posed by oral tolerance are mainly directed towards identifying newer and efficient antigen delivery systems capable of appropriately stimulating APCs in the mucosa

to drive a pro inflammatory response. While many hurdles must be overcome before newer mucosal vaccines become a practical reality, this is a potential area of research that has important implications in future oral vaccine development. An optimal vaccine against mucosally invasive pathogens may require the induction of different arms of immune responses in both mucosal and systemic sites.

Currently used mucosal vaccines

Polio vaccines

Poliovirus, the causative agent of poliomyelitis, is a human enterovirus that upon ingestion initiates infection at several sites, including the tonsils and PPs of the small intestine (Nathanson, 2008). From the primary sites of multiplication, virus drains into cervical and mesenteric lymph nodes. The virus then infects the cells of the GALT following transcytosis via M cells that express CD155, commonly known as poliovirus receptor (PVR) on their surface. The FAE lining over the PPs express the PVR, making them ideal candidates for viral entry into the deeper tissues of the mucosa (Nathanson, 2008). The virus then drains from its initial sites of infection in the mucosa to the local draining lymph nodes where it replicates further and spreads via the efferent lymphatic vessels and thoracic duct to enter into the bloodstream (Nathanson, 2008). In most instances, the natural infections in humans are cleared at this stage with mild symptoms. In rare instances (<1%) the virus spreads to the central nervous system (CNS) leading to permanent flaccid paralysis (Nathanson, 2008). Despite our understanding of the viral pathogenesis, there are huge gaps in our understanding of host responses to polio infection. Mucosal secretory IgA can block poliovirus replication in tonsils and intestinal tract. The circulating IgG and IgM antibodies can prevent the virus from spreading to the CNS (Mueller and Wimmer, 2003). In humans, the serum IgM responses are made transiently following vaccination, whereas the IgG response lasts for a longer time (Ogra and Karzon, 1969).

The polio vaccines are classical mucosal vaccines. Two different polio vaccines have been developed, the inactivated polio vaccine (IPV) by Salk and Youngner (Salk *et al.*, 1954) and the live, attenuated oral polio vaccine (OPV) of Sabin (Sabin *et al.*, 1960). The widespread use of these vaccines has led to near eradication of poliomyelitis. OPV is currently the vaccine of choice because it is inexpensive, easy to administer, and produces excellent immunity in the mucosal surfaces. Although OPV is safe and effective, in extremely rare cases the live attenuated vaccine virus can cause paralysis – either in the vaccinated child, or in a close contact. The Sabin OPV polio vaccine constitutes a model that has encouraged the development and use of other vaccines administered orally that have been licensed or presently in clinical trials.

While the polio vaccines are very effective and have nearly eradicated polio from the world, there is paucity in our understanding of both innate and cellular immune responses, involved in polio clearance. One reason for this is the lack of good animal models to study immunity associated with vaccination. The macaque model of polio infection or vaccination provides a unique opportunity for studying polio specific immune responses in non-human primates, which are naturally susceptible to poliovirus infection. In contrast, mice are not susceptible to poliomyelitis due to their lack of PVR expression (Crotty *et al.*, 2002). Researchers have however developed several lines of mice transgenics engineered to express human PVR (Ren *et al.*, 1990; Koike *et al.*, 1991; Horie *et al.*, 1994). These transgenic PVR (TgPVR) mice lines are highly susceptible to poliovirus when injected intravenously (i.v.) or intramuscularly (i.m.) and when injected intracranially (i.c.). Upon infection, TgPVR mice show signs of paralysis that closely resemble those in humans and monkeys (Ohka and Nomoto, 2001). Recently a new line of the TgPVR mice lacking the α/β interferon receptor (IFNAR) gene (hPVR-Tg/*Ifnar*KO) was generated, which were sensitive to the oral administration of polio virus (Ohka *et al.*, 2007). Thus hPVR-Tg/*Ifnar*KO can be considered as the first oral infection rodent model for poliovirus. Early studies performed on these mice imply that an interferon α and β response to poliovirus play a key role in the control and clearance of the infection (Ohka *et al.*, 2007). Future studies should be aimed at exploiting this transgenic line to study the anti viral responses at both mucosal and systemic sites. The lower costs relative to monkeys make this transgenic mouse line highly appealing for the neurovirulence and vaccine testing of OPV. Experiences with PVR transgenic mice may shed light on early events together with unravelling the complexities of the cellular responses that shape the host protective response. Additionally, studying transgenic mouse models of poliovirus infection would allow better visualization of viral pathogenesis and immunity *in vivo*.

Mucosal vaccines against enteric diseases

Enteric diseases take a heavy toll on the world's population, particularly among infants living in the developing world. Currently, it is estimated that 1.9 million deaths occur per year from diarrhoeal diseases (Girard *et al.*, 2006). Enteric diseases are caused by different viruses [adenoviruses, astroviruses, calicivirus (HuCV), rotavirus (RV)] and bacteria (*Campylobactor jejuni*, enterotoxigenic *E. coli* (ETEC), *Shigella*, *Salmonella typhi*, *Vibrio cholerae*). While several vaccines are now available for a few of these diseases (cholera, typhoid, rotavirus and influenza), none exists for the rest.

Cholera vaccine

Cholera is a disease of rapid onset, and leads to high lethality if left untreated. *Vibrio cholera* is the aetiological agent of cholera, and the pathogenicity of *V. cholerae* is mediated by the toxigenic effect of cholera toxin (CT), which disrupts ion transport of human intestinal epithelial cells (Sundaram *et al.*, 1991). When *V. cholerae* enters the small intestine, it colonizes the epithelium by means of vigorous flagellar activity, toxin-coregulated pili, and other virulence factors that aid in the attachment and colonization of bacteria in the mucus protected habitat. After successful colonization, *V. cholerae* secrete CT, a major enterotoxin responsible for the disease. CT triggers adenylate cyclase, subsequently increasing intracellular levels of cyclic AMP (cAMP), which in turn activates cAMP-dependent protein kinase affecting the normal ion transport mechanism of the cell, resulting in loss of water and electrolytes from the body.

Host immunity to cholera infection is primarily mediated by mucosal IgA antibodies directed against CT and LPS (Mestecky et al., 2008). These protective antitoxic antibodies in the gut prevent clinical manifestation by blocking toxin binding to epithelial GM1 ganglioside receptors (Sanchez and Holmgren, 2008). Because of its potent immunogenicity in mucosal tissues, CT has become the model mucosal immunogen and adjuvant. CT does not induce oral tolerance and can instead abrogate oral tolerance to unrelated proteins when administered simultaneously.

Presently three licensed oral cholera vaccines have been shown to be safe, immunogenic, and efficacious, at least for the induction of short term immunity that lasts for several months. The whole killed V. cholerae in combination with CT B subunit [whole cell/recombinant B subunit or WC/rBS (Dukoral)] confers a high level of protection (85–90%), but only for 6 months in all age groups (Black et al., 1987; Walker, 2005). Since CT cross-reacts to E. coli LT, the vaccine is capable of cross protecting against Enterotoxigenic Escherichia coli (ETEC) (Sundaram et al., 1991; Lopez-Gigosos et al., 2007). A variant of Dukoral vaccine containing no recombinant CT B-subunit has been produced and licensed in Vietnam (Trach et al., 2002). This vaccine (ORC-Vax) has shown an efficacy of 66% in all age groups in a clinical trial conducted in Vietnam (Trach et al., 1997, 2002). In addition, a bivalent O1/O139 whole cell vaccine developed in Vietnam is presently used in a large scale phase III trial in Kolkata, India, with active help from the International Vaccine Initiative (IVI) in Seoul. This vaccine has since been licensed for use in India and will be available shortly .

Compared to the killed whole-cell vaccine, a live oral attenuated vaccine offers great promise for preventing cholera because a single dose could cause active colonization, elicit high-titre serum antibodies, and stimulate mucosal immunity that is able to block bacterial adherence, kill the bacteria, and neutralize the toxin, thus providing excellent protection (Svennerholm et al., 1982; Butterton et al., 1993; Taylor et al., 1994). The CVD 103-HgR vaccine (Orochol, Berna Biotech,

Switzerland) is available for human use following successful trials in several countries. A single dose of this vaccine conferred robust protection (60–80%) in vaccinated volunteers challenged three months after vaccination (Hill et al., 2006). Although the vaccine has shown protective efficacy in multiple trials, the manufacturer Berna Biotech discontinued its production. Several other engineered live attenuated oral vaccine candidates have been developed, including CVD101, CVD103, CVD111, Peru-14, and Peru-15 (Rijpkema et al., 1992; Gotuzzo et al., 1993; Taylor et al., 1994; Kenner et al., 1995), and some are presently in late stages of clinical trials.

Typhoid vaccines

Typhoid fever is caused by Salmonella enterica serovar that infects humans and animals, causing a spectrum of disease ranging from systemic infection to gastroenteritis, depending on the particular bacterial serovar and the host species infected (Jones and Falkow, 1996). Typhoid fever is a systemic disease caused by Salmonella enterica, serovar typhi, a highly invasive enteric pathogen found almost exclusively in developing countries. Typhoid fever remains a serious public health problem throughout the world, especially in the developing world, with annual global incidence of around 16 million cases per year and accounts for 600, 000 deaths (2008). Salmonella enterica serovar typhimurium (S. typhimurium) infection of susceptible mouse strains is widely accepted as the best experimental model for studying human typhoid fever and has proved extremely valuable in uncovering mechanisms of innate and acquired immune resistance to intracellular pathogens (Ravindran and McSorley, 2005). Host defence against S. typhimurium infection requires significant contributions from both the innate and acquired arms of the immune system (Mittrucker and Kaufmann, 2000). The initial stages of infection are characterized by an innate immune response triggered by host recognition of several PAMPs such as flagellin, lipopolysaccharide (LPS), and lipoproteins. Each of these bacterial products can induce the production of inflammatory cytokines that are likely to contribute to the initial control of Salmonella infection (Gewirtz et al., 2001b; Sha et al., 2004). In addition to the innate protection, *Salmonella* infection also induces antigen-specific CD4[+], CD8[+], and B-cell responses, all of which can contribute to protective immunity (Ravindran and McSorley, 2005).

Currently there are two typhoid vaccines licensed for human use. These new-generation vaccines are the live, oral Ty21a and the injectable inactivated Vi polysaccharide.

The Vi polysaccharide vaccine This subunit Vi vaccine is a single dose injectable vaccine developed in the 1980s and licensed to Sanofi-Pasteur. The vaccine is composed of purified Vi capsular polysaccharide from the Ty2 *S. typhi* strain and elicits a T cell independent anti-Vi IgG response that is not boosted by additional doses. This vaccine is administered either subcutaneously (s.c.) or intramuscularly (i.m.). It is poorly immunogenic in infants and the serum antibody response in vaccine recipients is not boosted by additional doses of Vi antigen. The Vi vaccine provides protection in about 70% of the population and usually the immunity lasts for about 3 years (2008). The lack of long-term immunity has galvanized the development of newer variations of the Vi vaccine (Szu *et al.*, 1994; Kossaczka *et al.*, 1999; Lin *et al.*, 2001).

Ty21a live oral vaccine The attenuated *S. typhi* strain Ty21a was generated by chemical mutagenesis of wild type strain Ty2 and developed as the first live oral typhoid fever vaccine (Berna Biotech, Switzerland). This vaccine contains the ty21a strain of *S. typhi*, a mutant strain lacking UDP galactose epimerase enzyme implicated in completing the cell wall synthesis of bacilli (Gilman *et al.*, 1977). Once taken orally, these mutant bacilli colonize the mucosal surfaces but are unable to complete its cell wall synthesis. The bacilli are thus incapable of invading the mucosal cells and break open, liberating the antigens to the GALT. Host immunity is usually confined to the local mucosal sites without much systemic immunity. The oral Ty21a vaccine confers an efficacy of about 53–78% protection among vaccinated volunteers (2008).

Rotavirus vaccine Rotavirus (RV) is a genus of double-stranded RNA virus responsible for severe diarrhoeal disease and dehydration in infants. Worldwide, RV is estimated to account for 25–55% of all cases of severe diarrhoea, causing about 592,000 deaths mostly among infants (Parashar *et al.*, 2003). Rotavirus infections occur repeatedly throughout life. Most infections are asymptomatic or associated with mild enteric disease. Epidemiological studies of rotavirus have suggested that natural immunity is acquired following early exposure to the virus and that many children acquire immunity only following several rounds of infections (Mata *et al.*, 1983; Grinstein *et al.*, 1989; Linhares *et al.*, 1989; Raul Velazquez *et al.*, 1993). Rotavirus infection affects primarily the mature enterocytes on the tips of the small intestinal villi. Destruction of the affected cells reduces digestion and absorption of nutrients, resulting in diarrhoea. RV antigens are shed into the lumen of the intestine during an infection and can be detected by enzyme-linked immunosorbent assay (ELISA). The host immune response to rotavirus infection is partly defined. It is thought that control of RV infection is dependent upon mucosal IgA responses. However, work in animal models has shown that protection from reinfection can also be established in the absence of IgA (O'Neal *et al.*, 2000) and is thought to be primarily modulated by IgG.

Mice provide a good model for studying the immune response during a primary RV infection (Rose *et al.*, 1998). The mechanism of clearance of a RV infection in mouse has been difficult to define. Recombinase-activating gene-2[-/-] (RAG2[-/-]) and severe combined immunodeficiency (SCID) mice (deficient in both B and T lymphocytes) are unable to clear RV infection (Riepenhoff-Talty *et al.*, 1987; Franco and Greenberg, 1997), suggesting that either B or T lymphocytes mediate RV clearance. Further, mice that lack a functional T cell (TCR-deficient) are still capable of effectively clearing a primary infection, suggesting that clearance might be T cell independent (Franco and Greenberg, 1997).

Vaccine efforts have focused on the development of live, attenuated rotavirus strains of human and/or animal origin. In 1998, a rotavirus vaccine (Rotashield, Wyeth) was licensed for use in the United States. The vaccine was withdrawn following safety concerns after several cases of intussusception were reported following vaccination. Current vaccines RotaTeq, Merck (Matson, 2006) and Rotarix, GlaxoSmithKline (GSK) (O'Ryan, 2007) are safe and effective for reducing rotaviral gastroenteritis. In prelicensure trials, both vaccines showed high efficacy against severe RV gastroenteritis, and no side-effects (Ruiz-Palacios et al., 2006; Vesikari et al., 2006). Both vaccines are now included as part of the routine vaccination schedule for all infants in the United States and in other countries in Europe, Latin America and Australia.

RotaTeq, is a pentavalent live attenuated, three-dose oral vaccine developed by Merck Research Co. In a large clinical trial, the vaccine prevented 74% of cases of RV gastroenteritis, and 98% of cases of severe RV gastroenteritis in the first rotavirus season. Vaccine efficacy in the second season was 62.6% against any RV gastroenteritis and 88% against severe disease (Vesikari, 2008; Hyser and Estes, 2009). Vaccine efficacy was about 94.5% up to 2 years after vaccination. Results from a large clinical trial, as well as ongoing surveillance, indicate that the vaccine is safe and well tolerated, without any evidence of intussusception (Hyser and Estes, 2009).

A monovalent, live attenuated, two-dose oral vaccine Rotarix, was developed by GSK from a human RV strain G1P(8) isolated originally in Cincinnati (Bernstein et al., 1998). Rotarix induces an effective anti RV IgA response (Bernstein et al., 1998) but a lower IgG1 specific neutralizing antibody, indicating that serotype specific neutralizing antibodies might not be that important in controlling RV infection. Further research is needed to understand the targets and correlates of antiRV immunity.

Even though the present available mucosal vaccines are effective, immunogenicity is still not optimal as administration of several doses is required to induce modest protection. The Holy Grail in this area is to develop a prophylactic intervention that can induce robust protection in a solo dose for a longer period. A future breakthrough in vaccine research would potentially involve the use of genetically defined mutants that are highly immunogenic and relative safe to use.

Emerging mucosal vaccines against enteric diseases

Enterotoxigenic Escherichia coli (ETEC)
ETEC caused by a variety of E. coli strains remains the major cause of diarrhoea in the developing world (Abu Elyazeed et al., 1999). Disease caused by enterotoxigenic E. coli (ETEC) accounts for approximately 400,000 deaths annually (Abu-Elyazeed et al., 1999). Repeated ETEC infection in children leads to acquired immunity as reflected by lower incidence of infections with increasing age. Thus, a prophylactic intervention in infants and travellers may be an efficient strategy to control ETEC infections.

Cholera toxin is homologous to LT and vaccination with CT-B subunit (CT-B) -based vaccines elicits a protective short-lived immune response against ETEC strains (Rol et al., 2007; Lundkvist et al., 2009). A potential mucosal ETEC vaccine should, likely, contain the prevalent fimbrial antigens and/or a LT toxoid component, which imparts the crucial adjuvanticity to the preperation. To date, multiple strategies have been taken to deliver ETEC fimbriae and toxin antigens to the human immune system to elicit strong mucosal, in particular, intestinal immune responses that are considered to be of prime importance for protection against ETEC disease (Verdonck et al., 2008). The use of live attenuated Shigella vectors for expression of ETEC fimbrial and LT antigens seems to be an efficient strategy to protect against both Shigella and ETEC (Girard et al., 2006).

The use of ETEC candidate vaccines for protection against diarrhoea in adult travellers has been effective but they have been largely ineffective in children growing up in endemic areas (Girard et al., 2006). These findings have intensified efforts to improve the immunogenicity of different candidate vaccines, as well as to develop newer types of ETEC vaccines.

Shigellosis
Shigellosis is endemic throughout the world where it is responsible for over 120 million cases of severe diarrhoea, with about 1.1 million deaths worldwide mainly among children in the developing world (Kotloff et al., 1999). In addition, about

500,000 cases of shigellosis are reported each year among travellers and military personnel from developed countries. World Health Organization (WHO) has identified *Shigella* along with ETEC as priority targets for vaccines, and this has recently emerged as a '*Shigella*-ETEC vaccine initiative' by the Bill and Melinda Gates Foundation (Kotloff *et al.*, 1999). Although several strategies have been used to develop vaccines targeting shigellosis, none has been licensed for human use (Kweon, 2008). Most candidate vaccines centre around the usage of multivalent vaccine comprising of either killed or live vaccines that represents a mixture of serotypes prevalent in the endemic region. The choice of candidate antigens is complicated due to the large antigenic diversity among *Shigella* strains. Serotypic diversity due to the variations in different antigens is seen among Shigella strains and such diversity facilitates microbial evasion from the host immune system.

The promising Shigellosis candidate vaccines are the attenuated strains of *Shigella* used as live oral vaccines, parenterally administered O polysaccharide-carrier protein conjugates (Cohen *et al.*, 1996) and the intranasally administered proteosomes consisting of outer membrane protein vesicles of group B *Neisseria meningitidis* to which *Shigella* lipopolysaccharide is non-covalently bound (Mallett *et al.*, 1995). Host immunity seems to be dependent on the ability of these vaccines to invade epithelial cells. Further, immunity elicited by these candidate vaccines are serotype specific. Therefore, boosted by the initial success, new prototype vaccines are been currently developed to incorporate broad-spectrum protection against the epidemiologically relevant *Shigella* serotypes.

Mucosal vaccines against acute respiratory infections

Live attenuated influenza vaccine (LAIV; marketed in the US as Flumist®)

Influenza, commonly referred to as 'flu,' is a transmittable disease caused by the virus influenza. Influenza virus types A and B are both common causes of acute respiratory illnesses (Ambrose *et al.*, 2008). The flu virus is airborne and can be easily transmitted from person to person. Once the virus enters the respiratory system, it invades the lining of the throat, nasal passages and sometimes the lungs. The virus then infects healthy cells, rapidly reproducing thousands of copies, which can, in turn, infect other cells throughout the respiratory tract. It is estimated that 5–15% of the population develop upper respiratory tract infections and 250,000–500,000 deaths occur worldwide every year during these epidemics, with most deaths and hospitalizations occurring in the elderly and chronically ill (Ambrose *et al.*, 2008). Vaccination with inactivated vaccines remains the principal intervention to prevent seasonal influenza. In the developed world, vaccination alone has contributed to 70–90% protection in healthy adults (Lynch and Walsh, 2007). Influenza vaccines are among the oldest active vaccines. Over the years, both inactivated and live attenuated vaccines have been used for human use. The vaccines, which were initially bivalent, containing representative types A and B, became trivalent with the addition after 1977 of the A (H1N1) subtype (La Montagne *et al.*, 1983). To strengthen the immunity to influenza, vaccinologists have since replaced whole viruses with subunit preparations. While most of these parentally injected vaccines were efficient in inducing mucosal protection, attempts were made to utilize the mucosal route as an efficient means of inducing protection in oral surfaces (Ambrose *et al.*, 2008).

FluMist® originally licensed in 2003 is the only nasal spray flu vaccine approved in the U.S. to help prevent influenza. The vaccine is used in individuals between the age of 2 and 49 years of age against influenza disease caused by influenza virus subtypes A and type B contained in the vaccine. FluMist is distinctive from the flu shot in that it uses live, attenuated viruses to help stimulate an immune response in mucosal surfaces that is designed to closely resemble the body's natural response to an influenza infection.

Flumist viruses replicate primarily in the ciliated epithelial cells of the nasopharyngeal mucosa to induce immune responses in the form of both antibodies, and cellular immunity (Ambrose *et al.*, 2008). During the course of replication, all the viral proteins would be presented to the immune system in their native conformation and in the context of histocompatibility proteins; resultant immune responses should mimic those of natural infection with influenza virus. Formulations of LAIV against pandemic influenza strains, including H5N1, H9N2, and H7N3, are currently being tested in preclinical and phase I clinical studies (Ambrose *et al.*, 2008).

Emerging mucosal vaccines against bacterial respiratory pathogens

Current vaccines against bacterial respiratory pathogens are systemic vaccines that are administered intramuscularly and elicit a systemic immune response. Despite this substantial progress made in the prevention of bacterial respiratory infections, there is still an urgent need for better protection. In this section we focus on potential of nasal and oral vaccination as key strategies to optimize future vaccines that have to meet the challenges still posed by bacterial respiratory pathogens.

Streptococcus pneumoniae Invasive pneumococcal disease causes an estimated 1.6 million deaths annually worldwide, the majority of which could have been prevented with vaccination (Baumann, 2008). The pneumococcal polysaccharide 23-valent vaccine (PPV23) is immunogenic against 80% of more than 90 serotypes that cause invasive disease and reduces invasive pneumococcal disease by around 50% (Christenson *et al.*, 2001). The licensed conjugate vaccines can afford high level of protection against invasive disease, but are ineffective against oral infections due to their inability to impart mucosal immunity (Baumann, 2008). The nasal vaccines based on the commercial PCV7 (Prevenar) are capable of imparting immunogenicity at both systemic and mucosal sites (Sabirov and Metzger, 2006, 2008). The nasal vaccine was more successful in preventing pneumococcal colonization on the nasal mucosa surface but also in attenuating experimental pneumonia and peritonitis (Wu *et al.*, 1997). Interestingly, nasal but not systemic (intramuscular), vaccination with Prevenar-protected mice from nasal colonization, support-

ing the concept of enhanced mucosal efficacy of mucosal vaccines (Wu *et al.*, 1997).

Emerging mucosal vaccines against Neisseria meningitidis Neisseria meningitidis is a major cause of meningitis and septicaemia in children and adolescents (Finne *et al.*, 1983; Jodar *et al.*, 2002). *N. meningitidis* commonly colonizes the nasal mucosa as a commensal but occasionally causes serious invasive disease (Baumann, 2008). The immunological reasons for *Neisseria* to change from a state of commensalism to pathogenicity are ill defined. Currently, polysaccharide vaccines against serotypes A, C, Y and W are available. Vaccine induced cellular immunity is primarily against immunodominant surface antigens like PorA (Martin *et al.*, 2006).

Nasal vaccination strategies aim to reduce the risk of invasion at the mucosal site by limiting nasal colonization. In a recent study, conjugated group C meningococcal vaccine (CRM-MenC) mixed with chitosan either as microparticles or as suspensions insufflated into the nasal cavity elicited high levels of nasal mucosal IgA as compared to parenteral immunization (Baudner *et al.*, 2005). These data show that meningococcal vaccines can elicit mucosal immunity if combined with appropriate adjuvants.

HIV vaccine

HIV is considered as a mucosal pathogen as its natural transmission occurs mainly via mucosal surfaces. The immune system in the mucosa represents a critical element in the interaction between HIV and the host immune system. However, the exact nature of the cellular and molecular interaction between HIV and the host mucosa remains poorly understood. This lack of basic knowledge is one of the main reasons why we still do not have an effective AIDS vaccine. Additionally, most vaccine interventions have focused on parenteral immunization inducing mostly systemic immunity. Unfortunately, these approaches have not yielded the desired results needed to control HIV/AIDS. Ongoing efforts have since re-focused on novel mucosal vaccine strategies, in spite of the difficulty associated with induction and evaluation of a protective immune response at mucosal surfaces. Novel approaches to generate a suitable vaccine for HIV have prompted the search for efficient vaccine carriers capable of inducing robust immune responses at mucosal surfaces. Efforts are also ongoing in identifying vaccine antigens and relevant immunomodulators for oral delivery.

Bacterial vectors are a class of very effective oral carriers well suited for mucosal delivery of HIV antigens (Paterson and Johnson, 2004). The ability of harmless bacterial vaccine strains, such as *Lactobacilli* to present components of HIV to the host cells provides additional support for the development of live oral bacterial vaccine vectors for the delivery of an HIV-1 vaccine (Darji *et al.*, 1997; Pascual *et al.*, 1997; Paglia *et al.*, 1998; Xiang *et al.*, 2000; Shata and Hone, 2001; Shata *et al.*, 2001; Vecino *et al.*, 2002; Joseph *et al.*, 2006; Mohamadzadeh *et al.*, 2008; Mohamadzadeh and Klaenhammer, 2008). Oral vaccination of mice with recombinant *Lactococcus lactis* expressing surface bound HIV envelope protein *env* together with CT led to local and systemic humoral as well as cellular immunity. The resultant immunity led to reduced viraemia following challenge with vaccinia virus expressing HIV *env* (Xin *et al.*, 2003). Recombinant strains of *Streptococcus gordonii* expressing the V3 domain of the gp120 protein of the HIV-1 virus or the E7 protein of human papillomavirus have been engineered for use as delivery system. The use of these delivery systems induced antigen-specific local and systemic responses in vaccinated monkeys (Di Fabio *et al.*, 1998; Oggioni *et al.*, 1999). Additionally, *Salmonella* vector system is also capable of inducing CD8+ T cell responses to immunogens when delivered orally (Shata *et al.*, 2001). Further, studies showed that the same DNA vaccine formulated as naked DNA when injected intramuscularly did not induce a measurable mucosal immune response (Shata *et al.*, 2001). These findings strongly suggest that mucosally delivered vaccines have the capacity to prime both mucosal and systemic anti-HIV immune responses, whereas parenterally delivered vaccines are poor inducers of mucosal immunity. An oral/intramuscular combinational immunization regimen using attenuated *Listeria monocytogenes* recombinants has also been evaluated in a macaque model (Jiang *et al.*, 2007). The study showed that administration of *Listeria monocytogenes*-Gag (*Lmdd-gag*) induced cellular and antibody immune responses to HIV Gag. Based on these encouraging findings, it is possible to develop bacterial HIV-1 DNA vaccine vectors that induce both humoral and cell-mediated immune responses to HIV-1 antigens locally and systemically.

Space constraints prevent an exhaustive survey of the large numbers of studies aimed at HIV mucosal vaccine. Instead, we summarize and discuss a few key important results (Table 7.2). In spite of the progress made in understanding HIV pathogenesis, the quest for the elusive HIV vaccine presents unprecedented scientific challenges. Efforts to identify 'correlates of immunity' in humans as well as the non-human primate model of AIDS have yielded conflicting results. However, there is general consensus that generating neutralizing antibodies and a robust CD8+ T cell response is important in controlling HIV. However, unlike other viral pathogens, HIV chronically replicates in the host and cleverly evades the host immune responses. The challenge to make an effective vaccine becomes daunting due to the fact that HIV selectively targets and rapidly eliminates CD4+ memory cells, which initiate protective immune responses. Research needs to be refocused to harness the enormous potential offered by adopting mucosal vaccine strategies. Further, advancements in the development of potent carrier systems should galvanize efforts directed at making the mucosal HIV vaccine. Similarly, there are vast areas unchartered in our understanding regarding the interplay of innate and adaptive cells at mucosal surfaces following HIV infection. A better understanding of innate immune mechanisms at the mucosal front will improve HIV prevention, and will facilitate the development of novel, effective, and prophylactic oral mucosal HIV vaccines.

Emerging mucosal vaccines against human papillomavirus (HPV) vaccines

Cervical cancer is a leading cause of cancer deaths in women worldwide, and virtually most of these malignancies are attributed to infection with HPV, especially HPV-16 (Walboomers *et al.*, 1999). Natural immunity to HPV infection

Table 7.2 Mucosal HIV vaccine strategies

Vaccine	Experimental model	Route	Immune response	Reference
Subunit vaccine				
Haemagglutinating virus of Japan (HVJ)-liposome containing gp160	Mice	Intranasal	Neutralizing IgG (serum), IgA (nasal wash), Systemic CTL	Sakaue et al. (2003)
Glycoprotein Gp41-CTB (Plant)	Mice	Intranasal followed by intraperitoneal boost	Systemic IgG and mucosal IgA	Matoba et al. (2004)
Gp120 + mCT	Non-human primate	Intranasal	Anti-gp120 specific systemic IgG and mucosal IgA	Yoshino et al. (2004)
Virus-like particles (VLP)				
Bovine papillomavirus (BPV)-L1-HIV1-gp41	Mice	Oral	Mucosal and systemic Gag specific CTLs, Serum IgG and mucosal IgA	Zhang et al. (2004)
Influenza haemagglutinin (HA)/HIV envelope protein (Env)	Mice	Intranasal	Systemic IgG and mucosal IgA, systemic CTL	Guo et al. (2003)
HIV Env-CTB	Mice	Intranasal	Systemic IgG and mucosal IgA, CTL	Kang et al. (2003)
Live recombinant vaccine				
Recombinant modified vaccinia Ankara (rMVA) (gp160)	Mice	Intra rectal	Strong gp160 specific systemic and mucosal antibodies, Systemic CTL	Belyakov et al. (1999)
rMVA (Env) + CT	Mice	Intranasal	Systemic IgG and mucosa IgA, CTL	Gherardi et al. (2004)
Recombiannt baccile Calmette–Guerin (rBCG)-V3J1	Mice	Intranasal	Systemic IgG response	Hiroi et al. (2001)
Listeria monocytogenes (LM)-gag	Mice	Oral	Mucosal CTLs	Peters et al. (2003)

Peptide-based vaccine

Mutant *E. coli* heat labile toxin (LT-G192)	Mice	Intrarectal and oral	Serum antibody response	Douce *et al.* (1999)
Gal/pol peptide +CpG/CT	Mice	Transcutaneous	Systemic and mucosal CTL	Belyakov *et al.* (2004)
Env peptide in PLG-UEA1	Mice	Intranasal	Systemic IgG and mucosal IgA	Manocha *et al.* (2005)
Mutant *E. coli* heat labile toxin LT (R192G)	Non-human primates	Intrarectal	Mucosal CTLs	Belyakov *et al.* (2001)
Tetragalloyl-D-lysine dendrimer (TGDK) conjugated with HIV antigens	Non-human primates	Oral	SIV specific antibody responses	Misumi *et al.* (2009)

DNA vaccine

Microencapsulated gp 160 DNA	Mice	Oral	Systemic and mucosal antibodies	Kaneko *et al.* (2000)
Glycoprotein Gp160 DNA	Mice	Intranasal	Ag-specific mucosal IgA and plasma IgG responses, CTL immunity	Wang *et al.* (2003)

is poor because of viral evasion mechanism exhibited by HPV. Nevertheless, there is clear evidence suggesting a role for both cellular as well as antibody (secretory IgA) in anti-HPV responses (Nardelli-Haefliger *et al.*, 2005). The leading vaccine candidate is a subunit based HPV virus-like particle (VLP) vaccine (Schiller and Davies, 2004). The injectable HPV-VLP vaccines are well tolerated and are highly immunogenic (Evans *et al.*, 2001). However, the requirement for multiple injections and the loss of protective antibodies during ovulation might represent potential gaps in the success of this vaccine (Nardelli-Haefliger *et al.*, 2003). Mucosal vaccination might represent a potential alternative to overcome these difficulties by inducing local immune responses for a prolonged period of time (Balmelli *et al.*, 1998; Nardelli-Haefliger *et al.*, 1999; Decroix *et al.*, 2001).

Mucosal immunization of mice with HPV16 VLP via the lower respiratory tract can induce high antibody responses (Balmelli *et al.*, 1998; Balmelli *et al.*, 2002). HPV16 L1 VLPs vaccination of mice leads to the induction of anti-VLP and HPV16-neutralizing antibodies in serum and in genital, rectal and oral secretions. Intranasal vaccination of women with the HPV16L1VLPs induced a mucosal immune response demonstrated by the induction of anti-HPV16 IgA in the PBMC and SigA in the secretions suggesting that intranasal administration of this vaccine preperation may represent a potential alternative to the present parenteral immunizations. Testing in additional trials will be valuable in understanding the efficacy of this preperation delivered via mucosal routes.

Mucosal delivery of vaccines

Mucosal delivery of vaccines is an ideal way to induce effective local immunity (Mestecky, 1987). Unfortunately, mucosal delivery of antigens can elicit weak and short lived antigen-specific immune responses. The ineffectiveness of orally delivered antigens in eliciting a robust immune response has inspired widespread strategies for delivery systems that would increase the antigen absorption, prevent its degradation and facilitate its uptake and transport to the lymphoid tissue for presentation to immunocompetent cells, and induce the appropriate type of immune response. Several immune-enhancing strategies have been developed to overcome suboptimal mucosal immune responses including delivery vehicles (Rol *et al.*, 2005). Mucosal delivery vehicles are largely designed to preserve antigenic structure and effectively deliver the antigen in an immunologically competent manner.

Mucosal adjuvants

Adjuvants are particularly important in mucosal immunization strategies since most antigens are poorly immunogenic when delivered per os. Although the immunomodulatory effects of soluble holotoxins, such as cholera toxin (CT) and heat-labile enterotoxin (LT) have been shown to be effective mucosal adjuvants for nasal/oral delivery of numerous antigens, their use is restricted because of their toxicity (Takahashi *et al.*, 1996; Rappuoli *et al.*, 1999; Williams *et al.*, 1999). Hence, current efforts to overcome obstacles to the development of effective mucosal vaccines are mainly directed towards finding safe and efficient mucosal adjuvants that provide protective immunity against infections.

Characteristics of an ideal mucosal adjuvant

An ideal mucosal carrier should be able to co-deliver the antigens and adjuvants across the epithelial barrier to target antigen-presenting cell subsets. It can be envisaged that they travel across the epithelial lining through the intracellular or the paracellular route. Adjuvants that may facilitate the transport of antigen through the intracellular route are the bacterial toxins cholera toxin (CT), heat-labile enterotoxin (LT) and *Pseudomonas aeruginosa* exotoxin (Dougan and Hormaeche, 2006). Polarized epithelial cells internalize these toxins and it is thought that the antigen may also be engulfed during this process. Furthermore, vaccine components could be internalized by specialized M cells that would transfer them to the underlying mucosal lymphoid follicles. This is more likely to occur with molecules that specifically target M cells. Another way by which antigen and adjuvant may gain access to submucosa is through the uptake by mucosal DCs that have dendrites extending outside the epithelial layer, however this phenomenon has been so far demonstrated only with whole bacteria (Fig. 7.2) (Niess et al., 2005). Finally, some substances may disrupt the integrity of mucosa or form pores (as in the case of saponins) allowing antigen passage in the underlying tissues. Once in the submucosa, antigen and adjuvant encounter immune cells locally or are transported to draining lymph nodes.

Nasal mucosa is very attractive for mucosal vaccination because of the absence of acidity and abundance of secreted enzymes. Furthermore, small surface of this mucosa requires a low amount of antigen and adjuvant. Many devices (mainly sprays) have been developed for nasal vaccines and bio-adhesive delivery system has been developed to retain the vaccine formulation in the nasal cavity. On the other hand, the oral route raises problems of antigen dilution because of the large surface of the gastrointestinal tract and requires higher amount of antigen/adjuvant formulation, with consequent high costs of production. Moreover, vaccines have to be targeted to specific sites of the epithelial cell surface to ensure their appropriate uptake. For these purposes, several delivery systems have been developed. Most of them are particulate structures that may facilitate targeting of the antigen to M cells. It is not the intent of this chapter to review mucosal adjuvants in detail. A few mucosal adjuvants have been selected to be reviewed in this chapter.

Bacterial toxins and their derivatives

Bacterial enterotoxins, CT produced by *Vibrio cholerae* and LT produced by *Escherichia coli* are used as gold standards in experiments measuring mucosal adjuvant efficacy. These enterotoxins are homologous in amino acid composition (Dallas and Falkow, 1980) and tertiary structure (Sixma et al., 1991). They are both composed of an A subunit that has enzymatic activity and a pentameric B subunit that is responsible for binding to the cell membrane. The mechanisms that enable CT or LT to overcome tolerance and potentiate both systemic and mucosal immune responses are still unknown. These are potent enterotoxins that are remarkably stable to gut proteases and bile salts. Further, the pentameric B subunit bind in high affinity to GM1 ganglioside receptors, present on epithelial cells thus facilitating the uptake and presentation of the toxins to the GALT effector system. The inherent immunomodulating activities have been attributed to the A subunit, and their enzymatic ADP-ribosylating activity (Sanchez and Holmgren, 2008). Both CT and LT induce long-term mucosal and systemic responses (Lycke, 1997). CT has been shown to induce primarily Th2 responses (Xu-Amano et al., 1994; Marinaro et al., 1995). In contrast, LT has been reported to induce a mixed Th1/Th2 response (Takahashi et al., 1996). These differences have been credited to changes in ganglioside binding specificities of their B subunit (Yamamoto et al., 2000).

Although both CT and LT are extensively used as mucosal adjuvants in animal models, the strong toxicity of these molecules does not allow their use in humans. Therefore, to avoid toxicity problems with the holotoxins, several mutants of CT and LT with reduced toxicity have been developed in the last decade. For this reason, numerous attempts have been made to identify non-toxic modifications of both CT and LT that retain adjuvanticity by either removing the toxic A domain (Tamura et al., 1989) or by rendering it enzymatically inactive by site-directed mutagenesis. Interestingly, studies addressing the adjuvant effects of recombinant B subunits of CT and LT have generated conflicting results. Several studies reported an adjuvant effect for recombinant Cholera toxin B (rCTB) or recombinant heat labile toxin (rLTB) (de Haan et al., 1998; Isaka et al., 1998; Wu and Russell, 1998), while others have reported no adjuvant activity (Vajdy and Lycke, 1992). Generally, both CTB and LTB are considered to be poor adjuvants but the adjuvanticity is improved following coupling with antigens (Sanchez and Holmgren, 2008). This is due to the increased uptake of the coupled protein across the mucosal barrier and more efficient antigen presentation in the organized lymphoid tissues.

Numerous innovative modifications have been made to both CT and LT in order to preserve the adjuvant activity and to remove the toxicity. A mutant of LT that is completely devoid of its enzymatic activity, LT-K63, has proven safe in extensive preclinical toxicology studies and has been recently delivered intranasally to humans in a phase I clinical trial (Stephenson et al., 2006). Another CT-derived molecule is CTA1-DD, a fusion protein composed of the A1 subunit of CT with a B cell targeting moiety derived from *Staphylococcus aureus* protein A (Agren et al., 1998; Eriksson et al., 2004). This protein retains the enzymatic activity of the A subunit of CT but lacks its binding domain. CTA1-DD is an effective adjuvant through different mucosal routes (Marinaro et al., 2003). In addition to CT and LT, other bacterial enterotoxins have been found to act as mucosal adjuvants such as Zot from *Vibrio cholerae* (Marinaro et al., 2003) and fragilysin from *Bacteroides fragilis* (Vines et al., 2000). The adjuvant effect of these toxins is characterized by the induction of antigen-specific serum IgG and mucosal IgA in mice upon mucosal delivery. However, the exact mechanism by which these toxins amplify immune responses to co-delivered antigens is not known.

Microparticles

A wide range of polymeric materials have been explored for their ability to enhance the immunogenicity of oral vaccines, such as poly (DL)-lactide-co-glycolic acid (PLGA) (Herrmann et al., 1999; Sharpe et al., 2003; Singh et al., 2006), chitosan and other cationic polymers (Illum et al., 2001; Twaites et al., 2004). Biodegradable biopolymers composed of either poly (lactide-co-glycolide) (PLG) or poly (lactic-co-glycolic acid)(PLGA) are promising delivery systems capable of protecting the peptide and antigens from degradation in gut, increase the bioavailability of the vaccine and target the antigens to inductive sites in the GALT (Eldridge et al., 1989; Klencke et al., 2002; O'Hagan, 1998; Okada and Toguchi, 1995). Further, it has been widely explored in several immunological studies due to their biodegradability, biocompatibility and slow release properties *in vivo* (O'Hagan et al., 2006). These biodegradable particulate microparticles favour uptake by M cells in the intestine following oral administration (Lavelle, 2001; Garinot et al., 2007) and they are also effective by the intranasal route (Marazuela et al., 2008).

Micoparticle-delivered antigens administered orally could induce potent responses at both systemic and mucosal sites (O'Hagan et al., 2006).

Entrapment in PLG microparticles polarizes immune responses induced by soluble antigens to a protective Th1 cell subtype, and also facilitates the induction of CD8+ CTLs (O'Hagan et al., 2006). In rodent studies, microparticles have been shown to be effective against challenge with various pathogens, including *Bordetella pertussis* (Cahill et al., 1995; Jones et al., 1996), *Chlamydia trachomatis* ((Whittum-Hudson et al., 1996), *Salmonella typhimurium* (Allaoui-Attarki et al., 1997) and *Streptococcus pneumoniae* (Seo et al., 2002). Early studies in primates have shown that mucosal immunization with microparticles induces protective immunity against challenge with SIV (Marx et al., 1993; Tseng et al., 1995). Microparticles have also shown promise for mucosal delivery of DNA vaccines (Mathiowitz et al., 1997; Singh et al., 2001). Nevertheless, the challenge lies in translating the observations from smaller animal models to higher primates and humans.

TLR9 ligand: CpG motifs

CpG is a class of unmethylated bacterial DNAs consisting of synthetic oligodeoxynucleotides (ODNs) containing CpG motifs that have immunostimulatory properties (Krieg, 2002). CpG ODNs induce innate immune responses by binding to Toll-like receptor 9 (TLR 9) on several cell types, including professional antigen-presenting cells (Krieg, 2002). Their resistance against intestinal proteases makes them well suited for surface protection in the gut (Brandt-zaeg, 1997; Farstad et al., 2000). CpG ODNs as mucosal adjuvants induce mucosal IgA and systemic responses following oral, intranasal, intrarectal and intravaginal delivery (McCluskie and Davis, 1999; Harandi and Holmgren, 2004). When administered mucosally, CpG-ODNs can enhance both local and systemic Th1 immunity in addition to antibody responses (Krieg, 2002). Furthermore, a synergistic adjuvant activity was reported when CpG ODN was mixed with CT or LT; however, a mixture of CpG ODN with CTB and the mutant of LT protein derivative LTK63, had no synergistic effects on the magnitude of the immune response (McCluskie and Davis, 1998; McCluskie et al., 2000). The relative Th-1 polarization associated with CpG-ODN may make it a suitable adjuvant for the development of vaccines against a vast majority of pathogens. Thus, one can conclude that the use of CpG ODN as a Th1-tilting immunostimulator either singly or in combination with other mucosal activators holds promise for future mucosal vaccine strategies.

TLR4 ligand: monophosphoryl lipid A

Monophosphoryl lipid (MPL)-A is derived from the LPS of *Salmonella minnesota* R595, a Gram-negative bacterium and, has immunomodulatory properties (Freytag and Clements, 2005). Similar to LPS, MPL-A is thought to interact with TLR4 on APCs, resulting in the release of proinflammatory cytokines. The adjuvant activity of MPL is primarily attributed to its ability to activate APCs and elicit production of proinflammatory cytokines. MPL in combination with delivery systems has been used as a mucosal adjuvant through the nasal and the oral routes (Baldridge *et al.*, 2000; Childers *et al.*, 2000; Clark *et al.*, 2001; Doherty *et al.*, 2002; Yang *et al.*, 2002). Vaccines formulated with MPL and several antigens including hepatitis B surface antigen, tetanus toxoid or influenza antigens when administered intranasally in mice resulted in enhanced IgA titres at mucosal surfaces. Furthermore, MPL enhanced IgA antibody responses to these antigens were detected in samples from mucosal sites (Baldridge *et al.*, 2000). Thus, on the basis of collective information regarding mucosal adjuvant activity of MPL, this may be used as a mucosal adjuvant for future vaccine strategies.

Transgenic plants

Transgenic plant derived vaccine antigens have been shown to be orally immunogenic in humans and several animal species (Mason *et al.*, 1992; Modelska *et al.*, 1998; Tackaberry *et al.*, 1999). Initially, these vaccines were used to target diarrhoeal diseases (Arakawa *et al.*, 1998; Tacket *et al.*, 1998, 2000). Several other vaccines have since been tested in animals against diseases such as hepatitis B (Richter *et al.*, 2000) and foot-and-mouth disease (Wigdorovitz *et al.*, 1999). Clinical trials have also begun to assess human immune responses to plant-produced recombinant enterotoxin (LT-B) from *E. coli* (Tacket *et al.*, 1998) and Norwalk virus capsid protein (NVCP) delivered

in (raw) transgenic potatoes (Walmsley and Arntzen, 2000).

Plant-based vaccines hold promise and will likely form an essential part of a future vaccine. Several issues remain to be addressed including potential risk factors associated with production. First, progress needs to be made in improving the yield of antigens in ensuring uniformity of yield, in enhancing immunogenicity of orally administered vaccines, and in assessing the potential for development of oral tolerance. Transgenic plant-derived vaccines offer a new strategy for development of safe, inexpensive vaccines against diarrhoeal diseases such as enterotoxigenic *E. coli* infection and norovirus.

Concluding remarks and future directions

The inability of current vaccine candidates and immunization strategies to generate mucosal and systemic protective immunity justifies the development of novel mucosal delivery systems. The generation of fully defined and safe mucosal adjuvants for humans could have huge impact on vaccine development and in the treatment of both mucosally and non-mucosally acquired diseases. However, despite intense research, development of practical mucosal vaccines has been slow. In response to these limitations, significant effort has been focused on the mucosal adjuvant activities of both LT and CT. Unfortunately, even though LT and CT are effective as mucosal adjuvants, their toxicity is a major constraint. Recombinant, enzymatically inactive forms of both LT and CT toxins have been generated and some of the mutant derivatives retain adjuvant activity while having greatly reduced toxicity. Understanding the adjuvant mechanism of the non-toxic mutants of CT and LT would be extremely important before they are potentially suitable for transition into clinical evaluation.

Although developments in mucosal vaccine research may look promising, protection elicited by the host to both mucosal and non-mucosal pathogens is complex and would require consideration from multiple factors. The fast progress in basic immunology will certainly lead in few years to the development of needle-free vaccines that will galvanize the production of newer prophylactic attempts. Research is needed to understand the mechanisms of adaptive B-cell and T-cell memory in the gut. A better knowledge of gut immune responsiveness during early life is required to establish the usefulness of mucosal vaccination against mucosal pathogens encountered by young infants. With a concerted research agenda, mucosal immunologists may accelerate development of life-saving mucosal vaccines that can reduce the burden of disease and mortality worldwide.

Acknowledgements

Supported by grants from the US National Institutes of Health (grants R01 DK057665, R01 AI048638, U19 AI057266, U54 AI057157, N01 AI50019 and N01 AI50025) and from the Bill and Melinda Gates Foundation to B.P. We thank Dr Helder Nakaya for his help with the figures. We apologize for the inadvertent omission of any published work pertinent to the topic of this chapter.

References

Abreu, M.T., Fukata, M., and Arditi, M. (2005). TLR signaling in the gut in health and disease. J. Immunol. *174*, 4453–4460.

Abreu, M.T., Vora, P., Faure, E., Thomas, L.S., Arnold, E.T., and Arditi, M. (2001). Decreased expression of Toll-like receptor-4 and MD-2 correlates with intestinal epithelial cell protection against dysregulated proinflammatory gene expression in response to bacterial lipopolysaccharide. J. Immunol. *167*, 1609–1616.

Abu-Elyazeed, R., Wierzba, T.F., Mourad, A.S., Peruski, L.F., Kay, B.A., Rao, M., Churilla, A.M., Bourgeois, A.L., Mortagy, A.K., Kamal, S.M., *et al.* (1999). Epidemiology of enterotoxigenic *Escherichia coli* diarrhea in a pediatric cohort in a periurban area of lower Egypt. J. Infect. Dis. *179*, 382–389.

Acosta-Rodriguez, E.V., Napolitani, G., Lanzavecchia, A., and Sallusto, F. (2007). Interleukins 1beta and 6 but not transforming growth factor-beta are essential for the differentiation of interleukin 17-producing human T helper cells. Nat. Immunol. *8*, 942–949.

Agren, L., Lowenadler, B., and Lycke, N. (1998). A novel concept in mucosal adjuvanticity: the CTA1-DD adjuvant is a B cell–targeted fusion protein that incorporates the enzymatically active cholera toxin A1 subunit. Immunol. Cell Biol. *76*, 280–287.

Allaoui-Attarki, K., Pecquet, S., Fattal, E., Trolle, S., Chachaty, E., Couvreur, P., and Andremont, A. (1997). Protective immunity against *Salmonella typhimurium* elicited in mice by oral vaccination with phosphorylcholine encapsulated in poly(DL-lactide-co-glycolide) microspheres. Infect. Immun. *65*, 853–857.

Ambrose, C.S., Luke, C., and Coelingh, K. (2008). Current status of live attenuated influenza vaccine in the United States for seasonal and pandemic influenza. Influenza Other Resp. Viruses *2*, 193–202.

Annacker, O., Coombes, J.L., Malmstrom, V., Uhlig, H.H., Bourne, T., Johansson-Lindbom, B., Agace, W.W., Parker, C.M., and Powrie, F. (2005). Essential role for CD103 in the T cell-mediated regulation of experimental colitis. J. Exp. Med. *202*, 1051–1061.

Arakawa, T., Chong, D.K., and Langridge, W.H. (1998). Efficacy of a food plant-based oral cholera toxin B subunit vaccine. Nat. Biotechnol. *16*, 292–297.

Ayabe, T., Satchell, D.P., Wilson, C.L., Parks, W.C., Selsted, M.E., and Ouellette, A.J. (2000). Secretion of microbicidal alpha-defensins by intestinal Paneth cells in response to bacteria. Nat. Immunol. *1*, 113–118.

Aziz, M.A., Midha, S., Waheed, S.M., and Bhatnagar, R. (2007). Oral vaccines: new needs, new possibilities. Bioessays *29*, 591–604.

Baldridge, J.R., Yorgensen, Y., Ward, J.R., and Ulrich, J.T. (2000). Monophosphoryl lipid A enhances mucosal and systemic immunity to vaccine antigens following intranasal administration. Vaccine *18*, 2416–2425.

Balmelli, C., Demotz, S., Acha-Orbea, H., De Grandi, P., and Nardelli-Haefliger, D. (2002). Trachea, lung, and tracheobronchial lymph nodes are the major sites where antigen-presenting cells are detected after nasal vaccination of mice with human papillomavirus type 16 virus-like particles. J. Virol. *76*, 12596–12602.

Balmelli, C., Roden, R., Potts, A., Schiller, J., De Grandi, P., and Nardelli-Haefliger, D. (1998). Nasal immunization of mice with human papillomavirus type 16 virus-like particles elicits neutralizing antibodies in mucosal secretions. J. Virol. *72*, 8220–8229.

Bambou, J.C., Giraud, A., Menard, S., Begue, B., Rakotobe, S., Heyman, M., Taddei, F., Cerf-Bensussan, N., and Gaboriau-Routhiau, V. (2004). *In vitro* and *ex vivo* activation of the TLR5 signaling pathway in intestinal epithelial cells by a commensal *Escherichia coli* strain. J. Biol. Chem. *279*, 42984–42992.

Baudner, B.C., Verhoef, J.C., Giuliani, M.M., Peppoloni, S., Rappuoli, R., Del Giudice, G., and Junginger, H.E. (2005). Protective immune responses to meningococcal C conjugate vaccine after intranasal immunization of mice with the LTK63 mutant plus chitosan or trimethyl chitosan chloride as novel delivery platform. J. Drug Target. *13*, 489–498.

Baumann, U. (2008). Mucosal vaccination against bacterial respiratory infections. Expert Rev. Vaccines *7*, 1257–1276.

Begue, B., Dumant, C., Bambou, J.C., Beaulieu, J.F., Chamaillard, M., Hugot, J.P., Goulet, O., Schmitz, J., Philpott, D.J., Cerf-Bensussan, N., *et al.* (2006). Microbial induction of CARD15 expression in intestinal epithelial cells via toll-like receptor 5 triggers an antibacterial response loop. J. Cell Physiol. *209*, 241–252.

Belyakov, I.M., Hammond, S.A., Ahlers, J.D., Glenn, G.M., and Berzofsky, J.A. (2004). Transcutaneous immunization induces mucosal CTLs and protective immunity by migration of primed skin dendritic cells. J. Clin. Invest. *113*, 998–1007.

Belyakov, I.M., Hel, Z., Kelsall, B., Kuznetsov, V.A., Ahlers, J.D., Nacsa, J., Watkins, D.I., Allen, T.M., Sette, A., Altman, J., et al. (2001). Mucosal AIDS vaccine reduces disease and viral load in gut reservoir and blood after mucosal infection of macaques. Nat. Med. 7, 1320–1326.

Belyakov, I.M., Moss, B., Strober, W., and Berzofsky, J.A. (1999). Mucosal vaccination overcomes the barrier to recombinant vaccinia immunization caused by preexisting poxvirus immunity. Proc. Natl. Acad. Sci. U.S.A. 96, 4512–4517.

Bendelac, A. (1995). CD1: presenting unusual antigens to unusual T lymphocytes. Science 269, 185–186.

Benson, M.J., Pino-Lagos, K., Rosemblatt, M., and Noelle, R.J. (2007). All-trans retinoic acid mediates enhanced T reg cell growth, differentiation, and gut homing in the face of high levels of co-stimulation. J. Exp. Med. 204, 1765–1774.

Bernstein, D.I., Smith, V.E., Sherwood, J.R., Schiff, G.M., Sander, D.S., DeFeudis, D., Spriggs, D.R., and Ward, R.L. (1998). Safety and immunogenicity of live, attenuated human rotavirus vaccine 89–12. Vaccine 16, 381–387.

Bienenstock, J., and McDermott, M.R. (2005). Bronchus- and nasal-associated lymphoid tissues. Immunol. Rev. 206, 22–31.

Black, R.E., Levine, M.M., Clements, M.L., Young, C.R., Svennerholm, A.M., and Holmgren, J. (1987). Protective efficacy in humans of killed whole-vibrio oral cholera vaccine with and without the B subunit of cholera toxin. Infect. Immun. 55, 1116–1120.

Bowman, E.P., Kuklin, N.A., Youngman, K.R., Lazarus, N.H., Kunkel, E.J., Pan, J., Greenberg, H.B., and Butcher, E.C. (2002). The intestinal chemokine thymus-expressed chemokine (CCL25) attracts IgA antibody-secreting cells. J. Exp. Med. 195, 269–275.

Brandtzaeg, P. (1997). Review article: Homing of mucosal immune cells–a possible connection between intestinal and articular inflammation. Aliment. Pharmacol. Ther. 11 (Suppl. 3), 24–37; discussion 37–29.

Brandtzaeg, P., Kiyono, H., Pabst, R., and Russell, M.W. (2008). Terminology: nomenclature of mucosa-associated lymphoid tissue. Mucosal Immunol. 1, 31–37.

Butterton, J.R., Boyko, S.A., and Calderwood, S.B. (1993). Use of the Vibrio cholerae irgA gene as a locus for insertion and expression of heterologous antigens in cholera vaccine strains. Vaccine 11, 1327–1335.

Cahill, E.S., O'Hagan, D.T., Illum, L., Barnard, A., Mills, K.H., and Redhead, K. (1995). Immune responses and protection against Bordetella pertussis infection after intranasal immunization of mice with filamentous haemagglutinin in solution or incorporated in biodegradable microparticles. Vaccine 13, 455–462.

Cario, E., Gerken, G., and Podolsky, D.K. (2002). 'For whom the bell tolls!' – innate defense mechanisms and survival strategies of the intestinal epithelium against luminal pathogens. Z. Gastroenterol. 40, 983–990.

Cario, E., and Podolsky, D.K. (2000). Differential alteration in intestinal epithelial cell expression of toll-like receptor 3 (TLR3) and TLR4 in inflammatory bowel disease. Infect. Immun. 68, 7010–7017.

Cario, E., Rosenberg, I.M., Brandwein, S.L., Beck, P.L., Reinecker, H.C., and Podolsky, D.K. (2000). Lipopolysaccharide activates distinct signaling pathways in intestinal epithelial cell lines expressing Toll-like receptors. J. Immunol. 164, 966–972.

Castigli, E., Scott, S., Dedeoglu, F., Bryce, P., Jabara, H., Bhan, A.K., Mizoguchi, E., and Geha, R.S. (2004). Impaired IgA class switching in APRIL-deficient mice. Proc. Natl. Acad. Sci. U.S.A. 101, 3903–3908.

Cerovic, V., Jenkins, C.D., Barnes, A.G., Milling, S.W., MacPherson, G.G., and Klavinskis, L.S. (2009). Hyporesponsiveness of intestinal dendritic cells to TLR stimulation is limited to TLR4. J. Immunol. 182, 2405–2415.

Cerutti, A. (2008). The regulation of IgA class switching. Nat. Rev. Immunol. 8, 421–434.

Chamaillard, M., Girardin, S.E., Viala, J., and Philpott, D.J. (2003). Nods, Nalps and Naip: intracellular regulators of bacterial-induced inflammation. Cell Microbiol. 5, 581–592.

Cheroutre, H., and Lambolez, F. (2008). The thymus chapter in the life of gut-specific intra epithelial lymphocytes. Curr. Opin. Immunol. 20, 185–191.

Childers, N.K., Miller, K.L., Tong, G., Llarena, J.C., Greenway, T., Ulrich, J.T., and Michalek, S.M. (2000). Adjuvant activity of monophosphoryl lipid A for nasal and oral immunization with soluble or liposome-associated antigen. Infect. Immun. 68, 5509–5516.

Chirdo, F.G., Millington, O.R., Beacock-Sharp, H., and Mowat, A.M. (2005). Immunomodulatory dendritic cells in intestinal lamina propria. Eur. J. Immunol. 35, 1831–1840.

Christenson, B., Lundbergh, P., Hedlund, J., and Ortqvist, A. (2001). Effects of a large-scale intervention with influenza and 23-valent pneumococcal vaccines in adults aged 65 years or older: a prospective study. Lancet 357, 1008–1011.

Chung, D.R., Kasper, D.L., Panzo, R.J., Chitnis, T., Grusby, M.J., Sayegh, M.H., and Tzianabos, A.O. (2003). CD4+ T cells mediate abscess formation in intra-abdominal sepsis by an IL-17-dependent mechanism. J. Immunol. 170, 1958–1963.

Chung, W.O., and Dale, B.A. (2004). Innate immune response of oral and foreskin keratinocytes: utilization of different signaling pathways by various bacterial species. Infect. Immun. 72, 352–358.

Clark, M.A., Blair, H., Liang, L., Brey, R.N., Brayden, D., and Hirst, B.H. (2001). Targeting polymerised liposome vaccine carriers to intestinal M cells. Vaccine 20, 208–217.

Cohen, D., Ashkenazi, S., Green, M., Lerman, Y., Slepon, R., Robin, G., Orr, N., Taylor, D.N., Sadoff, J.C., Chu, C., et al. (1996). Safety and immunogenicity of investigational Shigella conjugate vaccines in Israeli volunteers. Infect. Immun. 64, 4074–4077.

Coombes, J.L., Siddiqui, K.R., Arancibia-Carcamo, C.V., Hall, J., Sun, C.M., Belkaid, Y., and Powrie, F. (2007). A functionally specialized population of mucosal CD103+ DCs induces Foxp3+ regulatory T cells via a TGF-beta and retinoic acid-dependent mechanism. J. Exp. Med. 204, 1757–1764.

Crotty, S., Hix, L., Sigal, L.J., and Andino, R. (2002). Poliovirus pathogenesis in a new poliovirus receptor transgenic mouse model: age-dependent paralysis and a mucosal route of infection. J. Gen. Virol. 83, 1707–1720.

Dallas, W.S., and Falkow, S. (1980). Amino acid sequence homology between cholera toxin and Escherichia coli heat-labile toxin. Nature 288, 499–501.

Darji, A., Guzman, C.A., Gerstel, B., Wachholz, P., Timmis, K.N., Wehland, J., Chakraborty, T., and Weiss, S. (1997). Oral somatic transgene vaccination using attenuated S. typhimurium. Cell 91, 765–775.

de Haan, L., Verweij, W.R., Feil, I.K., Holtrop, M., Hol, W.G., Agsteribbe, E., and Wilschut, J. (1998). Role of GM1 binding in the mucosal immunogenicity and adjuvant activity of the Escherichia coli heat-labile enterotoxin and its B subunit. Immunology 94, 424–430.

Debertin, A.S., Tschernig, T., Tonjes, H., Kleemann, W.J., Troger, H.D., and Pabst, R. (2003). Nasal-associated lymphoid tissue (NALT): frequency and localization in young children. Clin. Exp. Immunol. 134, 503–507.

Decroix, N., Hocini, H., Quan, C.P., Bellon, B., Kazatchkine, M.D., and Bouvet, J.P. (2001). Induction in mucosa of IgG and IgA antibodies against parenterally administered soluble immunogens. Scand. J. Immunol. 53, 401–409.

Denning, T.L., Wang, Y.C., Patel, S.R., Williams, I.R., and Pulendran, B. (2007). Lamina propria macrophages and dendritic cells differentially induce regulatory and interleukin 17-producing T cell responses. Nat. Immunol. 8, 1086–1094.

Di Fabio, S., Medaglini, D., Rush, C.M., Corrias, F., Panzini, G.L., Pace, M., Verani, P., Pozzi, G., and Titti, F. (1998). Vaginal immunization of Cynomolgus monkeys with Streptococcus gordonii expressing HIV-1 and HPV 16 antigens. Vaccine 16, 485–492.

Doherty, T.M., Olsen, A.W., van Pinxteren, L., and Andersen, P. (2002). Oral vaccination with subunit vaccines protects animals against aerosol infection with Mycobacterium tuberculosis. Infect. Immun. 70, 3111–3121.

Douce, G., Giannelli, V., Pizza, M., Lewis, D., Everest, P., Rappuoli, R., and Dougan, G. (1999). Genetically detoxified mutants of heat-labile toxin from Escherichia coli are able to act as oral adjuvants. Infect. Immun. 67, 4400–4406.

Dougan, G., and Hormaeche, C. (2006). How bacteria and their products provide clues to vaccine and adjuvant development. Vaccine 24 (Suppl. 2), S2–13–19.

Eckmann, L., and Kagnoff, M.F. (2005). Intestinal mucosal responses to microbial infection. Springer Semin. Immunopathol. 27, 181–196.

Eldridge, J.H., Meulbroek, J.A., Staas, J.K., Tice, T.R., and Gilley, R.M. (1989). Vaccine-containing biodegradable microspheres specifically enter the gut-associated lymphoid tissue following oral administration and induce a disseminated mucosal immune response. Adv. Exp. Med. Biol. 251, 191–202.

Eriksson, A.M., Schon, K.M., and Lycke, N.Y. (2004). The cholera toxin-derived CTA1-DD vaccine adjuvant administered intranasally does not cause inflammation or accumulate in the nervous tissues. J. Immunol. 173, 3310–3319.

Evans, T.G., Bonnez, W., Rose, R.C., Koenig, S., Demeter, L., Suzich, J.A., O'Brien, D., Campbell, M., White, W.I., Balsley, J., et al. (2001). A Phase 1 study of a recombinant viruslike particle vaccine against human papillomavirus type 11 in healthy adult volunteers. J. Infect. Dis. 183, 1485–1493.

Fagarasan, S., Kinoshita, K., Muramatsu, M., Ikuta, K., and Honjo, T. (2001). In situ class switching and differentiation to IgA-producing cells in the gut lamina propria. Nature 413, 639–643.

Farstad, I.N., Carlsen, H., Morton, H.C., and Brandtzaeg, P. (2000). Immunoglobulin A cell distribution in the human small intestine: phenotypic and functional characteristics. Immunology 101, 354–363.

Finne, J., Leinonen, M., and Makela, P.H. (1983). Antigenic similarities between brain components and bacteria causing meningitis. Implications for vaccine development and pathogenesis. Lancet 2, 355–357.

Franchi, L., Amer, A., Body-Malapel, M., Kanneganti, T.D., Ozoren, N., Jagirdar, R., Inohara, N., Vandenabeele, P., Bertin, J., Coyle, A., et al. (2006). Cytosolic flagellin requires Ipaf for activation of caspase-1 and interleukin 1beta in Salmonella-infected macrophages. Nat. Immunol. 7, 576–582.

Franco, M.A., and Greenberg, H.B. (1997). Immunity to rotavirus in T cell deficient mice. Virology 238, 169–179.

Freytag, L.C., and Clements, J.D. (2005). Mucosal adjuvants. Vaccine 23, 1804–1813.

Ganz, T. (2003). Defensins: antimicrobial peptides of innate immunity. Nat. Rev. Immunol. 3, 710–720.

Garinot, M., Fievez, V., Pourcelle, V., Stoffelbach, F., des Rieux, A., Plapied, L., Theate, I., Freichels, H., Jerome, C., Marchand-Brynaert, J., et al. (2007). PEGylated PLGA-based nanoparticles targeting M cells for oral vaccination. J. Control Release 120, 195–204.

George, M., and Abraham, T.E. (2006). Polyionic hydrocolloids for the intestinal delivery of protein drugs: alginate and chitosan – a review. J. Control Release 114, 1–14.

GeurtsvanKessel, C.H., and Lambrecht, B.N. (2008). Division of labor between dendritic cell subsets of the lung. Mucosal Immunol. 1, 442–450.

Gewirtz, A.T., Navas, T.A., Lyons, S., Godowski, P.J., and Madara, J.L. (2001a). Cutting edge: bacterial flagellin activates basolaterally expressed TLR5 to induce epithelial proinflammatory gene expression. J. Immunol. 167, 1882–1885.

Gewirtz, A.T., Simon, P.O., Jr., Schmitt, C.K., Taylor, L.J., Hagedorn, C.H., O'Brien, A.D., Neish, A.S., and Madara, J.L. (2001b). Salmonella typhimurium translocates flagellin across intestinal epithelia, inducing a proinflammatory response. J. Clin. Invest. 107, 99–109.

Gherardi, M.M., Perez-Jimenez, E., Najera, J.L., and Esteban, M. (2004). Induction of HIV immunity in the genital tract after intranasal delivery of a MVA vector: enhanced immunogenicity after DNA prime-modified vaccinia virus Ankara boost immunization schedule. J. Immunol. *172*, 6209–6220.

Gilman, R.H., Hornick, R.B., Woodard, W.E., DuPont, H.L., Snyder, M.J., Levine, M.M., and Libonati, J.P. (1977). Evaluation of a UDP-glucose-4-epimeraseless mutant of *Salmonella typhi* as a liver oral vaccine. J. Infect. Dis. *136*, 717–723.

Girard, M.P., Steele, D., Chaignat, C.L., and Kieny, M.P. (2006). A review of vaccine research and development: human enteric infections. Vaccine *24*, 2732–2750.

Girardin, S.E., Tournebize, R., Mavris, M., Page, A.L., Li, X., Stark, G.R., Bertin, J., DiStefano, P.S., Yaniv, M., Sansonetti, P.J., et al. (2001). CARD4/Nod1 mediates NF-kappaB and JNK activation by invasive *Shigella flexneri*. EMBO Rep. *2*, 736–742.

Gotuzzo, E., Butron, B., Seas, C., Penny, M., Ruiz, R., Losonsky, G., Lanata, C.F., Wasserman, S.S., Salazar, E., Kaper, J.B., et al. (1993). Safety, immunogenicity, and excretion pattern of single-dose live oral cholera vaccine CVD 103-HgR in Peruvian adults of high and low socioeconomic levels. Infect. Immun. *61*, 3994–3997.

Grinstein, S., Gomez, J.A., Bercovich, J.A., and Biscotti, E.L. (1989). Epidemiology of rotavirus infection and gastroenteritis in prospectively monitored Argentine families with young children. Am. J. Epidemiol. *130*, 300–308.

Guo, L., Lu, X., Kang, S.M., Chen, C., Compans, R.W., and Yao, Q. (2003). Enhancement of mucosal immune responses by chimeric influenza HA/SHIV virus-like particles. Virology *313*, 502–513.

Haller, D., Russo, M.P., Sartor, R.B., and Jobin, C. (2002). IKK beta and phosphatidylinositol 3-kinase/Akt participate in non-pathogenic Gram-negative enteric bacteria-induced RelA phosphorylation and NF-kappa B activation in both primary and intestinal epithelial cell lines. J. Biol. Chem. *277*, 38168–38178.

Harandi, A.M., and Holmgren, J. (2004). CpG DNA as a potent inducer of mucosal immunity: implications for immunoprophylaxis and immunotherapy of mucosal infections. Curr. Opin. Investig. Drugs *5*, 141–145.

Hausmann, M., Kiessling, S., Mestermann, S., Webb, G., Spottl, T., Andus, T., Scholmerich, J., Herfarth, H., Ray, K., Falk, W., et al. (2002). Toll-like receptors 2 and 4 are up-regulated during intestinal inflammation. Gastroenterology *122*, 1987–2000.

Heritage, P.L., Underdown, B.J., Arsenault, A.L., Snider, D.P., and McDermott, M.R. (1997). Comparison of murine nasal-associated lymphoid tissue and Peyer's patches. Am. J. Respir. Crit. Care Med. *156*, 1256–1262.

Herrmann, J.E., Chen, S.C., Jones, D.H., Tinsley-Bown, A., Fynan, E.F., Greenberg, H.B., and Farrar, G.H. (1999). Immune responses and protection obtained by oral immunization with rotavirus VP4 and VP7 DNA vaccines encapsulated in microparticles. Virology *259*, 148–153.

Higgins, S.C., Jarnicki, A.G., Lavelle, E.C., and Mills, K.H. (2006). TLR4 mediates vaccine-induced protective cellular immunity to *Bordetella pertussis*: role of IL-17-producing T cells. J. Immunol. *177*, 7980–7989.

Hill, D.R., Ford, L., and Lalloo, D.G. (2006). Oral cholera vaccines: use in clinical practice. Lancet Infect. Dis. *6*, 361–373.

Hiroi, T., Goto, H., Someya, K., Yanagita, M., Honda, M., Yamanaka, N., and Kiyono, H. (2001). HIV mucosal vaccine: nasal immunization with rBCG-V3J1 induces a long term V3J1 peptide-specific neutralizing immunity in Th1- and Th2-deficient conditions. J. Immunol. *167*, 5862–5867.

Honda, K., and Takeda, K. (2009). Regulatory mechanisms of immune responses to intestinal bacteria. Mucosal Immunol. *2*, 187–196.

Horie, H., Koike, S., Kurata, T., Sato-Yoshida, Y., Ise, I., Ota, Y., Abe, S., Hioki, K., Kato, H., Taya, C., et al. (1994). Transgenic mice carrying the human poliovirus receptor: new animal models for study of poliovirus neurovirulence. J. Virol. *68*, 681–688.

Hornef, M.W., Frisan, T., Vandewalle, A., Normark, S., and Richter-Dahlfors, A. (2002). Toll-like receptor 4 resides in the Golgi apparatus and colocalizes with internalized lipopolysaccharide in intestinal epithelial cells. J. Exp. Med. *195*, 559–570.

Hou, Y., Hu, W.G., Hirano, T., and Gu, X.X. (2002). A new intra-NALT route elicits mucosal and systemic immunity against *Moraxella catarrhalis* in a mouse challenge model. Vaccine *20*, 2375–2381.

Huang, W., Na, L., Fidel, P.L., and Schwarzenberger, P. (2004). Requirement of interleukin-17A for systemic anti-*Candida albicans* host defense in mice. J. Infect. Dis. *190*, 624–631.

Husby, S., Mestecky, J., Moldoveanu, Z., Holland, S., and Elson, C.O. (1994). Oral tolerance in humans. T cell but not B cell tolerance after antigen feeding. J. Immunol. *152*, 4663–4670.

Hussell, T., and Humphreys, I.R. (2002). Nasal vaccination induces protective immunity without immunopathology. Clin. Exp. Immunol. *130*, 359–362.

Hyser, J.M., and Estes, M.K. (2009). Rotavirus vaccines and pathogenesis: 2008. Curr. Opin. Gastroenterol. *25*, 36–43.

Iijima, N., Linehan, M.M., Saeland, S., and Iwasaki, A. (2007). Vaginal epithelial dendritic cells renew from bone marrow precursors. Proc. Natl. Acad. Sci. U.S.A. *104*, 19061–19066.

Iijima, N., Thompson, J.M., and Iwasaki, A. (2008). Dendritic cells and macrophages in the genitourinary tract. Mucosal Immunol. *1*, 451–459.

Iliev, I.D., Mileti, E., Matteoli, G., Chieppa, M., and Rescigno, M. (2009). Intestinal epithelial cells promote colitis-protective regulatory T-cell differentiation through dendritic cell conditioning. Mucosal Immunol. *2*, 340–350.

Illum, L., Church, A.E., Butterworth, M.D., Arien, A., Whetstone, J., and Davis, S.S. (2001). Development of systems for targeting the regional lymph nodes for diagnostic imaging: *in vivo* behaviour of colloidal PEG-coated magnetite nanospheres in the rat following interstitial administration. Pharm. Res. *18*, 640–645.

Imaoka, K., Miller, C.J., Kubota, M., McChesney, M.B., Lohman, B., Yamamoto, M., Fujihashi, K., Someya,

K., Honda, M., McGhee, J.R., *et al.* (1998). Nasal immunization of nonhuman primates with simian immunodeficiency virus p55gag and cholera toxin adjuvant induces Th1/Th2 help for virus-specific immune responses in reproductive tissues. J. Immunol. *161*, 5952–5958.

Inohara, N., Ogura, Y., Fontalba, A., Gutierrez, O., Pons, F., Crespo, J., Fukase, K., Inamura, S., Kusumoto, S., Hashimoto, M., *et al.* (2003). Host recognition of bacterial muramyl dipeptide mediated through NOD2. Implications for Crohn's disease. J. Biol. Chem. *278*, 5509–5512.

Isaka, M., Yasuda, Y., Kozuka, S., Miura, Y., Taniguchi, T., Matano, K., Goto, N., and Tochikubo, K. (1998). Systemic and mucosal immune responses of mice to aluminium-adsorbed or aluminium-non-adsorbed tetanus toxoid administered intranasally with recombinant cholera toxin B subunit. Vaccine *16*, 1620–1626.

Ismail, A.S., and Hooper, L.V. (2005). Epithelial cells and their neighbors. IV. Bacterial contributions to intestinal epithelial barrier integrity. Am. J. Physiol. Gastrointest. Liver Physiol. *289*, G779–784.

Ivanov, I.I., Frutos Rde, L., Manel, N., Yoshinaga, K., Rifkin, D.B., Sartor, R.B., Finlay, B.B., and Littman, D.R. (2008). Specific microbiota direct the differentiation of IL-17-producing T-helper cells in the mucosa of the small intestine. Cell Host Microbe *4*, 337–349.

Iwasaki, A. (2007). Mucosal dendritic cells. Annu. Rev. Immunol. *25*, 381–418.

Iwasaki, A., and Kelsall, B.L. (1999). Mucosal immunity and inflammation. I. Mucosal dendritic cells: their specialized role in initiating T cell responses. Am. J. Physiol. *276*, G1074–1078.

Iwasaki, A., and Kelsall, B.L. (2000). Localization of distinct Peyer's patch dendritic cell subsets and their recruitment by chemokines macrophage inflammatory protein (MIP)-3alpha, MIP-3beta, and secondary lymphoid organ chemokine. J. Exp. Med. *191*, 1381–1394.

Iwasaki, A., and Kelsall, B.L. (2001). Unique functions of CD11b+, CD8 alpha+, and double-negative Peyer's patch dendritic cells. J. Immunol. *166*, 4884–4890.

Jepson, M.A., Clark, M.A., and Hirst, B.H. (2004). M cell targeting by lectins: a strategy for mucosal vaccination and drug delivery. Adv. Drug Deliv. Rev. *56*, 511–525.

Jiang, S., Rasmussen, R.A., Nolan, K.M., Frankel, F.R., Lieberman, J., McClure, H.M., Williams, K.M., Babu, U.S., Raybourne, R.B., Strobert, E., *et al.* (2007). Live attenuated Listeria monocytogenes expressing HIV Gag: immunogenicity in rhesus monkeys. Vaccine *25*, 7470–7479.

Jodar, L., Feavers, I.M., Salisbury, D., and Granoff, D.M. (2002). Development of vaccines against meningococcal disease. Lancet *359*, 1499–1508.

Johansson, C., and Kelsall, B.L. (2005). Phenotype and function of intestinal dendritic cells. Semin. Immunol. *17*, 284–294.

Johansson-Lindbom, B., Svensson, M., Pabst, O., Palmqvist, C., Marquez, G., Forster, R., and Agace, W.W. (2005). Functional specialization of gut CD103+ dendritic cells in the regulation of tissue-selective T cell homing. J. Exp. Med. *202*, 1063–1073.

Johansson-Lindbom, B., Svensson, M., Wurbel, M.A., Malissen, B., Marquez, G., and Agace, W. (2003). Selective generation of gut tropic T cells in gut-associated lymphoid tissue (GALT): requirement for GALT dendritic cells and adjuvant. J. Exp. Med. *198*, 963–969.

Jones, B.D., and Falkow, S. (1996). Salmonellosis: host immune responses and bacterial virulence determinants. Annu. Rev. Immunol. *14*, 533–561.

Jones, D.H., McBride, B.W., Thornton, C., O'Hagan, D.T., Robinson, A., and Farrar, G.H. (1996). Orally administered microencapsulated *Bordetella pertussis* fimbriae protect mice from *B. pertussis* respiratory infection. Infect. Immun. *64*, 489–494.

Joseph, J., Saubi, N., Pezzat, E., and Gatell, J.M. (2006). Progress towards an HIV vaccine based on recombinant bacillus Calmette-Guerin: failures and challenges. Expert Rev. Vaccines *5*, 827–838.

Kagnoff, M.F. (1996). Oral tolerance: mechanisms and possible role in inflammatory joint diseases. Baillieres Clin. Rheumatol. *10*, 41–54.

Kaneko, H., Bednarek, I., Wierzbicki, A., Kiszka, I., Dmochowski, M., Wasik, T.J., Kaneko, Y., and Kozbor, D. (2000). Oral DNA vaccination promotes mucosal and systemic immune responses to HIV envelope glycoprotein. Virology *267*, 8–16.

Kang, S.M., Yao, Q., Guo, L., and Compans, R.W. (2003). Mucosal immunization with virus-like particles of simian immunodeficiency virus conjugated with cholera toxin subunit B. J. Virol. *77*, 9823–9830.

Kelsall, B. (2008). Recent progress in understanding the phenotype and function of intestinal dendritic cells and macrophages. Mucosal Immunol. *1*, 460–469.

Kenner, J.R., Coster, T.S., Taylor, D.N., Trofa, A.F., Barrera-Oro, M., Hyman, T., Adams, J.M., Beattie, D.T., Killeen, K.P., Spriggs, D.R., *et al.* (1995). Peru-15, an improved live attenuated oral vaccine candidate for *Vibrio cholerae* O1. J. Infect. Dis. *172*, 1126–1129.

Keshav, S. (2006). Paneth cells: leukocyte-like mediators of innate immunity in the intestine. J. Leukoc. Biol. *80*, 500–508.

Khader, S.A., Bell, G.K., Pearl, J.E., Fountain, J.J., Rangel-Moreno, J., Cilley, G.E., Shen, F., Eaton, S.M., Gaffen, S.L., Swain, S.L., *et al.* (2007). IL-23 and IL-17 in the establishment of protective pulmonary CD4+ T cell responses after vaccination and during *Mycobacterium tuberculosis* challenge. Nat. Immunol. *8*, 369–377.

Kim, J.G., Lee, S.J., and Kagnoff, M.F. (2004). Nod1 is an essential signal transducer in intestinal epithelial cells infected with bacteria that avoid recognition by toll-like receptors. Infect. Immun. *72*, 1487–1495.

Kiyono, H., and Fukuyama, S. (2004). NALT- versus Peyer's-patch-mediated mucosal immunity. Nat. Rev. Immunol. *4*, 699–710.

Klencke, B., Matijevic, M., Urban, R.G., Lathey, J.L., Hedley, M.L., Berry, M., Thatcher, J., Weinberg, V., Wilson, J., Darragh, T., *et al.* (2002). Encapsulated plasmid DNA treatment for human papillomavirus 16-associated anal dysplasia: a Phase I study of ZYC101. Clin. Cancer Res. *8*, 1028–1037.

Kobayashi, K.S., Chamaillard, M., Ogura, Y., Henegariu, O., Inohara, N., Nunez, G., and Flavell, R.A. (2005). Nod2-dependent regulation of innate and adaptive immunity in the intestinal tract. Science *307*, 731–734.

Koike, S., Ise, I., and Nomoto, A. (1991). Functional domains of the poliovirus receptor. Proc. Natl. Acad. Sci. U.S.A. *88*, 4104–4108.

Kossaczka, Z., Lin, F.Y., Ho, V.A., Thuy, N.T., Van Bay, P., Thanh, T.C., Khiem, H.B., Trach, D.D., Karpas, A., Hunt, S., *et al.* (1999). Safety and immunogenicity of Vi conjugate vaccines for typhoid fever in adults, teenagers, and 2- to 4-year-old children in Vietnam. Infect. Immun. *67*, 5806–5810.

Kotloff, K.L., Winickoff, J.P., Ivanoff, B., Clemens, J.D., Swerdlow, D.L., Sansonetti, P.J., Adak, G.K., and Levine, M.M. (1999). Global burden of Shigella infections: implications for vaccine development and implementation of control strategies. Bull. World Health Org. *77*, 651–666.

Kraus, T.A., Toy, L., Chan, L., Childs, J., and Mayer, L. (2004). Failure to induce oral tolerance to a soluble protein in patients with inflammatory bowel disease. Gastroenterology *126*, 1771–1778.

Krieg, A.M. (2002). CpG motifs in bacterial DNA and their immune effects. Annu. Rev. Immunol. *20*, 709–760.

Kuklin, N.A., Rott, L., Feng, N., Conner, M.E., Wagner, N., Muller, W., and Greenberg, H.B. (2001). Protective intestinal anti-rotavirus B cell immunity is dependent on alpha 4 beta 7 integrin expression but does not require IgA antibody production. J. Immunol. *166*, 1894–1902.

Kurono, Y., Yamamoto, M., Fujihashi, K., Kodama, S., Suzuki, M., Mogi, G., McGhee, J.R., and Kiyono, H. (1999). Nasal immunization induces *Haemophilus influenzae*-specific Th1 and Th2 responses with mucosal IgA and systemic IgG antibodies for protective immunity. J. Infect. Dis. *180*, 122–132.

Kweon, M.N. (2008). Shigellosis: the current status of vaccine development. Curr. Opin. Infect. Dis. *21*, 313–318.

La Montagne, J.R., Noble, G.R., Quinnan, G.V., Curlin, G.T., Blackwelder, W.C., Smith, J.I., Ennis, F.A., and Bozeman, F.M. (1983). Summary of clinical trials of inactivated influenza vaccine – 1978. Rev. Infect. Dis. *5*, 723–736.

Lambrecht, B.N., De Veerman, M., Coyle, A.J., Gutierrez-Ramos, J.C., Thielemans, K., and Pauwels, R.A. (2000). Myeloid dendritic cells induce Th2 responses to inhaled antigen, leading to eosinophilic airway inflammation. J. Clin. Invest. *106*, 551–559.

Lambrecht, B.N., Prins, J.B., and Hoogsteden, H.C. (2001). Lung dendritic cells and host immunity to infection. Eur. Respir. J. *18*, 692–704.

Lambrecht, B.N., Salomon, B., Klatzmann, D., and Pauwels, R.A. (1998). Dendritic cells are required for the development of chronic eosinophilic airway inflammation in response to inhaled antigen in sensitized mice. J. Immunol. *160*, 4090–4097.

Lamm, M.E. (1997). Interaction of antigens and antibodies at mucosal surfaces. Annu. Rev. Microbiol. *51*, 311–340.

Lavelle, E.C. (2001). Targeted delivery of drugs to the gastrointestinal tract. Crit. Rev. Ther. Drug Carrier Syst. *18*, 341–386.

Lavelle, E.C., and O'Hagan, D.T. (2006). Delivery systems and adjuvants for oral vaccines. Expert Opin. Drug Deliv. *3*, 747–762.

Levine, M.M. (2000). Immunization against bacterial diseases of the intestine. J. Pediatr. Gastroenterol. Nutr. *31*, 336–355.

Lightfield, K.L., Persson, J., Brubaker, S.W., Witte, C.E., von Moltke, J., Dunipace, E.A., Henry, T., Sun, Y.H., Cado, D., Dietrich, W.F., *et al.* (2008). Critical function for Naip5 in inflammasome activation by a conserved carboxy-terminal domain of flagellin. Nat. Immunol. *9*, 1171–1178.

Lin, F.Y., Ho, V.A., Khiem, H.B., Trach, D.D., Bay, P.V., Thanh, T.C., Kossaczka, Z., Bryla, D.A., Shiloach, J., Robbins, J.B., *et al.* (2001). The efficacy of a *Salmonella typhi* Vi conjugate vaccine in two-to-five-year-old children. N. Engl. J. Med. *344*, 1263–1269.

Linhares, A.C., Gabbay, Y.B., Freitas, R.B., da Rosa, E.S., Mascarenhas, J.D., and Loureiro, E.C. (1989). Longitudinal study of rotavirus infections among children from Belem, Brazil. Epidemiol. Infect. *102*, 129–145.

Lopez-Gigosos, R., Garcia-Fortea, P., Reina-Dona, E., and Plaza-Martin, E. (2007). Effectiveness in prevention of travellers' diarrhoea by an oral cholera vaccine WC/rBS. Travel Med. Infect. Dis. *5*, 380–384.

Luhrmann, A., Tschernig, T., and Pabst, R. (2002). Stimulation of bronchus-associated lymphoid tissue in rats by repeated inhalation of aerosolized lipopeptide MALP-2. Pathobiology *70*, 266–269.

Lund, J.M., Linehan, M.M., Iijima, N., and Iwasaki, A. (2006). Cutting edge: plasmacytoid dendritic cells provide innate immune protection against mucosal viral infection in situ. J. Immunol. *177*, 7510–7514.

Lundkvist, J., Steffen, R., and Jonsson, B. (2009). Cost–benefit of WC/rBS oral cholera vaccine for vaccination against ETEC-caused travelers' diarrhea. J. Travel Med. *16*, 28–34.

Luzza, F., Parrello, T., Monteleone, G., Sebkova, L., Romano, M., Zarrilli, R., Imeneo, M., and Pallone, F. (2000). Up-regulation of IL-17 is associated with bioactive IL-8 expression in *Helicobacter pylori*-infected human gastric mucosa. J. Immunol. *165*, 5332–5337.

Lycke, N. (1997). The mechanism of cholera toxin adjuvanticity. Res. Immunol. *148*, 504–520.

Lynch, J.P., 3rd, and Walsh, E.E. (2007). Influenza: evolving strategies in treatment and prevention. Semin. Respir. Crit. Care. Med. *28*, 144–158.

Macpherson, A.J., McCoy, K.D., Johansen, F.E., and Brandtzaeg, P. (2008). The immune geography of IgA induction and function. Mucosal Immunol. *1*, 11–22.

Mallett, C.P., Hale, T.L., Kaminski, R.W., Larsen, T., Orr, N., Cohen, D., and Lowell, G.H. (1995). Intransal or intragastric immunization with proteosome-Shigella lipopolysaccharide vaccines protects against lethal pneumonia in a murine model of Shigella infection. Infect. Immun. *63*, 2382–2386.

Manicassamy, S., and Pulendran, B. (2009). Retinoic acid-dependent regulation of immune responses by dendritic cells and macrophages. Semin. Immunol. *21*, 22–27.

Manicassamy, S., Ravindran, R., Deng, J., Oluoch, H., Denning, T.L., Kasturi, S.P., Rosenthal, K.M., Evavold, B.D., and Pulendran, B. (2009). Toll-like receptor 2-dependent induction of vitamin A-metabolizing enzymes in dendritic cells promotes T regulatory responses and inhibits autoimmunity. Nat. Med. *15*, 401–409.

Manocha, M., Pal, P.C., Chitralekha, K.T., Thomas, B.E., Tripathi, V., Gupta, S.D., Paranjape, R., Kulkarni, S., and Rao, D.N. (2005). Enhanced mucosal and systemic immune response with intranasal immunization of mice with HIV peptides entrapped in PLG microparticles in combination with *Ulex Europaeus*-I lectin as M cell target. Vaccine *23*, 5599–5617.

Marazuela, Prado, E.G., N., Moro, E., Fernandez-Garcia, H., Villalba, M., Rodriguez, R., and Batanero, E. (2008). Intranasal vaccination with poly(lactide-co-glycolide) microparticles containing a peptide T of Ole e 1 prevents mice against sensitization. Clin. Exp. Allergy *38*, 520–528.

Marinaro, M., Fasano, A., and De Magistris, M.T. (2003). Zonula occludens toxin acts as an adjuvant through different mucosal routes and induces protective immune responses. Infect. Immun. *71*, 1897–1902.

Marinaro, M., Staats, H.F., Hiroi, T., Jackson, R.J., Coste, M., Boyaka, P.N., Okahashi, N., Yamamoto, M., Kiyono, H., Bluethmann, H., et al. (1995). Mucosal adjuvant effect of cholera toxin in mice results from induction of T helper 2 (Th2) cells and IL-4. J. Immunol. *155*, 4621–4629.

Martin, D.R., Ruijne, N., McCallum, L., O'Hallahan, J., and Oster, P. (2006). The VR2 epitope on the PorA P1.7–2, 4 protein is the major target for the immune response elicited by the strain-specific group B meningococcal vaccine MeNZB. Clin. Vaccine Immunol. *13*, 486–491.

Marx, P.A., Compans, R.W., Gettie, A., Staas, J.K., Gilley, R.M., Mulligan, M.J., Yamshchikov, G.V., Chen, D., and Eldridge, J.H. (1993). Protection against vaginal SIV transmission with microencapsulated vaccine. Science *260*, 1323–1327.

Mason, H.S., Lam, D.M., and Arntzen, C.J. (1992). Expression of hepatitis B surface antigen in transgenic plants. Proc. Natl. Acad. Sci. U.S.A. *89*, 11745–11749.

Mason, K.L., Huffnagle, G.B., Noverr, M.C., and Kao, J.Y. (2008). Overview of gut immunology. Adv. Exp. Med. Biol. *635*, 1–14.

Masopust, D., Ha, S.J., Vezys, V., and Ahmed, R. (2006). Stimulation history dictates memory CD8 T cell phenotype: implications for prime-boost vaccination. J. Immunol. *177*, 831–839.

Mata, L., Simhon, A., Urrutia, J.J., Kronmal, R.A., Fernandez, R., and Garcia, B. (1983). Epidemiology of rotaviruses in a cohort of 45 Guatamalan Mayan Indian children observed from birth to the age of three years. J. Infect. Dis. *148*, 452–461.

Mathiowitz, E., Jacob, J.S., Jong, Y.S., Carino, G.P., Chickering, D.E., Chaturvedi, P., Santos, C.A., Vijayaraghavan, K., Montgomery, S., Bassett, M., et al. (1997). Biologically erodable microspheres as potential oral drug delivery systems. Nature *386*, 410–414.

Matoba, N., Magerus, A., Geyer, B.C., Zhang, Y., Muralidharan, M., Alfsen, A., Arntzen, C.J., Bomsel, M., and Mor, T.S. (2004). A mucosally targeted subunit vaccine candidate eliciting HIV-1 transcytosis-blocking Abs. Proc. Natl. Acad. Sci. U.S.A. *101*, 13584–13589.

Matson, D.O. (2006). The pentavalent rotavirus vaccine, RotaTeq. Semin. Pediatr. Infect. Dis. *17*, 195–199.

Matsui, M., Hafler, D.A., and Weiner, H.L. (1996). Pilot study of oral tolerance to keyhole limpet hemocyanin in humans: down-regulation of KLH-reactive precursor-cell frequency. Ann. N. Y. Acad. Sci. *778*, 398–404.

McCluskie, M.J., and Davis, H.L. (1998). CpG DNA is a potent enhancer of systemic and mucosal immune responses against hepatitis B surface antigen with intranasal administration to mice. J. Immunol. *161*, 4463–4466.

McCluskie, M.J., and Davis, H.L. (1999). CpG DNA as mucosal adjuvant. Vaccine *18*, 231–237.

McCluskie, M.J., Weeratna, R.D., and Davis, H.L. (2000). Intranasal immunization of mice with CpG DNA induces strong systemic and mucosal responses that are influenced by other mucosal adjuvants and antigen distribution. Mol. Med. *6*, 867–877.

McGhee, J.R., Kunisawa, J., and Kiyono, H. (2007). Gut lymphocyte migration: we are halfway 'home'. Trends Immunol. *28*, 150–153.

Melmed, G., Thomas, L.S., Lee, N., Tesfay, S.Y., Lukasek, K., Michelsen, K.S., Zhou, Y., Hu, B., Arditi, M., and Abreu, M.T. (2003). Human intestinal epithelial cells are broadly unresponsive to Toll-like receptor 2-dependent bacterial ligands: implications for host–microbial interactions in the gut. J. Immunol. *170*, 1406–1415.

Mestecky, J. (1987). The common mucosal immune system and current strategies for induction of immune responses in external secretions. J. Clin. Immunol. *7*, 265–276.

Mestecky, J., Nguyen, H., Czerkinsky, C., and Kiyono, H. (2008). Oral immunization: an update. Curr. Opin. Gastroenterol. *24*, 713–719.

Misumi, S., Masuyama, M., Takamune, N., Nakayama, D., Mitsumata, R., Matsumoto, H., Urata, N., Takahashi, Y., Muneoka, A., Sukamoto, T., et al. (2009). Targeted delivery of immunogen to primate m cells with tetragalloyl lysine dendrimer. J. Immunol. *182*, 6061–6070.

Mittrucker, H.W., and Kaufmann, S.H. (2000). Immune response to infection with *Salmonella* typhimurium in mice. J. Leukoc. Biol. *67*, 457–463.

Modelska, A., Dietzschold, B., Sleysh, N., Fu, Z.F., Steplewski, K., Hooper, D.C., Koprowski, H., and Yusibov, V. (1998). Immunization against rabies with plant-derived antigen. Proc. Natl. Acad. Sci. U.S.A. *95*, 2481–2485.

Modlin, R.L., and Cheng, G. (2004). From plankton to pathogen recognition. Nat. Med. *10*, 1173–1174.

Mohamadzadeh, M., Duong, T., Hoover, T., and Klaenhammer, T.R. (2008). Targeting mucosal dendritic cells with microbial antigens from probiotic lactic acid bacteria. Expert Rev. Vaccines 7, 163–174.

Mohamadzadeh, M., and Klaenhammer, T.R. (2008). Specific Lactobacillus species differentially activate Toll-like receptors and downstream signals in dendritic cells. Expert Rev. Vaccines 7, 1155–1164.

Mora, J.R., Iwata, M., Eksteen, B., Song, S.Y., Junt, T., Senman, B., Otipoby, K.L., Yokota, A., Takeuchi, H., Ricciardi-Castagnoli, P., *et al.* (2006). Generation of gut-homing IgA-secreting B cells by intestinal dendritic cells. Science *314*, 1157–1160.

Mora, J.R., and von Andrian, U.H. (2008). Differentiation and homing of IgA-secreting cells. Mucosal Immunol. *1*, 96–109.

Mowat, A.M., Parker, L.A., Beacock-Sharp, H., Millington, O.R., and Chirdo, F. (2004). Oral tolerance: overview and historical perspectives. Ann. N. Y. Acad. Sci. *1029*, 1–8.

Moyron-Quiroz, J.E., Rangel-Moreno, J., Kusser, K., Hartson, L., Sprague, F., Goodrich, S., Woodland, D.L., Lund, F.E., and Randall, T.D. (2004). Role of inducible bronchus associated lymphoid tissue (iBALT) in respiratory immunity. Nat. Med. *10*, 927–934.

Mueller, S., and Wimmer, E. (2003). Recruitment of nectin-3 to cell–cell junctions through trans-heterophilic interaction with CD155, a vitronectin and poliovirus receptor that localizes to alpha(v)beta3 integrin-containing membrane microdomains. J. Biol. Chem. *278*, 31251–31260.

Nardelli-Haefliger, D., Lurati, F., Wirthner, D., Spertini, F., Schiller, J.T., Lowy, D.R., Ponci, F., and De Grandi, P. (2005). Immune responses induced by lower airway mucosal immunisation with a human papillomavirus type 16 virus-like particle vaccine. Vaccine *23*, 3634–3641.

Nardelli-Haefliger, D., Roden, R., Balmelli, C., Potts, A., Schiller, J., and De Grandi, P. (1999). Mucosal but not parenteral immunization with purified human papillomavirus type 16 virus-like particles induces neutralizing titers of antibodies throughout the estrous cycle of mice. J. Virol. 73, 9609–9613.

Nardelli-Haefliger, D., Wirthner, D., Schiller, J.T., Lowy, D.R., Hildesheim, A., Ponci, F., and De Grandi, P. (2003). Specific antibody levels at the cervix during the menstrual cycle of women vaccinated with human papillomavirus 16 virus-like particles. J. Natl. Cancer Inst. 95, 1128–1137.

Nathanson, N. (2008). The pathogenesis of poliomyelitis: what we don't know. Adv. Virus Res. 71, 1–50.

Niess, J.H., Brand, S., Gu, X., Landsman, L., Jung, S., McCormick, B.A., Vyas, J.M., Boes, M., Ploegh, H.L., Fox, J.G., *et al.* (2005). CX3CR1-mediated dendritic cell access to the intestinal lumen and bacterial clearance. Science *307*, 254–258.

O'Hagan, D.T. (1998). Microparticles and polymers for the mucosal delivery of vaccines. Adv. Drug Deliv. Rev. *34*, 305–320.

O'Hagan, D.T., Singh, M., and Ulmer, J.B. (2006). Microparticle-based technologies for vaccines. Methods *40*, 10–19.

O'Neal, C.M., Harriman, G.R., and Conner, M.E. (2000). Protection of the villus epithelial cells of the small intestine from rotavirus infection does not require immunoglobulin A. J. Virol. 74, 4102–4109.

O'Ryan, M. (2007). Rotarix (RIX4414): an oral human rotavirus vaccine. Expert Rev. Vaccines *6*, 11–19.

Oggioni, M.R., Medaglini, D., Romano, L., Peruzzi, F., Maggi, T., Lozzi, L., Bracci, L., Zazzi, M., Manca, F., Valensin, P.E., *et al.* (1999). Antigenicity and immunogenicity of the V3 domain of HIV type 1 glycoprotein 120 expressed on the surface of Streptococcus gordonii. AIDS Res. Hum. Retroviruses *15*, 451–459.

Ogra, P.L., Faden, H., and Welliver, R.C. (2001). Vaccination strategies for mucosal immune responses. Clin. Microbiol. Rev. *14*, 430–445.

Ogra, P.L., and Karzon, D.T. (1969). Distribution of poliovirus antibody in serum, nasopharynx and alimentary tract following segmental immunization of lower alimentary tract with poliovaccine. J. Immunol. *102*, 1423–1430.

Ohka, S., Igarashi, H., Nagata, N., Sakai, M., Koike, S., Nochi, T., Kiyono, H., and Nomoto, A. (2007). Establishment of a poliovirus oral infection system in human poliovirus receptor-expressing transgenic mice that are deficient in alpha/beta interferon receptor. J. Virol. *81*, 7902–7912.

Ohka, S., and Nomoto, A. (2001). Recent insights into poliovirus pathogenesis. Trends Microbiol. *9*, 501–506.

Okada, H., and Toguchi, H. (1995). Biodegradable microspheres in drug delivery. Crit. Rev. Ther. Drug Carrier Syst. *12*, 1–99.

Opitz, B., Forster, S., Hocke, A.C., Maass, M., Schmeck, B., Hippenstiel, S., Suttorp, N., and Krull, M. (2005). Nod1-mediated endothelial cell activation by *Chlamydophila pneumoniae*. Circ. Res. *96*, 319–326.

Otte, J.M., Cario, E., and Podolsky, D.K. (2004). Mechanisms of cross hyporesponsiveness to Toll-like receptor bacterial ligands in intestinal epithelial cells. Gastroenterology *126*, 1054–1070.

Paglia, P., Medina, E., Arioli, I., Guzman, C.A., and Colombo, M.P. (1998). Gene transfer in dendritic cells, induced by oral DNA vaccination with *Salmonella* typhimurium, results in protective immunity against a murine fibrosarcoma. Blood *92*, 3172–3176.

Parashar, U.D., Hummelman, E.G., Bresee, J.S., Miller, M.A., and Glass, R.I. (2003). Global illness and deaths caused by rotavirus disease in children. Emerg. Infect. Dis. *9*, 565–572.

Park, J.H., Kim, Y.G., McDonald, C., Kanneganti, T.D., Hasegawa, M., Body-Malapel, M., Inohara, N., and Nunez, G. (2007). RICK/RIP2 mediates innate immune responses induced through Nod1 and Nod2 but not TLRs. J. Immunol. *178*, 2380–2386.

Pascual, D.W., Powell, R.J., Lewis, G.K., and Hone, D.M. (1997). Oral bacterial vaccine vectors for the delivery of subunit and nucleic acid vaccines to the organized lymphoid tissue of the intestine. Behring Inst. Mitt. 143–152.

Paterson, Y., and Johnson, R.S. (2004). Progress towards the use of Listeria monocytogenes as a live bacterial vaccine vector for the delivery of HIV antigens. Expert Rev. Vaccines 3, S119–134.

Perry, M., and Whyte, A. (1998). Immunology of the tonsils. Immunol. Today 19, 414–421.

Peters, C., Peng, X., Douven, D., Pan, Z.K., and Paterson, Y. (2003). The induction of HIV Gag-specific CD8+ T cells in the spleen and gut-associated lymphoid tissue by parenteral or mucosal immunization with recombinant Listeria monocytogenes HIV Gag. J. Immunol. 170, 5176–5187.

Porter, E.M., Bevins, C.L., Ghosh, D., and Ganz, T. (2002). The multifaceted Paneth cell. Cell Mol. Life Sci. 59, 156–170.

Pulendran, B. (2005). Variegation of the immune response with dendritic cells and pathogen recognition receptors. J. Immunol. 174, 2457–2465.

Pulendran, B., Tang, H., and Denning, T.L. (2008). Division of labor, plasticity, and crosstalk between dendritic cell subsets. Curr. Opin. Immunol. 20, 61–67.

Querec, T., Bennouna, S., Alkan, S., Laouar, Y., Gorden, K., Flavell, R., Akira, S., Ahmed, R., and Pulendran, B. (2006). Yellow fever vaccine YF-17D activates multiple dendritic cell subsets via TLR2, 7, 8, and 9 to stimulate polyvalent immunity. J. Exp. Med. 203, 413–424.

Querec, T.D., Akondy, R.S., Lee, E.K., Cao, W., Nakaya, H.I., Teuwen, D., Pirani, A., Gernert, K., Deng, J., Marzolf, B., et al. (2009). Systems biology approach predicts immunogenicity of the yellow fever vaccine in humans. Nat. Immunol. 10, 116–125.

Querec, T.D., and Pulendran, B. (2007). Understanding the role of innate immunity in the mechanism of action of the live attenuated Yellow Fever Vaccine 17D. Adv. Exp. Med. Biol. 590, 43–53.

Quiding-Jarbrink, M., Nordstrom, I., Granstrom, G., Kilander, A., Jertborn, M., Butcher, E.C., Lazarovits, A.I., Holmgren, J., and Czerkinsky, C. (1997). Differential expression of tissue-specific adhesion molecules on human circulating antibody-forming cells after systemic, enteric, and nasal immunizations. A molecular basis for the compartmentalization of effector B cell responses. J. Clin. Invest. 99, 1281–1286.

Raffatellu, M., Santos, R.L., Verhoeven, D.E., George, M.D., Wilson, R.P., Winter, S.E., Godinez, I., Sankaran, S., Paixao, T.A., Gordon, M.A., et al. (2008). Simian immunodeficiency virus-induced mucosal interleukin-17 deficiency promotes Salmonella dissemination from the gut. Nat. Med. 14, 421–428.

Rappuoli, R., Pizza, M., Douce, G., and Dougan, G. (1999). Structure and mucosal adjuvanticity of cholera and Escherichia coli heat-labile enterotoxins. Immunol. Today 20, 493–500.

Raul Velazquez, F., Calva, J.J., Lourdes Guerrero, M., Mass, D., Glass, R.I., Pickering, L.K., and Ruiz-Palacios, G.M. (1993). Cohort study of rotavirus serotype patterns in symptomatic and asymptomatic infections in Mexican children. Pediatr. Infect. Dis. J. 12, 54–61.

Ravindran, R., and McSorley, S.J. (2005). Tracking the dynamics of T-cell activation in response to Salmonella infection. Immunology 114, 450–458.

Ren, R.B., Costantini, F., Gorgacz, E.J., Lee, J.J., and Racaniello, V.R. (1990). Transgenic mice expressing a human poliovirus receptor: a new model for poliomyelitis. Cell 63, 353–362.

Richter, L.J., Thanavala, Y., Arntzen, C.J., and Mason, H.S. (2000). Production of hepatitis B surface antigen in transgenic plants for oral immunization. Nat. Biotechnol. 18, 1167–1171.

Riepenhoff-Talty, M., Dharakul, T., Kowalski, E., Michalak, S., and Ogra, P.L. (1987). Persistent rotavirus infection in mice with severe combined immunodeficiency. J. Virol. 61, 3345–3348.

Rijpkema, S.G., Bik, E.M., Jansen, W.H., Gielen, H., Versluis, L.F., Stouthamer, A.H., Guinee, P.A., and Mooi, F.R. (1992). Construction and analysis of a Vibrio cholerae delta-aminolevulinic acid auxotroph which confers protective immunity in a rabbit model. Infect. Immun. 60, 2188–2193.

Roland, K.L., Cloninger, C., Kochi, S.K., Thomas, L.J., Tinge, S.A., Rouskey, C., and Killeen, K.P. (2007). Construction and preclinical evaluation of recombinant Peru-15 expressing high levels of the cholera toxin B subunit as a vaccine against enterotoxigenic Escherichia coli. Vaccine 25, 8574–8584.

Roland, K.L., Tinge, S.A., Killeen, K.P., and Kochi, S.K. (2005). Recent advances in the development of live, attenuated bacterial vectors. Curr. Opin. Mol. Ther. 7, 62–72.

Rose, J., Franco, M., and Greenberg, H. (1998). The immunology of rotavirus infection in the mouse. Adv. Virus Res. 51, 203–235.

Ruiz-Palacios, G.M., Perez-Schael, I., Velazquez, F.R., Abate, H., Breuer, T., Clemens, S.C., Cheuvart, B., Espinoza, F., Gillard, P., Innis, B.L., et al. (2006). Safety and efficacy of an attenuated vaccine against severe rotavirus gastroenteritis. N. Engl. J. Med. 354, 11–22.

Sabin, A.B., Ramos-Alvarez, M., Alvarez-Amezquita, J., Pelon, W., Michaels, R.H., Spigland, I., Koch, M.A., Barnes, J.M., and Rhim, J.S. (1960). Live, orally given poliovirus vaccine. Effects of rapid mass immunization on population under conditions of massive enteric infection with other viruses. JAMA 173, 1521–1526.

Sabirov, A., and Metzger, D.W. (2006). Intranasal vaccination of neonatal mice with polysaccharide conjugate vaccine for protection against pneumococcal otitis media. Vaccine 24, 5584–5592.

Sabirov, A., and Metzger, D.W. (2008). Mouse models for the study of mucosal vaccination against otitis media. Vaccine 26, 1501–1524.

Sakaue, G., Hiroi, T., Nakagawa, Y., Someya, K., Iwatani, K., Sawa, Y., Takahashi, H., Honda, M., Kunisawa, J., and Kiyono, H. (2003). HIV mucosal vaccine: nasal immunization with gp160-encapsulated hemagglutinating virus of Japan-liposome induces antigen-specific CTLs and neutralizing antibody responses. J. Immunol. 170, 495–502.

Salazar-Gonzalez, R.M., Niess, J.H., Zammit, D.J., Ravindran, R., Srinivasan, A., Maxwell, J.R., Stoklasek, T., Yadav, R., Williams, I.R., Gu, X., *et al.* (2006). CCR6-mediated dendritic cell activation of pathogen-specific T cells in Peyer's patches. Immunity 24, 623–632.

Salk, J.E., Bazeley, P.L., Bennett, B.L., Krech, U., Lewis, L.J., Ward, E.N., and Youngner, J.S. (1954). Studies in human subjects on active immunization against poliomyelitis. II. A practical means for inducing and maintaining antibody formation. Am. J. Public Health Nations Health 44, 994–1009.

Sanchez, J., and Holmgren, J. (2008). Cholera toxin structure, gene regulation and pathophysiological and immunological aspects. Cell Mol. Life Sci. 65, 1347–1360.

Sansonetti, P.J. (2004). War and peace at mucosal surfaces. Nat. Rev. Immunol. 4, 953–964.

Santacroce, L., Buonfantino, M., and Santacroce, S. (1997). Surgical treatment and prognostic factors in colon cancer. J. Chemother. 9, 144–145.

Sato, A., and Iwasaki, A. (2004). Induction of antiviral immunity requires Toll-like receptor signaling in both stromal and dendritic cell compartments. Proc. Natl. Acad. Sci. U.S.A. 101, 16274–16279.

Schiller, J.T., and Davies, P. (2004). Delivering on the promise: HPV vaccines and cervical cancer. Nat. Rev. Microbiol. 2, 343–347.

Seo, J.Y., Seong, S.Y., Ahn, B.Y., Kwon, I.C., Chung, H., and Jeong, S.Y. (2002). Cross-protective immunity of mice induced by oral immunization with pneumococcal surface adhesin a encapsulated in microspheres. Infect. Immun. 70, 1143–1149.

Sha, J., Fadl, A.A., Klimpel, G.R., Niesel, D.W., Popov, V.L., and Chopra, A.K. (2004). The two murein lipoproteins of *Salmonella* enterica serovar Typhimurium contribute to the virulence of the organism. Infect. Immun. 72, 3987–4003.

Sharpe, S., Hanke, T., Tinsley-Bown, A., Dennis, M., Dowall, S., McMichael, A., and Cranage, M. (2003). Mucosal immunization with PLGA-microencapsulated DNA primes a SIV-specific CTL response revealed by boosting with cognate recombinant modified vaccinia virus Ankara. Virology 313, 13–21.

Shata, M.T., and Hone, D.M. (2001). Vaccination with a *Shigella* DNA vaccine vector induces antigen-specific CD8(+) T cells and antiviral protective immunity. J. Virol. 75, 9665–9670.

Shata, M.T., Reitz, M.S., Jr., DeVico, A.L., Lewis, G.K., and Hone, D.M. (2001). Mucosal and systemic HIV-1 Env-specific CD8(+) T-cells develop after intragastric vaccination with a *Salmonella* Env DNA vaccine vector. Vaccine 20, 623–629.

Shimoda, M., Nakamura, T., Takahashi, Y., Asanuma, H., Tamura, S., Kurata, T., Mizuochi, T., Azuma, N., Kanno, C., and Takemori, T. (2001). Isotype-specific selection of high affinity memory B cells in nasal-associated lymphoid tissue. J. Exp. Med. 194, 1597–1607.

Sierro, F., Dubois, B., Coste, A., Kaiserlian, D., Kraehenbuhl, J.P., and Sirard, J.C. (2001). Flagellin stimulation of intestinal epithelial cells triggers CCL20-mediated migration of dendritic cells. Proc. Natl. Acad. Sci. U.S.A. 98, 13722–13727.

Singh, J., Pandit, S., Bramwell, V.W., and Alpar, H.O. (2006). Diphtheria toxoid loaded poly-(epsilon-caprolactone) nanoparticles as mucosal vaccine delivery systems. Methods 38, 96–105.

Singh, M., Vajdy, M., Gardner, J., Briones, M., and O'Hagan, D. (2001). Mucosal immunization with HIV-1 gag DNA on cationic microparticles prolongs gene expression and enhances local and systemic immunity. Vaccine 20, 594–602.

Sixma, T.K., Pronk, S.E., Kalk, K.H., Wartna, E.S., van Zanten, B.A., Witholt, B., and Hol, W.G. (1991). Crystal structure of a cholera toxin-related heat-labile enterotoxin from *E. coli*. Nature 351, 371–377.

Sminia, T., van der Brugge-Gamelkoorn, G.J., and Jeurissen, S.H. (1989). Structure and function of bronchus-associated lymphoid tissue (BALT). Crit. Rev. Immunol. 9, 119–150.

Smith, P.D., Smythies, L.E., Mosteller-Barnum, M., Sibley, D.A., Russell, M.W., Merger, M., Sellers, M.T., Orenstein, J.M., Shimada, T., Graham, M.F., *et al.* (2001). Intestinal macrophages lack CD14 and CD89 and consequently are down-regulated for LPS- and IgA-mediated activities. J. Immunol. 167, 2651–2656.

Smythies, L.E., Sellers, M., Clements, R.H., Mosteller-Barnum, M., Meng, G., Benjamin, W.H., Orenstein, J.M., and Smith, P.D. (2005). Human intestinal macrophages display profound inflammatory anergy despite avid phagocytic and bactericidal activity. J. Clin. Invest. 115, 66–75.

Steinman, R.M., and Banchereau, J. (2007). Taking dendritic cells into medicine. Nature 449, 419–426.

Stephenson, I., Zambon, M.C., Rudin, A., Colegate, A., Podda, A., Bugarini, R., Del Giudice, G., Minutello, A., Bonnington, S., Holmgren, J., *et al.* (2006). Phase I evaluation of intranasal trivalent inactivated influenza vaccine with nontoxigenic *Escherichia coli* enterotoxin and novel biovector as mucosal adjuvants, using adult volunteers. J. Virol. 80, 4962–4970.

Sun, C.M., Hall, J.A., Blank, R.B., Bouladoux, N., Oukka, M., Mora, J.R., and Belkaid, Y. (2007). Small intestine lamina propria dendritic cells promote de novo generation of Foxp3 T reg cells via retinoic acid. J. Exp. Med. 204, 1775–1785.

Sundaram, U., Knickelbein, R.G., and Dobbins, J.W. (1991). Mechanism of intestinal secretion. Effect of serotonin on rabbit ileal crypt and villus cells. J. Clin. Invest. 87, 743–746.

Svennerholm, A.M., Sack, D.A., Holmgren, J., and Bardhan, P.K. (1982). Intestinal antibody responses after immunisation with cholera B subunit. Lancet 1, 305–308.

Szu, S.C., Taylor, D.N., Trofa, A.C., Clements, J.D., Shiloach, J., Sadoff, J.C., Bryla, D.A., and Robbins, J.B. (1994). Laboratory and preliminary clinical characterization of Vi capsular polysaccharide–protein conjugate vaccines. Infect. Immun. 62, 4440–4444.

Tackaberry, E.S., Dudani, A.K., Prior, F., Tocchi, M., Sardana, R., Altosaar, I., and Ganz, P.R. (1999). Development of biopharmaceuticals in plant expression systems: cloning, expression and immunological reactivity of human cytomegalovirus glycoprotein B (UL55) in seeds of transgenic tobacco. Vaccine 17, 3020–3029.

Tacket, C.O., Mason, H.S., Losonsky, G., Clements, J.D., Levine, M.M., and Arntzen, C.J. (1998). Immunogenicity in humans of a recombinant bacterial antigen delivered in a transgenic potato. Nat. Med. 4, 607–609.

Tacket, C.O., Mason, H.S., Losonsky, G., Estes, M.K., Levine, M.M., and Arntzen, C.J. (2000). Human immune responses to a novel Norwalk virus vaccine delivered in transgenic potatoes. J. Infect. Dis. 182, 302–305.

Takahashi, I., Marinaro, M., Kiyono, H., Jackson, R.J., Nakagawa, I., Fujihashi, K., Hamada, S., Clements, J.D., Bost, K.L., and McGhee, J.R. (1996). Mechanisms for mucosal immunogenicity and adjuvancy of Escherichia coli labile enterotoxin. J. Infect. Dis. 173, 627–635.

Takeda, K., and Akira, S. (2007). Toll-like receptors. Curr. Protoc. Immunol. Chapter 14, Unit 14, 12.

Takeda, K., Kaisho, T., and Akira, S. (2003). Toll-like receptors. Annu. Rev. Immunol. 21, 335–376.

Tamura, S., Funato, H., Nagamine, T., Aizawa, C., and Kurata, T. (1989). Effectiveness of cholera toxin B subunit as an adjuvant for nasal influenza vaccination despite pre-existing immunity to CTB. Vaccine 7, 503–505.

Taylor, D.N., Killeen, K.P., Hack, D.C., Kenner, J.R., Coster, T.S., Beattie, D.T., Ezzell, J., Hyman, T., Trofa, A., Sjogren, M.H., et al. (1994). Development of a live, oral, attenuated vaccine against El Tor cholera. J. Infect. Dis. 170, 1518–1523.

Tirouvanziam, R., Khazaal, I., N'Sonde, V., Peyrat, M.A., Lim, A., de Bentzmann, S., Fournie, J.J., Bonneville, M., and Peault, B. (2002). Ex vivo development of functional human lymph node and bronchus-associated lymphoid tissue. Blood 99, 2483–2489.

Toyoshima, M., Chida, K., and Sato, A. (2000). Antigen uptake and subsequent cell kinetics in bronchus-associated lymphoid tissue. Respirology 5, 141–145.

Trach, D.D., Cam, P.D., Ke, N.T., Rao, M.R., Dinh, D., Hang, P.V., Hung, N.V., Canh, D.G., Thiem, V.D., Naficy, A., et al. (2002). Investigations into the safety and immunogenicity of a killed oral cholera vaccine developed in Vietnam. Bull. World Health Organ. 80, 2–8.

Trach, D.D., Clemens, J.D., Ke, N.T., Thuy, H.T., Son, N.D., Canh, D.G., Hang, P.V., and Rao, M.R. (1997). Field trial of a locally produced, killed, oral cholera vaccine in Vietnam. Lancet 349, 231–235.

Tseng, J., Komisar, J.L., Trout, R.N., Hunt, R.E., Chen, J.Y., Johnson, A.J., Pitt, L., and Ruble, D.L. (1995). Humoral immunity to aerosolized staphylococcal enterotoxin B (SEB), a superantigen, in monkeys vaccinated with SEB toxoid-containing microspheres. Infect. Immun. 63, 2880–2885.

Twaites, B.R., de las Heras Alarcon, C., Cunliffe, D., Lavigne, M., Pennadam, S., Smith, J.R., Gorecki, D.C., and Alexander, C. (2004). Thermo and pH responsive polymers as gene delivery vectors: effect of polymer architecture on DNA complexation in vitro. J. Control Release 97, 551–566.

Uematsu, S., Fujimoto, K., Jang, M.H., Yang, B.G., Jung, Y.J., Nishiyama, M., Sato, S., Tsujimura, T., Yamamoto, M., Yokota, Y., et al. (2008). Regulation of humoral and cellular gut immunity by lamina propria dendritic cells expressing Toll-like receptor 5. Nat. Immunol. 9, 769–776.

Uren, T.K., Wijburg, O.L., Simmons, C., Johansen, F.E., Brandtzaeg, P., and Strugnell, R.A. (2005). Vaccine-induced protection against gastrointestinal bacterial infections in the absence of secretory antibodies. Eur. J. Immunol. 35, 180–188.

Vajdy, M., and Lycke, N.Y. (1992). Cholera toxin adjuvant promotes long-term immunological memory in the gut mucosa to unrelated immunogens after oral immunization. Immunology 75, 488–492.

van Wijk, F., and Cheroutre, H. (2009). Intestinal T cells: Facing the mucosal immune dilemma with synergy and diversity. Semin. Immunol. 21, 130–138.

Vecino, W.H., Morin, P.M., Agha, R., Jacobs, W.R., Jr., and Fennelly, G.J. (2002). Mucosal DNA vaccination with highly attenuated Shigella is superior to attenuated Salmonella and comparable to intramuscular DNA vaccination for T cells against HIV. Immunol. Lett. 82, 197–204.

Verdonck, F., Joensuu, J.J., Stuyven, E., De Meyer, J., Muilu, M., Pirhonen, M., Goddeeris, B.M., Mast, J., Niklander-Teeri, V., and Cox, E. (2008). The polymeric stability of the Escherichia coli F4 (K88) fimbriae enhances its mucosal immunogenicity following oral immunization. Vaccine 26, 5728–5735.

Vesikari, T. (2008). Rotavirus vaccines. Scand. J. Infect. Dis. 40, 691–695.

Vesikari, T., Giaquinto, C., and Huppertz, H.I. (2006). Clinical trials of rotavirus vaccines in Europe. Pediatr. Infect. Dis. J. 25, S42–47.

Vines, R.R., Perdue, S.S., Moncrief, J.S., Sentz, D.R., Barroso, L.A., Wright, R.L., and Wilkins, T.D. (2000). Fragilysin, the enterotoxin from Bacteroides fragilis, enhances the serum antibody response to antigen co-administered by the intranasal route. Vaccine 19, 655–660.

Walboomers, J.M., Jacobs, M.V., Manos, M.M., Bosch, F.X., Kummer, J.A., Shah, K.V., Snijders, P.J., Peto, J., Meijer, C.J., and Munoz, N. (1999). Human papillomavirus is a necessary cause of invasive cervical cancer worldwide. J. Pathol. 189, 12–19.

Waldo, F.B., van den Wall Bake, A.W., Mestecky, J., and Husby, S. (1994). Suppression of the immune response by nasal immunization. Clin. Immunol. Immunopathol. 72, 30–34.

Walker, R.I. (2005). Considerations for development of whole cell bacterial vaccines to prevent diarrheal diseases in children in developing countries. Vaccine 23, 3369–3385.

Walmsley, A.M., and Arntzen, C.J. (2000). Plants for delivery of edible vaccines. Curr. Opin. Biotechnol. 11, 126–129.

Wang, X., Hone, D.M., Haddad, A., Shata, M.T., and Pascual, D.W. (2003). M cell DNA vaccination for CTL immunity to HIV. J. Immunol. *171*, 4717–4725.

Whittum-Hudson, J.A., An, L.L., Saltzman, W.M., Prendergast, R.A., and MacDonald, A.B. (1996). Oral immunization with an anti-idiotypic antibody to the exoglycolipid antigen protects against experimental Chlamydia trachomatis infection. Nat. Med. *2*, 1116–1121.

Wigdorovitz, A., Perez Filgueira, D.M., Robertson, N., Carrillo, C., Sadir, A.M., Morris, T.J., and Borca, M.V. (1999). Protection of mice against challenge with foot and mouth disease virus (FMDV) by immunization with foliar extracts from plants infected with recombinant tobacco mosaic virus expressing the FMDV structural protein VP1. Virology *264*, 85–91.

Wijburg, O.L., Uren, T.K., Simpfendorfer, K., Johansen, F.E., Brandtzaeg, P., and Strugnell, R.A. (2006). Innate secretory antibodies protect against natural *Salmonella* typhimurium infection. J. Exp. Med. *203*, 21–26.

Williams, N.A., Hirst, T.R., and Nashar, T.O. (1999). Immune modulation by the cholera-like enterotoxins: from adjuvant to therapeutic. Immunol. Today *20*, 95–101.

World Health Organization (WHO) (2008). Typhoid vaccines: WHO position paper. Wkly Epidemiol. Rec. *83*, 49–59.

Wu, H.Y., Nahm, M.H., Guo, Y., Russell, M.W., and Briles, D.E. (1997). Intranasal immunization of mice with PspA (pneumococcal surface protein A) can prevent intranasal carriage, pulmonary infection, and sepsis with *Streptococcus pneumoniae*. J. Infect. Dis. *175*, 839–846.

Wu, H.Y., and Russell, M.W. (1998). Induction of mucosal and systemic immune responses by intranasal immunization using recombinant cholera toxin B subunit as an adjuvant. Vaccine *16*, 286–292.

Wu, Q., Martin, R.J., Rino, J.G., Breed, R., Torres, R.M., and Chu, H.W. (2007). IL-23-dependent IL-17 production is essential in neutrophil recruitment and activity in mouse lung defense against respiratory Mycoplasma pneumoniae infection. Microbes Infect. *9*, 78–86.

Xiang, R., Lode, H.N., Chao, T.H., Ruehlmann, J.M., Dolman, C.S., Rodriguez, F., Whitton, J.L., Overwijk, W.W., Restifo, N.P., and Reisfeld, R.A. (2000). An autologous oral DNA vaccine protects against murine melanoma. Proc. Natl. Acad. Sci. U.S.A. *97*, 5492–5497.

Xin, K.Q., Hoshino, Y., Toda, Y., Igimi, S., Kojima, Y., Jounai, N., Ohba, K., Kushiro, A., Kiwaki, M., Hamajima, K., et al. (2003). Immunogenicity and protective efficacy of orally administered recombinant Lactococcus lactis expressing surface-bound HIV Env. Blood *102*, 223–228.

Xu, W., He, B., Chiu, A., Chadburn, A., Shan, M., Buldys, M., Ding, A., Knowles, D.M., Santini, P.A., and Cerutti, A. (2007). Epithelial cells trigger frontline immunoglobulin class switching through a pathway regulated by the inhibitor SLPI. Nat. Immunol. *8*, 294–303.

Xu-Amano, J., Jackson, R.J., Fujihashi, K., Kiyono, H., Staats, H.F., and McGhee, J.R. (1994). Helper Th1 and Th2 cell responses following mucosal or systemic immunization with cholera toxin. Vaccine *12*, 903–911.

Yamamoto, M., Kiyono, H., Kweon, M.N., Yamamoto, S., Fujihashi, K., Kurazono, H., Imaoka, K., Bluethmann, H., Takahashi, I., Takeda, Y., et al. (2000). Enterotoxin adjuvants have direct effects on T cells and antigen-presenting cells that result in either interleukin-4-dependent or -independent immune responses. J. Infect. Dis. *182*, 180–190.

Yang, D., Chertov, O., Bykovskaia, S.N., Chen, Q., Buffo, M.J., Shogan, J., Anderson, M., Schroder, J.M., Wang, J.M., Howard, O.M., et al. (1999). Beta-defensins: linking innate and adaptive immunity through dendritic and T cell CCR6. Science *286*, 525–528.

Yang, Q.B., Martin, M., Michalek, S.M., and Katz, J. (2002). Mechanisms of monophosphoryl lipid A augmentation of host responses to recombinant HagB from *Porphyromonas gingivalis*. Infect. Immun. *70*, 3557–3565.

Yoshino, N., Lu, F.X., Fujihashi, K., Hagiwara, Y., Kataoka, K., Lu, D., Hirst, L., Honda, M., van Ginkel, F.W., Takeda, Y., et al. (2004). A novel adjuvant for mucosal immunity to HIV-1 gp120 in nonhuman primates. J. Immunol. *173*, 6850–6857.

Zasloff, M. (2002). Antimicrobial peptides of multicellular organisms. Nature *415*, 389–395.

Zelante, T., De Luca, A., Bonifazi, P., Montagnoli, C., Bozza, S., Moretti, S., Belladonna, M.L., Vacca, C., Conte, C., Mosci, P., et al. (2007). IL-23 and the Th17 pathway promote inflammation and impair antifungal immune resistance. Eur J. Immunol. *37*, 2695–2706.

Zhang, H., Fayad, R., Wang, X., Quinn, D., and Qiao, L. (2004). Human immunodeficiency virus type 1 gag-specific mucosal immunity after oral immunization with papillomavirus pseudoviruses encoding gag. J. Virol. *78*, 10249–10257.

Zhao, X., Deak, E., Soderberg, K., Linehan, M., Spezzano, D., Zhu, J., Knipe, D.M., and Iwasaki, A. (2003a). Vaginal submucosal dendritic cells, but not Langerhans cells, induce protective Th1 responses to herpes simplex virus-2. J. Exp. Med. *197*, 153–162.

Zhao, X., Sato, A., Dela Cruz, C.S., Linehan, M., Luegering, A., Kucharzik, T., Shirakawa, A.K., Marquez, G., Farber, J.M., Williams, I., et al. (2003b). CCL9 is secreted by the follicle-associated epithelium and recruits dome region Peyer's patch CD11b+ dendritic cells. J. Immunol. *171*, 2797–2803.

Zuercher, A.W., Jiang, H.Q., Thurnheer, M.C., Cuff, C.F., and Cebra, J.J. (2002). Distinct mechanisms for cross-protection of the upper versus lower respiratory tract through intestinal priming. J. Immunol. *169*, 3920–3925.

Intralymphatic Vaccination

Thomas M. Kündig, Pål Johansen and Gabriela Senti

8

Abstract

The immune response is initiated by dendritic cells (DCs) and other antigen-presenting cells. These cells are present in nearly all organs and tissues of the body, so that theoretically any organ or tissue could serve as a route for vaccine administration. The choice of route is therefore mainly based on practical aspects. Using conventional needle and syringe the subcutaneous or intramuscular route are standard. The dermis and especially the epidermis are technically more difficult to target, but are likely to gain more interest due to the recent development of micro-needle patches and needle free injection devices. Vaccine administration via mucosal surfaces such as nasal or oral vaccination represents another option for needle free vaccine administration.

While all the above-mentioned routes of administration have been proven to work and protect against childhood diseases, influenza and many other infectious agents, the discussion and comparison of these different routes usually focuses on patient convenience, reduction of pain and distress for children, cost, and on the possibility for mass vaccination. In this review, however, we would like to focus on how the route of administration can enhance the efficacy of vaccination.

Especially in therapeutic vaccination, i.e. in a smaller patient number that already suffers from a disease, vaccination efficiency rather than convenience is the main issue. This is particularly the case in therapeutic cancer vaccines and in allergen specific immunotherapy. Intralymphatic vaccination is a strategy to maximize immunogenicity and therefore vaccine efficacy. The main part of this review will discuss this long known vaccination route and its clinical applicability in therapeutic vaccination.

Introduction

The widespread introduction of vaccines approximately one century ago likely represents the most significant success in medicine, which together with hygiene and antibiotics have essentially alleviated the modern world from morbidity and mortality caused by infectious diseases. Despite this success, a medical need for improving vaccination remains. For mass vaccination campaigns it would be desirable to increase ease and speed of vaccination. For developing countries the main goal is reduction of cost and elimination of the cold chain. In children, reduction or absence of pain is an important aim. Such improvement of speed, convenience, and cost are the main issues when trying to improve vaccines which already work well, which are generally prophylactic vaccines against infectious agents. For another group of vaccines, e.g. for therapeutic vaccines to treat cancer or allergies, the main goal is to improve their efficacy. Most cancer vaccines so far produced disappointing clinical results (Rosenberg et al., 2004). Allergen specific immunotherapy, also called desensitization or allergy shots, has been proven to work, but only after dozens of vaccine doses and years of treatment (Bousquet et al., 1998). Hence, in both cancer and allergies, a marked improvement of vaccination efficacy is urgently needed. This review will focus on how the route of vaccine administration can enhance the immune response to the vaccine.

Subcutaneous and intramuscular vaccine administration

When administering vaccines using conventional needles and syringes, subcutaneous and intramuscular injections are the most practical routes. With a conventional needle and syringe, the human epidermis with a thickness of merely of 50–200 μm is difficult to target, and also the epidermal compartment cannot be injected with the volumes that are deliverable by conventional syringes. Even the dermis with a thickness of a few millimetres is technically not easy to inject and already the administration of smallest volumes such as 50 or 100 μl produces pain and a burning sensation. The next reasonably injectable anatomical layers therefore are the subcutaneous fat or the underlying muscles. Although over the last 100 years most inhabitants of this planet, including farm animals and pets has likely received several subcutaneous or intramuscular vaccinations, probably amounting to tens of billions vaccine administrations, it is surprising that only a small number of studies have compared these two routes of vaccination. The immunogenicity of the same vaccine administered either subcutaneously or intramuscularly has been compared for a multivalent pneumococcal vaccine (Cook *et al.*, 2007), influenza vaccines (Cook *et al.*, 2006; Delafuente *et al.*, 1998), a polyvalent meningococcal vaccine (Ruben *et al.*, 2001), a two component diphtheria tetanus vaccine (Rothstein *et al.*, 1996), booster diphtheria toxin vaccination (Mark *et al.*, 1999), varicella vaccination (Dennehy *et al.*, 1991), and hepatitis B vaccination (de Lalla *et al.*, 1988). Such comparative trials found that the intramuscular route appears to induce fewer local reactions (Cook *et al.*, 2006, 2007; Delafuente *et al.*, 1998; Dennehy *et al.*, 1991; Mark *et al.*, 1999; Rothstein *et al.*, 1996; Ruben *et al.*, 2001; Scheifele *et al.*, 1994), and these reactions were found to be even further reduced when longer needles were used (Diggle *et al.*, 2006). Likely the deeper location of the reaction makes the latter less visible

clinically. Also, injection of the vaccine into more vascularized muscle tissue may rapidly reduce local vaccine concentrations and therefore reduce the local reaction (Ajana *et al.*, 2008). In terms of efficacy intramuscular vaccine administration appears to be at least as good as the subcutaneous route (Ajana *et al.*, 2008).

Mucosal vaccination

Mucosal vaccination is considered the most effective route to produce an immune response protective against infections that are either limited to the mucosal surfaces or against pathogens that invade the host via the mucosa (Dietrich *et al.*, 2003). Mucosal vaccines are transported through microfold M cells into the mucosa-associated lymphoid tissue (MALT) where dendritic cells present them to T cells, which then induce B cells to produce IgA. As antibody forming cells preferentially return to their site of induction, in this case to the MALT, mucosal vaccination is an effective means to produce mucosal IgA mediated immunity. Aggregates of MALT are present in the mucosa of the respiratory, conjunctival, gastrointestinal, and genitourinary tract, and all these routes are theoretically possible for vaccine application. The attenuated Sabin oral poliovirus vaccine was introduced in 1961 and turned into a global success, followed by oral typhoid (Levine *et al.*, 1990) and cholera vaccines (Thompson, 2002). So far oral vaccines are live vaccines and therefore a safety concern remains. For the oral polio vaccine, for example, reversion to virulence has caused outbreaks of paralytic polio (Fine *et al.*, 2004; Kew *et al.*, 2004). A promising development are therefore live attenuated bacteria expressing foreign antigens (DiGiandomenico *et al.*, 2004; Garmory *et al.*, 2003; Kang *et al.*, 2002) and 'edible vaccines' from transgenic plants (Rigano *et al.*, 2003).

While an important problem with oral vaccination is the degradation and enzymatic digestion of protein antigens in the stomach and intestine, vaccine delivery via respiratory surfaces does not face this issue. The efficacy of influenza vaccination via nasal spray is well established (Belshe *et al.*, 1998; King *et al.*, 1998). Also, the efficacy of aerosol vaccination using nebulizers has been established (reviewed in Katz *et al.*, 2003), and nicely demonstrated in measles (Dilraj *et al.*, 2000; Sepulveda-Amor *et al.*, 2002).

Needle-free vaccination

The classical application of vaccines via needle and syringe poses not only the risk of needle stick injuries and transmission of infections from patient to health care providers (Dicko *et al.*, 2000; Miller and Pisani, 1999; Simonsen *et al.*, 1999), but also reduces distress or fear from the pain of injection especially in children (Jacobson *et al.*, 2001), but also in young adults (Nir *et al.*, 2003). This creates a strong interest in replacing injections with less painful and needle-free applications. Another goal of needle-free vaccination is to increase the speed of vaccination. Multi-use jet injectors have been used for mass vaccination campaigns in the developing world allowing vaccinating up to 1000 subjects per hour. However, reports of transmission of hepatitis B by multi-use jet injectors due to blood contamination of the nozzle (Canter *et al.*, 1990; Hoffman *et al.*, 2001) lead to the realization that also needle-free injection devices must be made disposable (Giudice and Campbell, 2006). Jet injections have been shown to induce at least comparable or even enhanced immune responses when compared to vaccine delivery via conventional needle and syringe (Jackson *et al.*, 2001; Parent du Chatelet *et al.*, 1997). One especially promising development is that of dry powder injection devices (Chen *et al.*, 2000, 2001), which also eliminate the need to maintain a product cold chain.

Transcutaneous immunization is another interesting alternative to classical needle-based vaccine applications (Frech *et al.*, 2008; Glenn *et al.*, 2000a; Bins *et al.*, 2005; Kenney *et al.*, 2004; Klimuk *et al.*, 2004; Partidos *et al.*, 2001; Seo *et al.*, 2000; Takigawa *et al.*, 2001; Zhao *et al.*, 2006). Mostly because the epidermis contains a high density (Bergstresser *et al.*, 1980) of potent antigen-presenting cells in the epidermis, i.e. Langerhans cells (Celluzzi and Falo, 1997; Romani *et al.*, 1989). The epidermis is of particularly interest in allergen immunotherapy, were subcutaneous allergen injections are associated with a high risk of systemic allergic adverse events, ranging from skin rash to asthma, but also to allergic shock or even death (Lockey *et al.*, 1987; Lockey *et al.*, 1990; Stewart and Lockey, 1992). These systemic adverse reactions are caused by allergen leaking into the circulation. Hence, the epidermis, which is not vascularized, is ideally suited to pulse antigen-presenting cells with allergens. In fact, the French allergologist Blamoutier in 1956 performed allergen-specific immunotherapy by needle scarification of the volar forearm in an area of 4×4 cm (Blamoutier *et al.*, 1959). These early results have recently been successfully reproduced in a double blind controlled clinical trial setting (Senti *et al.*, 2009).

One remaining problem in transcutaneous immunization is that the intact stratum corneum is practically impermeable for proteins. Vaccine delivery therefore requires the disruption of this outer skin layer and can be achieved chemically, mechanically or by electroporation. Hydration by occlusion of the skin changes the composition of the intercellular fluid such that antigen penetration is enhanced (Glenn *et al.*, 2000b). The outer corneal layers can also be removed by repeated stripping with an adhesive tape (Breternitz *et al.*, 2007) or by abrasive devices (Frerichs *et al.*, 2008).

However, stratum corneum disruption is not only required to enable vaccine penetration. Also, it appears to be important that such disruption induces a bystander effect mediated by keratinocyte derived proinflammatory cytokines. Antigen presentation and T cell stimulation depend on these bystander reactions, inducing secretion of interleukin (IL)-1, IL-6, IL-8 and TNF-α (Camp *et al.*, 1990; Corsini and Galli, 2000; Hauser *et al.*, 1986; Wang *et al.*, 2002). It is also reported that tape stripping induced production of IL-12 and interferon IFN-γ in the skin (Inoue *et al.*, 2005), and increased expression of MHC class II, CD86, CD40, CD54 and CD11c on Langerhans cells (Strid *et al.*, 2004). Tape stripping also stimulated expression of Toll-like receptor 9 (TLR9) in keratinocytes (Inoue and Aramaki, 2007). Trancutaneous immunization can further also be enhanced by adjuvants and carrier proteins. In the latter respect, cholera toxin, heat-labile enterotoxin from *E. coli*, and their mutants have been used (Frech *et al.*, 2008; Frerichs *et al.*, 2008; Guerena-Burgueno *et al.*, 2002).

Intralympatic vaccination

Function of lymph nodes

In a series of elegant skin flap experiments Frey and Wenk proved in 1957 (Frey and Wenk, 1957) that antigen, in order to induce a T-cell response, needs to reach lymph nodes via afferent lymph vessels. More recent experiments in alymphoplastic (aly/aly) and spleenless (Hox11–/–) mutant mice confirm the importance of secondary lymphoid organs, or neo-lymphoid aggregates (Greter *et al.*, 2009) in generating immune responses (Karrer *et al.*, 1997).

T- and B-cell receptors are randomly rearranged early in lymphocyte development generating T and B cells carrying a diverse repertoire of receptors. While this ensures that all possible antigens can be specifically recognized, it also requires that antigens must be presented to approximately 10^7 T and B cells in order to elicit an immune response. Therefore, only antigens that are drained into secondary lymphatic organs, where they can be presented to high numbers of T and B cells will induce an immune response, whereas antigens staying outside of secondary lymphoid organs have little chance to meet with specific T or B cells and are largely ignored, a phenomenon termed the 'geographic concept of immunogenicity' (Kundig *et al.*, 1995; Zinkernagel, 2000; Zinkernagel *et al.*, 1997). In light of the current understanding of immune regulation by dendritic cells and T cells, this concept may sound rather simplistic, but it remains valid nevertheless. Today's complexity of immune regulation should not make us forget that the key trigger and regulator of the immune response is the antigen.

Lymphatic drainage

Lymph vessels have evolved to drain pathogens into lymph nodes so that the immune system can generate an immune response as early as possible. Therefore, small particles sized 20–200 nm, i.e. the size of viruses, drain from peripheral injection sites into lymph nodes quite efficiently and in a free from, but even in this case usually only a few percent of the injected particles drain into the lymph nodes (Manolova *et al.*, 2008). Larger particles sized 500–2000 nm are mostly transported into lymph nodes by DCs (Manolova *et al.*, 2008).

Drainage from periphery to lymph nodes of non-particulate antigens, however, can be much less efficient and only very small fractions – between 10^{-3} and 10^{-6} – of the injected doses reach lymph nodes, as summarized in Table 8.1. As many of today's vaccines and immunotherapeutic agents are non-particulate, their direct administration into a lymph node may therefore enhance antigen presentation in that lymph node and therefore the immune response.

Intralymphatic vaccinations in research and clinics

The first review on intralymphatic vaccination was already written in 1977 (Juillard and Boyer, 1977). In the early 1970s Juillard and others aimed at enhancing tumour cell based cancer vaccines in dogs using this method. Ten years later, when researchers wanted to produce antibodies against proteins of which they had been able to purify only very small amounts, they were looking for the most efficient route of immunization. It was already reported in the 1980s that nanogram quantities of protein can elicit immune responses when injected into lymph nodes (Sigel *et al.*, 1983; Nilsson *et al.*, 1987). Thereafter intralymphatic vaccination was performed in various fields where conventional routes of administration produced insufficient results, or where the goal was to maximize the immune response, such as in cancer vaccines. In the following we will list the types of vaccines that are reported to have been used via the intralymphatic route.

Intralympatic vaccination with DCs

After subcutaneous or intradermal injection, the main fraction of DCs remains at the injection site (Barratt-Boyes *et al.*, 1997, 2000; De Vries *et al.*, 2003; Lesimple *et al.*, 2003; Morse *et al.*, 1999; Quillien *et al.*, 2005; Thomas *et al.*, 1999). Therefore, intralymphatic and intranodal delivery of DC-based vaccines have been attempted (Grover *et al.*, 2006; Lesimple *et al.*, 2006; Mackensen *et al.*, 1999) in order to enhance the immune response. Antigen-pulsed DCs injected into the lymph node localize to the paracortex (Brown *et al.*, 2003; De Vries *et al.*, 2003, 2005), and clinical trials suggested that intralymphatic administration of DC vaccines enhanced the immune

Table 8.1 Intralymphatic vaccination strongly enhances unstable vaccines

Type of vaccine	Dose required s.c./intralymphatic	Stability *in vivo* and drainage into lymph node
Naked RNA*	>10^6	–
Oligopeptide (Johansen *et al.*, 2005)	10^6	+
Naked DNA (Maloy *et al.*, 2001)	10^4	++
Protein (Martinez-Gomez *et al.*, 2009a)	10^3	+++
Live virus**	1	++++

responses(Lesimple *et al.*, 2003). However, in other clinical trials, no advantage of intranodal delivery over intradermal delivery of DCs was found (Brown *et al.*, 2003; Fong *et al.*, 2001).

Intralymphatic vaccination with tumour cells
Non-professional APC, such as a fibrosarcoma cell line efficiently induced antigen-specific CD8+ T-cell responses when injected directly into lymph nodes, but not if injected subcutaneously (Kundig *et al.*, 1995) (Ochsenbein *et al.*, 2001). The CD8+ T-cell response was found to be induced via direct antigen presentation on the MHC class I molecule of the fibrosarcoma (Kundig *et al.*, 1995) (Ochsenbein *et al.*, 2001). As DCs and T cells are present at very high densities in lymph nodes, costimulatory signals for T and B cell induction may be provided as a bystander effect. Intralymphatic immunization using tumour cells has been tried in both human cancer patients and dogs with indication of success (Juillard and Boyer, 1977; Juillard *et al.*, 1976, 1977, 1978, 1979).

Intralymphatic vaccination with MHC class I binding peptides
Aiming at enhancing CD8+ T-cell mediated cancer immunotherapy direct administration of MHC class I-binding peptide vaccines into lymph nodes or spleen was tried and found to dramatically enhance CD8+ T-cell responses in mice (Johansen *et al.*, 2005). Intralymphatic priming with a naked DNA vaccine followed by a peptide boost produced the strongest CD8+ T-cell responses that can be observed in mice, the frequencies of peptide specific CD8+ T cells being as high as 80% (Smith *et al.*, 2009).

Intralymphatic vaccination with naked DNA vaccines
Intralymphatic injection is shown to also markedly enhance naked DNA vaccines in mice (Heinzerling *et al.*, 2006; Maloy *et al.*, 2001; Smith *et al.*, 2009). These observations were confirmed in clinical trials on intralymphatic immunotherapy using a Melan-A/MART-1 DNA plasmid vaccine in stage IV melanoma patients, where immune responses that correlated with the tumour responses were generated (Weber *et al.*, 2008; Tagawa *et al.*, 2003). Similar clinical trials are now extended to include various forms of cancer (www.clinicaltrials.gov).

Intralymphatic vaccination with immunostimulating complexes (ISCOMS)
The intranasal route was compared to targeted lymph node immunization using HIV formulated in PR8-Flu ISCOM adjuvant in rhesus macaques. Targeting the vaccine to lymph nodes generated significantly stronger T- and B-cell responses (Koopman *et al.*, 2007).

Targeted lymph node immunization
This form of intralymphatic vaccination is extensively documented to be the most efficient means to immunized macaques against SIV. Vaccination using envelope gp120 and core p27 by targeted lymph node immunization was compared to the intradermal, nasal, or intramuscular route. Only targeted lymph node immunization induced protection against the SIV challenge, whereas none of the other routes induced a protective immune response (Lehner *et al.*, 1996). Similar results were obtained in macaques vaccinated with p27 alone (Kawabata *et al.*, 1998), with HSP70 linked

either to SIVgp120 or p27 (Lehner *et al.*, 2000), with HSP70 conjugated to HIV gp120, SIV p27 and CCR5 peptides (Bogers *et al.*, 2004a; Bogers *et al.*, 2004b), a particulate SIVp27 protein vaccine(Klavinskis *et al.*, 1996), or SIVp27 virus like particles (Lehner *et al.*, 1994).

Targeted lymph node immunization also proved to be the most efficient route to immunize cats against feline immunodeficiency virus using a protein based vaccine (Finerty *et al.*, 2001). The superior efficacy of lymph node targeting was also confirmed for protein vaccines in cows (Guidry *et al.*, 1994).

Intralymphatic vaccination with BCG
Julliard *et al.* report a comparison of intralymphatic versus intradermal vaccination with BCG. It was found that the tuberculin test became positive at an earlier time point (11–23 days) after intralymphatic than after intradermal vaccination (45 days) (Juillard and Boyer, 1977).

Intralymphatic administration of adjuvants
Lymph node targeting can also be used to enhance the efficacy of adjuvants. Intralymphatic administration of CpG required 100 times lower doses than subcutaneous administration, thus avoiding unwanted systemic adverse effects of the adjuvant (von Beust *et al.*, 2005). This is in line with reports of better safety profiles and enhanced efficacy of CpG when targeted to lymph nodes using particles (Storni *et al.*, 2004; Bourquin *et al.*, 2008).

Intralymphatic immunotherapy with allergens
IgE-mediated allergies, such as allergic rhino-conjunctivitis and asthma today affect up to 35% of the population in westernized countries (Beasley *et al.*, 1998; Arbes *et al.*, 2005; Verlato *et al.*, 2003; Wuthrich *et al.*, 1996). The gold standard treatment is subcutaneous allergen-specific immunotherapy (SIT), the administration of gradually increasing quantities of an allergen (Bousquet *et al.*, 1998; Lockey, 2001; Varney *et al.*, 1991). Immunotherapy confers long-term benefit (Durham *et al.*, 1999; Golden *et al.*, 1996; Pajno *et al.*, 2001; Moller *et al.*, 2002), but the assumed required doctor visits over 3–5 years are compromising patient compliance. Immunotherapy is

also associated with frequent allergic side-effects including a risk of anaphylaxis and even death (Lockey *et al.*, 1987; Lockey *et al.*, 1990; Stewart and Lockey, 1992).

Allergen immunotherapy shifts the T-cell response from a Th2 towards a Th1 phenotype (Norman, 2004; Till *et al.*, 2004), and it also stimulates the production of allergen-specific T-regulatory cells (Norman, 2004; Till *et al.*, 2004; Vissers *et al.*, 2004). In the serum, titres of allergen-specific IgG antibodies, especially IgG4, increase (Pierson-Mullany *et al.*, 2000). Which of these immunological mediators is ultimately responsible for ameliorating the allergy symptoms is a matter of debate.

In mice, intralymphatic administration of allergens was shown to enhance significantly the efficiency of immunization, inducing allergen-specific IgG2a antibody responses 10–20 times higher with only 0.1% of the allergen dose (Martinez-Gomez *et al.*, 2009a). Also, intralymphatic injection of allergens enhanced IL-2, IFN-γ, IL-4 and IL-10 secretion when compared to subcutaneous injection, suggesting that intralymphatic vaccination does not polarize the response to the allergen, but generated overall stronger Th1, Th2, and T-regulatory responses (Martinez-Gomez *et al.*, 2009a).

Meanwhile, the feasibility, safety and efficacy of intralymphatic allergen immunotherapy have been demonstrated in four separate clinical trials. In a first trial, eight bee-venom allergic patients, who would normally receive 70 subcutaneous injections of bee venom, were given only three low-dose injections of bee venom directly into their inguinal lymph nodes. In this proof of concept trial seven out of the eight treated patients were protected against a subsequent bee sting challenge (Senti *et al.*, manuscript in preparation). Similar results were obtained in a larger multicentre clinical trial with 66 bee venom-allergic patients (Senti *et al.*, manuscript in preparation). In a randomized controlled trial, 165 patients with grass pollen-induced hay fever received either 54 high-dose subcutaneous injections with pollen extract over 3 years or three low-dose intralymphatic injections over 2 months. The three low-dose intralymphatic allergen administrations enhanced safety and efficacy of immunotherapy and reduced treatment time from 3 years to 8 weeks (Senti *et al.*, 2008). These data have meanwhile been confirmed in a similar double-blind placebo-controlled trial using intralymphatic vaccination with a modified recombinant cat hair allergen in allergic patients (www.imvision-therapeutics.com).

Targeting intralymphatic vaccines to the MHC class II pathway

As intralymphatic vaccination brings the antigen directly to the DCs of the lymph node, intracellular translocation sequences and sequences further targeting the antigen to the MHC class II pathway may enhance the CD4+ T cell response. By fusing allergens to a tat-translocation peptide derived from HIV and to a part of the invariant chain, such allergy vaccines can be targeted to MHC class II molecules located in the endoplasmic reticulum. In several experimental studies, it has been shown that such targeting circumvents the inefficient pinocytosis process as well as enzymatic degradation in phagolysosomes, and that by such means the immunogenicity could be significantly enhanced (Martinez-Gomez *et al.*, 2009a,b; Rhyner *et al.*, 2007; Crameri *et al.*,

2007). This concept has meanwhile also been proven in a first clinical trial (www.imvision-therapeutics.com).

Biodistribution after intralymphatic delivery

Biodistribution studies in mice revealed that 100-fold higher antigen doses reached the lymph nodes after direct lymph-node injection than after subcutaneous injection in the proximity of a draining lymph node (Martinez-Gomez *et al.*, 2009a). Similar results were obtained in humans when proteins were radio-traced after intralymphatic and subcutaneous injection. On the right abdominal side, a ^{99m}Tc-labelled protein was injected directly into a superficial inguinal lymph node. On the left side, the same dose was injected subcutaneously, but 10 cm above the inguinal lymph nodes. As shown in Fig. 8.1, only a small fraction of the subcutaneously administered protein reached the lymph nodes after 4 hours and this fraction did not increase after 25 hours. In contrast, intralymphatic injection caused the protein to drain into the deep subcutaneous lymph nodes and further into one pelvic lymph node already within 20 min. Thus, intralymphatic

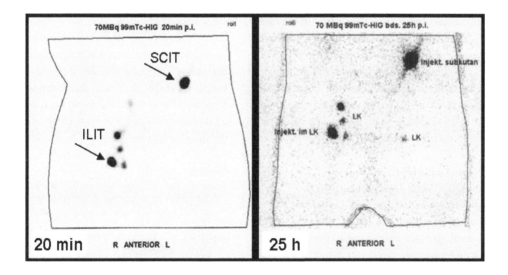

Figure 8.1 Biodistribution after intralymphatic administration: Biodistribution of 99mTc-labelled human IgG after intralymphatic (left abdominal side) and subcutaneous (right abdominal side) injections. Radio tracing was made by gamma-imaging 20 min (left panel) and 25 hours (right panel) after injection. Arrows indicate the site of injection (SCIT, subcutaneous, ILIT, intralymphatic).

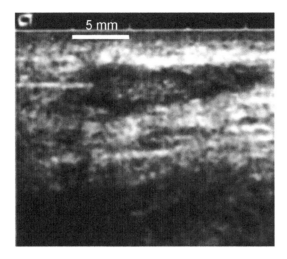

Figure 8.2 Intralymphatic injection. A sand blasted needle, being inserted into the lymph node from the left was used for better reflection and therefore visibility in the ultrasound. The dark, hypoechoic area represents the paracortex of the lymph node, which is approx. 15 mm long and 5 mm under the skin surface.

injection efficiently pulsed five lymph nodes with the full amount of the protein.

Is intralymphatic vaccination difficult or painful?

Subcutaneous lymph nodes are readily visible by ultrasound, as their paracortical area is hypoechoic (Fig. 8.2). Injection of a superficial lymph node in the groin area can be performed within minutes, even by doctors that have little experience in ultrasound. The pain of intralymphatic injection arises solely from penetrating the skin, whereas lymph nodes are poorly innervated. The pain of an intralymphatic injection is comparable to the subcutaneous injections. In fact, patients rated intralymphatic injection less painful than venous puncture (Senti *et al.*, 2008).

Which vaccines profit the most from intralymphatic administration?

Those vaccines which are the least stable and do not efficiently drain into lymph nodes after subcutaneous or intramuscular injection, appear to profit the most from direct intralymphatic administration. Using intralymphatic administration, we could induce specific CD8+ T cell responses even by injection of naked mRNA encoding for a viral glycoprotein, which we could not achieve even with the highest doses of mRNA by subcutaneous

vaccination (unpublished data), obviously because RNA in tissues is rapidly digested by RNase. On the other end of the spectrum, injection of a lymphocytic choriomeningitis virus (LCMV) into a subcutaneous lymph node of a mouse was found not to further enhance the T cell response over intravenous infection (TMK unpublished). Obviously, the virus particles found their way into secondary lymphatic organs efficiently.

Overall, vaccines that work only poorly by subcutaneous or intramuscular administration profit the most from intralymphatic delivery, while vaccines which already induce strong immune responses after subcutaneous of intramuscular administration may not be further enhanced by the intralymphatic route.

References

Ajana, F., Sana, C., and Caulin, E. (2008). [Are there differences in immunogenicity and safety of vaccines according to the injection method?]. Med. Mal. Infect. *38*, 648–657.

Arbes, S.J., Jr., Gergen, P.J., Elliott, L., and Zeldin, D.C. (2005). Prevalences of positive skin test responses to 10 common allergens in the US population: results from the third National Health and Nutrition Examination Survey. J. Allergy Clin. Immunol. *116*, 377–383.

Barratt-Boyes, S.M., Watkins, S.C., and Finn, O.J. (1997). Migration of cultured chimpanzee dendritic cells following intravenous and subcutaneous injection. Adv. Exp. Med. Biol. *417*, 71–75.

Barratt-Boyes, S.M., Zimmer, M.I., Harshyne, L.A., Meyer, E.M., Watkins, S.C., Capuano, S., 3rd, Murphey-Corb, M., Falo, L.D., Jr., and Donnenberg, A.D. (2000). Maturation and trafficking of monocyte-derived dendritic cells in monkeys: implications for dendritic cell-based vaccines. J. Immunol. *164*, 2487–2495.

Beasley, R., and The International Study of Asthma and Allergies in Childhood (I.S.A.A.C.) Steering Committee. (1998). Worldwide variation in prevalence of symptoms of asthma, allergic rhinoconjunctivitis, and atopic eczema: ISAAC. Lancet *351*, 1225–1232.

Belshe, R.B., Mendelman, P.M., Treanor, J., King, J., Gruber, W.C., Piedra, P., Bernstein, D.I., Hayden, F.G., Kotloff, K., Zangwill, K., *et al.* (1998). The efficacy of live attenuated, cold-adapted, trivalent, intranasal influenzavirus vaccine in children. N. Engl. J. Med. *338*, 1405–1412.

Bergstresser, P.R., Fletcher, C.R., and Streilein, J.W. (1980). Surface densities of Langerhans cells in relation to rodent epidermal sites with special immunologic properties. J. Invest. Dermatol. *74*, 77–80.

Bins, A.D., Jorritsma, A., Wolkers, M.C., Hung, C.F., Wu, T.C., Schumacher, T.N., and Haanen, J.B. (2005). A rapid and potent DNA vaccination strategy defined by *in vivo* monitoring of antigen expression. Nat. Med. *11*, 899–904.

Blamoutier, P., Blamoutier, J., and Guibert, L. (1959). [Treatment of pollinosis with pollen extracts by the method of cutaneous quadrille ruling.]. Presse Med. *67*, 2299–2301.

Bogers, W.M., Bergmeier, L.A., Ma, J., Oostermeijer, H., Wang, Y., Kelly, C.G., Ten Haaft, P., Singh, M., Heeney, J.L., and Lehner, T. (2004a). A novel HIV-CCR5 receptor vaccine strategy in the control of mucosal SIV/HIV infection. Aids *18*, 25–36.

Bogers, W.M., Bergmeier, L.A., Oostermeijer, H., ten Haaft, P., Wang, Y., Kelly, C.G., Singh, M., Heeney, J.L., and Lehner, T. (2004b). CCR5 targeted SIV vaccination strategy preventing or inhibiting SIV infection. Vaccine *22*, 2974–2984.

Bourquin, C., Anz, D., Zwiorek, K., Lanz, A.L., Fuchs, S., Weigel, S., Wurzenberger, C., von der Borch, P., Golic, M., Moder, S., *et al.* (2008). Targeting CpG oligonucleotides to the lymph node by nanoparticles elicits efficient antitumoral immunity. J. Immunol. *181*, 2990–2998.

Bousquet, J., Lockey, R., and Malling, H.J. (1998). Allergen immunotherapy: therapeutic vaccines for allergic diseases. A WHO position paper. J. Allergy Clin. Immunol. *102*, 558–562.

Breternitz, M., Flach, M., Prassler, J., Elsner, P., and Fluhr, J.W. (2007). Acute barrier disruption by adhesive tapes is influenced by pressure, time and anatomical location: integrity and cohesion assessed by sequential tape stripping. A randomized, controlled study. Br. J. Dermatol. *156*, 231–240.

Brown, K., Gao, W., Alber, S., Trichel, A., Murphey-Corb, M., Watkins, S.C., Gambotto, A., and Barratt-Boyes, S.M. (2003). Adenovirus-transduced dendritic cells

injected into skin or lymph node prime potent simian immunodeficiency virus-specific T cell immunity in monkeys. J. Immunol. *171*, 6875–6882.

Camp, R., Fincham, N., Ross, J., Bird, C., and Gearing, A. (1990). Potent inflammatory properties in human skin of interleukin-1 alpha-like material isolated from normal skin. J. Invest. Dermatol. *94*, 735–741.

Canter, J., Mackey, K., Good, L.S., Roberto, R.R., Chin, J., Bond, W.W., Alter, M.J., and Horan, J.M. (1990). An outbreak of hepatitis B associated with jet injections in a weight reduction clinic. Arch. Intern. Med. *150*, 1923–1927.

Celluzzi, C.M., and Falo, L.D., Jr. (1997). Epidermal dendritic cells induce potent antigen-specific CTL-mediated immunity. J. Invest. Dermatol. *108*, 716–720.

Chen, D., Endres, R.L., Erickson, C.A., Weis, K.F., McGregor, M.W., Kawaoka, Y., and Payne, L.G. (2000). Epidermal immunization by a needle-free powder delivery technology: immunogenicity of influenza vaccine and protection in mice. Nat. Med. *6*, 1187–1190.

Chen, D., Weis, K.F., Chu, Q., Erickson, C., Endres, R., Lively, C.R., Osorio, J., and Payne, L.G. (2001). Epidermal powder immunization induces both cytotoxic T-lymphocyte and antibody responses to protein antigens of influenza and hepatitis B viruses. J. Virol. *75*, 11630–11640.

Cook, I.F., Barr, I., Hartel, G., Pond, D., and Hampson, A.W. (2006). Reactogenicity and immunogenicity of an inactivated influenza vaccine administered by intramuscular or subcutaneous injection in elderly adults. Vaccine *24*, 2395–2402.

Cook, I.F., Pond, D., and Hartel, G. (2007). Comparative reactogenicity and immunogenicity of 23 valent pneumococcal vaccine administered by intramuscular or subcutaneous injection in elderly adults. Vaccine *25*, 4767–4774.

Corsini, E., and Galli, C.L. (2000). Epidermal cytokines in experimental contact dermatitis. Toxicology *142*, 203–211.

Crameri, R., Fluckiger, S., Daigle, I., Kundig, T., and Rhyner, C. (2007). Design, engineering and *in vitro* evaluation of MHC class-II targeting allergy vaccines. Allergy *62*, 197–206.

de Lalla, F., Rinaldi, E., Santoro, D., and Pravettoni, G. (1988). Immune response to hepatitis B vaccine given at different injection sites and by different routes: a controlled randomized study. Eur. J. Epidemiol. *4*, 256–258.

de Vries, I.J., Krooshoop, D.J., Scharenborg, N.M., Lesterhuis, W.J., Diepstra, J.H., Van Muijen, G.N., Strijk, S.P., Ruers, T.J., Boerman, O.C., Oyen, W.J., *et al.* (2003). Effective migration of antigen-pulsed dendritic cells to lymph nodes in melanoma patients is determined by their maturation state. Cancer Res. *63*, 12–17.

de Vries, I.J., Lesterhuis, W.J., Barentsz, J.O., Verdijk, P., van Krieken, J.H., Boerman, O.C., Oyen, W.J., Bonenkamp, J.J., Boezeman, J.B., Adema, G.J., *et al.* (2005). Magnetic resonance tracking of dendritic cells in melanoma patients for monitoring of cellular therapy. Nat. Biotechnol. *23*, 1407–1413.

Delafuente, J.C., Davis, J.A., Meuleman, J.R., and Jones, R.A. (1998). Influenza vaccination and warfarin anticoagulation: a comparison of subcutaneous and intramuscular routes of administration in elderly men. Pharmacotherapy *18*, 631–636.

Dennehy, P.H., Reisinger, K.S., Blatter, M.M., and Veloudis, B.A. (1991). Immunogenicity of subcutaneous versus intramuscular Oka/Merck varicella vaccination in healthy children. Pediatrics *88*, 604–607.

Dicko, M., Oni, A.Q., Ganivet, S., Kone, S., Pierre, L., and Jacquet, B. (2000). Safety of immunization injections in Africa: not simply a problem of logistics. Bull. World Health Organ. *78*, 163–169.

Dietrich, G., Griot-Wenk, M., Metcalfe, I.C., Lang, A.B., and Viret, J.F. (2003). Experience with registered mucosal vaccines. Vaccine *21*, 678–683.

Diggle, L., Deeks, J.J., and Pollard, A.J. (2006). Effect of needle size on immunogenicity and reactogenicity of vaccines in infants: randomised controlled trial. BMJ *333*, 571.

DiGiandomenico, A., Rao, J., and Goldberg, J.B. (2004). Oral vaccination of BALB/c mice with *Salmonella enterica* serovar Typhimurium expressing *Pseudomonas aeruginosa* O antigen promotes increased survival in an acute fatal pneumonia model. Infect. Immun. *72*, 7012–7021.

Dilraj, A., Cutts, F.T., de Castro, J.F., Wheeler, J.G., Brown, D., Roth, C., Coovadia, H.M., and Bennett, J.V. (2000). Response to different measles vaccine strains given by aerosol and subcutaneous routes to schoolchildren: a randomised trial. Lancet *355*, 798–803.

Durham, S.R., Walker, S.M., Varga, E.M., Jacobson, M.R., O'Brien, F., Noble, W., Till, S.J., Hamid, Q.A., and Nouri-Aria, K.T. (1999). Long-term clinical efficacy of grass-pollen immunotherapy. N. Engl. J. Med. *341*, 468–475.

Fine, P.E., Oblapenko, G., and Sutter, R.W. (2004). Polio control after certification: major issues outstanding. Bull. World Health Organ. *82*, 47–52.

Finerty, S., Stokes, C.R., Gruffydd-Jones, T.J., Hillman, T.J., Barr, F.J., and Harbour, D.A. (2001). Targeted lymph node immunization can protect cats from a mucosal challenge with feline immunodeficiency virus. Vaccine *20*, 49–58.

Fong, L., Brockstedt, D., Benike, C., Wu, L., and Engleman, E.G. (2001). Dendritic cells injected via different routes induce immunity in cancer patients. J. Immunol. *166*, 4254–4259.

Frech, S.A., Dupont, H.L., Bourgeois, A.L., McKenzie, R., Belkind-Gerson, J., Figueroa, J.F., Okhuysen, P.C., Guerrero, N.H., Martinez-Sandoval, F.G., Melendez-Romero, J.H., *et al.* (2008). Use of a patch containing heat-labile toxin from *Escherichia coli* against travellers' diarrhoea: a phase II, randomised, double-blind, placebo-controlled field trial. Lancet *371*, 2019–2025.

Frerichs, D.M., Ellingsworth, L.R., Frech, S.A., Flyer, D.C., Villar, C.P., Yu, J., and Glenn, G.M. (2008). Controlled, single-step, stratum corneum disruption as a pretreatment for immunization via a patch. Vaccine *26*, 2782–2787.

Frey, J.R., and Wenk, P. (1957). Experimental studies on the pathogenesis of contact eczema in the guinea-pig. Int. Arch. Allergy Appl. Immunol. *11*, 81–100.

Garmory, H.S., Leary, S.E., Griffin, K.F., Williamson, E.D., Brown, K.A., and Titball, R.W. (2003). The use of live attenuated bacteria as a delivery system for heterologous antigens. J. Drug Target *11*, 471–479.

Giudice, E.L., and Campbell, J.D. (2006). Needle-free vaccine delivery. Adv. Drug Deliv. Rev. *58*, 68–89.

Glenn, G.M., Taylor, D.N., Li, X., Frankel, S., Monemarano, A., and Alving, C.R. (2000a). Transcutaneous immunization: A human vaccine delivery strategy using a patch. Nat. Med. *6*, 1403–1406.

Glenn, G.M., Taylor, D.N., Li, X., Frankel, S., Montemarano, A., and Alving, C.R. (2000b). Transcutaneous immunization: a human vaccine delivery strategy using a patch. Nat. Med. *6*, 1403–1406.

Golden, D.B., Kwiterovich, K.A., Kagey-Sobotka, A., Valentine, M.D., and Lichtenstein, L.M. (1996). Discontinuing venom immunotherapy: outcome after five years. J. Allergy Clin. Immunol. *97*, 579–587.

Greter, M., Hofmann, J., and Becher, B. (2009). Neo-lymphoid aggregates in the adult liver can initiate potent cell-mediated immunity. PLoS Biol. *7*, e1000109.

Grover, A., Kim, G.J., Lizee, G., Tschoi, M., Wang, G., Wunderlich, J.R., Rosenberg, S.A., Hwang, S.T., and Hwu, P. (2006). Intralymphatic dendritic cell vaccination induces tumor antigen-specific, skin-homing T lymphocytes. Clin. Cancer Res. *12*, 5801–5808.

Guerena-Burgueno, F., Hall, E.R., Taylor, D.N., Cassels, F.J., Scott, D.A., Wolf, M.K., Roberts, Z.J., Nesterova, G.V., Alving, C.R., and Glenn, G.M. (2002). Safety and immunogenicity of a prototype enterotoxigenic *Escherichia coli* vaccine administered transcutaneously. Infect. Immun. *70*, 1874–1880.

Guidry, A.J., O'Brian, C.N., Oliver, S.P., Dowlen, H.H., and Douglass, L.W. (1994). Effect of whole *Staphylococcus aureus* and mode of immunization on bovine opsonizing antibodies to capsule. J. Dairy Sci. *77*, 2965–2974.

Hauser, C., Saurat, J.H., Schmitt, A., Jaunin, F., and Dayer, J.M. (1986). Interleukin 1 is present in normal human epidermis. J. Immunol. *136*, 3317–3323.

Heinzerling, L., Basch, V., Maloy, K., Johansen, P., Senti, G., Wuthrich, B., Storni, T., and Kundig, T.M. (2006). Critical role for DNA vaccination frequency in induction of antigen-specific cytotoxic responses. Vaccine *24*, 1389–1394.

Hoffman, P.N., Abuknesha, R.A., Andrews, N.J., Samuel, D., and Lloyd, J.S. (2001). A model to assess the infection potential of jet injectors used in mass immunisation. Vaccine *19*, 4020–4027.

Inoue, J., and Aramaki, Y. (2007). Toll-like receptor-9 expression induced by tape-stripping triggers on effective immune response with CpG-oligodeoxynucleotides. Vaccine *25*, 1007–1013.

Inoue, J., Yotsumoto, S., Sakamoto, T., Tsuchiya, S., and Aramaki, Y. (2005). Changes in immune responses to antigen applied to tape-stripped skin with CpG-oligodeoxynucleotide in mice. J. Control Release *108*, 294–305.

Jackson, L.A., Austin, G., Chen, R.T., Stout, R., DeStefano, F., Gorse, G.J., Newman, F.K., Yu, O., and Weniger, B.G. (2001). Safety and immunogenicity of varying dosages of trivalent inactivated influenza vaccine administered by needle-free jet injectors. Vaccine *19*, 4703–4709.

Jacobson, R.M., Swan, A., Adegbenro, A., Ludington, S.L., Wollan, P.C., and Poland, G.A. (2001). Making vaccines more acceptable–methods to prevent and minimize pain and other common adverse events associated with vaccines. Vaccine *19*, 2418–2427.

Johansen, P., Haffner, A.C., Koch, F., Zepter, K., Erdmann, I., Maloy, K., Simard, J.J., Storni, T., Senti, G., Bot, A., *et al.* (2005). Direct intralymphatic injection of peptide vaccines enhances immunogenicity. Eur. J. Immunol. *35*, 568–574.

Juillard, G.J., and Boyer, P.J. (1977). Intralymphatic immunization: current status. Eur. J. Cancer *13*, 439–440.

Juillard, G.J., Boyer, P.J., Niewisch, H., and Hom, M. (1979). Distribution and consequences of cell suspensions following intralymphatic infusion. Bull. Cancer *66*, 217–228.

Juillard, G.J., Boyer, P.J., and Snow, H.D. (1976). Intralymphatic infusion of autochthonous tumor cells in canine lymphoma. Int. J. Radiat. Oncol. Biol. Phys. *1*, 497–503.

Juillard, G.J., Boyer, P.J., and Yamashiro, C.H. (1978). A phase I study of active specific intralymphatic immunotherapy (ASILI). Cancer *41*, 2215–2225.

Juillard, G.J., Boyer, P.J., Yamashiro, C.H., Snow, H.D., Weisenburger, T.H., McCarthy, T., and Miller, R.J. (1977). Regional intralymphatic infusion (ILI) of irradiated tumor cells with evidence of distant effects. Cancer *39*, 126–130.

Kang, H.Y., Srinivasan, J., and Curtiss, R., 3rd (2002). Immune responses to recombinant pneumococcal PspA antigen delivered by live attenuated *Salmonella enterica* serovar typhimurium vaccine. Infect. Immun. *70*, 1739–1749.

Karrer, U., Althage, A., Odermatt, B., Roberts, C.W., Korsmeyer, S.J., Miyawaki, S., Hengartner, H., and Zinkernagel, R.M. (1997). On the key role of secondary lymphoid organs in antiviral immune responses studied in alymphoplastic (aly/aly) and spleenless (Hox11(–)/–) mutant mice. J. Exp. Med. *185*, 2157–2170.

Katz, D.E., DeLorimier, A.J., Wolf, M.K., Hall, E.R., Cassels, F.J., van Hamont, J.E., Newcomer, R.L., Davachi, M.A., Taylor, D.N., and McQueen, C.E. (2003). Oral immunization of adult volunteers with microencapsulated enterotoxigenic *Escherichia coli* (ETEC) CS6 antigen. Vaccine *21*, 341–346.

Kawabata, S., Miller, C.J., Lehner, T., Fujihashi, K., Kubota, M., McGhee, J.R., Imaoka, K., Hioi, T., and Kiyono, H. (1998). Induction of Th2 cytokine expression for p27-specific IgA B-cell responses after targeted lymph node immunization with simian immunodeficiency virus in rhesus macaques. J. Infect. Dis. *177*, 26–33.

Kenney, R.T., Yu, J., Guebre-Xabier, M., Frech, S.A., Lambert, A., Heller, B.A., Ellingsworth, L.R., Eyles, J.E., Williamson, E.D., and Glenn, G.M. (2004). Induction of protective immunity against lethal anthrax challenge with a patch. J. Infect. Dis. *190*, 774–782.

Kew, O.M., Wright, P.F., Agol, V.I., Delpeyroux, F., Shimizu, H., Nathanson, N., and Pallansch, M.A. (2004). Circulating vaccine-derived polioviruses: current state of knowledge. Bull. World Health Organ. *82*, 16–23.

King, J.C., Jr., Lagos, R., Bernstein, D.I., Piedra, P.A., Kotloff, K., Bryant, M., Cho, I., and Belshe, R.B. (1998). Safety and immunogenicity of low and high doses of trivalent live cold-adapted influenza vaccine administered intranasally as drops or spray to healthy children. J. Infect. Dis. *177*, 1394–1397.

Klavinskis, L.S., Bergmeier, L.A., Gao, L., Mitchell, E., Ward, R.G., Layton, G., Brookes, R., Meyers, N.J., and Lehner, T. (1996). Mucosal or targeted lymph node immunization of macaques with a particulate SIVp27 protein elicits virus-specific CTL in the genito-rectal mucosa and draining lymph nodes. J. Immunol. *157*, 2521–2527.

Klimuk, S.K., Najar, H.M., Semple, S.C., Aslanian, S., and Dutz, J.P. (2004). Epicutaneous application of CpG oligodeoxynucleotides with peptide or protein antigen promotes the generation of CTL. J. Invest. Dermatol. *122*, 1042–1049.

Koopman, G., Bogers, W.M., van Gils, M., Koornstra, W., Barnett, S., Morein, B., Lehner, T., and Heeney, J.L. (2007). Comparison of intranasal with targeted lymph node immunization using PR8-Flu ISCOM adjuvanted HIV antigens in macaques. J. Med. Virol. *79*, 474–482.

Kundig, T.M., Bachmann, M.F., DiPaolo, C., Simard, J.J., Battegay, M., Lother, H., Gessner, A., Kuhlcke, K., Ohashi, P.S., Hengartner, H., *et al.* (1995). Fibroblasts as efficient antigen-presenting cells in lymphoid organs. Science *268*, 1343–1347.

Lehner, T., Bergmeier, L.A., Tao, L., Panagiotidi, C., Klavinskis, L.S., Hussain, L., Ward, R.G., Meyers, N., Adams, S.E., Gearing, A.J., *et al.* (1994). Targeted lymph node immunization with simian immunodeficiency virus p27 antigen to elicit genital, rectal, and urinary immune responses in nonhuman primates. J. Immunol. *153*, 1858–1868.

Lehner, T., Mitchell, E., Bergmeier, L., Singh, M., Spallek, R., Cranage, M., Hall, G., Dennis, M., Villinger, F., and Wang, Y. (2000). The role of gammadelta T cells in generating antiviral factors and beta-chemokines in protection against mucosal simian immunodeficiency virus infection. Eur. J. Immunol. *30*, 2245–2256.

Lehner, T., Wang, Y., Cranage, M., Bergmeier, L.A., Mitchell, E., Tao, L., Hall, G., Dennis, M., Cook, N., Brookes, R., *et al.* (1996). Protective mucosal immunity elicited by targeted iliac lymph node immunization with a subunit SIV envelope and core vaccine in macaques. Nat. Med. 2, 767–775.

Lesimple, T., Moisan, A., Carsin, A., Ollivier, I., Mousseau, M., Meunier, B., Leberre, C., Collet, B., Quillien, V., Drenou, B., *et al.* (2003). Injection by various routes of melanoma antigen-associated macrophages: biodistribution and clinical effects. Cancer Immunol. Immunother. *52*, 438–444.

Lesimple, T., Neidhard, E.M., Vignard, V., Lefeuvre, C., Adamski, H., Labarriere, N., Carsin, A., Monnier, D., Collet, B., Clapisson, G., *et al.* (2006). Immunologic and clinical effects of injecting mature peptide-loaded dendritic cells by intralymphatic and intranodal routes in metastatic melanoma patients. Clin. Cancer Res. *12*, 7380–7388.

Levine, M.M., Ferreccio, C., Cryz, S., and Ortiz, E. (1990). Comparison of enteric-coated capsules and liquid formulation of Ty21a typhoid vaccine in randomised controlled field trial. Lancet *336*, 891–894.

Lockey, R.F. (2001). 'ARIA': global guidelines and new forms of allergen immunotherapy. J. Allergy Clin. Immunol. *108*, 497–499.

Lockey, R.F., Benedict, L.M., Turkeltaub, P.C., and Bukantz, S.C. (1987). Fatalities from immunotherapy (IT) and skin testing (ST). J. Allergy Clin. Immunol. *79*, 660–677.

Lockey, R.F., Turkeltaub, P.C., Olive, E.S., Hubbard, J.M., Baird-Warren, I.A., and Bukantz, S.C. (1990). The Hymenoptera venom study. III: Safety of venom immunotherapy. J. Allergy Clin. Immunol. *86*, 775–780.

Mackensen, A., Krause, T., Blum, U., Uhrmeister, P., Mertelsmann, R., and Lindemann, A. (1999). Homing of intravenously and intralymphatically injected human dendritic cells generated *in vitro* from CD34+ hematopoietic progenitor cells. Cancer Immunol. Immunother. *48*, 118–122.

Maloy, K.J., Erdmann, I., Basch, V., Sierro, S., Kramps, T.A., Zinkernagel, R.M., Oehen, S., and Kundig, T.M. (2001). Intralymphatic immunization enhances DNA vaccination. Proc. Natl. Acad. Sci. U.S.A. *98*, 3299–3303.

Manolova, V., Flace, A., Bauer, M., Schwarz, K., Saudan, P., and Bachmann, M.F. (2008). Nanoparticles target distinct dendritic cell populations according to their size. Eur J. Immunol. *38*, 1404–1413.

Mark, A., Carlsson, R.M., and Granstrom, M. (1999). Subcutaneous versus intramuscular injection for booster DT vaccination of adolescents. Vaccine *17*, 2067–2072.

Martinez-Gomez, J.M., Johansen, P., Erdmann, I., Senti, G., Crameri, R., and Kundig, T.M. (2009a). Intralymphatic injections as a new administration route for allergen-specific immunotherapy. Int. Arch. Allergy Immunol. *150*, 59–65.

Martinez-Gomez, J.M., Johansen, P., Rose, H., Steiner, M., Senti, G., Rhyner, C., Crameri, R., and Kundig, T.M. (2009b). Targeting the MHC class II pathway of antigen presentation enhances immunogenicity and safety of allergen immunotherapy. Allergy *64*, 172–178.

Miller, M.A., and Pisani, E. (1999). The cost of unsafe injections. Bull. World Health Organ. *77*, 808–811.

Moller, C., Dreborg, S., Ferdousi, H.A., Halken, S., Host, A., Jacobsen, L., Koivikko, A., Koller, D.Y., Niggemann, B., Norberg, L.A., *et al.* (2002). Pollen immunotherapy reduces the development of asthma in children with seasonal rhinoconjunctivitis (the PAT-study). J. Allergy Clin. Immunol. *109*, 251–256.

Morse, M.A., Coleman, R.E., Akabani, G., Niehaus, N., Coleman, D., and Lyerly, H.K. (1999). Migration of human dendritic cells after injection in patients with metastatic malignancies. Cancer Res. *59*, 56–58.

Nilsson, B.O., Svalander, P.C., and Larsson, A. (1987). Immunization of mice and rabbits by intrasplenic deposition of nanogram quantities of protein attached to Sepharose beads or nitrocellulose paper strips. J. Immunol. Methods 99, 67–75.

Nir, Y., Paz, A., Sabo, E., and Potasman, I. (2003). Fear of injections in young adults: prevalence and associations. Am. J. Trop. Med. Hyg. *68*, 341–344.

Norman, P.S. (2004). Immunotherapy: 1999–2004. J. Allergy Clin. Immunol. *113*, 1013–1023; quiz 1024.

Ochsenbein, A.F., Sierro, S., Odermatt, B., Pericin, M., Karrer, U., Hermans, J., Hemmi, S., Hengartner, H., and Zinkernagel, R.M. (2001). Roles of tumour localization, second signals and cross priming in cytotoxic T-cell induction. Nature *411*, 1058–1064.

Pajno, G.B., Barberio, G., De Luca, F., Morabito, L., and Parmiani, S. (2001). Prevention of new sensitizations in asthmatic children monosensitized to house dust mite by specific immunotherapy. A six-year follow-up study. Clin. Exp. Allergy *31*, 1392–1397.

Parent du Chatelet, I., Lang, J., Schlumberger, M., Vidor, E., Soula, G., Genet, A., Standaert, S.M., and Saliou, P. (1997). Clinical immunogenicity and tolerance studies of liquid vaccines delivered by jet-injector and a new single-use cartridge (Imule): comparison with standard syringe injection. Imule Investigators Group. Vaccine *15*, 449–458.

Partidos, C.D., Beignon, A.S., Semetey, V., Briand, J.P., and Muller, S. (2001). The bare skin and the nose as noninvasive routes for administering peptide vaccines. Vaccine *19*, 2708–2715.

Pierson-Mullany, L.K., Jackola, D., Blumenthal, M., and Rosenberg, A. (2000). Altered allergen binding capacities of Amb a 1-specific IgE and IgG4 from ragweed-sensitive patients receiving immunotherapy. Ann. Allergy Asthma Immunol. *84*, 241–243.

Quillien, V., Moisan, A., Carsin, A., Lesimple, T., Lefeuvre, C., Adamski, H., Bertho, N., Devillers, A., Leberre, C., and Toujas, L. (2005). Biodistribution of radiolabelled human dendritic cells injected by various routes. Eur. J. Nucl. Med. Mol. Imaging *32*, 731–741.

Rhyner, C., Kundig, T., Akdis, C.A., and Crameri, R. (2007). Targeting the MHC II presentation pathway in allergy vaccine development. Biochem. Soc. Trans. *35*, 833–834.

Rigano, M.M., Sala, F., Arntzen, C.J., and Walmsley, A.M. (2003). Targeting of plant-derived vaccine antigens to immunoresponsive mucosal sites. Vaccine *21*, 809–811.

Romani, N., Koide, S., Crowley, M., Witmer-Pack, M., Livingstone, A.M., Fathman, C.G., Inaba, K., and Steinman, R.M. (1989). Presentation of exogenous protein antigens by dendritic cells to T cell clones. Intact protein is presented best by immature, epidermal Langerhans cells. J. Exp. Med. *169*, 1169–1178.

Rosenberg, S.A., Yang, J.C., and Restifo, N.P. (2004). Cancer immunotherapy: moving beyond current vaccines. Nat. Med. *10*, 909–915.

Rothstein, E.P., Kamiya, H., Nii, R., Matsuda, T., Bernstein, H.H., Long, S.S., Hosbach, P.H., and Meschievitz, C.K. (1996). Comparison of diphtheria-tetanus-two component acellular pertussis vaccines in United States and Japanese infants at 2, 4, and 6 months of age. Pediatrics *97*, 236–242.

Ruben, F.L., Froeschle, J.E., Meschievitz, C., Chen, K., George, J., Reeves-Hoche, M.K., Pietrobon, P., Bybel, M., Livingood, W.C., and Woodhouse, L. (2001). Choosing a route of administration for quadrivalent meningococcal polysaccharide vaccine: intramuscular versus subcutaneous. Clin. Infect. Dis. *32*, 170–172.

Scheifele, D.W., Bjornson, G., and Boraston, S. (1994). Local adverse effects of meningococcal vaccine. CMAJ *150*, 14–15.

Senti, G., Graf, N., Haug, S., Ruedi, N., von Moos, S., Sonderegger, T., Johansen, P., and Kundig, T.M. (2009). Epicutaneous allergen administration as a novel method of allergen-specific immunotherapy. J. Allergy Clin. Immunol. *124*, 997–1002.

Senti, G., Prinz Vavricka, B.M., Erdmann, I., Diaz, M.I., Markus, R., McCormack, S.J., Simard, J.J., Wuthrich, B., Crameri, R., Graf, N., et al. (2008). Intralymphatic allergen administration renders specific immunotherapy faster and safer: a randomized controlled trial. Proc. Natl. Acad. Sci. U.S.A. *105*, 17908–17912.

Seo, N., Tokura, Y., Nishijima, T., Hashizume, H., Furukawa, F., and Takigawa, M. (2000). Percutaneous peptide immunization via corneum barrier-disrupted murine skin for experimental tumor immunoprophylaxis. Proc. Natl. Acad. Sci. U.S.A. *97*, 371–376.

Sepulveda-Amor, J., Valdespino-Gomez, J.L., Garcia-Garcia Mde, L., Bennett, J., Islas-Romero, R., Echaniz-Aviles, G., and de Castro, J.F. (2002). A randomized trial demonstrating successful boosting responses following simultaneous aerosols of measles and rubella (MR) vaccines in school age children. Vaccine *20*, 2790–2795.

Sigel, M.B., Sinha, Y.N., and VanderLaan, W.P. (1983). Production of antibodies by inoculation into lymph nodes. Methods Enzymol. *93*, 3–12.

Simonsen, L., Kane, A., Lloyd, J., Zaffran, M., and Kane, M. (1999). Unsafe injections in the developing world and transmission of bloodborne pathogens: a review. Bull. World Health Organ. *77*, 789–800.

Smith, K.A., Tam, V.L., Wong, R.M., Pagarigan, R.R, Meisenburg, B.L., Joea, D.K., Liu, X., Sanders, C., Diamond, D., Kundig, T.M., et al. (2009). Enhancing DNA vaccination by sequential injection of lymph nodes with plasmid vectors and peptides. Vaccine *27*, 2603–2615.

Stewart, G.E., 2nd, and Lockey, R.F. (1992). Systemic reactions from allergen immunotherapy. J. Allergy Clin. Immunol. *90*, 567–578.

Storni, T., Ruedl, C., Schwarz, K., Schwendener, R.A., Renner, W.A., and Bachmann, M.F. (2004). Nonmethylated CG motifs packaged into virus-like particles induce protective cytotoxic T cell responses in the absence of systemic side-effects. J. Immunol. *172*, 1777–1785.

Strid, J., Hourihane, J., Kimber, I., Callard, R., and Strobel, S. (2004). Disruption of the stratum corneum allows potent epicutaneous immunization with protein antigens resulting in a dominant systemic Th2 response. Eur. J. Immunol. *34*, 2100–2109.

Tagawa, S.T., Lee, P., Snively, J., Boswell, W., Ounpraseuth, S., Lee, S., Hickingbottom, B., Smith, J., Johnson, D., and Weber, J.S. (2003). Phase I study of intranodal delivery of a plasmid DNA vaccine for patients with Stage IV melanoma. Cancer *98*, 144–154.

Takigawa, M., Tokura, Y., Hashizume, H., Yagi, H., and Seo, N. (2001). Percutaneous peptide immunization via corneum barrier-disrupted murine skin for experimental tumor immunoprophylaxis. Ann. N.Y. Acad. Sci. *941*, 139–146.

Thomas, R., Chambers, M., Boytar, R., Barker, K., Cavanagh, L.L., MacFadyen, S., Smithers, M., Jenkins, M., and Andersen, J. (1999). Immature human monocyte-derived dendritic cells migrate rapidly to draining lymph nodes after intradermal injection for melanoma immunotherapy. Melanoma Res. *9*, 474–481.

Thompson, M.J. (2002). Immunizations for international travel. Prim. Care *29*, 787–814.

Till, S.J., Francis, J.N., Nouri-Aria, K., and Durham, S.R. (2004). Mechanisms of immunotherapy. J. Allergy Clin. Immunol. *113*, 1025–1034; quiz 1035.

Varney, V.A., Gaga, M., Frew, A.J., Aber, V.R., Kay, A.B., and Durham, S.R. (1991). Usefulness of immunotherapy in patients with severe summer hay fever uncontrolled by antiallergic drugs. BMJ *302*, 265–269.

Verlato, G., Corsico, A., Villani, S., Cerveri, I., Migliore, E., Accordini, S., Carolei, A., Piccioni, P., Bugiani, M., Lo Cascio, V., et al. (2003). Is the prevalence of adult asthma and allergic rhinitis still increasing? Results of an Italian study. J. Allergy Clin. Immunol. *111*, 1232–1238.

Vissers, J.L., van Esch, B.C., Hofman, G.A., Kapsenberg, M.L., Weller, F.R., and van Oosterhout, A.J. (2004). Allergen immunotherapy induces a suppressive memory response mediated by IL-10 in a mouse asthma model. J. Allergy Clin. Immunol. *113*, 1204–1210.

von Beust, B.R., Johansen, P., Smith, K.A., Bot, A., Storni, T., and Kundig, T.M. (2005). Improving the therapeutic index of CpG oligodeoxynucleotides by intralymphatic administration. Eur. J. Immunol. *35*, 1869–1876.

Wang, B., Feliciani, C., Howell, B.G., Freed, I., Cai, Q., Watanabe, H., and Sauder, D.N. (2002). Contribution of Langerhans cell-derived IL-18 to contact hypersensitivity. J. Immunol. *168*, 3303–3308.

Weber, J., Boswell, W., Smith, J., Hersh, E., Snively, J., Diaz, M., Miles, S., Liu, X., Obrocea, M., Qiu, Z., et al. (2008). Phase 1 trial of intranodal injection of a Melan-A/MART-1 DNA plasmid vaccine in patients with stage IV melanoma. J. Immunother. *31*, 215–223.

Wuthrich, B., Schindler, C., Medici, T.C., Zellweger, J.P., and Leuenberger, P. (1996). IgE levels, atopy markers and hay fever in relation to age, sex and smoking status in a normal adult Swiss population. SAPALDIA (Swiss

Study on Air Pollution and Lung Diseases in Adults) Team. Int. Arch. Allergy Immunol. *111*, 396–402.

www.clinicaltrials.gov Safety and Immune Response to a Multi-component Immune Based Therapy (MKC1106-MT) for Patients With Melanoma.

www.clinicaltrials.gov Safety and Immune Response to a Multi-component Immune Based Therapy (MKC1106-PP) for Patients With Advanced Cancer.

www.imvision-therapeutics.com Positive Phase I Clinical Results for Treatment of Cat Dander Allergy.

Zhao, Y.L., Murthy, S.N., Manjili, M.H., Guan, L.J., Sen, A., and Hui, S.W. (2006). Induction of cytotoxic T-lymphocytes by electroporation-enhanced needle-free skin immunization. Vaccine 24, 1282–1290.

Zinkernagel, R.M. (2000). Localization dose and time of antigens determine immune reactivity. Semin. Immunol. *12*, 163–171; discussion 257–344.

Zinkernagel, R.M., Ehl, S., Aichele, P., Oehen, S., Kundig, T., and Hengartner, H. (1997). Antigen localisation regulates immune responses in a dose- and time-dependent fashion: a geographical view of immune reactivity. Immunol. Rev. *156*, 199–209.

The First Vaccine Obtained Through Reverse Vaccinology: the Serogroup B Meningococcus Vaccine

Jeannette Adu-Bobie, Beatrice Aricò, Marzia M. Giuliani and Davide Serruto

Abstract

Neisseria meningitidis was isolated over 100 years ago when Anton Weicshelbaum identified the causative agent of cerebrospinal meningitis. Since its isolation in 1887, *N. meningitidis* has been recognized to cause endemic cases, case clusters, epidemics and pandemics of meningitis and devastating septicaemia. Despite over one century since its discovery, scientists have yet to identify a universal vaccine for this deadly bacterium. Although vaccines exist for several serogroups of pathogenic *N. meningitidis*, serotype B (MenB) has eluded scientists for decades, until the advent of genomics. The genome era has completely changed the way to design vaccines. The availability of the complete genome of microorganisms combined with a novel advanced technology has introduced a new prospective in vaccine research. This novel approach is now known as 'reverse vaccinology' and *N. meningitidis* can be considered the first successful example of its application. This chapter will describe the successful story of the development of the serogroup B vaccine, starting from the analysis of genome and finishing with the results obtained in clinical trials.

Introduction: meningococcal biology and epidemiology

N. meningitidis, also known as meningococcus, is a Gram-negative β-proteobacterium member of the family Neisseriaceae and is considered a primary cause of bacterial meningitis and other invasive bacterial infections worldwide. Since the introduction of capsular polysaccharide vaccines against infections from *Haemophilus influenzae* type b and *Streptococcus pneumoniae,* the role of meningococcus as a cause of bacterial meningitis and meningococcal septicaemia has become more prominent.

N. meningitidis is a strictly human pathogen and the nasopharynx is the only known reservoir. Transmission of the bacterium occurs via droplets. The bacterium colonizes a substantial proportion of the population 3–30%, where it resides as a commensal (Caugant and Maiden, 2009). A major step in the pathogenesis of *N. meningitidis* is colonization and adhesion of the nasopharynx. In particular occasions, for reasons yet unknown and in susceptible individuals, meningococcus is able to actively invade the respiratory tract epithelia, reach and multiply within the bloodstream, causing a fulminant sepsis. From the bloodstream, the bacterium can cross the blood–brain barrier and spread to the cerebrospinal fluid, causing meningitis. In spite of intensive research, little is yet known on molecular mechanisms of pathogenesis of meningococcal diseases.

The adhesion of *N. meningitidis* to the nasopharyngeal mucosa is the first step of the infection and part of a multistage process consisting of first localized and then intimate attachment (Carbonnelle *et al.,* 2009). Initial adhesion is mediated by meningococcal type IV pili (Nassif *et al.,* 1994; Virji, 2009; Virji *et al.,* 1991). Subsequently retraction of pili allows the intimate adhesion of the bacteria (Hauck and Meyer, 2003; Pujol *et al.,* 1999). Several factors such as Opc and Opa opacity proteins (Virji, 2009; Virji *et al.,* 1993), have been shown to mediate the adhesion and invasion of non-capsulate meningococci to both endothelial and epithelial cells by interacting with CD66 and heparan sulphate proteoglycan receptors (Hauck and Meyer, 2003; Virji *et al.*, 1993; Virji *et al.*, 1992). Recently new adhesins have been described: App (NMB1985), NhhA (NMB0992), NadA (NMB1194) and MspA (NMB1998) (Capecchi *et al.*, 2005; Scarselli *et al.*, 2006; Serruto *et al.*, 2003; Turner *et al.*, 2006; van Ulsen *et al.*, 2006). These proteins, identified using a genomic approach, have been found to mediate interaction of meningococcus to host cells. App (Adhesion and penetration protein) is highly homologous to the *Haemophilus* adhesion and penetration protein (Hap) from *Haemophilus influenzae*. App was shown to promote adherence of meningococcus to epithelial cells. App and Hap belong to the family of autotransporter proteins with a serine protease activity, which is responsible of the processing of the membrane-associated protein and extracellular release of the N-terminal portion (Serruto *et al.*, 2003). The function of this secreted fragment is yet unknown. NhhA (*Neisseria* hia/hsf homologue, also known as GNA0992) is an oligomeric outer membrane protein specific to *N. meningitidis*, homologue to Hia/Hsf adhesins of *H. influenzae* (Scarselli *et al.*, 2001) able to mediate adhesion to epithelial cells, heparain sulphate and laminin (Scarselli *et al.*, 2006). Hia/ Hsf and NhhA have been recently included in a trimeric autotransporter adhesin (TAA) family, together with other relevant virulence factors such as YadA of *Yersinia* spp., Uspa1 and Uspa2 of *Moraxella catarrhalis*, Omp100 of *Actinobacillus actinomycetemcomitans* as well as NadA of *N. meningitidis*. NadA is described in detail later in the chapter.

MspA, Meningococcal serine protease A is expressed by several but not all virulent *Neisseria* strains. Although its role in the context of meningococcus is not well elucidated, it is reported to mediate binding of *Escherichia coli* expressing the protein to both epithelial and endothelial cells (Turner *et al.*, 2006; van Ulsen *et al.*, 2006).

The role of *Neisseria* two-partner secretion system (TPS) proteins in bacterial interactions with target cells has been also suggested. The functionality of TPS in *N. meningitidis* has been recently described: these systems (also designated HrpA/HrpB in analogy with haemagglutin/ haemolysin-related proteins) may facilitate bacterial adhesion and intracellular survival (Schmitt *et al.*, 2007; Tala *et al.*, 2008).

The next steps, after bacterial adhesion, are the invasion of the cells, intracellular persistence and transcytosis. These sequences of events are modulated by the interaction of virulence factors with their host cell receptors, and signals are sent from pathogen to host and vice versa in the adhesion cascade (Merz and So, 2000). Although proteins with the capability of mediating *in vitro* meningococcal entry into host cells were identified (Opc and Opa are the main factors), proteins allowing the bacterial entry into the bloodstream are unknown. Recently, Type IV pili-mediated adhesion has been shown to be required for *N. meningitidis* to cross the blood–brain barrier through the intercellular junctions (Coureuil *et al.* 2009). Several factors allowing meningococcal survival in the blood have been identified (Geoffroy *et al.*, 2003; Mackinnon *et al.*, 1993) and recently fHBP was shown to play an essential role (a description of this antigen is provided below).

A capsule made up of complex polysaccharide surrounds all disease-causing meningococci. This is one of the essential meningococcal attributes for pathogenesis, which negatively affects bacterial adhesion and, consequently entry, but in contrast it is known to be important for bacterial survival in extracellular fluids and fundamental for the intracellular survival of this microorganism (Spinosa *et al.*, 2007). It is the structure of the polysaccharide that defines the highest order in meningococcal serological typing, the serogroup. Based on the chemical composition of the capsule, 13 structurally distinct capsular polysaccharides have been identified. Six *N. meningitidis* serogroups, designated A, B, C, Y, W135, and X account for virtually all disease-causing isolates globally (Stephens *et al.*, 2007). There are significant geographic differences in the distribution of the different disease-causing serogroups. Serogroup A is mainly responsible for the large epidemics observed in sub-Saharan Africa, in a region known as the 'meningitis belt'. Endemic meningococcal disease due to mainly serogroups B and C is observed in Europe and the US. Epidemics of serogroup B disease have occurred in Norway, Cuba, Brazil, and Chile and recently in

New Zealand. Surveillance of serogroup distribution shows that prevalence can change over time. For example, serogroup Y has in recent years accounted for a substantial proportion of cases in the US.

A further classification uses serological methods and is based on the antigenicity of the most abundant membrane proteins PorA and PorB, whose sequence defines the serotype and serosubtype, respectively. Moreover, on the basis of the lipopolysaccharide structure is possible to define 12 immunotypes (Morley and Pollard, 2001). New methods of classification have been developed more recently. Multilocus enzyme electrophoresis (MLEE) uses the electrophoretic mobility of various cytoplasmic enzymes to classify meningococci into clonal families with similar characteristics (Caugant *et al.*, 1986a,b). A genetic-based innovation involving the direct sequencing of seven genes that encode intracellular, cytoplasmic enzymes (multilocus sequence typing, MLST) is presently the 'gold standard' for classifying the meningococci as well as many other bacteria. Unlike antigens [such as surface-localized outer membrane proteins (COMPs)], these enzymes are not subject to immunological selective pressure, but are under stabilizing selection for conservation of their metabolic 'house-keeping' function (Maiden *et al.*, 1998; Maiden and Feavers, 2000). Strains that harbour identical sequences of these seven MLST genes are grouped within sequence types (or STs). Sequence types are grouped into clonal complexes by their similarity to the 'founding member' of each complex (the central genotype). More than 7000 STs have been identified today (http://pubmlst.org/neisseria/). Although there are currently 43 neisserial clonal complexes, the majority of strains that presently cause disease worldwide are grouped within four complexes, the so-called 'hypervirulent lineages': the ST-32, ST41/44, ST-8 and ST-11 complexes. The distribution of these hypervirulent complexes, like serogroups, is temporally and geographically variable (Caugant *et al.*, 2007). MLST is crucial in understanding the structure and dynamics within populations of *N. meningitidis* and aids in the development and evaluation for vaccine formulations. However, a limitation of MLST for vaccine development might be its reliance on housekeeping genes that are not exposed to the host immune system. Surface-exposed antigens show high levels of sequence diversity among strains due to evolutionary pressure. As a consequence, the antigenic repertoire is not predictable on the basis of MLST alone. Such antigens might be included in new typing methods to investigate population structures in a way that is relevant for vaccine development (Bambini *et al.*, 2009).

What is devastating about septicaemia and meningitis is that the death rate is still high (5–15%), and death can occur within hours despite the availability of effective antibiotics. In addition, depending on the different clinical presentations of infection, mortality rates can be as high as 40% in the case of septic shock. Moreover, up to 25% of survivors are left with neurological sequelae, limb loss or hearing loss.

The incidence of meningococcal disease ranges from very rare to 1000 cases per 100,000 population every year, and despite severe antibiotic therapies, the mortality rates remain high (Stephens *et al.*, 2007). Many *N. meningitidis* strains have reduced susceptibility to penicillins, but high levels of resistance are rarely found. Therefore penicillin remains a suitable treatment for meningococcal disease. Other antibiotics such as rifampin, ciprofloxacin or ceftriaxone are the antibiotics mostly used against meningococci (Stephens *et al.*, 2007). Although antibiotic resistance is not a major problem, the main concern is the rapid progression of the disease: mortality risk is high, and permanent sequelae are common among survivors. As a result vaccines are the only effective means to control this devastating disease.

Rates of meningococcal disease are highest among young children and increase again in adolescents and young adults. Neonates are relatively protected against meningococcal disease as a result of passive acquisition of transplacental maternal antibodies. With increasing age during the first months of life, maternal antibodies decrease and infants become increasingly susceptible to meningococcal disease. The lowest protective antibody levels are generally found in infants and toddlers between 6 months and 2 years of age, with increases in natural immunity beginning to appear during the second year of life.

Meningococcal disease is, therefore, a significant problem for infants and young children, although disease also occurs in adolescents and adults.

Meningococcal vaccine development

Vaccines offering protection against meningococcal disease have been available for more than 30 years but even now there is no a formulation that offers comprehensive protection against strains from all of the pathogenic capsular groups. The development of meningococcal vaccines has been hampered by the biology of *N. meningitidis*. This pathogen has evolved an array of sophisticated mechanisms to evade host defences. These mechanisms include production of poorly immunogenic polysaccharide capsules and lipopolysaccharides, some of which mimic glycosylated structures on human tissue. Surface-exposed proteins that elicit immune responses exhibit variability in both their antigenic structure and expression. In addition, the meningococcus is naturally able to acquire DNA, providing the organism with a mechanism for genetic exchange and almost continuous possibilities for changing of nucleic acid sequences including those encoding the principal antigens.

Several combinations of serogroup A, C, W-135 and Y polysaccharides have been used in vaccines over the years in different countries. The polysaccharide vaccines are safe and immunogenic and efficacious in children older than 2 years and adults (Al-Mazrou *et al.*, 2005; Gotschlich *et al.*, 1969). However, these vaccines are typically only used on people in high-risk groups and in response to epidemics due to several limitations: polysaccharide vaccines are poorly immunogenic in young infants and children less than 2 years old, the major group at risk of meningococcal disease. Polysaccharide vaccines do not reliably induce immunological memory due to their failure to generate memory T-cells. Many of the problems of simple polysaccharide vaccines were overcome with the discovery that the chemical conjugation of the capsular polysaccharide to an immunogenic protein carrier greatly increased its immunogenicity. This led to the development of several successful meningococcal monovalent group C and tetravalent A, C, Y, W-135 conjugate

vaccines (Costantino *et al.*, 1992; Granoff *et al.*, 2008).

Since the widespread introduction of the conjugate MenC vaccine (against Serogroup C) into infant immunization schedules, and the increased use of tetravalent vaccines in the US, serogroup B remains the largest challenge in confronting meningococcal disease in the developed world. Serogroup B (Men B) strains are responsible for the majority of endemic and epidemic meningococcal disease in developed countries; one-third of disease in the US (Rosenstein *et al.*, 1999) and more than half in Europe (Noah and Henderson, 2002). The global incidence has been estimated at between 20,000 and 80,000 cases per year, accounting for 2000–8000 deaths annually. The highest rate of MenB disease occurs in infants under the age of 1 year, as a result of waning maternal antibodies (Cartwright *et al.*, 2001).

A polysaccharide-based vaccine approach cannot be used for group B meningococcus because the MenB capsular polysaccharide is a polymer of a(2–8)-linked N-acetylneuranimic acid that is also present in mammalian tissues. This means that it is almost completely non-immunogenic and does not induce a protective response (Finne *et al.*, 1987; Nedelec *et al.*, 1990). This mimicry has also raised concerns about autoimmunity in vaccinated individuals, although this has not been observed to date.

An alternative approach to MenB vaccine development is based on the use of surface-exposed proteins contained in outer membrane preparations (OMVs, outer membrane vesicles). A variety of 'tailor-made' MenB OMV vaccines have been developed and licensed to control epidemics dominated by a single clone. OMV vaccines have been used in Norway (Fredriksen *et al.*, 1991), Cuba (Sierra *et al.*, 1991), Chile (Boslego *et al.*, 1995) and New Zealand (Oster *et al.*, 2005). They are able to induce protective antibodies against the homologous strain in all age groups, and have proved successful in controlling epidemic disease. The main limitation of OMV vaccines is that they are strain-specific and do not provide protection against heterologous strains due to the antigenic diversity of their protein antigens.

Although OMV components contain more than seventy proteins (Uli *et al.*, 2006; Vipond *et al.*, 2006), it is generally accepted that the major immunodeterminant of the OMV vaccines is the PorA outer membrane protein. Studies have provided evidence that in adults, OMV vaccination results in limited serological responses to heterologous strains that have different PorA types. Several studies have demonstrated that the immune response in infants to OMV vaccination is more strain-specific than in older children and adults. It is therefore clear that although OMV vaccines have a proven track record in combating serogroup B epidemics caused by a single strain, they have a limited value in providing protection against endemic strains that harbour a different PorA protein.

The homologous immune response to MenB OMV-based vaccines has led to efforts directed during the last three decades at identification and development of monovalent surface-exposed proteins that induce protective immune responses capable of providing protection against heterologous strains. However, until recently, all of the surface-exposed proteins described have had significant antigenic variability, and therefore these efforts have been generally unsuccessful.

The reverse vaccinology applied to MenB

Until recently, the development of vaccines has classically relied almost exclusively on bio-chemical and immunological methods. These approaches were indeed successful in certain cases but were hampered by the limited number of candidate antigens that could be identified as well as the time required for their identification.

The availability of whole genome sequences has entirely changed the approach to vaccine development. The genome represents a list of virtually all the protein antigens that the pathogen can express at any time. It becomes possible to choose potentially surface-exposed proteins in a reverse manner, starting from the genome rather than from the microorganism (Rappuoli, 2001).

In 1998, the research team at Novartis Vaccines (formerly Chiron Vaccines) embarked on a large-scale genome project. To develop a universal vaccine against serogroup B, the genome of a MenB isolate (MC58 strain) has been sequenced and used to discover novel antigens (Pizza *et al.*, 2000; Tettelin *et al.*, 2000). This approach was termed reverse vaccinology and *N. meningitidis* serogroup B became the prototype for the use of genomics in vaccine development.

The identification of new previously unidentified antigens was a process that took the research team 18 months to achieve. The sequence of the virulent strain MC58 was determined by the shotgun strategy and while the sequencing was still in progress, the unassembled DNA fragments were analysed to identify the open reading frames (ORFs). The MC58 genome consists of 2,272,352 base pairs with an average of G+C content of 53%. The 83% of the genome codes for 2158 ORFs. Out of these, 1158 have a putative biological role assigned on the basis of their similarity with known proteins (Tettelin *et al.*, 2000). To identify novel vaccine antigens a strategy has been aimed to select, among the more than 2000 predicted proteins, those which were predicted to be surface-exposed or secreted and tested them for their potential to induce protection against disease. *N. meningitidis* is essentially an extracellular pathogen and the major protective response relies on circulating antibody: complement-mediated bactericidal activity is, in fact, the accepted correlate for the *in vivo* protection and as such is the surrogate endpoint in clinical trials of potential meningococcal vaccines. On the basis of this evidence, the group worked on the assumption that protective antigens are more likely to be found among surface-exposed or secreted proteins. Hence the initial selection of candidate analysis is based on computer predictions of secretion or surface location. In the preliminary selection, all *N. meningitidis* predicted ORFs were searched using computer programs such as PSORT and SignalP, which predict signal peptide sequences. Moreover, proteins containing predicted membrane spanning regions (using TMPRED), lipoprotein signature, and proteins homologous to surface-exposed proteins in other microorganisms were also selected. Finally, proteins with homology to known virulence factors or protective antigens from other pathogens were added to the list. Of the 2158 predicted ORFs in the *N. meningitidis* genome, 570 were selected by these criteria and could therefore represent new potential vaccine candidates. The selected ORFs were amplified, cloned and analysed for expression in a heterologous system as either C-terminal His-tag or N-terminal glutathione S-transferase (GST) fusion proteins. These two expression systems were chosen to achieve the highest level of expression and the easiest purification procedure by a single chromatography step. Of the 570, 350 ORFs were successfully cloned in *E. coli* and purified in a sufficient amount for mice immunizations. Most of the failures, both in cloning and in expression, were related to proteins with more than one transmembrane spanning region. This is likely to be due to toxicity for *E. coli* or to their intrinsic insolubility.

Each purified recombinant protein was used to immunize CD1 mice in the presence of Freund's adjuvant. Immune response was analysed by Western blot analysis on both total cell extracts and on purified OMVs to verify whether the protein was expressed and localized on the membrane, and by enzyme-linked immunosorbent assay (ELISA) and flow cytometry on whole cells to verify whether the antigen was surface-exposed in meningococcus. Finally, the bactericidal assay was used to evaluate the complement-mediated killing activity of the antibodies (SBA, serum bactericidal activity), since this property correlates with vaccine efficacy in humans (Borrow *et al.*, 2005; Goldschneider *et al.*, 1969).

Of the 91 proteins found to be positive in at least one of these assays, 28 were able to induce antibodies with bactericidal activity (Pizza *et al.*, 2000) (Fig. 9.1). Several of the antigens previously identified using conventional approaches showed strain variability or were only expressed in some strains, and most of them are effective only against the homologous strains. Therefore, the potential vaccine candidates identified were evaluated for degree of sequence variability among multiple isolates and serogroups of *N. meningitidis*, three stains of *N. gonorrhoeae*, as well as one strain each of *N. cinerea* and *N. lactamica*. The genome analysis identified antigens that were quite different from those identified using the conventional approaches. The majority of them are present in all strains tested and are conserved in sequence. Many of the newly-identified serogroup B antigens included surface-exposed protein or lipoproteins with a globular structure and without membrane spanning domains and many of them are not abundant on the bacterial surface. Reverse vaccinology has therefore proven to be a rapid and reliable approach to identifying vaccine candidates. In the case of serogroup B, these potential vaccine candidates, able to induce broad strain coverage, were subjected to further evaluation and characterization. The candidates were gradually funnelled down, a process that took a further 24 months. Finally, the three most immunogenic antigens on the basis of their ability to induce bactericidal activity or *in vivo* passive protection were selected to be used in a multicomponent vaccine (Fig. 9.1). They were NHBA (GNA2132), fHBP (GNA1870) and NadA (GNA1994). Other two antigens (named GNA2091 and GNA1030) were also selected. To further enhance their immunogenicity and facilitate large-scale manufacturing of the vaccine, four of the selected antigens were combined into two fusion proteins so that the resulting protein vaccine contained three recombinant proteins. The antigen NHBA was fused to GNA1030 while GNA2091 was fused to fHBP. NadA was included as single antigen as it did not perform well when fused to a partner. It is thought that this may be due to the fact the protein may lose its native trimeric organization (Capecchi *et al.*, 2005). Results showed that the two fusion proteins formulated with aluminium hydroxide induced immunes in both FACS and bactericidal assays. These antibodies were more potent than those induced by the individual antigens. Twenty micrograms of each of the two fusion proteins and of the NadA antigen were adsorbed to aluminium hydroxide, an adjuvant suitable for human use, to make the vaccine formulation that was used in subsequent studies. This multicomponent vaccine was named 5CVMB (5-component vaccine against MenB) (Giuliani *et al.*, 2006). The rationale behind combining antigen was to increase the spectrum of vaccine coverage, minimizing the possibility of bacterial evasion and development of selection mutants.

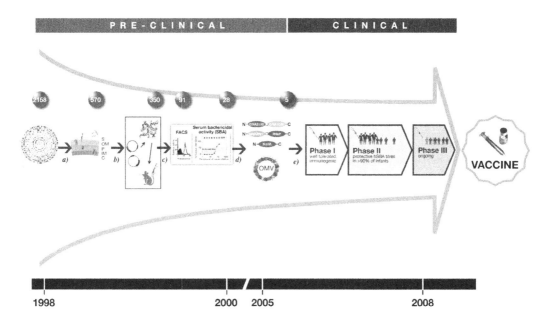

Figure 9.1 Reverse vaccinology applied to *Neisseria meningitidis* B. Based on the complete genome sequence of *N. meningitidis* strain MC58 (2158 ORFs), genetic sequences potentially encoding for novel surface exposed proteins were identified (a). DNA sequences encoding 570 potential surface-exposed antigens were amplified by PCR and cloned into an *Escherichia coli* expression vectors. 350 recombinant proteins were successfully produced, purified and used to immunize mice (b). The optimal recombinant protein candidates were then selected based on their surface expression, and ability to induce serum bactericidal antibodies: 91 new surface-exposed proteins have been identified and 28 novel protein antigens were able to induce antibodies with bactericidal activity were identified (c). The antigens selected by reverse vaccinology were prioritized based on these criteria: (i) the protein has to be surface-exposed; (ii) the protein has to be conserved in sequence across a range of different MenB strains; (iii) the protein must induce a broad bactericidal antibody response. The three top antigens that met the prioritization criteria were: NadA, fHBP and GNA2132 (NHBA), they gave high bactericidal titres and were bactericidal against most of the strain tested. The recombinant antigens were formulated with OMV-NZ (d) and tested in preclinical studies and clinical trials in adults, adolescents and infants (e).

Preclinical evaluation of the Novartis MenB vaccine

In order to properly assess whether the 5CVMB vaccine formulation was able to induce protection against most of the meningococcus B strains, a large panel of clinical isolates representing as much as possible the diversity of the bacterial population has been collected. Sera obtained by immunizing mice with the 5CVMB vaccine were tested in a bactericidal assay against a panel of 85 meningococcal strains. Almost all clonal complexes mainly described to be associated with disease were present in this panel. The four meningococcal hypervirulent clusters ST 32 complex (cpx) or ET5, ST41/44 cpx or Lineage 3, ST8 cpx or cluster A4, and ST11 cpx or ET37 represent 56/85 (65.9%) of the total strains; 29/85

(34.1%) of the strains grouped as 'others' include recently emerged ST-213, ST-269 clonal complexes and other complexes and some STs not yet assigned to any clonal complex (Bambini *et al.*, 2009; Giuliani *et al.*, 2006). The vaccine induced bactericidal antibodies against 78% of the strains. This percentage of killing is very promising, especially when compared with the protection offered by two separate OMV vaccines: Norwegian H44/76 strain-based vaccine 20% and New Zealand NZ98/254 strain-based vaccine 21%. In a closer observation the protection induced by 5CVMB formulated with aluminium was complete (100%) against all strains of the panel from the hypervirulent complexes ST32 and ST8, almost complete (95%) against the hypervirulent complex ST41/44 and substantial against strains of hypervirulent cluster ST11 and those classified as 'others' (65% and 59% respectively). The different percentage could be explained by the presence of different combinations of the subvariants of the antigens in strains belonging to the same clonal complex (Giuliani *et al.*, 2006). The poor coverage of strains and hypervirulent lineages provided by OMV vaccines validates the observations obtained so far in preclinical and clinical studies showing that OMV vaccines induce immunity that is mostly PorA specific. In fact the Norwegian vaccine (OMV-Nw) showed high SBA titres (> 1:8000) against all nine strains with PorA P1.16, identical to that of the vaccine strain and titres of 1:512 or lower against eight strains carrying a different PorA. Similarly the New Zealand vaccine (OMV-NZ) induced high SBA titres against the 11 strains with PorA P1.4 identical to the vaccine strain and low SBA titres against seven strains with different PorA.

Additional preclinical studies using different adjuvants have been performed in order to investigate whether the coverage of the 5CVMB vaccine could be increased. SBA induced in mice by different formulations has been measured against the same panel of 85 strains. Results showed that 5CVMB adjuvanted with aluminium hydroxide combined with CpG oligonucleotides or MF59, an adjuvant which is licensed for human use in Europe, showed coverage by bactericidal antibodies increased to about 92% and 94% respectively. Interestingly, with Freund's adjuvant, which is the most potent adjuvant for mouse immunization but unsuitable for human use, coverage by bactericidal antibodies reached 97.6% (Giuliani *et al.*, 2006). These data show that in mice the 5CVMB vaccine can induce immunity against nearly all MenB strains when the immune response is optimized.

In order to increase the immunogenicity of the vaccine, further investigations have been done using the combination of 5CVMB and the New Zealand vaccine based on the OMV from the NZ98/254 strain. The OMV-NZ was added in order to provide broader serogroup B strain coverage of the PorA variant contained in the OMV. The coverage against the panel of 85 strains increased from 78% to 85% (Rappuoli, 2008) (Fig. 9.2A).

During research phase and preclinical development the vaccine composed by the three recombinant proteins was named 5CVMB. In clinical

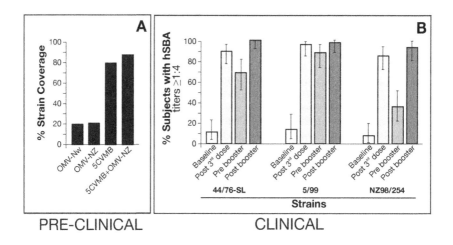

Figure 9.2 (A) Preclinical evaluation in mice of protective effect of 5CVMB, 5CVMB+OMV-NZ and OMV-based vaccines (OMV-NW and OMV-NZ). Results are expressed as percentage of strains killed in bactericidal assay with titres equal or higher than 1:128. The analysis has been performed against 85 *N. meningitidis* strains using rabbit serum as complement source. (B) Immunogenicity of the 'MenB vaccine' (rMenB+OMV-NZ) administered in infants. Data are represented as the percentage of subjects in the 'MenB vaccine' group achieving hSBA titres equal or greater than 1:4 after the third vaccination, pre- and post-booster against each of the meningococcal reference strains used (44/76-SL, 5/99 and NZ98/254).

development 5CVMB was renamed 'rMenB' and when combined with OMV-NZ became 'MenB vaccine'.

The Novartis MenB vaccine in clinical trials

The strategy to use reverse vaccinology to discover antigens for novel vaccines and test them against a large collection of clinical isolates representative of the bacterial population allowed the development of a vaccine against meningococcus B. The proteins were expressed in a system appropriate for industrial manufacturing and formulated with an adjuvant suitable for human use such as aluminium hydroxide. The final vaccine formulation consists of 50µg each recombinant protein GNA2132-GNA1030, NadA and GNA2091-fHBP (vaccine named 'rMenB'). In addition, the 3 recombinant proteins were also formulated with 25µg of OMV-based vaccine MeNZB (from *N. meningitidis* strain NZ98/254, expressing PorA serosubtype P1.4) in order to provide broader serogroup B strain coverage of the PorA variant contained in the OMV vaccine (named MenB vaccine). Both vaccines, with or without OMV, were formulated with aluminium hydroxide. The addition of MeNZB component was based on the experience from the use of the OMV-based vaccine, which was shown to be safe and efficacious in the control of the clonal meningococcal serogroup B epidemic in New Zealand (Oster *et al.*, 2005).

The recombinant vaccine with or without OMV was tested in clinical studies conducted in adults and demonstrated good safety and immunogenicity. The vaccines immunogenicity in the clinical trial was measured by serum bactericidal assay using human complement. It has been extensively demonstrated that bactericidal antibodies induced by meningococcal polysaccharide, polysaccharide–protein conjugates and outer membrane protein vaccines protect against meningococcal disease (Hols *et al.*, 2005; Perrett and Pollard, 2005; Stephens, 2007). Thus immunogenicity based on functional SBA activity will be the primary end-point for evaluating vaccines (Borrow *et al.*, 2006).

In order to evaluate safety and immunogenicity of the investigational vaccines, a phase II clinical study of the meningococcal B recombinant vaccine with or without OMV has been conducted in infants. Vaccines were administered at 2, 4, 6 and 12 months of age. Results demonstrated satisfactory safety, tolerability and immunogenicity (Miller *et al.*, 2008). Local and systemic reactions of the investigational MenB vaccines were in general similar in frequency and intensity to routine vaccinations.

The reference strains used in the infant trial were selected to assess the immunogenicity of the key antigens contained in the vaccines: fHBP (strain 44/76-SL), NadA (strain 5/99) and OMV (strain NZ98/254).

The immune response was measured by the percentage of subjects with serum bactericidal titres greater than 1:4. Immunogenicity analysis shows similar results for reference strains 44/76-SL and 5/99 in the rMenB alone and in the MenB vaccine groups. Titres in the MenB vaccine recipients were higher for NZ98/254 compared with MenB alone. Analysis of this vaccine group shows 89%, 96% and 85% with hSBA (human SBA) titres greater than 1:4 post 3rd dose against 44/76-SL (ST-32), 5/99 (ST-8) and NZ98/254 (ST41/44) respectively. The immune response further increase with the booster dose of the vaccine to 100% (44/76-SL), 98% (5/99) and 93% (NZ98/254). Persistence of bactericidal antibodies was demonstrated at 12 months for all reference strains (Fig. 9.2B).

The results from this study show that the MenB vaccine induces a robust immune response and good tolerability profile in infants. This vaccine is now being evaluated in larger phase III studies that will provide important further information on the safety profile and breadth of protection afforded by this vaccine.

Functional characterization of the vaccine antigens

While the vaccine is in clinical development, research efforts are focused on the functional and immunological characterization of the main vaccine antigens (NadA, fHBP and GNA2132), looking also at their potential role as virulence factors in meningococcal pathogenesis.

NadA

NadA belongs to the 'Oca' (oligomeric coiled-coil adhesin) family of bacterial trimeric autotransporter adhesins (Comanducci *et al.*, 2002), which are characterized by the ability to form trimers on the bacterial surface and by a common mechanism of secretion, which is linked to their trimerization (Cotter *et al.*, 2005; Linke *et al.*, 2006; Surana *et al.*, 2004) (Fig. 9.3). Oca proteins share a similar topology consisting of a conserved C-terminal membrane anchor (β-domain) through which the protein is translocated to the cell surface, a central alpha helical domain (stalk) with high propensity to form coiled-coil structures and a N-terminal globular 'head' that has been associated with receptor binding capabilities. Oca family includes well-known proteins such as YadA of *Yersinia* spp (Bliska *et al.*, 1993; Iriarte and Cornelis, 1996; Skurnik *et al.*, 1994) and UspAs proteins of *Moraxella catarrhalis* (Hill and Virji, 2003; Lafontaine *et al.*, 2000). Canonical architecture of these proteins has been proposed where the extracellular moiety, the passenger domain, consisting of the elongated stalk, neck and head, confers to the prototypic member of the family its characteristic drumstick appearance on the bacterial cell surface with the specific adhesive capabilities localized within the N-terminal globular head (Desvaux *et al.*, 2004; Hoiczyk *et al.*, 2000; Roggenkamp *et al.*, 2003).

NadA forms stable trimers on the bacterial surface, which mediate adhesion to and entry into epithelial cells (Capecchi *et al.*, 2005). Moreover a trimeric protein and a properly folded N-terminal 'head' domain are necessary to NadA-cell binding *in vitro*. Furthermore, a protein receptor molecule which is differentially expressed by different human epithelial cell lines seems to mediate the binding of the trimeric NadA (Capecchi *et al.*, 2005).

In order to preserve its functional organization and conformational structure, which could potentially also be implicated in an efficacious immune response, NadA was included as single trimeric soluble protein (NadA$_{\Delta 351-405}$, devoid of the membrane anchor domain) in the MenB vaccine. The presence of NadA seems to be more frequently associated with disease isolates than with carriage isolates. The gene is found in three out of the four known hypervirulent lineages of serogroup B and C strains, whereas is mostly absent from carrier strains and not found in *N. gonorrhoeae* and in the commensal species *N. lactamica* and *N. cinerea* (Comanducci *et al.*, 2002). NadA

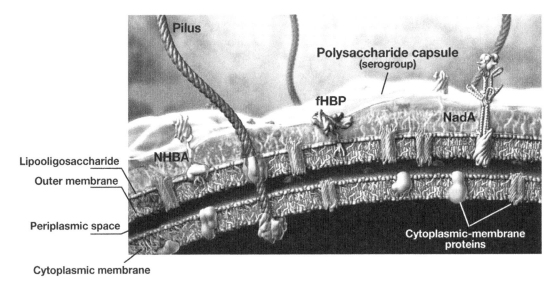

Figure 9.3 Schematic representation of the different bacterial compartments (outer membrane, periplasmic space, cytoplasmic membrane) and the main antigens identified through reverse vaccinology approach: GNA2132 (NHBA), fHBP and NadA. Other components of the meningococcal membranes are indicated (Pilus, polysaccharide capsule and lipooligosaccharide).

is expressed at different levels during growth, reaching the maximum in stationary phase. Expression of NadA is phase variable because of a tetranucleotide tract (TAAA) located upstream of the *nadA* promoter (Comanducci *et al.*, 2002; Martin *et al.*, 2005). Moreover, it has recently been shown that expression of NadA is controlled by a transcriptional regulator of the MarR family (Schielke *et al.*, 2009).

The gene clusters in three well-conserved alleles, whose overall identity ranges from 96% to 99%. Its low GC content suggests gene acquisition by horizontal gene transfer and subsequent limited evolution to generate the three well-conserved alleles. However, soluble NadA induces high level of bactericidal antibodies and the bactericidal activity is not influenced by allele diversity (Comanducci *et al.*, 2002). Interestingly, it is shown that NadA was able to induce bactericidal antibodies also when administered intranasally with mucosal adjuvants (Bowe *et al.*, 2004) or when expressed on the surface of the oral commensal bacterium *Streptococcus gordonii* (Ciabattini *et al.*, 2008). In addition, NadA is recognized by children convalescent sera suggesting that it is expressed and immunogenic *in vivo* (Litt *et al.*, 2004). Finally, NadA binds to and activates human monocyte-derived dendritic cells as well as monocytes/macrophages (Franzoso *et al.*, 2008; Mazzon *et al.*, 2007). Taken together these data support the hypothesis that NadA is an important vaccine immunogen involved in meningococcal colonization.

fHBP (GNA1870)

GNA1870 antigen (also named fHBP, factor H binding protein) is a well-studied membrane-anchored lipoprotein, which induces high levels of bactericidal antibodies and confers protection *in vivo* in the infant rat model. It is expressed by nearly all pathogenic isolates of *N. meningitidis* but the level of expression varies between strains, which have been therefore classified as high, intermediate and low expressors (Masignani *et al.*, 2003). The protein can be classified into three main distinct sequence variants: conservation within each variant ranges between 91.6 to 100%, while between the variants the conservation can be as low as 62.8%. This diversity has an

important impact on the immunological properties of GNA1870 as members of each variant induce a strong protective immune response against meningococcal strains carrying homologous alleles but are ineffective against strains that express distantly related variants (Masignani *et al.*, 2003).

Sequence analysis on a large panel of *N. meningitidis* clinical isolates showed that while the 120 N-terminal amino acids of the protein are well conserved, the last 154 C-terminal residues show significant diversity. Epitope mapping of the protein encoded by the pathogenic strain MC58 has shown that the 154 C-terminal residues contain most of the protective epitopes (Giuliani *et al.*, 2005; Welsch *et al.*, 2004). The solution structure of this C-terminal immunodominant domain has been determined by nuclear magnetic resonance (NMR) spectroscopy. This domain forms an eight-stranded β-barrel whose strands are connected by loops of variable lengths. The barrel is preceded by a short α-helix and by a flexible N-terminal tail. Sequence alignment of the three representative variants showed that variable residues were not exclusively located in correspondence of loops, but were spread on the entire surface of the β-barrel suggesting that it is fully exposed on the bacterium surface and accessible to the immune system (Cantini *et al.*, 2006). In addition, a number of studies using monoclonal antibodies have identified residues involved in protective epitopes (Beernink *et al.*, 2008; Giuliani *et al.*, 2005; Scarselli *et al.*, 2009; Welsch *et al.*, 2004). Recently, the structure of the full-length protein was determined by NMR, which improves the knowledge about the distribution of protective epitopes on the protein surface (Cantini *et al.*, 2009) (Fig. 9.3). The protein is composed of two independent barrels connected by a short link. Combining functional and structural data was possible to identify the location of the variable amino acids, the residues involved in binding protective monoclonal antibodies and the regions bound by antibodies against the three variants. The variable residues cluster in the upper part of the molecule. This is consistent with a model where the protein is anchored to the bacterial cell wall through the lipid moiety and exposes the upper part to the outside, where it is under the selective pressure of the immune system. Accordingly, the amino acids known to be part of epitopes are all localized in correspondence of the region of higher variability (Beernink *et al.*, 2008; Giuliani *et al.*, 2005; Scarselli *et al.*, 2009; Welsch *et al.*, 2004). The distribution of these residues on the full-length protein supports the hypothesis that the C-terminal domain of fHBP contains the major part of the epitopes. Therefore, the structure provides the basis for designing improved vaccine molecules.

It was recently discovered that GNA1870 binds specifically the human complement factor H (fH), a negative regulator of the alternative complement activation pathway and for this reason has been renamed fHBP, factor H binding protein (Madico *et al.*, 2006). Many pathogens have evolved the ability to evade killing by the innate immune system by recruiting host complement regulators (Lambris *et al.*, 2008) and same of them have adapted to avoid complement-mediated lysis by mimicking host surfaces, sequestering fH to their surface. By binding human factor H this protein allows the bacterium to escape the alternative pathway of complement activation and survive and grow in human blood and cause a devastating disease (Madico *et al.*, 2006; Seib *et al.*, 2009). Interestingly, fHBP binds only human fH and does not bind mouse, rat and lower primates fH (Granoff *et al.*, 2009) suggesting that *N. meningitidis* evolved to survive and growth only in human blood and explaining why this pathogen is strictly human specific and the lack of an animal model for meningococcus.

In a recent study the crystal structure of the fH in complex with fHBP has been solved (Schneider *et al.*, 2009). This reveals how *N. meningitidis* uses fHBP protein instead of charged-carbohydrate chemistry to recruit the host complement regulator fH. The structure also indicates the molecular basis of the host specificity of the interaction between fH and the meningococcus.

In addition, fHBP increases meningococcal survival in the presence of the antimicrobial peptide LL-37 most likely due to electrostatic interactions between fHBP and the cationic LL-37 molecule at the cell surface (Seib *et al.*, 2009). Since LL-37 is produced by cells that interact with *N. meningitidis* during infection, including the nasopharyngeal epithelia and phagocytic cells of the blood, it may be involved in innate host defences against meningococcal disease. Hence, the expression of fHBP by *N. meningitidis* strains is important for survival in human blood and human serum and in the presence of LL-37.

NHBA (GNA2132)

GNA2132 is a surface-exposed lipoprotein expressed by genetically diverse *N. meningitidis* strains (Fig. 9.3). The N-terminal region of GNA2132 is variable among different *Neisseria* strains, whereas the carboxyl-terminal region is highly conserved (Pizza *et al.*, 2000). No homologous proteins were found by searching non-redundant prokaryotic databases, suggesting that this protein is specific for *Neisseria* species.

Serum antibodies from mice immunized with recombinant GNA2132 are able to bind to the surface of diverse Nm strains and elicit complement-mediated bactericidal activity (Giuliani *et al.*, 2006; Pizza *et al.*, 2000). Moreover, anti-GNA2132 antibody elicited deposition of human C3b on the bacterial surface and passively protected infant rats against meningococcal bacteraemia after challenge with different *N. meningitidis* strains (Welsch *et al.*, 2003). GNA2132 is able to bind heparin and heparan sulphate (Serruto *et al.*, 2010). In order to indicate its functional binding properties we renamed the GNA2132 as Neisseria Heparin Binding Antigen (NHBA). Heparin binding is a common feature of several bacterial virulence factors and vaccine components (Rostand and Esko, 1997). The binding of heparin to bacteria has been reported to increase resistance to the bactericidal activity of normal human serum (Chen *et al.*, 1995). The mechanisms involved in increased serum resistance have not been well elucidated, but it has been suggested that heparin may function to recruit complement regulatory proteins that in turn act to prevent complement activation. Thus, the establishment of a NHBA–heparin complex on the meningococcus cell surface could lead to the heparin-mediated immobilization of complement regulatory proteins that in turn act to prevent complement activation. *In vivo*, NHBA is likely to bind glysoaminoglycans (such as heparan sulfate) that are present in mucosal secretion or on the surface of host cells.

Thus, while one of the five antigens present in MenB vaccine could improve the fitness of *N. meningitidis* contributing to colonization of the nasopharyngeal mucosa by strains that belong to the hypervirulent lineages, the other two might enhanced serum resistance by recruiting factor H or other complement regulatory proteins, increasing the ability of meningococcus to survive and multiply in human blood, which is a key factor in the development of fulminant meningococcal disease. These data suggest that antibodies to the MenB vaccine proteins could have two modes of action: (i) directly by activating classical complement pathway and (ii) indirectly by interfering with colonization of invasive strains and/or by preventing binding of fH on the bacterial surface, which increase the susceptibility of the bacteria to killing by the alternative pathway.

Conclusions

The quest for a universal vaccine against MenB has been a long and difficult one. Forty years of studies based on classical approaches to vaccine research failed to provide an efficacious solution. At the end of the last century the way in which biological research, clinical diagnostics and vaccine development was carried out was drastically redefined by the development of powerful tools such as genomic-based technologies, bioinformatics and proteomics. The availability of the genomic sequence of a pathogen provides the possibility to identify *in silico* vaccine candidates as well as important new virulence factors, irrespective of the protein abundance or cultivation of the bacteria. The MenB example shows the successful application of the reverse vaccinology in vaccine design and development. The use of this novel approach resulted in the identification of a large number of previously unidentified antigens and has radically changed the landscape of antigen discovery. More is known about *N. meningitidis* pathogen since the use of reverse vaccinology than in the previous forty years. Reverse vaccinology is now a routine approach that has subsequently been used in the development of vaccines against several bacterial pathogens including group B streptococcus, group A

streptococcus and pneumococcus (Serruto *et al.*, 2009).

References

Al-Mazrou, Y., Khalil, M., Borrow, R., Balmer, P., Bramwell, J., Lal, G., Andrews, N., and Al-Jeffri, M. (2005). Serologic responses to ACYW135 polysaccharide meningococcal vaccine in Saudi children under 5 years of age. Infect. Immun. 73, 2932–2939.

Bambini, S., Muzzi, A., Olcen, P., Rappuoli, R., Pizza, M., and Comanducci, M. (2009). Distribution and genetic variability of three vaccine components in a panel of strains representative of the diversity of serogroup B meningococcus. Vaccine 27, 2794–2803.

Beernink, P.T., Welsch, J.A., Bar-Lev, M., Koeberling, O., Comanducci, M., and Granoff, D.M. (2008). Fine antigenic specificity and cooperative bactericidal activity of monoclonal antibodies directed at the meningococcal vaccine candidate, factor H-binding protein. Infect. Immun. 76, 4232–4240.

Bliska, J.B., Copass, M.C., and Falkow, S. (1993). The *Yersinia pseudotuberculosis* adhesin YadA mediates intimate bacterial attachment to and entry into HEp-2 cells. Infect. Immun. 61, 3914–3921.

Borrow, R., Balmer, P., and Miller, E. (2005). Meningococcal surrogates of protection–serum bactericidal antibody activity. Vaccine 23, 2222–2227.

Borrow, R., Carlone, G.M., Rosenstein, N., Blake, M., Feavers, I., Martin, D., Zollinger, W., Robbins, J., Aaberge, I., Granoff, D.M., *et al.* (2006). *Neisseria meningitidis* group B correlates of protection and assay standardization–international meeting report Emory University, Atlanta, Georgia, United States, 16–17 March 2005. Vaccine 24, 5093–5107.

Boslego, J., Garcia, J., Cruz, C., Zollinger, W., Brandt, B., Ruiz, S., Martinez, M., Arthur, J., Underwood, P., Silva, W., *et al.* (1995). Efficacy, safety, and immunogenicity of a meningococcal group B (15:P1.3) outer membrane protein vaccine in Iquique, Chile. Chilean National Committee for Meningococcal Disease. Vaccine 13, 821–829.

Bowe, F., Lavelle, E.C., McNeela, E.A., Hale, C., Clare, S., Arico, B., Giuliani, M.M., Rae, A., Huett, A., Rappuoli, R., *et al.* (2004). Mucosal vaccination against serogroup B meningococci: induction of bactericidal antibodies and cellular immunity following intranasal immunization with NadA of *Neisseria meningitidis* and mutants of *Escherichia coli* heat-labile enterotoxin. Infect. Immun. 72, 4052–4060.

Cantini, F., Savino, S., Scarselli, M., Masignani, V., Pizza, M., Romagnoli, G., Swennen, E., Veggi, D., Banci, L., and Rappuoli, R. (2006). Solution structure of the immunodominant domain of protective antigen GNA1870 of *Neisseria meningitidis*. J. Biol. Chem. 281, 7220–7227.

Cantini, F., Veggi, D., Dragonetti, S., Savino, S., Scarselli, M., Romagnoli, G., Pizza, M., Banci, L., and Rappuoli, R. (2009). Solution structure of the factor H binding protein, a survival factor and protective antigen of *Neisseria meningitidis*. J. Biol. Chem. 284, 9022–9026.

Capecchi, B., Adu-Bobie, J., Di Marcello, F., Ciucchi, L., Masignani, V., Taddei, A., Rappuoli, R., Pizza, M., and Arico, B. (2005). *Neisseria meningitidis* NadA is a new invasin which promotes bacterial adhesion to and penetration into human epithelial cells. Mol. Microbiol. *55*, 687–698.

Carbonnelle, E., Hill, D.J., Morand, P., Griffiths, N.J., Bourdoulous, S., Murillo, I., Nassif, X., and Virji, M. (2009). Meningococcal interactions with the host. Vaccine *27* (Suppl. 2), 887–889.

Cartwright, K., Noah, N., and Peltola, H. (2001). Meningococcal disease in Europe: epidemiology, mortality, and prevention with conjugate vaccines. Report of a European advisory board meeting Vienna, Austria, 6–8 October, 2000. Vaccine *19*, 4347–4356.

Caugant, D.A., Bovre, K., Gaustad, P., Bryn, K., Holten, E., Hoiby, E.A., and Froholm, L.O. (1986a). Multilocus genotypes determined by enzyme electrophoresis of *Neisseria meningitidis* isolated from patients with systemic disease and from healthy carriers. J. Gen. Microbiol. *132*, 641–652.

Caugant, D.A., Froholm, L.O., Bovre, K., Holten, E., Frasch, C.E., Mocca, L.F., Zollinger, W.D., and Selander, R.K. (1986b). Intercontinental spread of a genetically distinctive complex of clones of *Neisseria meningitidis* causing epidemic disease. Proc. Natl. Acad. Sci. U.S.A. *83*, 4927–4931.

Caugant, D.A., and Maiden, M.C. (2009). Meningococcal carriage and disease – Population biology and evolution. Vaccine *27*, B64–B70.

Caugant, D.A., Tzanakaki, G., and Kriz, P. (2007). Lessons from meningococcal carriage studies. FEMS Microbiol. Rev. *31*, 52–63.

Ciabattini, A., Giomarelli, B., Parigi, R., Chiavolini, D., Pettini, E., Arico, B., Giuliani, M.M., Santini, L., Medaglini, D., and Pozzi, G. (2008). Intranasal immunization of mice with recombinant *Streptococcus gordonii* expressing NadA of *Neisseria meningitidis* induces systemic bactericidal antibodies and local IgA. Vaccine *26*, 4244–4250.

Comanducci, M., Bambini, S., Brunelli, B., Adu-Bobie, J., Arico, B., Capecchi, B., Giuliani, M.M., Masignani, V., Santini, L., Savino, S., *et al.* (2002). NadA, a novel vaccine candidate of *Neisseria meningitidis*. J. Exp. Med. *195*, 1445–1454.

Costantino, P., Viti, S., Podda, A., Velmonte, M.A., Nencioni, L., and Rappuoli, R. (1992). Development and phase 1 clinical testing of a conjugate vaccine against meningococcus A and C. Vaccine *10*, 691–698.

Cotter, S.E., Surana, N.K., and St Geme, J.W., 3rd (2005). Trimeric autotransporters: a distinct subfamily of autotransporter proteins. Trends Microbiol. *13*, 199–205.

Desvaux, M., Parham, N.J., and Henderson, I.R. (2004). The autotransporter secretion system. Res. Microbiol. *155*, 53–60.

Finne, J., Bitter-Suermann, D., Goridis, C., and Finne, U. (1987). An IgG monoclonal antibody to group B meningococci cross-reacts with developmentally regulated polysialic acid units of glycoproteins in neural and extraneural tissues. J. Immunol. *138*, 4402–4407.

Franzoso, S., Mazzon, C., Sztukowska, M., Cecchini, P., Kasic, T., Capecchi, B., Tavano, R., and Papini, E. (2008). Human monocytes/macrophages are a target of *Neisseria meningitidis* Adhesin A (NadA). J. Leukoc. Biol. *83*, 1100–1110.

Fredriksen, J.H., Rosenqvist, E., Wedege, E., Bryn, K., Bjune, G., Froholm, L.O., Lindbak, A.K., Mogster, B., Namork, E., Rye, U., *et al.* (1991). Production, characterization and control of MenB-vaccine 'Folkehelsa': an outer membrane vesicle vaccine against group B meningococcal disease. NIPH Ann. *14*, 67–79; discussion 79–80.

Geoffroy, M.C., Floquet, S., Metais, A., Nassif, X., and Pelicic, V. (2003). Large-scale analysis of the meningococcus genome by gene disruption: resistance to complement-mediated lysis. Genome Res. *13*, 391–398.

Giuliani, M.M., Adu-Bobie, J., Comanducci, M., Arico, B., Savino, S., Santini, L., Brunelli, B., Bambini, S., Biolchi, A., Capecchi, B., *et al.* (2006). A universal vaccine for serogroup B meningococcus. Proc. Natl. Acad. Sci. U.S.A. *103*, 10834–10839.

Giuliani, M.M., Santini, L., Brunelli, B., Biolchi, A., Arico, B., Di Marcello, F., Cartocci, E., Comanducci, M., Masignani, V., Lozzi, L., *et al.* (2005). The region comprising amino acids 100 to 255 of *Neisseria meningitidis* lipoprotein GNA 1870 elicits bactericidal antibodies. Infect. Immun. *73*, 1151–1160.

Goldschneider, I., Gotschlich, E.C., and Artenstein, M.S. (1969). Human immunity to the meningococcus. I. The role of humoral antibodies. J. Exp. Med. *129*, 1307–1326.

Gotschlich, E.C., Goldschneider, I., and Artenstein, M.S. (1969). Human immunity to the meningococcus. IV. Immunogenicity of group A and group C meningococcal polysaccharides in human volunteers. J. Exp. Med. *129*, 1367–1384.

Granoff, D., Harrison, L.H., and Borrow, R. (2008). Meningococcal vaccines. In Vaccines, 5th edn, Plotkin, S.A., Offit, P.A., and Orenstein, W.A., eds (Philadelphia, Saunders).

Granoff, D.M., Welsch, J.A., and Ram, S. (2009). Binding of complement factor H (fH) to *Neisseria meningitidis* is specific for human fH and inhibits complement activation by rat and rabbit sera. Infect. Immun. *77*, 764–769.

Hauck, C.R., and Meyer, T.F. (2003). 'Small' talk: Opa proteins as mediators of Neisseria-host-cell communication. Curr. Opin. Microbiol. *6*, 43–49.

Hill, D.J., and Virji, M. (2003). A novel cell-binding mechanism of Moraxella catarrhalis ubiquitous surface protein UspA: specific targeting of the N-domain of carcinoembryonic antigen-related cell adhesion molecules by UspA1. Mol. Microbiol. *48*, 117–129.

Hoiczyk, E., Roggenkamp, A., Reichenbecher, M., Lupas, A., and Heesemann, J. (2000). Structure and sequence analysis of *Yersinia* YadA and *Moraxella* UspAs reveal a novel class of adhesins. EMBO J. *19*, 5989–5999.

Hols, P., Hancy, F., Fontaine, L., Grossiord, B., Prozzi, D., Leblond-Bourget, N., Decaris, B., Bolotin, A., Delorme, C., Dusko Ehrlich, S., *et al.* (2005). New insights in the molecular biology and physiology of *Streptococcus thermophilus* revealed by comparative genomics. FEMS Microbiol. Rev. *29*, 435–463.

Iriarte, M., and Cornelis, G.R. (1996). Molecular determinants of *Yersinia* pathogenesis. Microbiologia *12*, 267–280.

Lafontaine, E.R., Cope, L.D., Aebi, C., Latimer, J.L., McCracken, G.H., Jr., and Hansen, E.J. (2000). The UspA1 protein and a second type of UspA2 protein mediate adherence of *Moraxella catarrhalis* to human epithelial cells *in vitro*. J. Bacteriol. *182*, 1364–1373.

Lambris, J.D., Ricklin, D., and Geisbrecht, B.V. (2008). Complement evasion by human pathogens. Nat. Rev. Microbiol. *6*, 132–142.

Linke, D., Riess, T., Autenrieth, I.B., Lupas, A., and Kempf, V.A. (2006). Trimeric autotransporter adhesins: variable structure, common function. Trends Microbiol. *14*, 264–270.

Litt, D.J., Savino, S., Beddek, A., Comanducci, M., Sandiford, C., Stevens, J., Levin, M., Ison, C., Pizza, M., Rappuoli, R., *et al.* (2004). Putative vaccine antigens from *Neisseria meningitidis* recognized by serum antibodies of young children convalescing after meningococcal disease. J. Infect. Dis. *190*, 1488–1497.

Mackinnon, F.G., Borrow, R., Gorringe, A.R., Fox, A.J., Jones, D.M., and Robinson, A. (1993). Demonstration of lipooligosaccharide immunotype and capsule as virulence factors for *Neisseria meningitidis* using an infant mouse intranasal infection model. Microb. Pathog. *15*, 359–366.

Madico, G., Welsch, J.A., Lewis, L.A., McNaughton, A., Perlman, D.H., Costello, C.E., Ngampasutadol, J., Vogel, U., Granoff, D.M., and Ram, S. (2006). The meningococcal vaccine candidate GNA1870 binds the complement regulatory protein factor H and enhances serum resistance. J. Immunol. *177*, 501–510.

Maiden, M.C., Bygraves, J.A., Feil, E., Morelli, G., Russell, J.E., Urwin, R., Zhang, Q., Zhou, J., Zurth, K., Caugant, D.A., *et al.* (1998). Multilocus sequence typing: a portable approach to the identification of clones within populations of pathogenic microorganisms. Proc. Natl. Acad. Sci. U.S.A. *95*, 3140–3145.

Maiden, M.C., and Feavers, I.M. (2000). Meningococcal genomics: two steps forward, one step back. Nat. Med. *6*, 1215–1216.

Martin, P., Makepeace, K., Hill, S.A., Hood, D.W., and Moxon, E.R. (2005). Microsatellite instability regulates transcription factor binding and gene expression. Proc. Natl. Acad. Sci. U.S.A. *102*, 3800–3804.

Masignani, V., Comanducci, M., Giuliani, M.M., Dambini, S., Adu-Bobie, J., Arico, B., Brunelli, B., Pieri, A., Santini, L., Savino, S., *et al.* (2003). Vaccination against *Neisseria meningitidis* using three variants of the lipoprotein GNA1870. J. Exp. Med. *197*, 789–799.

Mazzon, C., Baldani-Guerra, B., Cecchini, P., Kasic, T., Viola, A., de Bernard, M., Arico, B., Gerosa, F., and Papini, E. (2007). IFN-{gamma} and R-848 Dependent Activation of Human Monocyte-Derived Dendritic Cells by *Neisseria meningitidis* Adhesin A. J. Immunol. *179*, 3904–3916.

Merz, A.J., and So, M. (2000). Interactions of pathogenic neisseriae with epithelial cell membranes. Annu. Rev. Cell Dev. Biol. *16*, 423–457.

Miller, E., Pollard, A.J., Borrow, R., Findlow, J., Dawson, T., Morant, A., John, T., Snape, M., Southern, J., Morris, R., *et al.* (2008). Safety and immunogenicity of Novartis meningococcal serogroup B baccine (MenB vaccine) after three doses administered in infancy. Paper presented at: 26th Annual Meeting of the European Society for Paediatric Infectious Diseases (Graz, Austria).

Morley, S.L., and Pollard, A.J. (2001). Vaccine prevention of meningococcal disease, coming soon? Vaccine *20*, 666–687.

Nassif, X., Beretti, J.L., Lowy, J., Stenberg, P., O'Gaora, P., Pfeifer, J., Normark, S., and So, M. (1994). Roles of pilin and PilC in adhesion of *Neisseria meningitidis* to human epithelial and endothelial cells. Proc. Natl. Acad. Sci. U.S.A. *91*, 3769–3773.

Nedelec, J., Boucraut, J., Garnier, J.M., Bernard, D., and Rougon, G. (1990). Evidence for autoimmune antibodies directed against embryonic neural cell adhesion molecules (N-CAM) in patients with group B meningitis. J. Neuroimmunol. *29*, 49–56.

Noah, N., and Henderson, B. (2002). Surveillance of Bacterial Meningitis in Europe 1997/1998 (Communicable Disease Surveillance Centre, London), pp. 1–16.

Oster, P., Lennon, D., O'Hallahan, J., Mulholland, K., Reid, S., and Martin, D. (2005). MeNZB: a safe and highly immunogenic tailor-made vaccine against the New Zealand *Neisseria meningitidis* serogroup B disease epidemic strain. Vaccine *23*, 2191–2196.

Perrett, K.P., and Pollard, A.J. (2005). Towards an improved serogroup B *Neisseria meningitidis* vaccine. Expert Opin. Biol. Ther. *5*, 1611–1625.

Pizza, M., Scarlato, V., Masignani, V., Giuliani, M.M., Arico, B., Comanducci, M., Jennings, G.T., Baldi, L., Bartolini, E., Capecchi, B., *et al.* (2000). Identification of vaccine candidates against serogroup B meningococcus by whole-genome sequencing. Science *287*, 1816–1820.

Pujol, C., Eugene, E., Marceau, M., and Nassif, X. (1999). The meningococcal PilT protein is required for induction of intimate attachment to epithelial cells following pilus-mediated adhesion. Proc. Natl. Acad. Sci. U.S.A. *96*, 4017–4022.

Rappuoli, R. (2001). Reverse vaccinology, a genome-based approach to vaccine development. Vaccine *19*, 2688–2691.

Rappuoli, R. (2008). The application of reverse vaccinology, Novartis MenB vaccine developed by design. Paper presented at: 16th International Pathogenic Neisseria Conference, 7–12 September 2008 (Rotterdam, the Netherlands).

Roggenkamp, A., Ackermann, N., Jacobi, C.A., Truelzsch, K., Hoffmann, H., and Heesemann, J. (2003). Molecular analysis of transport and oligomerization of the *Yersinia enterocolitica* adhesin YadA. J. Bacteriol. *185*, 3735–3744.

Rosenstein, N.E., Perkins, B.A., Stephens, D.S., Lefkowitz, L., Cartter, M.L., Danila, R., Cieslak, P., Shutt, K.A., Popovic, T., Schuchat, A., et al. (1999). The changing epidemiology of meningococcal disease in the United States, 1992–1996. J. Infect. Dis. *180*, 1894–1901.

Rostand, K.S., and Esko, J.D. (1997). Microbial adherence to and invasion through proteoglycans. Infect. Immun. *65*, 1–8.

Scarselli, M., Cantini, F., Santini, L., Veggi, D., Dragonetti, S., Donati, C., Savino, S., Giuliani, M.M., Comanducci, M., Di Marcello, F., et al. (2009). Epitope mapping of a bactericidal monoclonal antibody against the factor H binding protein of *Neisseria meningitidis*. J. Mol. Biol. *386*, 97–108.

Scarselli, M., Rappuoli, R., and Scarlato, V. (2001). A common conserved amino acid motif module shared by bacterial and intercellular adhesins: bacterial adherence mimicking cell cell recognition? Microbiology *147*, 250–252.

Scarselli, M., Serruto, D., Montanari, P., Capecchi, B., Adu-Bobie, J., Veggi, D., Rappuoli, R., Pizza, M., and Arico, B. (2006). *Neisseria meningitidis* NhhA is a multifunctional trimeric autotransporter adhesin. Mol. Microbiol. *61*, 631–644.

Schielke, S., Huebner, C., Spatz, C., Nagele, V., Ackermann, N., Frosch, M., Kurzai, O., and Schubert-Unkmeir, A. (2009). Expression of the meningococcal adhesin NadA is controlled by a transcriptional regulator of the MarR family. Mol. Microbiol. *72*, 1054–1067.

Schmitt, C., Turner, D., Boesl, M., Abele, M., Frosch, M., and Kurzai, O. (2007). A functional two-partner secretion system contributes to adhesion of *Neisseria meningitidis* to epithelial cells. J. Bacteriol. *189*, 7968–7976.

Schneider, M.C., Prosser, B.E., Caesar, J.J., Kugelberg, E., Li, S., Zhang, Q., Quoraishi, S., Lovett, J.E., Deane, J.E., Sim, R.B., et al. (2009). *Neisseria meningitidis* recruits factor H using protein mimicry of host carbohydrates. Nature *458*, 890–893.

Seib, K.L., Serruto, D., Oriente, F., Delany, I., Adu-Bobie, J., Veggi, D., Arico, B., Rappuoli, R., and Pizza, M. (2009). Factor H-binding protein is important for meningococcal survival in human whole blood and serum and in the presence of the antimicrobial peptide LL-37. Infect. Immun. *77*, 292–299.

Serruto, D., Adu-Bobie, J., Scarselli, M., Veggi, D., Pizza, M., Rappuoli, R., and Arico, B. (2003). *Neisseria meningitidis* App, a new adhesin with autocatalytic serine protease activity. Mol. Microbiol. *48*, 323–334.

Serruto, D., Serino, L., Masignani, V., and Pizza, M. (2009). Genome-based approaches to develop vaccines against bacterial pathogens. Vaccine *27*, 3245–3250.

Serruto, D., Spadafina, T., Ciucchi, L., Lewis, L.A., Ram, S., Tontini, M., Santini, L., Biolchi, A., Seib, K.L., Giuliani, M.M., et al. *Neisseria meningitidis* GNA2132, a heparin-binding protein that induces protective immunity in humans. Proc. Natl. Acad. Sci. U.S.A. *107*, 3770–3775.

Sierra, G.V., Campa, H.C., Varcacel, N.M., Garcia, I.L., Izquierdo, P.L., Sotolongo, P.F., Casanueva, G.V., Rico, C.O., Rodriguez, C.R., and Terry, M.H. (1991). Vaccine against group B *Neisseria meningitidis*: protection trial and mass vaccination results in Cuba. NIPH Ann. *14*, 195–207; discussion 208–110.

Skurnik, M., el Tahir, Y., Saarinen, M., Jalkanen, S., and Toivanen, P. (1994). YadA mediates specific binding of enteropathogenic *Yersinia enterocolitica* to human intestinal submucosa. Infect. Immun. *62*, 1252–1261.

Spinosa, M.R., Progida, C., Tala, A., Cogli, L., Alifano, P., and Bucci, C. (2007). The *Neisseria meningitidis* capsule is important for intracellular survival in human cells. Infect. Immun. *75*, 3594–3603.

Stephens, D.S. (2007). Conquering the meningococcus. FEMS Microbiol. Rev. *31*, 3–14.

Stephens, D.S., Greenwood, B., and Brandtzaeg, P. (2007). Epidemic meningitis, meningococcaemia, and *Neisseria meningitidis*. Lancet *369*, 2196–2210.

Surana, N.K., Cutter, D., Barenkamp, S.J., and St. Geme, J.W., I.I.I. (2004). The *Haemophilus influenzae* Hia autotransporter contains an unusually short trimeric translocator domain. J. Biol. Chem. *279*, 14679–14685.

Tala, A., Progida, C., De Stefano, M., Cogli, L., Spinosa, M.R., Bucci, C., and Alifano, P. (2008). The HrpB-HrpA two-partner secretion system is essential for intracellular survival of *Neisseria meningitidis*. Cell Microbiol. *10*, 2461–2482.

Tettelin, H., Saunders, N.J., Heidelberg, J., Jeffries, A.C., Nelson, K.E., Eisen, J.A., Ketchum, K.A., Hood, D.W., Peden, J.F., Dodson, R.J., et al. (2000). Complete genome sequence of *Neisseria meningitidis* serogroup B strain MC58. Science *287*, 1809–1815.

Turner, D.P., Marietou, A.G., Johnston, L., Ho, K.K., Rogers, A.J., Wooldridge, K.G., and Ala'Aldeen, D.A. (2006). Characterization of MspA, an immunogenic autotransporter protein that mediates adhesion to epithelial and endothelial cells in *Neisseria meningitidis*. Infect. Immun. *74*, 2957–2964.

Uli, L., Castellanos-Serra, L., Betancourt, L., Dominguez, F., Barbera, R., Sotolongo, F., Guillen, G., and Pajon Feyt, R. (2006). Outer membrane vesicles of the VA-MENGOC-BC vaccine against serogroup B of *Neisseria meningitidis*: Analysis of protein components by two-dimensional gel electrophoresis and mass spectrometry. Proteomics *6*, 3389–3399.

van Ulsen, P., Adler, B., Fassler, P., Gilbert, M., van Schilfgaarde, M., van der Ley, P., van Alphen, L., and Tommassen, J. (2006). A novel phase-variable autotransporter serine protease, AusI, of *Neisseria meningitidis*. Microbes Infect. *8*, 2088–2097.

Vipond, C., Suker, J., Jones, C., Tang, C., Feavers, I.M., and Wheeler, J.X. (2006). Proteomic analysis of a meningococcal outer membrane vesicle vaccine prepared from the group B strain NZ98/254. Proteomics *6*, 3400–3413.

Virji, M. (2009). Pathogenic neisseriae: surface modulation, pathogenesis and infection control. Nat. Rev. Microbiol. *7*, 274–286.

Virji, M., Kayhty, H., Ferguson, D.J., Alexandrescu, C., Heckels, J.E., and Moxon, E.R. (1991). The role of pili in the interactions of pathogenic *Neisseria* with cultured human endothelial cells. Mol. Microbiol. *5*, 1831–1841.

Virji, M., Makepeace, K., Ferguson, D.J., Achtman, M., and Moxon, E.R. (1993). Meningococcal Opa and Opc proteins: their role in colonization and invasion of human epithelial and endothelial cells. Mol. Microbiol. *10*, 499–510.

Virji, M., Makepeace, K., Ferguson, D.J., Achtman, M., Sarkari, J., and Moxon, E.R. (1992). Expression of the Opc protein correlates with invasion of epithelial and endothelial cells by *Neisseria meningitidis*. Mol. Microbiol. *6*, 2785–2795.

Welsch, J.A., Moe, G.R., Rossi, R., Adu-Bobie, J., Rappuoli, R., and Granoff, D.M. (2003). Antibody to genome-derived neisserial antigen 2132, a *Neisseria meningitidis* candidate vaccine, confers protection against bacteremia in the absence of complement-mediated bactericidal activity. J. Infect. Dis. *188*, 1730–1740.

Welsch, J.A., Rossi, R., Comanducci, M., and Granoff, D.M. (2004). Protective activity of monoclonal antibodies to genome-derived neisserial antigen 1870, a *Neisseria meningitidis* candidate vaccine. J. Immunol. *172*, 5606–5615.

Vaccines for Neglected Diseases

10

Allan Saul

Abstract

Infectious diseases exert a major burden of disease in developing countries. While better use of existing vaccines would make an appreciable difference, the greatest burden is caused by diseases for which we currently have no vaccines. The picture, especially in children, is dominated by diarrhoeal and respiratory diseases. Paradoxically diseases have relatively low priority for funding in absolute terms, and especially in relationship to the burden of disease. Thus, new vaccines for these neglected diseases need both innovative scientific solutions and innovative development schemes involving scientific institutes, public financing and industrial input. The industrial input is critical: not only will vaccine manufacture require an industrial partner, but the knowledge to efficiently undertake the technical and clinical development leading to vaccine production largely resides in industry. A potentially important development in this area has been the recent formation of Industry Linked Vaccine Institutes: for example, the Novartis Vaccines Institute for Global Health and the Hilleman Laboratories. These are an important conduit for applying industrial know how for developing commercial vaccines to the pressing need for vaccines for neglected diseases of developing countries.

Burden of infectious disease

Despite the extensive history of development of vaccines, antibiotics and antiviral drugs, infectious diseases continue to have a major impact on the global burden of disease. In 2008, the World Health Organization issued an update of 'The Global Burden of Disease' (WHO, 2008a). This contains sobering estimates of the mortality and morbidity classified according to country and disease type. Within the broad WHO classifications used in this report, infectious diseases[1] account for a high proportion of the estimated global burden of disease as measured by disability-adjusted life years (DALYs) (26%) or by death rate (23%).

As this report indicates, the bulk of the world's morbidity and mortality from infectious disease lies in the low- and medium-income countries (99.0% of DALYs and 96.5% deaths). Although not surprising, unfortunately, not only are the absolute numbers much higher in low and middle income countries, but for infectious diseases the crude (non-age adjusted) DALYs and death rate are also much higher with 17.5 times as many DALYs and five times the death rate from infectious diseases in low and middle incomes as compared with high-income countries. The data in this report suggest a markedly differential impact of infectious diseases in these low- and middle-income countries compared with non-infectious disease (1.1 times the DALYs and 0.9 times the

---

[1] In this review, the WHO categories for 'Infectious and parasitic diseases' and 'Respiratory infections' have been combined to give the totals for infectious diseases.

death rates from non-infectious diseases occur in low- and middle-income countries compared with high-income countries).

Again sad, but not surprising, children < 5 years old bear a highly disproportionate burden of the morbidity and mortality associated with 'infectious diseases'. Although under-5s constitute only 9.7% of the global population, 53% and 43% of the total global infectious disease DALYs and death rate, respectively, occur in this age group.

However, a closer inspection of the causes of morbidity and deaths due to infectious disease reveals a number of surprises. Shown in Fig. 10.1A and B are the DALYs estimated by WHO for 2004 globally for all ages and for children < 5 years old. Since 99% of the DALYs are in low and middle income populations, deleting the DALYs in high income countries makes no significant difference to these figures. Not shown are the death rates attributed to different diseases. However, since most of the DALYs are due to loss of years

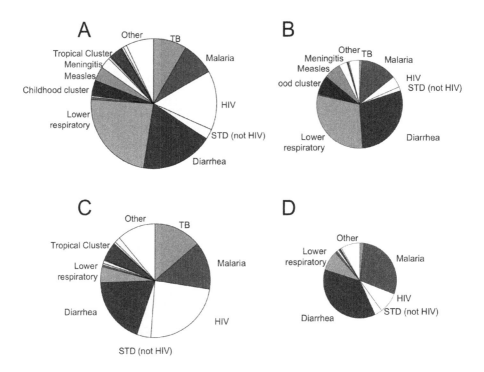

Figure 10.1 WHO estimates of DALYs due to infectious disease in 2004 for adults (A) and in children <5 years old (B). The estimates of DALY remaining after estimated DALYs due to diseases for which vaccines are available, or potentially available, have been deducted for adults (C) and children <5 years old (D). Area of each circle is proportional to the estimated total DALYs for each graph (A, 400 million; B, 214 million; C, 248 million; D, 100 million). All of the WHO report categories are plotted, but on this scale, several are not labelled or not visible. These are: upper respiratory and otitis media (between lower respiratory and childhood cluster); hepatitis B and hepatitis C (between meningitis and tropical cluster); leprosy, dengue, Japanese encephalitis, trachoma and intestinal nematodes (between tropical cluster and other). For these graphs, childhood cluster consists of pertussis, polio, diphtheria and tetanus. Most of the DALYs are due to pertussis and tetanus in children (9.9 million and 4.8 million, respectively). The WHO category includes measles, but measles has been plotted separately on these diagrams. Tropical cluster consists of trypanosomiasis, Chagas' disease, schistosomiasis, leishmaniasis, lymphatic filariasis and onchocerciasis. Intestinal nematodes consist of ascariasis, trichuriasis, hookworm disease and other intestinal infections.

of life rather than disability, the plots are qualitatively very similar. Nearly half of all deaths and disease is due to diarrhoea and lower respiratory infection. In young children, the proportion is even higher, with either lower respiratory disease or diarrhoea alone accounting for more than TB, malaria and HIV combined.

In considering the impact of infectious diseases, it is important to note that these WHO estimates only include the direct impact of the disease. For example, they do not include DALYs or deaths due to rheumatic heart disease as a consequence of infectious with group A streptococcus (estimated 233,000 deaths per year; Carapetis *et al.*, 2005), duodenal and gastric ulcers and gastric cancer due to *Helicobacter* (Mandeville *et al.*, 2009), cervical cancer due to papillomavirus (275,000 deaths per year; Parkin *et al.*, 2005) or hepatomas and liver cirrhosis due to hepatitis (estimated 600,000 deaths per year; WHO, 2008b).

In addition to these direct effects of infection, there is a growing awareness that many of the infectious diseases of developing countries have more wide- reaching consequences. For example, there is a growing body of evidence that repeated enteric infections, especially in areas with marginal nutrition, can have substantial impact on stunting and on retarding mental development (Petri, Jr. *et al.*, 2008; Oria *et al.*, 2009; Checkley *et al.*, 2008).

A tragic level of disease and deaths occurs from diseases for which there are effective vaccines, for example, from measles and other diseases in the 'childhood cluster', primarily pertussis and tetanus. WHO estimates in 2002 that about 1.4 million young children died from vaccine preventable diseases (WHO, 2008b) for which vaccines where then available. In addition another 1.1 million died from diseases for which licensed vaccines are available but were not available for use in developing countries in 2002 when these data were compiled. These include 716,000 from pneumococcus and 402,000 deaths from rotavirus (about 20% of deaths from diarrhoea) (WHO, 2008b). More recent figures suggest that the rotavirus burden is higher with 40% of hospital admissions due to diarrhoea and 611,000 deaths annually (Parashar *et al.*, 2006).

Need for new vaccines

Although better use of existing vaccines would have a major impact on the global burden of infectious diseases, there remains a high burden of diseases for which there are no current vaccines. Taking into account estimates of the disease burden due to childhood disease cluster, measles, and the proportion of pneumonia attributable to *Streptococcus pneumoniae*, *Haemophilus influenzae* B (Rudan *et al.*, 2008b) and meningitis attributable to these bacteria and *Neisseria meningitidis*; and the proportion of diarrhoeal disease due to Rotavirus (Parashar *et al.*, 2006) and cholera (Sack *et al.*, 2006) gives an estimate of the magnitude of the burden of diseases for which we have either no current vaccines or for which there are no new vaccine variants in advanced development that target the diseases found in the bulk of the world's at-risk populations (Fig. 10.1C and D). Together these diseases account for about 80% of the total population and about 70% of the <5 year old DALYs lost to infectious disease.

These figures highlight a number of areas for vaccine development: we have no vaccines for any sexually transmitted disease other than human papillomavirus (Starnbach and Roan, 2008), no vaccine for any of the parasitic diseases in the 'tropical disease cluster' nor any vaccine for intestinal worm infections (Hotez *et al.*, 2007).

In the case of respiratory disease, increasing deployment of Hib vaccines (Gessner, 2009) and pneumococcal vaccines, especially newer vaccines that cover the serotypes found in the developing countries where the majority of the population at risk live, should lead to a major reduction in disease (Scott, 2007) although even the 13 valent vaccine which has just (July 2009) been licensed in Chile (Wyeth, 2009) still leaves major gaps in the coverage of serotypes in some countries such as Bangladesh (Saha *et al.*, 2009).

The threat of H5N1 avian influenza, and more recently H1N1 swine flu has received considerable attention. However, the impact of seasonal influenza is unclear, although a recent study in Bangladesh highlighted both the impact of malaria in infants and the efficacy of immunization (in this case maternal immunization) (Zaman *et al.*, 2008) and raise the prospects that seasonal influenza is a more important cause of respiratory infections in developing countries than is widely appreciated.

Despite these caveats, there are real prospects that in the foreseeable future respiratory disease, especially in children in developing countries will see a substantial decline. There is however one major lower respiratory infection for which there is currently no vaccines: respiratory syncytial virus. This is a leading cause of morbidity in <5 year old children in developing countries being identified in 15–40% of pneumonia or bronchiolitis cases admitted to hospital (Rudan *et al.*, 2008a).

Although these diseases are individually important, the pattern of needed vaccines is dominated by four diseases in the all age DALY projections: HIV, diarrhoeal diseases, malaria and TB. In young children, just two disease categories, diarrhoeal disease and malaria account for about 70% of the unmet need.

Priorities for new vaccines and the global response

These figures show the need for new treatments. However, there are multiple other factors that impact on the priority for developing new vaccines for these diseases. These include the scientific probability that a safe, efficacious vaccine could be developed; the time taken for development; the likelihood that if an effective vaccine could be produced it would be widely deployed. The latter includes multiple components: the cost of the vaccine and its delivery, the perception of the disease importance; market schemes for promoting/subsidizing the vaccine; and the availability of other treatments for the disease in question at the time a vaccine could be launched.

In practice, such assessment has and is being made by national and transnational, philanthropic and industrial funding decisions. In theory, priorities could be assigned by systematically evaluating these parameters for each potential vaccine. Comparison of the two approaches highlights opportunities and gaps in the global response.

Recently, a study was undertaken by the George Institute for International Health documented investment in research and development of new therapies and prophylaxis for neglected diseases of developing countries, the G-FINDER

report (Moran and Guzman, 2009; Moran *et al.*, 2009). In this study 30 diseases were considered and estimates made of the expenditure on research and development specifically applicable to the developing world along with estimates taken from WHO figures and elsewhere of the burden of disease. As these authors comment

> Intuitively, there is a sense that the highest 'health return on investment' would result from investing in the highest burden diseases, as measured by DALYs (Disability Adjusted Life Years)…This cost/benefit ratio must then be discounted for risk, which will chiefly depend on the state of science and technology in the area of investment under consideration.

The G-FINDER study considered all expenditure including drugs, better diagnostics etc. In considering the case for new vaccines, in addition to the discounted cost/benefit ratio, the current availability of other treatments (e.g. drugs) or their likely future availability in the time taken to develop and deploy a vaccine needs to also be factored into the priority scheme.

From G-FINDER study, the expenditure/benefit (as $per DALY) can be calculated for the set of 'neglected diseases' for which vaccine R&D expenditure was documented (Table 10.1).

In this study there are three disease categories which by the cost/benefit ratio for vaccines are low: diarrhoeal disease, helminth infections and bacterial pneumonia and meningitis at $0.23, $0.01 and $0.26 per DALY, respectively, despite resulting in appreciable DALYs.

Because of the way in which the data were selected, for some diseases, these figures are only a portion of the total expenditure on vaccines. For example, for HIV, tuberculosis, pneumonia and meningitis, these figures do not include the costs of vaccine developments for developed country markets and so the total expenditure per DALY is substantially greater. However, for other diseases, e.g. diarrhoeal disease, as there is a negligible market for these vaccines in developed countries, the quoted expenditures represent close to total expenditure. Undoubtedly, as useful as this study is, it probably has not captured all of the expenditure on these vaccines. Indeed the report points out that expenditure by several vaccine manufacturers was not available.

The funding situation for diarrhoeal diseases is particularly pressing. As reviewed above, diarrhoeal diseases represent a major global health burden overall, and particularly in children, for which vaccines could make a major impact. So why is this area so far behind in the development and funding of vaccines? Some of the answers lie in a closer examination of the case for vaccines directed to the pathogens that cause the greatest burden of disease.

New vaccines for neglected diseases

Despite the constraints with funding for new vaccines for neglected diseases, there is significant progress towards identifying and testing new vaccines. There are many excellent recent reviews detailing of progress towards new vaccines for neglected diseases and the details of current vaccines research will not be addressed in this chapter.

Diarrhoeal diseases have been extensively reviewed for the field as a whole (Girard *et al.*, 2006; Nataro and Barry, 2008; Levine, 2006) as well as for individual vaccines: ETEC (Walker *et al.*, 2007), *Shigella* (Levine *et al.*, 2007; Kweon, 2008; Phalipon *et al.*, 2008), *Campylobacter* (Tribble *et al.*, 2004; Monteiro *et al.*, 2009), noroviruses (LoBue *et al.*, 2009).

As detailed in these reviews and elsewhere, there have been major advances in our understanding of the biology of these infectious agents. For many, the genome of the organisms have recently been sequenced, including complex organisms such as *Schistosoma* (Berriman *et al.*, 2009), and parasitic protozoa such as trypanosomes and *Leishmania* (El-Sayed *et al.*, 2005). For bacterial and viral pathogens, sequencing technology has advanced to allow comparison of multiple genomes as part of a directed approach to address antigenic diversity (Holt *et al.*, 2008). The power of these modern post genomic approaches are illustrated in Chapter 2 (Bagnoli *et al.*).

As important as these advances are, having good research does not make a vaccine. There are two critical processes that form a major rate limiting process for the development of vaccines for neglected diseases:

1 the translational research that takes a vaccine from the laboratory concept through proof of concept in humans;
2 the registration of a vaccine, its manufacture and introduction into public health programmes and the private market.

For the vaccines developed in the past 30 years, both of these steps have been undertaken by industry, almost always in western countries, and one of the critical drivers required to develop these vaccines is the expectation that these products will contribute to a profitable business. The profit motive has been important. It is estimated that new vaccines cost US\$231 to US821 million each (Gordon Douglas *et al.*, 2008) and not only does this investment have to be recovered for the particular vaccine under consideration and provide a sustainable income, the return has to cover the costs of the development of vaccines that do not make it to market. Western markets have supply the return on investment that makes this process feasible.

For many of the infectious diseases for which we urgently need vaccines for developing countries it is unlikely that this model of vaccine development will provide sufficient return to encourage pure private sector investment. Thus, to develop new vaccines for neglected diseases, not only will new methodologies for designing vaccines be important, but innovative and novel methods for their development and delivery will also be critical.

Two recent developments are providing an avenue for the translational research required to develop vaccines through proof of concept in humans:

• Product development partnerships, usually linked to large foundation such as the Bill and Melinda Gates Foundation (Ridley, 2004). Often, but not always, PDPs have grown out of a desire to make vaccines for a specific disease, e.g. the Hookworm Vaccine Initiative and the Malaria Vaccine Initiative. However, regardless of how they start, all of the vaccine orientated PDPs need access to the technical development capacity to covert an interesting idea in to a practical vaccine. Some of these, such as the Aeras Global TB Vaccine Foundation have developed extensive, state of the art, in-house technical development and pilot plant capacity. Others, for example, the PATH Malaria Vaccine Initiative and Enteric Vaccine Program works with private and public partners but has no internal capacity.

- Industry-linked vaccine institutes (ILVI) with a not for profit mission to develop vaccines for neglected diseases. At the time of writing, there are two such institutes, one operational, the Novartis Vaccines Institute for Global Health (NVGH) in Siena, Italy (Saul and Rappuoli, 2008) and the second announced but not yet functional the Hilleman Laboratories that will be based in India, linked to Merck and to the Wellcome Trust (Travis, 2009). In at least the NVGH model, the institute was created with direct access to the facilities of a major vaccine manufacturer thus providing the technical development expertise. Rather than the single disease model of several PDPs, NVGH and, as announced, the Hilleman Laboratories, operate in an industrial model, looking to find vaccines that will provide the best return in terms of public health benefit for the investment.

Both the PDPs and the industrial institutes share many attributes. Both seek to use philanthropic financing and access to expertise to get to the proof of principle stage for new vaccines. Like other pharmaceuticals, the majority of vaccine development programmes do not result in a successful product. Most fail during this technical development phase leading to demonstrated proof of principle in humans. Thus by taking on the risky part of the product development pathway, PDPs and ILVIs will make it more likely that manufacturers will take new vaccines through to market and distribution.

Even with a proof of principle in humans there are still major obstacles to the registration, manufacture and distribution of a vaccine intended for neglected diseases of low and middle income countries. Again, innovative and novel approaches will be required.

A major issue is the affordability of the vaccine. Reducing development costs and risk will go some way to lowering the cost. However, designing affordability into the vaccine's manufacture and its delivery are an important consideration.

'Affordability' has several components. One is the short term costs associated with introducing a new vaccine. In the longer term, vaccines have to be available at a cost that is sustainable both for the purchaser and the manufacturer. A critical part of this long term sustainability is the development of the high volume and consistent demand for a vaccine that enables economics of scale in the manufacture. Several innovative schemes are in place and are being trialled that address one or more aspects.

The most important of these has been the ability to fund vaccines through the GAVI Alliance (www.gavialliance.org). This has brought together donor countries, WHO and the World Bank, The Bill and Melinda Gates Foundation and other donors to provide a mechanism for subsidizing the introduction and subsequent purchase of vaccines by the poorest countries, usually through UNICEF.

More recently two innovative schemes have been set up to extend this programme: the International Finance Facility for Immunization (IFFIm) and the Advanced Market Commitment (AMC). The IFFIm enables GAVI to borrow funds against long term commitments by donor countries, thus allowing a more rapid access to existing vaccines. The AMC provides a mechanism for underwriting the development of new vaccines. The first vaccines to be funded under this programme launched in June 2009, are pneumococcal vaccines that cover serotypes of importance in developing countries (GAVI Alliance, 2009). The programme aims to provide manufacturers with an initial subsidized guaranteed price as an incentive to develop the vaccine, in return for a long-term commitment by the manufacturers for lower price when the market has been established.

Delivery costs can be particularly important: vaccines that can be delivered through existing

Table 10.1 Global expenditure on vaccines for neglected diseases

Disease	DALYs[a]	Total vaccine expenditure (US$)	US$ per DALY[b]
HIV	57,905,526	692,048,625	11.95
Malaria	33,973,675	55,745,483	1.64
TB	34,039,183	82,085,740	2.41
Kinetoplastids[c]	4,074,176	11,339,048	2.78
Diarrhoeal disease[d]	72,169,199	16,300,222	0.23
Dengue	663,938	36,004,100	54.23
Helminths[e]	12,017,750	98,824,000	0.01
Bacterial pneumonia and meningitis	104,631,161[f]	27,216,989	0.26

Data compiled from individual disease tables in Moran *et al.* (2009).

[a]Includes DALYs for diseases for which there are existing vaccines, especially for bacterial pneumonia and meningitis.
[b]Expenditure for vaccine R&D, G-FINDER also lists expenditure for other interventions, e.g. drugs.
[c]Approximately, the DALYs caused by kinetoplastids (trypanosomiais and leismaniasis) in the WHO 'Tropical Diseases Cluster'.
[d]includes typhoid and paratyphoid fever.
[e]Approximately DALYs due to WHO Intestinal worms with schistosomiasis and filariasis from 'Tropical Disease Cluster'.
[f]Approximately DALYs listed in GFINDER report are the sum of the WHO 'lower respiratory infections' and 'meningitis' categories. Presumably this also includes illness due to viral infections such as respiratory syncytial virus and influenza.

vaccination procedures such as the Extended Program of Immunization (EPI) are likely to incur lower delivery costs that those requiring a whole new infrastructure.

Within an EPI there are additional delivery costs in introducing new vaccines. For example, a study on the addition of Hib to the EPI in Ethiopia found that a major additional cost in substituting a pentavalent vaccine in a single dose vial for the existing 10 dose tetravalent vial was the extra cold storage required (Griffiths *et al.*, 2009). This amounted to US$0.62 (US$0.03 per cm³) of the $1.13 extra delivery costs per fully vaccinated child. In another example, in a study in India, substitution of a fully liquid pentavalent vaccine for a tetravalent liquid plus lyophilized Hib resulted in significant savings in distribution costs, in inventory control and two-thirds reduction in cold storage requirements (Wiedenmayer *et al.*, 2009).

The cost of the vaccine and its delivery are not the only driver of whether a vaccine is deployed especially for vaccines for 'neglected' diseases in developing countries. Almost by definition, their neglected status leads to a paucity of information on the epidemiology of the disease and the likely impact of a vaccine and this leads to major difficulties for counties in deciding priorities for introduction of new vaccines.

The global introduction of Hib conjugate vaccines is a good illustration of the problems and some of the innovative solutions that address uptake of a new vaccine in developing countries. The first conjugate vaccine (consisting of a surface polysaccharide, polyribosylribitol phosphate (PRP) from the Hib bacteria conjugated to diphtheria toxoid) was licensed in the USA in 1987 in 18- to 60-month-old children and a PRP–CRM197 conjugate for infants older than 2 months in 1990. High coverage rates were obtained and by 1993 (USA) and 1994 (UK)

Hib had essentially disappeared as a significant cause of disease (Wegner *et al.*, 2004). Nearly 20 years later, less than half the world's infants have access to this vaccine, despite a high burden of preventable disease(Gessner, 2009). However, that situation is rapidly changing with 85% of children living in countries that have vaccination are in the process of introducing Hib vaccines and almost all remaining children in countries that plan to introduce Hib. There are several factors that have made this difference. One is the subsidy of the vaccine through the Global Alliance for Vaccine Initiative (now the GAVI alliance). However, other important contributors are the development of tools for assessing impact of Hib (WHO, 2001a) and cost effectiveness of the vaccine (WHO, 2001b). Vaccine probe studies have demonstrated the burden of disease caused by Hib by measuring the reduction following vaccination (Adegbola *et al.*, 1996; Levine *et al.*, 1993, 1999; Gessner *et al.*, 2005) and effectiveness studies have documented to impact of the vaccine in a variety of settings (Baqui *et al.*, 2007). Since 2005, The Hib Initiative (www.hibaction.org) funded from GAVI, has coordinated and in part funded a programme for developing the country based evidence for decision making for the rapid increase in Hib vaccine introduction. The recent success of this HiB campaign can act as a model for helping the process for introduction of vaccines for other diseases of developing countries.

Conclusion

Analysis of the global disease burden caused by infectious diseases illustrates the need and opportunity for new vaccines, particularly for diarrhoeal diseases. Rapid advances in new technologies, e.g. genomic based methods for antigen discovery, provide tools for designing new vaccines. However, vaccine design is only part of the process: especially for vaccines for neglected diseases of the developing world, innovative and novel approaches to undertaking the translational research that leads to proof of principle in humans and for registering, manufacturing, introducing and distributing vaccines will play a vital role. The evolution of Product Development Partnerships, Industry Linked Vaccine Institutes will help meet the translational research gap and schemes such as the Advanced Market Commitment for providing incentives for the commercialization of vaccines will play a vital role. However, given the magnitude and urgency of the task, further innovations in these areas would make a big difference.

References

Adegbola, R.A., Mulholland, E.K., Falade, A.G., Secka, O., Sarge-Njai, R., Corrah, T., Palmer, A., Schneider, G., and Greenwood, B.M. (1996). *Haemophilus influenzae* type b disease in the western region of The Gambia: background surveillance for a vaccine efficacy trial. Ann. Trop. Paediatr. *16*, 103–111.

Baqui, A.H., El, A.S., Saha, S.K., Persson, L., Zaman, K., Gessner, B.D., Moulton, L.H., Black, R.E., and Santosham, M. (2007). Effectiveness of *Haemophilus influenzae* type B conjugate vaccine on prevention of pneumonia and meningitis in Bangladeshi children: a case-control study. Pediatr. Infect. Dis. J. *26*, 565–571.

Berriman, M., Haas, B.J., LoVerde, P.T., Wilson, R.A., Dillon, G.P., Cerqueira, G.C., Mashiyama, S.T., Al-Lazikani, B., Andrade, L.F., Ashton, P.D. *et al.* (2009). The genome of the blood fluke *Schistosoma mansoni*. Nature *460*, 352–358.

Carapetis, J.R., Steer, A.C., Mulholland, E.K., and Weber, M. (2005). The global burden of group A streptococcal diseases. Lancet Infect. Dis. *5*, 685–694.

Checkley, W., Buckley, G., Gilman, R.H., Assis, A.M., Guerrant, R.L., Morris, S.S., Molbak, K., Valentiner-Branth, P., Lanata, C.F., and Black, R.E. (2008). Multi-country analysis of the effects of diarrhoea on childhood stunting. Int. J. Epidemiol. *37*, 816–830.

El-Sayed, N.M., Myler, P.J., Blandin, G., Berriman, M., Crabtree, J., Aggarwal, G., Caler, E., Renauld, H., Worthey, E.A., Hertz-Fowler, C. *et al.* (2005). Comparative genomics of trypanosomatid parasitic protozoa. Science *309*, 404–409.

G.A.V.I Alliance (2009). The Pneumococcal AMC: The process. http://www.vaccineamc.org/files/AMC_ProcessSheet2009.pdf Accessed 5–10–2009.

Gessner, B.D. (2009). *Haemophilus influenzae* type b vaccine impact in resource-poor settings in Asia and Africa. Expert Rev. Vaccines *8*, 91–102.

Gessner, B.D., Sutanto, A., Linehan, M., Djelantik, I.G., Fletcher, T., Gerudug, I.K., Mercer, D., Mercer, D., Moniaga, V., Moulton, L.H. *et al.* (2005). Incidences of vaccine-preventable *Haemophilus influenzae* type b pneumonia and meningitis in Indonesian children: hamlet-randomised vaccine-probe trial. Lancet *365*, 43–52.

Girard, M.P., Steele, D., Chaignat, C.L., and Kieny, M.P. (2006). A review of vaccine research and development: human enteric infections. Vaccine *24*, 2732–2750.

Gordon Douglas, R., Sadoff, J., and Samant, V. (2008). The vaccine industry. In Vaccines, Plotkin, S.A., Orenstein, W.A., and Offit, P.A., eds (Elsevier), pp. 37–44.

Griffiths, U.K., Korczak, V.S., Ayalew, D., and Yigzaw, A. (2009). Incremental system costs of introducing combined DTwP-hepatitis B-Hib vaccine into national immunization services in Ethiopia. Vaccine *27*, 1426–1432.

Holt, K.E., Parkhill, J., Mazzoni, C.J., Roumagnac, P., Weill, F.X., Goodhead, I., Rance, R., Baker, S., Maskell, D.J., Wain, J. *et al.* (2008). High-throughput sequencing provides insights into genome variation and evolution in *Salmonella Typhi*. Nat. Genet. *40*, 987–993.

Hotez, P.J., Molyneux, D.H., Fenwick, A., Kumaresan, J., Sachs, S.E., Sachs, J.D., and Savioli, L. (2007). Control of neglected tropical diseases. N. Engl. J. Med. *357*, 1018–1027.

Kweon, M.N. (2008). Shigellosis: the current status of vaccine development. Curr. Opin. Infect. Dis. *21*, 313–318.

Levine, M.M. (2006). Enteric infections and the vaccines to counter them: future directions. Vaccine *24*, 3865–3873.

Levine, M.M., Kotloff, K.L., Barry, E.M., Pasetti, M.F., and Sztein, M.B. (2007). Clinical trials of Shigella vaccines: two steps forward and one step back on a long, hard road. Nat. Rev. Microbiol. *5*, 540–553.

Levine, O.S., Lagos, R., Munoz, A., Villaroel, J., Alvarez, A.M., Abrego, P., and Levine, M.M. (1999). Defining the burden of pneumonia in children preventable by vaccination against *Haemophilus influenzae* type b. Pediatr. Infect. Dis. J. *18*, 1060–1064.

Levine, O.S., Ortiz, E., Contreras, R., Lagos, R., Vial, P., Misraji, A., Ferreccio, C., Espinoza, C., Adlerstein, L., and Herrera, P. (1993). Cost–benefit analysis for the use of *Haemophilus influenzae* type b conjugate vaccine in Santiago, Chile. Am. J. Epidemiol. *137*, 1221–1228.

LoBue, A.D., Thompson, J.M., Lindesmith, L., Johnston, R.E., and Baric, R.S. (2009). Alphavirus-adjuvanted norovirus-like particle vaccines: heterologous, humoral, and mucosal immune responses protect against murine norovirus challenge. J. Virol. *83*, 3212–3227.

Mandeville, K.L., Krabshuis, J., Ladep, N.G., Mulder, C.J., Quigley, E.M., and Khan, S.A. (2009). Gastroenterology in developing countries: issues and advances. World J. Gastroenterol. *15*, 2839–2854.

Monteiro, M.A., Baqar, S., Hall, E.R., Chen, Y.H., Porter, C.K., Bentzel, D.E., Applebee, L., and Guerry, P. (2009). Capsule polysaccharide conjugate vaccine against diarrheal disease caused by *Campylobacter jejuni*. Infect. Immun. *77*, 1128–1136.

Moran, M., and Guzman, J. (2009). Neglected disease research and development: how much are we really spending? (The George Institute for International Health).

Moran, M., Guzman, J., Ropas, A.L., McDonald, A., Sturm, T., Jameson, N., Wu, L., Ryan, S., and Omune, B. (2009). G-FINDER 2008. http://www.thegeorgeinstitute.org/shadomx/apps/fms/fmsdownload.cfm?file_uuid=409D1EFD-BF15-8C94-E71C-288DE35DD0B2&siteName=iih. Accessed 20/8/2009.

Nataro, J.P., and Barry, E.M. (2008). Diarrheal disease vaccines. In Vaccines, Plotkin, S.A., Orenstein, W.A., and Offit, P.A., eds (Elsevier), pp. 1163–1171.

Oria, R.B., Costa, C.M., Lima, A.A., Patrick, P.D., and Guerrant, R.L. (2009). Semantic fluency: a sensitive marker for cognitive impairment in children with heavy diarrhea burdens? Med. Hypotheses. *73*, 682–686.

Parashar, U.D., Gibson, C.J., Bresse, J.S., and Glass, R.I. (2006). Rotavirus and severe childhood diarrhea. Emerg. Infect. Dis. *12*, 304–306.

Parkin, D.M., Bray, F., Ferlay, J., and Pisani, P. (2005). Global cancer statistics, 2002. CA Cancer J. Clin. *55*, 74–108.

Petri, W.A., Jr., Miller, M., Binder, H.J., Levine, M.M., Dillingham, R., and Guerrant, R.L. (2008). Enteric infections, diarrhea, and their impact on function and development. J. Clin. Invest. *118*, 1277–1290.

Phalipon, A., Mulard, L.A., and Sansonetti, P.J. (2008). Vaccination against shigellosis: is it the path that is difficult or is it the difficult that is the path? Microbes Infect. *10*, 1057–1062.

Ridley, R.G. (2004). Product Development, Private Public Partnerships for Diseases of Poverty. http://www.who.int/intellectualproperty/documents/en/R.Ridley.pdf. Accessed 10/4/2009.

Rudan, I., Boschi-Pinto, C., Biloglav, Z., Mulholland, K., and Campbell, H. (2008a). Epidemiology and etiology of childhood pneumonia. Bull. World Health Organ. *86*, 408–416.

Rudan, I., Boschi-Pinto, C., Biloglav, Z., Mulholland, K., and Campbell, H. (2008b). Epidemiology and etiology of childhood pneumonia. Bull. World Health Organ. *86*, 408–416.

Sack, D.A., Sack, R.B., and Chaignat, C.L. (2006). Getting serious about cholera. N. Engl. J. Med. *355*, 649–651.

Saha, S.K., Naheed, A., El, A.S., Islam, M., Al-Emran, H., Amin, R., Fatima, K., Brooks, W.A., Breiman, R.F., Sack, D.A. *et al.* (2009). Surveillance for invasive *Streptococcus pneumoniae* disease among hospitalized children in Bangladesh: antimicrobial susceptibility and serotype distribution. Clin. Infect. Dis. *48* (Suppl. 2), S75–S81.

Saul, A., and Rappuoli, R. (2008). The Novartis Vaccines Institute for Global Health: a new initiative for developing vaccines for neglected diseases in developing countries. J. Infect. Dev. Ctries. *2*, 154–155.

Scott, J.A. (2007). The preventable burden of pneumococcal disease in the developing world. Vaccine *25*, 2398–2405.

Starnbach, M.N., and Roan, N.R. (2008). Conquering sexually transmitted diseases. Nat. Rev. Immunol. *8*, 313–317.

Travis, J. (2009). A boost for vaccine development. Science *325*, 1489.

Tribble, D.R., Shahida, S., and Scott, D.A. (2004). Vaccines against *Campylobacter jejuni*. In New Generation Vaccines, Levine, M.M., Kaper, J.B., Rappuoli, R., Liu, M.A., and Good, M.F., eds (New York, Marcel Dekker, Inc), pp. 775–785.

Walker, R.I., Steele, D., and Aguado, T. (2007). Analysis of strategies to successfully vaccinate infants in developing countries against enterotoxigenic *E. coli* (ETEC) disease. Vaccine *25*, 2545–2566.

Wegner, J., Campbell, H., Miller, E., and Salisbury, D. (2004). Post Licensure impact of Hib and SCM conjugate vaccines. In New Generation Vaccines, Levine, M.M., Kaper, J.B., Rappuoli, R., Liu, M.A., and Good, M.F., eds (New York, Marcel Dekker Ltd), pp. 427–441.

W.H.O. (2001a). Estimating the local burden of *Haemophilus influenzae* type b (Hib) disease preventable by vaccination. Accessed 5/10/2009.

W.H.O. (2001b). Estimating the potential cost-effectiveness of using *Haemophilus influenzae* type b (Hib) vaccine. http://www.who.int/vaccines-documents/DocsPDF01/www654.pdf. Accessed 5/10/2009.

W.H.O. (2008a). The Global Burden of Disease 2004 Update. http://www.who.int/healthinfo/global_burden_disease/en/index.html. Accessed 5/10/2009.

W.H.O. (2008b). WHO vaccine preventable diseases – Monitoring system 2008 Global Update (WHO/IVB/2006). Accessed 5/10/20009.

Wiedenmayer, K.A., Weiss, S., Chattopadhyay, C., Mukherjee, A., Kundu, R., Aye, R., Tediosi, F., Hetzel, M.W., and Tanner, M. (2009). Simplifying paediatric immunization with a fully liquid DTP–HepB-Hib combination vaccine: evidence from a comparative time-motion study in India. Vaccine 27, 655–659.

Wyeth (14–7–2009). Wyeth's Prevenar 13 Receives First Approval. http://www.wyeth.com/news?nav=display&navTo=/wyeth_html/home/news/pressreleases/2009/1247572971477.html.

Zaman, K., Roy, E., Arifeen, S.E., Rahman, M., Raqib, R., Wilson, E., Omer, S.B., Shahid, N.S., Breiman, R.F., and Steinhoff, M.C. (2008). Effectiveness of maternal influenza immunization in mothers and infants. N. Engl. J. Med. 359, 1555–1564.

Vaccines to Combat *Pseudomonas aeruginosa* Infections in Immunocompromised Patients

11

Jennifer M. Scarff and Joanna B. Goldberg

Abstract

Pseudomonas aeruginosa is an important opportunistic pathogen that causes an array of nosocomial infections, such as ventilator-associated pneumonia and infections in cancer patients. *P. aeruginosa* infections are difficult to treat with antibiotics, making the need for other therapeutic options, such as vaccination, critical. Animal models, such as immunocompromised mice and dogs, have been used to investigate the efficacy of *P. aeruginosa*-specific vaccines. The main target antigen for these vaccines has been the lipopolysaccharide (LPS) of *P. aeruginosa*. These animal models have demonstrated that vaccination may be partially protective, but that a combination of vaccination with either antibiotic treatment or cell transfusion protocols typically works best. The efficacy of vaccination, particularly against LPS, has been investigated in human cancer patients. These patients were capable of mounting an immune response, but it was often short-lived or accompanied by severe side-effects. An anti-*Pseudomonas* vaccine could be beneficial to aid in treatment of nosocomial infections caused by this bacterium, but would need optimization for better efficacy.

Introduction

Pseudomonas aeruginosa is a Gram-negative bacterium found ubiquitously in the environment. It can cause both community-acquired and hospital-acquired infections. Community-acquired infections include bacterial keratitis following eye injury or contact-wear use, otitis externa, and skin and soft tissue infections, including hot tub folliculitis. *P. aeruginosa* is a predominant pathogen responsible for community-acquired pneumonia and both early- and late-onset ventilator-associated pneumonia. In the hospital setting, it is an important pathogen, mainly infecting compromised individuals, such as AIDS patients, as well as those undergoing chemotherapy or with other underlying conditions. *P. aeruginosa* can also cause serious life-threatening infections after burns by dissemination from the initial site of infection. It is also the leading pathogen in patients with cystic fibrosis (CF). In CF, the infection is localized to the lung and chronic infection at this site is the major cause of morbidity and mortality in this patient population.

P. aeruginosa is naturally antibiotic resistant and can easily develop new resistances, as well as multi-drug resistance, making infections by this bacterium very difficult to treat. Vaccination directed at *P. aeruginosa* would appear to be an appropriate response to this problem. In fact, a number of different vaccines are being developed to combat *P. aeruginosa* infections. These vaccines generally target cell surface antigens, including polysaccharides, such as lipopolysaccharide (LPS) and alginate, attachment or motility organelles, such as pili or flagella, or secretion apparatus, including the type III secretion component, PcrV. These antigens can be purified, conjugated to other antigens, expressed in heterologous and attenuated hosts, or delivered as DNA sequences. In addition, passive administration of antibodies can be effective immunotherapies. The target populations for most of these vaccines have been CF patients and burn patients. Even for these specific patient populations, very few vaccines have progressed to clinical trials and none has reached clinical approval (recently reviewed in Doring and Pier, 2008; Holder, 2004). In the case of immunocompromised patients, the precondition that makes them susceptible to infection in the first place would appear to interfere with active vaccination approaches. This chapter will summarize the limited literature on *P. aeruginosa* vaccines that combat infections in immunocompromised patients and address how recent efforts may be applicable to this at-risk group.

P. aeruginosa infections in the hospital setting

P. aeruginosa is an important pathogen in hospitalized patients. Nosocomial infections can include pneumonia following mechanical ventilation, trauma, or viral infection, urinary tract infections due to indwelling urinary catheters, bloodstream infections, surgical site infections, and skin infections after burns (Driscoll *et al.*, 2007). Only about 4% of the general population carry intestinal *P. aeruginosa*, while up to 18% of hospitalized patients can carry this bacterium (Bergogne-Berezin, 2004). The intestine, as well as the respiratory tract, the genitourinary tract, and the skin, can be the source of *P. aeruginosa*. Infection can also be due to contamination from the external environment. In addition, broad-spectrum antibiotic use can select for the dissemination of resistant *P. aeruginosa* from these sites (Sadikot *et al.*, 2005).

Those at particular risk are immunosuppressed individuals, such as cancer patients, transplants recipients, AIDS patients, and neutropenic patients (Rolston and Bodey, 1992). In the case of the immunocompromised host, increased susceptibility to infection combined with ineffective antibiotic treatment can lead to life-threatening pneumonia and bacteraemia. *P. aeruginosa* is one of the most prevalent bacterial causes of lung infections in immunocompromised patients. This is particularly true in those patients who have recently finished chemotherapy or are currently undergoing treatment (Vento *et al.*, 2008). In a study of patients with acute leukaemia, 14–21% of bacteraemic infections were caused by *P. aeruginosa* infection (Driscoll *et al.*, 2007). Also,

P. aeruginosa was isolated from 28% of children with cancer (Vento *et al.*, 2008). Neutropenic patients or patients with deficient neutrophil function tend to be susceptible to Gram-negative infections, in general (Vento *et al.*, 2008). *P. aeruginosa* was the cause of infection in 5–12% of these patients and specifically in 1–2.5% of febrile neutropenic patients. The mortality rate among immunocompromised patients with *P. aeruginosa* pneumonia is increased by 40% (Sadikot *et al.*, 2005).

Pneumonia-related deaths, including those caused by *P. aeruginosa*, have been increasing in HIV patients. HIV patients are 10 times more likely than the general population to acquire a *P. aeruginosa* infection (Driscoll *et al.*, 2007). *P. aeruginosa* infections seem to occur in patients with previous treatment with cephalosporins, low CD4 counts, and neutropenia. Recent bone marrow transplant recipients also have an increased risk of becoming infected with *P. aeruginosa*. These infections in immunocompromised patients can be severe, as patients with *P. aeruginosa* pneumonia had increased time of hospitalization compared to patients with pneumonia caused by other pathogens (Sadikot *et al.*, 2005).

Current antibiotic therapy

The current method of treatment of either pneumonia or bacteraemia in immunocompromised patients involves the immediate empirical treatment with one or two broad-spectrum antibiotics known to be effective against *P. aeruginosa*. After the antibiotic susceptibility of the cultured infecting organism is known, the appropriate effective antibiotics can be tailored to more specifically treat the infection.

Current antibiotics used to treat *P. aeruginosa* infections in neutropenic patients include β-lactam antibiotics such as cephalosporin and carbapenem. Aminoglycosides, such as tobramycin, are generally used in conjunction with another antibiotic. Fluoroquinolones such as ciprofloxacin can have anti-pseudomonal activity. Peptide antibiotics such as polymixins [polymixin B and polymixin E (colistin)] may also be effective, but are extremely toxic and thus are used as a last resort. Prophylactic antibiotic treatment is also administered upon admittance of immunocompromised patients to the hospital, as this environment increases the chance of infection (Pier and Ramphal, 2004). The challenges of this approach, especially related to extended hospitalizations and, therefore, long-term antibiotic treatment, include the development of antibiotic resistance in this organism.

One key factor to the difficulty of treating *P. aeruginosa* infections is the bacterium's innate antibiotic resistance. The bacterium can also decrease porin expression on the surface, which reduces the entry sites for some antibiotics. The bacterium contains multiple efflux pumps as well as enzymes, such as β-lactamases, capable of degrading antibiotics. It can also harbour enzymes to inactivate antibiotics, including quinolones. *P. aeruginosa* is also highly capable of developing antibiotic resistance as a result of previous exposure or treatment. Convergence of multiple resistance mechanisms within a single strain can result in multidrug resistance, which is being seen with increasing frequency in the hospital setting.

P. aeruginosa has minimal growth requirements and is able to grow in almost any moist environment. While it prefers growth in aerobic conditions, *P. aeruginosa* is a facultative anaerobe and can grow anaerobically in the presence of nitrate or arginine as a terminal electron acceptor. It can grow at a range of temperatures (from 4°C to 42°C). It has a broad environmental distribution, and can be found in soil, vegetation, and water. Because of this, it can also persist in the hospital environment in food, sinks, cut flowers, respiratory equipment, and even disinfectant solutions, increasing the exposure of hospitalized patients to infection.

Immunocompromised animal models and effects of vaccination

Even with the prevalence, persistence, and minimal growth requirements, as well as the myriad of degradative enzymes, as well as adherence, toxic, and signalling virulence factors released from this opportunisitic pathogen (reviewed in Van Delden, 2004), healthy humans are generally able to control infections by *P. aeruginosa*. However, compromise of anatomical barriers, such as in burned or wounded skin, eye injury, or indwelling devices can lead to increased susceptibility. And while all known monocytic, granulocytic, and lymphocytic cell types mediate some resistance to *P. aeruginosa*, the most important cellular mediator of resistance is polymorphonuclear leucocytes (PMN) (reviewed in Pier and Ramphal, 2004). In humans, the absence of neutrophils, increases the susceptibility to *P. aeruginosa* infection by orders of magnitude (Ramphal, 2004). The development of animal models of immunocompromise with respect to *P. aeruginosa* infection and vaccination has focused on both neutropenic and leucopenic treatments.

Irradiation of dogs as a model of infection

Irradiation of dogs has been used as a model of leucopenia. Dale *et al.* (1974) showed that irradiated dogs had an immediate decline in lymphocytes and a slower decline in neutrophils following treatment. This decrease over time was reflected in the outcome after infection: dogs infected intrabronchially with *P. aeruginosa* 4–5 days after irradiation fared better than dogs infected 6–7 days post irradiation. It was also noted that this correlated with the number of neutrophils in the dogs at the time of infection.

This irradiated dog model was used to study the efficacy of granulocyte transfusions, as well as the efficacy of combining granulocyte transfusion with antibiotic therapy (Dale *et al.*, 1974). Irradiated beagles were infected intrabronchially with 5×10^8 colony-forming units (CFU) of *P. aeruginosa* and then treated with either gentamicin alone or with granulocyte transfer. The dogs that received gentamicin and granulocytes fared better than the dogs that received the gentamicin alone, with less *P. aeruginosa* in the lungs at autopsy and no endotoxaemia. Using the same infection model, the effect of antibiotics with granulocyte transfer on infection was further investigated. Treatment with gentamicin alone or carbenicillin with gentamicin resulted in better survival than control dogs, 10% survival compared to 0% survival. Treating the dogs with carbenicillin alone resulted in an increase to 30% survival, but the treatment of granulocytes and gentamicin showed the best (57%) survival.

Cyclophosphamide treatment of dogs

A more common method of eliciting immuno-compromise involves treatment with the drug cyclophosphamide (Cy) (Cytoxan®). This drug is a chemotherapy agent, which can affect both lymphocytes and neutrophils. However, at lower doses, it causes a significant decrease specifically in the levels of neutrophils. Since not all studies using Cy report both neutrophil and lymphocyte counts, in this review we will refer to the effects of this treatment as 'neutropenic' or 'leucopenic' as defined in the source.

Epstein et al. (1974) described a neutropenic dog model using animals treated with Cy. They showed that normal dogs were able to clear an intravenous inoculum of 1–2×10^7 CFU of P. aeruginosa, but that neutropenic dogs were not. The neutropenic dogs had an initial reduction in bacteria in the blood, but by 18 hours after infection, the number of bacteria started to rise in the neutropenic dogs, which was concurrent with a fever. There was no fever seen in untreated dogs given P. aeruginosa. The susceptibility seen in the neutropenic dogs was dose dependent, as dogs given lower intravenous doses of 10^4 or 10^5 CFU of P. aeruginosa had longer time until death (7 days) compared with the dogs receiving 10^7 CFU (2–3 days).

Efforts to improve the outcome of infection by boosting the immune system have also been tested in this model (Epstein et al., 1974). Granulocytes were administered to neutropenic dogs to determine whether replenishing the neutrophils would be able to protect the dogs from P. aeruginosa infection. Dogs were intravenously inoculated with 1–2×10^7 CFU of P. aeruginosa and then received at least 10^9 granulocytes 24 hours after infection. At the time of infusion, the dogs already had detectable bacteria in blood, which initially declined after the transfusion. However, the single granulocyte transfusion was unable to protect the dogs, as they all succumbed to infection by 18 hours post infusion (42 hours post infection), some suffering from bacteraemia.

Vaccination in immunocompromised dogs

LPS is an immunodominant protective antigen and it has been one of the major targets for vaccine development for many P. aeruginosa infections (Pier, 2003). There are 20 International Antigenic Typing System (IATS) serotypes of P. aeruginosa that differ in their monosaccharide composition and/or linkages between sugar residues of the O antigen portion of LPS; additional variability and modification of the O antigen results in 31 subtypes (Liu et al., 1983). Antibodies to the LPS provide serogroup-specific protection (Cryz et al., 1984; Fomsgaard et al., 1989; Knirel, 1990; Pollack and Young, 1979). Fewer than 10 of these serotypes are prominent in infection (Faure et al., 2003; Pier and Thomas, 1982), making the possibility of developing a multivalent vaccine to all of the clinically relevant types feasible.

Studies by Harvath et al. (Harvath and Andersen, 1976; Harvath et al., 1976a,b) used the neutropenic dog model and tested the P. aeruginosa LPS vaccine that was available at that time, which was composed of semi-purified LPS (developed by Parke-Davis Co.). Dogs received either placebo or one of two different serotypes of LPS intramuscularly. These animals were then made neutropenic by treatment with Cy. Vaccinated dogs had a longer mean survival time following intravenous infection with 10^7 CFU of P. aeruginosa compared to unvaccinated dogs. As anticipated, this protection was serotype-specific, as challenge with a strain of a different serotype resulted in no difference compared to control unvaccinated dogs. The vaccinated dogs also had a slower onset of fever and lower bacterial numbers in blood and organs compared to control dogs (Harvath and Andersen, 1976).

This group further investigated a combination therapy using vaccination followed by granulocyte transfusion (Harvath et al., 1976b). In this study, dogs were vaccinated prior to being rendered neutropenic as described above. Both vaccinated and unvaccinated dogs received granulocyte transfusions 24 and 48 hours after intravenous infection with 0.9 to 3.5×10^7 CFU of P. aeruginosa. Vaccinated and transfused animals had increased survival compared to dogs that received only granulocyte transfusions. Only half of dogs receiving the vaccine succumbed to the infection and none were septicaemic, while 83% of unvaccinated dogs succumbed to infection and all were septicaemic.

Monoclonal antibodies have been investigated as a potential treatment for *P. aeruginosa* using immunosuppressed dogs. Harvath *et al.* (1976a) investigated the efficacy of serum transfer to protect Cy-treated mongrel dogs given 10^7 CFU of *P. aeruginosa* intravenously. They showed that giving serum from dogs immunized with an LPS vaccine as compared to control plasma one day prior to intravenous *P. aeruginosa* challenge resulted in no difference in survival. No significant difference in blood cultures, temperature, or bacterial levels in the lungs were detected, suggesting a lack of efficacy of this approach.

Kazmierowski *et al.* (1977) also treated leucopenic dogs with three different doses of monoclonal antibody to LPS prior to intrabronchial inoculation with 5×10^8 CFU of *P. aeruginosa*. The presence of antibody in the leucopenic dogs did not change overall survival of these dogs, but mean survival time was directly proportional to the dose of antibody given. They also tested granulocyte transfusion combined with monoclonal antibody treatment. Granulocyte transfusion alone, which began 24 hours after *P. aeruginosa* infection, was unable to protect more than 27% of dogs. However, granulocyte transfusion combined with the monoclonal antibody treatment was able to protect 67% of dogs. Also, fewer dogs that received the antibody and granulocytes together became septicaemic compared to the other groups, indicating that a single treatment after *P. aeruginosa* infection in immunosuppressed individuals may not be effective in clearing the infection. The dogs that fared best after infection had received combination treatments, whether it was granulocytes with antibiotics or with antibodies.

Mice as a model system

Cryz *et al.* (1983b) described a leucopenic infection model in mice for *P. aeruginosa* infection; mice were treated with Cy and then infected with *P. aeruginosa* in an incision made on their backs. These mice were extremely susceptible to *P. aeruginosa*: less than 200 CFU was lethal, with mean time to death around 2 days. These mice had rapid bacterial growth at the infected incision that eventually disseminated to the liver.

In an attempt to prevent mortality from *P. aeruginosa* infection in neutropenic mice, Matsumoto *et al.* (1987) treated animals with human granulocyte colony-stimulating factor (hG-CSF). Mice were rendered neutropenic by intraperitoneal treatment with a single dose of Cy. The neutropenia was at its most severe at 4 days post-Cy treatment with a rapid improvement to normal levels by day 7. When mice were infected intraperitoneally with *P. aeruginosa* at day 4, the neutropenic mice showed increased susceptibility to infection compared to control non-neutropenic mice ($\log_{10} LD_{50}$ of 2.64 compared to 5.47, respectively). If mice were given 4 daily injections of $1.0\,\mu g$ of hG-CSF starting day 1 post-Cy treatment prior to infection, the LD_{50} returned to that seen in wild-type non-neutropenic mice ($\log_{10} LD_{50}$ of 5.39). This is likely due to the fact treatment with hG-CSF after Cy injection abrogates the induction of neutropenia. In fact, this treatment has also been shown to increase the number of neutrophils in normal mice, making the mice neutrophilic (Tamura *et al.*, 1987), leading to an enhanced cellular influx at the site of infection.

BitMansour *et al.* (2002) transplanted haematopoietic stem cells (HSC) alone or with common myeloid progenitors (CMP) and granulocyte-monocyte progenitors (GMP) into irradiated mice. CMPs give rise to either GMPs or megakaryocyte/erythrocyte progenitors. The blood of mice that received the HSC with CMP/GMP transplant still remained neutropenic; however, an increase of myeloid cells was observed in organs, such as the spleen. Consequently, mice receiving CMP/GMP with the HSC transplant had better survival (75%) after intraperitoneal infection with 300–500 CFU of *P. aeruginosa* than the mice receiving only HSC (10% survival). These investigators also used luciferase-expressing *P. aeruginosa* and *in vivo* imaging to locate the bacteria after infection. In mice receiving only HSC, the bacteria were found in the gastrointestinal (GI) area and dissemination to the lungs could be seen. Conversely, in the mice receiving CMP/GMP with the HSC, there was low bacterial burden seen in the GI tract with little to no dissemination to other organs. Quantification of the bacteria in these mice showed significantly less bacteria in the spleen, lungs and heart of these mice, confirming the *in vivo* imaging data.

The neutropenic model most used in mice involves removal of neutrophils with a specific antibody, RB6–8C5 (also known as anti-Gr1). This antibody is specific for a cell surface marker found in high levels on neutrophils, so administration of the antibody selectively depletes neutrophils (Hestdal *et al.*, 1991), but may also affect dendritic cells, and subsets of monocytes, macrophages, and lymphocytes (Daley *et al.*, 2008).

Koh *et al.* (2005) monitored the spread of *P. aeruginosa* from the GI tract to the blood to mimic the dissemination of infection seen in cancer patients. They gave mice streptomycin and penicillin in their drinking water for 4 days. Subsequently, they were given 10^7 CFU/ml of *P. aeruginosa* per ml of drinking water for 5 days. After GI colonization was achieved, these investigators used either Cy or RB6-8C5 to induce immunocompromise in these mice. These workers found that the outer core polysaccharide portion of LPS and the LPS O antigen side chains were critical for establishing an infection. An *aroA P. aeruginosa* mutant that has been investigated as an attenuated vaccine strain (Priebe *et al.*, 2002, 2003) was able to establish GI colonization, but did not disseminate, suggesting it may be useful in this patient population.

LPS-based vaccines in immunocompromised mice

Cryz *et al.* (1983a) tested the efficacy of immunoglobulins against three known virulence factors of *P. aeruginosa*: exotoxin A, elastase, or LPS. Exotoxin A is a secreted protein that inhibits protein synthesis of host cells by ADP-ribosylating eukaryotic elongation factor-2, which leads to cell death. Elastase is a protease, which degrades a broad range of host proteins, including elastin. Elastase disrupts the tight junctions in the lung epithelium, increasing lung permeability, which can aid in the establishment of infection. Both elastase and other bacterial proteases have been shown to degrade immune effectors including antibodies, complement, and surfactant proteins.

These immunoglobulins were given intravenously and tested for their ability to protect leucopenic mice from a low, but lethal, dose (about 50 CFU) of *P. aeruginosa* that were instilled into an incision made on the back. Neither the anti-exotoxin A nor the anti-elastase antibodies protected better than control rabbit antibodies, while the anti-LPS antibodies protected in a serotype-specific manner, leading to 90% survival. The anti-LPS antibody was able to protect these mice when administered up to 6 hours post infection, and even at 24 hours post infection it increased survival time compared to control mice. In the mice receiving control serum, a sharp increase in the amount of bacteria in the skin occurred post infection, with bacteria disseminating to the blood and liver. In the mice receiving the anti-LPS antibody, the bacteria in the skin grew slower and the number started to decrease by 48 hours post infection; no bacteria were detectable in the blood and liver. Also, giving these antibodies in concurrence with gentamicin increased survival even further in the anti-LPS group, but had no effect on the other antibody-treated groups.

We have recently assessed the efficacy of an LPS-based vaccine that consists of the O antigen from *P. aeruginosa* serotype O11 expressed on an attenuated strain of *Salmonella Typhimurium*. This serotype is commonly found in infections and strains of this serotype have a much higher prevalence of cytotoxicity than other serotypes and are also associated with a high mortality rate (Faure *et al.*, 2003). Prior work from our laboratory showed that this vaccine could induce serotype-specific antibodies and that oral immunization provided greater protection from acute *P. aeruginosa* pneumonia than intraperitoneal immunization (DiGiandomenico *et al.*, 2004). We subsequently showed that intranasal administration of this vaccine was even more effective against acute pneumonia and could additionally protect against *P. aeruginosa* infections after burns and eye injury (DiGiandomenico *et al.*, 2007).

With this as a backdrop, we tested whether intranasal immunization could provide protection from infection in immunocompromised mice. Mice were vaccinated and an immune response was noted prior to immunocompromise. Mice were then treated with either Cy or RB6–8C5 to induce leucopenia or neutropenia, respectively.

We did not observe any decrease in the specific antibody response of these mice after either of these treatments. In both models, prior vaccination was able to reduce susceptibility to low doses (<2000 CFU) of *P. aeruginosa* given intranasally (Scarff and Goldberg, 2008). This protection was characterized by a longer time until death and lower bacterial load in respiratory tract as well as less dissemination to liver and spleen at 24 hours post-infection. Also, administration of immune sera to immunocompromised mice, via the intranasal route at the time of infection led to complete survival from *P. aeruginosa* doses of 200 CFU, while mice given either PBS or control serum died around 21 or 24.5 hours, respectively (Scarff and Goldberg, 2008). This study provides evidence that antibodies alone, generated by active immunization or given directly at the site of infection can slow the progress of acute pneumonia in these immunocompromised mice, allowing them to survive longer. We envisage that combining either active or passive vaccination with antibiotic therapy could be useful in clinical settings.

Passive immunotherapies may be the most effective means of controlling *P. aeruginosa* infections in immunocompromised hosts. Consequently, humanized antibodies against *P. aeruginosa* have been generated in mice by immunizing XenoMouse mice. The XenoMouse contains the human immunoglobulin locus instead of the mouse locus, allowing for production of humanized antibodies in mice. The IgG2 antibodies made in these mice strongly resemble the antibodies seen in humans in response to polysaccharide antigens. The antibody produced by immunizing these mice with heat-killed *P. aeruginosa* was found to be specific for the *P. aeruginosa* LPS. This antibody was opsonic for killing the *P. aeruginosa* by PMNs in the presence of human complement. In addition, this monoclonal antibody protected 100% of Cy-induced neutropenic mice from fatal *P. aeruginosa* sepsis (Hemachandra *et al.*, 2001).

More recently, a panel of humanized monoclonal antibodies to ten different *P. aeruginosa* serotypes has been generated. The majority of these antibodies were highly opsonic for uptake and killing of the homologous *P. aeruginosa* strains by human PMNs in the presence of human complement. All the monoclonal antibodies protected Cy-induced neutropenic mice from fatal *P. aeruginosa* sepsis with homologous serotypes (Lai *et al.*, 2005).

Outer membrane proteins as vaccines in immunocompromised mice

To address the serotype-specificity of LPS-based vaccines, outer membrane proteins, which are conserved among all serotypes, have been investigated as vaccine candidates (Gilleland *et al.*, 1984; Hancock *et al.*, 1985; Lee *et al.*, 1999). Fusion proteins comprised of epitopes to both OprI and OprF have also been used for vaccination studies (von Specht *et al.*, 1995, 1996). The rationale is that that immunization with both proteins would lead to greater efficacy, as there could be more opsonizing antibodies on the surface of the bacteria than from a vaccine composed of a single protein. The more immunoreactive portions of both OprF and OprI were fused to glutathione S-transferase (GST) both individually and in tandem. Mice were vaccinated with these constructs, treated with Cy, and then given an intraperitoneal challenge of *P. aeruginosa*. A mixture of GST-OprI and GST–OprF constructs was able to increase the LD_{50}, although the change was not significant. Interestingly, the GST–OprI-OprF construct saw no increase in the LD_{50} compared to mice given GST alone, while the mice receiving the GST–OprF–OprI construct saw the largest increase in LD_{50}. Serum from rabbits vaccinated with this GST–OprF–OprI construct was tested in severe combined immunodeficient (SCID) mice intraperitoneally infected with *P. aeruginosa*. The SCID mice receiving the vaccine sera were fully protected from a dose of 5×10^2 CFU and partially protected (40%) from a dose of 5×10^3 CFU. This demonstrates an ability of this passive reagent to protect against low doses of *P. aeruginosa*, which might be relevant early in infection.

In order to deliver proteins as they might be encountered during natural infection, OprI was expressed in an attenuated *Salmonella dublin* strain and tested in mice (Toth *et al.*, 1994). Mice were given an eight dose regimen, which consisted either entirely of oral doses of *Salmonella*-OprI or a mixture of two doses of *Salmonella*-OprI given orally followed by six intraperitoneal boosters with recombinant OprI protein. Both of these vaccinations resulted in specific IgG and IgA antibody responses that, surprisingly, declined within 1 week following immunocompromise via Cy treatment. While the mice receiving the combination vaccine strategy did have better survival compared vector-immunized mice following oral challenge with *P. aeruginosa*, the mice receiving only oral vaccination had the best survival. In fact, the mice receiving only oral vaccination had 100% survival following an oral challenge with of 4.7×10^4 CFU of *P. aeruginosa*. While this level of protection may seem promising, the eight vaccinations and boosters would appear to be a difficult regimen to use in the clinical setting. Also, the decline in antibody titers following immunocompromise does not bode well for efficacy of this vaccine in cancer patients undergoing chemotherapy.

The effect of passive administration of monoclonal antibodies specific for OprI has also been tested in immunocompromised mice (Rahner *et al.*, 1990). Mice were rendered leucopenic by treatment with Cy, given either antibody alone or antibody with irradiated human leucocytes, and then infected subcutaneously with *P. aeruginosa*. There was differential survival depending on the monoclonal antibody, suggesting some epitope specificity of these reagents. Giving the mice irradiated leucocytes with the antibodies gave a slight, but significant, increase in this survival. Surprisingly, treatment with only the irradiated leucocytes conferred some degree of protection, with 70–80% of mice surviving the *P. aeruginosa* infection.

A combination therapy involving vaccination and treatment with specific antibody prior to infection has been tested in mice. Intraperitoneal vaccination was performed with OprI expressed and purified from *Escherichia coli*; immunization yielded specific antibodies in mice prior to immunocompromise (Finke *et al.*, 1991). This vaccination was able to raise the LD_{50} following a subcutaneous infection to *P. aeruginosa* from 78 CFU in the control mice to 2900 CFU in the vaccinated mice. Treatment with one of the OprI-specific monoclonal antibodies in the vaccinated mice one day prior to infection saw a further increase in LD_{50} to 3400–7500 CFU.

Immunotherapies to other antigens in immunocompromised mice

Martinez *et al.* (Martinez and Callahan, 1985) investigated an immunization–antibody combination using the toxoid of *P. aeruginosa* exotoxin A in conjunction with human post-immune serum from a heat-killed *E. coli* vaccine (J5) as a potential therapy against *P. aeruginosa*. Mice were vaccinated with toxoid, and then treated with a single dose of Cy, one day prior to infection. After vaccination with the toxoid, 90% of mice had detectable toxin A-specific circulating antibodies. Vaccination with the toxoid alone had no effect on survival compared to controls, while administration of the human post-immune serum raised survival in these mice to about 50%. However, this survival was not much different from that derived from the human pre-immune serum. When mice were vaccinated and also received the post-immune serum, survival was raised to 91%, while the survival of the toxoid-immunized given pre-immune serum remained near 50%.

Passive administration of antibodies to the motility organelle flagella, have also been investigated in the neutropenic mouse model of pneumonia (Oishi *et al.*, 1993). Human IgG1 monoclonal antibodies to *P. aeruginosa* LPS and flagella were compared for their protective activities. The activity of the anti-flagella antibody was as effective as the anti-LPS antibody at protecting against *P. aeruginosa* pneumonia in neutropenic mice. The mechanism of action of the anti-flagella antibody was shown to be due to inhibition of bacterial motility rather than the typical opsonophagocytic killing seen for the anti-LPS antibody. These workers also showed a synergistic effect of the anti-flagella antibody and antibiotic treatment, leading to increased survival compared to either treatment alone.

The component PcrV of the *P. aeruginosa* type III secretion apparatus has also been investigated as a vaccine candidate. The type III secretion system acts as a 'needle' or 'syringe' and injects proteins (effectors) directly from the bacteria into eukaryotic host cells. The *P. aeruginosa* effectors include four toxins: ExoS, ExoT, ExoU, and ExoY. ExoS and ExoT both have both ADP-ribosyltransferase and Rho GTPase-activating protein domains. ExoY is an adenylate cyclase, which increases the cyclic-AMP in the cytosol of the target cell. ExoU is a cytotoxic phospholipase. PcrV is one of three components used for translocation of these effectors and is located at the tip of the needle complex (Hauser, 2009). Both active and passive vaccines based on PcrV have been shown to be protective in animal models of infection (reviewed by Lynch and Wiener-Kronish, 2008). These vaccines were also tested in immunocompromised mice. Vaccinated Cy-treated mice were infected intra-abdominally with *P. aeruginosa*. This study revealed specific epitopes of PcrV that were critical for protection (Moriyama *et al.*, 2009). Interestingly, the anti-PcrV Fab fragments themselves appear to directly block cytotoxicity and can be given to therapeutically inhibit *P. aeruginosa*-induced pulmonary damage (Baer *et al.*, 2009).

Summary of animal data

Overall, *P. aeruginosa* vaccinations in immunocompromised animals have tested a variety of different antigens and strategies, including active immunization and passive administration of antibodies, as well as immune system alterations. While some single strategy methods were capable of reducing the susceptibility, most animals ultimately succumbed to the *P. aeruginosa* infection. The most effective treatments seemed to involve combinations of immune system boosts, like granulocyte transfusion, with either active or passive antibody administration. Also, combining vaccine strategies with antibiotic therapy seemed to have a positive effect on the outcome of the infection.

P. aeruginosa vaccines in cancer patients

A study was carried out to analyse the immune response to *P. aeruginosa* in cancer patients, most of whom had either leukaemia or carcinoma (Crowe *et al.*, 1982). It was found that the patients were able to make antibodies to exotoxin A and two different bacterial proteases. The antibody response was higher in patients who were infected with *P. aeruginosa* (>10^5 CFU/ml in lungs, blood or urine) compared to those who were only colonized (<10^5 CFU/ml). A patient infected with other species of *Pseudomonas* had much lower antibody levels than those groups infected or colonized with *P. aeruginosa*. The sera from these patients, was transferred to mice to see if it provided protection against a *P. aeruginosa* challenge. Mice were protected based upon the level of response of the patient; the higher-responder sera (associated with high level of infection) completely protected mice, while the intermediate-responder sera (associated with lower level of infection) only protected some of the mice. Pre-immune sera or the serum from the patient infected with other *Pseudomonas* species was unable to protect these mice. This result indicates that patients are capable of forming antibodies against *P. aeruginosa* during the course of natural infection. Additionally, these antibodies are functional, as they are able to mediate clearance of bacteria in a mouse model of infection. However, this also suggests that this level of antibody in the context of cancer is unable to protect against infection.

A heptavalent vaccine against LPS (Pseudogen®, developed by Parke-Davis Co.) was tested in patients with acute leukaemia by Pennington *et al.* (1975). The test groups consisted of patients in bone marrow remission (<5% blast cells), not currently receiving chemotherapy or adrenal corticosteroids during vaccination. The patients receiving the vaccine had a strong, but short-lived, haemagglutinating antibody response. It was demonstrated that patients who had been on methotrexate had a lower antibody response. Unfortunately, those individuals receiving the vaccine actually had more bacterial and fungal infections than the control group, although less infections by *P. aeruginosa*. In a study by Young *et al.* (1973) investigating the efficacy of this same vaccine using a larger number of cancer patients (both leukaemias and solid tumours), vaccination was shown to significantly reduce mortality compared to control patients. Although not statistically significant, there were also less fatal bacteraemic (19 control vs. 10 vaccine) and non-bacteraemic (12 control vs. 3 vaccine) infections in these patients. For all groups, death was associated with severe leucopenia and/or low titres of opsonizing antibodies.

Pseudogen® was also tested in children with acute leukaemia without alteration to their chemotherapy regimens by Haghbin *et al.* (1973). They found that 85% of the patients responded to the vaccine, but predominately an IgM response was elicited and the titres were not maintained. In all three studies, severe reactions to the vaccination were common (high fevers, etc.), leading to reduction in the amount of antigen in booster vaccinations. The side-effects were severe enough in the Pennington *et al.* (1975) study that corticosteroids were mixed with the vaccine to reduce the side-effects. Although there is some evidence that vaccination could prevent *P. aeruginosa* infection in cancer patients, the vaccine used in these studies required multiple administrations, increasing the possibility for severe side-effects.

The ability to transfer immunity to bone marrow transplant recipients was tested by Gottlieb *et al.* (1990). Donor or recipients were vaccinated with Aerugen®, an octavalent LPS-based vaccine composed of the O antigen of eight different serotypes devoid of lipid A, and conjugated to *P. aeruginosa* exotoxin A (originally developed by the Swiss Serum and Vaccine Institute) (Cryz *et al.*, 1987a). When either the donor or recipient alone was vaccinated 7–10 days prior to bone marrow transplant, the serum antibody response to the polysaccharides was low, whereas vaccination of both the recipient and donor prior to transplant resulted in a rise in antibody titre to the polysaccharides. The peak antibody titre occurred at 2–4 weeks and consisted of high levels of both IgG1 and IgG2 subtypes. The presence of IgG2 is important, as IgG2 levels have been shown to decline in patients after bone marrow transplant (Velardi *et al.*, 1988).

P. aeruginosa vaccines in other susceptible populations

While *P. aeruginosa* remains a prominent pathogen, especially among susceptible individuals, there is currently no vaccine available for at-risk patients. Immunotherapies targeting *P. aeruginosa* infections in CF and burned patients have shown only moderate success (recently reviewed in Doring and Pier, 2008). In fact, a recent report reviewing the literature on the state-of-the-art concerning vaccines to prevent infections with *P. aeruginosa* in CF states, 'vaccines against *Pseudomonas aeruginosa* cannot be recommended' (Johansen and Gotzsche, 2008). However, these studies provide lessons and give insights into the challenges in developing a vaccine for immuno-compromised patients.

The heptavalent vaccine Pseudogen® was tested in patients with CF and similar problems were noted as were found in immunocompromised patients: not only was the organism not eliminated from the airways, but febrile responses were observed in 20–40% of patients (Pennington *et al.*, 1975). Similarly in burn patients, Pseudogen® administration led to increased survival, but there was also a high adverse reaction rate (Alexander and Fisher, 1974). This response was likely due to the presence of the lipid A (endotoxin) portion of LPS in this preparation. In the case of CF patients, it has been speculated that this vaccine may have led to increased immune complex formation, as they had already been infected with *P. aeruginosa* (Doring and Pier, 2008). Another LPS vaccine trial in CF patients that were not colonized showed that neither the acquisition nor the course of disease was altered compared to the non-vaccinated control group (Langford and Hiller, 1984).

A small open study with Aerugen® showed this vaccine to be safe in healthy volunteers and CF patients (Cryz *et al.*, 1987b; Schaad *et al.*, 1991). It induced antibodies to the O antigens included in the vaccine; these antibodies were opsonic and promoted *P. aeruginosa* killing by human neutrophils (Cryz Jr. *et al.*, 1994). In CF patients, after 6 years and 10 years, there was more infection noted in the non-vaccinated group compared to the vaccinated group. Also, lung function and weight was higher in the immunized CF patients compared to the non-immunized group (Lang *et al.*, 2004). Unfortunately, a larger, placebo-controlled trial with this same vaccine was discontinued by the manufacturer, Crucell, as there was no significant difference between the two groups in the clinical parameters chosen to measure outcomes.

In 1980, Jones *et al.* (1980) reported the results of passive immunization in burned patients. Immunoglobulin derived from healthy volunteers who were immunized with a 16-valent LPS based vaccine was given to burned patients. Those patients given the immune serum showed better survival compared to controls, although no further studies on this approach have been reported.

Vaccine studies using outer membrane protein have been shown to be safe and immunogenic in healthy volunteers (Jang *et al.*, 1999; Mansouri *et al.*, 1999) and to confer protection against *P. aeruginosa* bacteraemia in burned patients (Kim *et al.*, 2000) compared to placebo control patients. Patients with chronic lung disease [including those with CF and chronic obstructive pulmonary disease (COPD)] were intranasally immunized with a hybrid OprF-OprI recombinant protein expressed and purified from *E. coli* followed by a systemic booster vaccination. Antibody levels persisted in the airway mucosa of these patients after 6 months, while serum antibody levels were not maintained (Sorichter *et al.*, 2009). More recently, human volunteers were immunized with either a systemic, nasal, or oral vaccine, based on attenuated live *Salmonella* strains expressing an OprF–OprI fusion protein followed by a systemic booster vaccination. In this study, a nasal and oral vaccination appeared to provide the highest titre of specific IgA and IgG antibody response in the lung suggesting the potential for this delivery system for enhanced immunogenicity against lower airway infection with *P. aeruginosa* especially in patients with CF as well as COPD (Bumann *et al.*, 2010).

Alginate is a linear polymer of D-mannuronic acid and L-guluronic acid that gives *P. aeruginosa* strains isolated from chronic lung infections in CF their characteristic mucoid appearance. Since alginate is a prominent extracellular antigen, it would appear to represent a viable target for vaccine development for CF. However, problems have been found when trying to use alginate as an immunogen. Opsonic, but not non-opsonic, antibodies to alginate protect animals against chronic endobronchial infection (Garner *et al.*, 1990). However, when purified alginate was injected into healthy volunteers only a small proportion produced an increased level of opsonic antibodies to alginate (Pier, 2005). It has also been shown that antibodies to specific epitopes can mediate killing of mucoid strains *in vitro* (Pier and Thomas, 1983), while antibodies to alginate produced during chronic lung infection do not (Meluleni *et al.*,

1995). Fully human IgG1 monoclonal antibodies to alginate (Aerucin®, Aridis Pharmaceuticals) that are opsonic (Pier *et al.*, 2004) may prove to be useful in passive administration to CF patients (Doring and Pier, 2008).

A recent clinical trial of a *P. aeruginosa* flagella-based vaccine has also met with limited success (Doring and Pier, 2008). This vaccine was well tolerated and CF patients developed high serum IgG to the flagella subtypes that were included in the vaccine. The degree of protection against *P. aeruginosa* that was calculated from the relative risk was 34%. Among the *P. aeruginosa* strains isolated from the infected vaccinated group, most were 'flagella-positive', but were expressing flagella subtypes that were not included in the vaccine, suggesting the efficacy of this approach. Unfortunately, the second primary endpoint, prevention of chronic *P. aeruginosa* infection, was not achieved, due to a much lower than expected rate of colonization observed in the placebo-control group (Doring and Pier, 2008). This is probably due to the initiation of antibiotic treatment following an initial *P. aeruginosa* exposure in the CF patients enrolled in this trial (Doring and Pier, 2008). At present, the manufacturer, IMMUNO, has terminated production of this vaccine.

Future challenges

Recent studies have begun to explore why *P. aeruginosa* is such a major pathogen in immunocompromised patients. We have shown that intranasal infection with a non-cytotoxic *P. aeruginosa* strain, which does not disseminate in normal mice, was found in the liver and spleen of Cy-treated mice. This was not due to increased lung permeability of the lung by the treatment itself, but depends on the Cy treatment and *P. aeruginosa* infection (Scarff and Goldberg, 2008). This suggests that the combination of immunocompromise and *P. aeruginosa* promotes the dissemination that is characteristic of infection in patients undergoing chemotherapy. This strain did not have the cytotoxic phospholipase, ExoU (Allewelt *et al.*, 2000) indicating that this factor was not required for the observed effect. On the other hand, ExoU- and the type III secretion system-expressing *P. aeruginosa* strains have been associated with poor outcomes in patients with hospital-acquired and ventilator-associated pneumonia (El Solh *et al.*, 2008; Hauser *et al.*, 2002; Schulert *et al.*, 2003). In addition, ExoU itself has been shown to interfere with the ability of recruited phagocytic cells to eradicate bacteria from the lung, leading to a local immunosuppression (Diaz *et al.*, 2008). To the best of our knowledge, whether there are certain *P. aeruginosa* genes or phenotypes specifically associated with infections in immunocompromised patients is not currently known. Recognizing particular *P. aeruginosa* components important in this patient population may help identify valuable new vaccine candidates and/or virulence factors to target for new anti-infective or antimicrobial therapies.

The problem inherent in developing vaccines based on purified components is that antigens may not be presented to the host's immune system in a manner analogous to how they are detected during infection. Live vaccines based on attenuated bacteria have been shown to elicit immune responses of greater magnitude and of longer duration than other types of vaccine constructs likely because the duration of the infection resembles the early stages of a natural infection.

Salmonella can be an effective vehicle to deliver antigens. It can stimulate humoral, cell-mediated, and secretory immune responses (Chatfield *et al.*, 1994) and can home to cells within the mucosal-associated lymphoid tissue; presentation of an antigen at one mucosal site can stimulate immunity at a distant site. Vaccine studies using attenuated *S. typhimurium* strains expressing heterologous antigens in mouse models of infection are generally performed prior to expressing antigens for delivery to humans via *Salmonella typhi* Ty21a (Germanier and Furer, 1975). Ty21a has been licensed as an oral vaccine against typhoid fever and is the only live-attenuated bacterial vector that is currently being marketed as a commercial vaccine (Vivotif®, Berna Biotech). It has an unrivalled efficacy and safety record with over 200 million doses administered to humans with no documented cases of bacteraemic dissemination. However, the disadvantage of using heterologous expression system for antigen delivery is that single vaccine candidates must be expressed and evaluated independently and a multivalent

vaccine approach is likely needed to provide sufficient protection.

Attenuated strains of *P. aeruginosa* may have an advantage in this regard, as multiple antigens are expressed. Studies in normal mice have shown that an attenuated *aroA* mutant of *P. aeruginosa* can protect against acute pneumonia by strains of both homologous (Priebe *et al.*, 2003) and heterologous serotypes (Priebe *et al.*, 2008). In addition, unlike wild-type *P. aeruginosa*, the *aroA* mutant does not disseminate from the GI tract of mice following immunocompromise (Koh *et al.*, 2005).

Distinct from problems associated with animal models of chronic *P. aeruginosa* infection in CF (Kukavica-Ibrulj and Levesque, 2008), treatment of mice with either Cy or the RB6-8C5 antibody to induce immunosuppression appears to adequately mimic the clinical conditions of patients undergoing chemotherapy (Koh *et al.*, 2009; Opal and Cross, 2005). Also the ability to monitor the level and location of *P. aeruginosa* during infection through *in vivo* imaging provides critical information on the progress of the disease and the vaccine-mediated clearance (BitMansour *et al.*, 2002; DiGiandomenico *et al.*, 2007).

Another difficulty in the development of any active vaccine strategy in this patient population is that they may not make an adequate immune response once they are immunocompromised. Even with respect to vaccination prior to immunocompromise, some animal studies have suggested that vaccine-induced antibody levels wane following treatment (Toth *et al.*, 1994), while work from our laboratory suggests that they are stable (Scarff and Goldberg, 2008).

One of the major issues related to the delivery of vaccines to immunocompromised patients is that it is difficult to recognize at-risk populations. Therefore, passive immunotherapy may be the most desirable approach. The ability to generate hybridomas secreting human antibodies from vaccinated XenoMouse mice is promising for mass production of *Pseudomonas*-specific antibodies that could be used either prophylactically or therapeutically against infections in immunocompromised patients. Preferably the delivery of such reagents should be directed at producing effective immunity at mucosal sites;

aerosolization may provide the best manner to deliver any vaccine.

Finally as with any vaccine, it is anticipated that there will be significant challenges in the design and implementation of clinical studies attempting to protect immunocompromised patients from *P. aeruginosa*. Initially it will be important to determine that the vaccine is safe in this highly susceptible patient population. In addition, it will be critical to carefully define the patient population to be vaccinated and the appropriate clinical outcomes and endpoints to determine efficacy. As mentioned, in the case of the flagella vaccine trial, antibiotic treatment of the CF patients enrolled in the study diminished differences in the number of *P. aeruginosa* infections between the vaccinated and placebo control groups (Doring and Pier, 2008). Finally, studies have suggested that antibiotic treatments and immune boosting augment the positive effects of vaccination in immunocompromised animals and patients. This combination therapy may provide the best outcome for this patient population.

Acknowledgements

This work was supported by grants from the Cystic Fibrosis Foundation and the National Institutes of Health (NIH; 1 R01 AI068112) to JBG. JMS was partly supported by the NIH through the University of Virginia Infectious Diseases Training Program (5 T32 AI007046). We apologize to our colleagues whose work has been omitted due to lack of space.

References

Alexander, J.W., and Fisher, M.W. (1974). Immunization against *Pseudomonas* in infection after thermal injury. J. Infect. Dis. *130*, S152–S158.

Allewelt, M., Coleman, F.T., Grout, M., Priebe, G.P., and Pier, G.B. (2000). Acquisition of expression of the *Pseudomonas aeruginosa* ExoU cytotoxin leads to increased bacterial virulence in a murine model of acute pneumonia and systemic spread. Infect. Immun. *68*, 3998–4004.

Baer, M., Sawa, T., Flynn, P., Luehrsen, K., Martinez, D., Wiener-Kronish, J.P., Yarranton, G., and Bebbington, C. (2009). An engineered human antibody fab fragment specific for *Pseudomonas aeruginosa* PcrV antigen has potent antibacterial activity. Infect. Immun. *77*, 1083–1090.

Bergogne-Berezin, E. (2004). *Pseudomonads* and miscellaneous Gram-negative bacilli. In Infectious Diseases, Cohen, J.M., and Powderly, W.G., eds (Mosby).

BitMansour, A., Burns, S.M., Traver, D., Akashi, K., Contag, C.H., Weissman, I.L., and Brown, J.M. (2002). Myeloid progenitors protect against invasive aspergillosis and *Pseudomonas aeruginosa* infection following hematopoietic stem cell transplantation. Blood *100*, 4660–4667.

Bumann, D., Behre, C., Behre, K., Herz, S., Gewecke, B., Gessner, J.E., von Specht, B.U., and Baumann, U. (2010). Systemic, nasal and oral live vaccines against *Pseudomonas aeruginosa*: a clinical trial of immunogenicity in lower airways of human volunteers. Vaccine *28*, 707–713.

Chatfield, S.N., Dougan, G., and Roberts, M. (1994). Progress in the development of multivalent oral vaccines based on live attenuated *Salmonella*. In Modern Vaccinology, Kurstak, E., ed. (New York, NY, Plenum Medical Book Company), pp. 55–86.

Crowe, K.E., Bass, J.A., Young, V.M., and Straus, D.C. (1982). Antibody response to *Pseudomonas aeruginosa* exoproducts in cancer patients. J. Clin. Microbiol. *15*, 115–122.

Cryz Jr., S.J., Wedgwood, J.L., A.B., Ruedeberg, A., Que, J.U., Furer, E., and Schaad, U.B. (1994). Immunization of noncolonized cystic fibrosis patients against *Pseudomonas aeruginosa*. J. Infect. Dis. *169*, 1159–1162.

Cryz, S., Pitt, T., Furer, E., and Germanier, R. (1984). Role of lipopolysaccharide in virulence of *Pseudomonas aeruginosa*. Infect. Immun. *44*, 508–513.

Cryz, S.J., Jr., Furer, E., and Germanier, R. (1983a). Passive protection against *Pseudomonas aeruginosa* infection in an experimental leukopenic mouse model. Infect. Immun. *40*, 659–664.

Cryz, S.J., Jr., Furer, E., and Germanier, R. (1983b). Simple model for the study of *Pseudomonas aeruginosa* infections in leukopenic mice. Infect. Immun. *39*, 1067–1071.

Cryz, S.J., Jr., Lang, A.B., Sadoff, J.C., Germanier, R., and Furer, E. (1987a). Vaccine potential of *Pseudomonas aeruginosa* O-polysaccharide-toxin A conjugates. Infect. Immun. *55*, 1547–1551.

Cryz, S.J.J., Furer, E., Cross, A.S., Wegmann, A., Germanier, R., and Sadoff, J.C. (1987b). Safety and immunogenicity of a *Pseuomonas aeruginosa* O-polysaccharide toxin A conjugate vaccine in humans. J. Clin. Invest. *80*, 51–60.

Dale, D.C., Reynolds, H.Y., Pennington, J.E., Elin, R.J., Pitts, T.W., and Graw, R.G., Jr. (1974). Granulocyte transfusion therapy of experimental *Pseudomonas* pneumonia. J. Clin. Invest. *54*, 664–671.

Daley, J.M., Thomay, A.A., Connolly, M.D., Reichner, J.S., and Albina, J.E. (2008). Use of Ly6G-specific monoclonal antibody to deplete neutrophils in mice. J. Leukoc. Biol. *83*, 64–70.

Diaz, M.H., Shaver, C.M., King, J.D., Musunuri, S., Kazzaz, J.A., and Hauser, A.R. (2008). *Pseudomonas aeruginosa* induces localized immunosuppression during pneumonia. Infect. Immun. *76*, 4414–4421.

DiGiandomenico, A., Rao, J., and Goldberg, J.B. (2004). Oral vaccination of BALB/c mice with *Salmonella enterica* serovar Typhimurium expressing *Pseudomonas aeruginosa* O antigen promotes increased survival in an acute fatal pneumonia model. Infect. Immun. *72*, 7012–7021.

DiGiandomenico, A., Rao, J., Harcher, K., Zaidi, T.S., Gardner, J., Neely, A.N., Pier, G.B., and Goldberg, J.B. (2007). Intranasal immunization with heterologously expressed polysaccharide protects against multiple *Pseudomonas aeruginosa* infections. Proc. Natl. Acad. Sci. U.S.A. *104*, 4624–4629.

Doring, G., and Pier, G.B. (2008). Vaccines and immunotherapy against *Pseudomonas aeruginosa*. Vaccine *26*, 1011–1024.

Driscoll, J.A., Brody, S.L., and Kollef, M.H. (2007). The epidemiology, pathogenesis and treatment of *Pseudomonas aeruginosa* infections. Drugs 67, 351–368.

El Solh, A.A., Akinnusi, M.E., Wiener-Kronish, J.P., Lynch, S.V., Pineda, L.A., and Szarpa, K. (2008). Persistent infection with *Pseudomonas aeruginosa* in ventilator-associated pneumonia. Am. J. Respir. Crit. Care Med. *178*, 513–519.

Epstein, R.B., Waxman, F.J., Bennett, B.T., and Andersen, B.R. (1974). *Pseudomonas* septicemia in neutropenic dogs. I. Treatment with granulocyte transfusions. Transfusion *14*, 51–57.

Faure, K., Shimabukuro, D., Ajayi, T., Allmond, L.R., Sawa, T., and Wiener-Kronish, J.P. (2003). O-antigen serotypes and type III secretory toxins in clinical isolates of *Pseudomonas aeruginosa*. J. Clin. Microbiol. *41*, 2158–2160.

Finke, M., Muth, G., Reichhelm, T., Thoma, M., Duchene, M., Hungerer, K.D., Domdey, H., and von Specht, B.U. (1991). Protection of immunosuppressed mice against infection with *Pseudomonas aeruginosa* by recombinant *P. aeruginosa* lipoprotein I and lipoprotein I-specific monoclonal antibodies. Infect. Immun. *59*, 1251–1254.

Fomsgaard, A., Dinesen, B., Shand, G.H., Pressler, T., and Hoiby, N. (1989). Antilipopolysaccharide antibodies and differential diagnosis of chronic *Pseudomonas aeruginosa* lung infection in cystic fibrosis. J. Clin. Microbiol. 27, 1222–1229.

Garner, C.V., DesJardins, D., and Pier, G.B. (1990). Immunogenic properties of *Pseudomonas aeruginosa* mucoid exopolysaccharide. Infect. Immun. *58*, 1835–1842.

Germanier, R., and Furer, E. (1975). Isolation and characterization of *galE* mutant Ty 21a of *Salmonella typhi*: a candidate strain for a live, oral typhoid vaccine. J. Infect. Dis. *131*, 553–558.

Gilleland, H.E., Jr., Parker, M.G., Matthews, J.M., and Berg, R.D. (1984). Use of a purified outer membrane protein F (porin) preparation of *Pseudomonas aeruginosa* as a protective vaccine in mice. Infect. Immun. *44*, 49–54.

Gottlieb, D.J., Cryz, S.J., Jr., Furer, E., Que, J.U., Prentice, H.G., Duncombe, A.S., and Brenner, M.K. (1990). Immunity against *Pseudomonas aeruginosa* adoptively transferred to bone marrow transplant recipients. Blood 76, 2470–2475.

Haghbin, M., Armstrong, D., and Murphy, M.L. (1973). Controlled prospective trial of *Pseudomonas aeruginosa* vaccine in children with acute leukemia. Cancer *32*, 761–766.

Hancock, R.E., Mutharia, L.M., and Mouat, E.C. (1985). Immunotherapeutic potential of monoclonal antibodies against *Pseudomonas aeruginosa* protein F. Eur. J. Clin. Microbiol. *4*, 224–227.

Harvath, L., and Andersen, B.R. (1976). Evaluation of type-specific and non-type-specific *Pseudomonas* vaccine for treatment of pseudomonas sepsis during granulocytopenia. Infect. Immun. *13*, 1139–1143.

Harvath, L., Andersen, B.R., and Amirault, H.J. (1976a). Passive immunity against *Pseudomonas* sepsis during granulocytopenia. Infect. Immun. *14*, 1151–1155.

Harvath, L., Andersen, B.R., Zander, A.R., and Epstein, R.B. (1976b). Combined pre-immunization and granulocyte transfusion therapy for treatment of *Pseudomonas* septicemia in neutropenic dogs. J. Lab. Clin. Med. *87*, 840–847.

Hauser, A.R. (2009). The type III secretion system of *Pseudomonas aeruginosa*: infection by injection. Nat. Rev. Microbiol. 7, 654–665.

Hauser, A.R., Cobb, E., Bodi, M., Mariscal, D., Valles, J., Engel, J.N., and Rello, J. (2002). Type III protein secretion is associated with poor clinical outcomes in patients with ventilator-associated pneumonia caused by *Pseudomonas aeruginosa*. Crit. Care Med. *30*, 521–528.

Hemachandra, S., Kamboj, K., Copfer, J., Pier, G., Green, L.L., and Schreiber, J.R. (2001). Human monoclonal antibodies against *Pseudomonas aeruginosa* lipopolysaccharide derived from transgenic mice containing megabase human immunoglobulin loci are opsonic and protective against fatal pseudomonas sepsis. Infect. Immun. *69*, 2223–2229.

Hestdal, K., Ruscetti, F.W., Ihle, J.N., Jacobsen, S.E., Dubois, C.M., Kopp, W.C., Longo, D.L., and Keller, J.R. (1991). Characterization and regulation of RB––8C5 antigen expression on murine bone marrow cells. J. Immunol. *147*, 22–28.

Holder, I.A. (2004). *Pseudomonas* immunotherapy: a historical overview. Vaccine *22*, 831–839.

Jang, I.J., Kim, I.S., Park, W.J., Yoo, K.S., Yim, D.S., Kim, H.K., Shin, S.G., Chang, W.H., Lee, N.G., Jung, S.B., et al. (1999). Human immune response to a *Pseudomonas aeruginosa* outer membrane protein vaccine. Vaccine 17, 158–168.

Johansen, H.K., and Gotzsche, P.C. (2008). Vaccines for preventing infection with *Pseudomonas aeruginosa* in cystic fibrosis. Cochrane Database Syst. Rev. CD001399.

Jones, R.J., Roe, E.A., and Gupta, J.L. (1980). Controlled trial of *Pseudomonas* immunoglobulin and vaccine in burn patients. Lancet 2, 1263–1265.

Kazmierowski, J.A., Reynolds, H.Y., Kauffman, J.C., Durbin, W.A., Graw, R.G., Jr., and Devlin, H.B. (1977). Experimental pneumonia due to *Pseudomonas aeruginosa* in leukopenic dogs: prolongation of survival by combined treatment with passive antibody to

Pseudomonas and granulocyte transfusions. J. Infect. Dis. *135*, 438–446.

Kim, D.K., Kim, J.J., Kim, J.H., Woo, Y.M., Kim, S., Yoon, D.W., Choi, C.S., Kim, I., Park, W.J., Lee, N., *et al.* (2000). Comparison of two immunization schedules for a *Pseudomonas aeruginosa* outer membrane proteins vaccine in burn patients. Vaccine *19*, 1274–1283.

Knirel, Y.A. (1990). Polysaccharide antigens of *Pseudomonas aeruginosa*. Microbiology *17*, 273–304.

Koh, A.Y., Priebe, G.P., and Pier, G.B. (2005). Virulence of *Pseudomonas aeruginosa* in a murine model of gastrointestinal colonization and dissemination in neutropenia. Infect. Immun. *73*, 2262–2272.

Koh, A.Y., Priebe, G.P., Ray, C., Van Rooijen, N., and Pier, G.B. (2009). Inescapable need for neutrophils as mediators of cellular innate immunity to acute *Pseudomonas aeruginosa* pneumonia. Infect. Immun. *77*, 5300–5310.

Kukavica-Ibrulj, I., and Levesque, R.C. (2008). Animal models of chronic lung infection with *Pseudomonas aeruginosa*: useful tools for cystic fibrosis studies. Lab. Anim. *42*, 389–412.

Lai, Z., Kimmel, R., Petersen, S., Thomas, S., Pier, G., Bezabeh, B., Luo, R., and Schreiber, J.R. (2005). Multi-valent human monoclonal antibody preparation against *Pseudomonas aeruginosa* derived from transgenic mice containing human immunoglobulin loci is protective against fatal pseudomonas sepsis caused by multiple serotypes. Vaccine *23*, 3264–3271.

Lang, A.B., Horn, M.P., Imboden, M.A., and Zuercher, A.W. (2004). Prophylaxis and therapy of *Pseudomonas aeruginosa* infection in cystic fibrosis and immunocompromised patients. Vaccine *22* (Suppl. 1), S44–48.

Langford, D.T., and Hiller, J. (1984). Prospective, controlled study of a polyvalent *Pseudomonas* vaccine in cystic fibrosis–three year results. Arch. Dis. Child. *59*, 1131–1134.

Lee, N.G., Ahn, B.Y., Jung, S.B., Kim, Y.G., Lee, Y., Kim, H.S., and Park, W.J. (1999). Human anti-*Pseudomonas aeruginosa* outer membrane proteins IgG cross-protective against infection with heterologous immunotype strains of *P. aeruginosa*. FEMS Immunol. Med. Microbiol. *25*, 339–347.

Liu, P.V., Matsumoto, H., Kusama, H., and Bergan, T. (1983). Survey of heat-stable major somatic antigens of *Pseudomonas aeruginosa*. Int. J. Syst. Bacteriol. *33*, 256–275.

Lynch, S.V., and Wiener-Kronish, J.P. (2008). Novel strategies to combat bacterial virulence. Curr. Opin. Crit. Care *14*, 593–599.

Mansouri, E., Gabelsberger, J., Knapp, B., Hundt, E., Lenz, U., Hungerer, K.D., Gilleland, H.E., Jr., Staczek, J., Domdey, H., and von Specht, B.U. (1999). Safety and immunogenicity of a *Pseudomonas aeruginosa* hybrid outer membrane protein F-I vaccine in human volunteers. Infect. Immun. *67*, 1461–1470.

Martinez, D., and Callahan, L.T., 3rd (1985). Prophylaxis of *Pseudomonas aeruginosa* infections in leukopenic mice by a combination of active and passive immunization. Eur. J. Clin. Microbiol. *4*, 186–189.

Matsumoto, M., Matsubara, S., Matsuno, T., Tamura, M., Hattori, K., Nomura, H., Ono, M., and Yokota, T. (1987). Protective effect of human granulocyte colony-stimulating factor on microbial infection in neutropenic mice. Infect. Immun. *55*, 2715–2720.

Meluleni, G.J., Grout, M., Evans, D.J., and Pier, G.B. (1995). Mucoid *Pseudomonas aeruginosa* growing in a biofilm *in vitro* are killed by opsonic antibodies to the mucoid exopolysaccharide capsule but not by antibodies produced during chronic lung infection in cystic fibrosis patients. J. Immun. *155*, 2029–2038.

Moriyama, K., Wiener-Kronish, J.P., and Sawa, T. (2009). Protective effects of affinity-purified antibody and truncated vaccines against *Pseudomonas aeruginosa* V-antigen in neutropenic mice. Microbiol. Immunol. *53*, 587–594.

Oishi, K., Sonoda, F., Iwagaki, A., Ponglertnapagorn, P., Watanabe, K., Nagatake, T., Siadak, A., Pollack, M., and Matsumoto, K. (1993). Therapeutic effects of a human antiflagella monoclonal antibody in a neutropenic murine model of *Pseudomonas aeruginosa* pneumonia. Antimicrob. Agents Chemother. *37*, 164–170.

Opal, S.M., and Cross, A.S. (2005). The use of immunocompromised animals as models for human septic shock. Shock *24* (Suppl. 1), 64–70.

Pennington, J.E., Reynolds, H.Y., Wood, R.E., Robinson, R.A., and Levine, A.S. (1975). Use of a *Pseudomonas aeruginosa* vaccine in patients with acute leukemia and cystic fibrosis. Am. J. Med. *58*, 629–636.

Pier, G. (2005). Application of vaccine technology to prevention of *Pseudomonas aeruginosa* infections. Expert Rev. Vaccines *4*, 645–656.

Pier, G.B. (2003). Promises and pitfalls of *Pseudomonas aeruginosa* lipopolysaccharide as a vaccine antigen. Carbohydr. Res. *338*, 2549–2556.

Pier, G.B., Boyer, D., Preston, M., Coleman, F.T., Llosa, N., Mueschenborn-Koglin, S., Theilacker, C., Goldenberg, H., Uchin, J., Priebe, G.P., *et al.* (2004). Human monoclonal antibodies to *Pseudomonas aeruginosa* alginate that protect against infection by both mucoid and nonmucoid strains. J. Immunol. *173*, 5671–5678.

Pier, G.B., and Ramphal, R.R. (2004). *Pseudomonas aeruginosa*. In Principles and Practices of Infectious Diseases, 6th edn, Mandell, G.L., Bennett, J.E., Dolin, R., eds (Philadelphia, Churchill Livingstone), pp. 2587–2614.

Pier, G.B., and Thomas, D.M. (1982). Lipopolysaccharide and high molecular weight polysaccharide serotypes of *Pseudomonas aeruginosa*. J. Infect. Dis. *148*, 217–223.

Pier, G.B., and Thomas, D.M. (1983). Characterization of the human immune response to a polysaccharide vaccine from *Pseudomonas aeruginosa*. J. Infect. Dis. *148*, 206–213.

Pollack, M., and Young, L.S. (1979). Protective activity of antibodies to exotoxin A and lipopolysaccharide at the onset of *Pseudomonas aeruginosa* septicemia in man. J. Clin. Invest. *63*, 276–286.

Priebe, G.P., Brinig, M.M., Hatano, K., Grout, M., Coleman, F.T., Pier, G.B., and Goldberg, J.B. (2002). Construction and characterization of a live, attenuated *aroA* deletion mutant of *Pseudomonas aeruginosa* as a candidate intranasal vaccine. Infect. Immun. *70*, 1507–1517.

Priebe, G.P., Meluleni, G.J., Coleman, F.T., Goldberg, J.B., and Pier, G.B. (2003). Protection against fatal *Pseudomonas aeruginosa* pneumonia in mice after nasal immunization with a live, attenuated *aroA* deletion mutant. Infect. Immun. *71*, 1453–1461.

Priebe, G.P., Walsh, R.L., Cederroth, T.A., Kamei, A., Coutinho-Sledge, Y.S., Goldberg, J.B., and Pier, G.B. (2008). IL-17 is a critical component of vaccine-induced protection against lung infection by lipopolysaccharide-heterologous strains of *Pseudomonas aeruginosa*. J. Immunol. *181*, 4965–4975.

Rahner, R., Eckhardt, A., Duchene, M., Domdey, H., and von Specht, B.U. (1990). Protection of immunosuppressed mice against infection with *Pseudomonas aeruginosa* by monoclonal antibodies to outer membrane protein OprI. Infection *18*, 242–245.

Ramphal, R. (2004). Changes in the etiology of bacteremia in febrile neutropenic patients and the susceptibilities of the currently isolated pathogens. Clin. Infect. Dis. *39* (Suppl. 1) S25–31.

Rolston, K.V., and Bodey, G.P. (1992). *Pseudomonas aeruginosa* infection in cancer patients. Cancer Invest. *10*, 43–59.

Sadikot, R.T., Blackwell, T.S., Christman, J.W., and Prince, A.S. (2005). Pathogen–host interactions in *Pseudomonas aeruginosa* pneumonia. Am. J. Respir. Crit. Care Med. *171*, 1209–1223.

Scarff, J.M., and Goldberg, J.B. (2008). Vaccination against *Pseudomonas aeruginosa* pneumonia in immunocompromised mice. Clin. Vaccine Immunol. *15*, 367–375.

Schaad, U.B., Lang, A.B., Wedgwood, J., Ruedeberg, A., Que, J.U., Furer, E., and Cryz Jr., S.J. (1991). Safety and immunogenicity of *Pseudomonas aeruginosa* conjugate-A vaccine in cystic fibrosis. Lancet *338*, 1236–1237.

Schulert, G.S., Feltman, H., Rabin, S.D., Martin, C.G., Battle, S.E., Rello, J., and Hauser, A.R. (2003). Secretion of the toxin ExoU is a marker for highly virulent *Pseudomonas aeruginosa* isolates obtained from patients with hospital-acquired pneumonia. J. Infect. Dis. *188*, 1695–1706.

Sorichter, S., Baumann, U., Baumgart, A., Walterspacher, S., and von Specht, B.U. (2009). Immune responses in the airways by nasal vaccination with systemic boosting against *Pseudomonas aeruginosa* in chronic lung disease. Vaccine *27*, 2755–2759.

Tamura, M., Hattori, K., Nomura, H., Oheda, M., Kubota, N., Imazeki, I., Ono, M., Ueyama, Y., Nagata, S., Shirafuji, N., *et al.* (1987). Induction of neutrophilic granulocytosis in mice by administration of purified human native granulocyte colony-stimulating factor (G-CSF). Biochem. Biophys. Res. Commun. *142*, 454–460.

Toth, A., Schodel, F., Duchene, M., Massarrat, K., Blum, B., Schmitt, A., Domdey, H., and von Specht, B.U. (1994). Protection of immunosuppressed mice against translocation of *Pseudomonas aeruginosa* from the gut by oral immunization with recombinant *Pseudomonas aeruginosa* outer membrane protein I expressing *Salmonella dublin*. Vaccine *12*, 1215–1221.

Van Delden, C. (2004). Virulence factors in *Pseudomonas aeruginosa*. In *Pseudomonas*, Ramos, J.-L., ed. (New York, Kluwer Academic/Plenum Publishers), pp. 3–45.

Velardi, A., Cucciaioni, S., Terenzi, A., Quinti, I., Aversa, F., Grossi, C.E., Grignani, F., and Martelli, M.F. (1988). Acquisition of Ig isotype diversity after bone marrow transplantation in adults. A recapitulation of normal B cell ontogeny. J. Immunol. *141*, 815–820.

Vento, S., Cainelli, F., and Temesgen, Z. (2008). Lung infections after cancer chemotherapy. Lancet Oncol. *9*, 982–992.

von Specht, B., Knapp, B., Hungerer, K., Lucking, C., Schmitt, A., and Domdey, H. (1996). Outer membrane proteins of *Pseudomonas aeruginosa* as vaccine candidates. J. Biotechnol. *44*, 145–153.

von Specht, B.U., Knapp, B., Muth, G., Broker, M., Hungerer, K.D., Diehl, K.D., Massarrat, K., Seemann, A., and Domdey, H. (1995). Protection of immunocompromised mice against lethal infection with *Pseudomonas aeruginosa* by active or passive immunization with recombinant *P. aeruginosa* outer membrane protein F and outer membrane protein I fusion proteins. Infect. Immun. *63*, 1855–1862.

Young, L.S., Meyer, R.D., and Armstrong, D. (1973). *Pseudomonas aeruginosa* vaccine in cancer patients. Ann. Intern. Med. *79*, 518–527.

Nosocomial Infections: *Staphylococcus aureus*

12

Alice G. Cheng, Olaf Schneewind and Dominique Missiakas

Abstract

Staphylococcus aureus is the most frequent cause of human skin and soft tissue, bloodstream and respiratory tract infections. Staphylococcal strains have acquired antibiotic resistance traits against available therapies and drug-resistant strains (MRSA, methicillin-resistant *S. aureus*) are currently isolated in up to 80% of hospital and 60% of community-acquired infections (CA-MRSA). Unlike pneumococci and group A streptococci; *S. aureus* infections do not raise immunity against subsequent infections. Consistent with this observation, early efforts to develop vaccines from whole-cell killed preparations of staphylococci have failed. More recent work characterized proteins and carbohydrates in the staphylococcal envelope and examined these molecules as protective antigens in vaccine studies. This article reviews the pathogenesis of *S. aureus* infections as well as past and current efforts that have been pursued to develop effective vaccines.

Isolation and identification of *S. aureus*

In his lecture at the Ninth Surgical Congress in Berlin in 1880, Sir Alexander Ogston reported his observation of the presence of 'Micrococci' associated with pus in surgical wound infections. Later, he would coin the word staphylococci to refer to these particular organisms (Ogston, 1883) and use eggs to isolate pure cultures. By inoculating pure cultures of staphylococci into rabbits, he was able to observe the development of abscesses thereby fulfilling Koch's postulates for the identification of the aetiological agent of suppurative abscesses (Ogston, 1883). *S. aureus* is a physiological commensal of the human skin, nares, and mucosal surfaces. In 1884, Rosenbach isolated two colony types of staphylococci found on humans and based on their pigmentation proposed the nomenclature *S. aureus* and *S. albus* for the yellow and white isolates, respectively (Rosenbach, 1884). The latter species is now named *S. epidermidis*. Genotypic properties and refined taxonomy have led to the distinction of over 30 species during the last four decades (Gotz *et al.*, 2006).

S. aureus and nosocomial burdens

S. aureus is the leading cause of hospital acquired infections (Lowy, 1998). The spectrum of human diseases caused by staphylococci ranges from soft tissue infections, abscesses in organ tissues, osteomyelitis, endocarditis, and toxic shock syndrome to necrotizing pneumonia (Archer, 1998; Lowy, 1998). Over the past forty years, *S. aureus* strains exhibiting multiple antibiotic resistances, methicillin-resistant *S. aureus* (MRSA), have evolved due to continued selective pressure (Kaplan *et al.*, 2005). MRSA are now isolated in up to 60% of community and 80% of hospital infections and represent a formidable challenge for antibiotic therapy, which in many instances fails because of rapidly evolving drug resistance traits (Kaplan *et al.*, 2005). On average, 4.6% of all individuals admitted into American hospitals, 10% of dialysis patients, 10% of individuals with ventriculoperitoneal shunts, 2% of intravenous drug abusers, and 3% of nursing home residents will suffer MRSA infections (Fridkin *et al.*, 2005; Kaplan *et al.*, 2005). Community-acquired MRSA infections (CA-MRSA) occur in children without predisposing risk factors, in young adults as well as healthy newborns (Projan *et al.*, 2006). As the rate of MRSA and CA-MRSA infections continues to rise, the United States is on the brink of a public health crisis, which can only be diverted by the development of new therapeutic strategies or by vaccines that protect humans against *S. aureus* infections (Projan *et al.*, 2006; Tenover *et al.*, 2001).

Epidemiology of *Staphylococcus* infections and genomic variability

Epidemiology of *Staphylococcus* infections

Because they colonize the human skin, staphylococcal strains are exposed to all antibiotic therapies (Neu, 1992). *S. aureus* is a major cause of nosocomial and community-acquired infections. Whenever drug-resistant microbes emerge, these strains can spread by direct contact very rapidly among human populations, as exemplified by the threat of MRSA and vancomycin-resistant *S. aureus* (VRSA) (Brumfitt and Hamilton-Miller, 1989; Chang *et al.*, 2003; Hiramatsu *et al.*, 1997). Remarkably, a rapid spread of highly virulent *S. aureus* strains has been observed worldwide (Ochoa *et al.*, 2005; Purcell and Fergie, 2005; Smith and Cook, 2005; Tristan *et al.*, 2007). These strains demonstrate a novel epidemiological pattern. They are frequently transmitted outside of the hospital environment, among otherwise healthy individuals. They have been designated community-associated methicillin-resistant *S. aureus* (CA-MRSA) (Chambers, 2001; Control, 1981; Herold *et al.*, 1998). Most isolates carry the SCC*mec* IV genetic element that confers resistance to β-lactam antimicrobials, rendering this entire class of antimicrobials obsolete (Ma *et al.*, 2002) and appear to encode unique virulence traits. For example, the Panton–Valentine leukocidin (PVL), a pore-forming cytotoxin with specificity for leucocytes (Gillet *et al.*, 2002; Gladstone and Van Heyningen, 1957; Issartel *et al.*, 2005; Lina *et al.*, 1999; Vandenesch *et al.*, 2003) shows

a high degree of epidemiological association with invasive *S. aureus* disease. The PVL-encoded gene is carried on a bacteriophage (Kaneko and Kamio, 2004). Additional phage-encoded proteins that may contribute increased virulence include the plasminogen activator staphylokinase (Sak) (Bokarewa *et al.*, 2006; Coleman *et al.*, 1989; Rooijakkers *et al.*, 2005a), as well as the immunomodulatory proteins CHIPS (chemotaxis inhibiting protein) and SCIN (staphylococcal complement inhibitor) (Postma *et al.*, 2004; Rooijakkers *et al.*, 2005b, 2006). In addition, a novel class of secreted staphylococcal peptides, termed phenol-soluble modulins (PSMs) appear to be highly expressed in current CA-MRSA isolates and has been shown to target human neutrophils for destruction (Wang *et al.*, 2007). Thus, the collective acquisition of pathogenic traits may render these strains more capable of causing significant infection in healthy hosts.

Genomic variability

Several different *S. aureus* and *S. epidermidis* strains have been sequenced and encompass between 2,550 and 2,870 genes (Baba *et al.*, 2002; Diep *et al.*, 2006; Fitzgerald *et al.*, 2001; Kuroda *et al.*, 2001) and display up to 22% of DNA sequence variability (Fitzgerald *et al.*, 2001). The vast majority of staphylococcal virulence functions (including surface proteins, sortase, α-haemolysin, exotoxins and *agr* regulon) are encoded on the bacterial chromosome (Kuroda *et al.*, 2001). These genes are not associated with mobile DNA elements and are found in all staphylococcal strains. Most of the genome variability is brought about by insertions of transposons and mobile elements, as well as prophages and plasmids (Novick, 2003b). For example, MRSA strains carry a known hotspot for the insertion of transposons and insertion sequences of the methicillin resistance cassette (Hanssen and Ericson Sollid, 2006; Katayama *et al.*, 2000). Genomic analysis of strain USA300 shows that the same hotspot acquired the ACME (arginine catabolic mobile element) gene cluster (Diep *et al.*, 2006). ACME may enable the strain to evade host immune responses. USA300 also encodes a pathogenicity island that carries two enterotoxins not found in other sequenced genomes except

S. aureus strain COL, and three plasmids with multiple antibiotic resistance determinants (Diep *et al.*, 2006). Pathogenicity islands are common in *S. aureus* and encode toxins associated with specific clinical conditions (toxinoses), some of which function as superantigens (Novick and Subedl, 2007). Secretion of superantigen stimulates large T-cell populations, and causes massive cytokine release that eventually leads to toxic shock. Pathogenicity islands are 15–20 kbp DNA elements that occupy constant positions in the chromosomes of toxigenic strains. Phage-related features and the presence of flanking direct repeats characterize these mobile elements (Novick and Subedi, 2007).

Recent work identified *S. aureus* bacteriophages as important contributors to pathogenesis and the evolution of staphylococcal genomes (Novick, 2003b; van Wamel *et al.*, 2006). All sequenced genomes of clinical *S. aureus* harbour at least one or up to three prophages (Brussow *et al.*, 2004). Phage excision or transduction has been observed during the course of chronic lung infections in cystic fibrosis patients and are cause of genomic variation (Goerke *et al.*, 2006). Phages have been shown to transfer pathogenicity islands between strains which is best exemplified by φ80a-mediated excision and transduction of SaPI1 (Ruzin *et al.*, 2001). Innate immune modulators such as chemotaxis inhibitory protein (CHIPS) and staphylococcal complement inhibitor (SCIN) have been shown to be encoded by β-haemolysin converting phages (van Wamel *et al.*, 2006). Prophages are likely to encode additional factors that favour their replication in infected hosts. Recently, a variant of the clinical isolate *S. aureus* Newman lacking all prophages was constructed and shown to be unable to cause disease in animals (Bae *et al.*, 2006).

The envelope of *S. aureus*, peptidoglycan biogenesis and antibiotic resistance

Staphylococci elaborate a thick (60–80 nm) continuous cell wall also called the murein sacculus. The cell wall is composed largely of peptidoglycan also known as mucopeptide or murein. Other essential cell wall polymers include wall teichoic acid and surface proteins both of which are covalently attached to the peptidoglycan or tethered to the membrane by a diglucosyl diacylglycerol modification, respectively.

Other polymers are located on the cell surface of staphylococci, and these include capsular polysaccharides (O'Riordan and Lee, 2004) as well as polysaccharide intercellular adhesin (PIA). The genes responsible for PIA synthesis were first revealed in *S. epidermidis* and designated *icaABC*. PIA is a polymer of *N*-acetylglucosamine, a surface carbohydrate involved in biofilm formation (Heilmann *et al.*, 1996) that contributes to the pathogenesis of biomaterial-associated infections (Rupp *et al.*, 2001). The *ica* genes are also found in other staphylococci including *S. aureus* where they contribute to the formation of biofilms on the surfaces of biomedical implants (Cramton *et al.*, 1999). Detailed information on capsular polysaccharides and PIA can be found in the several reviews (Gotz *et al.*, 2006; O'Riordan and Lee, 2004). Cell wall teichoic acid (WTA) and lipoteichoic acid (LTA) are secondary wall polymers that traverse the envelope (Archibald *et al.*, 1973). In *S. aureus* about 8% of peptidoglycan is covalently modified with WTA by phosphodiester linkages on the C6 carbon of muramic acid residues (Hay *et al.*, 1965a; Hay *et al.*, 1965b). TA (teichoic acids) consist of chains of as many as 30 glycerol or ribitol (a 5-C polyhydric alcohol) residues with phosphodiester links and with various substituents (sugars, choline, D-alanine) (Neuhaus and Baddiley, 2003). D-Alanyl esterification of TA is promoted by the Dlt locus and is dispensable for cell viability; however, *dlt* mutants of *S. aureus* are more susceptible to neutrophil killing during infection (Collins *et al.*, 2002). WTA mutants of *S. aureus* are viable *in vitro* but have been shown to be unable to colonize cotton rat nares (Weidenmaier *et al.*, 2004). Unlike WTA, LTA is a 1–3-linked glycerol-phosphate polymer linked to a glycolipid in the membrane. *S. aureus* LTA is retained by a glycolipid anchor [diglucosyl diacylglycerol (Glc$_2$-DAG)] in bacterial membranes (Duckworth *et al.*, 1975; Fischer, 1990) and synthesized as a polyglycerol-phosphate polymer on the outer surface of bacterial membranes from phosphatidyl glycerol substrate (Koch *et al.*, 1984). Staphylococci can produce polyglycerol-phosphate polymers even in the absence of glycolipids, and instead anchor LTA via diacylglycerol (Gründling and Schneewind, 2007b). In contrast, synthesis of polyglycerol-phosphate LTA is required for *S. aureus* growth and cell division, as *ltaS*, the gene responsible for LTA synthesis, cannot be deleted without loss of viability at 37°C (Gründling and Schneewind, 2007a; Oku *et al.*, 2009).

Peptidoglycan

The peptidoglycan consists of a backbone of glycan chains of the repeating disaccharides N-acetylglucosamine (NAG) and its lactic ether N-acetylmuramic acid (NAM). Each disaccharide carries a tetrapeptide substituent of alternating L and D amino acids. A peptide bridge links the terminal COOH of D-alanine of one tetrapeptide to an NH2 group of a tetrapeptide on a neighbouring glycan chain. In *S. aureus*, each interbridge peptide consists of a pentaglycine. *S. aureus* peptidoglycan is extensively modified on the *N*-acetylmuramic acid residues with *O*-acetyl substituents such as the 4-*N*, 6-*O*-diacetyl derivative (Tipper *et al.*, 1971). This modification renders the peptidoglycan resistant to egg-white lysozyme and other muramidases (Warren and Gray, 1965).

Resistance to vancomycin and methicillin

In response to vancomycin selective pressure, some staphylococcal strains acquired mutations that trigger altered cell wall envelope structures and low-level vancomycin resistance (Hiramatsu *et al.*, 1997; Weigel *et al.*, 2003). Other strains such as VRSA acquired the *vanA* resistance gene from enterococci (Chang *et al.*, 2003; Tenover *et al.*, 2001). VanA ligates β-hydroxy carboxylic acids, for example D-lactate, to D-Ala and the product, D-Ala-D-lactate is then incorporated into wall peptides and lipid II: L-Ala-D-iGln-(NH₂-Gly₅)-L-Lys-D-Ala-D-lactate-COOH as opposed to the wild-type wall peptide L-Ala-D-iGln-(NH₂-Gly₅)-L-Lys-D-Ala-D-Ala-COOH (Bugg *et al.*, 1991). The net result of staphylococcal *vanA* gene acquisition, vancomycin resistance, stems from the reduced affinity of the glycopeptide antibiotic to the depsipeptide bond between D-Ala-D-lactate (Walsh, 1993).

Resistance to methicillin by MRSA results from the acquisition of the oxacillin/methicillin-resistance gene *mecA*. *mecA* was possibly acquired from *S. sciuri* by an unknown mechanism (Wu *et al.*, 2001). MecA encodes a penicillin-binding protein (PBP) named PBP2a that displays unusual low affinity for all β-lactam antibiotics (Archer and Niemeyer, 1994; Hartman and Tomasz, 1984). The genetic determinant of methicillin resistance is located on a mobile genetic element called staphylococcal cassette chromosome (SCC) in *S. aureus* (21–67 kb fragment) (Katayama *et al.*, 2000). Because MRSA and VRSA strains have also acquired macrolide, tetracycline, aminoglyocoside, chloramphenicol and fluoroquinolone resistance mechanisms (Kuroda *et al.*, 2001), pharmaceutical, academic research and medical communities have suddenly begun to realize that their therapeutic arsenals are nearing depletion, posing the pre-antibiotic threat for infectious disease therapy (Projan, 2003).

Virulence strategies and diseases

Virulence determinants

S. aureus is a versatile pathogen that can be isolated from many different pathological-anatomical sites or disease entities (Kuehnert *et al.*, 2006). Research over the past several decades identified *S. aureus* cell wall anchored proteins, secreted toxins, capsular- and exo-polysaccharides, iron-transport systems, and modulators of host immune functions in addition to antibiotic-resistance genes as important virulence factors (Archer, 1998; Marraffini *et al.*, 2006; Novick and Jiang, 2003). In addition, the complex bacterial surface of *S. aureus* forms the organism's first line of defence against the host immune system. The cell wall of *S. aureus* is comprised of peptidoglycan, providing a rigid structure to the pathogen while serving as a scaffold for the attachment of surface proteins and carbohydrates. Together, these surface structures play an essential role in allowing the pathogen to gain access to host tissues.

Staphylococcal versatility and productive outcome of infection are enabled by complex regulation of virulence determinants. Of note, all virulence factors are secreted products or surface exposed molecules (teichoic acids; capsule). We will refer to the secreted products as the staphylococcal secretome. The secretome can be subdivided into factors ubiquitously present in all strains, and those carried on prophages or pathogenicity islands (prone to genetic variability). The latter factors are exemplified by prophage encoded PVL, Sak, CHIPS and SCIN (reviewed by (Foster, 2005; Rooijakkers *et al.*, 2005b)) as well as by pathogenicity island encoded superantigens (Novick, 2003b). Ubiquitous factors include lipoproteins, cell wall anchored proteins and extracellular proteins (reviewed by (Dinges *et al.*, 2000; Marraffini *et al.*, 2006; Navarre and Schneewind, 1999)). Further, we can distinguish between factors secreted by the Sec machinery and factors transported by separate pathways. In staphylococci, the ESAT-6 secretion system represents one of the Sec-independent secretion systems that contributes to persistence during infection (Burts *et al.*, 2008; Burts *et al.*, 2005; Missiakas and Schneewind, 2007). Likewise, the newly described phenol-soluble modulins (PSMs) highly expressed in CA-MRSA isolates (McNeil *et al.*, 2007), must also be secreted by a Sec-independent mechanism. The corresponding genes are, however, not yet known.

One important class of staphylococcal surface proteins is anchored to the cell wall through the activity of the transpeptidase sortase A (SrtA) (Mazmanian *et al.*, 2001). The substrates of SrtA contain an LPXTG motif sorting signal at the C-terminus; cleavage of the surface protein between the T and G residues of the LPXTG motif allows for the generation of an acyl enzyme, which is resolved upon nucleophilic attack by the lipid II moiety. The modified surface protein attached to lipid II is subsequently incorporated into the growing cell wall. Many SrtA anchored proteins bind extracellular matrix components and are known as microbial surface components recognizing adhesive matrix molecules, or MSCRAMMs (Patti *et al.*, 1994). These include fibronectin-binding proteins (FnbA and FnbB), fibrinogen binding proteins ClfA and ClfB, collagen adhesin (Cna) and a collection of Sdr proteins containing serine-aspartate repeats. A second group of SrtA substrates consists of the

iron-regulated surface determinants in particular IsdA and IsdB which have been shown to be critical for iron uptake during infection (Mazmanian *et al.*, 2003; Mazmanian *et al.*, 2002). Because they facilitate entry into the host tissues and nutrient acquisition by invading staphylococci, the substrates of SrtA have been considered attractive candidates for vaccine development (Rivas *et al.*, 2004; Stranger-Jones *et al.*, 2006).

Recently, it has been established that the increased virulence in the USA300/USA500 sublineage is afforded by differential expression of core virulence determinants, such as phenol-soluble modulins and alpha-toxin (Li *et al.*, 2009). Because, the virulence phenotype of USA300 was already established in its progenitor it was also suggested that acquisition of mobile genetic elements played a limited role in the evolution of USA300 virulence or a role not yet revealed (Li *et al.*, 2009). Thus, it is important to emphasize that differential gene expression appears to be key to the evolution of USA300 virulence as opposed to acquisition of new molecular mechanisms of virulence.

Regulation of virulence determinants

The pathogenesis of staphylococcal infections requires the complex regulation of virulence determinants by *agr*, the accessory gene regulator locus, in addition to other regulatory elements. During infection, staphylococci perform a bacterial census via AgrB-mediated secretion of autoinducing peptides that are derived from AgrD. Autoinducing peptides bind the sensory kinase AgrC at threshold concentration. Signal transduction involves phosphorelay with the response regulator AgrA and transcriptional regulation of a plethora of exotoxin genes (Novick, 2003a) (Fig. 12.1). This bacterial census ensures massive secretion of exotoxins when staphylococcal counts are high, increasing the likelihood of bacterial spread within infected tissues and systemic dissemination.

Additional transcription factors of the Sar family as well as two-component regulatory systems play important roles in regulating *agr* expression or modifying the production of toxin genes, albeit that the mechanisms of signal transduction or environmental signalling are not yet known in

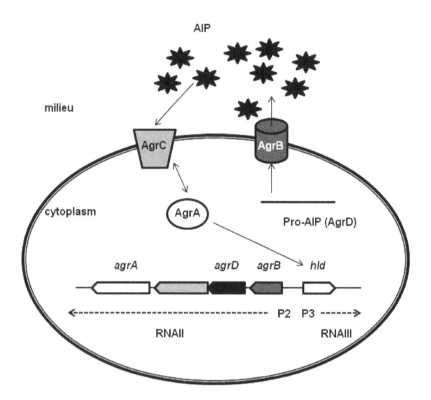

Figure 12.1 The Agr system. *agrD* encodes the pro-AIP peptide that is processed and secreted by AgrB. Mature extracellular AIP binds to the receptor-histidine kinase, AgrC. This activates a phosphorelay cascade between AgrC and the response regulator, AgrA, which in conjunction with SarA (not shown), activates the two *agr* promoters, P2 and P3, leading to the production of RNA III, which controls transcription of target genes through additional intracellular regulatory mediators, including a second two-component module, *saeRS* (see text for details).

molecular detail (Cheung *et al.*, 2004; Novick, 2003a; Pragman and Schlievert, 2004). SarA homologues represents a family of DNA binding proteins homologous to SarA a global regulator of virulence factors (Cheung *et al.*, 2004; Novick, 2003a; Pragman and Schlievert, 2004) (Fig. 12.2). The regulation of virulence determinants can be summarized as follows. The synthesis of cell surface adhesins (cell wall proteins) during exponential growth of staphylococci coincides with the expression of SarA and the two component system SaeRS (Giraudo *et al.*, 1997, 1999). Synthesis of cell wall proteins is disrupted upon transition from exponential to post-exponential growth and instead, synthesis and production of extracellular toxins such as haemolysin is observed. Maximal expression of SarA and the ensuing activation of Agr (Roberts *et al.*, 2006) are responsible for this transition (Dunman *et al.*,

2001). SarA expression is repressed by SarA and SarR but induced by SigB, a stress-response sigma factor (Bischoff *et al.*, 2001). The Agr locus is activated at threshold concentration of its secreted auto-inducing peptides AgrD (Novick, 2003a). Activation of the Agr locus is also regulated by SarA, MgrA, ArlRS, SarX and SarU (reviewed by Cheung *et al.*, 2007). Agr activation results in increased transcription at two divergent promoters yielding RNAII a transcript for the *agrBDCA* operon and RNAIII, a regulatory RNA that also leads to the synthesis of δ-haemolysin (Fig. 12.1) (Novick, 2003a). The molecular link between RNAIII and downstream gene regulation is unclear but coincides with activation of SaeRS (Novick and Jiang, 2003) and down-regulation of Rot (SarA homologue) (Said-Salim *et al.*, 2003). As a result, *sarT* and subsequently *sarS* expressions are repressed. SarT is an activator of SarS.

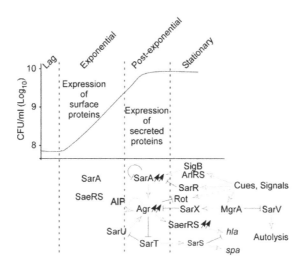

Figure 12.2 Regulation of virulence determinants in *S. aureus* (adapted from (Cheung *et al.*, 2007)). Synthesis of virulence determinants and activation/repression of regulatory factors are shown as a function of cell growth. Synthesis of surface proteins during exponential phase coincides with SarA and SaeRS expression. Instead, when cells enter post-exponential growth, production of extracellular toxins is observed. This transition corresponds to maximal expression of SarA and the ensuing activation of *agr*. A cross-talk between many regulatory factors has been implicated in the fine tuning of the Agr-SarA response, most of which are depicted on the figure (see details in text).

SarS is a repressor of alpha toxin production and an activator of protein A synthesis (Cheung *et al.*, 2004; Tegmark *et al.*, 2000). In addition, the Agr response appears to be amplified by SarU (Cheung *et al.*, 2007) (Fig. 12.2).

In response to environmental cues, this complex regulatory network is thought to coordinate expression of adhesins early during colonization and expression of toxins late during infection to facilitate bacterial spread in host tissues (Fig. 12.2). However, it is important to note that this assumption has not been validated in an animal model of infection and is derived solely from *in vitro* data (Chien *et al.*, 1998; Dunman *et al.*, 2001; Novick, 2003a). It is hypothesized that regulatory switches between exponential and post-exponential growth may mimic replication of bacteria and migration to tissues where abscess formation occurs followed by bacterial escape and spread.

Survival of staphylococci in blood

To survive in blood, *S. aureus* must escape a variety of innate immune mechanisms, such as antimicrobial peptides, complement, and phagocytic killing (Foster, 2005; Peschel and Sahl, 2006). The principal defence against *S. aureus* infection is provided by neutrophilic PMNs (neutrophils), which constitute 60–70% of human white blood cells (Voyich *et al.*, 2005). Patients afflicted with chronic granulomatous disease (CGD), a genetic disorder that impairs neutrophils from generating a cellular oxidative response, suffer from recurrent *S. aureus* infection. The disease can be explained by a mutation in the multi-subunit NADPH oxidase complex responsible for the generation of the superoxide radical in the phagocytic vacuole (Dinauer and Orkin, 1992). Reactive oxygen species, along with the acidic vacuolar environment, prove toxic to *S. aureus*, and help curtail bacterial spread. *S. aureus* forms a fairly large yellow colony on rich medium caused by the production of the orange carotenoid staphyloxanthin (Marshall and Wilmoth, 1981; Pelz *et al.*, 2005). Production of this membrane-bound pigment helps scavenge reactive oxygen species and protects *S. aureus* from neutrophil killing (Clauditz *et al.*, 2006).

This attribute seems important for subcutaneous abscess formation (Liu *et al.*, 2005).

Proteins of the complement cascade are also important in innate host defence against *S. aureus* (Neth *et al.*, 2002; Verbrugh *et al.*, 1979; Wilkinson *et al.*, 1978). These proteins may bind to the staphylococcal surface, thereby facilitating phagocytic uptake of the pathogen and by releasing proteolytic fragments of C3 and C5 serve as potent chemoattractant peptides for phagocytes. Complement depletion in experimental animals renders them more susceptible to septicaemia (Sakiniene *et al.*, 1999). Staphylococci subverts these mechanisms by secreting factors inhibitory for complement activation and neutrophil chemotaxis (de Haas *et al.*, 2004; Rooijakkers *et al.*, 2005a). They also secrete toxins that lyse neutrophils (Wang *et al.*, 2007) and superantigens to inappropriately activate the host's immune system (Jardetzky *et al.*, 1994).

A recent study examined the contribution of all proteins anchored by sortase A for survival in whole blood and identified ClfA and AdsA (Thammavongsa *et al.*, 2009). The phenotype of *clfA* mutants represented an expected result, as the encoded clumping factor A product is known to precipitate fibrin and interfere with macrophage and neutrophil phagocytosis (Higgins *et al.*, 2006; Palmqvist *et al.*, 2004). Mutations in *clfA* displayed defects of survival in blood that were associated with a reduction in staphylococcal load in kidney tissue on day 5 (Thammavongsa *et al.*, 2009). However, unlike *clfA* variants, *adsA* mutants displayed reduced ability to form abscess lesions (Thammavongsa *et al.*, 2009). AdsA is a cell wall-anchored protein of *S. aureus* with 5-nucleotidase signature sequences ILHTnDiH-GrL (residues 124–134) and YdamaVGNHEFD (residues 189–201), suggesting that the protein may catalyse the synthesis of adenosine from 5′-AMP. This conjecture was tested and confirmed by purifying recombinant AdsA and demonstrating that it indeed displays 5′-nucleotidase activity *in vitro*. Further, *adsA* was shown to be required for synthesis of adenosine during infection (Thammavongsa *et al.*, 2009). These data could be explained on the basis that in mammals, adenosine assumes an essential role in regulating innate and acquired immune responses (Thiel *et*

al., 2003). Strong or excessive host inflammatory responses, e.g. in response to bacterial infection, exacerbate the tissue damage inflicted by invading pathogens (Thiel *et al.*, 2003). Successful immune clearance of microbes therefore involves the balancing of pro- and anti-inflammatory mediators. In humans, generation of adenosine at sites of inflammation, hypoxia, organ injury, and traumatic shock is mediated by two sequential enzymes. CD39/ecto-nucleoside triphosphate diphosphohydrolase-type-1 (ENTPD1) is the dominant vascular ecto-nucleotidase that hydrolyses extracellular nucleotides. CD39 converts circulating ATP and ADP to AMP (Eltzschig *et al.*, 2003). CD73, expressed on the surface of endothelial cells and subsets of T cells, then converts 5′-AMP to adenosine (Zimmermann, 1992). Thus, production of AdsA allows staphylococci to generate adenosine at sites of infection to and dampen the host immune response favouring escape from phagocytic clearance and subsequent formation of staphylococcal organ abscesses. All of these activities are mediated by AdsA, a cell surface anchored 5′-nucleotidase that converts AMP to adenosine (Thammavongsa *et al.*, 2009).

Liquefaction necrosis and abscesses

Essentially every organ system and tissue of humans is susceptible to infection with *S. aureus*. The most common site of infection is the skin and soft tissues, however infection with this pathogen also results in frequent infection of deeper tissues, causing pneumonia upon replication in the lungs, osteomyelitis of the skeletal system, and endocarditis when affecting the lining of the heart (Lowy, 1998). Bloodstream infection, or septicaemia, is often related to seeding of these deeper organs, and accounts for approximately 75,000 cases of disease per year in the United States alone (Kuehnert *et al.*, 2005). The pathological consequence of seeding can be probed as the ability of the organism to form abscesses. Abscesses appear as early as day 2 post infection and mature to enclose a central population of staphylococci, surrounded by a layer of eosinophilic, amorphous material, and a large cuff of PMNs (Fig. 12.3; Cheng *et al.*, 2009). The contribution of each cell wall-anchored surface proteins, of envelope-associated proteins Eap and Emp and of two carbohydrate

day 2   day 5

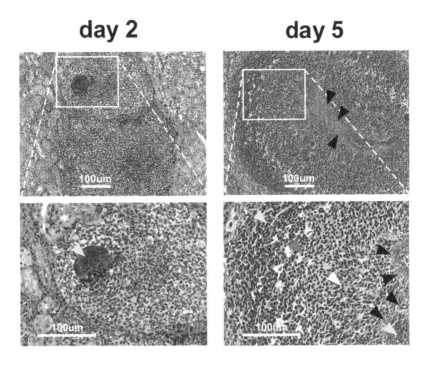

Figure 12.3 Histopathology of staphylococcal abscess communities, SACs. BALB/c mice were infected with *S. aureus* Newman *via* retro-orbital injection. Thin-sectioned, H&E-stained tissues of infected kidneys on day 2 and day 5 following infection were analysed by light microscopy. On day 2, an unorganized infiltrate of PMNs are characteristic of early infectious lesions, these lesions may be accompanied by bacterial communities (grey arrow). By day 5, the abscess structure matures where the central nidus of bacteria (grey arrow) is enclosed by an amorphous, eosinophilic pseudocapsule (black arrows) and surrounded by a zone of dead PMNs (white arrow), a zone of apparently healthy PMNs (red arrow), and a rim of necrotic PMNs (green arrow), separated by a layer of eosinophilic fibrin from healthy kidney tissue. A colour version of this figure can be located in the plate section at the back of the book.

structures, capsular polysaccharide and poly-*N*-acetylglucosamine has been examined in the renal abscess model of disease (Cheng *et al.*, 2009). This comparative analysis was used to define four discrete stages of a developmental programme for abscess formation and maturation as shown in Fig. 12.4. Stage I represents a step during which *S. aureus* survives in the bloodstream following intravenous inoculation and disseminates *via* the vasculature to peripheral organ tissues. During Stage II, staphylococci in renal tissues (or other organs) attract a massive infiltrate of polymorphonuclear leucocytes and other immune cells. In Stage III abscesses mature with a central accumulation of the pathogen, referred as a *staphylococcal abscess community* (SAC), enclosed by an eosinophilic pseudocapsule (Cheng *et al.*, 2009). Bacteria are surrounded by a zone of dead PMNs,

apparently healthy PMNs, and finally an outer zone of dead PMNs with a rim of eosinophilic material. In Stage IV, abscesses mature and rupture on the organ surface, thereby releasing staphylococci once again into circulation.

Mutations in genes affecting capsular polysaccharide or poly-*N*-acetylglucosamine biosyntheses do not affect the virulence of *S. aureus* (Cheng *et al.*, 2009). Mutations in cell wall anchored protein genes *sdrD*, *isdB*, *clfB*, *isdA*, *clfA*, and *isdC* caused reduced bacterial load. Significant defects in abscess formation were only observed for *sdrD*, *isdB*, and *isdA* (Cheng *et al.*, 2009). Interestingly, mutations in *clfA* and *clfB* exhibited defects in staphylococcal load but not in abscess formation in agreement with previous studies suggesting that clumping factor proteins mediate fibrinogen binding, as well as resistance to phagocytic

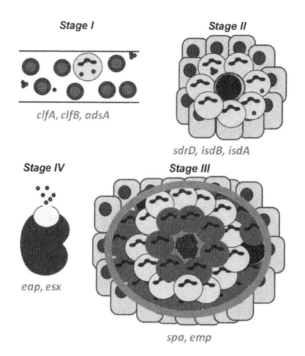

Stage I

clfA, clfB, adsA

Stage II

sdrD, isdB, isdA

Stage IV

eap, esx

Stage III

spa, emp

Figure 12.4 Working model for staphylococcal abscess formation and persistence in host tissues showing stage I–IV. See text for details.

clearance, attributes required for pathogen survival and dissemination in blood (McDevitt *et al.*, 1994; Ní Eidhin *et al.*, 1998). Thus, *clfA* and *clfB* along with *adsA* are crucial at Stage I (Fig. 12.4), while *isdA*, *isdB* and *sdrD* are important for Stage II of the developmental process. The function of SdrD remains unknown. Most certainly, staphylococci must require haem-iron scavenging *via* IsdA and IsdB for expansive growth during this stage. Protein A (*spa*) and *emp* mutations do not greatly affect the bacterial load on day 5 post infection and instead reduce the number of abscesses, suggesting that these genes function at a later time represented as Stage III on Fig. 12.4. Protein A impedes phagocytosis by binding the Fc component of immunoglobulin (Jensen, 1958; Uhlén *et al.*, 1984), activates platelet aggregation *via* the von Willebrand factor (Hartleib *et al.*, 2000), functions as a B-cell superantigen by capturing the Fab region of VH3 bearing IgM (Roben *et al.*, 1995), and, through its activation of TNFR1, can initiate staphylococcal pneumonia (Gomez *et al.*, 2004). Staphylococcal abscesses mature over weeks and, following rupture and release into the

peritoneal cavity, lead to new infectious lesions (Stage IV). Variants lacking *eap* or a functional type VII secretion system are defective in persistence and are, therefore, assigned to Stage IV (Burts *et al.*, 2008; Cheng *et al.*, 2009).

Hla and staphylococcal pneumonia

S. aureus is haemolytic on blood agar. All *S. aureus* sequenced genomes examined thus harbour *hla*, the structural gene for α-haemolysin, on their chromosome (O'Reilly *et al.*, 1986, 1990). Some strains of *S. aureus* appear non-haemolytic on blood agar, and this can be attributed to point mutations in either *hla* or the *agr* locus, the global regulator for the expression of secreted proteins. The *hla* gene encodes a 293 amino acid long protein secreted as a water-soluble monomer (Bhakdi and Tranum-Jensen, 1991). Staphylococcal α-haemolysin (Hla or α-toxin) is the founding member of bacterial pore-forming β-barrel toxins (Song *et al.*, 1996). Hla is thought to engage surface receptors of sensitive host cells, thereby promoting its oligomerization into a heptameric prepore and insertion of a β-barrel with 2 nm pore

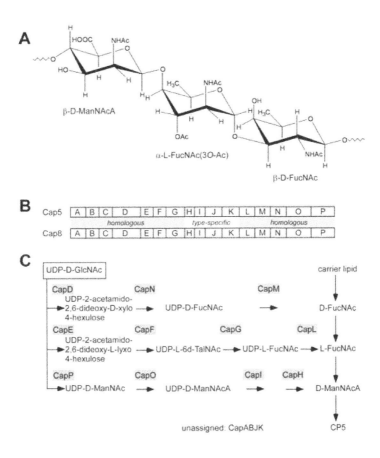

Figure 12.5 *S. aureus* capsular polysaccharide (CP). (A) Repeat structure of type 5 CP. (B) Schematic to illustrate the gene content of the capsular polysaccharide locus on the *S. aureus* chromosome, comparing type 5 (Cap5) and type 8 (Cap8) strains. (C) Schematic illustrating a putative biosynthesis pathway for CP5.

diameter into the plasma membrane (Gouaux *et al.*, 1997). Hla pores form in lymphocytes, macrophages, type I alveolar cells, pulmonary endothelium and erythrocytes, whereas granulocytes and fibroblasts appear resistant to lysis (Bhakdi and Tranum-Jensen, 1991; McElroy *et al.*, 1999). Instillation of purified Hla into rabbit or rat lung tissue triggers vascular leakage and pulmonary hypertension (McElroy *et al.*, 1999; Seeger *et al.*, 1984; Seeger *et al.*, 1990), which has been attributed to release of different signalling molecules, e.g. phosphatidyl inositol, nitric oxide, prostanoids and thromboxane A_2 (Rose *et al.*, 2002; Seeger *et al.*, 1984, 1990; Suttorp *et al.*, 1985).

S. aureus pneumonia is the second most common invasive disease caused by this pathogen, with mortality estimates cited in excess of 30%. In the setting of concomitant influenza A infection, mortality from *S. aureus* pneumonia is often between 50–60% (Finelli *et al.*, 2008; Frazee, 2007; King *et al.*, 2006). *S. aureus* pneumonia causes mortality not only in the hospital, but also in otherwise healthy individuals (Lowy, 1998). Recent increases in morbidity have been attributed to the rapid spread of CA-MRSA (Gillet *et al.*, 2002; Voyich *et al.*, 2006). Many CA-MRSA strains appear to be lysogenized with a bacteriophage that carries *lukS-PV* and *lukF-PV*, the structural genes for PVL (Gillet *et al.*, 2002). The epidemiological association of PVL bacteriophage with *S. aureus* strains from patients with necrotizing pneumonia has led to the hypothesis that PVL is a major contributor to invasive *S. aureus* disease. However, using a mouse model of staphylococcal pneumonia that mimics clinicopathological features of human disease, Bubeck Wardenburg and colleagues identified Hla, but not PVL, as an essential virulence factor in *S. aureus* Newman (Bubeck Wardenburg *et al.*, 2007). Like Hla, PVL is a member of the family of β-channel pore-forming toxins that assemble into heptameric structures that penetrate cell membranes (Gouaux *et al.*, 1997).

Vaccine approaches

Preventative measures to reduce the burden of *S. aureus* disease have been needed for many years, however an FDA licensed vaccine with proven clinical efficacy is still not available (Bubeck Wardenburg *et al.*, 2008). Whole cell vaccination with either killed or live-attenuated *S. aureus* strains did not yield protective immunity against highly virulent *S. aureus* (Rogers and Melly, 1965). Thus, subtractive vaccine development approaches, for example the deletion of genes to reveal crucial protective antigens in whole cell preparations, has not been feasible in the past.

Capsular polysaccharide-based vaccines

Capsular polysaccharides (CP) have played a dominant role in vaccine development, including vaccines that prevent *Haemophilus influenza* meningitis, meningococcal meningitis, or pneumococcal diseases (Austrian, 1977; Claesson *et al.*, 2005; Eskola *et al.*, 1990; Gotschlich *et al.*, 1969). Based on the significant burden of pneumococcal disease in infants and individuals with advancing age, the costs associated with *S. pneumoniae* infection, and the rising rates of *S. pneumoniae* drug resistance, the U.S. Department of Health declared vaccination against pneumococcal disease a public health priority (Barocchi *et al.*, 2007; Services, 2000). Two vaccines, Pneumovax® (Merck) and Prevnar® (Wyeth), were manufactured in response to this demand and both vaccines are based on capsular polysaccharides, as antibodies against the pneumococcal sugar-coat enable opsono-phagocytic clearance of the invading pathogen (Austrian, 1977; Avery and Goebel, 1931).

Not surprisingly then, when Karakawa and Vann discovered a capsule on the surface of *S. aureus* strains (Karakawa and Vann, 1982), significant efforts were directed towards discovering the chemical nature, genetic determinants, contributions to virulence in animal models and the use of capsular polysaccharides as staphylococcal vaccines (Fig. 12.5) (O'Riordan and Lee, 2004). Karakawa and Vann proposed a capsular polysaccharide typing scheme for *S. aureus* based upon the preparation of absorbed rabbit antiserum to prototype *S. aureus* strains (Karakawa and Vann, 1982). The heavily encapsulated strains M and Smith diffuse, which form mucoid colonies on agar, were assigned serotypes 1 and 2, respectively. However, serotype 1 and 2 strains are rarely encountered among clinical isolates (Kuehnert *et al.*, 2006). Serotypes 3–11 isolates produce non-mucoid colonies on most agar media; their colony morphology is indistinguishable from strains lacking capsule (Sompolinsky *et al.*, 1985). Serotyping of staphylococcal isolates from diverse strain collections revealed that serotype 5 and 8 isolates account for about 80% of strains associated with human disease (Arbeit *et al.*, 1984).

Lee and co-workers isolated a capsule mutant in the serotype 1 strain of *S. aureus* and reported a 1000-fold increase in LD_{50} over the wild-type parent (Lee *et al.*, 1987). In contrast, a capsule-deficient Tn918 insertion mutant of the CP5 prototype strain Reynolds displayed no virulence defect (measured as LD_{50} dose, replication in blood or abscess formation), as compared to its wild-type parent (Albus *et al.*, 1991). The capsule mutant phenotype was re-examined with strains that were grown on media where capsule synthesis is increased; under such conditions a 10-fold reduction in virulence was reported (Thakker *et al.*, 1998). Purified CP5 or CP8 failed to elicit serum antibodies when injected into mice, however coupling of polysaccharides to purified exotoxin A of *Pseudomonas aeruginosa* increased immunogenicity and T-cell-dependent properties of the vaccine (Fattom *et al.*, 1993). Differences in the carrier protein and the chemical methods used for coupling proteins to the polysaccharide affected the magnitude of the immune response in mice, but these variables did not affect the distribution of IgG subclasses detected in immune serum (Fattom *et al.*, 1995). Use of monophosphoryl lipid A as an adjuvant enhanced the immunogenicity of conjugate vaccines and induced a shift in the IgG subclass composition towards the more opsonic IgG2a and IgG2b subclasses (Fattom *et al.*). The protective efficacy of antibodies to CP5-exotoxin A conjugate vaccine was tested in a mouse model of lethality and disseminated infection (Fattom *et al.*, 1996). Immunization with CP5-exotoxin A provided a significant but moderate level of protection for mice against serotype 5 *S. aureus* strain challenge: 33 of 45 mice immunized with the conjugate survived, compared with 4 of 30 mice injected with phosphate-buffered saline. Passive immunization with immune IgG afforded similar moderate levels of protection for mice against *S. aureus* challenge. However, another study failed to detect protection with capsular antibodies (Nemeth and Lee, 1995).

Fattom and colleagues at NABI, Inc. (Boca Raton, FL, USA), prepared CP5 and CP8-exotoxin A conjugates for commercial use. Both antigens were combined into a bivalent vaccine (StaphVax) intended for immunization of individuals at risk for *S. aureus* infection. Clinical studies demonstrated safety in 70 healthy human volunteers (Fattom *et al.*, 1993). NABI conducted a phase II, double-blinded, placebo-controlled clinical study of StaphVax in about 200 chronic ambulatory peritoneal dialysis patients at high risk of staphylococcal disease. Patients were immunized with StaphVax, and antibody responses and infection rates monitored. The results showed that a vaccine dose of 25 μg of CP5 and CP8 conjugate generated weak antibody responses and that infection rates between immunized and non-immunized patients were similar. NABI conducted a double-blind clinical trial to evaluate safety, immunogenicity, and efficacy of StaphVax for prevention of bacteraemia in 1800 patients with end-stage renal disease receiving haemodialysis (Shinefield *et al.*, 2002). These patients are at high risk for staphylococcal infection, with 3 to 4 of every 100 patients infected with *S. aureus* per year. Half of the patients in the trial were administered placebo, and the other half were immunized with a single injection of StaphVax (100 μg CP5 and CP8 each conjugated to 100 μg

exotoxin A). Efficacy was estimated by comparing the incidence of *S. aureus* bacteraemia in patients who received vaccine with control patients. For 3–54 weeks following immunization, the vaccine reduced the incidence of bacteraemia by 26% (not significant, $P = 0.23$), with no significant differences in the number of deaths for vaccine and control groups (Shinefield *et al.*, 2002). A subsequent phase III trial failed also, calling into question whether CP5/CP8 conjugate vaccines alone can generate protective immune responses (Fattom *et al.*, 2004; Projan *et al.*, 2006).

Protein-based vaccines

Several surface proteins have been tested as vaccine candidates (Stranger-Jones *et al.*, 2006). One promising antigen, IsdB, binds haemoglobin on the staphylococcal surface (Fig. 12.2) (Kuklin *et al.*, 2006). IsdB functions as a haemophore and removes haem from haemoglobin for subsequent passage of the iron compound across the cell wall and cytoplasmic membrane (Torres *et al.*, 2006). Antibodies against IsdB generate a significant level of protection against staphylococcal infection and *isdB* mutants displayed a small reduction in virulence (Torres *et al.*, 2006). Other surface proteins under investigation include ClfA and ClfB, the clumping factor of *S. aureus* (McDevitt *et al.*, 1994; Ní Eidhin *et al.*, 1998). Inhibitex Inc. developed an experimental hyperimmune IgG preparation called Veronate, which was produced by selecting sera from patients with high titres to clumping factor A (ClfA) (Rivas *et al.*, 2004; Vernachio *et al.*, 2003). Positive phase II data had indicated the potential for protecting low birth-weight babies from staphylococcal infections, however a phase III study failed to reach its target endpoints for protection (Projan *et al.*, 2006).

Acknowledgements

The authors wish to acknowledge funding by grants from the National Institute of Allergy and Infectious Diseases, Infectious Diseases Branch (AI52474 to O.S.; AI75258 to D.M.) and by Novartis Vaccines and Diagnostics (Siena, Italy). A.G.C. was a trainee of the National Institutes of Health Medical Scientist Training Program at The University of Chicago (GM07281). O.S. and D.M. acknowledge membership within and support from the Region V (Great Lakes) Regional Centre of Excellence in Biodefense and Emerging Infectious Diseases Consortium (GLRCE, National Institute of Allergy and Infectious Diseases award U54 AI057153).

References

Albus, A., Arbeit, R.D., and Lee, J.C. (1991). Virulence studies of *Staphylococcus aureus* mutants altered in type 5 capsule production. Infect. Immun. *59*, 1008–1014.

Arbeit, R.D., Karakawa, W.W., Vann, W.F., and Robbins, J.B. (1984). Predominance of two newly described capsular polysaccharide types among clinical isolates of *Staphylococcus aureus*. Diagn. Microbiol. Infect. Dis. *2*, 85–91.

Archer, G.L. (1998). *Staphylococcus aureus*: a well-armed pathogen. Clin. Infect. Dis. *26*, 1179–1181.

Archer, G.L., and Niemeyer, D.M. (1994). Origin and evolution of DNA associated with resistance to methicillin in staphylococci. Trends Microbiol. *2*, 343–347.

Archibald, A.R., Baddiley, J., and Heckels, J.E. (1973). Molecular arrangment of teichoic acid in the cell wall of *Staphylococcus aureus*. Nature New Biol. *241*, 29–31.

Austrian, R. (1977). Prevention of pneumococcal infection by immunization with capsular polysaccharides of *Streptococcus pneumoniae*: current status of polyvalent vaccines. J. Infect. Dis. S38–42.

Avery, O.T., and Goebel, W.F. (1931). Chemo-immunological studies on conjugated carbohydrate-protein. V. The immunological specificity of an antigen prepared by combining the capsular polysaccharide of type III pneumococcus with foreign protein. J. Exp. Med. *54*, 437–447.

Baba, T., Takeuchi, F., Kuroda, M., Yuzawa, H., Aoki, K., Oguchi, A., Nagai, Y., Iwama, N., Asano, K., Naimi, T., *et al.* (2002). Genome and virulence determinants of high virulence community-acquired MRSA. Lancet *359*, 1819–1827.

Bae, T., Baba, T., Hiramatsu, K., and Schneewind, O. (2006). Prophages of *Staphylococcus aureus* Newman and their contribution to virulence. Mol. Microbiol. *62*, 1035–1047.

Barocchi, M.A., Censini, S., and Rappuoli, R. (2007). Vaccines in the era of genomics: the pneumococcal challenge. Vaccine *25*, 2963–2973.

Bhakdi, S., and Tranum-Jensen, J. (1991). Alpha-toxin of *Staphylococcus aureus*. Microbiol. Rev. *55*, 733–751.

Bischoff, M., Entenza, J.M., and Giachino, P. (2001). Influence of a functional sigB operon on the global regulators sar and agr in *Staphylococcus aureus*. J. Bacteriol. *183*, 5171–5179.

Bokarewa, M.I., Jin, T., and Tarkowski, A. (2006). *Staphylococcus aureus*: Staphylokinase. Int. J. Biochem. Cell Biol. *38*, 504–509.

Brumfitt, W., and Hamilton-Miller, J. (1989). Methicillin-resistant *Staphylococcus aureus*. N. Engl. J. Med. *320*, 1188–1199.

Brussow, H., Canchaya, C., and Hardt, W.D. (2004). Phages and the evolution of bacterial pathogens: from genomic rearrangements to lysogenic conversion. Microbiol. Mol. Biol. Rev. *68*, 560–602.

Bubeck Wardenburg, J., Missiakas, D.M., and Schneewind, O. (2008). Vaccines for *Staphylococcus aureus* Infections. In New Generation Vaccines, Levine, M.M., Kaper, J.B., Rappuoli, R., Liu, M.F., and Good, A.L., eds (Washington DC, Informa Health Care).

Bubeck Wardenburg, J., Patel, R.J., and Schneewind, O. (2007). Surface proteins and exotoxins are required for the pathogenesis of *Staphylococcus aureus* pneumonia. Infect. Immun. *75*, 1040–1044.

Bugg, T.D.H., Wright, G.D., Dutka-Malen, S., Arthur, M., Courvalin, P., and Walsh, C.T. (1991). Molecular basis for vancomycin resistance in *Enterococcus faecium* BM4147: biosynthesis of a depsipeptide peptidoglycan precursor by vancomycin resistance proteins VanH and VanA. Biochemistry *30*, 10408–10415.

Burts, M.L., DeDent, A.C., and Missiakas, D.M. (2008). EsaC substrate for the ESAT-6 secretion pathway and its role in persistent infections of *Staphylococcus aureus*. Mol. Microbiol. *69*, 736–746.

Burts, M.L., Williams, W.A., DeBord, K., and Missiakas, D.M. (2005). EsxA and EsxB are secreted by an ESAT-6-like system that is required for the pathogenesis of *Staphylococcus aureus* infections. Proc. Natl. Acad. Sci. U.S.A. *102*, 1169–1174.

Chambers, H.F. (2001). The changing epidemiology of *Staphylococcus aureus*? Emerg. Infect. Dis. *7*, 178–182.

Chang, S., Sievert, D.M., Hageman, J.C., Boulton, M.L., Tenover, F.C., Downes, F.P., Shah, S., Rudrik, J.T., Pupp, G.R., Brown, W.J., *et al.* (2003). Infection with vancomycin-resistant *Staphylococcus aureus* containing the vanA resistance gene. N. Engl. J. Med. *348*, 1342–1347.

Cheng, A.G., Kim, H.K., Burts, M.L., Krausz, T., Schneewind, O., and Missiakas, D.M. (2009). Genetic requirements for *Staphylococcus aureus* abscess formation and persistence in host tissues. Faseb J. *23*, 3393–3404.

Cheung, A.L., Bayer, A.S., Zhang, G., Gresham, H., and Xiong, Y.Q. (2004). Regulation of virulence determinants *in vitro* and *in vivo* in *Staphylococcus aureus*. FEMS Immunol. Med. Microbiol. *40*, 1–9.

Cheung, A.L., Nishina, K.A., Pous, M.P., and Tamber, S. (2007). The SarA protein family of *Staphylococcus aureus*. Int. J. Biochem. Cell Biol. *40*, 355–361.

Chien, Y., Manna, A.C., and Cheung, A.L. (1998). SarA level is a determinant of *agr* activation in *Staphylococcus aureus*. Mol. Microbiol. *30*, 991–1001.

Claesson, B.A., Trollfors, B., Lagergård, T., Knutsson, N., Schneerson, R., and Robbins, J.B. (2005). Antibodies against *Haemophilus influenzae* type b capsular polysaccharide and tetanus toxoid before and after a booster dose of the carrier protein nine years after primary vaccination with a protein conjugate vaccine. Pediatr. Infect. Dis. J. *24*, 463–464.

Clauditz, A., Resch, A., Wieland, K.P., Peschel, A., and Gotz, F. (2006). Staphyloxanthin plays a role in the fitness of *Staphylococcus aureus* and its ability to cope with oxidative stress. Infect. Immun. *74*, 4950–4953.

Coleman, D.C., Sullivan, D.J., Russell, R.J., Arbuthnott, J.P., Carey, B.F., and Pomeroy, H.M. (1989). *Staphylococcus aureus* bacteriophages mediating the simultaneous lysogenic conversion of beta-lysin, staphylokinase and enterotoxin A: molecular mechanism of triple conversion. J. Gen. Microbiol. *135*, 1679–1697.

Collins, L.V., Kristian, S.A., Weidenmaier, C., Faigle, M., Van Kessel, K.P., Van Strijp, J.A., Gotz, F., Neumeister, B., and Peschel, A. (2002). *Staphylococcus aureus* strains lacking D-alanine modifications of teichoic acids are highly susceptible to human neutrophil killing and are virulence attenuated in mice. J. Infect. Dis. *186*, 214–219.

Control, C.f.D. (1981). Methicillin-resistant *Staphylococcus aureus* – United States. Morb. Mortal Wkly Rep. *30*, 557–559.

Cramton, S.E., Gerke, C., Schnell, N.F., Nichols, W.W., and Gotz, F. (1999). The intercellular adhesion (ica) locus is present in *Staphylococcus aureus* and is required for biofilm formation. Infect. Immun. *67*, 5427–5433.

de Haas, C.J., Veldkamp, K.E., Peschel, A., Weerkamp, F., Van Wamel, W.J., Heezius, E.C., Poppelier, M.J., Van Kessel, K.P., and van Strijp, J.A. (2004). Chemotaxis inhibitory protein of *Staphylococcus aureus*, a bacterial antiinflammatory agent. J. Exp. Med. *199*, 687–695.

Diep, B.A., Gill, S.R., Chang, R.F., Phan, T.H., Chen, J.H., Davidson, M.G., Lin, F., Lin, J., Carleton, H.A., Mongodin, E.F., *et al.* (2006). Complete genome sequence of USA300, an epidemic clone of community-acquired meticillin-resistant *Staphylococcus aureus*. Lancet *367*, 731–739.

Dinauer, M.C., and Orkin, S.H. (1992). Chronic granulomatous disease. Annu. Rev. Med. *43*, 117–124.

Dinges, M.M., Orwin, P.M., and Schlievert, P.M. (2000). Exotoxins of *Staphylococcus aureus*. Clin. Microbiol. Rev. *13*, 16–34, table of contents.

Duckworth, M., Archibald, A.R., and Baddiley, J. (1975). Lipoteichoic acid and lipoteichoic acid carrier in *Staphylococcus aureus* H. FEBS Lett. *53*, 176–179.

Dunman, P.M., Murphy, E., Haney, S., Palacios, D., Tucker-Kellogg, G., Wu, S., Brown, E.L., Zagursky, R.J., Shlaes, D., and Projan, S.J. (2001). Transcription profiling-based identification of *Staphylococcus aureus* genes regulated by the agr and/or sarA loci. J. Bacteriol. *183*, 7341–7353.

Eltzschig, H.K., Ibla, J.C., Furuta, G.T., Leonard, M.O., Jacobson, K.A., Enjyoji, K., Robson, S.C., and Colgan, S.P. (2003). Coordinated adenine nucleotide phosphohydrolysis and nucleoside signaling in posthypoxic endothelium: role of ectonucleotidases and adenosine A2B receptors. J. Exp. Med. *198*, 783–796.

Eskola, J., Kayhty, H., Takala, A.K., Peltola, H., Ronnberg, P.R., Kela, E., Pekkanen, E., McVerry, P.H., and Makela, P.H. (1990). A randomized, prospective field trial of a conjugate vaccine in the protection of infants and young children against invasive *Haemophilus influenzae* type b disease. N. Engl. J. Med. *323*, 1381–1387.

Fattom, A., Li, X., Cho, Y.H., Burns, A., Hawwari, A., Shepherd, S.E., Coughlin, R., Winston, S., and Naso, R. (1995). Effect of conjugation methodology, carrier protein, and adjuvants on the immune response to *Staphylococcus aureus* capsular polysaccharides. Vaccine *13*, 1288–1293.

Fattom, A., Schneerson, R., Watson, D.C., Karakawa, W.W., FitzGerald, D., Pastan, I., Li, X., Shiloach, J., Bryla, D.A., and Robbins, J.B. (1993). Laboratory and clinical evaluation of conjugate vaccines composed of *Staphylococcus aureus* type 5 and type 8 capsular polysaccharides bound to *Pseudomonas aeruginosa* recombinant exoprotein A. Infect. Immun. *61*, 1023–1032.

Fattom, A.I., Horwith, G., Fuller, S., Propst, M., and Naso, R. (2004). Development of StaphVAX, a polysaccharide conjugate vaccine against *S. aureus* infection: from the lab bench to phase III clinical trials. Vaccine *22*, 880–887.

Fattom, A.I., Sarwar, J., Ortiz, A., and Naso, R. (1996). A *Staphylococcus aureus* capsular polysaccharide (CP) vaccine and CP-specific antibodies protect mice against bacterial challenge. Infect. Immun. *64*, 1659–1665.

Finelli, L., Fiore, A., Dhara, R., Brammer, L., Shay, D.K., Kamimoto, L., Fry, A., Hageman, J., Gorwitz, R., Bresee, J., *et al.* (2008). Influenza-associated pediatric mortality in the United States: increase of *Staphylococcus aureus* coinfection. Pediatrics *122*, 805–811.

Fischer, W. (1990). In Glycolipids, Phosphoglycolipids and Sulfoglycopids, Morris, K., ed. (New York, Plenum Press), pp. 123–234.

Fitzgerald, J.R., Sturdevant, D.E., Mackie, S.M., Gill, S.R., and Musser, J.M. (2001). Evolutionary genomics of *Staphylococcus aureus*: insights into the origin of methicillin-resistant strains and the toxic shock syndrome peidemic. Proc. Natl. Acad. Sci. U.S.A. *98*, 8821–8826.

Foster, T.J. (2005). Immune evasion by staphylococci. Nat. Rev. Microbiol. *3*, 948–958.

Frazee, B.W. (2007). Update on emerging infections: news from the Centers for Disease Control and Prevention. Severe methicillin-resistant *Staphylococcus aureus* community-acquired pneumonia associated with influenza–Louisiana and Georgia, December 2006–January 2007. Ann. Emerg. Med. *50*, 612–616.

Fridkin, S.K., Hageman, J.C., Morrison, M., Sanza, L.T., Como-Sabetti, K., Jernigan, J.A., Harriman, K., Harrison, L.H., Lynfield, R., and Farley, M.M. (2005). Methicillin-resistant *Staphylococcus aureus* disease in three communities. N. Engl. J. Med. *352*, 1436–1444.

Gillet, Y., Issartel, B., Vanhems, P., Fournet, J.C., Lina, G., Bes, M., Vandenesch, F., Piemont, Y., Brousse, N., Floret, D., *et al.* (2002). Association between *Staphylococcus aureus* strains carrying gene for Panton-Valentine leukocidin and highly lethal necrotising pneumonia in young immunocompetent patients. Lancet *359*, 753–759.

Giraudo, A.T., Calzolari, A., Cataldi, A.A., Bogni, C., and Nagel, R. (1999). The sae locus of *Staphylococcus aureus* encodes a two-component regulatory system. FEMS Microbiol. Lett. *177*, 15–22.

Giraudo, A.T., Cheung, A.L., and Nagel, R. (1997). The sae locus of *Staphylococcus aureus* controls exoprotein synthesis at the transcriptional level. Arch. Microbiol. *168*, 53–58.

Gladstone, G.P., and Van Heyningen, W.E. (1957). Staphylococcal leucocidins. Br. J. Exp. Pathol. *38*, 123–137.

Goerke, C., Wirtz, C., Fluckiger, U., and Wolz, C. (2006). Extensive phage dynamics in *Staphylococcus aureus* contributes to adaptation to the human host during infection. Mol. Microbiol. *61*, 1673–1685.

Gomez, M.I., Lee, A., Reddy, B., Muir, A., Soong, G., Pitt, A., Cheung, A., and Prince, A. (2004). *Staphylococcus aureus* protein A induces airway epithelial inflammatory responses by activating TNFR1. Nat. Med. *10*, 842–848.

Gotschlich, E.C., Liu, T.Y., and Artenstein, M.S. (1969). Human immunity to the meningococcus. IV. Immunogenicity of group A and group C meningococcal polysaccharides in human volunteers. J. Exp. Med. *129*, 1367–1384.

Gotz, F., Bannerman, T., and Schleifer, K.-H. (2006). The Genera *Staphylococcus* and *Macrococcus*. In The Prokaryotes, Dworkin, M., Falkow, S., Rosenberg, E., Schleifer, K.-H., and Stackebrandt, E., eds (New York, Springer), pp. 5–75.

Gouaux, E., Hobaugh, M., and Song, L. (1997). alpha-Hemolysin, gamma-hemolysin, and leukocidin from *Staphylococcus aureus*: distant in sequence but similar in structure. Protein Sci. *6*, 2631–2635.

Gründling, A., and Schneewind, O. (2007a). Genes required for glycolipid synthesis and lipoteichoic acid anchoring in *Staphylococcus aureus*. J. Bacteriol. *189*, 2521–2530.

Gründling, A., and Schneewind, O. (2007b). Synthesis of glycerol phosphate lipoteichoic acid in *Staphylococcus aureus*. Proc. Nat. Acad. Sci. U.S.A. *104*, 8478–8483.

Hanssen, A.M., and Ericson Sollid, J.U. (2006). SCCmec in staphylococci: genes on the move. FEMS Immunol. Med. Microbiol. *46*, 8–20.

Hartleib, J., Kohler, N., Dickinson, R., Chhatwal, G., Sixma, J., Hartford, O., Foster, T.J., Peters, G., Kehrl, B., and Herrmann, M. (2000). Protein A is the von Willebrand factor binding protein of *Staphylococcus aureus*. Blood *96*, 2149–2156.

Hartman, B.J., and Tomasz, A. (1984). Low affinity penicillin binding protein associated with β-lactam resistance in *Staphylococcus aureus*. J. Bacteriol. *158*, 513–516.

Hay, J.B., Archibald, A.R., and Baddiley, J. (1965a). The molecular structure of bacterial walls. The size of ribitol teichoic acids and the nature of their linkage to glycosaminopeptides. Biochem. J. *97*, 723–730.

Hay, J.B., Davey, N.B., Archibald, A.R., and Baddiley, J. (1965b). The Chain length of ribitol teichoic acids and the nature of their association with bacterial cell walls. Biochem. J. *94*, 7C–9C.

Heilmann, C., Schweitzer, O., Gerke, C., Vanittanakom, N., Mack, D., and Gotz, F. (1996). Molecular basis of intercellular adhesion in the biofilm-forming Staphylococcus epidermidis. Mol. Microbiol. *20*, 1083–1091.

Herold, B.C., Immergluck, L.C., Maranan, M.C., Lauderdale, D.S., Gaskin, R.E., Boyle-Vavra, S., Leitch, C.D., and Daum, R.S. (1998). Community-acquired methicillin-resistant *Staphylococcus aureus* in children with no identified predisposing risk. JAMA *279*, 593–598.

Higgins, J., Loughman, A., van Kessel, K.P., van Strijp, J.A., and Foster, T.J. (2006). Clumping factor A of *Staphylococcus aureus* inhibits phagocytosis by human polymorphonuclear leucocytes. FEMS Microbiol. Lett. *258*, 290–296.

Hiramatsu, K., Aritaka, N., Hanaki, H., Kawasaki, S., Hosoda, Y., Hori, S., Fukuchi, Y., and Kobayashi, I. (1997). Dissemination in Japanese hospitals of strains of *Staphylococcus aureus* heterogeneously resistant to vancomycin. Lancet *350*, 1670–1673.

Issartel, B., Tristan, A., Lechevallier, S., Bruyere, F., Lina, G., Garin, B., Lacassin, F., Bes, M., Vandenesch, F., and Etienne, J. (2005). Frequent carriage of Panton-Valentine leucocidin genes by *Staphylococcus aureus* isolates from surgically drained abscesses. J. Clin. Microbiol. *43*, 3203–3207.

Jardetzky, T.S., Brown, J.H., Gorga, J.C., Stern, L.J., Urban, R.G., Chi, Y.I., Stauffacher, C., Strominger, J.L., and Wiley, D.C. (1994). Three-dimensional structure of a human class II histocompatibility molecule complexed with superantigen. Nature *368*, 711–718.

Jensen, K. (1958). A normally occurring staphylococcus antibody in human serum. Acta Pathol. Microbiol. Scand. *44*, 421–428.

Kaneko, J., and Kamio, Y. (2004). Bacterial two-component and hetero-heptameric pore-forming cytolytic toxins: structures, pore-forming mechanism, and organization of the genes. Biosci. Biotechnol. Biochem. *68*, 981–1003.

Kaplan, S.L., Hulten, K.G., Gonzalez, B.E., Hammerman, W.A., Lamberth, L., Versalovic, J., and Mason, E.O., Jr. (2005). Three-year surveillance of community-acquired *Staphylococcus aureus* infections in children. Clin. Infect. Dis. *40*, 1785–1791.

Karakawa, W.W., and Vann, W.F. (1982). Capsular polysaccharides of *Staphylococcus aureus*. Semin. Infect. Dis. *4*, 285–293.

Katayama, Y., Ito, T., and Hiramatsu, K. (2000). A new class of genetic element, staphylococcus cassette chromosome mec, encodes methicillin resistance in *Staphylococcus aureus*. Antimicrob. Agents Chemother. *44*, 1549–1555.

King, M.D., Humphrey, B.J., Wang, Y.F., Kourbatova, E.V., Ray, S.M., and Blumberg, H.M. (2006). Emergence of community-acquired methicillin-resistant *Staphylococcus aureus* USA 300 clone as the predominant cause of skin and soft-tissue infections. Ann. Intern. Med. *144*, 309–317.

Koch, H.U., Haas, R., and Fischer, W. (1984). The role of lipoteichoic acid biosynthesis in membrane lipid metabolism of growing *Staphylococcus aureus*. Eur. J. Biochem. *138*, 357–363.

Kuehnert, M.J., Hill, H.A., Kupronis, B.A., Tokars, J.I., Solomon, S.L., and Jernigan, D.B. (2005). Methicillin-resistant-*Staphylococcus aureus* hospitalizations, United States. Emerg. Infect. Dis. *11*, 868–872.

Kuehnert, M.J., Kruszon-Moran, D., Hill, H.A., McQuillan, G., McAllister, S.K., Fosheim, G., McDougal, L.K., Chaitram, J., Jensen, B., Fridkin, S.K., *et al.* (2006).

Prevalence of *Staphylococcus aureus* nasal colonization in the United States, 2001–2002. J. Infect. Dis. *193*, 172–179.

Kuklin, N.A., Clark, D.J., Secore, S., Cook, J., Cope, L.D., McNeely, T., Noble, L., Brown, M.J., Zorman, J.K., Wang, X.M., *et al.* (2006). A novel *Staphylococcus aureus* vaccine: iron surface determinant B induces rapid antibody responses in rhesus macaques and specific increased survival in a murine *S. aureus* sepsis model. Infect. Immun. *74*, 2215–2223.

Kuroda, M., Ohta, T., Uchiyama, I., Baba, T., Yuzawa, H., Kobayashi, L., Cui, L., Oguchi, A., Aoki, K., Nagai, Y., *et al.* (2001). Whole genome sequencing of methicillin-resistant *Staphylococcus aureus*. Lancet *357*, 1225–1240.

Lee, J.C., Betley, M.J., Hopkins, C.A., Perez, N.E., and Pier, G.B. (1987). Virulence studies, in mice, of transposon-induced mutants of *Staphylococcus aureus* differing in capsule size. J. Infect. Dis. *156*, 741–750.

Li, M., Diep, B.A., Villaruz, A.E., Braughton, K.R., Jiang, X., DeLeo, F.R., Chambers, H.F., Lu, Y., and Otto, M. (2009). Evolution of virulence in epidemic community-associated methicillin-resistant *Staphylococcus aureus*. Proc. Natl. Acad. Sci. U.S.A. *106*, 5883–5888.

Lina, G., Piemont, Y., Godail-Gamot, F., Bes, M., Peter, M.O., Gauduchon, V., Vandenesch, F., and Etienne, J. (1999). Involvement of Panton-Valentine leukocidin-producing *Staphylococcus aureus* in primary skin infections and pneumonia. Clin. Infect. Dis. *29*, 1128–1132.

Liu, G.Y., Essex, A., Buchanan, J.T., Datta, V., Hoffman, H.M., Bastian, J.F., Fierer, J., and Nizet, V. (2005). *Staphylococcus aureus* golden pigment impairs neutrophil killing and promotes virulence through its antioxidant activity. J. Exp. Med. *202*, 209–215.

Lowy, F.D. (1998). *Staphylococcus aureus* infections. N. Engl. J. Med. *339*, 520–532.

Ma, X.X., Ito, T., Tiensasitorn, C., Jamklang, M., Chongtrakool, P., Boyle-Vavra, S., Daum, R.S., and Hiramatsu, K. (2002). Novel type of staphylococcal cassette chromosome mec identified in community-acquired methicillin resistant *Staphylococcus aureus* strains. Antimicrob. Agents Chemother. *46*, 1147–1152.

Marraffini, L.A., Dedent, A.C., and Schneewind, O. (2006). Sortases and the art of anchoring proteins to the envelopes of gram-positive bacteria. Microbiol. Mol. Biol. Rev. *70*, 192–221.

Marshall, J.H., and Wilmoth, G.J. (1981). Proposed pathway of triterpenoid carotenoid biosynthesis in *Staphylococcus aureus*: evidence from a study of mutants. J. Bacteriol. *147*, 914–919.

Mazmanian, S.K., Skaar, E.P., Gasper, A.H., Humayun, M., Gornicki, P., Jelenska, J., Joachimiak, A., Missiakas, D.M., and Schneewind, O. (2003). Passage of heme-iron across the envelope of *Staphylococcus aureus*. Science *299*, 906–909.

Mazmanian, S.K., Ton-That, H., and Schneewind, O. (2001). Sortase-catalysed anchoring of surface proteins to the cell wall of *Staphylococcus aureus*. Mol. Microbiol. *40*, 1049–1057.

Mazmanian, S.K., Ton-That, H., Su, K., and Schneewind, O. (2002). An iron-regulated sortase anchors a class of surface protein during *Staphylococcus aureus* pathogenesis. Proc. Natl. Acad. Sci. U.S.A. *99*, 2293–2298.

McDevitt, D., Francois, P., Vaudaux, P., and Foster, T.J. (1994). Molecular characterization of the clumping factor (fibrinogen receptor) of *Staphylococcus aureus*. Mol. Microbiol. *11*, 237–248.

McElroy, M.C., Harty, H.R., Hosford, G.E., Boylan, G.M., Pittet, J.F., and Foster, T.J. (1999). Alpha-toxin damages the air–blood barrier of the lung in a rat model of *Staphylococcus aureus*-induced pneumonia. Infect. Immun. *67*, 5541–5544.

McNeil, L.K., Reich, C., Aziz, R.K., Bartels, D., Cohoon, M., Disz, T., Edwards, R.A., Gerdes, S., Hwang, K., Kubal, M., *et al.* (2007). The National Microbial Pathogen Database Resource (NMPDR): a genomics platform based on subsystem annotation. Nucleic Acids Res. *35*, D347–353.

Missiakas, D., and Schneewind, O. (2007). What has genomics taught us about Gram-positive protein secretion and targeting? In Bacterial Pathogenomics, Pallen, M.J., Nelson, K.E., Preston, G.M., eds (Washington DC, ASM Press), pp. 301–326.

Navarre, W.W., and Schneewind, O. (1999). Surface proteins of gram-positive bacteria and mechanisms of their targeting to the cell wall envelope. Microbiol. Mol. Biol. Rev. *63*, 174–229.

Nemeth, J., and Lee, J.C. (1995). Antibodies to capsular polysaccharides are not protective against experimental *Staphylococcus aureus* endocarditis. Infect. Immun. *63*, 375–380.

Neth, O., Jack, D.L., Johnson, M., Klein, N.J., and Turner, M.W. (2002). Enhancement of complement activation and opsonophagocytosis by complexes of mannose-binding lectin with mannose-binding lectin-associated serine protease after binding to *Staphylococcus aureus*. J. Immunol. *169*, 4430–4436.

Neu, H.C. (1992). The crisis in antibiotic resistance. Science *257*, 1064–1073.

Neuhaus, F.C., and Baddiley, J. (2003). A continuum of anionic charge: structures and functions od D-alanyl-teichoic acids in gram-positive bacteria. Microbiol. Mol. Biol. Rev. *67*, 686–723.

Ní Eidhin, D., Perkins, S., Francois, P., Vaudaux, P., Höök, M., and Foster, T.J. (1998). Clumping factor B (ClfB), a new surface-located fibrinogen-binding adhesin of *Staphylococcus aureus*. Mol. Microbiol. *30*, 245–257.

Novick, R.P. (2003a). Autoinduction and signal transduction in the regulation of staphylococcal virulence. Mol. Microbiol. *48*, 1429–1449.

Novick, R.P. (2003b). Mobile genetic elements and bacterial toxinoses: the superantigen-encoding pathogenicity islands of *Staphylococcus aureus*. Plasmid *49*, 93–105.

Novick, R.P., and Jiang, D. (2003). The staphylococcal saeRS system coordinates environmental signals with agr quorum sensing. Microbiology *149*, 2709–2717.

Novick, R.P., and Subedi, A. (2007). The SaPIs: mobile pathogenicity islands of staphylococcus. Chem. Immunol. Allergy *93*, 42–57.

O'Reilly, M., de Azavedo, J.C., Kennedy, S., and Foster, T.J. (1986). Inactivation of the alpha-haemolysin gene of *Staphylococcus aureus* 8325–4 by site-directed mutagenesis and studies on the expression of its haemolysins. Microb. Pathog. *1*, 125–138.

O'Reilly, M., Kreiswirth, B.N., and Foster, T.J. (1990). Cryptic alpha-toxin gene in toxic shock syndrome and septicemia strains of *Staphylococcus aureus*. Mol. Microbiol. *4*, 1947–1955.

O'Riordan, K., and Lee, J.C. (2004). *Staphylococcus aureus* capsular polysaccharides. Clin. Microbiol. Rev. *17*, 218–234.

Ochoa, T.J., Mohr, J., Wanger, A., Murphy, J.R., and Heresi, G.P. (2005). Community-associated methicillin-resistant *Staphylococcus aureus* in pediatric patients. Emerg. Infect. Dis. *11*, 966–968.

Ogston, A. (1883). Micrococcus poisoning. J. Anat. Physiol. *17*, 24–58.

Oku, Y., Kurokawa, K., Matsuo, M., Yamada, S., Lee, B.L., and Sekimizu, K. (2009). Pleiotropic roles of polyglycerolphosphate synthase of lipoteichoic acid in growth of *Staphylococcus aureus* cells. J. Bacteriol. *191*, 141–151.

Palmqvist, N., Patti, J.M., Tarkowski, A., and Josefsson, E. (2004). Expression of staphylococcal clumping factor A impedes macrophage phagocytosis. Microbes and infection/Institut Pasteur *6*, 188–195.

Patti, J.M., Allen, B.L., McGavin, M.J., and Höök, M. (1994). MSCRAMM-mediated adherence of microorganisms to host tissues. Annu. Rev. Microbiol. *48*, 89–115.

Pelz, A., Wieland, K.P., Putzbach, K., Hentschel, P., Albert, K., and Gotz, F. (2005). Structure and biosynthesis of staphyloxanthin from *Staphylococcus aureus*. J. Biol. Chem. *280*, 32493–32498.

Peschel, A., and Sahl, H.G. (2006). The co-evolution of host cationic antimicrobial peptides and microbial resistance. Nat. Rev. Microbiol. *4*, 529–536.

Postma, B., Poppelier, M.J., van Galen, J.C., Prossnitz, E.R., van Strijp, J.A., de Haas, C.J., and van Kessel, K.P. (2004). Chemotaxis inhibitory protein of *Staphylococcus aureus* binds specifically to the C5a and formylated peptide receptor. J. Immunol. *172*, 6994–7001.

Pragman, A.A., and Schlievert, P.M. (2004). Virulence regulation in *Staphylococcus aureus*: the need for *in vivo* analysis of virulence factor regulation. FEMS Immunol. Med. Microbiol. *42*, 147–154.

Projan, S.J. (2003). Why is big Pharma getting out of antibacterial drug discovery? Curr. Opin. Microbiol. *6*, 427–430.

Projan, S.J., Nesin, M., and Dunman, P.M. (2006). Staphylococcal vaccines and immunotherapy: to dream the impossible dream? Curr. Opin. Pharmacol. *6*, 473–479.

Purcell, K., and Fergie, J. (2005). Epidemic of community-acquired methicillin-resistant *Staphylococcus aureus* infections: a 14-year study at Driscoll Children's Hospital. Arch. Pediatr. Adolesc. Med. *159*, 980–985.

Rivas, J.M., Speziale, P., Patti, J.M., and Hook, M. (2004). MSCRAMM–targeted vaccines and immunotherapy for staphylococcal infection. Curr. Opin. Drug Discov. Dev. *7*, 223–227.

Roben, P.W., Salem, A.N., and Silverman, G.J. (1995). VH3 family antibodies bind domain D of staphylococcal protein A. J. Immunol. *154*, 6437–6445.

Roberts, C., Anderson, K.L., Murphy, E., Projan, S.J., Mounts, W., Hurlburt, B., Smeltzer, M., Overbeek, R., Disz, T., and Dunman, P.M. (2006). Characterizing the effect of the *Staphylococcus aureus* virulence factor regulator, SarA, on log-phase mRNA half-lives. J. Bacteriol. *188*, 2593–2603.

Rogers, D.E., and Melly, M.A. (1965). Speculations on the immunology of staphylococcal infections. Ann. NY Acad. Sci. *128*, 274–284.

Rooijakkers, S.H., Ruyken, M., Roos, A., Daha, M.R., Presanis, J.S., Sim, R.B., van Wamel, W.J., van Kessel, K.P., and van Strijp, J.A. (2005a). Immune evasion by a staphylococcal complement inhibitor that acts on C3 convertases. Nat. Immunol. *6*, 920–927.

Rooijakkers, S.H., Ruyken, M., van Roon, J., van Kessel, K.P., van Strijp, J.A., and van Wamel, W.J. (2006). Early expression of SCIN and CHIPS drives instant immune evasion by *Staphylococcus aureus*. Cell. Microbiol. *8*, 1282–1293.

Rooijakkers, S.H., van Kessel, K.P., and van Strijp, J.A. (2005b). Staphylococcal innate immune evasion. Trends Microbiol. *13*, 596–601.

Rose, F., Dahlem, G., Guthmann, B., Grimminger, F., Maus, U., Hanze, J., Duemmer, N., Grandel, U., Seeger, W., and Ghofrani, H.A. (2002). Mediator generation and signaling events in alveolar epithelial cells attacked by *S. aureus* alpha-toxin. Am. J. Physiol. Lung Cell Mol. Physiol. *282*, L207–L214.

Rosenbach, F.J. (1884). Mikroorganismen bei den Wund-infections-Krankheiten des Menschen (Wiesbaden, Germany).

Rupp, M.E., Fey, P.D., Heilmann, C., and Gotz, F. (2001). Characterization of the importance of *Staphylococcus epidermidis* autolysin and polysaccharide intercellular adhesin in the pathogenesis of intravascular catheter-associated infection in a rat model. J. Infect. Dis. *183*, 1038–1042.

Ruzin, A., Lindsay, J., and Novick, R.P. (2001). Molecular genetics of SaPI1–a mobile pathogenicity island in *Staphylococcus aureus*. Mol. Microbiol. *41*, 365–377.

Said-Salim, B., Dunman, P.M., McAleese, F.M., Macapagal, D., Murphy, E., McNamara, P.J., Arvidson, S., Foster, T.J., Projan, S.J., and Kreiswirth, B.N. (2003). Global regulation of *Staphylococcus aureus* genes by rot. J. Bacteriol. *185*, 610–619.

Sakiniene, E., Bremell, T., and Tarkowski, A. (1999). Complement depletion aggravates *Staphylococcus aureus* septicaemia and septic arthritis. Clin. Exp. Immunol. *115*, 95–102.

Seeger, W., Bauer, M., and Bhakdi, S. (1984). Staphylococcal alpha-toxin elicits hypertension in isolated rabbit lungs. Evidence for thromboxane formation and the role of extracellular calcium. J. Clin. Invest. *74*, 849–858.

Seeger, W., Birkemeyer, R.G., Ermert, L., Suttorp, N., Bhakdi, S., and Duncker, H.R. (1990). Staphylococcal alpha-toxin-induced vascular leakage in isolated perfused rabbit lungs. Lab. Invest. *63*, 341–349.

Services, U.S.D.o.H.a.H. (2000). Healthy people 2010. US Department of Health and Human Services, Washington, DC.

Shinefield, H., Black, S., Fattom, A., Horwith, G., Rasgon, S., Ordonez, J., Yeoh, H., Law, D., Robbins, J.B., Schneerson, R., *et al.* (2002). Use of a *Staphylococcus aureus* conjugate vaccine in patients receiving hemodialysis. N. Engl. J. Med. *346*, 491–496.

Smith, J.M., and Cook, G.M. (2005). A decade of community MRSA in New Zealand. Epidemiol. Infect. *133*, 899–904.

Sompolinsky, D., Samra, Z., Karakawa, W.W., Vann, W.F., Schneerson, R., and Malik, Z. (1985). Encapsulation and capsular types in isolates of *Staphylococcus aureus* from different sources and relationship to phage types. J. Clin. Microbiol. *22*, 828–834.

Song, L., Hobaugh, M.R., Shustak, C., Cheley, S., Bayley, H., and Gouaux, J.E. (1996). Structure of staphylococcal alpha-hemolysin, a heptameric transmembrane pore. Science *274*, 1859–1866.

Stranger-Jones, Y.K., Bae, T., and Schneewind, O. (2006). Vaccine assembly from surface proteins of *Staphylococcus aureus*. Proc. Natl. Acad. Sci. U.S.A. *103*, 16942–16947.

Suttorp, N., Seeger, W., Dewein, E., Bhakdi, S., and Roka, L. (1985). Staphylococcal alpha-toxin-induced PGI2 production in endothelial cells: role of calcium. Am. J. Physiol. *248*, C127–C134.

Tegmark, K., Karlsson, A., and Arvidson, S. (2000). Identification and characterization of SarH*1*, a new global regulator of virulence gene expression in *Staphylococcus aureus*. Mol. Microbiol. *37*, 398–409.

Tenover, F.C., Biddle, J.W., and Lancaster, M.V. (2001). Increasing resistance to vancomycin and other glycopeptides in *Staphylococcus aureus*. Emerg. Infect. Dis. *7*, 327–332.

Thakker, M., Park, J.S., Carey, V., and Lee, J.C. (1998). *Staphylococcus aureus* serotype 5 capsular polysaccharide is antiphagocytic and enhances bacterial virulence in a murine bacteremia model. Infect. Immun. *66*, 5183–5189.

Thammavongsa, V., Kern, J.W., Missiakas, D.M., and Schneewind, O. (2009). *Staphylococcus aureus* synthesizes adenosine to escape host immune responses. J. Exp. Med. *206*, 2417–2427.

Thiel, M., Caldwell, C.C., and Sitkovsky, M.V. (2003). The critical role of adenosine A2A receptors in downregulation of inflammation and immunity in the pathogenesis of infectious diseases. Microbes Infect. *5*, 515–526.

Tipper, D.J., Tomoeda, M., and Strominger, J.L. (1971). Isolation and characterization of –1, 4-N-acetyl-muramyl-N-acetylglucosamine and its O-acetyl derivative. Biochemistry *10*, 4683–4690.

Torres, V.J., Pishchany, G., Humayun, M., Schneewind, O., and Skaar, E.P. (2006). *Staphylococcus aureus* IsdB is a hemoglobin receptor required for heme-iron utilization. J. Bacteriol. *188*, 8421–8429.

Tristan, A., Bes, M., Meugnier, H., Lina, G., Bozdogan, B., Courvalin, P., Reverdy, M.E., Enright, M.C., Vandenesch, F., and Etienne, J. (2007). Global distribution of Panton-Valentine leukocidin–positive methicillin-resistant *Staphylococcus aureus*, 2006. Emerg. Infect. Dis. *13*, 594–600.

Uhlén, M., Guss, B., Nilsson, B., Gatenbeck, S., Philipson, L., and Lindberg, M. (1984). Complete sequence of the staphylococcal gene encoding protein A. A gene evolved through multiple duplications. J. Biol. Chem. *259*, 1695–1702 and 13628 (Corr.).

van Wamel, W.J., Rooijakkers, S.H., Ruyken, M., van Kessel, K.P., and van Strijp, J.A. (2006). The innate immune modulators staphylococcal complement inhibitor and chemotaxis inhibitory protein of *Staphylococcus aureus* are located on beta-hemolysin-converting bacteriophages. J. Bacteriol. *188*, 1310–1315.

Vandenesch, F., Naimi, T., Enright, M.C., Lina, G., Nimmo, G.R., Heffernan, H., Liassine, N., Bes, M., Greenland, T., Reverdy, M.E., *et al.* (2003). Community-acquired methicillin-resistant *Staphylococcus aureus* carrying Panton-Valentine leukocidin genes: worldwide emergence. Emerg. Infect. Dis. *9*, 978–984.

Verbrugh, H.A., Van Dijk, W.C., Peters, R., Van Der Tol, M.E., and Verhoef, J. (1979). The role of *Staphylococcus aureus* cell-wall peptidoglycan, teichoic acid and protein A in the processes of complement activation and opsonization. Immunology 37, 615–621.

Vernachio, J., Bayer, A.S., Le, T., Chai, Y.L., Prater, B., Schneider, A., Ames, B., Syribeys, P., Robbins, J., and Patti, J.M. (2003). Anti-clumping factor A immunoglobulin reduces the duration of methicillin-resistant *Staphylococcus aureus* bacteremia in an experimental model of infective endocarditis. Antimicrob. Agents Chemother. *47*, 3400–3406.

Voyich, J.M., Braughton, K.R., Sturdevant, D.E., Whitney, A.R., Said-Salim, B., Porcella, S.F., Long, R.D., Dorward, D.W., Gardner, D.J., Kreiswirth, B.N., *et al.* (2005). Insights into mechanisms used by *Staphylococcus aureus* to avoid destruction by human neutrophils. J. Immunol. *175*, 3907–3919.

Voyich, J.M., Otto, M., Mathema, B., Braughton, K.R., Whitney, A.R., Welty, D., Long, R.D., Dorward, D.W., Gardner, D.J., Lina, G., *et al.* (2006). Is Panton-Valentine leukocidin the major virulence determinant in community-associated methicillin-resistant *Staphylococcus aureus* disease? J. Infect. Dis. *194*, 1761–1770.

Walsh, C.T. (1993). Vancomycin resistance: decoding the molecular logic. Science *261*, 308–309.

Wang, R., Braughton, K.R., Kretschmer, D., Bach, T.H., Queck, S.Y., Li, M., Kennedy, A.D., Dorward, D.W., Klebanoff, S.J., Peschel, A., *et al.* (2007). Identification of novel cytolytic peptides as key virulence determinants for community-associated MRSA. Nat. Med. *13*, 1510–1514.

Warren, G.H., and Gray, J. (1965). Effect of sublethal concentrations of penicillins on the lysis of bacteria by lysozyme and trypsin. Proc. Soc. Exp. Biol. Med. *120*, 504–511.

Weidenmaier, C., Kokai-Kun, J.F., Kristian, S.A., Chanturiya, T., Kalbacher, H., Gross, M., Nicholson, G.,

Neumeister, B., Mond, J.J., and Peschel, A. (2004). Role of teichoic acids in *Staphylococcus aureus* nasal colonization, a major risk factor in nosocomial infections. Nat. Med. *10*, 243–245.

Weigel, L.M., Clewell, D.B., Gill, S.R., Clark, N.C., McDougal, L.K., Flannagan, S.E., Kolonay, J.F., Shetty, J., Killgore, G.E., and Tenover, F.C. (2003). Genetic analysis of a high-level vancomycin-resistant isolate of *Staphylococcus aureus*. Science *302*, 1569–1571.

Wilkinson, B.J., Kim, Y., Peterson, P.K., Quie, P.G., and Michael, A.F. (1978). Activation of complement by cell surface components of *Staphylococcus aureus*. Infect. Immun. *20*, 388–392.

Wu, S.W., de Lencastre, H., and Tomasz, A. (2001). Recruitment of the mecA gene homologue of Staphylococcus sciuri into a resistance determinant and expression of the resistant phenotype in *Staphylococcus aureus*. J. Bacteriol. *183*, 2417–2424.

Zimmermann, H. (1992). 5′-Nucleotidase: molecular structure and functional aspects. Biochem. J. *285*, 345–365.

Towards the Development of a Universal Vaccine Against Group B *Streptococcus*

13

Roberta Cozzi, John L. Telford and Domenico Maione

Abstract

Group B *Streptococcus* (GBS) is one of the most common cause of life-threatening bacterial infections in infants and is also an emerging pathogen among adult humans, especially in the elderly, immunocompromised and diabetic adults.

Capsular polysaccharide based vaccines of the most common serotypes present in the United States and Europe are in an advanced stage of development but they are not effective against serotypes present in other parts of the world.

Many protein antigens have been studied for the discovery of an effective universal vaccine that could overcome serotype specificity. Thanks to reverse vaccinology and new technologies, a vaccine combination based on the pilus proteins has been discovered for the development of a universal GBS vaccine that is potentially capable of preventing all GBS infections.

Introduction

Group B *Streptococcus* (GBS), also referred to as *Streptococcus agalactiae*, is one of the most common causes of life-threatening bacterial infections in infants. GBS is a Gram-positive pathogen that colonizes the urogenital and the gastrointestinal tracts of more than 30% of the healthy population and, in particular, it colonizes the vagina of 25–40% of healthy women (Dillon *et al.*, 1982; Schuchat, 1998; Hansen *et al.*, 2004). Neonatal GBS infections can result in pneumonia, sepsis, meningitis and in some cases, death (McCracken, 1973; Ferrieri, 1985; Gibbs *et al.*, 2004). Moreover, GBS infections are increasing also among adults, especially in the elderly,

immunocompromised and diabetic adults (Farley *et al.*, 1993; Schrag *et al.*, 2000; Blancas *et al.*, 2004; Edwards *et al.*, 2005; Skoff *et al.*, 2009).

GBS, first recognized as a pathogen in bovine mastitis, is distinguished from other pathogenic streptococci by the cell wall-associated group B carbohydrate. The microorganism also expresses a capsular polysaccharide (CPS) that allows GBS isolates to be classified in 10 different serotypes based on the distinct structure and antigenicity of the capsule (Kong *et al.*, 2002). However, around 8–14% of the clinical isolates in Europe and in the USA are non-typeable strains because they cannot be distinguished on the basis of CPS antigenicity (Bisharat *et al.*, 2005; Gherardi *et al.*, 2007; Skoff *et al.*, 2009).

GBS disease in newborns has been divided in early-onset disease (EOD) and late-onset disease (LOD) depending on the infants' age and disease manifestation. Early-onset disease manifests in the first week of life and the neonate is usually infected by exposure to GBS during birth. The transmission from mothers to newborns usually occurs when the neonate aspirates contaminated amniotic and vaginal fluids. Early onset disease can progress as pneumonia and the bacteria can spread into the bloodstream resulting in septicaemia, meningitis and osteomyelitis (Rubens *et al.*, 1991; Edwards, 2001; Puopolo *et al.*, 2005). Infants who present with late-onset disease do not show signs of infection in the first 6 days of life. LOD (7–90 days) is less frequent than EOD and the mortality rate is lower but morbidity is high, as around 50% of neonates that survive to GBS infection suffer complications, including mental retardation, hearing loss and speech and language delay (Schuchat, 1998; Schrag *et al.*, 2000; Edwards, 2001).

The introduction in the US of national guidelines for GBS disease prevention, first issued in 1996 and updated in 2002, recommending universal screening of pregnant women for rectovaginal GBS colonization at 35–37 weeks' gestation and administering intrapartum antimicrobial prophylaxis to carriers, was associated with a decline in the incidence of EOD in the United States (Boyer *et al.*, 1983; Gibbs *et al.*, 1994; Baker, 1997; CDC, 2002; Moore *et al.*, 2003; Law *et al.*, 2005). But EOD still occurred with an incidence of 0.34 per 1000 live births in 2003–2005 in the USA (Phares *et al.*, 2008). Not surprisingly, late-onset GBS infections did not decline despite the implementation of prophylactic measures and occurred in 1999–2005 with an incidence averaging 0.34 per 1000 live births in USA (Phares *et al.*, 2008).

Hence, GBS is still a public health concern for human infants and adults and the introduction of additional prevention and therapeutic strategies against GBS infection is highly desirable. Vaccination represents the most attractive strategy for GBS disease prevention. An April 1999 NIAID commissioned study from the Institute of Medicine cited GBS as one of the four most favourable infectious disease vaccine targets. Effective vaccines would stimulate the production of functionally active antibodies that could cross the placenta and provide protection against neonatal GBS infection.

During the last two decades, polysaccharide based vaccines against GBS have been extensively studied but also several promising protein antigens have been identified leading to the development of universal protein-based vaccines (Korzeniowska-Kowal *et al.*, 2001; Martin *et al.*, 2002; Maione *et al.*, 2005; Margarit *et al.*, 2009). In this work we review the different approaches undertaken to develop effective vaccines against GBS infections.

Capsular polysaccharide-based vaccines

Most of the work on GBS vaccines focused on capsular polysaccharides. Evidence of the protective nature of polysaccharide-specific antibodies came from the studies of Rebecca Lancefield in the 1930s. She demonstrated that, using CPS-specific polyclonal rabbit serum, mice could be protected against GBS infections (Lancefield, 1934; Lancefield, 1938).

During the 1970s, Baker *et al.* showed that there is a correlation between a low level of maternal antibodies against type III CPS and susceptibility of infants to GBS disease. These studies established that CPS-specific antibodies are able to be transferred from the mother to the newborn, increasing in the neonates the amount of protective antibodies present at the time of delivery. These findings showed that vaccination of pregnant women could be a prophylactic strategy to avoid GBS disease in newborns, providing the rationale for the development of a vaccine against GBS based on the CPS antigens (Baker and Kasper, 1976; Lin *et al.*, 2004).

The first human clinical trials took place in the 1980s and were conducted with purified native CPS. Even if these trials demonstrated the safety and the immunogenicity of these antigens, it was evident that these preparations needed to be improved because the vaccine based on the CPS III was able to induce a significant IgG response only in 60% of the recipients (Baker *et al.*, 1981, 1988; Baker and Kasper, 1985). As a result of these findings the necessity to improve the immunogenicity of the CPS antigens to induce an higher immune response was recognized. To this purpose, the first GBS conjugate vaccine was prepared using the CPS serotype III coupled to tetanus toxoid (TT) (Paoletti *et al.*, 1992; Wessels *et al.*, 1993, 1995; Paoletti *et al.*, 2001).

Since then, CPS-tetanus toxoid conjugates vaccines based on nine GBS serotypes have been produced and tested pre-clinically (Paoletti *et al.*, 1994; Baker *et al.*, 2003b). All the GBS conjugated vaccines prepared with TT were more immunogenic than uncoupled CPS in mice and rabbits. These studies showed that these antigens were able to induce functionally active CPS-specific IgG, that, in the presence of complement, were able to opsonize and induce killing of GBS by human peripheral blood leucocytes in *in vitro* assays. The success of the preclinical tests in animals constituted the rationale to proceed with the clinical studies of the CPS–TT conjugated in humans. Clinical phase 1 of conjugate vaccines prepared with GBS types Ia, Ib, II, III, and V and the clinical phase 2 of conjugate vaccines prepared with GBS types Ia, Ib, II, and III CPS demonstrated that these preparations were safe, well-tolerated and highly immunogenic in humans (Paoletti and Kasper, 2003).

Although the immunogenicity in humans of the CPS antigens was successfully increased through conjugation to tetanus toxoid, cross-protection between serotypes was still lacking, representing the main obstacle to the development of an effective GBS vaccine. (Lin *et al.*, 1998; Hickman *et al.*, 1999; Berg *et al.*, 2000; Davies *et al.*, 2001). For this reason the need to create a multivalent vaccine in order to obtain a broad coverage of the vaccine against the prevalent GBS serotypes was evident. A combination of four TT-conjugated serotypes (Ia, Ib, II and III) was successfully tested in a mouse infection model (Paoletti *et al.*, 1994) and, following these results, also human trials were set up using the combination of two TT-conjugated serotypes (II and III) (Baker *et al.*, 2003a). These studies have shown that the combinations have the same immunogenicity and reactogenicity of each monovalent CPS vaccine. However, recent epidemiological studies (Harrison *et al.*, 1998; Berg *et al.*, 2000; Fluegge *et al.*, 2005; Barcaite *et al.*, 2008; Phares *et al.*, 2008) suggest that in order to achieve a coverage against the majority of GBS strains circulating in Europe and North America, it is necessary to combine at least four serotypes (Ia, Ib, III and V) in a multivalent vaccine but there are other regions where such a combination would not be effective, owing to a different distribution of serotypes, for example serotypes VI and VIII, which are predominant in Japan (Lachenauer *et al.*, 1999). Moreover, the CPS-conjugated vaccines would not protect against all the non-typeable isolates.

An additional obstacle to the licensure of vaccines against GBS is the difficulty of conducting clinical efficacy trials. Efficacy studies in humans are very complicated to carry out, because of the low incidence of neonatal GBS disease. A possible solution to overcome this difficulty came from the studies of Feng-Ying C. Lin and coworkers (Lin *et al.*, 2001). The authors carried out a prospective study to estimate the level of maternal antibodies necessary to protect newborns against EOD caused by GBS type Ia infection. The levels of maternal antibodies against GBS type Ia were measured by ELISA in 45 case patients neonates born after 34 weeks of gestation who developed EOD (caused by GBS serotype Ia) and 319 controls (neonates colonized by GBS Ia but without EOD). From this comparison it was demonstrated that the probability of developing EOD declined with increasing maternal levels of anti GBS Ia IgG. In particular, newborns whose mothers had anti-GBS Ia IgG levels > 5 mg/ml had an 88% lower risk of developing EOD. They demonstrated through this study that it is possible to predict if a vaccine can induce type-specific immunity and prevent EOD in infants by evaluating if it is able to elicit GBS Ia specific IgG levels above the defined threshold in their mothers (Lin *et al.*, 2001).

More recently, the same authors have estimated the level of maternal IgG anti-GBS type III required to protect newborns from EOD caused by GBS serotype III. Using a similar case–control trial they demonstrated that the probability of developing EOD decreased with increasing levels of maternal IgG anti-GBS type III. In particular, infants whose mothers had >10 mg/ml IgG anti-GBS type III had 91% lower risk for EOD, thus showing that a vaccine that induces IgG levels above that threshold in mothers can be predicted to confer effective protection against EOD caused by type III GBS strains (Lin *et al.*, 2004). The results of Lin's studies suggest a new strategy that may lead to the registration of a vaccine against GBS based on the quantification of specific antibodies induced by the vaccines, as a correlate of protection, instead of more complex efficacy clinical trials.

Presently, although great progress have been made towards the development of polysaccharide based vaccines, further clinical studies are still required to fulfil this urgent medical need.

Protein-based vaccine

In order to develop a serotype-independent universal GBS vaccine, several efforts have been directed to the identification of highly protective protein antigens. In contrast to the CPS antigens, proteins able to induce protective T-cell-dependent antibody responses and long-lasting immunity so that conjugation to other molecules (as the TT for the CPS) is not necessary. Conserved surface proteins are considered the best candidates for the development of a vaccine against GBS infection, because antibodies directed against surface antigens can interfere with bacterial virulence factors and can promote complement dependent opsonophagocytosis. Before 2005, only a few GBS protein antigens had been identified as potential vaccine candidates; these include Rib, the alpha and beta subunits of the C protein, Sip, and the C5a peptidase proteins (Table 13.1). However, these proteins, with the exception of C5a peptidase and Sip, are either not expressed by all strains or are highly variable in different isolates (Madoff *et al.*, 1992; Stalhammar-Carlemalm *et al.*, 1993; Brodeur *et al.*, 2000; Cheng *et al.*, 2001; Lindahl *et al.*, 2005; Santillan *et al.*, 2008).

The C protein complex was able to induce passive protection against GBS infections in an animal model (Madoff *et al.*, 1992; Larsson *et al.*, 1996; Gravekamp *et al.*, 1999; Lachenauer *et al.*, 1999; Yang *et al.*, 2008). Further studies showed also that the C protein complex could be one of the factors that confer resistance to opsonization, allowing GBS strains expressing the complex to escape from opsonophagocytosis (Payne and Ferrieri, 1985; Payne *et al.*, 1987). Rib is a surface protein with a similar structure and sequence to the alpha subunit of the C protein and it is also able to induce protective immunity (Musser *et al.*, 1989). Unfortunately, the alpha subunit of the C protein is present in the genome of only approximately 50% of clinical isolates, while Rib is present in the genome of all serotype III strains but not in other serotypes. Moreover, both these proteins contain repeated sequences that show strain-to-strain variations and that can affect their protective efficacy (Musser *et al.*, 1989).

Sip (surface immunogenic protein) is a surface GBS protein that was identified after immunological screening of a genomic library. Sip has been identified in GBS strains of every serotype and the *sip* gene is highly conserved among GBS isolates (Brodeur *et al.*, 2000). The protection conferred by the Sip protein was determined by a mouse neonatal infection model. Newborn mice were protected against infections of GBS strains of serotypes Ia/c, Ib, II, III, and V (Brodeur *et al.*, 2000). Furthermore, it has been observed that sera collected from pregnant women and their healthy newborns have anti-Sip antibody (Martin *et al.*, 2002).

C5a peptidase is a serine protease that inactivates human C5a, a factor produced during complement activation. C5a peptidase is highly conserved surface-bound protein that is expressed on the surface of all serotypes of both group A streptococcus (GAS) and group B streptococcus (Wexler *et al.*, 1985; Chmouryguina *et al.*, 1996). The GBS C5a peptidase is 98% identical in sequence to the GAS homologue protein sequence. It has been shown that C5a peptidase is a protective antigen (Cheng *et al.*, 2001; Santillan *et al.*, 2008). Moreover, Cheng and coworkers have shown that it could be also used as a carrier protein for type III polysaccharide vaccine (Cheng *et al.*, 2002).

Reverse vaccinology: a new approach to identify potential GBS vaccine candidates

For more than a century, vaccines were developed by isolating, inactivating and injecting the cause of the infection. This traditional approach is time-consuming and expensive. Moreover, it usually identifies only abundant antigens that are expressed under *in vitro* culture conditions (Andre, 2003). When the first complete microbial genome sequences became available in 1995, a new era, 'the genomic era', began and it changed completely the approach to vaccine development (Rappuoli, 2001; De Groot and Rappuoli, 2004). This new approach, named reverse vaccinology (Pizza *et al.*, 2000; Rappuoli, 2000; Mora *et al.*, 2003), has provided a new impulse to the vaccinology field because the vaccine research starts from the genome and not from the pathogen itself (Capecchi *et al.*, 2004). In reverse vaccinology, antigen discovery is achieved by using the integration of several techniques such as genomics, bioinformatics, and molecular biology. Reverse vaccinology shows several advantages respect to the conventional approach. In fact, it permits the identification also of less common, low expressed

Table 13.1 Protein antigens described so far for the development of a serotype independent vaccine

Protein vaccine candidates	Features	References
C5a peptidase	Serine-protease, highly conserved and present in all GBS strains. GBS C5α peptidase is 98% identical in aa sequence to that expressed by group A *Streptococcus*. The protein has been used also as a carrier for CPS polysaccharides III	Cheng *et al.* (2001)
C protein (a and b)	Presents in many strains but not in the clinically relevant serotype III strains. It is composed of two unrelated proteins, the trypsin-resistant α protein and the trypsin-sensitive β protein. Strains reported to carry the c antigen may express either or both the α and the β protein. Each of the two components of the c antigen, elicits protective immunity	Bevanger and Naess (1985)
Rib	Similar structure and sequence to the alpha subunit of the C protein Confers protection in mucosal challenge and systemic heterologous challenge	Musser *et al.* (1989)
Sip	Surface protein identified after immunological screening of a genomic library. Found in all serotypes and present in all GBS strains. Function unknown	Brodeur *et al.* (2000)
OCT and PGK	Ornithine carbamoyltransferase and Phosphoglycerate kinase. Surface proteins identified by proteomic approach predicted to be cytoplasmic and missed by the genomic screening	Hughes *et al.* (2001)
Srr-2	High-molecular-mass, serine-rich repeat protein (Srr-2), contains internal repeats and LPXTG motif. Present in the ST-17 strains, a virulent lineage of serotype III GBS	Doro *et al.* (2009)
BP-2a (GBS59)	Present in all GBS strains carrying PI-2a in 6 variants. Backbone protein of Pilus 2a, involved in biofilm formation. Each variant protects only against homologous strains	Margarit *et al.* (2009), Rinaudo *et al.* (2010), Rosini *et al.* (2006)
AP-2a (GBS67)	Present in all GBS strains carrying PI-2a in 2 variants, protects against heterologous challenge. Is the major ancillary protein of Pilus 2a. Contains a Von Willebrand (VWA) domain involved in cells adhesion	Maione *et al.* (2005), Rosini *et al.* (2006), Konto-Ghiorghi *et al.* (2009)
BP-1 (GBS80)	Pilus 1 backbone, contributes to GBS paracellular translocation through epithelial cells	Maione *et al.* (2005), Pezzicoli *et al.* (2008)
AP-1 (GBS104)	Major ancillary protein of the pilus 1, dispensable for backbone protein polymerization	Maione *et al.* (2005), Rosini *et al.* (2006)
BP-2b	Backbone protein of Pilus 2b. Role of this pilus type is under investigation	Margarit *et al.* (2009), Rosini *et al.* (2006)

and/or not expressed *in vitro* antigens. Moreover, the reverse vaccinology approach can also be applied to non-cultivable microorganisms. One of the major disadvantages is that the reverse vaccinology can be applied only for the discovery of proteins antigens and not for other antigens like lipopolysaccharides and glycolipids (Rappuoli, 2000; Serruto and Rappuoli, 2006).

The reverse vaccinology approach was applied to the development of a vaccine against GBS starting from the sequencing of the complete genome of a virulent GBS strain (2603v/r, serotype V). But a comparative genome hybridization (CGH) analysis showed that the genetic variability within the GBS isolates was too high, suggesting that more genome sequences were necessary for the identification of vaccine candidates (Glaser *et al.*, 2002; Herbert *et al.*, 2005; Tettelin *et al.*, 2005). From this analysis it was clear the need to include genome sequences of more serotypes for the selection of protein antigens.

In order to study the genome variability in GBS, Tettelin and coworkers sequenced the genome of six GBS strains that represent the most frequent disease-causing serotypes (serotype Ia strains A909 and 515, type Ib strain H36B, type II strain 18RS21, type III strain COH1 and type V strain CJB111). By a comparative analysis of all available genomes, it was possible to identify two subgenomes: the 'core genome' and the 'variable genome', together defined as 'pan-genome' (Maione *et al.*, 2005; Medini *et al.*, 2005; Mora *et al.*, 2006; Telford, 2008). The 'core genome' includes genes present in all the strains and constitutes around 80% of each genome. It contains all genes necessary for the basic biology of the bacteria. The 'variable genome' is responsible for strain diversity and comprises genes that are dispensable and unique to each strain. The introduction of the concept of the 'pan-genome' represented a huge potential for the application of reverse vaccinology to the identification of novel vaccine candidates.

The availability of the antigens on the bacterial surface for antibody recognition is a prerequisite for a protective immune response. Maione and coworkers, by using modern computer algorithms and bioinformatic software, selected within the GBS pan-genome the genes coding for putative surface-associated and secreted proteins. Different bioinformatic tools were used in order to identify the presence of signal peptides (Signal IP, PSORT), transmembrane domains (TMPRED), lipoproteins and cell-wall anchored proteins (motifs), and homology to other bacterial surface proteins (FastA). Around 589 putative surface proteins were selected, 396 belonged to the core genome and 193 were variable genes (Maione *et al.*, 2005). The proteins predicted to contain more than three transmembrane domains were excluded from the selection because they are difficult to produce in *E. coli*. By

using an high-throughput cloning and expression approach, 312 of the selected GBS genes were successfully produced in *E. coli*. Each of the genes was cloned with either an N-terminal 6XHis Tag or a C-terminal GST tag, and the expressed proteins were purified by affinity chromatography (Montigiani *et al.*, 2002; Maione *et al.*, 2005).

All the 312 purified recombinant GBS antigens were tested by an active maternal immunization/neonatal pup challenge mouse model of GBS infection. Briefly, the antigens were used to immunize intraperitoneally groups of 6–8 CD-1 female mice (6–8 weeks of age) with a three-dose immunization schedule. After the last immunization, mice were mated and their pups were challenged, within 48 h after birth, with a lethal dose of GBS. The survival of the neonates was monitored for 3 days (Maione *et al.*, 2005). Immune sera were also collected from immunized mice for *in vitro* analysis. Immunoblot assays were used for the identification of the protein in GBS total protein extracts, while flow cytometry assays were carried out to confirm the surface exposure of the antigens.

From this first systematic screening, four antigens were identified as capable of significantly increasing the survival rate of challenged infant mice. When the four antigens were mixed and administered simultaneously, an almost universal protection was achieved against challenge model using a panel of strains comprehensive of the most pathogenic GBS serotypes. In particular, the levels of protection reached were similar to those achieved using the polysaccharides-based vaccines. Only one (SAG0032) of these four antigens was part of the 'core genome', and this protein was the already described Sip protein. The other three antigens – GBS67 (SAG1408), GBS80 (SAG0645), and GBS104 (SAG0649) – were present in the variable portion of the subgenome.

The major outcome of this study was that while none of the protective antigens was able to confer wide protection against GBS infections alone, the combination of the four antigens, each effective against overlapping populations of isolates, was able to confer broad serotype-independent protection. Moreover, despite the levels of expression of the protective antigens were variable from strain to strain, the authors found that protection mediated by these antigens was strictly correlated to their surface exposure, as determined by FACS analysis. This allowed to estimate a coverage of 87% for a vaccine based on the combination of the four selected proteins, based on the FACS analysis performed on a panel of clinical isolates (Maione *et al.*, 2005).

Pilus-based vaccine

Characterization studies of the antigens selected in the work of Maione and colleagues revealed that three of them (GBS67, GBS80 and GBS104) were components of pilus-like structures (Lauer *et al.*, 2005; Rosini *et al.*, 2006) (Fig. 13.1).

Pili have now been identified in many Gram-positive pathogens, even if they had been overlooked by conventional molecular microbiology for a century (for a detailed review on Gram-positive pili, refer to Telford *et al.* (2006). The mechanism of pilus assembly and function of Gram-positive pili are just beginning to be unravelled (Ton-That and Schneewind, 2004; Lauer *et al.*, 2005; Mora *et al.*, 2005). Pili of Gram-positive bacteria are composed of three proteins that are substrates of specific sortase enzymes and assembled into high molecular weight covalently linked structures which are subsequently covalently attached to the peptidoglycan cell wall (Ton-That and Schneewind, 2004).

Thanks to comparative analyses of the available GBS genomes, it has been possible to identify in GBS three pilus genomic islands named pilus island 1 (PI-1), pilus island 2a (PI-2a) and pilus island 2b (PI-2b), which are localized in two distinct loci (PI-1 and PI-2, respectively). In fact, pilus island 2b (PI-2b) is an allelic variant of PI-2a that has a similar genetic organization to PI-1 and PI-2a, but varies substantially in gene sequence (Rosini *et al.*, 2006). Each genomic pilus island codes for the main structural protein, known as the backbone protein (BP), constituting the pilus shaft, two accessory proteins, known as ancillary proteins (AP), and two sortases that are required for pilus polymerization (Fig. 13.2).

GBS pilus proteins as vaccine candidates

The studies by Maione *et al.* and by Rosini *et al.* demonstrated that at least two of the three pilus structural components, the BP and the AP1, encoded by the PI-1 (respectively GBS80 and GBS104) and by the PI-2a (respectively GBS59 and GBS67) were able to induce protective immunity against GBS infection in mice. More

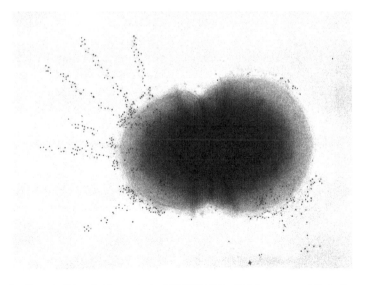

Figure 13.1 Visualization of pili in wild-type strain JM1300013 by EM: Immunogold labelling and transmission electron microscopy of BP of PI-1 (GBS80) in strain JM9130013, showing the presence of long pilus-like structures.

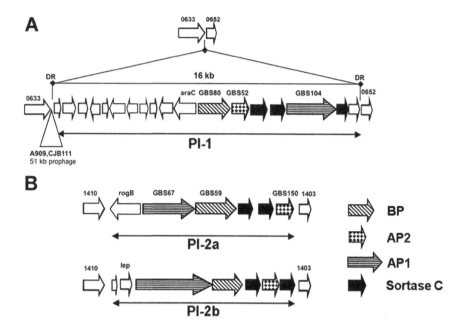

Figure 13.2 Schematic representation of the GBS pilus island operons (A. pilus island 1; B. pilus island 2a and 2b) containing the protective antigens. Genes coding for LPXTG-containing proteins (BP; AP1 and AP2) are represented with arrows with different filling motives as in the legend. Each PI contains at least two sortase genes (black arrows), transcriptional regulators and conserved flanking genes. A signal peptidase is also present in PI-2b (lep).

recently, Margarit and coworkers tested by the active maternal immunization/neonatal pup challenge mouse model already described the three structural proteins belonging to PI-2b, in order to establish if also the components of the pilus island 2b were able to induce protection *in vivo*. The results confirmed, also for the PI-2b, that the backbone (BP-2b) and the main ancillary (AP–2b) were able to confer protection against GBS strains carrying the homologous proteins.

These results paved the way to the development of a vaccine exclusively composed of pilus components. Strain-specific pili components have been tested as vaccine candidates against different pathogens; however, gene variability has, until now, represented the major obstacle to the development of pilus-based vaccines (Margarit *et al.*, 2009).

To verify whether the GBS pili antigens could be considered suitable candidates for the development of an effective vaccine against GBS, Margarit and colleagues conducted a series of studies to verify the distribution of each pilus island, the

sequence conservation and the surface exposure of each pilus protein. A collection of 289 GBS clinical isolates was analysed, by PCR, to detect the presence of the three pilus islands and it was shown that all the investigated GBS strains carried at least one of the three pilus islands. In many strains the concurrent presence of two pilus islands was observed, with the PI-1 and PI-2a being the most represented combination (46% of the analysed strains) (Margarit *et al.*, 2009).

Based on the analysis of genomic sequence analysis of 186 GBS isolates it was also demonstrated that pili structures are well conserved within each of the islands. In particular, all three structural proteins of PI-1 and PI-2b were highly conserved and differed by very few amino acids whereas the backbone (BP-2a) and the ancillary protein 1 (AP1–2a) of pilus island 2a were found to be more variable, as seven alleles were found for both proteins with sequence identities ranging from 48% to 98% and from 87% to 98%, respectively (Margarit *et al.*, 2009). It was also demonstrated, by FACS analysis on 289 GBS clinical isolates, that 94% of strains expressed at least one pilus at high level on their surface. Since it was already demonstrated that antigen protection strictly correlates with surface antigen exposure (Maione *et al.*, 2005), it was estimated that a pilus based vaccine may confer protection almost against 100% of circulating GBS strains (Margarit *et al.*, 2009).

All studies confirmed that GBS pili protein components could represent valid candidates for the development of a vaccine against GBS. Hence, based on *in vivo* protection experiments, antigen conservation and antigen expression, a combination of three proteins representing one protein from each of the 3 pilus island was selected. The combination of the backbone proteins from PI-1 and PI-2b and the ancillary protein 1 from PI-2a conferred protection in the mouse maternal immunization model against a panel of GBS strains representing the most common serotypes (Margarit *et al.*, 2009).

The results of these studies provided strong support to the development of a pilus-based vaccine against GBS infection. An additional rationale to support the development of a pilus-based vaccine is the fact that it could also be useful in preventing GBS colonization, as it is known that pili are involved in bacterial adhesion to the host (Maisey *et al.*, 2007; Manetti *et al.*, 2007; Bagnoli *et al.*, 2008; Pezzicoli *et al.*, 2008), and that antibodies against pili can inhibit GBS biofilm formation (Rinaudo *et al.*, 2010).

Pilus proteins as vaccine candidates in other pathogenic streptococci

The GBS pilus-based vaccine can be considered the prototype of vaccines against pili of other Gram-positive pathogens. In fact, multiple genome analysis was applied to identify pilus-like structures in two other major human streptococcal pahtogens: group A *Streptococcus* (GAS) and *S. pneumoniae* (Mora *et al.*, 2005; Barocchi *et al.*, 2006). In these three pathogenic streptococci, pilus genomic islands have a similar organization (Hava and Camilli, 2002; Mora *et al.*, 2005) and therefore there is reason to believe that pilus-like structures are important virulence factors and may represent potential vaccine candidates.

Mora and coworkers described four variants of pili in GAS and showed that immunization of mice with a combination of recombinant pilus proteins conferred protection against mucosal challenge with virulent GAS bacteria (Mora *et al.*, 2005). Also in the case of *S. pneumoniae*, even if piliated strains constitute only a subset of clinical isolates, it has been demonstrated that recombinant pilus antigens are able to confer protection against pneumococcal *strains expressing homologous proteins* in a mouse model of infection (Gianfaldoni *et al.*, 2007).

Proteomic approaches for GBS vaccine development

Proteomic strategies are valuable methods to identify proteins present in particluar subcellular compartments. A new proteomic approach is currently being applied to discover surface proteins of a pathogen aimed at the identification of vaccine candidates and to the understanding of bacterial pathogenesis.

This strategy has the strength to identify anchorless cell surface proteins potentially overlooked with a computational genome-wide screening approach (Rodriguez-Ortega *et al.*, 2006). The first proteomic analysis to identify the most abundant proteins present on the surface of GBS was performed by Hughes and coworkers and led to the identification of 27 highly expressed proteins, two of which, the ornithine carbamoyltransferase (OCT) and the phosphoglycerate kinase (PGK), were shown to confer protection in a mouse model of infection (Hughes *et al.*, 2002).

More recently, Doro and coworkers applied a fast and efficient proteomic approach for the identification of GBS surface-exposed proteins to be used as vaccine candidates. In this study the authors utilized a surface digestion procedure based on proteolytic enzymes (trypsin and Proteinase K) used to 'shave' protruding protein domains off the bacterial surface and peptides released from the bacterial surface. The peptides released were identified by mass spectrometry. Moreover, the protocol included a second proteolytic digestion in order to allow the identification of all the surface proteins particularly resistant to proteolysis, that otherwise would be lost by mass spectrometry. Using this approach with the GBS strain COHI serotype III, they identified 43 GBS surface-exposed or secreted proteins some of which were missed by the genomic approach performed by Maione *et al.* (2005). The effectiveness of this approach is supported by the fact that most of the identified proteins were extracellular proteins, cell-wall anchored proteins, lipoproteins and membrane proteins (Doro *et al.*, 2009). Also from this study the new protective antigen SAN_1485, a serine-rich repeat protein (Srr-2), was identified. This antigen was particularly interesting because is associated with a hypervirulent lineage of serotype III GBS (Seifert *et al.*, 2006).

Conclusions

After years of intensive studies aiming at the selection of protective antigens, today we may be close to the development of an effective vaccine to prevent GBS diseases. Capsular polysaccharide based vaccines are in an advanced stage of development. Many protein antigens are currently available for the development of a serotype independent vaccine (Table 13.1), with the pilus proteins being the most interesting ones. As the majority of the studies have demonstrated that no single antigen can work alone as a universal vaccine, the main goal of the ongoing studies is to select the best combination of antigens to achieve a broad vaccine coverage. A combination of glycoconjugates representing the most diffused serotypes plus a mix of highly protective pilus proteins seems to be the most promising vaccine.

One of the major obstacles to the licensure of a GBS vaccine including protein antigens remains the difficulty of conducting clinical efficacy trials. Since little is known about the level of maternal antibodies against protective proteins, it is hoped that studies similar to those conducted by Feng-Ying C. Lin for the capsular polysaccharides will be soon carried out also for protein antigens.

Today, even if further studies are still required, clinical trials are ongoing, and it can be predicted that in the next few years an effective vaccine to prevent GBS diseases will be finally available to the market to fulfil this very urgent medical need.

References

Andre, F.E. (2003). Vaccinology: past achievements, present roadblocks and future promises. Vaccine *21*, 593–595.

Bagnoli, F., Moschioni, M., Donati, C., Dimitrovska, V., Ferlenghi, I., Facciotti, C., Muzzi, A., Giusti, F., Emolo, C., Sinisi, A., *et al.* (2008). A second pilus type in *Streptococcus pneumoniae* is prevalent in emerging serotypes and mediates adhesion to host cells. J. Bacteriol. *190*, 5480–5492.

Baker, C.J. (1997). Group B streptococcal infections. Clin. Perinatol. *24*, 59–70.

Baker, C.J., Edwards, M.S., and Kasper, D.L. (1981). Role of antibody to native type III polysaccharide of group B Streptococcus in infant infection. Pediatrics *68*, 544–549.

Baker, C.J., and Kasper, D.L. (1976). Correlation of maternal antibody deficiency with susceptibility to neonatal group B streptococcal infection. N. Engl. J. Med. *294*, 753–756.

Baker, C.J., and Kasper, D.L. (1985). Group B streptococcal vaccines. Rev. Infect. Dis. *7*, 458–467.

Baker, C.J., Rench, M.A., Edwards, M.S., Carpenter, R.J., Hays, B.M., and Kasper, D.L. (1988). Immunization of pregnant women with a polysaccharide vaccine of group B streptococcus. N. Engl. J. Med. *319*, 1180–1185.

Baker, C.J., Rench, M.A., Fernandez, M., Paoletti, L.C., Kasper, D.L., and Edwards, M.S. (2003a). Safety and immunogenicity of a bivalent group B streptococcal conjugate vaccine for serotypes II and III. J. Infect. Dis. *188*, 66–73.

Baker, C.J., Rench, M.A., and McInnes, P. (2003b). Immunization of pregnant women with group B streptococcal type III capsular polysaccharide-tetanus toxoid conjugate vaccine. Vaccine *21*, 3468–3472.

Barcaite, E., Bartusevicius, A., Tameliene, R., Kliucinskas, M., Maleckiene, L., and Nadisauskiene, R. (2008). Prevalence of maternal group B streptococcal colonisation in European countries. Acta Obstet. Gynecol. Scand. *87*, 260–271.

Barocchi, M.A., Ries, J., Zogaj, X., Hemsley, C., Albiger, B., Kanth, A., Dahlberg, S., Fernebro, J., Moschioni, M., Masignani, V., *et al.* (2006). A pneumococcal pilus influences virulence and host inflammatory responses. Proc. Natl. Acad. Sci. U.S.A. *103*, 2857–2862.

Berg, S., Trollfors, B., Lagergard, T., Zackrisson, G., and Claesson, B.A. (2000). Serotypes and clinical manifestations of group B streptococcal infections in western Sweden. Clin. Microbiol. Infect. *6*, 9–13.

Bisharat, N., Jones, N., Marchaim, D., Block, C., Harding, R.M., Yagupsky, P., Peto, T., and Crook, D.W. (2005). Population structure of group B streptococcus from a low-incidence region for invasive neonatal disease. Microbiology *151*, 1875–1881.

Blancas, D., Santin, M., Olmo, M., Alcaide, F., Carratala, J., and Gudiol, F. (2004). Group B streptococcal disease in nonpregnant adults: incidence, clinical characteristics, and outcome. Eur. J. Clin. Microbiol. Infect. Dis. *23*, 168–173.

Boyer, K.M., Gadzala, C.A., Kelly, P.D., and Gotoff, S.P. (1983). Selective intrapartum chemoprophylaxis of neonatal group B streptococcal early-onset disease. III. Interruption of mother-to-infant transmission. J. Infect. Dis. *148*, 810–816.

Brodeur, B.R., Boyer, M., Charlebois, I., Hamel, J., Couture, F., Rioux, C.R., and Martin, D. (2000). Identification of group B streptococcal Sip protein, which elicits cross-protective immunity. Infect. Immun. *68*, 5610–5618.

Capecchi, B., Serruto, D., Adu-Bobie, J., Rappuoli, R., and Pizza, M. (2004). The genome revolution in vaccine research. Curr. Issues Mol. Biol. *6*, 17–27.

Cheng, Q., Carlson, B., Pillai, S., Eby, R., Edwards, L., Olmsted, S.B., and Cleary, P. (2001). Antibody against surface-bound C5a peptidase is opsonic and initiates macrophage killing of group B streptococci. Infect. Immun. *69*, 2302–2308.

Cheng, Q., Debol, S., Lam, H., Eby, R., Edwards, L., Matsuka, Y., Olmsted, S.B., and Cleary, P.P. (2002). Immunization with C5a peptidase or peptidase-type III polysaccharide conjugate vaccines enhances clearance of group B Streptococci from lungs of infected mice. Infect. Immun. *70*, 6409–6415.

Chmouryguina, I., Suvorov, A., Ferrieri, P., and Cleary, P.P. (1996). Conservation of the C5a peptidase genes in group A and B streptococci. Infect. Immun. *64*, 2387–2390.

Davies, H.D., Raj, S., Adair, C., Robinson, J., and McGeer, A. (2001). Population-based active surveillance for neonatal group B streptococcal infections in Alberta, Canada: implications for vaccine formulation. Pediatr. Infect. Dis. J. *20*, 879–884.

De Groot, A.S., and Rappuoli, R. (2004). Genome-derived vaccines. Expert Rev. Vaccines *3*, 59–76.

Dillon, H.C., Jr., Gray, E., Pass, M.A., and Gray, B.M. (1982). Anorectal and vaginal carriage of group B streptococci during pregnancy. J. Infect. Dis. *145*, 794–799.

Doro, F., Liberatori, S., Rodriguez-Ortega, M.J., Rinaudo, C.D., Rosini, R., Mora, M., Scarselli, M., Altindis, E., D'Aurizio, R., Stella, M., *et al.* (2009). Surfome analysis as a fast track to vaccine discovery: identification of a novel protective antigen for Group B Streptococcus hypervirulent strain COH1. Mol. Cell Proteomics *8*, 1728–1737.

Edwards, M.S., and Baker, C.J. (2001). Group B streptococcal infections. In Infectious Diseases of the Fetus and Newborn Infant, 5th edn, Remington, J.S., and Klein, J.O., eds (Philadelphia, WB Saunders), pp. 1091–1156.

Edwards, M.S., Rench, M.A., Palazzi, D.L., and Baker, C.J. (2005). Group B streptococcal colonization and serotype-specific immunity in healthy elderly persons. Clin. Infect. Dis. *40*, 352–357.

Farley, M.M., Harvey, R.C., Stull, T., Smith, J.D., Schuchat, A., Wenger, J.D., and Stephens, D.S. (1993). A population-based assessment of invasive disease due to group B Streptococcus in nonpregnant adults. N. Engl. J. Med. *328*, 1807–1811.

Ferrieri, P. (1985). GBS infections in the newborn infant: diagnosis and treatment. Antibiot. Chemother. *35*, 211–224.

Fluegge, K., Supper, S., Siedler, A., and Berner, R. (2005). Serotype distribution of invasive group B streptococcal isolates in infants: results from a nationwide active laboratory surveillance study over 2 years in Germany. Clin. Infect. Dis. *40*, 760–763.

Gherardi, G., Imperi, M., Baldassarri, L., Pataracchia, M., Alfarone, G., Recchia, S., Orefici, G., Dicuonzo, G., and Creti, R. (2007). Molecular epidemiology and distribution of serotypes, surface proteins, and antibiotic resistance among group B streptococci in Italy. J. Clin. Microbiol. *45*, 2909–2916.

Gianfaldoni, C., Censini, S., Hilleringmann, M., Moschioni, M., Facciotti, C., Pansegrau, W., Masignani, V., Covacci, A., Rappuoli, R., Barocchi, M.A., *et al.* (2007). *Streptococcus pneumoniae* pilus subunits protect mice against lethal challenge. Infect. Immun. *75*, 1059–1062.

Gibbs, R.S., McDuffie, R.S., Jr., McNabb, F., Fryer, G.E., Miyoshi, T., and Merenstein, G. (1994). Neonatal group B streptococcal sepsis during 2 years of a universal screening program. Obstet. Gynecol. *84*, 496–500.

Gibbs, R.S., Schrag, S., and Schuchat, A. (2004). Perinatal infections due to group B streptococci. Obstet. Gynecol. *104*, 1062–1076.

Glaser, P., Rusniok, C., Buchrieser, C., Chevalier, F., Frangeul, L., Msadek, T., Zouine, M., Couve, E., Lalioui, L., Poyart, C., *et al.* (2002). Genome sequence of *Streptococcus agalactiae*, a pathogen causing invasive neonatal disease. Mol. Microbiol. *45*, 1499–1513.

Gravekamp, C., Kasper, D.L., Paoletti, L.C., and Madoff, L.C. (1999). Alpha C protein as a carrier for type III capsular polysaccharide and as a protective protein in group B streptococcal vaccines. Infect. Immun. *67*, 2491–2496.

Hansen, S.M., Uldbjerg, N., Kilian, M., and Sorensen, U.B.R. (2004). Dynamics of *Streptococcus agalactiae* colonization in women during and after pregnancy and in their infants. J. Clin. Microbiol. *42*, 83–89.

Harrison, L.H., Elliott, J.A., Dwyer, D.M., Libonati, J.P., Ferrieri, P., Billmann, L., and Schuchat, A. (1998). Serotype distribution of invasive group B streptococcal isolates in Maryland: implications for vaccine formulation. Maryland Emerging Infections Program. J. Infect. Dis. *177*, 998–1002.

Hava, D.L., and Camilli, A. (2002). Large-scale identification of serotype 4 *Streptococcus pneumoniae* virulence factors. Mol. Microbiol. *45*, 1389–1406.

Herbert, M.A., Beveridge, C.J., McCormick, D., Aten, E., Jones, N., Snyder, L.A., and Saunders, N.J. (2005). Genetic islands of *Streptococcus agalactiae* strains NEM316 and 2603VR and their presence in other Group B streptococcal strains. BMC Microbiol. *5*, 31.

Hickman, M.E., Rench, M.A., Ferrieri, P., and Baker, C.J. (1999). Changing epidemiology of group B streptococcal colonization. Pediatrics *104*, 203–209.

Hughes, M.J., Moore, J.C., Lane, J.D., Wilson, R., Pribul, P.K., Younes, Z.N., Dobson, R.J., Everest, P., Reason, A.J., Redfern, J.M., *et al.* (2002). Identification of major outer surface proteins of *Streptococcus agalactiae*. Infect. Immun. *70*, 1254–1259.

Kong, F., Gowan, S., Martin, D., James, G., and Gilbert, G.L. (2002). Serotype identification of group B streptococci by PCR and sequencing. J. Clin. Microbiol. *40*, 216–226.

Korzeniowska-Kowal, A., Witkowska, D., and Gamian, A. (2001). [Molecular mimicry of bacterial polysaccharides and their role in etiology of infectious and autoimmune diseases]. Postepy Hig. Med. Dosw. *55*, 211–232.

Lachenauer, C.S., Kasper, D.L., Shimada, J., Ichiman, Y., Ohtsuka, H., Kaku, M., Paoletti, L.C., Ferrieri, P., and Madoff, L.C. (1999). Serotypes VI and VIII predominate among group B streptococci isolated from pregnant Japanese women. J. Infect. Dis. *179*, 1030–1033.

Lancefield, R.C. (1934). Serological differentiation of specific types of bovine haemolytic streptococci (group B). J. Exp. Med. *59*, 441–458.

Lancefield, R.C. (1938). Two serological types of group B hemolytic streptococci with related, but not identical, type-specific substances. J. Exp. Med. *67*, 25–40.

Larsson, C., Stalhammar-Carlemalm, M., and Lindahl, G. (1996). Experimental vaccination against group B streptococcus, an encapsulated bacterium, with highly purified preparations of cell surface proteins Rib and alpha. Infect. Immun. *64*, 3518–3523.

Lauer, P., Rinaudo, C.D., Soriani, M., Margarit, I., Maione, D., Rosini, R., Taddei, A.R., Mora, M., Rappuoli, R., Grandi, G., et al. (2005). Genome analysis reveals pili in group B Streptococcus. Science *309*, 105.

Law, M.R., Palomaki, G., Alfirevic, Z., Gilbert, R., Heath, P., McCartney, C., Reid, T., and Schrag, S. (2005). The prevention of neonatal group B streptococcal disease: a report by a working group of the Medical Screening Society. J. Med. Screen. *12*, 60–68.

Lin, F.Y., Clemens, J.D., Azimi, P.H., Regan, J.A., Weisman, L.E., Philips, J.B., 3rd, Rhoads, G.G., Clark, P., Brenner, R.A., and Ferrieri, P. (1998). Capsular polysaccharide types of group B streptococcal isolates from neonates with early-onset systemic infection. J. Infect. Dis. *177*, 790–792.

Lin, F.Y., Philips, J.B., 3rd, Azimi, P.H., Weisman, L.E., Clark, P., Rhoads, G.G., Regan, J., Concepcion, N.F., Frasch, C.E., Troendle, J., et al. (2001). Level of maternal antibody required to protect neonates against early-onset disease caused by group B Streptococcus type Ia: a multicenter, seroepidemiology study. J. Infect. Dis. *184*, 1022–1028.

Lin, F.Y., Weisman, L.E., Azimi, P.H., Philips, J.B., 3rd, Clark, P., Regan, J., Rhoads, G.G., Frasch, C.E., Gray, B.M., Troendle, J., et al. (2004). Level of maternal IgG anti-group B streptococcus type III antibody correlated with protection of neonates against early-onset disease caused by this pathogen. J. Infect. Dis. *190*, 928–934.

Lindahl, G., Stalhammar-Carlemalm, M., and Areschoug, T. (2005). Surface proteins of *Streptococcus agalactiae* and related proteins in other bacterial pathogens. Clin. Microbiol. Rev. *18*, 102–127.

Madoff, L.C., Michel, J.L., Gong, E.W., Rodewald, A.K., and Kasper, D.L. (1992). Protection of neonatal mice from group B streptococcal infection by maternal immunization with beta C protein. Infect. Immun. *60*, 4989–4994.

Maione, D., Margarit, I., Rinaudo, C.D., Masignani, V., Mora, M., Scarselli, M., Tettelin, H., Brettoni, C., Iacobini, E.T., Rosini, R., et al. (2005). Identification of a universal group B streptococcus vaccine by multiple genome screen. Science *309*, 148–150.

Maisey, H.C., Hensler, M., Nizet, V., and Doran, K.S. (2007). Group B streptococcal pilus proteins contribute to adherence to and invasion of brain microvascular endothelial cells. J. Bacteriol. *189*, 1464–1467.

Manetti, A.G., Zingaretti, C., Falugi, F., Capo, S., Bombaci, M., Bagnoli, F., Gambellini, G., Bensi, G., Mora, M., Edwards, A.M., et al. (2007). *Streptococcus pyogenes* pili promote pharyngeal cell adhesion and biofilm formation. Mol. Microbiol. *64*, 968–983.

Margarit, I., Rinaudo, C.D., Galeotti, C.L., Maione, D., Ghezzo, C., Buttazzoni, E., Rosini, R., Runci, Y., Mora, M., Buccato, S., et al. (2009). Preventing bacterial infections with pilus-based vaccines: the group B streptococcus paradigm. J. Infect. Dis. *199*, 108–115.

Martin, D., Rioux, S., Gagnon, E., Boyer, M., Hamel, J., Charland, N., and Brodeur, B.R. (2002). Protection from group B streptococcal infection in neonatal mice by maternal immunization with recombinant Sip protein. Infect. Immun. *70*, 4897–4901.

McCracken, G.H., Jr. (1973). Group B streptococci: the new challenge in neonatal infections. J. Pediatr. *82*, 703–706.

Medini, D., Donati, C., Tettelin, H., Masignani, V., and Rappuoli, R. (2005). The microbial pan-genome. Curr. Opin. Genet. Dev. *15*, 589–594.

Montigiani, S., Falugi, F., Scarselli, M., Finco, O., Petracca, R., Galli, G., Mariani, M., Manetti, R., Agnusdei, M., Cevenini, R., et al. (2002). Genomic approach for analysis of surface proteins in *Chlamydia pneumoniae*. Infect. Immun. *70*, 368–379.

Moore, M.R., Schrag, S.J., and Schuchat, A. (2003). Effects of intrapartum antimicrobial prophylaxis for prevention of group-B-streptococcal disease on the incidence and ecology of early-onset neonatal sepsis. Lancet Infect. Dis. *3*, 201–213.

Mora, M., Bensi, G., Capo, S., Falugi, F., Zingaretti, C., Manetti, A.G., Maggi, T., Taddei, A.R., Grandi, G., and Telford, J.L. (2005). Group A Streptococcus produce pilus-like structures containing protective antigens and Lancefield T antigens. Proc. Natl. Acad. Sci. U.S.A. *102*, 15641–15646.

Mora, M., Donati, C., Medini, D., Covacci, A., and Rappuoli, R. (2006). Microbial genomes and vaccine design: refinements to the classical reverse vaccinology approach. Curr. Opin. Microbiol. *9*, 532–536.

Mora, M., Veggi, D., Santini, L., Pizza, M., and Rappuoli, R. (2003). Reverse vaccinology. Drug Discov. Today *8*, 459–464.

Musser, J.M., Mattingly, S.J., Quentin, R., Goudeau, A., and Selander, R.K. (1989). Identification of a high-virulence clone of type III *Streptococcus agalactiae* (group B Streptococcus) causing invasive neonatal disease. Proc. Natl. Acad. Sci. U.S.A. *86*, 4731–4735.

Paoletti, L.C., and Kasper, D.L. (2003). Glycoconjugate vaccines to prevent group B streptococcal infections. Expert Opin. Biol. Ther. *3*, 975–984.

Paoletti, L.C., Peterson, D.L., Legmann, R., and Collier, R.J. (2001). Preclinical evaluation of group B streptococcal polysaccharide conjugate vaccines prepared with a modified diphtheria toxin and a recombinant duck hepatitis B core antigen. Vaccine *20*, 370–376.

Paoletti, L.C., Wessels, M.R., Michon, F., DiFabio, J., Jennings, H.J., and Kasper, D.L. (1992). Group B Streptococcus type II polysaccharide-tetanus toxoid conjugate vaccine. Infect. Immun. *60*, 4009–4014.

Paoletti, L.C., Wessels, M.R., Rodewald, A.K., Shroff, A.A., Jennings, H.J., and Kasper, D.L. (1994). Neonatal mouse protection against infection with multiple group B streptococcal (GBS) serotypes by maternal immunization with a tetravalent GBS polysaccharide-tetanus toxoid conjugate vaccine. Infect. Immun. *62*, 3236–3243.

Payne, N.R., and Ferrieri, P. (1985). The relation of the Ibc protein antigen to the opsonization differences between strains of type II group B streptococci. J. Infect. Dis. *151*, 672–681.

Payne, N.R., Kim, Y.K., and Ferrieri, P. (1987). Effect of differences in antibody and complement requirements on phagocytic uptake and intracellular killing of 'c' protein-positive and -negative strains of type II group B streptococci. Infect. Immun. *55*, 1243–1251.

Pezzicoli, A., Santi, I., Lauer, P., Rosini, R., Rinaudo, D., Grandi, G., Telford, J.L., and Soriani, M. (2008). Pilus backbone contributes to group B Streptococcus paracellular translocation through epithelial cells. J. Infect. Dis. *198*, 890–898.

Phares, C.R., Lynfield, R., Farley, M.M., Mohle-Boetani, J., Harrison, L.H., Petit, S., Craig, A.S., Schaffner, W., Zansky, S.M., Gershman, K., et al. (2008). Epidemiology of invasive group B streptococcal disease in the United States, 1999–2005. JAMA *299*, 2056–2065.

Pizza, M., Scarlato, V., Masignani, V., Giuliani, M.M., Arico, B., Comanducci, M., Jennings, G.T., Baldi, L., Bartolini, E., Capecchi, B., et al. (2000). Identification of vaccine candidates against serogroup B meningococcus by whole-genome sequencing. Science *287*, 1816–1820.

Puopolo, K.M., Madoff, L.C., and Eichenwald, E.C. (2005). Early-onset group B streptococcal disease in the era of maternal screening. Pediatrics *115*, 1240–1246.

Rappuoli, R. (2000). Reverse vaccinology. Curr. Opin. Microbiol. *3*, 445–450.

Rappuoli, R. (2001). Reverse vaccinology, a genome-based approach to vaccine development. Vaccine *19*, 2688–2691.

Rinaudo, C.D., Rosini, R., Galeotti, C.L., Berti, F., Necchi, F., Reguzzi, V., Ghezzo, C., Telford, J.L., Grandi, G., and Maione, D. (2010). Specific involvement of pilus type 2a in biofilm formation in group B Streptococcus. PLoS One *5*, e9216.

Rodriguez-Ortega, M.J., Norais, N., Bensi, G., Liberatori, S., Capo, S., Mora, M., Scarselli, M., Doro, F., Ferrari, G., Garaguso, I., et al. (2006). Characterization and identification of vaccine candidate proteins through analysis of the group A Streptococcus surface proteome. Nat. Biotechnol. *24*, 191–197.

Rosini, R., Rinaudo, C.D., Soriani, M., Lauer, P., Mora, M., Maione, D., Taddei, A., Santi, I., Ghezzo, C., Brettoni, C., et al. (2006). Identification of novel genomic islands coding for antigenic pilus-like structures in *Streptococcus agalactiae*. Mol. Microbiol. *61*, 126–141.

Rubens, C.E., Raff, H.V., Jackson, J.C., Chi, E.Y., Bielitzki, J.T., and Hillier, S.L. (1991). Pathophysiology and histopathology of group B streptococcal sepsis in Macaca nemestrina primates induced after intraamniotic inoculation: evidence for bacterial cellular invasion. J. Infect. Dis. *164*, 320–330.

Santillan, D.A., Andracki, M.E., and Hunter, S.K. (2008). Protective immunization in mice against group B streptococci using encapsulated C5a peptidase. Am. J. Obstet. Gynecol. *198*, 114, e111–116.

Schrag, S., Gorwitz, R., Fultz-Butts, K., and Schuchat, A. (2002). Prevention of perinatal group B streptococcal disease: revised guidelines from CDC. MMWR Morb. Mortal. Wkly Rep. *51*, 1–22.

Schrag, S.J., Zywicki, S., Farley, M.M., Reingold, A.L., Harrison, L.H., Lefkowitz, L.B., Hadler, J.L., Danila, R., Cieslak, P.R., and Schuchat, A. (2000). Group B streptococcal disease in the era of intrapartum antibiotic prophylaxis. N. Engl. J. Med. *342*, 15–20.

Schuchat, A. (1998). Epidemiology of group B streptococcal disease in the United States: shifting paradigms. Clin. Microbiol. Rev. *11*, 497–513.

Seifert, K.N., Adderson, E.E., Whiting, A.A., Bohnsack, J.F., Crowley, P.J., and Brady, L.J. (2006). A unique serine-rich repeat protein (Srr-2) and novel surface antigen (epsilon) associated with a virulent lineage of serotype III *Streptococcus agalactiae*. Microbiology *152*, 1029–1040.

Serruto, D., and Rappuoli, R. (2006). Post-genomic vaccine development. FEBS Lett. *580*, 2985–2992.

Skoff, T.H., Farley, M.M., Petit, S., Craig, A.S., Schaffner, W., Gershman, K., Harrison, L.H., Lynfield, R., Mohle-Boetani, J., Zansky, S., et al. (2009). Increasing burden of invasive group B streptococcal disease in nonpregnant adults, 1990–2007. Clin. Infect. Dis. *49*, 85–92.

Stalhammar-Carlemalm, M., Stenberg, L., and Lindahl, G. (1993). Protein rib: a novel group B streptococcal cell surface protein that confers protective immunity and is expressed by most strains causing invasive infections. J. Exp. Med. *177*, 1593–1603.

Telford, J.L. (2008). Bacterial genome variability and its impact on vaccine design. Cell Host Microbe *3*, 408–416.

Telford, J.L., Barocchi, M.A., Margarit, I., Rappuoli, R., and Grandi, G. (2006). Pili in gram-positive pathogens. Nat. Rev. Microbiol. *4*, 509–519.

Tettelin, H., Masignani, V., Cieslewicz, M.J., Donati, C., Medini, D., Ward, N.L., Angiuoli, S.V., Crabtree, J., Jones, A.L., Durkin, A.S., et al. (2005). Genome analysis of multiple pathogenic isolates of *Streptococcus agalactiae*: Implications for the microbial 'pan-genome'. Proc. Natl. Acad. Sci. U.S.A. *102*, 13950–13955.

Ton-That, H., and Schneewind, O. (2004). Assembly of pili in Gram-positive bacteria. Trends Microbiol. *12*, 228–234.

Wessels, M.R., Paoletti, L.C., Pinel, J., and Kasper, D.L. (1995). Immunogenicity and protective activity in animals of a type V group B streptococcal polysaccharide-tetanus toxoid conjugate vaccine. J. Infect. Dis. *171*, 879–884.

Wessels, M.R., Paoletti, L.C., Rodewald, A.K., Michon, F., DiFabio, J., Jennings, H.J., and Kasper, D.L. (1993).

Stimulation of protective antibodies against type Ia and Ib group B streptococci by a type Ia polysaccharide-tetanus toxoid conjugate vaccine. Infect. Immun. *61*, 4760–4766.

Wexler, D.E., Chenoweth, D.E., and Cleary, P.P. (1985). Mechanism of action of the group A streptococcal C5a inactivator. Proc. Natl. Acad. Sci. U.S.A. *82*, 8144–8148.

Yang, H.H., Mascuch, S.J., Madoff, L.C., and Paoletti, L.C. (2008). Recombinant group B Streptococcus alpha-like protein 3 is an effective immunogen and carrier protein. Clin. Vaccine Immunol. *15*, 1035–1041.

Vaccines Against *Streptococcus pneumoniae*

14

James C. Paton

Abstract

Existing vaccines against *Streptococcus pneumoniae* are targeted at the capsular polysaccharide (PS) of which there are 93 distinct serotypes. Polyvalent purified PS vaccines are immunogenic in healthy adults, but not in high risk groups such as young children and the elderly. Development of PS–protein conjugate vaccines has overcome the poor immunogenicity of PS in children, but the protection imparted is strictly serotype-specific, and the number of included serotypes is even more restricted than in the PS vaccine formulations. Widespread introduction of conjugate vaccines in developed countries has dramatically reduced the incidence of invasive pneumococcal disease due to serotypes included in the vaccine. However, these benefits are being eroded by increases in the incidence of disease caused by non-vaccine serotypes. Conjugate vaccines are also expensive, limiting their use in developing countries, where the burden of pneumococcal disease is greatest. Clearly, there is an urgent need to develop alternative pneumococcal vaccines that are (i) inexpensive, (ii) immunogenic in young children, and (iii) provide protection against all pneumococci regardless of serotype. Advances towards this goal are discussed herein, with particular emphasis on vaccines comprising pneumococcal proteins that contribute to virulence and are common to all serotypes.

The need for new pneumococcal vaccines

The potential contribution of vaccines to the control of pneumococcal disease has been recognized for nearly a century. The first experimental vaccines, comprising killed whole cells, were tested in the early 1900s, albeit with inconclusive results. Although interest fluctuated in the decades that followed, the steady increase in knowledge of the immunobiology of pneumococcal disease enabled a rational approach to vaccine design. This has resulted in the licensing of polyvalent capsular polysaccharide vaccines, and more recently, the polysaccharide–protein conjugates. These vaccines are aimed at preventing invasive diseases such as pneumonia, meningitis and bacteraemia, as well as less serious, but highly prevalent infections such as otitis media. They are being targeted principally at specific groups at high-risk of pneumococcal disease, particularly children under 2 years and adults over 65 years of age.

Research aimed at the development of more effective vaccines against *Streptococcus pneumoniae* is being driven by several factors. First amongst these is the fact that the pneumococcus continues to cause high morbidity and mortality throughout the world, even in regions where antibiotics are readily available. *S. pneumoniae* is the single commonest cause of community-acquired pneumonia, and has become the commonest cause of meningitis in many regions. Although determining the true burden of pneumococcal disease is complicated by difficulties in establishing an aetiological diagnosis, particularly in cases of non-bacteraemic pneumonia, the pneumococcus is conservatively estimated to kill more than a million children under the age of 5 years each year in developing countries. This accounts for 20–25% of all deaths in this age group (Williams *et al.*, 2002; WHO, 2007). Even in developed countries, where effective antimicrobial drugs are readily available, morbidity and mortality from pneumococcal disease is significant. For example, in the United States there are approximately 500,000 cases of pneumococcal pneumonia, 50,000 cases of bacteraemia and 3000 cases of pneumococcal meningitis each year, collectively resulting in an estimated 40,000 deaths (Centers for Disease Control, 1997; Klein, 2000). *S. pneumoniae* is also the single commonest cause of otitis media, which in the USA results in over 24 million visits to paediatricians each year, and more prescriptions for antibiotics than any other infectious disease (Klein, 2001). Thus, otitis media has a significant impact on health-care costs in developed countries, and estimates for USA exceed $5 billion per annum (Brixner, 2005).

The second major driver for vaccine development has been the increasing threat posed by antibiotic-resistant pneumococci. In the first two decades after introduction of penicillin, clinical isolates of *S. pneumoniae* were universally and exquisitely sensitive this drug. However, strains with reduced susceptibility to penicillin (MIC > 0.1 µg/ml) were detected in the late 1960s and since then have steadily increased in prevalence throughout the world. The problem is greatest in areas in which the use of antibiotics has been poorly regulated, and rates of resistance to β-lactams may exceed 50% of isolates (Klugman, 1990). Moreover, the degree of resistance has been increasing as a consequence of accumulation of multiple mutations in penicillin binding protein genes, resulting in strains with high-level penicillin resistance (MIC ≥ 2 µg/ml). Penicillin-resistant pneumococci are also often resistant to one or more other classes of antibiotics, and multiply-resistant clones of *S. pneumoniae* have spread globally (Klugman, 1996; McGee *et al.*, 2000). The increasing prevalence of penicillin- and multiply resistant pneumococci is complicating management of patients with suspected pneumococcal disease, particularly those with meningitis. In developed countries this is necessitating the use of more expensive alternative antimicrobials, but this option is not available in poorer parts of the world.

Purified polysaccharide vaccines

The protection imparted by pneumococcal polysaccharide (PS) vaccines is largely a result of binding of specific antibody to the capsule, resulting in opsonization and rapid clearance of invading pneumococci. A well-constructed trial of a tetravalent PS vaccine in US military recruits during World War II demonstrated a high degree of protection against pneumonia caused by types included in the vaccine (McLeod *et al.*, 1945). However, the resultant commercial production of two hexavalent PS vaccines coincided with the introduction of penicillin and other antibiotics, which at the time appeared to be spectacularly effective against the pneumococcus. As a result the vaccines were not utilized to any great extent and were eventually withdrawn from the market (Austrian, 2001). Nevertheless, it soon became clear that prompt and appropriate antibiotic therapy could not be relied upon to prevent death from invasive pneumococcal disease in certain high-risk patient groups. The continued high morbidity and mortality rekindled interest in PS vaccines and further trials were conducted in healthy young adults (South African miners) who had high attack rates of pneumococcal pneumonia and bacteraemia. Multivalent formulations were shown to be approximately 80% effective in preventing invasive pneumococcal disease caused by serotypes contained in the vaccine (Austrian, 2001). A 14-valent PS vaccine was licensed in 1977 and coverage was expanded to 23 types in 1983. This latter formulation includes types 1, 2, 3, 4, 5, 6B, 7F, 8, 9N, 9V, 10A, 11A, 12F, 14, 15B, 17F, 18C, 19A, 19F, 20, 22F, 23F and 33F (Centers for Disease Control, 1997).

A number of additional trials have been conducted in older adults since the introduction of the PS vaccines (reviewed by Briles *et al.*, 1999a). Comparisons of the results of the different studies are complicated by differences in study design and clinical criteria, but they have tended to show somewhat lower efficacy rates (usually of the order of 60%), particularly in elderly recipients, the immunocompromised, and those with underlying chronic diseases. Nevertheless, controversy surrounding the efficacy of PS vaccines for prevention of pneumococcal pneumonia has continued, particularly since some prospective randomized trials in older adults failed to demonstrate any protection whatsoever (Simberkoff *et al.*, 1986; Ortqvist *et al.*, 1998). In spite of this, the vaccine is currently recommended for all persons aged over 65 years, and those under 65 belonging to other high-risk groups (Centers for Disease Control, 1997). PS vaccines are not recommended for children aged < 2 years, even though they are a particularly high-risk group, because efficacy has not been demonstrated in clinical trials.

The likely explanation for the failure of the vaccine in young children is the poor immunogenicity of many of the component PS antigens. Children < 2 years of age can mount an adequate antibody response to some types (e.g. type 3), but responses are particularly poor for the PS types which most commonly cause invasive disease in children, namely 6A/B, 14, 18C, 19F and 23F (Douglas *et al.*, 1983). Indeed, responses to these types are weak up to the age of 5 years and do not reach adult levels until 8–10 years of age (Paton *et al.*, 1986). Elderly adults also exhibit weaker and more transient antibody responses to some of these PS serotypes compared with younger adults, and this undoubtedly accounts for the poorer clinical efficacy in this age group referred to above (Briles *et al.*, 1999a). PSs are referred to as 'thymus independent type 2' antigens and activate B lymphocytes independently of CD4+ cells by directly binding and cross-linking antigen receptors on the B cell surface. This process is distinct from that induced by protein antigens and involves co-stimulation by CD21 (type 2 complement receptor) after binding of C3d (generated by activation of the alternative complement pathway by PS). Neonatal B lymphocytes express low levels of CD21, and this may be one explanation for the hyporesponsiveness to PS during infancy (Rijkers *et al.*, 2001). PS antigens do not induce immunological memory and antibodies produced are mainly of the IgG_2 subclass. Even in healthy adults, antibody levels begin to decline about 1 year post-vaccination and for many types return to preimmunization levels after about 5 years (Briles *et al.*, 1999a). Interestingly, Musher *et al.* (2001b) have reported that healthy individuals

varied markedly in their capacity to mount an antibody response to various pneumococcal PS serotypes, and that this was controlled genetically and inherited in a codominant pattern. Thus, a subset of the adult population may be refractory to immunization with pneumococcal PS vaccines.

The other principal weakness of PS vaccines is that the protection they elicit is strictly serotype dependent. As mentioned above, the current vaccine formulation includes PS purified from 23 of the 93 known serotypes of *S. pneumoniae*, and clinical trials have confirmed that it provides negligible cover against serotypes not included in the formulation. Fortunately, not all types are equally prevalent, and the formulation was determined with reference to available data on the distribution of types causing invasive disease in adults and children (Robbins *et al.*, 1983). Most of these data emanated from the US or Europe, and the 23 included serotypes currently account for about 90% of invasive pneumococcal infections in these regions. However, there are geographical and temporal differences in the serotype distribution of disease-causing pneumococci, and the existing formulation may cover as little as 60% of strains in parts of Asia (Lee *et al.*, 1991). Moreover, serotype prevalence data are scanty for many developing countries and vaccine coverage in these regions is uncertain.

Some PS types included in the vaccine (for example type 6B) are known to elicit antibodies that cross-react with structurally related PS types that are not included (in this example type 6A). Types 6A and 6B pneumococci are both important causes of invasive disease in children, but the vaccine was formulated with the expectation that the cross-reacting antibodies would provide cross-protection (Robbins *et al.*, 1983). However, this assumption may be incorrect, as the cross-reacting antibodies appear to be of low avidity and function poorly in *in vitro* opsonophagocytic assays against the heterologous type. More recent clinical trials are also strongly suggestive of weaker than expected cross-protection (Briles *et al.*, 1999a).

The inevitable conclusion that must be drawn from the above is that notwithstanding the high protective efficacy in healthy adults, existing PS vaccines have suboptimal efficacy in groups who are most at risk from life-threatening invasive pneumococcal disease.

Polysaccharide–protein conjugate vaccines

A solution to the poor immunogenicity of PS in young children emanated from the seminal work of Avery and Goebel (1931), who reported that chemical conjugation of type 3 pneumococcal PS to a protein carrier massively increased its immunogenicity in rabbits. This approach was subsequently used against *Haemophilus influenzae* type b and the Hib polysaccharide–protein conjugate vaccine has been spectacularly successful (Robbins *et al.*, 1996). This has encouraged development of multivalent pneumococcal PS–protein conjugate vaccines, the first generation of which was introduced in USA in 2000, and is now licensed in many countries. Conjugation to a protein carrier converts the PS into a T-cell dependent antigen. The PS component is thought to react with receptors on B-cells, which then internalize the conjugate, process it, and present peptide fragments in association with class II MHC molecules to peptide-specific T-cells. Memory responses are generated and primed B-cells can be boosted either with conjugate or free PS (Briles *et al.*, 1999a).

Development of pneumococcal PS–protein conjugate vaccines has been considerably more complex than was the case with Hib, owing to the multiplicity of disease-causing serotypes. A number of parameters which influence immunogenicity of conjugate antigens need to be optimized for each type, including the molecular size of the PS component, the carrier protein, the PS:carrier ratio, and the method used to covalently link the two components. In view of this developmental complexity, the number of serotypes that can be included is by necessity less than in the PS vaccine. However, the conjugate vaccines were principally designed to prevent invasive disease and otitis media in young children, for whom the range of disease-causing serotypes is more restricted than in adults.

Currently–licensed conjugate vaccines or those under development cover from between 7 and 13 serotypes, use different cross-linking chemistries, and employ a range of carriers such as tetanus or diphtheria toxoids, the diphtheria toxin derivative CRM_{197}, or outer membrane proteins from *Neisseria meningitidis* group B or non-typable *H. influenzae*. The 7-valent formulation introduced in 2000 (PCV7; Prevnar/Prevenar™) includes types 4, 6B, 9V, 14, 18C, 19F and 23F, and it is estimated that this would cover 60–90% of paediatric infections based on North American and European seroprevalence studies (Eskola, 2001). A recently licensed 10-valent vaccine (PCV10; Synflorix™) includes these same types with the addition of types 1, 5 and 7F. Type 7F is an increasingly frequent cause of invasive disease, and although types 1 and 5 are uncommon in Europe and North America, they are important causes of invasive paediatric disease in other geographic regions. Indeed, a ten-year study of the seroprevalence of pneumococci causing invasive disease in children in Southern Israel indicated that inclusion of these two types would increase coverage of PCV7 from 41% to 67% in Jewish children and from 22% to 63% in Bedouin children (Fraser *et al.*, 2001).

These conjugate vaccines are typically administered as a course of three injections at 2, 4 and 6 months of age, followed by a booster of either conjugate or PS vaccine at 12–15 months. Several clinical studies have demonstrated that they are well tolerated by infants and elicit strong, boostable antibody responses (reviewed in Wuorimaa and Kayhty, 2002). A large study of PCV7 in Northern California demonstrated 97% protection against invasive (bacteraemic) disease caused by vaccine types (Black *et al.*, 2000). A Finish study designed to test the protective efficacy of the same vaccine against otitis media, which included microbiological analysis of middle ear fluid from all suspected cases, demonstrated a 57% reduction in infections caused by vaccine types (Eskola *et al.*, 2001). This figure is similar to the 67% reduction in otitis caused by vaccine types reported in the Californian study, although microbiological analysis had been confined to spontaneously draining ears (Black *et al.*, 2000).

Although the degree of type-specific protection imparted by PCV7 was less spectacular against otitis media as compared to invasive disease, this outcome was not unexpected, as higher antibody concentrations are probably required for prevention of the former. Nevertheless, the prevalence of pneumococcal otitis media is such that even a partially protective vaccine would prevent a very large number of cases. Interestingly, in the Finish study the vaccine also reduced the number of otitis episodes caused by pneumococci belonging to non-included types such as 6A and 19A, which cross-react with vaccine types (6B and 19F, respectively) by 51% (Eskola *et al.*, 2001). This occurred even though antibodies to type 6B PS elicited by the conjugate have been shown to have weaker *in vitro* opsonophagocytic activity against type 6A pneumococci relative to type 6B strains (Vakevainen *et al.*, 2001).

A major concern emanating from the Finish study, however, was the finding that otitis media caused by all other non-included *S. pneumoniae* serotypes increased by 33% (Eskola *et al.*, 2001). This finding was also not unexpected. In previous trials conducted in the Gambia (Obaro *et al.*, 1996), South Africa (Mbelle *et al.*, 1999) and Israel (Dagan *et al.*, 2003), the conjugate vaccine significantly reduced nasopharyngeal carriage of vaccine types in children, but this was off-set at least partially by an increase in carriage of non-vaccine types, many of which were known to be capable of causing disease.

Nasopharyngeal carriage of *S. pneumoniae* is generally accepted as a prerequisite for pneumococcal disease, and there is a degree of correlation between serotypes being carried and those causing disease in a community. Carriers are the major source for transmission of pneumococci, and communities with high rates of carriage also have high attack rates of pneumococcal disease. Carriage rates are high in young children, particularly in developing countries, where infants acquire pneumococci (presumably from their colonized mothers) in the first few days of life. Individuals may be colonized by multiple strains or serotypes of *S. pneumoniae*, and these presumably compete (with each other and perhaps with other microflora) for occupation of the nasopharyngeal niche (Lipsitch *et al.*, 2000). Detection of multiple *S. pneumoniae* serotypes in nasopharyngeal cultures is technically difficult if one strain is significantly outnumbered. Thus, it is hard to determine the extent to which 'replacement carriage' observed

in a conjugate vaccine recipient is a result of acquisition of new *S. pneumoniae* types not previously present, or facilitated detection of a pre-existing non-vaccine serotype whose numbers have increased after elimination of interference from vaccine types. However, regardless of the mechanism, replacement carriage does appear to translate into increased disease caused by non-vaccine serotypes (Eskola *et al.*, 2001).

S. pneumoniae strains undoubtedly differ in their capacity to colonize the nasopharynx, as well as in their capacity to cause either otitis media or invasive disease once carriage has been established. These differences have a multifactorial basis and depend upon capsular serotype as well as upon other ill-defined virulence traits. This accounts for the non-uniform distribution and relative prevalence of the 93 serotypes of *S. pneumoniae*, as well as for the existence of highly successful, widely distributed clones, some of which are resistant to multiple antibiotics. Molecular analysis of one such highly transmissible, multiply resistant strain (the so-called Spanish type 23F clone) demonstrated that pneumococci are capable of switching serotype by recombinational exchange of capsule biosynthesis loci *in vivo* (Coffey *et al.*, 1998). It is easy to see how antibiotic therapy would facilitate such exchanges between co-colonizing sensitive and resistant strains; DNA released from the sensitive strain would directly transform the resistant type, enabling it to assume the serotype of the donor.

Most, but by no means all, serotype exchanges in resistant pneumococci detected to date have been from one vaccine type to another vaccine type, presumably because the other vaccine types are also commonly carried (Spratt and Greenwood, 2000). However, increased colonization by non-vaccine types due to use of the conjugate vaccines increases the likelihood of *in vivo* transformation of multiply resistant pneumococci to non-vaccine serotypes. Widespread use of the conjugate vaccines is also providing direct selective pressure for acquisition of non-vaccine serotype capsule loci by highly virulent *S. pneumoniae* clones, which had hitherto expressed a vaccine type capsule (Jefferies *et al.*, 2004).

The impact of widespread use of conjugate vaccines on the complex biology of pneumococcal disease is difficult to predict. The full effect of the vaccine on serotype prevalence may take many years to become apparent, and will vary from region to region depending upon levels of endemic carriage, baseline seroprevalence, and rates of vaccine utilization. USA probably represents a best-case scenario; PCV7 has caused a dramatic reduction in invasive disease caused by vaccine types, with significant concomitant homotypic herd immunity, including in the older population. However, there has been a significant increase in invasive disease due to non-vaccine types notably 3, 6C, 15, 19A, 22F and 33F, with 19A now the predominant cause of invasive disease in children (Hicks *et al.*, 2007; Jacobs *et al.*, 2008). To date, the extent of replacement disease in USA due to non-vaccine types has been modest compared with the overall reduction in disease burden. However, as use of PCV grows, so too will the rate of nasopharyngeal carriage of non-vaccine types. This, in turn, will increase the likelihood of transmission of non-vaccine serotype pneumococci, inevitably increasing rates of non-vaccine type disease. Capsule type switching may also facilitate vaccine escape, as well as enabling spread of antibiotic resistance to a broader range of serotypes than is currently the case. Moreover, in regions where carriage rates and/or baseline prevalence on non-vaccine types are much higher, the initial impact on pneumococcal disease burden is likely to be lower, and replacement by non-vaccine serotypes more rapid. On-going surveillance of the serotype distribution of pneumococci causing disease or being carried will therefore be essential. Inclusion of additional conjugated polysaccharides in the formulation may be required if particular non-vaccine types become too prevalent.

There are limits, however, on just how many capsular types can be accommodated. Polyvalent PS–protein conjugate vaccines are very expensive to produce, and addition of further PS types or periodic reformulation to take account of altered serotype prevalence will add further to this cost. This may place the vaccine even further out of the reach of many developing countries, whose need for effective pneumococcal vaccines is greatest. In countries that can afford them, conjugate vaccines are likely to have a major impact upon the burden of pneumococcal disease in the short term, but their overall efficacy is likely to diminish with time, necessitating on-going investment in development of alternative vaccination strategies capable of eliciting more broad-based and affordable protection.

Pneumococcal common protein vaccines

The established and potential limitations of both the PS and PS–protein conjugate vaccines have sparked efforts to identify protein components of the pneumococcus that are common to all or the vast majority of strains worldwide. The observation that age-related reduction in incidence of invasive pneumococcal disease in children does not correlate with natural acquisition of antibodies to PS provides support for the possibility that other immune mechanisms, perhaps including antibody responses to protein antigens, may play a key role in natural resistance to infection (Lipsitch *et al.*, 2005). If such 'common' protein vaccines are successfully developed, they could have a number of advantages over the currently available PS-based vaccines. Similar to other protein vaccines, they should induce high concentrations of antibody and immunological memory in infants and children, who respond well to T cell-dependent protein antigens. Common protein vaccines will probably be composed of several distinct protein components, which will direct the immune system to attack the organism at various stages of its pathogenesis. Indeed, if protein components associated with nasopharyngeal colonization are included and are effective in blocking such colonization, the vaccine could potentially result in a significant reduction in transmission and a large herd immunity effect. An effective common protein vaccine would reduce or eliminate the threat of serotype replacement inherent to the capsular PS-based vaccines and provide broad coverage worldwide. Finally, efficient and high-level expression of proteins can be engineered in recombinant systems enabling large-scale production at potentially very low cost, resulting in vaccines that are more affordable in developing countries. A number of candidate pneumococcal protein antigens have been examined for vaccine potential, and these are discussed below.

Pneumolysin

Pneumolysin was the first pneumococcal protein to be proposed as a vaccine antigen (Paton *et al.*, 1983). It is a potent 53 kDa thiol-activated pore-forming cytolysin produced by virtually all strains of *S. pneumoniae*. It is a bifunctional toxin and in addition to its cytotoxic properties, it is capable of directly activating the classical complement pathway (with a concomitant reduction in serum opsonic activity) (Paton *et al.*, 1984). *In vitro* studies using purified toxin have demonstrated that pneumolysin has a variety of detrimental effects on cells and tissues, which undoubtedly contribute to the pathogenesis of disease (reviewed by Paton, 1996). These properties include inhibition of the bactericidal activity of leucocytes, blockade of proliferative responses and Ig production by lymphocytes, reduction of ciliary beating of human respiratory epithelium, and direct cytotoxicity for respiratory endothelial and epithelial cells. Thus, pneumolysin may function in pathogenesis by interfering with both phagocytic and ciliary clearance of pneumococci, by blocking humoral immune responses, and by aiding penetration of host tissues.

Pneumolysin is also capable of direct induction of inflammatory responses (Houldsworth *et al.*, 1994), and injection of purified pneumolysin into rat lungs induces severe lobar pneumonia, indistinguishable histologically from that seen when virulent pneumococci are injected (Feldman *et al.*, 1991). Additional insights into the role of pneumolysin in pathogenesis have been gained by studies of the behaviour of defined pneumolysin-negative mutants of *S. pneumoniae* in a number of animal models. Such strains have significantly reduced virulence in mouse models of sepsis and pneumonia (Berry *et al.*, 1989b). Intranasal challenge with these mutants results in a less severe inflammatory response, a reduced rate of multiplication within the lung, a reduced capacity to injure the alveolar–capillary barrier and a delayed onset of bacteraemia, compared with the wild type strain (Paton, 1996). Additional site-directed mutagenesis studies have shown that both the cytotoxic and complement activation properties of the toxin contribute to the pathogenesis of pneumococcal pneumonia (Berry *et al.*, 1996; Rubins *et al.*, 1996).

Although native pneumolysin is a protective immunogen in mice, it is not suitable as a human vaccine antigen, because of its toxicity. To overcome this, mutations have been introduced into the pneumolysin gene in regions essential for its cytotoxic and/or complement activation properties, resulting in expression of non-toxic but immunogenic 'pneumolysoids', which are easily purified from recombinant *E. coli* expression systems (Paton *et al.*, 1991). Pneumolysin is a highly conserved protein and extensive analysis of genes from a wide range of *S. pneumoniae* serotypes has detected negligible variation in deduced amino acid sequence, auguring well for broad coverage. Indeed, immunization of mice with a pneumolysoid carrying a Trp_{433}-Phe mutation resulting in >99% reduction in cytotoxicity (designated PdB) provided a significant degree of protection against all nine serotypes of *S. pneumoniae* that were tested (Alexander *et al.*, 1994).

Humans are known to mount an antibody response to pneumolysin as a result of natural exposure to *S. pneumoniae*, and purified human anti-pneumolysin IgG also passively protects mice from challenge with virulent pneumococci (Musher *et al.*, 2001a). Thus, it is anticipated that the various pneumolysoids will be immunogenic in humans. However, pneumolysoid may not provide a sufficient degree of protection to be an effective stand-alone human vaccine antigen. Pneumolysin is not displayed on the surface of the pneumococcus. Rather, it is located in the cytoplasm and is released into the external milieu when pneumococci undergo spontaneous autolysis in some strains (Berry *et al.*, 1989a), as well as by an as yet uncharacterized export mechanism in others (Balachandran *et al.*, 2001). Antibodies to pneumolysin are presumed to impart protection by neutralization of the biological properties of the toxin, thereby impeding the kinetics of infection, rather than by stimulating opsonophagocytic clearance of the invading bacteria. Thus, protein-based vaccines combining pneumolysoid with pneumococcal surface proteins capable of eliciting opsonic antibodies would be expected to be more effective.

Choline-binding surface proteins

The choline binding proteins (CBPs) are a diverse family of proteins, which are tethered to the pneumococcal surface via non–covalent interactions with phosphoryl choline moieties on cell wall teichoic acid and membrane lipoteichoic acid. This binding is mediated by a domain comprising up to ten highly-conserved 20 amino acid repeats. For most of the CBPs, these repeats are located near their respective C-termini (Yamamoto et al., 1997; Lopez et al., 2004). Up to 15 proven or putative CBPs have been identified by examination of S. pneumoniae genome sequences, but some of these are not present in all strains. In spite of the similarity of the choline binding repeat domains, the remainder of the CBP molecules are structurally and functionally dissimilar. CBPs have diverse functions, including cell wall modification, adherence to host cell surface molecules, and modulation of complement activation (Swiatlo et al., 2004). Those with potential as vaccine antigens are discussed below.

Pneumococcal surface protein A (PspA)

PspA is one of the best-characterized members of the CBP family and has strong credentials as a vaccine antigen. It is found on the surface of all pneumococci (Crain et al., 1990) and has a proven role in the pathogenesis of disease, as evidenced by significantly reduced virulence of defined PspA-negative pneumococci in animal models (McDaniel et al., 1987; Tu et al., 1999). Its principal function appears to be inhibition of complement-dependent host defences mediated by factor B. This results in reduced deposition of C3b on the pneumococcal surface and concomitant impairment of complement receptor-mediated clearance (Tu et al., 1999). PspA also binds lactoferrin (Hammerschmidt et al., 1999; Hakansson et al., 2001), and this may aid colonization of host mucosae by protecting the pneumococcus from the bactericidal effects of apolactoferrin (Shaper et al., 2004). The biological properties of PspA reside in the N-terminal portion of the molecule, which forms a highly charged, largely alpha-helical anti-parallel coiled-coil structure (Hollingshead et al., 2000; Jedrzejas

et al., 2000). This region of PspA is exposed on the surface of S. pneumoniae and the presence of the capsule does not impede accessibility to exogenous antibodies (Gor et al., 2005; Daniels et al., 2006), suggesting that anti-PspA should also be opsonophagocytic.

Immunization of mice with a soluble 43 kDa N-terminal PspA fragment has been shown to be highly protective against systemic challenge (Talkington et al., 1991) and pneumonia (Briles et al., 2003). This region of the molecule is variable in terms of its amino acid sequence, although this does not appear to impact upon biological function (Ren et al., 2003). PspA proteins produced by various S. pneumoniae strains have been grouped into three families, with 95% of isolates producing PspA belonging to families 1 or 2 (Hollingshead et al., 2000; Vela Coral et al., 2001). In spite of the significant variation in amino acid sequence, the helical domain of PspA contains epitopes that elicit antibodies that are highly protective against challenge with S. pneumoniae strains producing heterologous PspA types (McDaniel et al., 1994).

Studies on the vaccine potential of PspA have extended to human trials, and immune sera from volunteers immunized with a family 1 PspA fragment reacted with 37 different S. pneumoniae strains belonging to diverse capsular and PspA types (Nabors et al., 2000). Moreover the sera passively protected mice against challenge with S. pneumoniae strains of three different capsular types expressing either family 1 or family 2 PspAs (Briles et al., 2000b). Thus, it appears likely that a human vaccine may only need to include two or three different PspA types in order to provide near species-wide protection. A theoretical concern that arose in a subset of patients during the human trials was that the PspA antigen, which has a coiled-coil structure, elicited antibodies capable of cross-reacting with human cardiac myosin. However, the significance of this finding is questionable. Both nasopharyngeal carriage and infection with S. pneumoniae elicits antibody to PspA and can even boost the low levels of antibody present in 10–20% of human sera that is reactive with human myosin. Such natural exposure to pneumococci is of course very common, but it has never been shown to be linked to any autoimmunity.

In addition to its promise as a parenteral vaccine antigen capable of preventing systemic disease, PspA exhibits considerable promise as a mucosal vaccine antigen for prevention of nasopharyngeal carriage. As described previously, a vaccine capable of preventing carriage is likely to impart substantial herd immunity. Intranasal immunization with full-length native PspA, using cholera toxin B subunit (CTB) as an adjuvant, has been shown to elicit significant mucosal and serum antibody responses and to protect mice against both nasal carriage of *S. pneumoniae* and systemic disease (Wu *et al.*, 1997; Yamamoto *et al.*, 1997). However, a recombinant N-terminal PspA fragment (also administered with CTB) appeared to be less effective, reducing the level of colonization after intranasal challenge, but not preventing it altogether (Briles *et al.*, 2000a). On the other hand, parenteral immunization with PspA elicits negligible levels of mucosal antibody and no detectable protection against carriage (Wu *et al.*, 1997). The potential efficacy of mucosal immunization with PspA for prevention of carriage in humans is also supported by the findings of a human volunteer study, which demonstrated that pre-existing (naturally acquired) antibody to PspA prevented colonization after intranasal administration of *S. pneumoniae* (McCool *et al.*, 2002).

PspC/CbpA

Several other members of the choline binding protein family have been proposed as vaccine antigens. The best characterized of these was isolated independently in three laboratories and is referred to as either pneumococcal surface protein C (PspC) (Brooks-Walter *et al.*, 1997), choline binding protein A (CbpA) (Rosenow *et al.*, 1997) or SpsA (Hammerschmidt *et al.*, 1997) (the foremost terminology will be used here). Its choline-binding repeat region is approximately 95% identical to that of PspA, while its N-terminal helical portion, like PspA, is highly variable. The N-terminal half mediates binding to the secretory component of IgA (Hammerschmidt *et al.*, 1997; 2000), as well as to C3 (Cheng *et al.*, 2000) and factor H (Dave *et al.*, 2001). PspC has been shown to be involved in adherence of pneumococci to cytokine-activated lung epithelial cells

in vitro, as well as to glycoconjugates previously identified as pneumococcal binding ligands. Furthermore, PspC-deficient pneumococci have a reduced capacity to colonize the nasopharynx of infant rats and mice (Rosenow *et al.*, 1997; Balachandran *et al.*, 2002). Interestingly, PspC appears to be expressed at greater levels in transparent phase pneumococci, which are favoured in the nasopharynx over opaque phase variants (Rosenow *et al.*, 1997). PspC may also be directly involved in invasion of nasopharyngeal cells, through interaction with the secretory component associated with the polymeric immunoglobulin receptor pIgR (Zhang *et al.*, 2000). The fact that PspC can interact with C3 and factor H is also strongly suggestive of a role in systemic disease, and significant differences in virulence between *pspC*-negative and otherwise isogenic wild type pneumococci have been demonstrated in mouse models of lung infection and bacteraemia (Balachandran *et al.*, 2002). An earlier study did not detect differences in virulence between wild type and *pspC*-negative pneumococci in an intraperitoneal challenge model, but mutation of both *pspC* and the pneumolysin gene had an additive attenuating effect (Berry and Paton, 2000).

Immunization of mice with PspC is highly protective against intravenous or intraperitoneal challenge with *S. pneumoniae* (Brooks-Walter *et al.*, 1999; Ogunniyi *et al.*, 2001). Theoretically, the suitability of PspC as a vaccine antigen might be diminished to some extent by the fact that it is present on only about 75% of *S. pneumoniae* strains (Brooks-Walter *et al.*, 1999; Hammerschmidt *et al.*, 1997). However, immunization with PspC was shown to provide significant protection against a strain that did not produce the protein (Brooks-Walter *et al.*, 1999). This can be explained by the fact that polyclonal antibodies to PspC cross-react with PspA as well as other protein species (Brooks-Walter *et al.*, 1999). Furthermore, in those strains that lack *pspC*, the locus is occupied by an allele *hic*, which encodes a protein with a high degree of similarity to PspC in the N-terminal half (Janulczyk *et al.*, 2000; Ianelli *et al.*, 2002), and this might also be recognized by PspC antibodies. Interestingly, the C-terminal portion of Hic does not have a choline-binding domain, but instead has a typical Gram-positive sortase-dependent cell wall anchorage domain, including a LPXTG motif (Janulczyk *et al.*, 2000). Like PspC, Hic can bind to factor H, and thereby interfere with complement activation (Jarva *et al.*, 2002). However, there are differences in the precise nature of this interaction, as Hic binds to short consensus repeats 8–11 of factor H, whereas PspC binds to repeats 13–15 (Duthy *et al.*, 2002).

Other choline-binding proteins

As mentioned previously, access to the pneumococcal genome sequence has facilitated the search for additional vaccine antigens, because it enables entire families of genes encoding proteins with recognizable structural features to be targeted (Paton and Giammarinaro, 2001). The choline binding proteins are good examples of this approach. Although several members of this family were previously identified by conventional techniques, such as elution from the cell surface with choline, a search of the genome sequence identified a dozen or so functional genes encoding proteins with choline-binding motifs. Site-specific mutagenesis was then used to demonstrate that five of the novel choline binding proteins (CbpD, CbpE, CbpG, LytB and LytC) were involved in *in vitro* adherence to epithelial cells, nasopharyngeal colonization or sepsis, thereby identifying them as vaccine candidates (Gosink *et al.*, 2000). LytB and LytC are unusual in that their choline-binding domains are located in the N-terminal part of the molecule, while the C-terminal portions have murein hydrolase activity (Lopez *et al.*, 2000). Purified recombinant LytB and LytC were subsequently tested for protective efficacy as part of another large-scale study. Immunization with these proteins conferred significant protection against intraperitoneal challenge in mice, although the degree of protection observed was marginally less than that observed using PspA, which was used as a control antigen (Wizemann *et al.*, 2001).

Another CBP with cell wall modification (in this case amidase) activity is the major pneumococcal autolysin LytA. Mutatgenesis of the *lytA* gene prevents the autolysis of pneumococci that occurs spontaneously in stationary phase cultures, or on addition of deoxycholate, and also attenuates virulence in mouse models of sepsis. It might seem paradoxical that inactivation of what is essentially a suicide gene could have such an effect. However, LytA is largely responsible for release of intracellular pneumolysin, inflammatory cell wall degradation products and other cell-associated virulence factors, and so prevention of autolysis might be of considerable benefit to the host (Berry *et al.*, 1989a, 1992; Canvin *et al.*, 1995). Exogenous antibody to LytA is capable of penetrating the surface layers of the pneumococcus and inhibiting autolysis and release of pneumolysin *in vitro*. Active immunization of mice with purified LytA also elicited a similar degree of protection as pneumolysoid against challenge with fully virulent pneumococci, but it conferred no significant protection against challenge with high doses of a pneumolysin-negative strain. This suggested that the LytA-induced protection is mediated largely through blockade of pneumolysin release (Lock *et al.*, 1992).

Lipoproteins

The pneumococcal genome includes over 30 putative lipoproteins, with prolipoprotein signal peptidase recognition sequences (LXXC) (Tettelin *et al.*, 2001). This so-called lipobox motif directs covalent attachment of a diacyl glycerol moiety to the N-terminal Cys residue of the mature protein, anchoring it to the outer face of the plasma membrane. Thus, they are located beneath the cell wall and the capsule in *S. pneumoniae*. These lipoproteins have diverse functions, the commonest being substrate binding components of ATP binding cassette (ABC) transport systems, and many are important for growth and survival of the pneumococcus *in vitro* and *in vivo*. Their cellular location suggests that they are not exposed on the cell surface to any significant extent, which in turn implies that they are unlikely to elicit opsonic antibodies. However, this does not necessarily preclude their utility as vaccine targets, since exogenous antibody may diffuse through the capsule and cell wall layers and inhibit the biological function of the lipoprotein. Indeed, several pneumococcal lipoproteins have been shown to have potential as vaccine antigens, as discussed below.

Pneumococcal surface antigen A (PsaA)

PsaA is a highly conserved 37-kDa lipoprotein produced by all pneumococci. It was initially thought to be an adhesin based on sequence homology with putative lipoprotein adhesins of oral streptococci, but it is actually the binding component of a Mn^{2+}-specific ABC transport system (Dintilhac et al., 1997). Defined psaA-negative mutants of S. pneumoniae are virtually avirulent for mice and exhibit markedly reduced adherence in vitro to human type II pneumocytes (Berry and Paton, 1996; McAllister et al., 2004). This is presumed to be a consequence of growth retardation due to an inability to scavenge Mn^{2+} in vivo, as well as pleiotropic effects on expression of a range of cellular processes or virulence factors. Intracellular Mn^{2+} appears to play a critical role in regulation of expression of oxidative stress response enzymes and intracellular redox homeostasis, and psaA-negative pneumococci exhibit hypersensitivity to superoxide and hydrogen peroxide (Tseng et al., 2002; McAllister et al., 2004).

One study has shown that parenteral immunization of mice with purified PsaA in the presence of strong adjuvants elicits significant protection against systemic challenge with S. pneumoniae (Talkington et al., 1996). However, in other studies immunization with PsaA elicited only marginal protection and was less efficacious than pneumolysoid in an intraperitoneal challenge model (Ogunniyi et al., 2000; Gor et al., 2002). The dimensions of PsaA (approximately 7 nm at its longest axis) (Lawrence et al., 1998) are such that if it is indeed anchored to the outer face of the cell membrane via its N-terminal lipid moiety, it is unlikely to be exposed on the outer surface of the pneumococcus. This is consistent with the fact that whereas the known surface-exposed domains of PspA and PspC are variable, the amino acid sequence of PsaA is highly conserved (Sampson et al., 1997). Gor et al. (2005) used flow cytometry to compare surface accessibility of PsaA and PspA to exogenous specific antibodies in 12 S. pneumoniae strains. PspA was readily detectable on the surface of all strains, whereas PsaA was not. This directly correlated with the protective efficacy of either active or passive immunization with the respective protein or antibody; significant protection against systemic challenge was achieved using PspA or anti-PspA, but not using PsaA or anti-PsaA. Given the virtual absence of surface exposure, any protection elicited by immunization with PsaA is unlikely to be a consequence of enhanced opsonophagocytic clearance. Rather, it is presumably due to in vivo blockade of ion transport, which necessitates diffusion of antibody through the capsule and cell wall layers. Such penetration of antibody is likely to be concentration-dependent and so high anti-PsaA titres may be required for protection. Moreover, accessibility of PsaA to exogenous antibody may well be influenced by the thickness of the capsule, which may vary from strain to strain. Expression of pneumococcal capsule biosynthesis genes has also been shown to be up-regulated during invasive infection (Ogunniyi et al., 2002). In contrast, pneumococci colonizing the nasopharynx are thought to down-regulate capsule expression, thereby facilitating interaction between surface adhesins and the host mucosa. Consistent with this hypothesis, several studies have shown that intranasal immunization of mice with PsaA in the presence of strong mucosal adjuvants such as cholera toxin B subunit (CTB) significantly reduces the level of nasopharyngeal carriage of S. pneumoniae (Briles et al., 2000a; De et al., 2000). A lesser, but still significant reduction in susceptibility to carriage was also achieved by subcutaneous immunization of mice with synthetic lipidated multi-antigenic PsaA peptides (Johnson et al., 2002). Thus, at least in the nasopharynx, PsaA appears to be accessible to exogenous antibody. Immunization with a PsaA–CTB fusion protein also significantly reduced carriage of S. pneumoniae in mice without significantly disturbing the oropharyngeal microflora (Pimenta et al., 2006).

Iron transporter lipoproteins PiuA and PiaA

Two other metal-binding lipoproteins have been proposed as pneumococcal vaccine antigens. These proteins, designated PiuA and PiaA, are components of two separate ABC iron transport systems. At least one of these proteins is required for optimal growth of pneumococci in iron-depleted media, and they are capable of acquiring iron from haemoglobin (Brown *et al.*, 2001a). Indeed, PiuA has been shown to be capable of directly binding both haemin and haemoglobin (Tai *et al.*, 2003). PiuA and PiaA are produced by all pneumococci and their genes are highly conserved (Jomaa *et al.*, 2005). Mutagenesis studies have shown that both proteins contribute to virulence in mice using both lung and intraperitoneal models of infection (Brown *et al.*, 2001a). They are immunologically cross-reactive, and immunization of mice with either protein conferred a similar degree of protection against intraperitoneal challenge to that elicited by the pneumolysoid PdB. Moreover, immunization with a combination of PiuA and PiaA resulted in additive protection (Brown *et al.*, 2001b). Although a direct comparison has not been conducted, immunization with either PiuA or PiaA provided a higher degree of protection against systemic disease than that previously published for PsaA, using the same mouse model and *S. pneumoniae* challenge strain (Ogunniyi *et al.*, 2000). Like PsaA, PiuA and PiaA are predicted to be attached to the outer face of the plasma membrane (Tai *et al.*, 2003), and so the superior protective efficacy of the latter proteins ought not be due to a difference in accessibility to exogenous antibody. However, Jomaa *et al.* (2005) have shown by flow cytometry that both PiaA and PiuA are accessible to exogenous antibody in intact bacteria, and that these antibodies stimulate *in vitro* opsonophagocytic activity, particularly in the presence of complement. They also reported that the antibodies did not appear to interfere with iron uptake *in vitro*. Mucosal immunization with PiuA and PiaA has also been shown to elicit antibody responses both in serum and respiratory secretions, which protected mice against intranasal challenge (Jomaa *et al.*, 2006). The reason for the apparent difference in surface accessibility between PsaA and the two iron-binding lipoproteins is unclear, given their predicted location. One possibility is that in the latter two cases, at least a proportion of the proteins are released from the membrane and are then able to bind to more exposed domains on the pneumococcal surface, where they can interact with exogenous antibody more freely. Regardless of the underlying mechanism, available data suggest that PiaA and PiuA have more

promise than PsaA as vaccine antigens, as least for prevention of systemic disease.

Pneumococcal histidine triad proteins

The pneumococcal histidine triad proteins are a recently recognized family of surface proteins, that have an unusual polyhistidine motif HXX-HXH, repeated five or six times in their amino acid sequences. The prototype, PhtA, was discovered as part of a genome-wide screen for potential vaccine antigens (Wizemann *et al.*, 2001). Over 100 proteins were expressed and tested for efficacy in a mouse model, and PhtA was one of only 5 that were protective. The others were the choline binding proteins LytB and LytC (discussed previously), a cell wall-associated serine protease PrtA, and another protein of unknown function designated PvaA. Further examination of the pneumococcal genome sequence revealed three additional related open reading frames (designated PhtB, PhtD and PhtE), each with five or six copies of the histidine triad motif. The four proteins range in size from 91 to 114 kDa, and are closely related at the amino acid sequence level, exhibiting 32–87% identity; this similarity is strongest in the N-terminal regions (Adamou *et al.*, 2001). Although their signal peptides all contain a LXXC motif, this does not appear to function as a true lipobox, as they are not labelled by [3H]-palmitate (Hamel *et al.*, 2004). Thus, the manner in which the Pht proteins are attached to the pneumococcal cell surface and the precise site of attachment is uncertain. Nevertheless, flow cytometric analyses have shown that the C-terminal regions are more readily accessible to exogenous antibodies (Adamou *et al.*, 2001; Hamel *et al.*, 2004), suggesting that they are tethered via their N-termini. The histidine triads are believed to form a novel Zn^{2+}-binding motif (Riboldi-Tunnicliffe *et al.*, 2005), and the *pht* genes are regulated by the Zn^{2+}dependent repressor AdcR (Ogunniyi *et al.*, 2008). The Pht proteins act by inhibiting complement deposition on the pneumococcal surface through binding of factor H. They exhibit substantial functional redundancy and significant effects on virulence in mouse models of pneumococcal sepsis and on complement deposition *in vitro* were only observed when all four *pht* genes were deleted (Ogunniyi *et al.*, 2008).

There is a high degree of protein sequence conservation amongst individual Pht proteins from diverse *S. pneumoniae* serotypes (Hamel *et al.*, 2004), which combined with a degree of immunological cross-reactivity between the proteins, augers well for broad strain coverage for these candidate vaccine antigens. Immunization with purified PhtA, PhtB or PhtD has been shown to confer significant protection against i.p. challenge with type 3, 6A, 6B and one of two type 4 *S. pneumoniae* strains (Adamou *et al.*, 2001; Wizemann *et al.*, 2001). PhtD has also been shown to protect against intranasal challenge with a type 3 strain (Zhang *et al.*, 2001), while immunization with either PhtB or PhtE also protects against type 3 pneumococci in models of sepsis and pneumonia (Hamel *et al.*, 2004). In this latter study, immunization experiments with truncated PhtE fragments localized the protective epitopes to the more surface-exposed C-terminal region of the molecule. However, notwithstanding these promising results, the only direct comparative studies of protective efficacy of Pht proteins with other well-characterized antigens indicate that the level of protection elicited is no better than that achieved by either PspA or pneumolysoid (Wizemann *et al.*, 2001; Ogunniyi *et al.*, 2007b).

Sortase-dependent surface proteins

Sortase-dependent surface proteins of Gram-positive bacteria are identifiable by the presence of a C-terminal anchoring motif, which consists of a conserved LPXTG sequence, followed by a hydrophobic domain and usually a tail of positively charged residues. This motif is recognized by sortase, a membrane-localized cysteine protease, which cleaves between the T and G residues, and covalently links the processed protein to the peptidoglycan cross-bridges (Schneewind *et al.*, 1993). In pneumococci, inactivation of the sortase gene releases known sortase-dependent surface proteins such as the major pneumococcal neuraminidase NanA, and reduces adherence to pharyngeal cells (Kharat and Tomasz, 2003). NanA-deficient mutants of *S. pneumoniae* have been shown to have a reduced

capacity to colonize the upper and lower respiratory tract of mice (Orihuela *et al.*, 2004; Manco *et al.*, 2006) and the nasopharynx and middle ear cleft of chinchillas (Tong *et al.*, 2000). An early study indicated that purified NanA had modest but significant protective efficacy (weaker than pneumolysin) in a mouse sepsis model (Lock *et al.*, 1988). More recently, it has shown to be protective against both carriage and otitis media in chinchillas (Long *et al.*, 2004; Tong *et al.*, 2005).

In recent years, significant attention has been paid to pilus-like structures on the surface of *S. pneumoniae*, which contribute to virulence (Barocchi *et al.*, 2006). These are encoded on the *rlrA* pathogenicity islet, which is present in some, but not all clinical isolates. The *rlrA* islet encodes a transcriptional regulator, three pilus structural components (RrgA, RrgB and RrgC), and three sortases (SrtB, SrtC and SrtD) which are required for pilus assembly (Barocchi *et al.*, 2006; LeMieux *et al.*, 2006). RrgB is the major pilin, while RrgA and RrgC are ancillary pilin subunits decorating the shaft and tip (Fälker *et al.*, 2008). Immunization of mice with purified RrgA, RrgB, or the combination of all three pilus proteins elicited significant protection against intraperitoneal challenge with the *S. pneumoniae* strain from which the proteins originated (Gianfaldoni *et al.*, 2007). However, the utility of pilus proteins as stand-alone vaccine antigens is limited by the fact that there are three different sequence clades of the *rlrA* islet, each associated with distinct *S. pneumoniae* clonal groups. At the amino acid sequence level, RrgB is the most divergent, with only about 50% identity between clades (Moschioni *et al.*, 2008), and the extent of cross-protection is unknown. Moreover, *rlrA*-positive pneumococci account for only about 30% of strains, and the majority of these belong to serotypes covered by PCV7 (Basset *et al.*, 2007; Moschioni *et al.*, 2008). Interestingly, a second pneumococcal pilus locus has recently been described, and this was present in about 16% of strains, again belonging to discrete clonal groups, all but one of which lacked *rlrA* (Bagnoli *et al.*, 2008). However, the vaccine potential of these distinct pilus proteins has not been investigated.

Other sortase-dependent pneumococcal surface proteins have been proposed as vaccine candidates, including hyaluronidase (Hyl) and the IgA1 protease (Iga). The latter is of interest as its sortase motif is located in the N-terminal region of the molecule, but is nevertheless essential for proper function and surface localization (Bender and Weiser, 2006). However, although these proteins have been shown to contribute to pathogenesis using either *in vitro* or *in vivo* models (Berry and Paton, 1996; Weiser *et al.*, 2003), Hyl is at best a weak protective antigen (Paton *et al.*, 1997), while Iga is yet to be tested for protective efficacy individually.

Other protein vaccine candidates

A number of other apparently surface-associated pneumococcal proteins with at least theoretical vaccine potential have been identified, in most cases using immuno-proteomic approaches. This list includes proteins that lack export or anchorage signals and would have been predicted to be cytoplasmic and hence dismissed by previous motif-based searches. Examples include metabolic enzymes such as enolase (which also binds plasminogen) (Bergmann *et al.*, 2001), 6-phosphogluconate dehydrogenase (a putative adhesin) (Daniely *et al.*, 2006), fructose-biphosphate aldolase and glyceraldehyde-3-phosphate dehydrogenase (Ling *et al.*, 2004), as well as the heat-shock protease ClpP (Kwon *et al.*, 2004). Another candidate is the putative proteinase maturation protein PpmA (Overweg *et al.*, 2000), although its degree of surface exposure and protective efficacy has been questioned (Gor *et al.*, 2005).

Giefing *et al.* (2008) conducted a comprehensive study of the antibody repertoire induced in humans during natural pneumococcal infection, by extensive screening of *E. coli* display libraries representing the entire *S. pneumoniae* proteome with carefully selected panels of convalescent patient sera. Sequence analysis of reactive clone inserts identified not only proteins that reacted consistently with the sera, but also the key epitopes therein. Over 140 immunoreactive proteins were identified, and many of the vaccine candidates referred to above, most notably PspA, PspC, PhtA/B/D/E, NanA, Iga, LytC and LytA, were detected with high frequency. Interestingly, about 20% of the proteins represented unknown

gene products, and so may have been missed in previous motif-based in silico screens. A subset of the identified proteins were selected on the basis of further seroreactivity screens of synthetic peptides, surface localization and opsonophagocytic studies using epitope-specific hyperimmune mouse sera, and PCR-based gene distribution analyses. Twenty selected proteins were then purified and tested for protective immunogenicity in mouse models of sepsis and pneumonia. Significantly, two proteins (PcsB and StkP) provided a degree of protection similar to that seen for PspA. Both are highly conserved and appear to play an important role in cell separation and/or formation of division septa. PcsB and StkP both protected against a variety of challenge strains and were superior to PspA when the challenge strain expressed a dissimilar family/clade of PspA (Giefing *et al.*, 2008). A Phase I clinical trials of these two antigens commenced in April, 2009.

From the above it can be seen that there are a number of pneumococcal proteins that exhibit potential as vaccine antigens. However, assessment of their protective efficacy has generally been carried out in different laboratories, using a variety of animal models and challenge strains. Clearly, a comprehensive series of direct comparative protection studies needs to be performed, in order to determine which of these proteins provides the strongest protection against the widest variety of *S. pneumoniae* strains.

Combination protein vaccines

The vast majority of the pneumococcal proteins under consideration as vaccine antigens are directly or indirectly involved in the pathogenesis of pneumococcal disease. Mutagenesis of some combinations of virulence factor genes, for example those encoding pneumolysin and either PspA or PspC, PspA and PspC, or all three genes, has been shown to synergistically attenuate pneumococcal virulence in animal models, implying that the respective proteins function independently in the pathogenic process (Berry and Paton, 2000; Balachandran *et al.*, 2002; Ogunniyi *et al.*, 2007b). This strongly suggests that immunization with combinations of these antigens might provide additive protection. Moreover, there may be differences in the relative protective capacities of the individual antigens against particular *S. pneumoniae* strains, particularly for surface-exposed antigens that exhibit some degree of sequence variation. Thus, a combined pneumococcal protein vaccine may elicit a higher degree of protection against a wider variety of strains than any single antigen.

To date only a limited number of combination experiments have been performed. Immunization of mice with a combination of the pneumolysoid PdB and PspA provided significantly increased protection against intraperitoneal challenge than immunization with either protein alone. However, combining either protein with PsaA did not result in enhanced protection (Ogunniyi *et al.*, 2000). The potential benefits of combination protein vaccines are also well illustrated using a mouse model of non-bacteraemic pneumonia, which closely reflects the commonest form of pneumococcal respiratory disease in humans (Briles *et al.*, 2003). In this system, subcutaneous immunization (using alum adjuvant) with either PdB or PspA, but not PsaA, significantly reduced numbers of *S. pneumoniae* in the lungs 7 days after challenge. A significant additional reduction in bacterial load was achieved by immunization with a combination of PdB and PspA, but not when either protein was combined with PsaA (Briles *et al.*, 2003). These findings contrast with those obtained using mucosal (intranasal) immunization with the same proteins with CTB as adjuvant. As discussed previously, immunization with either PspA or PsaA, but not PdB, reduced the level of carriage of *S. pneumoniae* after intranasal challenge, and the combination of PspA and PsaA was more effective than either antigen alone (Briles *et al.*, 2000a). These findings imply that PsaA is either more important for survival of *S. pneumoniae* in the nasopharynx than in the lung, or that it is more accessible to exogenous antibody in the former niche. On the other hand, pneumolysin appears to play only a minor role during the colonization phase, but is clearly important once the organism has been aspirated into the lungs. Thus, optimum vaccine formulation will be dependent upon the mode of vaccine delivery and the stage of the pathogenic process being targeted for immunoprophylaxis.

Few other pneumococcal protein combinations have been tested for additive protective immunogenicity. Immunization with both the iron transporters PiuA and PiaA was more effective than either antigen alone (Brown *et al.*, 2001b). A combination of Pneumolysoid (PdB), PspA and PspC has also shown stronger protection than single or paired antigens (Ogunniyi *et al.*, 2007a). In that study, combinations of PhtB and PhtE were less protective than those involving two or more of PdB, PspA and PspC, and inclusion of PhtB did not improve the protection imparted by either PdB/PspA or PdB/PspC (Ogunniyi *et al.*, 2007a). Clearly, additional comparative studies of the protective efficacy of the better-characterized proteins, as well as the more recently identified vaccine candidates (both singly and in combination), are required to enable informed decisions on the formulation of a protein-based pneumococcal vaccine.

Pneumococcal proteins as supplements or carriers for PS-conjugate vaccines

Consideration should also be given to using pneumococcal protein antigens as supplements to PS–protein conjugate vaccines. Incorporation of one or more proteins common to all *S. pneumoniae* serotypes may significantly reduce the problems associated with limited serotype cover and replacement carriage associated with the conjugate vaccines, although the problem of high cost remains. Pneumolysoid has also been proposed as an alternative carrier in PS–protein conjugate vaccines, and conjugates of PdB with type 19F PS have been shown to be highly immunogenic and protective in mice (Paton *et al.*, 1991; Lee *et al.*, 1994). In a later study, a similar detoxified pneumolysin derivative was shown to be a very effective carrier protein in a tetravalent conjugate vaccine formulation including PS types 6B, 14, 19F and 23F (Michon *et al.*, 1998). Use of pneumolysoid, or other suitable pneumococcal proteins, as carriers for PS in conjugate vaccines may also minimize any problems associated with overuse of existing carrier proteins.

An interesting variation to the above approach has recently been described by Lu *et al.* (2009), who immunized mice with a PsaA–pneumolysoid fusion protein conjugated to the cell wall polysaccharide. This trivalent conjugate of species-wide antigens elicited stronger antibody responses to each component than simple mixtures of the three antigens, or various bivalent conjugates. Moreover, the trivalent conjugate was effective when administered either intranasally or subcutaneously, and elicited strong protection in models of colonization and fatal pneumonia.

Mucosal vaccination strategies

Given the pivotal role of nasopharyngeal colonization in the transmission of S. pneumoniae and as a precursor of pneumococcal disease, vaccination strategies specifically designed to elicit mucosal immune responses may be more efficacious than parenteral immunization for certain antigens, particularly those implicated in colonization. To date this has been examined in animal models using direct intranasal administration of vaccine formulations with a strong mucosal adjuvants, or cytokines (Wortham et al., 1998; Arulanandam et al., 2001). Anti-PS responses in mice have been achieved using purified PS conjugated to either cholera toxin B subunit (CTB) or an E. coli labile enterotoxin (LT) derivative (Jakobsen et al., 1999; Seong et al., 1999). The protective efficacy against carriage of mucosal immunization with purified PspA and/or PsaA in the presence of CTB has already been discussed (Briles et al., 2000a). Co-encapsulation of PsaA and CTB within alginate microspheres was also shown to elicit higher serum and mucosal immune responses after oral immunization of mice than did non-encapsulated antigens, and this translated into stronger protection against intranasal challenge (Seo et al., 2002).

An alternative means of eliciting mucosal immune responses involves oral administration of live recombinant carrier bacteria expressing pneumococcal antigens. Recombinant attenuated Salmonellae expressing PdB (Paton et al., 1993) or PspA (Nayak et al., 1998; Kang et al., 2002) have been constructed and shown to elicit mucosal and humoral antibody responses in mice after oral immunization. Furthermore, for strains expressing PspA, protection against systemic challenge was also demonstrated (Nayak et al., 1998; Kang et al., 2002). Recent studies have further refined

the Salmonella delivery vectors such that they exhibit regulated, delayed attenuation in vivo, resulting in stronger antibody responses and enhanced protection against challenge (Li et al., 2009).

Expression of type 3 PS has also been achieved in Lactococcus lactis, which has been proposed as an alternative carrier for vaccine antigens (Gilbert et al., 2000). However, the mechanism of biosynthesis of this PS serotype is much simpler than all the other clinically significant pneumococcal PS types, and requires expression of only a small number of genes (Paton and Morona, 2006). Expression of the other much larger PS biosynthesis loci in heterologous bacteria may be extremely difficult, and any such live vaccines would also suffer from the disadvantages of serotype-dependent protection and poor immunogenicity of PS antigens in high risk groups. Recently, Lactobacillus casei has been used to express PspA and PspC, inducing protective Th-1 type immune responses (Ferreira et al., 2009). Protection against lung infection in mice has also been demonstrated by intranasal immunization with a combination of adenoviral vectors expressing pneumolysoid PdB, PspA and PsaA (Arévalo et al., 2009).

Killed whole cell vaccines

Development of killed whole cell vaccines (WCVs) may seem old-fashioned in the era of well-characterized, purified subunit antigens, particularly for parenteral immunization. However, they have considerable potential when delivered via the mucosal route, and if they can be proven to be safe and effective, WCVs have cost and stability advantages over more sophisticated alternatives, which would facilitate deployment in Third World settings. Immunogenicity of WCVs may be enhanced by co-presentation of antigens and molecules that engage toll-like receptors TLR2 and TLR4 (Blander and Medzhitov, 2006). Intranasal administration of heat-killed type 4 pneumococci with or without CT adjuvant has been shown to elicit strong humoral and mucosal responses to type 4 PS in mice and protection from homologous challenge (Hvalbye et al., 1999). On the other hand, intranasal or oral immunization with a WCV comprising ethanol-inactivated,

LytA-negative, non-encapsulated pneumococci with CTB or CT adjuvant conferred non-serotype-dependent protection against nasopharyngeal colonization, middle ear infection, and fatal pneumonia (Malley *et al.*, 2001, 2004). Interestingly, passive immunization studies indicated that protection against fatal pneumonia was antibody dependent (Malley *et al.*, 2001), whereas protection against colonization was dependent on CD4+ T cells (Malley *et al.*, 2005). The latter phenomenon was subsequently shown to be a function of IL-17A production, which enhanced clearance of pneumococci from the nasopharynx by neutrophils (Lu *et al.*, 2008). Interestingly, antibody-independent, antigen-specific, CD4+-dependent protection against colonization has also been observed in mice immunized with a mixture of purified PspC, PsaA and pneumolysoid with CTB adjuvant (Basset *et al.*, 2007).

A variation of the killed whole cell vaccine approach has been described by Audouy *et al.* (2007), involving use of heat-killed trichloroacetic acid-treated *Lactococcus lactis* as a display vehicle for purified pneumococcal proteins (in this case a combination of PpmA, SlrA and Iga) via a protein anchor. Intranasal immunization of mice with this construct elicited significant protection against pneumonia, albeit no better than that achieved using ethanol-killed non-encapsulated pneumococci.

DNA vaccines

A further strategy under consideration for prevention of pneumococcal disease is the use of DNA vaccines. This involves introduction of naked plasmid DNA carrying genes encoding protective antigens under the control of a eukaryotic promoter, either by intramuscular injection or transdermally using a 'gene gun'. The naked DNA is taken up by host cells and the antigens are expressed *in vivo*. This approach has been used for a variety of pathogens, and is attractive because DNA vaccines are potentially cheap to produce on a large scale. They usually elicit both humoral and cell-mediated immune responses. One study reported construction of a DNA vaccine plasmid encoding the α-helical N-terminal half of PspA (the region which contains the cross-protective epitopes). This induced

strong antibody responses in mice and conferred long-lasting protection against both homologous and heterologous *S. pneumoniae* challenge strains (Bosarge *et al.*, 2001). However, in another study, immunization with constructs directing expression of similar regions of PspA from different *S. pneumoniae* strains resulted in production of cross-reacting antibodies, but protection against challenge was confined to strains expressing related PspA types (Miyaji *et al.*, 2002). The same group also examined DNA vaccine constructs directing expression of either the C-terminal two-thirds of PspA or full-length PsaA, which elicited significant antibody responses in mice to the respective protein (Miyaji *et al.*, 2001). However, protection against challenge was not elicited by the C-terminal PspA-expressing construct in a subsequent study (Miyaji *et al.*, 2003), while that for the PsaA-expressing construct has not been examined.

Use of DNA vaccine delivery systems for PS antigens is extremely problematic, not only because of the multiplicity of serotypes, but also because the genetic loci encoding PS biosynthesis are very large, comprising up to 20 or more genes for each PS type (Paton and Morona, 2006). An innovative solution to the latter problem has been achieved using an anti-idiotype approach. Firstly, a monoclonal antibody to type 4 PS was used to screen a phage display library, and this identified a peptide mimic capable of eliciting an anti-type 4 PS response. An oligonucleotide encoding this peptide was then inserted into a DNA vaccine vector and this elicited an anti-type 4 antibody response in mice (Lesinski *et al.*, 2001). It remains to be seen whether such antibodies are protective against challenge with type 4 pneumococci, and whether peptide mimics can be developed for a sufficient number of the other PS serotypes.

Conclusions

The ongoing high global morbidity and mortality associated with pneumococcal disease, and the complications caused by increasing rates of resistance to antimicrobials has underpinned extensive efforts in recent years to develop more effective vaccination strategies against *S. pneumoniae*. These efforts have benefited from a better understanding of the mechanisms of pathogenesis of pneumococcal disease, and the advances made possible by the advent of recombinant DNA technology and access to genome sequence data. The polyvalent PS vaccines have doubtless prevented a large number of deaths from invasive disease in recipients belonging to those patient groups for whom this vaccine is currently recommended. The newer PS–protein conjugate formulations also confer a very high degree of protection on young children against included serotypes, with significant herd immunity as well. However, there is now general acceptance that this vaccination approach is not without its drawbacks, and as explained above, the initially substantial clinical benefits of widespread use of conjugate vaccines may diminish with time. It will take many years for the overall impact of conjugate vaccines on disease burden and the population biology of *S. pneumoniae* to become apparent. At the very least, use of the conjugate vaccines will buy time for development of cheaper, non-serotype-specific vaccines based on combinations of protein antigens. It must be emphasized, however, that the success of these protein vaccines is not dependent upon real or perceived failure of the conjugates. Rather, the two approaches should be viewed as complementary, each having an important role to play in global prevention of pneumococcal disease. Neither should development of parenteral protein vaccines impede future research on mucosal- or DNA-based delivery systems, which may further improve presentation of protective antigens to the immune system, thereby optimizing host responses.

References

Adamou, J.E., Heinrichs, J.H., Erwin, A.L., Walsh, W., Gayle, T., Dormitzer, M., Dagan, R., Brewah, Y.A., Barren, P., Lathigra, R., *et al.* (2001). Identification and characterization of a novel family of pneumococcal proteins that are protective against sepsis. Infect. Immun. *69*, 949–958.

Alexander, J.E., Lock, R.A., Peeters, C.C.A.M., Poolman, J.T., Andrew, P.W., Mitchell, T.J., Hansman, D., and Paton, J.C. (1994). Immunization of mice with pneumolysin toxoid confers a significant degree of protection against at least nine serotypes of *Streptococcus pneumoniae*. Infect. Immun. *62*, 5683–5688.

Arévalo, M.T., Xu, Q., Paton, J.C., Hollingshead, S., Pichichero, M.E., Briles, D.E., Girgis, N., and Zeng, M. (2009). Mucosal vaccination with a multicomponent adenovirus-vectored vaccine protects against *Streptococcus pneumoniae* infection in the lung. FEMS Immunol. Med. Microbiol. *55*, 346–351.

Arulanandam, B.P., Lynch, J.M., Briles, D.E., Hollingshead, S., and Metzger, D.W. (2001). Intranasal vaccination with pneumococcal surface protein A and IL-12 augments antibody-mediated opsonization and protective immunity against *Streptococcus pneumoniae* infection. Infect. Immun. *69*, 6718–6724.

Audouy, S.A., van Selm, S., van Roosmalen, M.L., Post, E., Kanninga, R., Neef, J., Estevão, S., Nieuwenhuis, E.E., Adrian, P.V., Leenhouts, K., *et al.* (2007). Development of lactococcal GEM-based pneumococcal vaccines. Vaccine *25*, 2497–2506.

Austrian, R. (2001). Pneumococcal otitis media and pneumococcal vaccines, a historical perspective. Vaccine *19*, S71–S77.

Avery, O.T., and Goebel, W.F. (1931). Chemo-immunological studies on conjugated carbohydrate-proteins. V. The immunological specificity of an antigen prepared by combining the capsular polysaccharide of type III pneumococcus with foreign protein. J. Exp. Med. *54*, 437–447.

Bagnoli, F., Moschioni, M., Donati, C., Dimitrovska, V., Ferlenghi, I., Facciotti, C., Muzzi, A., Giusti, F., Emolo, C., Sinisi, A., *et al.* (2008). A second pilus type in *Streptococcus pneumoniae* is prevalent in emerging serotypes and mediates adhesion to host cells. J. Bacteriol. *190*, 5480–5492.

Balachandran, P., Brooks-Walter, A., Virolainen-Julkunen, A., Hollingshead, S.K., and Briles, D.E. (2002). Role of pneumococcal surface protein C in nasopharyngeal carriage and pneumonia and its ability to elicit protection against carriage of *Streptococcus pneumoniae*. Infect. Immun. *70*, 2526–2534.

Balachandran, P., Hollingshead, S.K., Paton, J.C., and Briles, D.E. (2001). The autolytic enzyme LytA of *Streptococcus pneumoniae* is not responsible for releasing pneumolysin. J. Bacteriol. *183*, 3108–3116.

Barocchi, M.A., Ries, J., Zogaj, X., Hemsley, C., Albiger, B., Kanth, A., Dahlberg, S., Fernebro, J., Moschioni, M., Masignani, V., *et al.* (2006). A pneumococcal pilus influences virulence and host inflammatory responses. Proc. Natl. Acad. Sci. U.S.A. *103*, 2857–2862.

Basset, A., Thompson, C.M., Hollingshead, S.K., Briles, D.E., Ades, E.W., Lipsitch, M., and Malley, R. (2007). Antibody-independent, CD4+ T-cell-dependent protection against pneumococcal colonization elicited by intranasal immunization with purified pneumococcal proteins. Infect. Immun. *75*, 5460–5464.

Basset, A., Trzcinski, K., Hermos, C., O'Brien, K.L., Reid, R., Santosham, M., McAdam, A.J., Lipsitch, M., and Malley, R. (2007). Association of the pneumococcal pilus with certain capsular serotypes but not with increased virulence. J. Clin. Microbiol. *45*, 1684–1689.

Bender, M.H., and Weiser, J.N. (2006). The atypical amino-terminal LPNTG-containing domain of the pneumococcal human IgA1-specific protease is required for proper enzyme localization and function. Mol. Microbiol. *61*, 526–543.

Bergmann, S., Rohde, M., Chhatwal, G.S., and Hammerschmidt, S. (2001). Alpha-Enolase of *Streptococcus pneumoniae* is a plamin(ogen)-binding protein displayed on the bacterial cell surface. Mol. Microbiol. *40*, 1273–1287.

Berry, A.M., Alexander, J.E., Mitchell, T.J., Andrew, P.W., Hansman, D., and Paton, J.C. (1995). Effect of defined point mutations in the pneumolysin gene on the virulence of *Streptococcus pneumoniae*. Infect. Immun. *63*, 1969–1974.

Berry, A.M., Lock, R.A., Hansman, D., and Paton, J.C. (1989a). Contribution of autolysin to the virulence of *Streptococcus pneumoniae*. Infect. Immun. *57*, 2324–2330.

Berry, A.M., and Paton, J.C. (1996). Sequence heterogeneity of PsaA, a 37-kDa putative adhesin essential for virulence of *Streptococcus pneumoniae*. Infect. Immun. *64*, 5255–5262.

Berry, A.M., and Paton, J.C. (2000). Additive attenuation of virulence of *Streptococcus pneumoniae* by mutation of the genes encoding pneumolysin and other putative pneumococcal virulence proteins. Infect. Immun. *68*, 133–140.

Berry, A.M., Paton, J.C., and Hansman, D. (1992). Effect of insertional inactivation of the genes encoding pneumolysin and autolysin on the virulence of *Streptococcus pneumoniae* type 3. Microb. Pathog. *45*, 87–93.

Berry, A.M., Yother, J., Briles, D.E., Hansman, D., and Paton, J.C. (1989b). Reduced virulence of a defined pneumolysin-negative mutant of *Streptococcus pneumoniae*. Infect. Immun. *57*, 2037–2042.

Black, S., Shinefield, H., Fireman, B., Lewis, E., Ray, P., Hansen, J.R., Elvin, L., Ensor, K.M., Hackell, J., Siber, G., *et al.* (2000). Efficacy, safety and immunogenicity of heptavalent pneumococcal conjugate vaccine in children. Northern California Kaiser Permanente Vaccine Study Center Group. Pediatr. Infect. Dis. J. *19*, 187–195.

Blander, J.M., and Medzhitov, R. (2006). Toll-dependent selection of microbial antigens for presentation by dendritic cells. Nature *440*, 808–812.

Bosarge, J.R., Watt, J.M., McDaniel, D.O., Swiatlo, E., and McDaniel, L.S. (2001). Genetic immunization with the region encoding the alpha-helical domain of PspA elicits protective immunity against *Streptococcus pneumoniae*. Infect. Immun. *69*, 5456–5463.

Briles, D.E., Ades, E., Paton, J.C., Sampson, J.S., Carlone, G.M., Huebner, R.C., Virolainen, A., Swiatlo, E., and Hollingshead, S.K. (2000a). Intranasal immunization of mice with a mixture of the pneumococcal proteins PsaA and PspA is highly protective against nasopharyngeal carriage of *Streptococcus pneumoniae*. Infect. Immun. *68*, 796–800.

Briles, D.E., Hollingshead, S.K., King, J., Swift, A., Braun, P.A., Park, M.K., Ferguson, L.M., Nahm, M.H., and Nabors, G.S. (2000b). Immunization of humans with recombinant pneumococcal surface protein A (rPspA) elicits antibodies that passively protect mice from fatal infection with *Streptococcus pneumoniae* bearing heterologous PspA. J. Infect. Dis. *182*, 1694–1701.

Briles, D.E., Hollingshead, S.K., Paton, J.C., Ades, E.W., Novak, L., van Ginkel, F.W., and Benjamin, W.H., Jr. (2003). Immunizations with pneumococcal surface protein A and pneumolysin are protective against pneumonia in a murine model of pulmonary infection with *Streptococcus pneumoniae*. J. Infect. Dis. *188*, 339–348.

Briles, D.E., Paton, J.C., Nahm, M.H., and Swiatlo, E. (1999). Immunity to *Streptococcus pneumoniae*. In Effects of Microbes on the Immune System, Cunningham, M.W., and Fujinami, R.S., eds (Philadelphia, Lippincott Williams & Wilkins), pp. 263–280.

Brixner, D.I. (2005). Improving acute otitis media outcomes through proper antibiotic use and adherence. Am. J. Manag. Care *11*, S202–S210.

Brooks-Walter, A., Briles, D.E., and Hollingshead, S.K. (1999). The pspC gene of *Streptococcus pneumoniae* encodes a polymorphic protein PspC, which elicits cross-reactive antibodies to PspA and provides immunity to pneumococcal bacteremia. Infect. Immun. *67*, 6533–6542.

Brooks-Walter, A., Tart, R.C., Briles, D.E., and Hollingshead, S.K. (1997). The pspC gene encodes a second pneumococcal surface protein homologous to the gene encoding the protection-eliciting PspA protein of *Streptococcus pneumoniae*. Abstr. 35. In Abstracts of the 97th General Meeting of the American Society for Microbiology 1997.

Brown, J.S., Gilliland, S.M., and Holden, D.W. (2001a). A *Streptococcus pneumoniae* pathogenicity island encoding an ABC transporter involved in iron uptake and virulence. Mol. Microbiol. *40*, 572–585.

Brown, J.S., Ogunniyi, A.D., Woodrow, M.C., Holden, D.W., and Paton, J.C. (2001b). Immunization with components of two iron-uptake ABC transporters protects mice against systemic *Streptococcus pneumoniae* infection. Infect. Immun. *69*, 6702–6706.

Canvin, J.R., Marvin, A.P., Sivakumaran, M., Paton, J.C., Boulnois, G.J., Andrew, P.W., and Mitchell, T.J. (1995). The role of pneumolysin and autolysin in the pathology of pneumonia and septicemia in mice infected with a type 2 pneumococcus. J. Infect. Dis. *172*, 119–123.

Centers for Disease Control and Prevention. (1997). Prevention of pneumococcal disease: recommendations of the Advisory Committee on Immunization Practices (ACIP). MMWR 46(RR-8), 1–24.

Cheng, Q., Finkel, D., and Hostetter, M.K. (2000). Novel purification scheme and functions for a C3-binding protein from *Streptococcus pneumoniae*. Biochemistry *39*, 5450–5457.

Coffey, T.J., Enright, M.C., Daniels, M., Morona, J.K., Morona, R., Hryniewicz, W., Paton, J.C., and Spratt, B.G. (1998). Recombinational exchanges at the capsular polysaccharide biosynthetic locus lead to frequent serotype changes among natural isolates of *Streptococcus pneumoniae*. Mol. Microbiol. *27*, 73–84.

Crain, M.J., Waltman, W.D., 2nd, Turner, J.S., Yother, J., Talkington, D.F., McDaniel, L.S., Gray, B.M., and Briles, D.E. (1990). Pneumococcal surface protein A (PspA) is serologically highly variable and is expressed by all clinically important capsular serotypes of *Streptococcus pneumoniae*. Infect. Immun. 58, 3293–3299.

Dagan, R., Givon-Lavi, N., Zamir, O., and Fraser, D. (2003). Effect of a nonavalent conjugate vaccine on carriage of antibiotic-resistant *Streptococcus pneumoniae* in day-care centers. Pediatr. Infect. Dis. J. *22*, 532–540.

Daniels, C.C., Briles, T.C., Mirza, S., Hakansson, A.P., and Briles, D.E. (2006). Capsule does not block antibody binding to PspA, a surface virulence protein of *Streptococcus pneumoniae*. Microb. Pathog. *40*, 228–233.

Daniely, D., Portnoi, M., Shagan, M., Porgador, A., Givon-Lavi, N., Ling, E., Dagan, R., and Mizrachi Nebenzahl, Y. (2006). Pneumococcal 6-phosphogluconate-dehydrogenase, a putative adhesin, induces protective immune response in mice. Clin. Exp. Immunol. *144*, 254–263.

Dave, S., Brooks-Walter, A., Pangburn, M.K., and McDaniel, L.S. (2001). PspC, a pneumococcal surface protein, binds human factor H. Infect. Immun. *69*, 3435–3437.

De, B.K., Sampson, J.S., Ades, E.W., Huebner, R.C., Jue, D.L., Johnson, S.E., Espina, M., Stinson, A.R., Briles, D.E., and Carlone, G.M. (2000). Purification and characterization of *Streptococcus pneumoniae* palmitoylated pneumococcal surface adhesin A expressed in *Escherichia coli*. Vaccine *18*, 1811–1821.

Dintilhac, A., Alloing, G., Granadel, C., and Claverys, J.P. (1997). Competence and virulence of S. pneuminiae: Adc and PsaA mutants exhibit a requirement for Zn and Mn resulting from inactivation of metal permeases. Mol. Microbiol. *25*, 727–739.

Douglas, R.M., Paton, J.C., Duncan, S.J., and Hansman, D. (1983). Antibody response to pneumococcal vaccination in children younger than five years of age. J. Infect. Dis. *148*, 131–137.

Duthy, T.G., Ormsby, R.J., Giannakis, E., Ogunniyi, A.D., Stroeher, U.H., Paton, J.C., and Gordon, D.L. (2002). The human complement regulator factor H binds pneumococcal surface protein PspC via short consensus repeat domains 13–15. Infect. Immun. *70*, 5604–5611.

Eskola, J. (2001). Polysaccharide-based pneumococcal vaccines in the prevention of acute otitis media. Vaccine *19*, S78–S82.

Eskola, J., Kilpi, T., Palmu, A., Jokinen, J., Haapakoski, J., Herva, E., Takala, A., Kayhty, H., Karma, P., Kohberger, R., et al., and the Finnish Otitis Media Study Group. (2001). Efficacy of a pneumococcal conjugate vaccine against acute otitis media. N. Engl. J. Med. *344*, 403–409.

Fälker, S., Nelson, A.L., Morfeldt, E., Jonas, K., Hultenby, K., Ries, J., Melefors, O., Normark, S., and Henriques-Normark, B. (2008). Sortase-mediated assembly and surface topology of adhesive pneumococcal pili. Mol. Microbiol. *70*, 595–607.

Feldman, C., Munro, N.C., Jeffery, P.K., Mitchell, T.J., Andrew, P.W., Boulnois, G.J., Guerreiro, D., Rohde, J.A., Todd, H.C., Cole, P.J., et al. (1991). Pneumolysin induces the salient histological features of pneumococcal infection in the rat lung in vivo. Am. J. Respir. Cell Mol. Biol. *5*, 416–423.

Ferreira, D.M., Darrieux, M., Silva, D.A., Leite, L.C., Ferreira, J.M., Jr., Ho, P.L., Miyaji, E.N., and Oliveira, M.L. (2009). Characterization of protective mucosal and systemic immune responses elicited by pneumococcal surface protein PspA and PspC nasal vaccines against a respiratory pneumococcal challenge in mice. Clin. Vaccine Immunol. *16*, 636–645.

Fraser, D., Givon-Lavi, N., Bilenko, N., and Dagan, R. (2001). A decade (1989–1998) of pediatric invasive pneumococcal disease in 2 populations residing in 1 geographical location: implications for vaccine choice. Clin. Infect. Dis. *33*, 421–427.

Gianfaldoni, C., Censini, S., Hilleringmann, M., Moschioni, M., Facciotti, C., Pansegrau, W., Masignani, V., Covacci, A., Rappuoli, R., Barocchi, M.A., et al. (2007). *Streptococcus pneumoniae* pilus subunits protect mice against lethal challenge. Infect. Immun. *75*, 1059–1062.

Giefing, C., Meinke, A.L., Hanner, M., Henics, T., Bui, M.D., Gelbmann, D., Lundberg, U., Senn, B.M., Schunn, M., Habel, A., et al. (2008). Discovery of a novel class of highly conserved vaccine antigens using genomic scale antigenic fingerprinting of pneumococcus with human antibodies. J. Exp. Med. *205*, 117–131.

Gilbert, C., Robinson, K., Le Page, R.W., and Wells, J.M. (2000). Heterologous expression of an immunogenic pneumococcal type 3 capsular polysaccharide in Lactococcus lactis. Infect. Immun. *68*, 3251–3260.

Gor, D.O., Ding, X., Briles, D.E., Jacobs, M.R., and Greenspan, N.S. (2005). Relationship between surface acessibility for PpmA, PsaA, and PspA and antibody-mediated immunity to systemic infection by *Streptococcus pneumoniae*. Infect. Immun. *73*, 1304–1312.

Gor, D.O., Ding, X., Li, Q., Schreiber, J.R., Dubinsky, M., and Greenspan, N.S. (2002). Enhanced immunogenicity of pneumococcal surface adhesin A by genetic fusion to cytokines and evaluation of protective immunity in mice. Infect. Immun. *70*, 5589–5595.

Gosink, K.K., Mann, E.R., Guglielmo, C., Tuomanen, E.I., and Masure, H.R. (2000). Role of novel choline binding proteins in virulence of *Streptococcus pneumoniae*. Infect. Immun. *68*, 5690–5695.

Hakansson, A., Roche, H., Mirza, S., McDaniel, L.S., Brooks-Walter, A., and Briles, D.E. (2001). Characterization of the binding of human lactoferrin to pneumococcal surface protein A (PspA). Infect. Immun. *69*, 3372–3381.

Hamel, J., Charland, N., Pineau, I., Ouellet, C., Rioux, S., Martin, D., and Brodeur, B.R. (2004). Prevention of pneumococcal disease in mice immunized with conserved surface-accessible proteins. Infect. Immun. *72*, 2659–2670.

Hammerschmidt, S., Bethe, G., Remanen, P., and Chhatwal, G.S. (1999). Identification of pneumococcal surface protein A as a lactoferrin-binding protein of *Streptococcus pneumoniae*. Infect. Immun. *67*, 1683–1687.

Hammerschmidt, S., Talay, S., Brandtzaeg, P., and Chhatwal, G.S. (1997). SpsA, a novel pneumococcal surface protein with specific binding to secretory immunoglobulin A and secretory component. Mol. Microbiol. 25, 1113–1124.

Hammerschmidt, S., Tillig, M.P., Wolff, S., Vaerman, J.P., and Chhatwal, G.S. (2000). Species-specific binding of human secretory component to SpsA protein of *Streptococcus pneumoniae* via a hexapeptide motif. Mol. Microbiol. 36, 726–736.

Hicks, L.A., Harrison, L.H., Flannery, B., Hadler, J.L., Schaffner, W., Craig, A.S., Jackson, D., Thomas, A., Beall, B., Lynfield, R., *et al.* (2007). Incidence of pneumococcal disease due to non–pneumococcal conjugate vaccine (PCV7) serotypes in the United States during the era of widespread PCV7 vaccination, 1998–2004. J. Infect. Dis. 196, 1346–1354.

Hollingshead, S.K., Becker, R.S., and Briles, D.E. (2000). Diversity of PspA: mosaic genes and evidence for past recombination in *Streptococcus pneumoniae*. Infect. Immun. 68, 5889–5900.

Houldsworth, S., Andrew, P.W., and Mitchell, T.J. (1994). Pneumolysin stimulates production of TNFa and IL-1β by human mononuclear phagocytes. Infect. Immun. 62, 1501–1503.

Hvalbye, B.K., Aaberge, I.S., Lovik, M., and Haneberg, B. (1999). Intranasal immunization with heat-inactivated *Streptococcus pneumoniae* protects mice against systemic pneumococcal infection. Infect. Immun. 67, 4320–4325.

Iannelli, F., Oggioni, M.R., and Pozzi, G. (2002). Allelic variation in the highly polymorphic locus pspC of *Streptococcus pneumoniae*. Gene 284, 63–71.

Jacobs, M.R., Good, C.E., Bajaksouzian, S., and Windau, A.R. (2008). Emergence of *Streptococcus pneumoniae* serotypes 19A, 6C, and 22F and serogroup 15 in Cleveland, Ohio, in relation to introduction of the protein-conjugated pneumococcal vaccine. Clin. Infect. Dis. 47, 1388–1395.

Jakobsen, H., Schulz, D., Pizza, M., Rappuoli, R., and Jonsdottir, I. (1999). Intranasal immunization with pneumococcal polysaccharide conjugate vaccines with non-toxic mutants of *Escherichia coli* heat-labile enterotoxins as adjuvants protects mice against invasive pneumococcal infections. Infect. Immun. 67, 5892–5897.

Janulczyk, R., Iannelli, F., Sjoholm, A.G., Pozzi, G., and Bjorck, L. (2000). Hic, a novel surface protein of *Streptococcus pneumoniae* that interferes with complement function. J. Biol. Chem. 275, 37257–37263.

Jarva, H., Janulczyk, R., Hellwage, J., Zipfel, P.F., Bjorck, L., and Meri, S. (2002). *Streptococcus pneumoniae* evades complement attack and opsonophagocytosis by expressing the pspC locus-encoded Hic protein that binds to short consensus repeats 8–11 of factor H. J. Immunol. 168, 1886–1894.

Jedrzejas, M.J., Hollingshead, S.K., Lebowitz, J., Chantalat, L., Briles, D.E., and Lamani. E.J. (2000). Production and characterization of the functional fragment of pneumococcal surface protein A. Arch. Biochem. Biophys. 373, 116–125.

Jefferies, J.M., Smith, A., Clarke, S.C., Dowson, C., and Mitchell, T.J. (2004). Genetic analysis of diverse disease-causing pneumococci indicates high levels of diversity within serotypes and capsule switching. J. Clin. Microbiol. 42, 5681–5688.

Johnson, S.E., Dykes, J.K., Jue, D.L., Sampson, J.S., Carlone, G.M., and Ades, E.W. (2002). Inhibition of pneumococcal carriage in mice by subcutaneous immunization with peptides from the common surface protein pneumococcal surface adhesin A. J. Infect. Dis. 185, 489–496.

Jomaa, M., Terry, S., Hale, C., Jones, C., Dougan, G., and Brown, J. (2006). Immunization with the iron uptake ABC transporter proteins PiA and PiuA prevents respiratory infection with *Streptococcus pneumoniae*. Vaccine 245, 133–5139.

Jomaa, M., Yuste, J., Paton, J.C., Jones, C., Dougan, C., and Brown, J.S. (2005). Antibodies to the iron uptake ABC transporter lipoproteins PiaA and PiuA promote opsonophagocytosis of *Streptococcus pneumoniae*. Infect. Immun. 73, 6852–6859.

Kang, H.Y., Srinivasan, J., and Curtiss, R., 3rd. (2002). Immune responses to recombinant pneumococcal PspA antigen delivered by live attenuated *Salmonella enterica* serovar typhimurium vaccine. Infect. Immun. 70, 1739–1749.

Kharat, A.S., and Tomasz, A. (2003). Inactivation of the srtA gene affects localization of surface proteins and decreases adhesion of *Streptococcus pneumoniae* to human pharyngeal cells in vitro. Infect. Immun. 71, 2758–2765.

Klein, D.L. (2000). Pneumococcal disease and the role of conjugate vaccines. In *Streptococcus pneumoniae*: Molecular Biology and Mechanisms of Disease, Tomasz, A., ed. (New York, Mary Ann Liebert Inc.), pp. 467–477.

Klein, J.O. (2001). The burden of otitis media. Vaccine 19, S2–S8.

Klugman, K.P. (1990). Pneumococcal resistance to antibiotics. Clin. Microbiol. Rev. 3, 171–196.

Klugman, K.P. (1996). Epidemiology, control and treatment of multiresistant pneumococci. Drugs 52 (Suppl. 2), 42–46.

Kwon, H.Y., Ogunniyi, A.D., Choi, M.H., Pyo, S.N., Rhee, D.K., and Paton, J.C. (2004). The ClpP protease of *Streptococcus pneumoniae* modulates virulence gene expression and protects against fatal pneumococcal challenge. Infect. Immun. 72, 5646–5653.

Lawrence, M.C., Pilling, P.A., Ogunniyi, A.D., Berry, A.M., and Paton, J.C. (1998). The crystal structure of pneumococcal surface antigen PsaA reveals a metal-binding site and a novel structure for a putative ABC-type binding protein. Structure 6, 1553–1561.

Lee, C.J., Banks, S.D., and Li, J.P. (1991). Virulence, immunity and vaccine related to S. pneumoniae. Crit. Rev. Microbiol. 18, 89–114.

Lee, C.J., Lock, R.A., Mitchell, T.J., Andrew, P.W., Boulnois, G.J., and Paton, J.C. (1994). Protection of infant mice from challenge with *Streptococcus pneumoniae* type 19F by immunization with a type 19F polysaccharide–pneumolysoid conjugate. Vaccine 12, 875–878.

LeMieux, J., Hava, D.L., Basset, A., and Camilli, A. (2006). RrgA and RrgB are components of a multisubunit pilus encoded by the *Streptococcus pneumoniae* rlrA pathogenicity islet. Infect. Immun. *74*, 2453–2456.

Lesinski, G.B., Smithson, S.L., Srivastava, N., Chen, D., Widera, G., and Westerink, M.A. (2001). A DNA vaccine encoding a peptide mimic of *Streptococcus pneumoniae* serotype 4 capsular polysaccharide induces specific anti-carbohydrate antibodies in Balb/c mice. Vaccine *19*, 1717–1726.

Li, Y., Wang, S., Scarpellini, G., Gunn, B., Xin, W., Wanda, S.Y., Roland, K.L., and Curtiss, R., 3rd. (2009). Evaluation of new generation *Salmonella* enterica serovar Typhimurium vaccines with regulated delayed attenuation to induce immune responses against PspA. Proc. Natl. Acad. Sci. U.S.A. *106*, 593–598.

Ling, E., Feldman, G., Portnoi, M., Dagan, R., Overweg, K., Mulholland, F., Chalifa-Caspi, V., Wells, J., and Mizrachi-Nebenzahl, Y. (2004). Glycolytic enzymes associated with the cell surface of *Streptococcus pneumoniae* are antigenic in humans and elicit protective immune responses in the mouse. Clin. Exp. Immunol. *138*, 290–298.

Lipsitch, M., Dykes, J.K., Johnson, S.E., Ades, E.W., King, J., Briles, D.E., and Carlone. G.M. (2000). Competition among *Streptococcus pneumoniae* for intranasal colonization in a mouse model. Vaccine *18*, 2895–2901.

Lipsitch, M., Whitney, C.G., Zell, E., Kaijalainen, T., Dagan, R., and Malley, R. (2005). Are anticapsular antibodies the primary mechanism of protection against invasive pneumococcal disease? PLoS Med. *2*, e15.

Lock, R.A., Hansman, D., and Paton, J.C. (1992). Comparative efficacy of autolysin and pneumolysin as immunogences protecting mice against infection by *Streptococcus pneumoniae*. Microb. Pathog. *12*, 137–143.

Lock, R.A., Paton, J.C., and Hansman, D. (1988). Comparative efficacy of pneumococcal neuraminidase and pneumolysin as immunogens protective against *Streptococcus pneumoniae*. Microb. Pathog. *5*, 461–467.

Long, J.P., Tong, H.H., and DeMaria, T.F. (2004). Immunization with native or recombinant *Streptococcus pneumoniae* neuraminidase affords protection in the chinchilla otitis media model. Infect. Immun. *72*, 4309–4313.

Lopez, R., Garcia, E., Garcia, P., and Garcia, J.L. (2004). Cell wall hydrolases. In The Pneumococcus, E.I Tuomanen, Mitchell, T.J., Morrison, D.A., and Spratt, B.G., eds (Washington DC, ASM Press), pp. 75–88.

Lopez, R., Gonzalez, M.P., Garcia, E., Garcia, J.L., and Garcia, P. (2000). Biological roles of two new murein hydrolases of *Streptococcus pneumoniae* representing examples of module shuffling. Res. Microbiol. *151*, 437–443.

Lu, Y.J., Forte, S., Thompson, C.M., Anderson, P.W., and Malley, R. (2009). Protection against pneumococcal colonization and fatal pneumonia by a trivalent conjugate of a fusion protein with the cell wall polysaccharide. Infect. Immun. *77*, 2076–2083.

Lu, Y.J., Gross, J., Bogaert, D., Finn, A., Bagrade, L., Zhang, Q., Kolls, J.K., Srivastava, A., Lundgren, A., Forte, S., *et*

al. (2008). Interleukin-17A mediates acquired immunity to pneumococcal colonization. PLoS Pathog. *4*, e1000159.

MacLeod, C.M., Hodges, R.G., Heidelberger, M., and Bernhard, W.G. (1945). Prevention of pneumococcal pneumonia by vaccination. J. Exp. Med. *82*, 45–465.

Malley, R., Lipsitch, M., Stack, A., Saladino, R., Fleisher, G., Pelton, S., Thompson, C., Briles, D., and Anderson, P. (2001). Intranasal immunization with killed unencapsulated whole cells prevents colonization and invasive disease by capsulated pneumococci. Infect. Immun. *69*, 4870–4873.

Malley, R., Morse, S.C., Leite, L.C., Areas, A.P.M., Ho, P.L., Kunrusly, F.S., Almeida, I.C., and Anderson, P.W. (2004). Multi-serotype protection of mice against pneumococcal colonization of the nasopharynx and middle ear by killed nonencapsulated cells given intranasally with a non-toxic adjuvant. Infect. Immun. *72*, 4290–4292.

Malley, R., Trzcinski, K., Srivastava, A., Thompson, C.M., Anderson, P.W., and Lipsitch, M. (2005). CD4+ T cells mediate antibody-independent acquired immunity to pneumococcal colonization. Proc. Natl. Acad. Sci. U.S.A. *102*, 4848–4853.

Manco, S., Hernon, F., Yesilkaya, H., Paton, J.C., Andrew, P.W., and Kadioglu, A. (2006). Pneumococcal neuraminidases A and B both have essential roles during infection of the respiratory tract and sepsis. Infect. Immun. *74*, 4014–4020.

Mbelle, N., Huebner, R.E., Wasas, A.D., Kimura, A., Chang, I., and Klugman, K.P. (1999). Immunogenicity and impact on nasopharyngeal carriage of a nonavalent pneumococcal conjugate vaccine. J. Infect. Dis. *180*, 1171–1176.

McAllister, L.J., Tseng, H., Ogunniyi, A.D., Jennings, M.P., McEwan, A.G., and Paton, J.C. (2004). Molecular analysis of the psa permease complex of *Streptococcus pneumoniae*. Mol. Microbiol. *53*, 889–901.

McCool, T.L., Cate, T.R., Moy, G., and Weiser, J.N. (2002). The immune response to pneumococcal proteins during experimental human carriage. J. Exp. Med. *195*, 359–365.

McDaniel, L.S., Ralph, B.A., McDaniel, D.O., and Briles, D.E. (1994). Localization of protection-eliciting epitopes on PspA of *Streptococcus pneumoniae* between amino acid residues 192 and 260. Microb. Pathog. *17*, 323–337.

McDaniel, L.S., Yother, J., Vijayakumar, M., McGarry, L., Guild, W.R., and Briles, D.E. (1987). Use of insertional inactivation to facilitate studies of biological properties of pneumococcal surface protein A (PspA). J. Exp. Med. *165*, 381–394.

McGee, L., Klugman, K.P., and Tomasz, A. (2000). Serotypes and clones of antibiotic-resistant pneumococci. In *Streptococcus pneumoniae*: Molecular Biology and Mechanisms of Disease, Tomasz, A., ed. (New York, Mary Ann Liebert Inc.), pp. 375–379.

Michon, F., Fusco, P.C., Minetti, C.A., Laude-Sharp, M., Uitz, C., Huang, C.H., D'Ambra, A.J., Moore, S., Remeta, D.P., Heron, I., *et al.* (1998). Multivalent pneumococcal capsular polysaccharide conjugate vaccines employing genetically detoxified pneumolysin as a carrier protein. Vaccine *16*, 1732–1741.

Miyaji, E.N., Dias, W.O., Gamberini, M., Gebara, V.C., Schenkman, R.P., Wild, J., Riedl, P., Reimann, J., Schirmbeck, R., and Leite, L.C. (2001). PsaA (pneumococcal surface adhesin A) and PspA (pneumococcal surface protein A) DNA vaccines induce humoral and cellular immune responses against *Streptococcus pneumoniae*. Vaccine *20*, 805–812.

Miyaji, E.N., Dias, W.O., Tanizaki, M.M., and Leite, L.C. (2003). Protective efficacy of PspA (pneumococcal surface protein A)-based DNA vaccines: contribution of both humoral and cellular immune responses. FEMS Immunol. Med. Microbiol. *37*, 53–57.

Miyaji, E.N., Ferreira, D.M., Lopes, A.P., Brandileone, M.C., Dias, W.O., and Leite, L.C. (2002). Analysis of serum cross-reactivity and cross-protection elicited by immunization with DNA vaccines against *Streptococcus pneumoniae* expressing PspA fragments from different clades. Infect. Immun. *70*, 5086–5090.

Moschioni, M., Donati, C., Muzzi, A., Masignani, V., Censini, S., Hanage, W.P., Bishop, C.J., Reis, J.N., Normark, S., Henriques-Normark, B., *et al.* (2008). *Streptococcus pneumoniae* contains 3 rlrA pilus variants that are clonally related. J. Infect. Dis. *197*, 888–896.

Musher, D.M., Phan, H.M., and Baughn, R.E. (2001). Protection against bacteremic pneumococcal infection by antibody to pneumolysin. J. Infect. Dis. *183*, 827–830.

Musher, D.M., Watson, D.A., and Baughn, R.E. (2001). Genetic control of the immunological response to pneumococcal capsular polysaccharides. Vaccine *19*, 623–627.

Nabors, G.S., Braun, P.A., Herrmann, D.J., Heise, M.L., Pyle, D.J., Gravenstein, S., Schilling, M., Ferguson, L.M., Hollingshead, S.K., Briles, D.E., *et al.* (2000). Immunization of healthy adults with a single recombinant pneumococcal surface protein A (PspA) variant stimulates broadly cross-reactive antibodies. Vaccine *18*, 1743–1754.

Nayak, A.R., Tinge, S.A., Tart, R.C., McDaniel, L.S., Briles, D.E., and Curtiss, R., 3rd. (1998). A live recombinant oral *Salmonella* vaccine expressing pneumococcal surface protein A induces protective responses against *Streptococcus pneumoniae*. Infect. Immun. *66*, 3744–3751.

Obaro, S.K., Adegbola, R.A., Banya, W.A., and Greenwood, B.M. (1996). Carriage of pneumococci after pneumococcal vaccination. Lancet *348*, 271–272.

Ogunniyi, A.D., Folland, R.L., Hollingshead, S., Briles, D.E., and Paton, J.C. (2000). Immunization of mice with combinations of pneumococcal virulence proteins elicits enhanced protection against challenge with *Streptococcus pneumoniae*. Infect. Immun. *68*, 3028–3033.

Ogunniyi, A.D., Giammarinaro, P., and Paton, J.C. (2002). The genes encoding virulence-associated proteins and the capsule of *Streptococcus pneumoniae* are upregulated and differentially expressed in vivo. Microbiology *148*, 2045–2053.

Ogunniyi, A.D., Grabowicz, M., Briles, D.E., Cook, J., and Paton, J.C. (2007a). Development of a vaccine against invasive pneumococcal disease based on combinations of virulence proteins of *Streptococcus pneumoniae*. Infect. Immun. *75*, 350–357.

Ogunniyi, A.D., LeMessurier, K.S., Graham, R.M.A., Watt, J.M., Briles, D.E., Stroeher, U.H., and Paton, J.C. (2007b). Contributions of pneumolysin, pneumococcal surface protein A (PspA) and PspC to pathogenicity of *Streptococcus pneumoniae* D39 in a mouse model. Infect. Immun. *75*, 1843–1851.

Ogunniyi, A.D., Woodrow, M.C., Poolman, J.T., and Paton, J.C. (2001). Protection against *Streptococcus pneumoniae* elicited by immunization with pneumolysin and CbpA. Infect. Immun. *69*, 5997–6003.

Orihuela, C.J., Gao, G., Francis, K.P., Yu, J., and Tuomanen, E.I. (2004). Tissue-specific contributions of pneumococcal virulence factors to pathogenesis. J. Infect. Dis. *190*, 1661–1669.

Ortqvist, A., Hedlund, J., Burman, L.A., Elbel, E., Hofer, M., Leinonen, M., Lindblad, I., Sundelof, B., and Kalin, M. (1998). Randomised trial of 23-valent pneumococcal capsular polysaccharide vaccine in prevention of pneumonia in middle-aged and elderly people: Swedish pneumococcal vaccine group study. Lancet *351*, 399–403.

Overweg, K., Kerr, A., Sluijter, M., Jackson, M.H., Mitchell, T.J., de Jong, A.P., de Groot, R., and Hermans, P.W. (2000). The putative proteinase maturation protein A of *Streptococcus pneumoniae* is a conserved surface protein with potential to elicit protective immune responses. Infect. Immun. *68*, 4180–4188.

Paton, J.C. (1996). The contribution of pneumolysin to the pathogenicity of *Streptococcus pneumoniae*. Trends Microbiol. *4*, 103–106.

Paton, J.C., Berry, A.M., and Lock, R.A. (1997). Molecular analysis of putative pneumococcal virulence proteins. Microb. Drug Resist. *3*, 1–10.

Paton, J.C., and Giammarinaro, P. (2001). Genome-based analysis of pneumococcal virulence factors: the quest for novel vaccine antigens and drug targets. Trends Microbiol. *9*, 515–518.

Paton, J.C., Lock, R.A., and Hansman, D.J. 1983. Effect of immunization with pnuemolysin on survival time of mice challenged with *Streptococcus pneumoniae*. Infect. Immun. *40*, 548–552.

Paton, J.C., Lock, R.A., Lee, C.J., Li, J.P., Berry, A.M., Mitchell, T.J., Andrew, P.W., Hansman, D., and Boulnois, G.J. (1991). Purification and immunogenicity of genetically obtained pneumolysin toxoids and their conjugation to *Streptococcus pneumoniae* type 19F polysaccharide. Infect. Immun. *59*, 2297–2304.

Paton, J.C., and Morona, J.K. (2006). *Streptococcus pneumoniae* capsular polysaccharide. In Gram-Positive Pathogens, 2nd edn, Fischetti, V., Novick, R., Ferretti, J., Portnoy, D., and Rood, J., eds (Washington DC, USA, ASM Press), pp. 241–252.

Paton, J.C., Morona, J.K., Harrer, S., Hansman, D., and Morona, R. (1993). Immunization of mice with *Salmonella* typhimurium C5 aroA expressing a genetically toxoided derivative of the pneumococcal toxin pneumolysin. Microb. Pathog. *14*, 95–102.

Paton, J.C., Rowan-Kelly, B., and Ferrante, A. (1984). Activation of human complement by the pneumococcal toxin, pneumolysin. Infect. Immun. *43*, 1085–1087.

Paton, J.C., Toogood, I.R., Cockington, R., and Hansman, D. (1986). Antibody response to pneumococcal vaccine in children aged 5 to l5 years. Am. J. Dis. Child. *140*, 135–138.

Pimenta, F.C., Miyaji, E.N., Arêas, A.P., Oliveira, M.L., de Andrade, A.L., Ho, P.L., Hollingshead, S.K., and Leite, L.C. (2006). Intranasal immunization with the cholera toxin B subunit-pneumococcal surface antigen A fusion protein induces protection against colonization with *Streptococcus pneumoniae* and has negligible impact on the nasopharyngeal and oral microbiota of mice. Infect. Immun. *74*, 4939–4944.

Ren, B., Szalai, A.J., Thomas, O., Hollingshead, S.K., and Briles, D.E. (2003). Both family 1 and family 2 PspA proteins can inhibit complement deposition and confer virulence to a capsular serotype 3 strain of *Streptococcus pneumoniae*. Infect. Immun. *71*, 75–85.

Riboldi-Tunnicliffe, A., Isaacs, N.W., and Mitchell, T.J. (2005). 1.2 Angstroms crystal structure of the S. pneumoniae PhtA histidine triad domain a novel zinc binding fold. FEBS Lett. *579*, 5353–5360.

Rijkers, G.T., Sanders, E.A., Breukels, M.A., and Zegers, B.J. (2001). Infant B cell responses to polysaccharide determinants. Vaccine *16*, 1396–1400.

Robbins, J.B., Austrian, R., Lee, C.J., Rastogi, S.C., Schiffman, G., Henrichsen, J., Makela, P.H., Broome, C.V., Facklam, R.R., Tiesjema, R.H., *et al.* (1983). Considerations for formulating the second-generation pneumococcal capsular polysaccharide vaccine with emphasis on the cross-reactive types within groups. J. Infect. Dis. *148*, 1136–1159.

Robbins, J.B., Schneerson, R., Anderson, P., and Smith, D.H. (1996). Prevention of systemic infections, especially meningitis, caused by Haemophilus influenzae type b. Impact on public health and implications for other polysaccharide-based vaccines. J. Am. Med. Assoc. *276*, 1181–1185.

Rosenow, C., Ryan, P.J., Weiser, J.N., Johnson, S., Fontan, P., Ortqvist, A., and Masure, H.R. (1997). Contribution of novel choline-binding proteins to adherence, colonization and immunogenicity of *Streptococcus pneumoniae*. Mol. Microbiol. *25*, 819–829.

Rubins, J.B., Charboneau, D., Fasching, C., Berry, A.M., Paton, J.C., Alexander, J.E., Andrew, P.W., Mitchell, T.J., and Janoff, E.N. (1996). Distinct roles for pneumolysin's cytotoxic and complement activities in the pathogenesis of pneumococcal pneumonia. Am. J. Respir. Crit. Care Med. *153*, 1339–1346.

Sampson, J.S., Furlow, Z., Whitney, A.M., Williams, D., Facklam, R., and Carlone, G.M. (1997). Limited diversity of *Streptococcus pneumoniae* psaA among pneumococcal vaccine serotypes. Infect. Immun. *65*, 1967–1971.

Schneewind, O., Mihaylova-Petkov, D., and Model, P. (1993). Cell wall sorting signals in surface proteins of Gram-positive bacteria. EMBO J. *12*, 4803–4811.

Seo, J.Y., Seong, S.Y., Ahn, B.Y., Kwon, I.C., Chung, H., and Jeong, S.Y. (2002). Cross-protective immunity of mice induced by oral immunization with pneumococcal surface adhesin a encapsulated in microspheres. Infect. Immun. *70*, 1143–1149.

Seong, S.Y., Cho, N.H., Kwon, I.C., and Jeong, S.Y. (1999). Protective immunity of microsphere-based mucosal vaccines against lethal intranasal challenge with *Streptococcus pneumoniae*. Infect. Immun. *67*, 3587–3592.

Shaper, M., Hollingshead, S.K., Benjamin, W.H., Jr., and Briles, D.E. (2004). PspA protects *Streptococcus pneumoniae* from killing by apolactoferrin, and antibody to PspA enhances killing of pneumococci by apolactoferrin. Infect. Immun. *72*, 5031–5040.

Simberkoff, M.S., Cross, A.P., Al-Ibrahim, M., Baltch, A.L., Geiseler, P.J., Nadler, J., Richmond, A.S., Smith, R.P., Schiffman, G., Shepard, D.S., *et al.* (1986). Efficacy of pneumococcal vaccine in high-risk patients: results of a Veterans Administration cooperative study. N. Engl. J. Med. *315*, 1318–1327.

Spratt, B.G., and Greenwood, B.M. (2000). Prevention of pneumococcal disease by vaccination: does serotype replacement matter. Lancet *356*, 1210–1211.

Swiatlo, E., McDaniel, L.S., and Briles, D.E. (2004). Choline-binding proteins. In The Pneumococcus, E.I Tuomanen, Mitchell, T.J., Morrison, D.A., and Spratt, B.G., eds (Washington DC, ASM Press), pp. 49–60.

Tai, S.S., Yu, C., and Lee, J.K. (2003). A solute binding protein of *Streptococcus pneumoniae* iron transport. FEMS Microbiol. Lett. *220*, 303–308.

Talkington, D.F., Brown, B.G., Tharpe, J.A., Koenig, A., and Russell, H. (1996). Protection of mice against fatal pneumococcal challenge by immunization with pneumococcal surface adhesin A (PsaA). Microb. Pathog. *21*, 17–22.

Talkington, D.F., Crimmins, D.L., Voellinger, D.C., Yother, J., and Briles, D.E. (1991). A 43-kilodalton pneumococcal surface protein, PspA: isolation, protective abilities, and structural analysis of the amino-terminal sequence. Infect. Immun. *59*, 1285–1289.

Tettelin, H., Nelson, K.E., Paulsen, I.T., Eisen, J.A., Read, T.D., Peterson, S., Heidelberg, J., DeBoy, R.T., Haft, D.H., Dodson, R.J., *et al.* (2001). Complete genome sequence of a virulent isolate of *Streptococcus pneumoniae*. Science *293*, 498–506.

Tong, H.H., Blue, L.E., James, M.A., and DeMaria, T.F. (2000). Evaluation of the virulence of a *Streptococcus pneumoniae* neuraminidase-deficient mutant in nasopharyngeal colonization and development of otitis media in the chinchilla model. Infect. Immun. *68*, 921–924.

Tong, H.H., Li, D., Chen, S., Long, J.P., and DeMaria, T.F. (2005). Immunization with recombinant *Streptococcus pneumoniae* neuraminidase NanA protects chinchillas against nasopharyngeal colonization. Infect. Immun. *73*, 7775–7778.

Tseng, H.J., McEwan, A.G., Paton, J.C., and Jennings, M.P. (2002). Virulence of *Streptococcus pneumoniae*: PsaA mutants are hypersensitive to oxidative stress. Infect. Immun. *70*, 1635–1639.

Tu, A.H.T., Fulgham, R.L., McCory, M.A., Briles, D.E., and Szalai, A.J. (1999). Pneumococcal surface protein A (PspA) inhibits complement activation by *Streptococcus pneumoniae*. Infect. Immun. *67*, 4720–4724.

Vakevainen, M., Eklund, C., Eskola, J., and Kayhty, H. (2001). Cross-reactivity of antibodies to type 6B and 6A polysaccharides of *Streptococcus pneumoniae* evoked by pneumococcal conjugate vaccine in infants. J. Infect. Dis. *184*, 789–793.

Vela Coral, M.C., Fonseca, N., Castaneda, E., Di Fabio, J.L., Hollingshead, S.K., and Briles, D.E. (2001). Families of pneumococcal surface protein A (PspA) of *Streptococcus pneumoniae* invasive isolates recovered from Colombian children. Emerg. Infect. Dis. *7*, 832–836.

Weiser, J.N., Bae, D., Fasching, C., Scamurra, R.W., Ratner, A.J., and Janoff, E.N. (2003). Antibody-enhanced pneumococcal adherence requires IgA1 protease. Proc. Natl. Acad. Sci. USA. *100*, 4215–4220.

Williams, B.G., Gouws, E., Boschi-Pinto, C., Bryce, J., and Dye, C. (2002). Estimates of world-wide distribution of child deaths from acute respiratory infections. Lancet Infect. Dis. *2*, 25–32.

Wizemann, T.M., Heinrichs, J.H., Adamou, J.E., Erwin, A.L., Kunsch, C., Choi, G.H., Barash, S.C., Rosen, C.A., Masure, H.R., Tuomanen, E., *et al.* (2001). Use of a whole genome approach to identify vaccine molecules affording protection against *Streptococcus pneumoniae* infection. Infect. Immun. *69*, 1593–1598.

World Health Organization. (2007). Pneumococcal conjugate vaccine for childhood immunization – WHO position paper. Wkly. Epidemiol. Record *82*, 93–104.

Wortham, C., Grinberg, L., Kaslow, D.C., Briles, D.E., McDaniel, L.S., Lees, A., Flora, M., Snapper, C.M., and Mond, J.J. (1998). Enhanced protective antibody responses to PspA after intranasal or subcutaneous injections of PspA genetically fused to granulocyte-macrophage colony-stimulating factor or interleukin-2. Infect. Immun. *66*, 1513–1520.

Wu, H.Y., Nahm, M.H., Guo, Y., Russell, M.W., and Briles, D.E. (1997). Intranasal immunization of mice with PspA (pneumococcal surface protein A) can prevent intranasal carriage and infection with *Streptococcus pneumoniae*. J. Infect. Dis. *175*, 839–846.

Wuorimaa, T., and Kayhty, H. (2002). Current state of pneumococcal vaccines. Scand. J. Immunol. *56*, 111–129.

Yamamoto, M., McDaniel, L.S., Kawabata, K., Briles, D.E., Jackson, R.J., McGhee, J.R., and Kiyono, H. (1997). Oral immunization with PspA elicits protective humoral immunity against *Streptococcus pneumoniae* infection. Infect. Immun. *65*, 640–644.

Zhang, J.R., Mostov, K.E., Lamm, M.E., Nanno, M., Shimida, S., Ohwaki, M., and Tuomanen, E. (2000). The polymeric immunoglobulin receptor translocates pneumococci across human nasopharyngeal epithelial cells. Cell *102*, 827–837.

Zhang, Y., Masi, A.W., Barniak, V., Mountzouros, K., Hostetter, M.K., and Green, B.A. (2001). Recombinant PhpA protein, a unique histidine motif-containing protein from *Streptococcus pneumoniae*, protects mice against intranasal pneumococcal challenge. Infect. Immun. *69*, 3827–3836.

15

Veterinary Vaccines with a Focus on Bovine Mastitis

John R. Middleton

Abstract

While novel approaches to vaccination against diseases of veterinary importance are being explored, currently marketed products, in general, employ old technology with the majority of products still being killed, modified live, or toxoid preparations. Due to the breadth of diseases encountered in veterinary medicine and the large number of vaccines marketed and under development, this chapter will focus on vaccines aimed at preventing bovine mastitis with a particular focus on *Staphylococcus aureus*, a bacterium that not only causes mastitis in cattle, but is a leading cause of human infection. Vaccine developments for *S. aureus* in cattle will be compared with research aimed at preventing staphylococcal infection in humans. The remainder of the chapter will discuss other available vaccines aimed at preventing bovine mastitis and serve to illustrate that the goals of vaccination may differ depending on the type of infection being prevented.

Introduction

Veterinary medicine covers a broad range of animal species and their associated diseases. There are, therefore, a wide variety of veterinary vaccines available for use and in the pipeline. Success of currently marketed vaccines in preventing or ameliorating disease is not only dependent on inciting cause, but the desired type of protection afforded by the vaccine. Currently marketed products for livestock fall broadly into three categories: (1) killed or lysed whole organism, (2) modified live whole organism and (3) toxoids. Vaccines marketed as an aid to prevention of bacterial infections in cattle include killed or modified live bacterins with or without toxoids against organisms such as *Escherichia coli*, *Salmonella enterica* subsp. *typhimurium*, clostridial organisms, brucellosis, *Staphylococcus aureus*, *Moraxella bovis*, *Histophilus somni*, leptospirosis, vibriosis (*Campylobacter foetus*), *Mycoplasma bovis*, *Mannheimia haemolytica*, *Pasteurella multocida*, and *Fusobacterium necrophorum*. Common vaccines aimed at preventing viral diseases in livestock include killed or modified live vaccines against bovine herpes virus-1 (BHV-1; infectious bovine rhinotracheitis), bovine viral diarrhoea virus (BVDV), bovine respiratory syncytial virus, parainfluenza-3, rotavirus, coronavirus, and rabies.

Alternative strategies to whole or lysed whole organism vaccines or toxoids have or are being explored in veterinary medicine. These approaches include subunit and subunit conjugate vaccines, new approaches to modified live vaccines including gene deletion or mutation, vector vaccines that employ specific antigen expression by an attenuated virus or bacteria or through gene recombination, DNA vaccines that utilize host cells to express target antigens and stimulate immunity, or plant vector vaccines (Shams, 2005). While a gene deletion vaccine was licensed by the United States Department of Agriculture for pseudorabies virus control in pigs in 1986 (Kit, 1990), there do not appear to be many, if any, vaccines on the market for livestock that utilize such alternative strategies for delivering antigen(s) and stimulating immunity. A gene deletion vaccine for BHV-1 (Belknap *et al.*, 1999) showed promise in experimental trials for preventing infectious bovine rhinotracheitis, but does not appear to have been commercialized. Others have more recently studied viral and bacterial vector vaccines and DNA vaccines for important diseases of cattle including tuberculosis, foot and mouth disease (FMD), BHV-1, BVDV, and *Neospora caninum*, but only preliminary evaluations are published (Donofrio *et al.*, 2008; Hema *et al.*, 2009; Liu *et al.*, 2008; Momtaz *et al.*, 2008; Ren *et al.*, 2009; Vordermeier *et al.*, 2009; Zhao *et al.*, 2009). Due to the cost of research, development, and manufacture, newer vaccine technologies are unlikely to replace existing vaccines for livestock diseases unless they are cheaper, safer, and/or demonstrate enhanced efficacy over the currently marketed products.

Owing to the breadth of diseases encountered in veterinary medicine and the large number of vaccines marketed and under development, this chapter will focus on vaccines aimed at preventing bovine mastitis with a particular focus on *Staphylococcus aureus*, a bacterium that not only causes mastitis in livestock, but is a leading cause of human infection. Vaccine developments for *S. aureus* in cattle will be compared with research aimed at preventing staphylococcal infection in humans. The remainder of the chapter will discuss other available vaccines for bovine mastitis pathogens and serve to illustrate that the goals of vaccination and therefore determination of efficacy may differ depending on the type of infection being prevented.

Bovine mastitis

Mastitis, inflammation of the mammary gland, is regarded as the most costly infectious disease on dairy farms worldwide. The vast majority of mastitis is caused by an intramammary infection (IMI) with a bacterial organism (Bramley *et al.*, 1998). A wide variety of bacterial pathogens can cause mastitis with varying clinical outcomes in the cow. Disease manifestations range from subclinical to death (Bramley *et al.*, 1998). Mastitis can be classified based on the origin of the inciting pathogen as either contagious or environmental (Bramley *et al.*, 1998). Contagious pathogens are usually harboured in the mammary gland of infected cattle and spread cow-to-cow during milking, while environmental pathogens are found in the cow's environment and are acquired between milkings. Common contagious pathogens include *Staphylococcus aureus*, *Mycoplasma* spp., *Streptococcus agalactiae* and *Corynebacterium bovis* (Bramley *et al.*, 1998). Contagious pathogens are of concern because they tend to cause chronic subclinical infections and hence infected cattle become a reservoir for infection of other cattle. Common environmental pathogens include faecal coliforms, *Streptococcus dysgalactiae*, *Streptococcus uberis* and *Enterococcus* spp. (Bramley *et al.*, 1998). While many environmental pathogens cause subclinical or mild clinical infections that spontaneously cure, the coliforms, as a group, cause the greatest proportion of severe clinical mastitis that can, in some cases, result in death of the cow.

Costs associated with mastitis include milk production losses, pharmaceuticals, discarded milk, veterinary services, labour, milk quality deficits, investment in mastitis management materials and infrastructure, diagnostic testing, and cattle replacement (Halasa *et al.*, 2007). Estimates of the actual costs associated with cases of clinical and subclinical mastitis vary widely, are limited in the literature, and are somewhat herd and pathogen specific. A recent comprehensive review on the economic effects of bovine mastitis and mastitis management, however, helps provide some generalities regarding the economic impact of bovine mastitis on the dairy operation (Halasa *et al.*, 2007). The cost per case of clinical mastitis was estimated at €287 and €102 per case of subclinical mastitis. In a UK study, the estimated annual output losses, treatment costs, and costs of prevention for mastitis were £197.9 million, £79.8 million, and £9.3 million, respectively (Bennett *et al.*, 1999). Earlier estimates from the United States suggest that mastitis costs the US dairy industry in excess of $2 billion per annum, with the cost of an individual case of mastitis being about $185 (Yancey, 1999). These data underline the important economic impact of mastitis on the dairy industry, but when considering the costs of treating cases of mastitis, it must also be considered that treatment is not uniformly efficacious.

Efficacy of antimicrobial treatment varies by bacterial cause with Gram-positive pathogens being more responsive to antibiotic treatment than Gram-negative pathogens (Smith *et al.*, 1993). Of the Gram-positive pathogens, *S. aureus* is generally the most refractory to antimicrobial treatment. Mastitis caused by *Mycoplasma* spp. is considered untreatable and cattle afflicted with mycoplasma mastitis are therefore culled from the herd (Fox *et al.*, 2005). It is generally accepted that antimicrobial treatment of Gram-negative intramammary infection is not efficacious and mastitis caused by these pathogens is usually treated with supportive care. However, antimicrobials may be systemically administered to cattle with severe Gram-negative clinical mastitis to alleviate secondary bacteraemia (Morin *et al.*, 1998; Wenz *et al.*, 2001). Hence, not only are there significant costs associated with treatment, but for some pathogens treatment efficacy is not satisfactory.

For those pathogens that are refractory to antimicrobial treatment, immunization is an attractive alternative control and prevention measure. Vaccination has been adopted for a broad range of animal and human diseases. While efficacy of vaccination tends to vary according to the disease of interest due to factors such as pathogen virulence and immunogenicity, if vaccination is shown efficacious, immunization is generally cheaper than trying to treat an established infection with antimicrobial therapy that may require 2–8 days of therapy and may not be uniformly effective. By example, the cost of three doses of a currently available commercial coliform mastitis vaccine (Upjohn J5 Bacterin, Pfizer, Inc.) is around $US4.00, whereas the cost of treating a case of clinical coliform mastitis was reported to be US$378.15 in one study (Degraves and Fetrow, 1991). Furthermore, Degraves and Fetrow (1991) used partial budget analysis and estimated that if >1% of cow lactations result in clinical coliform mastitis, profits could be increased by $US57.00/cow lactation by using a Gram-negative core antigen coliform mastitis vaccine. Unfortunately, economic analyses of the benefits of vaccination against other mastitis pathogens are lacking, in part due to marginal efficacy.

Vaccines against *S. aureus*, *Mycoplasma* spp., streptococci, and coliforms have been studied for the prevention of IMI, and vaccines against *S. aureus*, *Mycoplasma* spp., and coliforms have

been commercialized. This chapter will discuss these vaccines, their reported efficacy, and possible future developments of vaccines for bovine mastitis.

Contagious mastitis vaccines

The major economic impact of contagious mastitis pathogens is due to increased milk somatic cell count and lost milk production. Because the cow is usually the reservoir of infection, the primary goal of a contagious mastitis pathogen vaccine should be to prevent infection and colonization. Vaccines that decrease the severity of clinical or subclinical mastitis may aid in increasing cow and herd productivity, but ultimately if the vaccine does not prevent infection the pathogen will remain in the herd and spread cow-to-cow.

Mycoplasma mastitis vaccines

Mycoplasma mastitis

Mycoplasma mastitis in cattle can be caused by a variety of mycoplasmal species including *Mycoplasma bovis*, *M. alkelescens*, *M. bovigenitalium*, *M. californicum*, *M. canadense*, *M. bovoculi*, *M. gallinaarum*, *M. canis*, and *M. bovirhinis* (Fox *et al.*, 2005). While mastitis can be manifest by a range of clinical signs including subclinical disease demonstrated by an elevated milk somatic cell count, clinical mastitis, and agalactia, the exact pathogenesis of mycoplasma intramammary infection is not well described. Prevalence of mycoplasma mastitis is hard to estimate because routine culture of milk is not always performed due to the unique growth conditions of these organisms (7–10 days on selective media at 37°C with 10% CO_2; Hogan *et al.*, 1999b). Published prevalence estimates suggest that 1–8% of herds have at least one affected cow (Fox *et al.*, 2005). Mycoplasmas are considered contagious mastitis pathogens, but the mode of infection is not completely understood. Common sites of infection other than the mammary gland include the joints (arthritis) and respiratory tract. Introduction of mastitis into a naïve herd has been assumed to occur through infected purchased cattle. However, evidence exists to suggest that cattle may be asymptomatically colonized with mycoplasma and acquire IMI from other body sites or via haematogenous spread (Fox *et al.*, 2005). Treatment is unrewarding and therefore usually not pursued. Control measures include strict milking time hygiene to aid in preventing contagious spread, segregation of affected animals into a separate milking group that is milked last, and culling of affected cattle (Fox *et al.*, 2005). While these procedures have been found efficacious in some herds they may not be universally successful. Hence, an effective vaccine that prevents intramammary infection and/or prevents asymptomatic colonization of other body sites would be a definite benefit for controlling bovine mastitis caused by mycoplasma as well as potentially other manifestations such as arthritis and respiratory disease.

Mycoplasma mastitis vaccines
One of the earliest applications of vaccination in veterinary medicine was inoculation against contagious bovine pleuropneumonia (CBPP) caused by *Mycoplasma mycoides* subsp. *mycoides*. Pleural fluid or lung tissue from cattle that had died of the disease was inserted subcutaneously in the distal tail of non-diseased cattle as a form of immunization. Inoculated cattle frequently shed the end of their tail, but were immune to CBPP. In their review on mycoplasma vaccines, Nicholas *et al.* (2009) state that 'surprisingly, little progress has been made in developing safe, defined and protective alternatives, the vaccines today still consisting of mildly attenuated strains serially passaged in eggs or in culture'.

Work on vaccines specifically targeted at preventing bovine mastitis is sparse. Boothby and co-workers published a series of studies evaluating a single strain *M. bovis* vaccine in dairy cattle (Boothby *et al.*, 1986a,b, 1987, 1988). Eight lactating dairy cattle were evenly divided into two groups, vaccinates and controls. Vaccinates received formalin killed *M. bovis* strain California 201 subcutaneously in Freund's complete adjuvant (0, 2, and 4 weeks) and subsequently without adjuvant by intramammary infusion (6 and 8 weeks). All cows (vaccinates and controls) were then challenged with live *M. bovis* of the same strain by intramammary infusion at week 12. All challenged quarters became infected postchallenge and the majority of unchallenged quarters in both groups of cattle also became infected

(Boothby *et al.*, 1986a). At the end of study (19.5 weeks), all vaccinates were mycoplasma culture negative, while the majority of controls remained culture positive. However, even though vaccinates became culture negative they had persistently high milk somatic cell counts and low milk yield (Boothby *et al.*, 1986b). Antibody responses were measured in serum and milk whey. The authors reported that *Mycoplasma bovis*-specific serum IgM, IgG, and IgG_2, but not IgA reactivities were markedly increased in vaccinates over control cattle, and challenge exposure with *M. bovis* resulted in high specific serum IgM, IgG1, and IgA responses in vaccinates and controls. The only response in milk whey was an elevated specific IgG_1 reactivity at the time of challenge (Boothby *et al.*, 1987). Likewise, lymphocytes from peripheral blood, but not from the mammary gland, had increased responsiveness to mitogens, and vaccination and challenge resulted in skin test reactivity (Boothby *et al.*, 1988). Hence, while evidence of immunity to vaccination and challenge was documented, immunity at the site of IMI was not robust and might explain the inability of the vaccine to prevent IMI. No statistical analyses of these data were reported.

Others have since studied formalin-inactivated *M. bovis* vaccines in cattle, but the major focus has been bovine respiratory disease (Nicholas *et al.*, 2002; Urbaneck *et al.*, 2000). These studies showed promise by reducing losses from pneumonia and treatment expense (Urbaneck *et al.*, 2000) or by decreasing incidence of pyrexia, lung lesions, weight loss, and systemic spread of the organism. In contrast, a more recent study used purified *M. bovis* antigens as a vaccine and showed an increased severity of pneumonia in vaccinated cattle (Bryson *et al.*, 2002). While these studies and the data generated by Boothby and co-workers (Boothby *et al.*, 1986a,b) illustrate that some *M. bovis* vaccines are immunogenic, they also demonstrate that the stimulated immunity may not protect from infection or potentially exacerbate the infection inflammatory response. Vaccines against mastitis caused by other species of mycoplasma do not appear to have been studied in cattle. However, mycoplasma vaccines, both live attenuated and formalin inactivated, have been studied and are used in various parts of the world for causes of contagious agalactia in sheep and goats. *Mycoplasma agalactiae* is the chief cause of contagious agalactia, but other species of mycoplasma including *M. capricolum* subsp. *capricolum* and *M. mycoides* subsp. *mycoides* can cause similar clinical signs. Vaccines for contagious agalactia appear to have variable efficacy (Nicholas *et al.*, 2009), and there is concern about live attenuated formulations leading to transient shedding of organism thus increasing the likelihood of contagious spread. In addition, shedding may present human health implications in regions were non-pasteurized milk is consumed.

In the USA, there are commercially available *M. bovis* vaccines for use in cattle as well as availability of autogenous preparations. One vaccine is specifically marketed for the prevention of bovine mastitis caused by *M. bovis* (Mycomune™, Bioimmune, Inc.), and is an inactivated bacterin containing multiple strains of *M. bovis*. However, no peer-reviewed efficacy data for this vaccine appears to have been published. Calloway and co-workers studied another commercially available *M. bovis* vaccine in dairy cattle during the non-lactating (dry) period to determine its ability to stimulate antibody responses in serum and colostrum (Calloway *et al.*, 2008). The vaccine used is marketed for prevention of *M. bovis* respiratory disease in cattle (Pulmo-Guard MpB™, Boehringer Ingelheim Vetmedica, Inc.) and according to the manufacturer's information contains antigens from two field isolates of *M. bovis*. Vaccinates had significantly higher serum anti-*M. bovis* IgG$_1$ concentrations after the initial vaccination (60 days prior to calving), booster vaccination (39 days prior to calving), and at calving than controls. Likewise, vaccinates had higher colostral concentrations of anti-*M. bovis* IgG$_1$ than controls. However, IgG$_1$ concentrations in milk did not differ between groups. Efficacy at preventing bovine mastitis was not an aim of the study, but the lack of a significant increase in antibody against *M. bovis* in milk further illustrates the difficulties of stimulating adequate immunity in the mammary gland against *M. bovis*.

Newer technologies for vaccination against bovine or caprine mastitis caused by mycoplasma such as subunit or DNA vaccines do not appear to have been studied, and although such approaches have been applied to other mycoplasmal diseases of livestock none of the studied vaccines have made it to clinical trials or market (Nicholas *et al.*, 2009). At least one study evaluating subunit mycoplasma vaccines in cattle for CBPP demonstrated an exacerbation of disease (Nicholas *et al.*, 2009). Similar findings have been reported with subunit vaccines for *M. pneumoniae*, the cause of atypical pneumonia in people (Nicholas *et al.*, 2009).

Commentary and outlook

Currently there is one commercially available vaccine marketed for mycoplasma mastitis in cattle although some companies offer autogenous vaccines and there is another vaccine marketed for bovine respiratory disease. There is apparently no published efficacy data for the commercially available mastitis bacterin. Efficacy of the other vaccines discussed above is variable and lack of complete prevention of infection seems to prevail. In addition, some of the vaccines discussed may exacerbate the inflammatory response (Boothby *et al.*, 1986a, b, 1987; Boothby *et al.*, 1988) even after clearing the bacteria, which would negatively impact milk quality and yield. Hence, while an efficacious vaccine for preventing mycoplasma mastitis and/or body site colonization would be desirable for the control of mycoplasma mastitis, there does not appear to be a promising candidate vaccine on the horizon.

Staphylococcus aureus vaccines

Staphylococcus aureus *mastitis*
Staphylococcus aureus is another contagious mastitis pathogen of dairy cattle that is primarily spread cow-to-cow during milking either through contact with fomites or the milker's hands (Bramley *et al.*, 1998). The main reservoir of infection is the cow's mammary gland. However, some studies report high prevalences of *S. aureus* IMI in heifers before their first lactation and the strains of *S. aureus* found in pre-partum heifers are frequently the same as those found in lactating cattle (Fox *et al.*, 1995; Middleton *et al.*, 2002; Roberson *et al.*, 1998). While clinical disease occurs, the most common form of *S. aureus* mastitis on modern dairies is chronic subclinical infection manifest

by elevations in milk somatic cell count and decreased milk production. Prevalence estimates suggest that about 10% of dairy cattle in the USA possess a *S. aureus* IMI (Makovec and Ruegg, 2003; Wilson *et al.*, 1997). Antimicrobial treatment outcomes for *S. aureus* mastitis vary by study, chronicity of disease, and length of treatment (Barkema *et al.*, 2006). Treatment success declines with increasing age of the cow, increasing milk somatic cell count, chronicity of infection, increasing bacteria count prior to treatment, and increasing number of quarters infected (Barkema *et al.*, 2006). Control and prevention of *S. aureus* mastitis has been focused on milking time hygiene (including post-milking teat disinfection, single-use udder clothes, and gloves worn by milkers), proper maintenance of milking equipment, culling of chronically infected cattle, appropriate treatment of clinical mastitis, and routine use of intramammary antimicrobials during the non-lactating (dry) period (Dodd *et al.*, 1969; Neave *et al.*, 1969). Segregation of infected cattle and milking them last has also been recommended (Middleton *et al.*, 2001). While these control measures have led to a decline in prevalence of contagious pathogens on many farms they are not always successful (Middleton *et al.*, 2001; Smith *et al.*, 1998). Vaccination has been studied as a control measure for *S. aureus* mastitis for many years and a vaccine for *S. aureus* mastitis has been marketed in the USA since the mid-1970s. However, universal efficacy at preventing IMI has not been demonstrated.

Unlike mycoplasma, the pathogenesis of *S. aureus* IMI has been extensively studied and this in depth understanding of *S. aureus* mastitis pathogenesis has been used to elucidate potential vaccine targets. Infection of the mammary gland begins with *S. aureus* entering the teat orifice, traversing the keratinized teat canal, and moving into the mammary gland parenchyma (Sutra and Poutrel, 1994). Once in the gland, successful infection requires adhesion of *S. aureus* to host cells or extracellular matrix proteins via expressed molecules such as fibronectin binding protein, clumping factor, collagen binding protein, and teichoic acid (Shinefield and Black, 2005). Neutrophils are the primary phagocyte responsible for clearing *S. aureus* from the mammary gland

(Paape *et al.*, 1981). However, in milk, neutrophils have impaired phagocytic function due to the presence of fat globules (Kent and Newbould, 1969; Paape *et al.*, 1975) that leads to impaired *S. aureus* clearance. Phagocytosis is facilitated by opsonization with immunoglobulins and complement, and studies in dairy cattle show that IgG_2 and IgM are the most important opsonins for protection against *S. aureus* IMI (Barrio *et al.*, 2003; Guidry *et al.*, 1993; Watson, 1976). Data from dairy cattle mastitis also demonstrates a delayed innate immune response, based on cytokine profiles, compared with *Escherichia coli* IMI suggesting that different bacteria elicit different *in vivo* responses and providing one explanation for chronic *S. aureus* IMI (Bannerman *et al.*, 2004).

Staphylococcus aureus produces several factors that allow it to evade host immunity including protein A, an extracellular polysaccharide capsule, pseudocapsule, and some exotoxins (e.g. TSST-1) (Hermans *et al.*, 2004; Shinefield and Black, 2005). Protein A disrupts antibody mediated opsonization and phagocytosis of *S. aureus* by binding the Fc rather than Fab portion of IgG (Hermans *et al.*, 2004). Extracellular polysaccharide capsule interferes with phagocytosis by masking opsonizing antibody bound to the bacterial cell wall and preventing the activation of complement (Peterson *et al.*, 1978a; Peterson *et al.*, 1978b). Watson and co-workers discovered that *S. aureus* strains cultured from bovine milk had the potential to produce a pseudocapsule that had antiphagocytic properties (Watson, 1992). Certain *S. aureus* exotoxins, such as TSST-1, have the potential to act as superantigens by directly binding major histocompatibility complex class II with T-cell receptors leading to a massive, uncoordinated proliferation of T-cells and release of proinflammatory cytokines (Shinefield and Black, 2005).

Over the years, many of the factors involved in *S. aureus* pathogenesis have been investigated as vaccine targets for both bovine mastitis and human infection.

Current state of S. aureus *mastitis vaccination*
The simplest *S. aureus* vaccines are whole cell or whole cell lysates with or without adjuvants. Whole cell lysate bacterins for the prevention

of bovine mastitis (Brock *et al.*, 1975; Lee *et al.*, 2005; Leitner *et al.*, 2003b,c; Luby *et al.*, 2007; Middleton *et al.*, 2006; Williams *et al.*, 1966, 1975) and autogenous bacterins of similar formulation (Sears and Belschner, 1999) have been extensively studied. One such vaccine (Lysigin™ Boehringer Ingelheim Vetmedica, Inc.) has been commercially available for use in dairy cattle since the mid-1970s, and is a polyvalent whole cell lysate vaccine containing five strains of *S. aureus* (one serotype 5, two serotype 8, and two serotype 336 strains, the three predominant serotypes isolated from cases of bovine mastitis; Ma *et al.*, 2004). Efficacy has been evaluated in a number of studies (Nickerson *et al.*, 1999; Pankey *et al.*, 1985; Williams *et al.*, 1966; Williams *et al.*, 1975), and results have demonstrated decreased clinical severity of mastitis, lower milk SCC, increased spontaneous cures, and in one study (Williams *et al.*, 1966) a significant reduction in new IMI in two of three herds studied.

More recently Lysigin has been evaluated for prevention of mastitis in dairy heifers as staphylococci are the most common pathogens isolated from the udder of primiparous dairy cattle in the peripartum period (Fox *et al.*, 1995). The efficacy of Lysigin was compared with two experimental vaccine formulations and unvaccinated controls in primiparous heifers (Luby *et al.*, 2007; Middleton *et al.*, 2006). Heifers were vaccinated twice, 28 days apart in late gestation with either a three-isolate experimental bacterin (Group I; $n = 11$), a five-isolate experimental bacterin (Group II; $n = 11$), or commercially available Lysigin (Group III; $n = 14$). Group IV consisted of 11 unvaccinated control cattle. All groups (vaccinates and controls) were challenged with a heterologous strain of *S. aureus* by intramammary infusion on days 6, 7, and 8 of lactation in a single infection-free mammary quarter. All cattle became infected with *S. aureus* after challenge. While three cattle in Group I, and one cow in Group III, cleared their *S. aureus* IMI by the end of the study, there were no significant differences in *S. aureus* clearance rates between groups. However, similar to previous studies, cattle vaccinated with Lysigin had a lower mean duration of clinical mastitis and lower total mastitis score post-challenge than controls. Overall, there was no evidence that any

of the vaccinated groups had a lower mean SCC than control, and no evidence that vaccinates had greater milk yield than controls post-challenge. Heifers vaccinated with Lysigin had higher mean serum IgG_1 and IgG_2 sample-to-positive (S:P) ratios than controls against *S. aureus* strains of serotypes 5, 8, and 336. Milk total anti-*S. aureus* IgG S:P ratios were only different from controls against strains of serotype 8 and 336, and there were no significant differences between groups for milk IgG_1, IgG_2, and IgM (Luby *et al.*, 2007). Very similar results were found in a follow-up study utilizing lactating dairy cattle receiving two doses of Lysigin according to the manufacturer's directions (Middleton *et al.*, 2009). No animals in either group, vaccinates ($n = 44$) or controls ($n = 46$) developed a new *S. aureus* IMI after vaccination. The numbers of mammary quarters that developed a new coagulase-negative staphylococcal (CNS) IMI, time to new CNS IMI, milk somatic cell count, and milk antibody isotype (IgM, IgG_1, IgG_2, IgA) s:p ratio did not significantly differ between groups. In a herd with a 3% prevalence of *S. aureus* IMI and a 30% prevalence of CNS IMI, the vaccine did not reduce the new staphylococcal IMI rate. Together, these data suggest that there may be insufficient vaccine-induced opsonizing antibody in milk to facilitate clearance of *S. aureus* from the mammary gland.

A similar approach to vaccination of dairy heifers was studied by Lee and co-workers (Lee *et al.*, 2005), who characterized the immune response of dairy heifers following immunization with a trivalent *S. aureus* mastitis bacterin containing formalin-inactivated *S. aureus* strains belonging to serotypes 5, 8, and 336 with and without adjuvants. All vaccine formulations stimulated antigen-specific IgG_1 and IgG_2 production in serum, and formulations with adjuvant, either alum or Freund's incomplete, stimulated more IgG_2 than those without adjuvant. In addition, immune sera from vaccinated cattle seemed to increase *in vitro* neutrophil phagocytosis of the 3 serotypes of killed *S. aureus*, although differences generally were not statistically significant. Unfortunately, Lee and co-workers did not characterize mammary immunity or efficacy against *S. aureus* IMI (Lee *et al.*, 2005).

Finally, Nickerson and co-workers (1999) vaccinated heifers with commercially available Lysigin at 6-months of age followed by a booster dose 2 weeks later and subsequent vaccinations every 6 months until calving. Vaccinates had a 45% reduction in both new *S. aureus* IMI during pregnancy and new *S. aureus* IMI at calving relative to controls. In addition, vaccinates had a 30% reduction in coagulase negative staphylococcal (CNS) IMI which became chronic and a 31% reduction in new CNS IMI at calving relative to controls, thus providing evidence that Lysigin may be of use in reducing staphylococcal mastitis in periparturient heifers (Nickerson *et al.*, 1999). These data suggest that multiple immunizations prior to first calving may be a useful adjunct in controlling staphylococcal mastitis in heifers, but larger field studies in multiple herds are needed.

Another *S. aureus* bacterin for dairy cattle mastitis has been patented in Israel (MASTIVAC) (Leitner *et al.*, 2003a–c). MASTIVAC comprises three field strains of *S. aureus* and contains bacterial homogenates and exosecretions. Experimental challenge trials showed that 19 of 21 mammary quarters in non-vaccinated cattle developed a persistent *S. aureus* IMI after challenge while only 9 of 17 mammary quarters developed an IMI post challenge in the vaccinated group (Leitner *et al.*, 2003a). The researchers did not report overall cow IMI rates between groups which seems most relevant because if the vaccine is administered to the cow not the quarter it should prevent the cow from becoming infected. If one quarter remains infected the cow can still be a contagious reservoir for infection of herdmates. Leitner and co-workers also studied MASTIVAC under field conditions and found that 3 of 228 cows had a *S. aureus* IMI in the vaccinated group versus 6 of 224 cows in the unvaccinated control group (Leitner *et al.*, 2003c). No statistical analyses were performed. Vaccinated cattle in the field trial had about a 50% reduction in milk SCC over non-vaccinates. Given the low prevalence of *S. aureus* IMI in the field trial study herd and the lack of individual cow data from the challenge study, it is difficult to draw any significant conclusions about MASTIVAC.

Hence, whole organism inactivated vaccines for bovine mastitis have met with varied success, and while some studies report reduced clinical severity of disease and/or reduction in new IMI rates, none of the vaccines studied to date have completely prevented new IMI in dairy cattle. In contrast, a recent case report in the human medical literature documented that autovaccination with a formalin-inactivated methicillin-resistant *S. aureus* (MRSA) isolated from a patient with post-operative mediastinitis that was unresponsive to antibiotic therapy resolved after repeated dosing with the autogenous bacterin (Rizzo *et al.*, 2007). Similar autogenous preparations have been shown to increase the efficacy of antibiotic therapy for mastitis in dairy cattle, but results tend to vary from herd-to-herd presumably due to differences in virulence of the endemic strains of *S. aureus* found in each herd (Sears and Belschner, 1999).

Recent and future developments in S. aureus vaccination

Given that *S. aureus* is an important pathogen of humans as well as being an important cause of bovine mastitis and the only commercially available vaccine has proven less than optimal at preventing infection, there have been a number of recent studies evaluating alternative strategies to *S. aureus* infection. Several strategies have been employed including evaluation of surface, capsule, and cell wall antigens and evaluation of toxoids as well as other novel strategies detailed below.

Surface, capsular, and cell wall-associated antigen vaccines As many as 12 polysaccharide serotypes have been described in the literature (Guidry *et al.*, 1998; Karakawa and Vann, 1982; Sompolinsky *et al.*, 1985), 11 capsular (1–11) and a surface polysaccharide (336). However, only polysaccharides of serotypes 1, 2, 5, and 8 have been purified and characterized (Jones, 2005; O'Riordan and Lee, 2004). The majority of *S. aureus* isolated from bovine mastitis cases belong to serotypes 5 and 8 with a proportion of isolates being non-typeable (Guidry *et al.*, 1997; Sompolinsky *et al.*, 1985). Guidry and co-workers reported that the non-typeable bovine isolates belonged to serotype 336 (Guidry *et al.*, 1998), but the uniqueness of the serotype 336 antigen has been questioned (Kenny and Tollersrud, 1999).

Recently, the 336 antigen was characterized as polyribitol-phosphate-N-acetylglucosamine, a component of cell wall teichoic acid (Verdier *et al.*, 2007). Furthermore, Cocchiaro and co-workers (Cocchiaro *et al.*, 2006) present data suggesting that the non-typeable isolates possess mutations in the *cap5(8)* locus and therefore do not express capsule antigens, and conclude that there are likely only four unique polysaccharide types of *S. aureus* (1, 2, 5, and 8).

Purified capsular polysaccharide antigens have been investigated as vaccines. By themselves capsular polysaccharide antigens are poorly immunogenic, and therefore recent investigations have evaluated these antigens conjugated to proteins. Tollersrud and co-workers compared antibody responses of a purified capsular polysaccharide (CP) 5 conjugated to human serum albumin (CP5-HSA) with a whole cell lysate bacterin in Norwegian Red Cattle between 3 and 11 months of age (Tollersrud *et al.*, 2001). Unfortunately, cattle vaccinated with the whole cell bacterin exhibited a greater mean level and duration of anti-CP5 antibody than those vaccinated with the CP5–HSA conjugate vaccine. A trademarked vaccine (StaphVAX™, Nabi Biopharmaceuticals, Inc.) containing purified CP5 and CP8 antigens conjugated to mutant non-toxic recombinant *Pseudomonas aeruginosa* exotoxin A (rEPA) has been studied in Phase III clinical trials for prevention of *S. aureus* infection in humans (Fattom *et al.*, 2004a; Fattom *et al.*, 2004b; Robbins *et al.*, 2004; Shinefield *et al.*, 2002). Similar to studies in the bovine using whole cell lysate vaccines, the CP5/CP8 conjugate vaccine was found to be safe and immunogenic, but overall a significant reduction in the occurrence of *S. aureus* bacteraemia in human haemodialysis patients was not detected (Fattom *et al.*, 2004b). Post-hoc analysis of the data demonstrated a significant difference between vaccinates and controls in the occurrence of bacteraemia through 40 weeks post-vaccination, but the importance of this difference in a clinical setting is questioned. Development of the vaccine seems to have stalled based on the unfavourable outcome of the Phase III clinical trials. Based on the company's website (http://www.nabi.com/pipeline/pipeline.php?id=1) further development of the vaccine to include

serotype 336 antigen along with detoxified toxins is being pursued, but to date a commercially available formulation has not been marketed. Given that the serotype 336 antigen is likely a cell wall component, the importance of its inclusion in a *S. aureus* vaccine is debatable especially in light of the fact the majority of *S. aureus* strains are encapsulated *in vivo* (Sompolinsky *et al.*, 1985) and thus cell wall antigens, even if bound by antibody, are likely masked from immune surveillance by capsule.

Surface proteins referred to as microbial surface components recognizing adhesive matrix molecules (MSCRAMMs) or extracellular matrix binding proteins (ECMBPs) also have been investigated as vaccine targets. Bacterial adhesion to host extracellular matrix components is a key component of *S. aureus* infection. Important factors that facilitate this binding include clumping factors A and B (ClfA, ClfB), fibronectin binding proteins A and B (FnBPA, FnBPB), and collagen binding protein (Cna). A recent study in dairy cattle evaluated a series of DNA vaccines directed against ClfA and demonstrated a strong and specific antibody response in vaccinated groups relative to control cattle that received only the expression vector (Nour El-Din *et al.*, 2006). When *S. aureus* were pre-incubated with sera or milk from vaccinated cattle there was a reduction in the ability of the *S. aureus* to adhere to MAC-T cells *in vitro* compared with preincubation with sera or milk from cattle that were only injected with the expression vector (Nour El-Din *et al.*, 2006). However, no *in vivo* efficacy data were reported. More recently the same group evaluated protective immune responses against *S. aureus* in a murine model using a multi-gene DNA vaccine (Gaudreau *et al.*, 2007). A multivalent polyprotein DNA vaccine was formulated against ClfA, FnBPA, and the enzyme sortase (Srt) and compared with monovalent DNA vaccines against each of the individual proteins and a plasmid vector control. When mice were challenged with a virulent strain of *S. aureus*, 55% of vaccinates survived compared with 15% of control mice. However, all surviving mice developed arthritis. Additionally, in the virulent *S. aureus* challenge trial, the monovalent vaccines were not compared with the multivalent vaccine making it difficult to determine the importance of each antigen. The authors did compare the monovalent and multivalent vaccines with control using a less virulent *S. aureus* challenge and found that all mice in the multivalent vaccine group were protected from arthritis, while at least some of the mice in the other groups developed arthritis, but there were no significant differences between any vaccinated group and control. Hence, this approach to vaccination does not appear promising especially in light of other studies by the same group that demonstrate a lack of protection against intraperitoneal challenge with *S. aureus* in mice using DNA vaccines against ClfA and Cna, respectively (Brouillette *et al.*, 2002; Therrien *et al.*, 2007).

Schaffer and co-workers evaluated mucosal immunization with killed *S. aureus* grown under conditions to maximally express ClfB and systemic and mucosal vaccination with recombinant ClfB (rClfB) versus control in mice and found that while there was a significant reduction in colony forming units (CFU) of *S. aureus* isolated from the nose, carriage rates did not differ between treated and control groups (Schaffer *et al.*, 2006). These data suggest antibodies targeting ClfB are not sufficient to prevent nasal colonization, but may reduce nasal shedding by reducing intranasal bacterial loads.

Zhou and co-workers evaluated the immunogenicity of a Cna–FnBP fusion protein against *S. aureus* in a mouse model (Zhou *et al.*, 2006). The Cna–FnBP fusion protein stimulated a stronger humoral immune response, enhanced *in vitro* opsonophagocytosis, and decreased postchallenge mortality in vaccinates versus controls. However, 3 of 12 vaccinated mice still succumbed following intraperitoneal challenge with *S. aureus*, and results were not different from mice that were vaccinated with Cna or FnBP alone. In a previous study, Nelson and co-workers vaccinated dairy cattle with a FnBP, staphylococcal protein A fusion protein (zz-FnBPA) twice during the non-lactating period and once during lactation one week before intramammary bacterial challenge with 1000 CFU of *S. aureus*, and found no difference between vaccinates and unvaccinated controls with regard to bacterial shedding and milk somatic cell count (SCC) (Nelson *et al.*, 1990). When the fusion protein was conjugated

with immunostimulating complexes (ISCOMs) they noted higher antibody titres following immunization and no cases of clinical mastitis post-intramammary-challenge with 10,000 CFU of *S. aureus* in vaccinates compared with 3 of 5 cases (60%) of clinical mastitis in unvaccinated controls. Milk SCC was also lower than control immediately following challenge. However, the authors did not report whether *S. aureus* were recovered from the challenged mammary quarters of vaccinates post-challenge and milk SCCs were similar in both groups by day 17 postchallenge. Hence, the ability of the vaccine to prevent infection cannot be determined from the data presented.

Other surface molecules recently investigated as vaccine targets include iron-regulated surface determinants (Isd) and serine-aspartate repeat (Sdr) proteins (Kuklin *et al.*, 2006; Stranger-Jones *et al.*, 2006). Kuklin and co-workers reported that IsdB, an iron sequestering protein, is highly conserved among a diverse population of MRSA and methicillin-susceptible (MSSA) *S. aureus* isolates (Kuklin *et al.*, 2006). Purified IsdB with an aluminium hydroxyphosphate sulphate adjuvant was highly immunogenic in mice and rhesus macaques. Additionally, in a mouse sepsis model, the vaccine conferred a significant reduction in mortality relative to control mice immunized with adjuvant alone (Kuklin *et al.*, 2006). However, approximately 30–70% of the vaccinated mice succumbed to sepsis by day 10 post-challenge depending on challenge strains and *S. aureus* dose (Kuklin *et al.*, 2006). A slightly more recent study evaluated a multivalent vaccine containing IsdA, IsdB, SdrD, and SdrE antigens in a mouse challenge model (Stranger-Jones *et al.*, 2006). The authors tested 19 conserved surface proteins (including surface protein A, FnBPs, Clf, Sdr, Sas, HarA, Isd, and Aap) found in the genome sequences of 8 *S. aureus* strains, and identified 4 polypeptides that produced the greatest protective immunity, i.e. IsdA, IsdB, SdrD, and SdrE. Purified proteins were mixed with adjuvant and administered to mice twice 11 days apart. Mice were challenged 10 days after the second vaccination with various *S. aureus* isolates by intraperitoneal injection. Mice receiving the combined vaccine had significantly greater survival at 7 days post challenge compared with saline-treated controls, and in three of the experiments (*S. aureus* strains Newman, NRS248 and USA 400) 100% of the mice in the combined vaccination group survived to 7 days post-challenge. Three of antigens in the vaccine (IsdB, SdrD, and SdrE) are known to be immunogenic in humans because antibodies to these antigens have been detected in healthy people and people with *S. aureus* disease (Stranger-Jones *et al.*, 2006). Hence, while these results need to be interpreted with caution because mouse models do not often correlate with human infection or bovine mastitis, combining a series of conserved *S. aureus* surface antigens with other conserved *S. aureus* virulence factors may prove useful in future vaccine development.

Methicillin-resistant *S. aureus* produce an additional penicillin binding protein (PBP) referred to a PBP2a (PBP2') that is coded by the *mec*A gene and has a low affinity for beta-lactam antibiotics. Researchers in Brazil have investigated a DNA vaccine containing a fragment of the *mec*A gene that generates antibodies specifically targeted at a portion of PBP2a (Roth *et al.*, 2006; Senna *et al.*, 2003). The vaccine caused a significant reduction in bacterial loads in the kidney of mice challenged with a sublethal dose of MRSA, while commensal Gram-positive flora were not affected by the vaccine. Similar results were noted previously by Ohwada and co-workers using a DNA vaccine containing the *mec*A sequence (Ohwada *et al.*, 1999). While MRSA is infrequently isolated from the cow's mammary gland, these data provide some evidence that DNA vaccines targeting specific virulence factors may be of use in future *S. aureus* vaccine development.

Toxoids Staphylococcus aureus produces a number of exotoxins that either directly cause disease or facilitate tissue penetration and immune cell recruitment. A few recent studies have focused on *S. aureus* toxins as vaccine targets (Adlam *et al.*, 1977; Bubeck Wardenburg *et al.*, 2007; Bubeck Wardenburg and Schneewind, 2008; DeLeo and Otto, 2008; Menzies and Kernodle, 1996).

Bubeck Wardenburg and Schneewind investigated the role of *S. aureus* α-haemolysin (α-toxin; Hla), a pore forming β-barrel toxin, in the pathogenesis of *S. aureus* pneumonia using mouse and *in vitro* models, and tested a non-pore-forming mutant (Hla$_{H35L}$) as an immunogen for the protection of *S. aureus* pneumonia (Bubeck Wardenburg and Schneewind, 2008). Their study demonstrated that Hla was required for the pathogenesis of pneumonia and demonstrated an association between Hla expression and virulence. Active immunization of mice with Hla$_{H35L}$ generated antigen-specific IgG and caused a significant decrease in mortality of mice post-intranasal-challenge relative to saline treated controls. Decreased mortality was correlated with a decrease in bacteria recovered from the lungs 24 h after infections and fewer lung lesions on histopathological examination. Furthermore, passive immunization of mice with rabbit anti-Hla significantly reduced mortality due to pneumonia compared with mice treated with control serum. As part of their work, Bubeck Wardenburg and Schneewind also found that rabbit antisera raised against Panton-Valentine leukocidin (PVL) did not confer protection against *S. aureus* pneumonia in mice suggesting that PVL is dispensable for bacterial virulence, and therefore may not be useful as a vaccine antigen (Bubeck Wardenburg and Schneewind, 2008). While these data suggest that, at least in the model used, Hla appears to be an important virulence determinant of mortality in cases of *S. aureus* pneumonia and mutant Hla presents a promising vaccine target for *S. aureus* pneumonia in humans (Bubeck Wardenburg and Schneewind, 2008; DeLeo and Otto, 2008), it should be noted that the same authors previously reported that *S. aureus* with mutations in the sortase A (*srtA*) and protein A (*spa*) genes also had attenuated virulence (Bubeck Wardenburg *et al.*, 2007). Earlier studies (Adlam *et al.*, 1977; Hume *et al.*, 2000; Menzies and Kernodle, 1996) using purified alpha-toxin in rabbit and mouse models either by passive or active immunization demonstrated that high circulating alpha-toxin antibody levels reduced the lethal effects of alpha-toxin, but did not prevent abscess formation (Adlam *et al.*, 1977; Menzies and Kernodle, 1996) or reduce bacterial loads (Hume *et al.*, 2000) suggesting that antibody blocking of alpha-toxin may be important in preventing severe disease, but less useful in preventing sublethal disease or colonization.

Another *S. aureus* toxin that has been the focus of renewed interest as a vaccine target is TSST-1. Hu and co-workers vaccinated mice with a non-toxic mutant TSST-1 (mTSST-1) and subsequently challenged the mice with live *S. aureus* (Hu *et al.*, 2003). Vaccinates had a higher survival rate and significantly lower bacterial counts in organs than controls (Hu *et al.*, 2003). However, about 40% of vaccinates still succumbed to infection by day 11 post-challenge. Passive transfer of rabbit serum containing mTSST-1-specific antibodies was also shown to confer protection against *S. aureus* sepsis in mice but, similar to actively immunized mice, about 80% of mice receiving anti-mTSST-1 antibodies died by day 12 post-challenge. Immunized mice had lower titres of interferon gamma (IFN-γ) and higher titres of interleukin-10 (IL-10) than control mice following *S. aureus* infection. Interferon gamma is thought to play an important role in the pathogenesis of *S. aureus* infection, whereas IL-10 is thought to play a role in protecting the host from *S. aureus* exotoxin-mediated shock. Hence, the authors suggest that vaccination with mTSST-1 may positively influence cytokine production aiding in the amelioration of disease (Hu *et al.*, 2003). In a more recent study the same group showed that a fusion protein composed of glutathione S-transferase and mTSST-1 (GST-mTSST-1) conferred protection against *S. aureus* challenge and immune sera from vaccinated mice inhibited IFN-γ and tumour necrosis factor alpha (TNFα) production by murine splenocytes after *in vitro* stimulation with TSST-1 (Cui *et al.*, 2005). Hu and co-workers have also investigated a double mutant staphylococcal enterotoxin C (dmSEC) that was devoid of superantigenic activity as an intranasal vaccine for protection against *S. aureus* challenge in mice (Hu *et al.*, 2006). Similar to their findings with mTSST-1 vaccination, vaccination with dmSEC reduced mortality and bacterial counts in organs over controls, but by day 13 post-challenge over 50% of the mice in the vaccinated group had succumbed. By comparison, however, 100% of the controls had died by day 13 post-challenge. Vaccinates exhibited significantly higher titres of IL-4 and IL-10 and reduced titres of IFN-γ also similar to the mTSST-1 studies. In this latter study, mice were only challenged with

an SEC secreting strain of *S. aureus* and hence it cannot be determined whether the vaccine would have similar protective effects against a non-exotoxin secreting strain of *S. aureus*. The importance of TSST-1 as a vaccine antigen for human toxic shock syndrome is questioned because a large proportion (85%) of the at risk population (menstruating women) have existent circulating antibody titres to TSST-1 (Parsonnet *et al.*, 2005). Furthermore, 98% of women between the ages of 13 and 40 that are carriers for *S. aureus* with TSST-1 have positive antibody titres (Parsonnet *et al.*, 2005).

Another group (Chang *et al.*, 2008), recently studied a recombinant SEC mutant in lactating dairy cattle and found that vaccinated cattle had lower milk SCC concentrations, and lower numbers of IMI post-challenge than unvaccinated controls. The authors stated that the difference in the rate of *S. aureus* isolation between vaccinated (0 of 9 quarters) and control (6 of 8 quarters) cattle was significant, but no *p*-value was stated. Critical evaluation of their data using a Fisher's exact test reveals $P = 0.048$, i.e. marginally significant. In addition, only three cattle were evaluated in each group, cattle were only studied for 13 days post challenge, and the challenge strain was fairly innocuous causing a milk SCC of 409,000 cells/ml in the cow of origin. Given the culture inoculum size (0.025 ml), it is plausible that vaccinates were infected with *S. aureus*, but the infectious burden was below the detection limit of the culture technique. Hence, if the cows had been followed for a longer period of time or a higher culture inoculum had been studied, recrudescent infections may have been documented as seen in another study (Middleton *et al.*, 2006). Vaccinates had higher serum anti-SEC antibody levels, but milk anti-SEC antibody levels were not quantified. Furthermore, similar to the study by Hu and others (2006) noted above, cattle were only challenged with an exotoxin-secreting strain of *S. aureus* making it impossible to determine the effect of the vaccine on other non-exotoxin-secreting strains of *S. aureus*. The negative controls were vaccinated with PBS, but the vaccine contained a carboxy-methyl cellulose-sodium carrier. Hence, the influence of this carrier on immunity cannot be assessed because a carrier inoculated control group was not studied. Therefore, data from a larger number of cattle with appropriate control groups, preferably in a field trial in herds with a high prevalence of *S. aureus* mastitis, will be needed before meaningful conclusions can be drawn about the efficacy of this vaccine.

Alternative vaccine targets

As illustrated thus far, most vaccine preparations have focused on a limited repertoire of antigens. Some recent studies have taken a more global approach to vaccination by targeting conserved antigens produced by multiple strains of *S. aureus* (Balaban *et al.*, 1998; Goji *et al.*, 2004; Kerro-Dego *et al.*, 2006; Perez-Casal *et al.*, 2006; Yang *et al.*, 2005).

Goji and co-workers isolated two surface-located proteins, GapB and GapC, which have homology to glyceraldehyde-3-phosphate dehydrogenase (GAPDH) and appear to be conserved across a number *S. aureus* strains (Goji *et al.*, 2004). Glyceraldehyde-3-phosphate dehydrogenase proteins have been investigated as antigens in vaccines against microbial infections including *Streptococcus uberis* IMI in dairy cattle (Fontaine *et al.*, 2002). Recently, Perez-Casal and co-workers (Perez-Casal *et al.*, 2006) reported that a *S. aureus* GapC/B chimera elicited strong humoral and cellular immune responses in mice based on elevated levels of antigen-specific IgG, IgG_1, IgG_{2a} and numbers of IL-4 and IFN-γ-secreting cells over placebo treated controls. The same group more recently explored the use of DNA vaccination against the GapB and GapC proteins isolated from a bovine mastitis *S. aureus* strain (Kerro-Dego *et al.*, 2006). Immunization with plasmids encoding either GapB, GapC, or GapC/B did not elicit a significant humoral immune response and only elicited a low cell-mediated immune response after three immunizations in mice (Kerro-Dego *et al.*, 2006). However, significant humoral immune responses were noted if mice were boost immunized with recombinant Gap proteins, but the authors concluded that the observed immune responses were probably the result of the recombinant protein alone and the plasmid vaccines likely had no priming effect (Kerro-Dego *et al.*, 2006). Hence, it remains to be determined whether GAPDH homologues will

be successful as conserved antigen vaccines for *S. aureus* infection.

Exoprotein virulence factor secretion by *S. aureus* occurs in the post-exponential growth phase and is regulated by a quorum-sensing mechanism coded for, in part, by the *agr* locus. Exoprotein virulence factor synthesis is regulated by an RNA, termed RNAIII. Several peptides have been discovered as part of this regulatory mechanism including RNAIII-activating protein (RAP), target of RAP (TRAP), and RNAIII inhibiting protein (RIP) (Balaban *et al.*, 1998; Yang *et al.*, 2005; Yang *et al.*, 2006), and they have been studied as vaccine antigens. Balaban and co-workers found that 28% of mice vaccinated with purified RAP developed post-challenge skin lesions versus 70% of control mice in a *S. aureus* cutaneous infection model (Balaban *et al.*, 1998). Additionally vaccinates that developed lesions had smaller lesions than controls and only 3% of vaccinates died versus 22% of controls. In the same study, they also found that mice treated with purified RIP and challenged with *S. aureus* were protected from infection relative to controls, but the results were dependent on the ratio of RIP to *S. aureus* challenge dose (Balaban *et al.*, 1998). Yang and co-workers have recently explored TRAP, as a *S. aureus* vaccine target (Yang *et al.*, 2005; Yang *et al.*, 2006). By screening a phage display library with anti-TRAP antibodies, Yang and co-workers identified a peptide (TA21) as a potential vaccine target (Yang *et al.*, 2005). Mice vaccinated with an *Escherichia coli* engineered to express TA21 (FTA21) on its surface were protected from *S. aureus* infection in a sepsis model (0% mortality in FTA21 vaccinated mice versus 100% mortality in the controls) and a cellulitis model. All groups developed *S. aureus* subcutaneous lesions, but FTA21 vaccinates had a 70% reduction in lesion size versus controls (Yang *et al.*, 2005). Furthermore, it appears that TA21 is highly conserved among *S. aureus* and *Staphylococcus epidermidis* isolates (Yang *et al.*, 2005). In a subsequent study, the same group identified a mimotope of RAP and evaluated it as an immunogen for the protection of *S. aureus* infection in mice (Yang *et al.*, 2006). Mice were vaccinated with *E. coli* that expressed the mimotope on their surface or wild-type *E. coli* and studied in both sepsis and cellulitis models. Similar to the previous study, 100% of the mimotope vaccinated mice survived and 100% of the controls died by day 7 in the sepsis model, whereas all mice regardless of group developed lesions in the cellulitis model but mimotope immunized mice had significantly smaller lesions than controls (Yang *et al.*, 2006). Hence, as with other antigens, the vaccine had efficacy in decreasing the severity of disease, but not in preventing infection.

Others have recently evaluated avirulent mutants of *S. aureus* as immunogens to protect against *S. aureus* infection (Bogni *et al.*, 1998; Buzzola *et al.*, 2006; Pellegrino *et al.*, 2008). Pellegrino and co-workers tested an avirulent *S. aureus* mutant (generated by chemical mutagenesis) in dairy heifers using an intramammary challenge model with a homologous virulent strain of *S. aureus* (Pellegrino *et al.*, 2008). While the avirulent strain was immunogenic, no significant differences were found between vaccinates and controls with regard to bacterial shedding in milk or milk somatic cell count. Interestingly, the same mutant was shown to increase the *S. aureus* LD_{50} over controls in a mouse sepsis model when mice were challenged with the homologous *S. aureus* strain (Bogni *et al.*, 1998). The same increase in LD_{50} was not, however, observed when the mice were challenged with heterologous strains (Bogni *et al.*, 1998). These findings not only illustrate the differential specificity of the vaccine, but also demonstrate that vaccine studies using mice models often do not reflect outcomes in larger mammals. Buzzola and co-workers studied an avirulent *aroA S. aureus* mutant in mice and found that the mutant had a higher LD_{50} (1×10^6 CFU/mouse) than the parent strain (4.3×10^4 CFU/mouse) and a decreased ability to persist in the lungs, spleens, and mammary glands of mice (Buzzola *et al.*, 2006). Intramammary immunization of mice stimulated Th1 and Th2 responses in the mammary gland and reduced bacterial shedding from glands challenged with either the parental strain or a heterologous strain (Buzzola *et al.*, 2006). While the vaccine did not prevent IMI in mice, mammary glands of immunized mice did not exhibit cellular infiltration or vascular congestion suggesting that the vaccine ameliorated the inflammatory response.

Commentary and outlook

While it is clear that there are technologies emerging for vaccinating against *S. aureus* infection, there are currently no commercially available vaccines for human infection and the available vaccine for prevention of bovine mastitis has inconsistent results. The complexity of the organism and diversity of strains which cause infection have made the discovery of a pluripotent staphylococcal vaccine difficult, if not impossible. Attempts at defining a 'core antigen' for *S. aureus* have not yet yielded a commercially viable target. It is likely that the best vaccine will need to possess an array of *S. aureus* virulence factors or block expression of major pathogenic determinants. That said, previous reviews on *S. aureus* vaccination have stated these same goals, but most research groups continue to focus efforts on one or two antigens. In this chapter, it is apparent that some novel approaches to *S. aureus* vaccination are being taken, but the majority of studies are still focused on antigens that have already received a great deal of attention without a fruitful outcome.

A recent focus of *S. aureus* vaccination has been generation of antibody to capsular or surface polysaccharide antigens including CP5, CP8, and antigen 336 with the concept of increasing opsonophagocytosis. DeLeo and Otto suggest that intracellular survival of *S. aureus* within phagocytes may be one reason that vaccines targeting capsular or surface polysaccharides fail to prevent infection (DeLeo and Otto, 2008). Furthermore, they provide evidence that phagocytes may be a vector for bacterial dissemination (DeLeo and Otto, 2008) thus calling into question whether vaccines that enhance opsonophagocytosis are the best approach to protecting against *S. aureus* infection. A large proportion of *S. aureus* strains seem to express capsule *in vivo*. If capsule masks cell wall antigens and thus interferes with complement-mediated opsonophagocytosis, then vaccine strategies targeting cell wall antigens such as Clf and FnBP have to be questioned as to their efficacy *in vivo*, at least against encapsulated strains. For example, Risley and co-workers recently demonstrated that expression of capsular polysaccharide inhibited ClfA-mediated binding of *S. aureus* to fibrinogen and platelets suggesting ClfA may not be a viable vaccine antigen when trying to prevent infection with *S. aureus* that express capsule *in vivo* (Risley *et al.*, 2007).

Some recent studies have taken a novel approach to identifying putative vaccine antigens by screening for antigens using sera from healthy colonized, healthy non-colonized, and ill humans. Dryla and co-workers compared anti-*S. aureus* antibody repertoires in healthy people who were either colonized or not colonized with *S. aureus* and people with acute infections and noted marked differences in antibody repertoires between the three populations (Dryla *et al.*, 2005). Infected patients had higher antibody levels in sera directed at SdrD, HarA, FnBPA, Enolase, EbpS, and SA0688 leading the authors to conclude that these antigens are expressed early in infection and responded to by the host's immune system. The authors also found that anti-staphylococcal antibody levels were stable over a 12-month period in healthy adults irrespective of *S. aureus* colonization status. Higher antistaphylococcal IgG levels were correlated with enhanced opsonophagocytosis by murine monocytic cells and human polymorphonuclear leucocytes and toxin neutralization *in vitro*. Finally, they demonstrated that infected patients were missing antibodies or had low levels of antibodies to certain antigenic proteins suggesting differential susceptibility to *S. aureus* infection in a given population (Dryla *et al.*, 2005). Similarly, Clarke and co-workers identified *in vivo*-expressed *S. aureus* antigens by probing bacteriophage expression libraries with sera from infected and uninfected individuals (Clarke *et al.*, 2006). After generating 11 recombinant antigenic proteins, they determined antigen specific antibody titres in a large collection of human serum samples and found significantly higher concentrations of reactive IgG to 4 antigens including IsdA and IsdII in healthy individuals without nasal *S. aureus* carriage compared with healthy nasal carriers. They also demonstrated that vaccination of cotton rats with IsdA and IsdH protected against nasal carriage. Weichhart and others used a similar approach to 'fingerprint the repertoire of immune reactive proteins serving as target candidates for active and passive immunization against *S. aureus*' (Weichhart *et al.*, 2003). These types of study

are likely the way forward in defining antigenic targets necessary for protection against *S. aureus* infection, but they also demonstrate that existent antibody titres to *S. aureus* virulence factors are quite common in humans. Hence, vaccines that solely stimulate humoral immunity will likely not be sufficient in preventing disease. Similar approaches aimed at defining important antigens involved in immunity to *S. aureus* mastitis in the bovine seem to be lacking.

Environmental mastitis vaccines

Environmental mastitis pathogens, as the name implies, are found in the cow's environment and can be shed in the cow's faeces. Environmental pathogens include coliforms (*Escherichia coli*, *Klebsiella* spp., and *Enterobacter* spp.), environmental streptococci (*Streptococcus dysgalactiae*, *Streptococcus uberis*, *Streptococcus equinus*, and other *Streptococcus* spp.), *Citrobacter* spp., *Proteus* spp., *Pseudomonas* spp., *Serratia* spp., and *Enterococcus* spp. Environmental pathogens are typically acquired between milkings, but can be acquired during milking. Cow-to-cow transmission is possible but does not usually play a significant role in the rate of new environmental IMIs. Environmental pathogens are ubiquitous and it is impossible to completely eliminate them from the cow's environment. Hence, control measures are aimed at preventing environmental pathogens from entering the mammary gland through the open teat sphincter and maintaining clean housing and bedding areas for cattle. Cows are most susceptible to environmental mastitis during the early dry period and the periparturient period (Bramley *et al.*, 1998). Most environmental pathogen IMIs are short-lived (<30 days) with 40–50% of streptococcal and 80% of coliform infections causing clinical mastitis (Bramley *et al.*, 1998). Coliforms can cause peracute mastitis in about 10% of cases (Bramley *et al.*, 1998). Peracute mastitis is characterized by mammary gland inflammation and systemic disease and is the result of endotoxaemia and bacteraemia (Cebra *et al.*, 1996; Cullor, 1992; Wenz *et al.*, 2001). Since most environmental pathogen IMIs are short-lived and the instigating pathogens are not harboured in the cow's mammary gland, the goal of an environmental mastitis pathogen vaccine should be to decrease the clinical severity of disease because it is clinical disease that has the greatest economic impact. Currently, there are several vaccines available that decrease the clinical severity of clinical coliform mastitis and there are some streptococcal bacterins under development.

Coliform mastitis vaccines

Coliform bacteria are highly heterogeneous with a large number of genera, species and serotypes causing disease. This heterogeneity hindered early research because of the difficulty of finding a vaccine antigen that was uniformly immunogenic against all aetiological agents. The oligosaccharide side chains, also termed somatic antigens, vary considerably among Gram-negative bacteria making them poor universal immunogens. However, Gram-negative bacteria share a conserved core antigen composed of lipid A, N-acetylglucosamine, 2-keto-3-deoxyoctonate, heptose, and glucose residues (Cullor, 1993). In the 1980s, rough ('R') mutant Gram-negative bacteria were identified that lacked enzymes needed to produce a complete lipopolysaccharide (LPS) and hence had an exposed LPS core that was conserved among many Gram-negative bacteria. Monoclonal antibodies specific for an R-mutant *E. coli* were cross-reactive with other Gram-negative bacterial species (Barclay and Scott, 1987; Mutharia *et al.*, 1984; Nelles and Niswander, 1984; Tomita *et al.*, 1995). Two R-mutant bacteria, *E. coli* 0111:B4 (strain J5) (Upjohn J5 Bacterin, Pfizer, Inc.; J-Vac, Merial, Inc.) and *Salmonella typhimurium* Re-17 (Endovac Bovi, Immvac, Inc.), have been employed in R-mutant coliform mastitis vaccines (Wilson and Gonzalez, 2003). The vaccines do not prevent intramammary infection (Hogan *et al.*, 1999a; Hogan *et al.*, 1992a,b, 1995; Tomita *et al.*, 1998). However, controlled field studies using *E. coli* core antigen vaccines (J5) have shown reductions in the rate of clinical coliform mastitis ranging from 65% to 80% (Wilson and Gonzalez, 2003). In contrast, experimental challenge trials using Gram-negative core antigen vaccines have generally shown only subtle or no differences in the rate of clinical mastitis between vaccinates and non-vaccinated control cattle suggesting that experimental challenge

models do not duplicate the conditions of a natural IMI (Hogan *et al.*, 1992b, 1995; Tomita *et al.*, 2000). Finally, Wilson and co-workers recently showed that cattle vaccinated with J5 that experienced clinical mastitis had a lower reduction in daily milk production following a clinical mastitis case than non-vaccinates experiencing a case of clinical mastitis regardless of whether the clinical mastitis was caused by a coliform (Wilson *et al.*, 2008).

The mechanism by which these vaccines confer protection from clinical disease has not been clearly understood, and it has been assumed that protection is conferred through increased opsonization of bacteria or LPS by anti-J5 antibodies. However, a recent study (Dosogne *et al.*, 2002) suggests that J5 vaccination may reduce the severity of coliform mastitis by enhancing neutrophil diapedesis into the mammary gland via a T-helper 1 response and mediated through memory cells in the mammary gland.

Lipopolysaccharide is poorly immunogenic stimulating poor IgG_1 and IgG_2 anamnestic responses (Kehrli and Harp, 2001). Hence, immunity following vaccination is generally short-lived and thus Gram-negative core antigen vaccines require multiple vaccinations immediately prior to the time when Gram-negative mastitis is most prevalent. A three-dose regimen starting at the beginning of the non-lactating (dry) period, 4 weeks later, and finally during the first week of lactation is generally recommended (Wilson and Gonzalez, 2003). This regimen should provide protection against Gram-negative pathogens during the first 90 days of lactation which is when cattle traditionally are most susceptible to clinical coliform mastitis. Recently, it has been recognized that clinical coliform mastitis is occurring later in lactation in vaccinated herds and therefore research on the use of hyperimmunization with multiple vaccinations has been investigated. In an initial study, Chaiyotwittayakun and co-workers hyperimmunized Holstein steers at time 0, 30 days later, and every 2 weeks for 10 subsequent vaccinations (Chaiyotwittayakun *et al.*, 2004). Two immunizations increased mean serum IgM and IgG_1 above pre-immunization levels, but at least five immunizations were required to detect IgG_2, the antibody thought to best facilitate opsonization of bacteria in the mammary gland, levels that exceeded pre-immunization levels. Cross-reactivity of antibody with heterologous Gram-negative bacteria was also enhanced by hyperimmunization. More recently the same group studied the effects of hyperimmunization on serum IgG$_2$, incidence of clinical mastitis, and rate of survival to the end of lactation in 1012 adult lactating cattle on two commercial dairy farms (Erskine *et al.*, 2007). Results of the study showed that cows receiving six doses of a commercial J5 bacterin (UpJohn J5 Bacterin, Pfizer, Inc.) had higher mean anti-J5 IgG$_2$ concentrations after administration of the 4th, 5th, and 6th doses of vaccine than control cows that only received 3 doses of vaccine. However, the difference in mean IgG$_2$ concentrations between groups did not persist into the following lactation. Likewise, the proportion of cattle that developed clinical coliform mastitis did not differ between groups. Control cows were more likely to develop severe clinical coliform mastitis and a lower percentage of control cows remained in the herd until the end of lactation.

Degraves and Fetrow used partial budget analysis to conclude that if >1% of cow lactations result in clinical coliform mastitis profits could be increased by $57.00/cow lactation by using a Gram-negative core antigen vaccine (DeGraves and Fetrow, 1991). However, on-farm management practices may minimize the usefulness of a J5 vaccination programme in some herds. The National Animal Health Monitoring System (NAHMS) in their Dairy 2003 survey reported that the majority of cows on 35% of dairy operations (57% of all cows) were vaccinated against coliform mastitis. In the 2007 NAHMS report, 34% of US dairy operations vaccinated cows against *E. coli* mastitis with larger herds using the vaccine more commonly than smaller herds (25% of small herds (<100 head), 50% of medium-sized herds (100–499 head), and 79% of large herds (≥500 head)).

Beyond the research on core antigen vaccines there is a relative paucity of research on vaccines for coliform mastitis with the exception of a series of studies on ferric citrate receptor (FecA) of *E. coli* (Lin *et al.*, 1999; Takemura *et al.*, 2002; Takemura *et al.*, 2004). Lin and co-workers tested

a series of *E. coli* and *Klebsiella pneumoniae* isolates from naturally occurring bovine IMI in five herds and found that anti-FecA polyclonal antisera reacted with all isolates in the presence of citrate. They concluded that ferric citrate iron transport may be induced in coliform bacteria and used to acquire iron in milk to support survival and growth (Lin *et al.*, 1999), and suggested that FecA might be a useful vaccine component for the prevention of coliform mastitis. Further work using an intramammary *E. coli* challenge model showed that cattle vaccinated with FecA had higher serum and mammary secretion anti-FecA IgG titres at calving, immediately before challenge, and 7 days post-challenge than cows immunized with *E. coli* J5 or non-vaccinates (Takemura *et al.*, 2002). Likewise, FecA vaccination conferred increased IgG titres in serum and mammary secretions against the challenge strain of *E. coli*. Bacterial counts in milk, duration of *E. coli* infection, rectal temperature, and milk somatic cell count following challenge did not differ between vaccinates and controls. Hence, while the FecA vaccine was immunogenic it did not confer a decrease in clinical severity of induced *E. coli* mastitis. *In vitro* examination of the effects of purified IgG from cows immunized with FecA, *E. coli* J5, or control showed that IgG from cows vaccinated with FecA had no greater inhibitory effect on the growth of *E. coli* isolates from bovine mastitis cases than IgG from cattle in the other groups (Takemura *et al.*, 2004). Together these data demonstrate that while FecA is immunogenic and anti-FecA antibody recognizes an array of coliform bacteria, it may not be wholly useful as a univalent antigen for vaccination to prevent coliform mastitis.

Streptococcal mastitis vaccines

Vaccines for protection against environmental streptococcal mastitis pathogens have been studied experimentally, but to date none are commercially available. Researchers in the United Kingdom have studied live and killed *S. uberis* bacterins (Finch *et al.*, 1994, 1997). Both vaccines were found to prevent clinical mastitis when vaccinates were challenged with the same strain of *S. uberis* that was used in the vaccine. However, protection against clinical mastitis with heterologous strains of *S. uberis* was not found.

More recently, research in Canada has focused on two antigenic proteins which are shared by a wide variety of streptococci, GapC and a chimeric CAMP antigen, in an attempt to overcome the lack of vaccine efficacy against heterologous strains of streptococci. Fontaine and co-workers vaccinated cattle with either the recombinant GapC of *S. uberis*, the recombinant GapC of *S. dysgalactiae*, or with a chimeric CAMP-factor antigen (CAMP-3) and subsequently challenged mammary quarters with a heterologous strain of *S. uberis* (Fontaine *et al.*, 2002). The inflammatory response following experimental intramammary challenge with *S. uberis* was significantly reduced in cattle vaccinated with the GapC of *S. uberis* or CAMP-3. The inflammatory response in cattle vaccinated with *S. dysgalactiae* GapC was not reduced following intramammary challenge with *S. uberis* suggesting that there is no cross-species protection between the GapC protein of *S. uberis* and *S. dysgalactiae*. In another study, Bolton and co-workers evaluated the ability of two purified recombinant *S. dysgalactiae* proteins (GapC and Mig) to protect against homologous bacterial challenge in non-lactating cattle (Bolton *et al.*, 2004). Intramammary infection rates were reduced in GapC vaccinated cattle, but not Mig vaccinated cattle. Mammary secretion somatic cell counts were significantly lower in both GapC and Mig vaccinated groups.

Finally, two separate research groups have evaluated PauA, a plasminogen activating protein secreted by *S. uberis*, as a candidate vaccine antigen (Leigh *et al.*, 1999; McVey *et al.*, 2005). Leigh and co-workers showed that bacterial culture filtrate containing PauA (whole antigen) mixed with adjuvant and subcutaneously administered to dairy cattle was partially protective against intramammary challenge with *S. uberis,* whereas the depleted antigen control preparation conferred no protection from IMI following challenge. Protection from infection corresponded with an anti-PauA antibody response. In a more recent study, McVey and others identified multiple linear epitopes of PauA that are recognized by mouse monoclonal antibodies to PauA. The authors suggest that characterization of these epitopes may facilitate the design of future vaccines for streptococcal mastitis (McVey *et al.*, 2005).

Commentary and outlook

While R-mutant coliform mastitis vaccines do not prevent IMI, they have proven a useful adjunct in helping prevent clinical coliform mastitis. Adoption of vaccination programmes for coliform mastitis is not universal and appears to be herd-size dependent. On some farms there is an economic benefit to using R-mutant vaccines. With the exception of some recent work on FecA, there appears to be little research and development for advancing vaccination against coliform mastitis.

Given that streptococci can be a significant cause of subclinical and clinical environmental mastitis, an efficacious vaccine would be of benefit to the dairy industry. While some vaccine formulations have been studied and shown immunogenic, there are obstacles to overcome as the studied vaccines only protect against homologous strains of streptococci.

Conclusions

In order to evaluate efficacy studies on mastitis pathogen vaccines we need to define the goals of vaccination. This chapter argues that the efficacy of a contagious pathogen vaccine should be based on prevention of IMI whereas as the efficacy of an environmental mastitis vaccine should be based on decreased severity of infection. The rationale for these efficacy goals is based on the economic impact of each type of mastitis pathogen. These goals may be an oversimplification because vaccines for contagious pathogens which decrease mastitis severity may have a modicum of economic benefit. Likewise, environmental mastitis vaccines which prevent IMI would certainly have benefit because if an IMI does not occur in the first place clinical severity of mastitis is not an issue. In addition, some environmental pathogen infections can cause chronic subclinical IMI. Currently available contagious mastitis pathogen bacterins do not completely prevent new IMIs, whereas currently available coliform mastitis bacterins have proven efficacy in decreasing the severity of mastitis. More research is needed on streptococcal vaccines. Mastitis vaccines and vaccination protocols continue to be studied and novel approaches to immunization for mastitis will continue to evolve as we understand more about the host–pathogen–environment interaction.

While this chapter has taken a relatively narrow focus by only examining vaccine technologies for bovine mastitis, the examples chosen illustrate that the vaccine technologies used in veterinary medicine are somewhat outdated, and, for at least two of the pathogen types discussed herein, little new research and development has occurred (mycoplasmas and coliforms). Novel approaches have been taken to vaccination against staphylococci and streptococci, but few, if any, of these approaches appear to be currently in product development. The same is generally true for vaccines against other important diseases of livestock. This is in part because if current technology has proven efficacious, e.g. modified live viral vaccines for use against bovine respiratory disease complex pathogens, there is little impetus for vaccine manufacturers to develop or adopt new approaches to vaccination unless they are cheaper, safer, do not require cold chain storage, and/or demonstrate enhanced efficacy over the currently marketed products. In addition the farmer will not spend money on vaccines unless they can be shown economically beneficial to livestock production.

References

Adlam, C., Ward, P.D., McCartney, A.C., Arbuthnott, J.P., and Thorley, C.M. (1977). Effect immunization with highly purified alpha- and beta-toxins on staphylococcal mastitis in rabbits. Infect. Immun. *17*, 250–256.

Balaban, N., Goldkorn, T., Nhan, R.T., Dang, L.B., Scott, S., Ridgley, R.M., Rasooly, A., Wright, S.C., Larrick, J.W., Rasooly, R., *et al.* (1998). Autoinducer of virulence as a target for vaccine and therapy against *Staphylococcus aureus*. Science *280*, 438–440.

Bannerman, D.D., Paape, M.J., Lee, J.W., Zhao, X., Hope, J.C., and Rainard, P. (2004). *Escherichia coli* and *Staphylococcus aureus* elicit differential innate immune responses following intramammary infection. Clin. Diagn. Lab. Immunol. *11*, 463–472.

Barclay, G.R., and Scott, B.B. (1987). Serological relationships between *Escherichia coli* and *Salmonella* smooth- and rough-mutant lipopolysaccharides as revealed by enzyme-linked immunosorbent assay for human immunoglobulin G antiendotoxin antibodies. Infect. Immun. *55*, 2706–2714.

Barkema, H.W., Schukken, Y.H., and Zadoks, R.N. (2006). Invited Review: The role of cow, pathogen, and treatment regimen in the therapeutic success of bovine *Staphylococcus aureus* mastitis. J. Dairy Sci. *89*, 1877–1895.

Barrio, M.B., Rainard, P., Gilbert, F.B., and Poutrel, B. (2003). Assessment of the opsonic activity of purified bovine sIgA following intramammary immunization of cows with *Staphylococcus aureus*. J. Dairy Sci. *86*, 2884–2894.

Belknap, E.B., Walters, L.M., Kelling, C., Ayers, V.K., Norris, J., McMillen, J., Hayhow, C., Cochran, M., Reddy, D.N., Wright, J., *et al.* (1999). Immunogenicity and protective efficacy of a gE, gG and US2 gene-deleted bovine herpesvirus-1 (BHV-1) vaccine. Vaccine *17*, 2297–2305.

Bennett, R.M., Christiansen, K., and Clifton-Hadley, R.S. (1999). Estimating the costs associated with endemic diseases of dairy cattle. J. Dairy Res. *66*, 455–459.

Bogni, C., Segura, M., Giraudo, J., Giraudo, A., Calzolari, A., and Nagel, R. (1998). Avirulence and immunogenicity in mice of a bovine mastitis *Staphylococcus aureus* mutant. Can. J. Vet. Res. *62*, 293–298.

Bolton, A., Song, X.M., Willson, P., Fontaine, M.C., Potter, A.A., and Perez-Casal, J. (2004). Use of the surface proteins GapC and Mig of *Streptococcus dysgalactiae* as potential protective antigens against bovine mastitis. Can. J. Microbiol. *50*, 423–432.

Boothby, J.T., Jasper, D.E., and Thomas, C.B. (1986a). Experimental intramammary inoculation with Mycoplasma bovis in vaccinated and unvaccinated cows: effect on milk production and milk quality. Can. J. Vet. Res. *50*, 200–204.

Boothby, J.T., Jasper, D.E., and Thomas, C.B. (1986b). Experimental intramammary inoculation with Mycoplasma bovis in vaccinated and unvaccinated cows: effect on the mycoplasmal infection and cellular inflammatory response. Cornell Vet. *76*, 188–197.

Boothby, J.T., Jasper, D.E., and Thomas, C.B. (1987). Experimental intramammary inoculation with Mycoplasma bovis in vaccinated and unvaccinated cows: effect on local and systemic antibody response. Can. J. Vet. Res. *51*, 121–125.

Boothby, J.T., Schore, C.E., Jasper, D.E., Osburn, B.I., and Thomas, C.B. (1988). Immune responses to Mycoplasma bovis vaccination and experimental infection in the bovine mammary gland. Can. J. Vet. Res. *52*, 355–359.

Bramley, A.J., Cullor, J.S., Erskine, R.J., Fox, L.K., Harmon, R.J., Hogan, J.S., Nickerson, S.C., Oliver, S.P., Smith, K.L., and Sordillo, L.M. (1998). Current Concepts of Bovine Mastitis, 4th edn (Verona, National Mastitis Council), pp. 1–19.

Some of the material included in the sections on *S. aureus* vaccination was previously published by the author in Expert Reviews of Vaccines (Middleton, 2008) and is included here with the permission of Expert Reviews, Ltd.

Brock, J.H., Steel, E.D., and Reiter, B. (1975). The effect of intramuscular and intramammary vaccination of cows on antibody levels and resistance to intramammary infection by *Staphylococcus aureus*. Res. Vet. Sci. *19*, 152–158.

Brouillette, E., Lacasse, P., Shkreta, L., Belanger, J., Grondin, G., Diarra, M.S., Fournier, S., and Talbot, B.G. (2002). DNA immunization against the clumping factor A (ClfA) of *Staphylococcus aureus*. Vaccine *20*, 2348–2357.

Bryson, T.D.G., Ball, H.J., Foster, F., and N., B. (2002). Enhanced severity of induced *Mycoplasma bovis* pneumonia in calves following immunization with different antigenic extracts. Res. Vet. Sci. *72*, 19.

Bubeck Wardenburg, J., Patel, R.J., and Schneewind, O. (2007). Surface proteins and exotoxins are required for the pathogenesis of *Staphylococcus aureus* pneumonia. Infect. Immun. *75*, 1040–1044.

Bubeck Wardenburg, J., and Schneewind, O. (2008). Vaccine protection against *Staphylococcus aureus* pneumonia. J. Exp. Med. *205*, 287–294.

Buzzola, F.R., Barbagelata, M.S., Caccuri, R.L., and Sordelli, D.O. (2006). Attenuation and persistence of and ability to induce protective immunity to a *Staphylococcus aureus* aroA mutant in mice. Infect. Immun. *74*, 3498–3506.

Calloway, C.D., Schultz, L.G., Chigerwe, M., Larson, R.L., Youngquist, R.S., and Steevens, B.J. (2008). Determination of serologic and colostral response in late-gestation cows vaccinated with a Mycoplasma bovis bacterin. Am. J. Vet. Res. *69*, 912–915.

Cebra, C.K., Garry, F.B., and Dinsmore, R.P. (1996). Naturally occurring acute coliform mastitis in Holstein cattle. J. Vet. Intern. Med. *10*, 252–257.

Chaiyotwittayakun, A., Burton, J.L., Weber, P.S., Kizilkaya, K., Cardoso, F.F., and Erskine, R.J. (2004). Hyperimmunization of steers with J5 *Escherichia coli* bacterin: effects on isotype-specific serum antibody responses and cross reactivity with heterogeneous gram-negative bacteria. J. Dairy Sci. *87*, 3375–3385.

Chang, B.S., Moon, J.S., Kang, H.M., Kim, Y.I., Lee, H.K., Kim, J.D., Lee, B.S., Koo, H.C., and Park, Y.H. (2008). Protective effects of recombinant staphylococcal enterotoxin type C mutant vaccine against experimental bovine infection by a strain of *Staphylococcus aureus* isolated from subclinical mastitis in dairy cattle. Vaccine *26*, 2081–2091.

Clarke, S.R., Brummell, K.J., Horsburgh, M.J., McDowell, P.W., Mohamad, S.A., Stapleton, M.R., Acevedo, J., Read, R.C., Day, N.P., Peacock, S.J., *et al.* (2006). Identification of *in vivo*-expressed antigens of *Staphylococcus aureus* and their use in vaccinations for protection against nasal carriage. J. Infect. Dis. *193*, 1098–1108.

Cocchiaro, J.L., Gomez, M.I., Risley, A., Solinga, R., Sordelli, D.O., and Lee, J.C. (2006). Molecular characterization of the capsule locus from non-typeable *Staphylococcus aureus*. Mol. Microbiol. *59*, 948–960.

Cui, J.C., Hu, D.L., Lin, Y.C., Qian, A.D., and Nakane, A. (2005). Immunization with glutathione S-transferase and mutant toxic shock syndrome toxin 1 fusion protein protects against *Staphylococcus aureus* infection. FEMS Immunol. Med. Microbiol. *45*, 45–51.

Cullor, J.S. (1992). Shock attributable to bacteremia and endotoxemia in cattle: clinical and experimental findings. J. Am. Vet. Med. Assoc. *200*, 1894–1902.

Cullor, J.S. (1993). J5 *Escherichia coli*: A core antigen vaccine for coliform mastitis. Paper presented at: Coliform Mastitis Symposium (Pullman, WA, Veterinary Learning Systems), pp. 30–35.

DeGraves, F.J., and Fetrow, J. (1991). Partial budget analysis of vaccinating dairy cattle against coliform mastitis with an *Escherichia coli* J5 vaccine. J. Am. Vet. Med. Assoc. *199*, 451–455.

DeLeo, F.R., and Otto, M. (2008). An antidote for *Staphylococcus aureus* pneumonia? J. Exp. Med. *205*, 271–274.

Dodd, F.H., Westgarth, D.R., Neave, F.K., and Kingwill, R.G. (1969). Mastitis – the strategy of control. J. Dairy Sci. *52*, 689–695.

Donofrio, G., Sartori, C., Franceschi, V., Capocefalo, A., Cavirani, S., Taddei, S., and Flammini, C.F. (2008). Double immunization strategy with a BoHV-4-vectorialized secreted chimeric peptide BVDV-E2/BoHV-1-gD. Vaccine *26*, 6031–6042.

Dosogne, H., Vangroenweghe, F., and Burvenich, C. (2002). Potential mechanism of action of J5 vaccine in protection against severe bovine coliform mastitis. Vet. Res. *33*, 1–12.

Dryla, A., Prustomersky, S., Gelbmann, D., Hanner, M., Bettinger, E., Kocsis, B., Kustos, T., Henics, T., Meinke, A., and Nagy, E. (2005). Comparison of antibody repertoires against *Staphylococcus aureus* in healthy individuals and in acutely infected patients. Clin. Diagn. Lab. Immunol. *12*, 387–398.

Erskine, R.J., VanDyk, E.J., Bartlett, P.C., Burton, J.L., and Boyle, M.C. (2007). Effect of hyperimmunization with an *Escherichia coli* J5 bacterin in adult lactating dairy cows. J. Am. Vet. Med. Assoc. *231*, 1092–1097.

Fattom, A., Fuller, S., Propst, M., Winston, S., Muenz, L., He, D., Naso, R., and Horwith, G. (2004a). Safety and immunogenicity of a booster dose of *Staphylococcus aureus* types 5 and 8 capsular polysaccharide conjugate vaccine (StaphVAX) in hemodialysis patients. Vaccine *23*, 656–663.

Fattom, A.I., Horwith, G., Fuller, S., Propst, M., and Naso, R. (2004b). Development of StaphVAX, a polysaccharide conjugate vaccine against *S. aureus* infection: from the lab bench to phase III clinical trials. Vaccine *22*, 880–887.

Finch, J.M., Hill, A.W., Field, T.R., and Leigh, J.A. (1994). Local vaccination with killed Streptococcus uberis protects the bovine mammary gland against experimental intramammary challenge with the homologous strain. Infect. Immun. *62*, 3599–3603.

Finch, J.M., Winter, A., Walton, A.W., and Leigh, J.A. (1997). Further studies on the efficacy of a live vaccine against mastitis caused by *Streptococcus uberis*. Vaccine *15*, 1138–1143.

Fontaine, M.C., Perez-Casal, J., Song, X.M., Shelford, J., Willson, P.J., and Potter, A.A. (2002). Immunisation of dairy cattle with recombinant *Streptococcus uberis*

GapC or a chimeric CAMP antigen confers protection against heterologous bacterial challenge. Vaccine 20, 2278–2286.

Fox, L.K., Chester, S.T., Hallberg, J.W., Nickerson, S.C., Pankey, J.W., and Weaver, L.D. (1995). Survey of intramammary infections in dairy heifers at breeding age and first parturition. J. Dairy Sci. 78, 1619–1628.

Fox, L.K., Kirk, J.H., and Britten, A. (2005). Mycoplasma mastitis: a review of transmission and control. J. Vet. Med. B. Infect. Dis. Vet. Public Health 52, 153–160.

Gaudreau, M.C., Lacasse, P., and Talbot, B.G. (2007). Protective immune responses to a multi-gene DNA vaccine against Staphylococcus aureus. Vaccine 25, 814–824.

Goji, N., Potter, A.A., and Perez-Casal, J. (2004). Characterization of two proteins of Staphylococcus aureus isolated from bovine clinical mastitis with homology to glyceraldehyde-3-phosphate dehydrogenase. Vet. Microbiol. 99, 269–279.

Guidry, A., Fattom, A., Patel, A., and O'Brien, C. (1997). Prevalence of capsular serotypes among Staphylococcus aureus isolates from cows with mastitis in the United States. Vet. Microbiol. 59, 53–58.

Guidry, A., Fattom, A., Patel, A., O'Brien, C., Shepherd, S., and Lohuis, J. (1998). Serotyping scheme for Staphylococcus aureus isolated from cows with mastitis. Am. J. Vet. Res. 59, 1537–1539.

Guidry, A.J., Berning, L.M., and Hambleton, C.N. (1993). Opsonization of Staphylococcus aureus by bovine immunoglobulin isotypes. J. Dairy Sci. 76, 1285–1289.

Halasa, T., Huijps, K., Osteras, O., and Hogeven, H. (2007). Economic effects of bovine mastitis and mastitis management: A review. Vet. Q. 29, 18–31.

Hema, M., Chandran, D., Nagendrakumar, S.B., Madhanmohan, M., and Srinivasan, V.A. (2009). Construction of an infectious cDNA clone of foot-and-mouth disease virus type O 1 BFS 1860 and its use in the preparation of candidate vaccine. J. Biosci. 34, 45–58.

Hermans, K., Devriese, L.A., and Haesebrouck, F. (2004). Staphylococcus. In Pathogenesis of Bacterial Infections in Animals, Gyles, C.L., Prescott, J.F., Songer, J.G., and Thoen, C.O., eds (Ames, Blackwell Publishing), pp. 43–55.

Hogan, J.S., Bogacz, V.L., Aslam, M., and Smith, K.L. (1999a). Efficacy of an Escherichia coli J5 bacterin administered to primigravid heifers. J. Dairy Sci. 82, 939–943.

Hogan, J.S., Gonzalez, R.N., Harmon, R.J., Nickerson, S.C., Oliver, S.P., Pankey, J.W., and Smith, K.L. (1999b). Laboratory Handbook on Bovine Mastitis, 3rd edn (Madison, National Mastitis Council, Inc.).

Hogan, J.S., Smith, K.L., Todhunter, D.A., and Schoenberger, P.S. (1992a). Field trial to determine efficacy of an Escherichia coli J5 mastitis vaccine. J. Dairy Sci. 75, 78–84.

Hogan, J.S., Weiss, W.P., Smith, K.L., Todhunter, D.A., Schoenberger, P.S., and Sordillo, L.M. (1995). Effects of an Escherichia coli J5 vaccine on mild clinical coliform mastitis. J. Dairy Sci. 78, 285–290.

Hogan, J.S., Weiss, W.P., Todhunter, D.A., Smith, K.L., and Schoenberger, P.S. (1992b). Efficacy of an Escherichia coli J5 mastitis vaccine in an experimental challenge trial. J. Dairy Sci. 75, 415–422.

Hu, D.L., Omoe, K., Narita, K., Cui, J.C., Shinagawa, K., and Nakane, A. (2006). Intranasal vaccination with a double mutant of staphylococcal enterotoxin C provides protection against Staphylococcus aureus infection. Microbes Infect. 8, 2841–2848.

Hu, D.L., Omoe, K., Sasaki, S., Sashinami, H., Sakuraba, H., Yokomizo, Y., Shinagawa, K., and Nakane, A. (2003). Vaccination with nontoxic mutant toxic shock syndrome toxin 1 protects against Staphylococcus aureus infection. J. Infect. Dis. 188, 743–752.

Hume, E.B., Dajcs, J.J., Moreau, J.M., and O'Callaghan, R.J. (2000). Immunization with alpha-toxin toxoid protects the cornea against tissue damage during experimental Staphylococcus aureus keratitis. Infect. Immun. 68, 6052–6055.

Jones, C. (2005). Revised structures for the capsular polysaccharides from Staphylococcus aureus Types 5 and 8, components of novel glycoconjugate vaccines. Carbohydr. Res. 340, 1097–1106.

Karakawa, W.W., and Vann, W.F. (1982). Capsular polysaccharide of Staphylococcus aureus. In Seminars in Infectious Disease, Neinstein, L., and Fields, B.N., eds (New York, Thieme-Stratton), pp. 285–293.

Kehrli, M.E., Jr., and Harp, J.A. (2001). Immunity in the mammary gland. Vet. Clin. North Am. Food Anim. Pract. 17, 495–516, vi.

Kenny, K., and Tollersrud, T. (1999). Questions uniqueness of surface polysaccharide. Am. J. Vet. Res. 60, 530.

Kent, G.M., and Newbould, F.H. (1969). Phagocytosis and related phenomena in polymorphonuclear leukocytes from cow's milk. Can. J. Comp. Med. 33, 214–219.

Kerro-Dego, O., Prysliak, T., Potter, A.A., and Perez-Casal, J. (2006). DNA-protein immunization against the GapB and GapC proteins of a mastitis isolate of Staphylococcus aureus. Vet. Immunol. Immunopathol. 113, 125–138.

Kit, S. (1990). Genetically engineered vaccines for control of Aujeszky's disease (pseudorabies). Vaccine 8, 420–424.

Kuklin, N.A., Clark, D.J., Secore, S., Cook, J., Cope, L.D., McNeely, T., Noble, L., Brown, M.J., Zorman, J.K., Wang, X.M., et al. (2006). A novel Staphylococcus aureus vaccine: iron surface determinant B induces rapid antibody responses in rhesus macaques and specific increased survival in a murine S. aureus sepsis model. Infect. Immun. 74, 2215–2223.

Lee, J.W., O'Brien, C.N., Guidry, A.J., Paape, M.J., Shafer-Weaver, K.A., and Zhao, X. (2005). Effect of a trivalent vaccine against Staphylococcus aureus mastitis lymphocyte subpopulations, antibody production, and neutrophil phagocytosis. Can. J. Vet. Res. 69, 11–18.

Leigh, J.A., Finch, J.M., Field, T.R., Real, N.C., Winter, A., Walton, A.W., and Hodgkinson, S.M. (1999). Vaccination with the plasminogen activator from Streptococcus uberis induces an inhibitory response and protects against experimental infection in the dairy cow. Vaccine 17, 851–857.

Leitner, G., Lubashevsky, E., Glickman, A., Winkler, M., Saran, A., and Trainin, Z. (2003a). Development of a *Staphylococcus aureus* vaccine against mastitis in dairy cows. I. Challenge trials. Vet. Immunol. Immunopathol. *93*, 31–38.

Leitner, G., Lubashevsky, E., and Trainin, Z. (2003b). *Staphylococcus aureus* vaccine against mastitis in dairy cows, composition and evaluation of its immunogenicity in a mouse model. Vet. Immunol. Immunopathol. *93*, 159–167.

Leitner, G., Yadlin, N., Lubashevsy, E., Ezra, E., Glickman, A., Chaffer, M., Winkler, M., Saran, A., and Trainin, Z. (2003c). Development of a *Staphylococcus aureus* vaccine against mastitis in dairy cows. II. Field trial. Vet. Immunol. Immunopathol. *93*, 153–158.

Lin, J., Hogan, J.S., and Smith, K.L. (1999). Antigenic homology of the inducible ferric citrate receptor (FecA) of coliform bacteria isolated from herds with naturally occurring bovine intramammary infections. Clin. Diagn. Lab. Immunol. *6*, 966–969.

Liu, S., Gong, Q., Wang, C., Liu, H., Wang, Y., Guo, S., Wang, W., Liu, J., Shao, M., Chi, L., *et al.* (2008). A novel DNA vaccine for protective immunity against virulent *Mycobacterium bovis* in mice. Immunol. Lett. *117*, 136–145.

Luby, C.D., Middleton, J.R., Ma, J., Rinehart, C.L., Bucklin, S., Kohler, C., and Tyler, J.W. (2007). Characterization of the antibody isotype response in serum and milk of heifers vaccinated with a *Staphylococcus aureus* bacterin (Lysigin). J. Dairy Res. *74*, 239–246.

Ma, J., Cocchiaro, J., and Lee, J.C. (2004). Evaluation of serotypes of *Staphylococcus aureus* strains used in the production of a bovine mastitis bacterin. J. Dairy Sci. *87*, 178–182.

Makovec, J.A., and Ruegg, P.L. (2003). Results of milk samples submitted for microbiological examination in Wisconsin from 1994 to 2001. J. Dairy Sci. *86*, 3466–3472.

McVey, D.S., Shi, J., Leigh, J.A., Rosey, E.L., Ward, P.N., Field, T.R., and Yancey, R.J. (2005). Identification of multiple linear epitopes of the plasminogen activator A (PauA) of Streptococcus uberis with murine monoclonal antibodies. Vet. Immunol. Immunopathol. *104*, 155–162.

Menzies, B.E., and Kernodle, D.S. (1996). Passive immunization with antiserum to a nontoxic alpha-toxin mutant from *Staphylococcus aureus* is protective in a murine model. Infect. Immun. *64*, 1839–1841.

Middleton, J.R. (2008). *Staphylococcus aureus* antigens and challenges in vaccine development. Expert Rev. Vaccines *7*, 805–815.

Middleton, J.R., Fox, L.K., Gay, J.M., Tyler, J.W., and Besser, T.E. (2002). Use of pulsed-field gel electrophoresis for detecting differences in *Staphylococcus aureus* strain populations between dairy herds with different cattle importation practices. Epidemiol. Infect. *129*, 387–395.

Middleton, J.R., Fox, L.K., and Smith, T.H. (2001). Management strategies to decrease the prevalence of mastitis caused by one strain of *Staphylococcus aureus* in a dairy herd. J. Am. Vet. Med. Assoc. *218*, 1615–1618, 1581–1612.

Middleton, J.R., Luby, C.D., and Adams, D.S. (2009). Efficacy of vaccination against staphylococcal mastitis: a review and new data. Vet. Microbiol. *134*, 192–198.

Middleton, J.R., Ma, J., Rinehart, C.L., Taylor, V.N., Luby, C.D., and Steevens, B.J. (2006). Efficacy of different Lysigin formulations in the prevention of *Staphylococcus aureus* intramammary infection in dairy heifers. J. Dairy Res. 73, 10–19.

Momtaz, H., Hemmatzadeh, F., and Keyvanfar, H. (2008). Expression of bovine leukemia virus p24 protein in bacterial cell. Pak. J. Biol. Sci. *11*, 2433–2437.

Morin, D.E., Shanks, R.D., and McCoy, G.C. (1998). Comparison of antibiotic administration in conjunction with supportive measures versus supportive measures alone for treatment of dairy cows with clinical mastitis. J. Am. Vet. Med. Assoc. *213*, 676–684.

Mutharia, L.M., Crockford, G., Bogard, W.C., Jr., and Hancock, R.E. (1984). Monoclonal antibodies specific for *Escherichia coli* J5 lipopolysaccharide: cross-reaction with other gram-negative bacterial species. Infect. Immun. *45*, 631–636.

Neave, F.K., Dodd, F.H., Kingwill, R.G., and Westgarth, D.R. (1969). Control of mastitis in the dairy herd by hygiene and management. J. Dairy Sci. *52*, 696–707.

Nelles, M.J., and Niswander, C.A. (1984). Mouse monoclonal antibodies reactive with J5 lipopolysaccharide exhibit extensive serological cross-reactivity with a variety of gram-negative bacteria. Infect. Immun. *46*, 677–681.

Nelson, L., Flock, J.I., and Hook, M. (1990). Adhesins in staphylococcal mastitis as vaccine components. Flem. Vet. J. *62*, 111–125.

Nicholas, R.A., Ayling, R.D., and McAuliffe, L. (2009). Vaccines for Mycoplasma diseases in animals and man. J. Comp. Pathol. *140*, 85–96.

Nicholas, R.A., Ayling, R.D., and Stipkovits, L.P. (2002). An experimental vaccine for calf pneumonia caused by *Mycoplasma bovis*: clinical, cultural, serological and pathological findings. Vaccine *20*, 3569–3575.

Nickerson, S.C., Owens, W.E., Tomita, G.M., and Widel, P. (1999). Vaccinating dairy heifers with a *Staphylococcus aureus* bacterin reduces mastitis at calving. Large An. Pract. *20*, 16–28.

Nour El-Din, A.N., Shkreta, L., Talbot, B.G., Diarra, M.S., and Lacasse, P. (2006). DNA immunization of dairy cows with the clumping factor A of *Staphylococcus aureus*. Vaccine *24*, 1997–2006.

O'Riordan, K., and Lee, J.C. (2004). *Staphylococcus aureus* capsular polysaccharides. Clin. Microbiol. Rev. *17*, 218–234.

Ohwada, A., Sekiya, M., Hanaki, H., Arai, K.K., Nagaoka, I., Hori, S., Tominaga, S., Hiramatsu, K., and Fukuchi, Y. (1999). DNA vaccination by mecA sequence evokes an antibacterial immune response against methicillin-resistant *Staphylococcus aureus*. J. Antimicrob. Chemother. *44*, 767–774.

Paape, M.J., Guidry, A.J., Kirk, S.T., and Bolt, D.J. (1975). Measurement of phagocytosis of ^{32}P-labeled *Staphylococcus aureus* by bovine leukocytes: lysostaphin digestion and inhibitory effect of cream. Am. J. Vet. Res. *36*, 1737–1743.

Paape, M.J., Wergin, W.P., Guidry, A.J., and Schultze, W.D. (1981). Phagocytic defense of the ruminant mammary gland. Adv. Exp. Med. Biol. *137*, 555–578.

Pankey, J.W., Boddie, N.T., Watts, J.L., and Nickerson, S.C. (1985). Evaluation of protein A and a commercial bacterin as vaccines against *Staphylococcus aureus* mastitis by experimental challenge. J. Dairy Sci. *68*, 726–731.

Parsonnet, J., Hansmann, M.A., Delaney, M.L., Modern, P.A., Dubois, A.M., Wieland-Alter, W., Wissemann, K.W., Wild, J.E., Jones, M.B., Seymour, J.L., et al. (2005). Prevalence of toxic shock syndrome toxin 1-producing *Staphylococcus aureus* and the presence of antibodies to this superantigen in menstruating women. J. Clin. Microbiol. *43*, 4628–4634.

Pellegrino, M., Giraudo, J., Raspanti, C., Nagel, R., Odierno, L., Primo, V., and Bogni, C. (2008). Experimental trial in heifers vaccinated with *Staphylococcus aureus* avirulent mutant against bovine mastitis. Vet. Microbiol. *127*, 186–190.

Perez-Casal, J., Prysliak, T., Kerro-Dego, O., and Potter, A.A. (2006). Immune responses to a *Staphylococcus aureus* GapC/B chimera and its potential use as a component of a vaccine for *S. aureus* mastitis. Vet. Immunol. Immunopathol. *109*, 85–97.

Peterson, P.K., Wilkinson, B.J., Kim, Y., Schmeling, D., Douglas, S.D., Quie, P.G., and Verhoef, J. (1978a). The key role of peptidoglycan in the opsonization of *Staphylococcus aureus*. J. Clin. Invest. *61*, 597–609.

Peterson, P.K., Wilkinson, B.J., Kim, Y., Schmeling, D., and Quie, P.G. (1978b). Influence of encapsulation on staphylococcal opsonization and phagocytosis by human polymorphonuclear leukocytes. Infect. Immun. *19*, 943–949.

Ren, X.G., Xue, F., Zhu, Y.M., Tong, G.Z., Wang, Y.H., Feng, J.K., Shi, H.F., and Gao, Y.R. (2009). Construction of a recombinant BHV-1 expressing the VP1 gene of foot and mouth disease virus and its immunogenicity in a rabbit model. Biotechnol. Lett. *31*, 1159–1165.

Risley, A.L., Loughman, A., Cywes-Bentley, C., Foster, T.J., and Lee, J.C. (2007). Capsular polysaccharide masks clumping factor A-mediated adherence of *Staphylococcus aureus* to fibrinogen and platelets. J. Infect. Dis. *196*, 919–927.

Rizzo, C., Brancaccio, G., De Vito, D., and Rizzo, G. (2007). Efficacy of autovaccination therapy on postcoronary artery bypass grafting methicillin-resistant *Staphylococcus aureus* mediastinitis. Interact. Cardiovasc. Thorac. Surg. *6*, 228–229.

Robbins, J.B., Schneerson, R., Horwith, G., Naso, R., and Fattom, A. (2004). *Staphylococcus aureus* types 5 and 8 capsular polysaccharide–protein conjugate vaccines. Am. Heart J. *147*, 593–598.

Roberson, J.R., Fox, L.K., Hancock, D.D., Gay, J.M., and Besser, T.E. (1998). Sources of intramammary infections from *Staphylococcus aureus* in dairy heifers at first parturition. J. Dairy Sci. *81*, 687–693.

Roth, D.M., Senna, J.P., and Machado, D.C. (2006). Evaluation of the humoral immune response in BALB/c mice immunized with a naked DNA vaccine

anti-methicillin-resistant *Staphylococcus aureus*. Genet. Mol. Res. *5*, 503–512.

Schaffer, A.C., Solinga, R.M., Cocchiaro, J., Portoles, M., Kiser, K.B., Risley, A., Randall, S.M., Valtulina, V., Speziale, P., Walsh, E., et al. (2006). Immunization with *Staphylococcus aureus* clumping factor B, a major determinant in nasal carriage, reduces nasal colonization in a murine model. Infect. Immun. *74*, 2145–2153.

Sears, P.M., and Belschner, A.P. (1999). Alternative management and economic consideration in *Staphylococcus aureus* eliminationprograms. Paper presented at: National Mastitis Council Annual Meeting (Arlington, VA, National Mastitis Council, Inc.).

Senna, J.P., Roth, D.M., Oliveira, J.S., Machado, D.C., and Santos, D.S. (2003). Protective immune response against methicillin resistant *Staphylococcus aureus* in a murine model using a DNA vaccine approach. Vaccine *21*, 2661–2666.

Shams, H. (2005). Recent developments in veterinary vaccinology. Vet. J. *170*, 289–299.

Shinefield, H., Black, S., Fattom, A., Horwith, G., Rasgon, S., Ordonez, J., Yeoh, H., Law, D., Robbins, J.B., Schneerson, R., et al. (2002). Use of a *Staphylococcus aureus* conjugate vaccine in patients receiving hemodialysis. N. Engl. J. Med. *346*, 491–496.

Shinefield, H.R., and Black, S. (2005). Prevention of *Staphylococcus aureus* infections: advances in vaccine development. Expert Rev. Vaccines *4*, 669–676.

Smith, K.L., Hogan, J.S., and Todhunter, D.A. (1993). Management strategies to control coliform mastitis. Paper presented at: Coliform Mastitis Symposium (Pullman, WA, Veterinary Learning Systems).

Smith, T.H., Fox, L.K., and Middleton, J.R. (1998). Outbreak of mastitis caused by one strain of *Staphylococcus aureus* in a closed dairy herd. J. Am. Vet. Med. Assoc. *212*, 553–556.

Sompolinsky, D., Samra, Z., Karakawa, W.W., Vann, W.F., Schneerson, R., and Malik, Z. (1985). Encapsulation and capsular types in isolates of *Staphylococcus aureus* from different sources and relationship to phage types. J. Clin. Microbiol. *22*, 828–834.

Stranger-Jones, Y.K., Bae, T., and Schneewind, O. (2006). Vaccine assembly from surface proteins of *Staphylococcus aureus*. Proc. Natl. Acad. Sci. U.S.A. *103*, 16942–16947.

Sutra, L., and Poutrel, B. (1994). Virulence factors involved in the pathogenesis of bovine intramammary infections due to *Staphylococcus aureus*. J. Med. Microbiol. *40*, 79–89.

Takemura, K., Hogan, J.S., Lin, J., and Smith, K.L. (2002). Efficacy of immunization with ferric citrate receptor FecA from *Escherichia coli* on induced coliform mastitis. J. Dairy Sci. *85*, 774–781.

Takemura, K., Hogan, J.S., and Smith, K.L. (2004). Growth responses of *Escherichia coli* to immunoglobulin G from cows immunized with ferric citrate receptor, FecA. J. Dairy Sci. *87*, 316–320.

Therrien, R., Lacasse, P., Grondin, G., and Talbot, B.G. (2007). Lack of protection of mice against *Staphylococcus aureus* despite a significant immune response to immunization with a DNA vaccine encoding collagen-binding protein. Vaccine 25, 5053–5061.

Tollersrud, T., Zernichow, L., Andersen, S.R., Kenny, K., and Lund, A. (2001). *Staphylococcus aureus* capsular polysaccharide type 5 conjugate and whole cell vaccines stimulate antibody responses in cattle. Vaccine 19, 3896–3903.

Tomita, G.M., Nickerson, S.C., Owens, W.E., and Wren, B. (1998). Influence of route of vaccine administration against experimental intramammary infection caused by *Escherichia coli*. J. Dairy Sci. 81, 2159–2164.

Tomita, G.M., Ray, C.H., Nickerson, S.C., Owens, W.E., and Gallo, G.F. (2000). A comparison of two commercially available *Escherichia coli* J5 vaccines against *E. coli* intramammary challenge. J. Dairy Sci. 83, 2276–2281.

Tomita, G.M., Todhunter, D.A., Hogan, J.S., and Smith, K.L. (1995). Antigenic crossreactivity and lipopolysaccharide neutralization properties of bovine immunoglobulin G. J. Dairy Sci. 78, 2745–2752.

Urbaneck, D., Leibig, F., Forbig, T.H., and Stache, B. (2000). Erfahrungsbericht zur Anwedung bestandsspezifischer impfstoffe gegen respiratorische infektionen mit beteiligung von *Mykoplasma bovis* in einem mastrindergrossbestand. Der Praktische Tierarzt 81, 13.

Verdier, I., Durand, G., Bes, M., Taylor, K.L., Lina, G., Vandenesch, F., Fattom, A.I., and Etienne, J. (2007). Identification of the capsular polysaccharides in *Staphylococcus aureus* clinical isolates by PCR and agglutination tests. J. Clin. Microbiol. 45, 725–729.

Vordermeier, H.M., Villarreal-Ramos, B., Cockle, P.J., McAulay, M., Rhodes, S.G., Thacker, T., Gilbert, S.C., McShane, H., Hill, A.V., Xing, Z., et al. (2009). Viral booster vaccines improve *Mycobacterium bovis* BCG-induced protection against bovine tuberculosis. Infect. Immun. 77, 3364–3373.

Watson, D.L. (1976). The effect of cytophilic IgG2 on phagocytosis by ovine polymorphonuclear leucocytes. Immunology 31, 159–165.

Watson, D.L. (1992). Vaccination against experimental staphylococcal mastitis in dairy heifers. Res. Vet. Sci. 53, 346–353.

Weichhart, T., Horky, M., Sollner, J., Gangl, S., Henics, T., Nagy, E., Meinke, A., von Gabain, A., Fraser, C.M., Gill, S.R., et al. (2003). Functional selection of vaccine candidate peptides from *Staphylococcus aureus* whole-genome expression libraries *in vitro*. Infect. Immun. 71, 4633–4641.

Wenz, J.R., Barrington, G.M., Garry, F.B., McSweeney, K.D., Dinsmore, R.P., Goodell, G., and Callan, R.J. (2001). Bacteremia associated with naturally occuring acute coliform mastitis in dairy cows. J. Am. Vet. Med. Assoc. 219, 976–981.

Williams, J.M., Mayerhofer, H.J., and Brown, R.W. (1966). Clinical evaluation of a *Staphylococcus aureus* bacterin (polyvalent somatic antigen). Vet. Med. Small Anim. Clin. 61, 789–793.

Williams, J.M., Shipley, G.R., Smith, G.L., and Gerber, D.L. (1975). A clinical evaluation of *Staphylococcus aureus* bacterin in the control of staphylococcal mastitis in cows. Vet. Med. Small Anim. Clin. 70, 587–594.

Wilson, D.J., and Gonzalez, R.N. (2003). Vaccination strategies for reducing clinical severity of coliform mastitis. Vet. Clin. North Am. Food Anim. Pract. 19, 187–197, vii–viii.

Wilson, D.J., Gonzalez, R.N., and Das, H.H. (1997). Bovine mastitis pathogens in New York and Pennsylvania: prevalence and effects on somatic cell count and milk production. J. Dairy Sci. 80, 2592–2598.

Wilson, D.J., Grohn, Y.T., Bennett, G.J., Gonzalez, R.N., Schukken, Y.H., and Spatz, J. (2008). Milk production change following clinical mastitis and reproductive performance compared among J5 vaccinated and control dairy cattle. J. Dairy Sci. 91, 3869–3879.

Yancey, R.J. (1999). Vaccines and diagnostic methods for bovine mastitis: fact and fiction. Adv. Vet. Med. 41, 257–273.

Yang, G., Gao, Y., Dong, J., Liu, C., Xue, Y., Fan, M., Shen, B., and Shao, N. (2005). A novel peptide screened by phage display can mimic TRAP antigen epitope against *Staphylococcus aureus* infections. J. Biol. Chem. 280, 27431–27435.

Yang, G., Gao, Y., Dong, J., Xue, Y., Fan, M., Shen, B., Liu, C., and Shao, N. (2006). A novel peptide isolated from phage library to substitute a complex system for a vaccine against staphylococci infection. Vaccine 24, 1117–1123.

Zhao, Z., Ding, J., Liu, Q., Wang, M., Yu, J., and Zhang, W. (2009). Immunogenicity of a DNA vaccine expressing the Neospora caninum surface protein NcSRS2 in mice. Acta Vet. Hung. 57, 51–62.

Zhou, H., Xiong, Z.Y., Li, H.P., Zheng, Y.L., and Jiang, Y.Q. (2006). An immunogenicity study of a newly fusion protein Cna-FnBP vaccinated against *Staphylococcus aureus* infections in a mice model. Vaccine 24, 4830–4837.

Vaccines Against Newly Emerging Viral Diseases: the Example of SARS

16

Bart L. Haagmans

Abstract

Several newly emerging viral diseases in humans have been reported recently. The ability to identify and characterize the relevant pathogen and develop safe and effective vaccines against these newly emerging pathogens in a timely manner is of utmost importance. In this respect, the global response to the SARS epidemic provided valuable experience which can be utilized to respond quickly to future emerging viral infections. In only few weeks time the nucleotide sequence of this virus was available and through computational analysis of gene sequences diagnostic tests and vaccine candidates were identified and subsequently developed. Eight years after the first SARS outbreak several candidate SARS-CoV vaccines are at various stages of pre-clinical and clinical development. The 'classical' inactivated whole virus vaccine as well as a DNA vaccine expressing the spike gene ultimately reached the phase 1 clinical trial testing. These vaccines induce neutralizing antibodies to SARS-CoV and protect against SARS-CoV challenge in diverse animal models. However, these vaccines still need to be further tested against viruses closely related to SARS-CoV that potentially may emerge and for the absence of significant side-effects. The lessons learned from this outbreak combined with more recently developed techniques may aid the development of effective vaccines against future emerging viral diseases.

Newly emerging viral diseases

In recent years, several outbreaks of infectious diseases in humans were linked to an initial zoonotic transmission. When viruses cross a species barrier, the infection can be devastating, causing pathology and mortality. Examples with dramatic implications are the introduction of HIV-1 as well as other primate viruses such as monkeypox virus and herpesvirus simiae into humans. In addition, the recently increased distribution of arthropod vectors like the mosquito *Aedes aegypti*, has led to massive outbreaks of Dengue in South America and South-East Asia. Bats represent another important reservoir for a plethora of zoonotic pathogens: two closely related paramyxoviruses – Hendra and Nipahvirus – that cause persistent infections in frugivorous bats have spread to horses and pigs, respectively. Whereas these viruses directly infect the new host species, other viruses like influenza A viruses and severe acute respiratory syndrome coronavirus (SARS-CoV) may need to adapt to acquire differential receptor usage, enhanced replication, evasion of innate and specific host immune defences, and/or efficient transmission, to successfully jump the species barrier. Factors that contribute to or predispose humans and animals for the apparent increase in interspecies transmissions over the last decades are related to our globalizing society with changes in human behaviour, increase in human mobility, demographic changes and human exploitation of the environment.

One other reason for the emergence of new viral disease may be the rapid evolution of pathogen genomes within the host. These changes may be caused by error-prone replication leading to mutations as well as recombination, reassortment and deletion of the viral genomes. At this moment only a fraction of pathogens infecting animals and insects has been identified. New molecular techniques such as MassTag PCR, microarrays and high throughput sequencing are now applied but contributed already significantly to the identification of newly emerging pathogens such as Nipah virus and SARS-CoV. Essential in the process of identification and characterization of genome sequences is the exploitation of extensive databases that allow the alignment of viral genome sequences and the linkage of data obtained by classical virus culture and serological techniques, clinical studies and pathological data.

Coronaviruses

Coronaviruses – a genus within the Coronaviridae family – are positive-stranded RNA viruses that can be found in a remarkably broad range of different mammals and bird species (Weiss and Navas-Martin, 2005). Five different CoVs infect humans and are believed to cause a significant percentage of all common colds in human adults. Other animal coronaviruses such as transmissible gastroenteritis, which causes diarrhoea in piglets, and infectious bronchitis virus, a cause of severe upper respiratory tract and kidney disease in chickens virus are known to cause life-threatening disease. Feline infectious peritonitis virus may cause an invariably fatal disease in domestic cats and other felines. More recently several CoVs have been identified in bats (Poon *et al.*, 2005; Woo *et al.*, 2006). These viruses are genetically diverse and cluster with existing known human CoV, while others formed a separate lineage of viruses exclusively found in bats.

At this moment several examples of coronavirus cross-species transmission exist (Weiss and Navas-Martin, 2005). Bovine coronavirus and human coronavirus OC43 for example are very similar and a zoonotic transmission probably occurred 100 years ago. Other bovine CoVs found in alpaca and wild ruminants share more than 99% sequence similarity. There is evidence that recombination of canine, feline and unknown CoV genomes caused the emergence of a novel canine and feline CoV that subsequently also spread to pigs. Recombination within the spike gene for example may result in viruses with altered tropism as shown for feline and ferret

CoVs. A more recent example of an emerging coronavirus is the CoV that causes SARS.

SARS-CoV

Severe acute respiratory syndrome coronavirus (SARS-CoV) first emerged in the human population in November 2002. Phylogenetic analysis of SARS-CoV isolates from animals indicated that this virus most probably originated from bats and was transmitted first to palm civets and subsequently to humans at the wet markets in southern China (Kan *et al.*, 2005). Subsequent outbreaks occurred early 2003 in Hong Kong, Hanoi, Toronto and Singapore and could be directly traced back to one index patient who acquired the infection in Guangdong and travelled to Hong Kong. A worldwide epidemic was halted through the efforts of the World Health Organization, which responded rapidly to this threat by issuing a global alert, rigorous local containment efforts, warning against unnecessary travel to affected areas and by creating a network of international experts to combat this virus. In the end only 8096 people became ill, and 774 people died in this first SARS epidemic. Because SARS-CoV could re-emerge and cause another epidemic at any time, development of effective vaccines remains of vital importance.

Although several infectious agents, including chlamydia, influenza A subtype H5N1 and human metapneumovirus, were considered as a possible cause of SARS, three groups independently reported the isolation of a not previously discovered CoV from clinical specimens of SARS patients (Ksiazek *et al.*, 2003; Marra *et al.*, 2003; Peiris *et al.*, 2003a; Rota *et al.*, 2003; Drosten *et al.*, 2003). Through electron microscopy (Fig. 16.1), serology and reverse-transcription PCR with consensus- and random-primers and subsequent sequencing of the replicase gene its identity could be revealed and consistently demonstrated in clinical specimens from patients with the disease but not in healthy controls. To conclusively establish a causal role for this CoV, cynomolgus macaques were inoculated with a SARS-CoV isolate. Because the disease in macaques caused by SARS-CoV infection was pathologically similar to that seen in human patients with SARS, and since the virus was successfully re-isolated from the

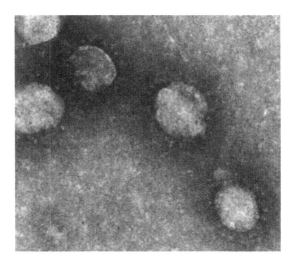

Figure 16.1 Negative-contrast electron microscopy of SARS-CoV virus particles.

nasal swabs and lung lesions of these animals and since a specific antibody response to the virus was shown in the infected animals, SARS-CoV proved to be the causative agent of this infectious disease (Fouchier *et al.*, 2003; Kuiken *et al.*, 2003).

SARS-CoV appears to have arisen from a recombination between two bat viruses and phylogenetic analysis of SARS-CoV isolates from animals indicated that the resulting bat virus was transmitted first to palm civets (*Paguma larvata*) and subsequently to humans at the wet markets in southern China (Chinese Consortium, 2004). Recent studies indicated that SARS-CoV like viruses clustered in a putative group consisting of two subgroups, one of bat CoVs and another of SARS-CoVs from humans and other mammalian hosts. Genome analyses have provided evidence that genetic variation in the spike gene of these viruses from civet cats is linked to an increased transmission capacity of the virus (Chinese Consortium, 2004). In addition, species to species variation in the sequence of the angiotensin-converting enzyme 2 (ACE2) gene, encoding the SARS-CoV receptor, also affects the efficiency by which the virus can enter cells (Li *et al.*, 2005a) Combined phylogenetic and bioinformatics analyses, chimeric gene design, and reverse genetics aided generation of viruses that encode spike proteins of diverse isolates allowed the reconstruction of events that led to the emergence of a virus able to spread efficiently in

humans. Structural modelling predicted that the SARS-CoV that caused the epidemic obtained an increased affinity for both civet and human ACE2 receptors through adaptation via repeated passages.

SARS-CoV is phylogenetically related to coronaviruses from group 2 (Fig. 16.2) despite the fact that it does not encode a haemagglutinin-esterase protein (Snijder *et al.*, 2003). The genome is packaged together with the nucleocapsid protein, at least five membrane proteins (M, E, 3a, 7a and 7b) and the spike protein (Fig. 16.3). The S1 region within the spike protein and more specifically a 193-amino acid fragment of the S protein (corresponding to residues 318–510) has been identified as the region that interacts with the viral receptor, angiotensin-converting enzyme 2 (Li *et al.*, 2005a). The majority of neutralizing antibodies are directed against this region of the spike protein. Antibodies raised against the N-terminal region of 3a protein or the M protein also inhibit SARS-CoV replication *in vitro* but their relevance in protection remains unclear. The genome also encodes two large poly-proteins with diverse enzymatic activities needed for efficient replication and several accessory proteins with unknown function (3b, 6, 8a, 8b and 9b).

Epidemiology

At the end of 2004, 30 countries reported a total of 8096 probable cases of SARS. Pathogenic

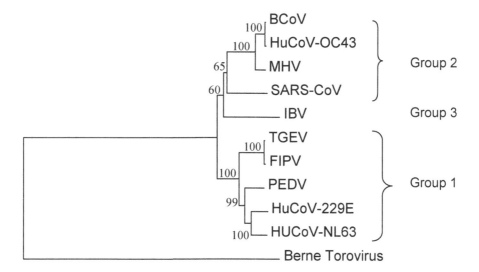

Figure 16.2 Phylogenetic tree based on deduced amino acid sequences of the coronavirus replicase ORF1b gene for bovine coronavirus (BCoV), human coronavirus 22E (HuCoV-OC43), mouse hepatitis virus (MHV), SARS-CoV, infectious bronchitis virus (IBV), transmissible gastroenteritis virus (TGEV), feline infectious peritonitis virus (FIPV), porcine epidemic diarrhoea virus (PEDV), human coronavirus 229E (HuCoV-229E), human coronavirus NL63 (HuCoV-NL63) and Berne Torovirus (used as an outgroup).

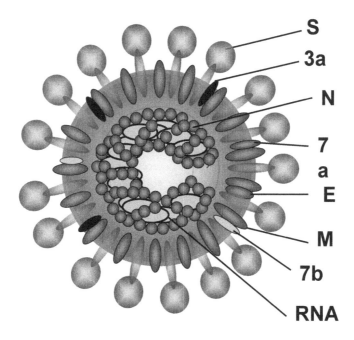

Figure 16.3 Schematic diagram of the SARS-CoV particle. S, spike protein; M, membrane protein, E, envelope protein; N, nucleocapsid protein; 3a, 7a and 7b; structural proteins of SARS-CoV.

SARS-CoVs do not circulate in the human population at the moment, but their re-emergence from animal reservoirs may likely occur in the future. Because many of the early SARS patients in Guandong had epidemiological links to the live-animal market trade, different animal species were tested for the presence of SARS like viruses. Soon after the outbreak, a SARS-like coronavirus, which had more than 99% homology with human SARS-CoV, was detected by RT-PCR in the nasal and faecal swabs of palm civets (*Paguma larvata*) and a raccoon dog (*Nyctereutes procyonoides*) (Guan *et al.*, 2003). More recent studies indicate that bats may potentially act as natural reservoirs for SARS-like CoVs (Li *et al.*, 2005b; Lau *et al.*, 2005). However, sequence comparison of the spike genes from bat SARS-like CoV and palm civet SARS-like CoV revealed only 64% genetic homology. Subsequent studies by Tang *et al.* (2006) have demonstrated that approximately 6% of bats sampled in China were positive for CoVs. Interestingly, these CoVs are genetically diverse and many bat CoVs clustered with existing group 1 viruses, while others formed a separate lineage that included only viruses from bats (putative group 5). Other SARS-CoV like viruses clustered in a putative group 4 consisting of two subgroups, one of bat CoVs and another of SARS-CoVs from humans and other mammalian hosts. However, from these studies the direct progenitor of the SARS-CoV isolated from palm civets has not been determined. Major genetic variations in the spike gene of these viruses from civet cats, seemed essential for the transition from animal-to-human transmission to human-to human transmission which eventually caused the SARS outbreak of 2002–2003.

There is at present no evidence for the virus persisting in the human population. Possible options for the re-emergence of SARS include the escape of the virus from laboratories, which already occurred on three occasions. The re-emergence of the virus from its animal reservoir remains possible, given that the virus is detectable in the faeces and respiratory secretions of some animals. Indeed, SARS-CoV re-emerged in four patients in Guangdong in December 2003, although these SARS-like CoVs caused milder clinical disease (Liang *et al.*, 2004). The National Institute of Allergy and Infectious Diseases Biodefense classified it as a category C priority pathogen pointing out that SARS-CoV can be used as a biological weapon.

Clinical disease

The clinical symptoms of SARS-CoV infection are those of lower respiratory tract disease and include fever, malaise, peripheral T-cell lymphocytopenia, decreased platelet counts, prolonged coagulation profiles and mildly elevated serum hepatic enzymes (Peiris *et al.*, 2004; Li *et al.*, 2004). Chest radiography reveals infiltrates with subpleural consolidation or 'ground glass' changes compatible with viral pneumonitis. Around 20–30% of individuals with SARS require management in intensive care units and the overall fatality rate reached approximately 10%.

Although the main clinical symptoms are those of severe respiratory illness, SARS-CoV actually also causes a gastrointestinal and urinary tract infection; SARS-CoV can be detected in the faeces and urine of patients and electron microscopic studies of biopsies of the upper and lower intestinal mucosae of patients with SARS confirmed the presence of the virus in these tissues (Peiris *et al.*, 2004). Faecal transmission proved to be important in at least one major community outbreak in Hong Kong (Amoy Gardens), in which over 300 patients were infected within a few days.

Three features of SARS may be relevant for intervention strategies. First, progressive age dependence in mortality and disease severity is observed in SARS patients. In fact, none of the SARS-CoV infected children aged below 12-year in Hong Kong required intensive care or mechanical ventilation. This is not totally explained by co-morbid factors but similar age dependence in mortality is seen in patients with other (non-viral) causes of acute respiratory stress syndrome (Rubenfeld *et al.*, 2005). Secondly, virus transmission is low in the first days of illness and peaks around day 10 after disease onset (Chu *et al.*, 2004). Finally, several studies revealed that high viral load in the nasopharyngeal aspirate was found to be an independent predictor of mortality (Hung *et al.*, 2004; Chu *et al.*, 2004). Therefore, vaccine strategies aimed at reducing the viral load may suffice to provide clinical benefit.

Pathogenesis

The major sources of transmission in humans are droplets that deposit on the respiratory epithelium. Unlike the situation in several other respiratory viral infections, viral load of SARS-CoV in the upper respiratory tract peaked around day 10 after disease onset (Peiris et al., 2004). Therefore, virus transmission may be less efficient in the first days of illness, a finding supported by epidemiological observations. Real-time PCR assays detect SARS-CoV during the first week in specimens of the lower respiratory tract (e.g. bronchoalveolar lavage, sputum, endotracheal aspirates), nasopharyngeal aspirate, throat swabs and/or serum (Chan et al., 2004), whereas faecal samples may show very high viral loads towards the end of the first week and second week of illness. In typical cases, which were largely confined to adult and elderly individuals, SARS presented with acute respiratory distress syndrome, characterized by the presence of diffuse alveolar damage and multiorgan dysfunction upon autopsy (Nicholls et al., 2003). The pathological changes in lung alveoli most likely follow a common pathway characterized by an acute phase of protein-rich alveolar fluid influx into the alveolar lumina as a consequence of the injury to the alveolar wall. Subsequently type-2 pneumocyte hyperplasia takes place to replace the loss of infected type 1 pneumocytes and to cover the denuded epithelial basement membrane, resulting in restoration of the normal alveolar architecture. Severe alveolar injury may lead to fibrosis with loss of alveolar function in more protracted cases.

Innate immune response to infection

It has been hypothesized that the pathological changes observed in the lungs are initiated by a disproportional innate immune response, illustrated by elevated levels of inflammatory cytokines and chemokines, such as CXCL10 (IP-10), CCL2 (MCP-1), IL-6, IL-8, IL-12, IL-1β and IFN-γ (Huang et al., 2005; Wong et al., 2004). These in vivo data have been confirmed in vitro, demonstrating that SARS-CoV infection induces

a range of cytokines and chemokines in diverse cell types (Cheung et al., 2005). Although in vitro studies argued that production of type I IFNs is inhibited or delayed by SARS-CoV, in SARS-CoV infected macaques but also early during SARS in humans, type I IFNs can be readily demonstrated (De Lang et al., 2007). Because prophylactic treatment of macaques with pegylated-IFN-α reduces SARS-CoV replication in the lungs, regulation of the production of IFNs may be important in controlling SARS (Haagmans et al., 2004). Subsequent functional genomics studies of these viruses in diverse species further provided insight in the role of specific host genes involved in the pathogenic response (De Lang et al., 2007; Smits et al., 2010).

Humoral immune response

Seroconversion against SARS-CoV usually occurred in weeks 2 or 3 of illness and virus neutralizing antibodies can be detected in convalescent human serum. In patients who had recovered from SARS titres peaked at month 4 but neutralizing antibodies were undetectable in 16% at month 36 (Cao et al., 2007). Most neutralizing antibodies are directed against different regions of the spike protein (S1 and S2) as monoclonal antibodies against these epitopes exert potent neutralization of SARS-CoV in vitro (Sui et al., 2004; Zhong et al., 2005; Lip et al., 2006). Conversely, peptides which are located in these regions were able to induce neutralizing antibodies (Bisht et al., 2005; Keng et al., 2005; Zhang et al., 2004). Neutralizing antibodies have also been detected against other viral proteins including 3a and M. SARS-CoV 3a protein consist of 274 amino acids, contains three putative transmembrane domains, and is expressed on the virus and cell surface (Ito et al., 2005; Tan et al., 2004b). In addition, it was recently reported that the N-terminal domain of 3a protein elicits strong and potentially protective humoral responses in infected patients (Zhong et al., 2006). Accordingly, rabbit polyclonal antibodies raised against a synthetic peptide corresponding to aa 15–28 of 3a protein inhibit SARS-CoV propagation in Vero E6 cells, in contrast to antibodies specific for the C-terminal domain of the protein (Akerstrom et al., 2006). It has been shown that SARS-CoV M protein also induced virus neutralizing antibodies in the absence of complement (Buchholz *et al.*, 2004). Nevertheless, expression of M, E, or N proteins in the absence of S protein did not confer detectable protection. These results identify S as a main SARS-CoV neutralization and protective antigen among the structural proteins, and confers a limited role to SARS-CoV M protein in protection.

In SARS patients that recover, high levels of neutralizing antibody responses are observed, suggesting that antibody responses play a role in determining the ultimate disease outcome of SARS-CoV-infected patients (Zhang *et al.*, 2004). Although attempts have been made to test the efficacy of serum preparations from seroconvalescent SARS patients in the acute phase of SARS, no conclusive evidence has been obtained regarding their efficacy. In mice, on the other hand, SARS-CoV infection is efficiently controlled upon passive transfer of convalescent immunoglobulines (Subbarao *et al.*, 2004). The concept that antibodies protect against SARS has been further explored through the generation of human monoclonal antibodies against SARS-CoV. Prophylactic administration of a human monoclonal antibody reduced replication of SARS coronavirus in the lungs of infected ferrets by 3 logs, completely prevented the development of SARS coronavirus-induced macroscopic lung pathology, and abolished shedding of virus in pharyngeal secretions (Ter Meulen *et al.*, 2004). In subsequent studies several other monoclonal antibodies were evaluated for their efficacy in mouse- and hamster- models (Sui *et al.*, 2005; Traggiai *et al.*, 2004).

Although experiments in diverse animal models have revealed that relatively low levels of neutralizing antibodies exert potent protection against lower respiratory tract infection, the neutralizing antibody titre necessary to achieve protection in humans exposed to SARS-CoV is not known. A concern in case of re-emergence of SARS is the possible absence of cross protection against these potential divergent viruses. However, studies by He *et al.* (2006) and Liu *et al.* (2007) have shown that the some neutralizing epitopes of SARS-CoV have been maintained during cross-species transmission, suggesting

that receptor binding domain-based vaccines may induce broad protection against both human and animal SARS-CoV variants.

Knowledge on the diversity of serotypes is essential information for vaccine design. Group 1 consists of viruses originating from animals isolated in 2003. A prototype of this group is the isolate SZ16 which primarily uses civet but not human angiotensin-converting enzyme 2 (ACE2) as a receptor. Group 2 are low pathogenic viruses originating from civets, raccoon dogs, or sporadic human cases, such as strain GD03 reported from a sporadic SARS on in 22 December 2003. This virus represented an independent introduction of a less pathogenic virus, having an S glycoprotein sequence that is the most divergent of all human strains (Chinese-Consortium, 2004). In general, group 2 isolates appear to have a receptor binding domain (RBD) that is capable of recognizing the human ACE2 receptor, and have been successfully cultured. GD03 S glycoprotein contains 18 amino acid substitutions relative to group 3 Urbani S protein, many of which map within neutralizing epitopes between amino acids 130–150 and 318–510, corresponding to the RBD. Recombinant viruses encoding the GD03 S glycoprotein have been isolated using reverse genetics (Baric *et al.*, 2006). Recombinant icGD03 virus replicates about 0.5–1.0 logs less efficiently in human airway epithelial cells as compared to *wt* Urbani. Group 3 are highly pathogenic viruses representing the 2002–2003 epidemic strains associated with the early, middle, or late phase. Prototypes of these viruses are the early isolate and middle isolates GZ02, and CUHK-W, respectively, and the late Urbani, FRA-1, or Toronto 2 (TOR-2) strains. Group 4, bat SARS-CoV strains have not been successfully cultured but were sequenced from samples taken from *Rhinolophus* spp. like the Chinese horseshoe bat. These viruses differ from Urbani by about 12–22% in amino acid sequence and generally have about 3–4 out of 13 contact interface residues with human ACE2 receptor. Using the S glycoprotein gene, an unrooted Bayesian analysis suggests that bat strains are most closely related to early phase human strains. Basically, all virus pseudotyped with S proteins from different strains were neutralized to the same extent, except the human GD03 and the two civet cat isolates SZ16 and SZ3, indicating that there were at least two human SARS-CoV serotypes, most likely originated from two independent transmissions of the virus from civet cat to human (Baric et al., 2006; Yang et al., 2005). Importantly, human convalescent sera had plaque reduction neutralization titre of 50% (PRNT50) values of about 1:1600 against late phase isolates like Urbani, yet were reduced about 10–15 fold against the heterologous icGD03 virus. Current evidence indicates that multiple overlapping neutralizing epitopes exist within the receptor binding domain of the SARS-CoV spike glycoprotein. Monoclonal antibodies that neutralize all human and zoonotic SARS-CoV strains have been identified and a combination of two of these antibodies that recognize different epitopes selects for escape mutants at different sites in the RBD that are severely attenuated in aged animals (Rockx et al., 2010).

Cellular immune response

The role of the cellular immune response in protection or exacerbation of disease in SARS patients is not well understood. The development of severe disease as the viraemia declines suggests that the adaptive response may contribute to progression of disease as has been observed in the pathogenesis of other coronaviruses. In most SARS autopsies, extensive necrosis of the spleen and atrophy of the white pulp with severe lymphocyte depletion have been observed. On the other hand, rapid peripheral lymphocyte recovery usually coincided with improved clinical conditions of SARS patients (Peiris et al., 2003b). However, despite potent immune responses and clinical recovery, peripheral lymphocyte counts in the recovered patients were not restored to normal levels one year after the infection (Li et al., 2006).

Long-lived memory T cell responses against SARS-CoV nucleocapsid and spike protein have been demonstrated in recovered SARS patients, although their relevance in antiviral protection is not well understood (Yang et al., 2006; Peng et al., 2006; Li et al., 2006). One in depth study explored the T cell response against peptides that cover the whole SARS-CoV genome in 128 patients one year post infection (Li et al., 2008).

Most responses were focused on the structural proteins (spike, envelope, membrane and nucleocapsid) whereas 30% were distributed in nonstructural proteins (replicase and Orfs). Overall the most frequently recognized T cell epitopes were located in Orf3 and the spike protein. It is important to note that none of the epitopes identified showed potential sequence homology with known human CoVs. Patients with more severe course of SARS had significantly stronger CD4+ T cell responses to the spike protein in terms of the number of epitopes recognized and the magnitude of the response. By contrast, the level of the memory CD8+ T cell response is independent of disease severity.

Studies in mice infected with SARS-CoV did not reveal a pivotal role for cell mediated immune responses in clearance of the virus. Interestingly, mice that lack NK-T cells, or NK cells or T and B cells all cleared the virus by day 9 after infection (Glass et al., 2004.). These data argue that cell-mediated immune responses are not essential to control virus clearance. However, it should be noted that differences in the pathogenesis of SARS-CoV in different animal species may certainly affect the role of different immune responses in resolution of the infection.

Animal models

In the preclinical development stage preferably different animal models are used to evaluate the safety and efficacy of candidate vaccines. Overall, a wide range of animal species, including rodents (mice and hamsters), carnivores (ferrets and cats) and non-human primates (cynomolgus and rhesus macaques, common marmosets and African green monkeys) can be experimentally infected with SARS-CoV (Subbarao et al., 2004; Roberts et al., 2005a; Roberts et al., 2005b; Martina et al., 2003; Kuiken et al., 2003; McAuliffe et al., 2004; Haagmans and Osterhaus 2007). Although the virus replicates efficiently in respiratory tissues, most species show no clinical signs of disease. In young mice, SARS-CoV replication has not been associated with overt signs of clinical illness or pronounced pathology. Young BALB/c mice displayed moderate interstitial pneumonitis on day 3 p.i., comparable to that seen in B6 and Rag1−/− mice (Glass et al., 2004). This pneumonitis is quickly resolved and is not observed on day 4 p.i. and the transient inflammation coincides with viral clearance since little viral antigen is observed at or after day 3 p.i., and virus titres in lungs also steadily decline from day 3 p.i. onwards (Subbarao *et al.*, 2004) Aged BALB/c mouse demonstrates clinical signs of illness including weight loss, dehydration, and ruffled fur, and several histopathological findings, i.e. diffuse alveolar damage including oedema, hyaline membrane formation, and pneumonitis, that correlate with those reported in autopsies of SARS patients (Roberts *et al.*, 2005a). The drawback to widespread utility of the aged mouse model is the difficulty in procuring large numbers of mice that are over 12 months of age. However, aged mice and ferrets show signs of clinical disease, albeit in the absence of the typical lung lesions seen in humans with SARS. SARS-CoV inoculation in the respiratory tract of cynomolgus macaques causes infection of bronchial epithelial cells and type-1 pneumocytes 1–4 days post infection followed by extensive type-2 pneumocyte hyperplasia in the lungs at 4–6 days post infection (Kuiken *et al.*, 2003; Haagmans and Osterhaus, 2007). The lesions, consisting of multiple foci of acute diffuse alveolar damage and characterized by flooding of alveoli with protein-rich oedema fluid mixed with variable numbers of neutrophils, are quite similar to those observed in humans in the acute stages of SARS. Host genomic profiling of aged macaques revealed the expression of several chemokines and cytokines including IL-1, IL-6, IL-8 and IP-10 at high levels in the lungs (De Lang *et al.*, 2007; Smits *et al.*, 2010). Comprehensive analysis of the genomic host response in different species using different SARS-CoV variants may be essential to select the appropriate animal model that ultimately reflects best the pathogenesis/host gene expression in humans (Haagmans *et al.*, 2009). Given the overlap of genes activated in cynomolgus macaques infected with SARS-CoV with host genes activated in SARS- and ARDS-patients, this model may be useful to ultimately analyse the efficacy of vaccines. Due to the lack of available immunological reagents (including microarray assays) for other NHPspecies, rhesus and cynomolgus macaques are likely to be the most informative of the non-human primate models for continued SARS research.

SARS-CoV vaccines

Based on the observation that SARS-CoV infection is efficiently controlled upon passive transfer of antibodies directed against the spike protein of SARS-CoV, a range of vaccines containing the spike protein/gene has been developed. Animals immunized with inactivated whole virus vaccines or live-recombinant vaccines expressing the SARS-CoV spike protein (e.g. using rabies virus, vesicular stomatitis virus, bovine parainfluenza virus type 3, adenovirus or attenuated vaccinia virus MVA as a vector), as well as mice immunized with DNA vaccines expressing the spike gene all developed neutralizing antibodies to SARS-CoV and were protected against SARS-CoV challenge. Table 16.1 displays an overview of vaccines that have been tested for efficacy in animal models.

Inactivated whole virus and subunit vaccines

Recombinant protein vaccines are well accepted as being among the safest vaccines although they are not always cost-effective. Recombinant protein vaccines are excellent at eliciting sterilizing immunity that can last up to several years depending upon the immunogenicity of the protein. These vaccines do not carry the risks of incomplete inactivation, genetic recombination with circulating viruses, or reversion to virulent phenotypes and are often also very stable. Furthermore, adjuvants such as aluminium salts $(Al(OH)_3$ and $Al_2(SO_4)_3$ or alum) or a squalene emulsion (MF59) (licensed for use in the US and Europe, respectively) may be added to the formulation of protein-based vaccines to increase immunogenicity.

Major disadvantages of inactivated, whole-virus vaccines would be the need to work with whole infectious virus in the preparation of the vaccine and the potential for incomplete inactivation in the processing and manufacturing of the vaccine.

Inactivated SARS vaccines have been reported to elicit high titres of spike-specific neutralizing antibodies. Few studies, however, have addressed whether inactivated whole SARS-CoV virions

Table 16.1 SARS-CoV vaccines[a]

Vaccine[b]	Animal species	Immunogenicity[c]	Protection	Author
Inactivated whole virus	Mice and macaques	Neutralizing Abs	Yes	Spruth et al. (2006)
	Mice	Neutralizing Abs	Yes	Stadler et al. (2005)
	Macaques	Neutralizing Abs	Yes	Qin et al. (2006)
	Macaques	Neutralizing Abs	Yes	Zhou et al. (2005)
Subunit	Mice	Neutralizing Abs	Yes	Bisht et al. (2005)
	Mice	Neutralizing Abs	Yes	Kam et al. (2007)
	Mice	Neutralizing Abs/lung IgA	Yes	Hu et al. (2007)
Plasmid DNA vector	Mice	Neutralizing Abs	Yes	Yang et al. (2004)
Adenovirus vector	Mice	Neutralizing Abs	Yes	See et al. (2006)
MVA vector	Mice	Neutralizing Abs	Yes	Bisht et al. (2004)
	Macaques	Neutralizing Abs	Yes	Chen et al., (2005)
	Ferrets	No neutralizing Abs	No	Weingartl et al. (2004)
Parainfluenza virus vector	Hamsters	Neutralizing Abs	Yes	Buchholz et al. (2004)
	Macaques	Neutralizing Abs	Yes	Bukreyev et al. (2004)
VSV vector	Mice	Neutralizing Abs	Yes	Kapadia et al. (2005)
	Aged mice	Neutralizing Abs	Yes	Vogel et al. (2007)
NDV vector	African green monkeys	Neutralizing Abs	Yes	DiNapoli et al. (2007)
VEEV vector	Aged mice	Neutralizing Abs	Yes	Deming et al. (2006)
	Aged mice and heterologous challenge	No neutralizing Abs	No	

[a]Only those SARS-CoV vaccines containing the spike protein/gene and tested for protection against a SARS-CoV challenge are listed.

[b]MVA, modified vaccinia virus Ankara; VSV, vesicular stomatitis virus; NDV, Newcastle disease virus; VEEV, Venezuelan equine encephalitis virus.

[c]Presence of neutralizing antibodies at time of challenge.

confer protection from virus challenge. Mice that were immunized twice with a candidate SARS-CoV vaccine, produced through a two-step inactivation procedure involving sequential formaldehyde and U.V. inactivation developed high antibody titres against the SARS-CoV spike protein and high levels of neutralizing antibodies (Spruth et al., 2006). Moreover, the vaccine conferred protective immunity as demonstrated by prevention of SARS-CoV replication in the respiratory tract of mice after intranasal challenge with SARS-CoV. Protection of mice was correlated to the antibody titre against the SARS-CoV S protein and neutralizing antibody titre. Similar results have been obtained using a beta-propiolactone inactivated SARS-CoV vaccine in mice (Stadler et al., 2005). In addition, two Chinese groups have demonstrated some protective efficacy of inactivated SARS vaccines in rhesus monkeys (Qin et al., 2006; Zhou et al., 2005). However, limited protection from SARS-CoV challenge was observed in ferrets vaccinated with inactivated whole virus (Darnell et al., 2007). Two out of four ferrets showed little or no neutralizing antibody even after the second immunization and all vaccinated ferrets were not able to reduce virus excretion at day 2 after challenge but subsequently cleared the virus more rapidly compared to control animals. Overall the efficacy and safety of these kind of vaccines needs to be evaluated further.

A soluble recombinant polypeptide containing the N-terminal segment of the spike glycoprotein have been produced in eukaryotic cells (Song et al., 2004) and plants (Pogrebnyak et al., 2005) and may suffice to induce neutralizing antibodies and protective immunity in mice (Bisht et al., 2005). In addition, a trimeric recombinant spike protein was able to elicit an efficacious protective immune response in hamsters (Kam et al., 2007).

One of the most promising vaccine candidates is based on the combination of recombinant spike protein with the Protollin adjuvant. In both young and aged mice an intranasal Protollin-formulated spike protein vaccine elicited high levels of antigen-specific IgG in serum and significant levels of antigen specific lung IgA (Hu et al., 2007). In contrast, mice immunized intramuscularly with alum absorbed spike protein did not develop detectable IgA responses. Following virus challenge of the aged mice no virus was detected in the lungs of mice vaccinated intranasally whereas intramuscularly immunized mice did not show significant control of virus replication compared to controls (Hu et al., 2007).

Although several types of vectored vaccines have been developed, several companies favoured the classical approach using inactivated whole virus to develop a vaccine to be used for preclinical testing. Methods in place for the production of available vaccines could be easily used using well-established technologies. Eight years after the first SARS outbreak a range of candidate vaccines has been developed and early 2006 some companies in China and the US initiated phase one trials and several other candidate SARS vaccines are at various stages of pre-clinical and clinical development. The first clinical trial has been initiated by a Chinese company, Sinovac Biotech of Beijing in collaboration with the Chinese academy of Medical Sciences using an inactivated whole virus vaccine. To evaluate the safety and immunogenicity of this vaccine, 36 subjects received two doses of vaccine or placebo control. On day 42, all individuals showed seroconversion and peak titres of neutralizing antibodies were reached 2 weeks after the second vaccination followed by a significant decline 4 weeks later (Lin et al., 2007).

DNA vaccines

A DNA vaccine encoding the spike glycoprotein of the SARS-CoV induces neutralizing antibody responses and protective immunity in a mouse model (Yang et al., 2004). These authors also demonstrated that antibody responses in mice vaccinated with an expression vector encoding a form of S that includes its transmembrane domain elicited neutralizing antibodies. Viral replication was reduced by more than six orders of magnitude in the lungs of mice vaccinated with these S plasmid DNA expression vectors, and protection was mediated by a humoral but not a T-cell-dependent immune mechanism. Subsequent studies using a prime–boost combination of DNA and whole killed SARS-CoV vaccines elicited higher antibody responses than DNA or whole killed virus vaccines alone (Kong et al., 2005). Apart from this study, several other groups have analysed the immunogenicity of SARS DNA vaccines but none of these challenged the vaccinated animals with SARS-CoV (Jin *et al.*, 2005; Kim *et al.*, 2004; Qu *et al.*, 2005; Wang *et al.*, 2005; Woo *et al.*, 2005; Zakhartchouk *et al.*, 2005; Zeng *et al.*, 2004; Zhao *et al.*, 2005; Zhu *et al.*, 2004). More recently, the results of an open-label Phase I clinical trial have been reported (Martin *et al.*, 2008). The vaccine was composed of a single closed circular plasmid that contained a codon-optimized deletion mutant of the SARS CoV spike protein with a cytoplasmic domain truncated. Three injections of vaccine at a dose of 4 mg each induced virus specific T cell responses and antibody responses that could be detected by ELISA and pseudovirus neutralization assay but not in the microneutralization plaque-reduction assay. Given the low levels of neutralization induced by this vaccine, prime boost protocols may be needed to achieve some level of protection in humans. Further evaluation of these kind of optimized vaccines may require use of the animal rule in models that reflect the pathogenesis of SARS-CoV in humans.

Viral expression vector vaccines

Adenovirus-vector based vaccination strategies against SARS-CoV were employed early on after the SARS outbreak to demonstrate that vaccinated rhesus macaques developed virus-neutralizing antibody responses against fragment S1 of spike and T-cell responses against the nucleocapsid (Gao *et al.*, 2003). More recently, See *et al.* (2006) demonstrated that vaccination of C57B/L6 mice with adenovirus type 5-expressing spike and nucleocapsid administered intranasally, but not intramuscularly, significantly limited SARS-CoV replication in the lungs. Adenovirus based vaccines tested in ferrets seemed more powerful as they protected the lower respiratory tract efficiently but had less effect on virus excretion in the upper respiratory tract (Kobinger *et al.*, 2007).

The highly attenuated modified vaccinia virus Ankara (MVA) has been used to express the spike glycoprotein of SARS-CoV in vaccination experiments using mouse, ferret and rhesus monkey models (Bisht *et al.*, 2004; Chen *et al.*, 2005; Weingartl *et al.*, 2004). Intranasal and intramuscular administration of MVA encoding the SARS-CoV spike protein led to the induction of a humoral immune response in BALB/c mice, as well as reduced viral titres in the respiratory tract (Bisht *et al.*, 2005).

Remarkably, vaccine candidates tested in ferrets showed reduced efficacy. Vaccination with MVA encoding the spike protein induced only moderate antibody responses and consequently did not protect against intranasal SARS-CoV infection and even resulted in an inflammatory response in the livers of the vaccinated ferrets (Czub *et al.*, 2005). Whether these aberrant responses resulted from immunopathological mechanisms, like antibody dependent enhancement of infection or represented recall responses to viral antigen in the liver is not clear at the moment but deserves further investigation.

Recombinant bovine-human parainfluenza virus type 3 vector (BHPIV3) is being developed as a live attenuated, intranasal paediatric vaccine against human parainfluenza virus type 3. Immunization of African green monkeys with a single dose of BHPIV3 expressing SARS-CoV spike protein administered via the respiratory tract induced the production of SARS-CoV neutralizing antibodies (Bukreyev *et al.*, 2004). A recombinant BHPIV3 expressing SARS-CoV structural protein (S, M, and N) individually or in combination has been evaluated for immunogenicity and protective efficacy in hamsters (Buchholz *et al.*, 2004). In the absence of spike, expression of M, N, or E did not induce a detectable serum SARS-CoV-neutralizing antibody response and no protection against SARS-CoV challenge in the respiratory tract, whereas the vectors expressing the S protein induced neutralizing antibody responses and protection.

Recombinant rabies virus expressing the S or the N protein of SARS-CoV induced a neutralizing antibody response in mice (Faber *et al.*, 2005). Similarly an attenuated vesicular stomatitis virus vector that encodes the SARS-CoV spike may be used to induce neutralizing antibody responses (Kapadia *et al.*, 2005). Mice vaccinated with vesicular stomatitis virus S developed SARS-CoV-neutralizing antibody and were able to control a challenge with SARS-CoV performed at 1 month or 4 months after a single vaccination. In addition, by passive antibody transfer experiments those authors demonstrated that the antibody response induced by the vaccine was sufficient for controlling SARS-CoV infection.

The efficacy of these vectors was further demonstrated in studies using aged mice. In aged mice vaccinated with vesicular stomatitis virus S, antibody titres induced were sufficient to protect them against subsequent challenge with SARS-CoV (Vogel *et al.*, 2007).

African green monkeys immunized through the respiratory tract with two doses of a recombinant Newcastle disease virus encoding the spike protein developed a relatively high titre of SARS-CoV neutralizing antibodies and upon challenge infection viral replication demonstrated a 1000-fold reduction in pulmonary SARS-CoV titre compared with control animals (DiNapoli *et al.*, 2007).

Finally, Venezuelan equine encephalitis virus based vaccines have been tested extensively in young and aged mice. Most importantly, different recombinant SARS-CoV bearing epidemic and zoonotic spike variants were used to challenge the vaccinated mice. Venezuelan equine encephalitis virus replicon particles expressing the 2003 epidemic Urbani SARS-CoV strain spike glycoprotein but not particles containing the nucleocapsid protein from the same strain provided complete short- and long-term protection against homologous strain challenge in young and senescent mice (Deming *et al.*, 2006). Although the spike encoding vaccine provided complete short-term protection against heterologous (strain GD03) challenge in young mice, only limited protection was seen in vaccinated senescent animals. Therefore, it is likely that declining immunity of senescent animals in combination with the reduced ability of antibody to neutralize heterologous challenge viruses resulted in vaccine failure in aged animals.

Interestingly, nucleocapsid encoding vaccines not only failed to protect from homologous or heterologous challenge, but resulted in enhanced immunopathology with eosinophilic infiltrates within the lungs of SARS-CoV–challenged mice (Deming *et al.*, 2006). In a recent study, Yasui *et al.* (2008) demonstrated that immunization with recombinant vaccinia viruses that encode the nuceocapsid protein in mice resulted in severe

pneumonia upon challenge with SARS-CoV. Furthermore, these mice exhibited up-regulation of both Th1 and Th2 cytokines and down-regulation of anti-inflammatory cytokines resulting in the infiltration of neutrophils, eosinophils and lymphocytes into the lung. Therefore skewing of the vaccine induced immune response dependent on the vaccine–adjuvant combination and/or the differences in host response in different hosts may determine the phenotype of the recall response upon challenge with the wild type virus. In the absence of effective neutralizing antibody responses these most likely ineffective T cell responses can exacerbate the infection. It is important to realize that ineffective neutralizing antibody responses may also be caused by challenge of the host with a closely related SARS-CoV that has maintained the nucleocapsid but not the spike homology to the vaccine strain. Other problems with weak antibody responses may arise when neutralizing antibody responses diminish in time more rapidly as compared to T cell responses to the nucleocapsid protein.

Live attenuated virus vaccines

Live attenuated virus vaccines face a series of potential concerns including reversion to *wt* and recombination repair with circulating heterogeneous human coronaviruses or zoonotic SARS strains. Consequently, live virus vaccine formulations should include rational approaches for minimizing the potential for reversion to *wt* phenotype and simultaneously resist recombination repair. One of them is the construction of replication-competent, propagation-defective viruses (pseudovirions) that are defective in one gene conferring an attenuated phenotype or even the ability for virus propagation (Enjuanes *et al.*, 2005). These viruses could be grown in packaging cell lines providing *in trans* the missing protein.

Given safety concerns it is often difficult to get regulatory approval of attenuated vaccines without strong proof that no other options exist to develop effective vaccines. However, because inactivated virus and DNA vaccines may induce low or transient levels of neutralizing antibodies, these kind of strategies need to be explored. New generation of vaccines may be obtained from further manipulating the full-length infectious cDNA clone of SARS-CoV. One approach would be to delete the ORFs 3a, 3b, 6, 7a, 7b, 8a, 8b or 9b similar to other coronavirus mutants generated previously; some mouse hepatitis or feline infectious peritonitis deletion viruses replicate to the same extent as wild type viruses *in vitro* but are severely attenuated *in vivo* making them interesting vaccine candidates. However, SARS-CoV deletion mutants lacking ORFs 3a, 3b, 6, 7a, or 7b, did grow similar to that of the parental wild type virus in the mouse model (Yount *et al.*, 2005). On the other hand, a recombinant SARS-CoV that lacks the E gene was attenuated *in vitro* and *in vivo* (DeDiego *et al.*, 2007). Viable recombinant virus with the E gene deleted was recovered in Vero cells with a titre around 10^6 pfu/ml but titres in the respiratory tract of hamsters were 100–1000 fold reduced compared to wildtype SARS-CoV replication, suggesting that this mutant is attenuated. Growing these viruses in packaging cell llines would provide the missing protein *in trans* and would make SARS-CoV that has the E gene deleted a promising vaccine candidate.

Live attenuated virus vaccines may revert to wildtype and recombine with other circulating human or zoonotic coronaviruses. In order to prevent this it has been proposed to delete an essential gene, located in a position distant from gene E, and the relocation of the deleted gene to the position previously occupied by gene E (Enjuanes *et al.*, 2007). A potential recombination leading to the rescue of gene E would lead to the loss of the essential gene. Alternatively, the transcriptional regulatory sequences (TRS) of a vaccine virus to a sequence incompatible with the TRS of any known circulating coronavirus could be performed as described by Yount *et al.* (2006). This virus could be further modified by building attenuating mutations on the genetic template of the recombination resistant TRS rewired virus either for use as a safe high titre seed stock for making killed vaccines or as a live virus vaccine. One such attenuating mutation could be targeted to the non-structural protein 1. Recombinant MHV encoding a deletion in the nsp1-coding sequence grew normally in tissue culture but was severely attenuated *in vivo* (Züst *et al.*, 2007). Low doses of nsp1 mutant MHV elicited potent cytotoxic T cell responses and protected mice against homologous and heterologous virus challenge. This attenuation strategy provides a new paradigm for the development of highly efficient coronavirus vaccines.

Future perspectives

In the past viruses have been grown *in vitro* and potential antigens identified using microbial, biochemical and serological methods. These approaches have provided vaccines such as inactivated viruses, life modified and subunit vaccines from purified parts of the virus. These methods require growth of the pathogen which may need serious bio-safety restrictions. Especially in the case of SARS-CoV safety issues were highly relevant and infections with SARS-CoV after the epidemic were mostly related to improper inactivation of the virus in the laboratory.

Whole genome sequencing and the prediction of open reading frames that encode proteins that theoretically are membrane proteins has provided candidate vaccine candidates. In the case of SARS-CoV several antibody accessible proteins have been identified that potentially mediate a protective response. However, most efforts have been focused on only one gene, the one that encodes the spike protein. A rigorous analysis of other potential vaccine candidates, e.g. the ORFs 3, 7a and 7b, has not been performed. More recently comparative analysis of multiple complete genome sequences combined with structural modelling of the relevant binding sites of the viral protein to antibodies and cellular receptor has been performed (Rockx *et al.*, 2010). These studies may aid in the identification of antigens able to induce cross reactive protective responses. In the case of re-emergence of SARS-CoV or a related virus, molecular epidemiological studies for the selection of representative strains will be fundamental in the design of such vaccines. Because viruses have relatively small genomes newly identified genes encoding the viral envelope proteins may be synthesized synthetically to develop new vaccine candidates in a timely way.

The importance of assessing immunogenicity of candidate SARS-CoV vaccines using virus neutralization assays is well acknowledged, but the variety of these tests in use is a significant problem since there is at this time no consensus on the most sensitive, specific, and reproducible assay system. To compare data from each of the candidate vaccines requires international standardization of the immunological assays and the availability of an antibody standard used for the evaluation of these vaccines.

Enhanced disease and mortality have been observed in kittens immunized against or infected with a type-I coronavirus, feline infectious peritonitis virus (FIPV), when subsequently exposed to FIPV infection (Weiss *et al.*, 1981; Vennema *et al.*, 1990; Olsen *et al.*, 1992). Macrophages are able to take up feline coronavirus–antibody complexes more efficiently causing the virus to replicate to higher titres. Interestingly, one study also demonstrated that antibodies against human SARS-CoV isolates enhance entry of pseudo-typed viruses expressing the civet cat SARS-like CoV-spike protein into cells but not replication (Yang *et al.*, 2005). However, so far there is no evidence for enhanced replication following SARS-CoV challenge in previously immunized animals. One other problem which may arise after vaccination with whole inactivated virus when absorbed with certain adjuvants such as alum, could relate to the induction of skewed Th2 recall responses similar to what has been observed in children vaccinated with inactivated respiratory syncytial and measles virus vaccines.

Although much effort has been focused on developing a SARS vaccine, the commercial viability of developing a vaccine for SARS-CoV will ultimately depend on whether the virus re-emerges in the near future. It is questionable whether possible future outbreaks will cause major outbreaks but vaccines, antivirals or passive immunization would be relevant in the context of protecting high-risk individuals such as laboratory and health-care workers.

References

Akerstrom, S., Tan, Y.J., and Mirazimi, A. (2006). Amino acids 15–28 in the ectodomain of SARS coronavirus 3a protein induces neutralizing antibodies. FEBS Lett. *580*, 3799–3803.

Baric, R.S., Sheahan, T., Deming, D., Donaldson, E., Yount, B., Sims, A.C., Roberts, R.S., Frieman, M., and Rockx, B. (2006). SARS coronavirus vaccine development. In The Nidovirus: Towards Control of SARS and other Nidovirus Diseases, vol. 581, Perlman, S., and Holmes, K., eds (New York, Springer), pp. 553–560.

Bisht, H., Roberts, A., Vogel, L., Bukreyev, A., Collins, P.L., Murphy, B.R., Subbarao, K., and Moss, B. (2004). Severe acute respiratory syndrome coronavirus spike protein expressed by attenuated vaccinia virus protectively immunizes mice. Proc. Natl. Acad. Sci. U.S.A. *101*, 6641–6646.

Bisht, H., Roberts, A., Vogel, L., Subbarao, K., and Moss, B. (2005). Neutralizing antibody and protective immunity to SARS coronavirus infection of mice induced by a soluble recombinant polypeptide containing an N-terminal segment of the spike glycoprotein. Virology *334*, 160–165.

Buchholz, U.J., Bukreyev, A., Yang, L., Lamirande, E.W., Murphy, B.R., Subbarao, K., and Collins, P.L. (2004). Contributions of the structural proteins of severe acute respiratory syndrome coronavirus to protective immunity. Proc. Natl. Acad. Sci. U.S.A. *101*, 9804–9809.

Bukreyev, A., Lamirande, E.W., Buchholz, U.J., Vogel, L.N., Elkins, W.R., St Claire, M., Murphy, B.R., Subbarao, K., and Collins, P.L. (2004). Mucosal immunization of African green monkeys (*Cercopithecus aethiops*) with an attenuated parainfluenza virus expressing the SARS coronavirus spike protein for the prevention of SARS. Lancet *363*, 2122–2127.

Cao, W.C., Liu, W., Zhang, P.H., Zhang, F., and Richardus, J.H. (2007). Disappearance of antibodies to SARS-associated coronavirus after recovery. N. Engl. J. Med. *357*, 1162–1163.

Chan, K.H., Poon, L.L., Cheng, V.C., Guan, Y., Hung, I.F., Kong, J., Yam, L.Y., Seto, W.H., Yuen, K.Y., and Peiris, J.S. (2004). Detection of SARS coronavirus in patients with suspected SARS. Emerg. Infect. Dis. *10*, 294–299.

Chen, Z., Zhang, L., Qin, C., Ba, L., Yi, C.E., Zhang, F., Wei, Q., He, T., Yu, W., Yu, J., *et al.* (2005). Recombinant modified vaccinia virus ankara expressing the spike glycoprotein of severe acute respiratory syndrome coronavirus induces protective neutralizing antibodies primarily targeting the receptor binding region. J. Virol. *79*, 2678–2688.

Cheung, C.Y., Poon, L.L., Ng, I.H., Luk, W., Sia, S.F., Wu, M.H., Chan, K.H., Yuen, K.Y., Gordon, S., Guan, Y., *et al.* (2005). Cytokine responses in severe acute respiratory syndrome coronavirus-infected macrophages *in vitro*: possible relevance to pathogenesis. J. Virol. *79*, 7819–7826.

Chinese-Consortium, S.M.E. (2004). Molecular evolution of the SARS coronavirus during the course of the SARS epidemic in China. Science *303*, 1666–1669.

Chu, C.M., Poon, L.L., Cheng, V.C., Chan, K.S., Hung, I.F., Wong, M.M., Chan, K.H., Leung, W.S., Tang, B.S., Chan, V.L., *et al.* (2004). Initial viral load and the outcomes of SARS. Can. Med. Assoc. J. *171*, 1349–1352.

Czub, M., Weingartl, H., Czub, S., He, R., and Cao, J. (2005). Evaluation of modified vaccinia virus Ankara based recombinant SARS vaccine in ferrets. Vaccine *23*, 2273–2279.

Darnell, M.E., Plant, E.P., Watanabe, H., Byrum, R., St Claire, M., Ward, J.M., and Taylor, D.R. (2007). Severe acute respiratory syndrome coronavirus infection in vaccinated ferrets. J. Infect. Dis. *196*, 1329–38. Epub 27 Sept 2007.

DeDiego, M.L., Alvarez, E., Almazan, F., Rejas, M.T., Lamirande, E., Roberts, A., Shieh, W.J., Saki, S., Subbarao, K., and Enjuanes, L. (2007). A SARS coronavirus that lacks the E gene is attenuated *in vitro* and *in vivo*. J. Virol. *81*, 1701–1713.

de Lang, A., Baas, T., Teal, T., Leijten, L.M., Rain, B., Osterhaus, A.D., Haagmans, B.L., and Katze, M.G. (2007). Functional genomics highlights differential induction of antiviral pathways in the lungs of SARS-CoV-infected macaques. PLoS Pathog. 3, e112.

Deming, D., Sheahan, T., Yount, B., Heise, M., Davis, N., Sims, A., Suthar, M., Pickles, R., Harkema, J., Wihitmore, A., *et al.* (2006). Vaccine efficacy in senescent mice challenged with recombinant SARS-CoV bearing epidemic and zoonotic spike variants. PLoS Med. 3, 2359–2374.

DiNapoli, J.M., Kotelkin, A., Yang, L., Elankumaran, S., Murphy, B.R., Samal, S.K., Collins, P.L., and Bukreyev, A. (2007). Newcastle disease virus, a host range-restricted virus, as a vaccine vector for intranasal immunization against emerging pathogens. Proc. Natl. Acad. Sci. U.S.A. *104*, 9788–93. Epub 29 May 2007.

Drosten, C., Gunther, S., Preiser, W., van der Werf, S., Brodt, H.R., Becker, S., Rabenau, H., Panning, M., Kolesnikova, L., Fouchier, R.A., *et al.* (2003). Identification of a novel coronavirus in patients with severe acute respiratory syndrome. N. Engl. J. Med. *348*, 67–76.

Enjuanes, L., Sola, I., Alonso, S., Escors, D., and Zúñiga, S. (2005). Coronavirus reverse genetics and development of vectors for gene expression. Curr. Top. Microbiol. Immunol. *287*, 161–197.

Enjuanes, L., Dediego, M.L., Alvarez, E., Deming, D., Sheahan, T., and Baric, R. (2007). Vaccines to prevent severe acute respiratory syndrome coronavirus-induced disease. Virus Res. *133*, 45–62. Epub 9 Apr 2007.

Faber, M., Lamirande, E.W., Roberts, A., Rice, A.B., Koprowski, H., Dietzschold, B., and Schnell, M.J. (2005). A single immunization with a rhabdovirus-based vector expressing severe acute respiratory syndrome coronavirus (SARS-CoV) S protein results in the production of high levels of SARS-CoV-neutralizing antibodies. J. Gen. Virol. *86*, 1435–1440.

Fouchier, R.A., Kuiken, T., Schutten, M., van Amerongen, G., van Doornum, G.J., van den Hoogen, B.G., Peiris, M., Lim, W., *et al.* (2003). Aetiology: Koch's postulates fulfilled for SARS virus. Nature *423*, 240.

Gao, W., Tamin, A., Soloff, A., D'Aiuto, L., Nwanegbo, E., Robbins, P.D., Bellini, W.J., Barratt-Boyes, S., and Gambotto, A. (2003). Effects of a SARSassociated coronavirus vaccine in monkeys. Lancet *362*, 1895–1896.

Glass, W.G., Subbarao, K., Murphy, B., and Murphy, P.M. (2004). Mechanisms of host defense following severe acute respiratory syndrome-coronavirus (SARSCoV) pulmonary infection of mice. J. Immunol. *173*, 4030–4039.

Guan, Y., Zheng, B.J., He, Y.Q., Liu, X.L., Zhuang, Z.X., Cheung, C.L., Luo, S.W., Li, P.H., Zhang, L.J., Guan, Y.J., *et al.* (2003). Isolation and characterization of viruses related to the SARS coronavirus from animals in southern China. Science *302*, 276–278.

Haagmans, B.L., Kuiken, T., Martina, B.E., Fouchier, R.A., Rimmelzwaan, G.F., van Amerongen, G., van Riel, D., de Jong, T., Itamura, S., Chan, K.H., *et al.* (2004). Pegylated interferon-alpha protects type 1 pneumocytes against SARS coronavirus infection in macaques. Nat. Med. *10*, 290–293. Epub 22 Feb 2004.

Haagmans, B.L., and Osterhaus, A.D. (2006). Nonhuman primate models for SARS. PLoS Med. 3, e194. Epub 18 Apr 2006.

Haagmans, B.L., Andeweg, A.C., and Osterhaus, A.D. (2009). The application of genomics to emerging zoonotic viral diseases. PLoS Pathog. 5, e1000557. Epub 26 Oct 2009.

He, Y., Li, J., Li, W., Lustigman, S., Farzan, M., and Jiang, S. (2006). Crossneutralization of human and palm civet severe acute respiratory syndrome coronaviruses by antibodies targeting the receptor-binding domain of spike protein. Immunol. J., *176*, 6085–6092.

Hu, M.C., Jones, T., Kenney, R.T., Barnard, D.L., Burt, D.S., and Lowell, G.H. (2007). Intranasal Protollin-formulated recombinant SARS S-protein elicits respiratory and serum neutralizing antibodies and protection in mice. Vaccine *25*, 6334–6340. Epub 29 Jun 2007.

Huang, K.J., Su, I.J., Theron, M., Wu, Y.C., Lai, S.K., Liu, C.C., and Lei, H.Y. (2005). An interferon gamma-related cytokine storm in SARS patients. J. Med. Virol. *75*, 185–194.

Hung, I.F., Cheng, V.C., Wu, A.K., Tang, B.S., Chan, K.H., Chu, C.M., Wong, M.M., Hui, W.T., Poon, L.L., Tse, D.M., *et al.* (2004). Viral loads in clinical specimens and SARS manifestations. Emerg. Infect. Dis. *10*, 1550–1557.

Ito, N., Mossel, E.C., Narayanan, K., Popov, V.L., Huang, C., Inoue, T., Peters, C.J., and Makino, S. (2005). Severe acute respiratory syndrome coronavirus 3a protein is a viral structural protein. J. Virol. *79*, 3182–3186.

Jin, H., Xiao, C., Chen, Z., Kang, Y., Ma, Y., Zhu, K., Xie, Q., Tu, Y., Yu, Y., and Wang, B. (2005). Induction of Th1 type response by DNA vaccinations with N, M, and E genes against SARS-CoV in mice. Biochem. Biophys. Res. Commun. *328*, 979–986.

Kam, Y.W., Kien, F., Roberts, A., Cheung, Y.C., Lamirande, E.W., Vogel, L., Chu, S.L., Tse, J., Guarner, J., Zaki, S.R. *et al.* (2007). Antibodies against trimeric S glycoprotein protect hamsters against SARS-CoV challenge despite their capacity to mediate Fcgamma-RII-dependent entry into B cells *in vitro*. Vaccine *25*, 729–740.

Kan, B., Wang, M., Jing, H., Xu, H., Jiang, X., Yan, M., Liang, W., Zheng, H., Wan, K., Liu, Q., *et al.* (2005). Molecular evolution analysis and geographic investigation of severe acute respiratory syndrome coronavirus-like virus in palm civets at an animal market and on farms. J. Virol. *79*, 11892–11900.

Kapadia, S.U., Rose, J.K., Lamirande, E., Vogel, L., Subbarao, K., and Roberts, A. (2005). Long-term protection from SARS coronavirus infection conferred by a single immunization with an attenuated VSV-based vaccine. Virology *340*, 174–182.

Keng, C.T., Zhang, A., and Shen, S. (2005). Amino acids 1055 to 1192 in the S2 region of severe acute respiratory syndrome coronavirus S protein induce neutralizing antibodies: implications for the development of vaccines and antiviral agents. J. Virol. *79*, 3289–3296.

Kim, T.W., Lee, J.H., Hung, C.F., Peng, S., Roden, R., Wang, M.C., Viscidi, R., Tsai, Y.C., He, L., Chen, P.J., *et al.* (2004). Generation and characterization of DNA vaccines targeting the nucleocapsid protein of severe acute respiratory syndrome coronavirus. J. Virol. *78*, 4638–4645.

Kobinger, G.P., Figueredo, J.M., Rowe, T., Zhi, Y., Gao, G., Sanmiguel, J.C., Bell, P., Wivel, N.A., Zitzow, L.A., Flieder, D.B., *et al.* (2007). Adenovirus-based vaccine prevents pneumonia in ferrets challenged with the SARS coronavirus and stimulates robust immune responses in macaques. Vaccine *25*, 5220–5231. Epub 7 May 2007.

Kong, W.P., Xu, L., Stadler, K., Ulmer, J.B., Abrignani, S., Rappuoli, R., and Nabel, G.J. (2005). Modulation of the immune response to the severe acute respiratory syndrome spike glycoprotein by gene-based and inactivated virus immunization. J. Virol. *79*, 13915–13923.

Ksiazek, T.G., Erdman, D., Goldsmith, C., Zaki, S., Peret, T., Emery, S., Tong, S., Urbani, C., Comer, J.A., Lim, W., *et al.* (2003). A novel coronavirus associated with severe acute respiratory syndrome. N. Engl. J. Med. *348*, 1953–1966.

Kuiken, T., Fouchier, R.A., Schutten, M., Rimmelzwaan, G.F., van Amerongen, G., van Riel, D., Laman, J.D., de Jong, T., van Doornum, G., Lim, W., *et al.* (2003). Newly discovered coronavirus as the primary cause of severe acute respiratory syndrome. Lancet *362*, 263–270.

Lau, S.K., Woo, P.C., Li, K.S., Huang, Y., Tsoi, H.W., Wong, B.H., Wong, S.S., Leung, S.Y., Chan, K.H., and Yuen, K.Y. (2005). Severe acute respiratory syndrome coronavirus-like virus in Chinese horseshoe bats. Proc. Natl. Acad. Sci. U.S.A. *102*, 14040–14045.

Li, T., Qiu, Z., Zhang, L., Han, Y., He, W., Liu, Z., Ma, X., Fan, H., Lu, W., Xie, J., *et al.* (2004). Significant changes of peripheral T lymphocyte subsets in patients with severe acute respiratory syndrome. J. Infect. Dis. *189*, 648–651.

Li, F., Li, W., Farzan, M., and Harrison, S.C. (2005a). Structure of SARS coronavirus spike receptor-binding domain complexed with receptor. Science *309*, 1864–1868.

Li, T., Xie, J., He, Y., Fan, H., Baril, L., Qiu, Z., Han, Y., Xu, W., Zhang, W., You, H., *et al.* (2006). Long-term persistence of robust antibody and cytotoxic T cell responses in recovered patients infected with SARS coronavirus. PLoS One *20*, e24.

Li, W., Shi, Z., Yu, M., Ren, W., Smith, C., Epstein, J.H., Wang, H., Crameri, G., Hu, Z., Zhang, H., *et al.* (2005b). Bats are natural reservoirs of SARS-like coronaviruses. Science *310*, 676–679.

Li, C.K., Wu, H., Yan, H., Ma, S., Wang, L., Zhang, M., Tang, X., Temperton, N.J., Weiss, R.A., Brenchley, J.M., *et al.* (2008). T cell responses to whole SARS coronavirus in humans. J. Immunol. *181*, 5490–5500.

Liang, G., Chen, Q., Xu, J., Liu, Y., Lim, W., Peiris, J.S., Anderson, L.J., Ruan, L., Li, H., Kan, B., *et al.* (2004). Laboratory diagnosis of four recent sporadic cases of community-acquired SARS, Guangdong province, China. Emerg. Infect. Dis. *10*, 1774–1781.

Lin, J.T., Zhang, J.S., Su, N., Xu, J.G., Wang, N., Chen, J.T., Chen, X., Liu, Y.X., Gao, H., Jia, Y.P., *et al.* (2007). Safety and immunogenicity from a phase I trial of inactivated severe acute respiratory syndrome coronavirus vaccine. Antivir. Ther. *12*, 1107–1113.

Lip, K.M., Shen, S., and Yang, X. (2006). Monoclonal antibodies targeting the HR2 domain and the region immediately upstream of the HR2 of the S protein neutralize *in vitro* infection of severe acute respiratory syndrome coronavirus. J. Virol. *80*, 941–50.

Liu, L., Fang, Q., Deng, F., Wang, H., Yi, C.E., Ba, L., Yu, W., Lin, R.D., Li, T., Hu, Z., *et al.* (2007). Natural mutations in the receptor binding domain of spike glycoprotein determine the reactivity of cross-neutralization between palm civet coronavirus and severe acute respiratory syndrome coronavirus. J. Virol. *81*, 4694–4700. Epub 21 Feb 2007.

Marra, M.A., Jones, S.J.M., Astell, C.R., Holt, R.A., Brooks-Wilson, A., Butterfield, Y.S.N., Khattra, J., Asano, J.K., Barber, S.A., and Chan, S.Y. (2003). The genome sequence of the SARS-associated coronavirus. Science *300*, 1399–1404.

Martin, J.E., Louder, M.K., Holman, L.A., Gordon, I.J., Enama, M.E., Larkin, B.D., Andrews, C.A., Vogel, L., Koup, R.A., Roederer, M., *et al.* (2008). A SARS DNA vaccine induces neutralizing antibody and cellular immune responses in healthy adults in a Phase I clinical trial. Vaccine *26*, 6338–6343. Epub 26 Sep 2008.

Martina, B.E., Haagmans, B.L., Kuiken, T., Fouchier, R.A., Rimmelzwaan, G.F., Van Amerongen, G., Peiris, J.S., Lim, W., and Osterhaus, A.D. (2003). SARS virus infection of cats and ferrets. Nature *425*, 915.

McAuliffe, J., Vogel, L., Roberts, A., Fahle, G., Fischer, S., Shieh, W.J., Butler, E., Zaki, S., St Claire, M., Murphy, B., *et al.* (2004). Replication of SARS coronavirus administered into the respiratory tract of African Green, rhesus and cynomolgus monkeys. Virology *330*, 8–15.

Nicholls, J.M., Poon, L.L., Lee, K.C., Ng, W.F., Lai, S.T., Leung, C.Y., Chu, C.M., Hui, P.K., Mak, K.L., Lim, W., *et al.* (2003). Lung pathology of fatal severe acute respiratory syndrome. Lancet *361*, 1773–1778.

Olsen, C.W., Corapi, W.V., Ngichabe, C.K., Baines, J.D., and Scott, F.W. (1992). Monoclonal antibodies to the spike protein of feline infectious peritonitis virus mediate antibody-dependent enhancement of infection of feline macrophages. J. Virol. *66*, 956–965.

Peiris, J.S., Lai, S.T., Poon, L.L., Guan, Y., Yam, L.Y., Lim, W., Nicholls, J., Yee, W.K., Yan, W.W., Cheung, M.T., *et al.* (2003a) Coronavirus as a possible cause of severe acute respiratory syndrome. Lancet *361*, 1319–1325.

Peiris, J.S., Yuen, K.Y., Osterhaus, A.D., and Stohr, K. (2003b). The severe acute respiratory syndrome. N. Engl. J. Med. *349*, 2431–2441.

Peiris, J.S., Guan, Y., and Yuen, K.Y. (2004). Severe acute respiratory syndrome. Nat. Med. *10*, S88–S97.

Peng, H., Yang, L.T., Wang, L.Y., Li, J., Huang, J., Lu, Z.Q., Koup, R.A., Bailer, R.T., and Wu, C.Y. (2006). Long-lived memory T lymphocyte responses against SARS coronavirus nucleocapsid protein in SARS-recovered patients. Virology. *351*, 466–475. Epub 11 May 2006.

Pogrebnyak, N., Golovkin, M., Andrianov, V., Spitsin, S., Smirnov, Y., Egolf, R., and Koprowski, H. (2005). Severe acute respiratory syndrome (SARS) S protein production in plants: development of recombinant vaccine. Proc. Natl. Acad. Sci. U.S.A. *102*, 9062–9067.

Poon, L.L., Chu, D.K., Chan, K.H., Wong, O.K., Ellis, T.M., Leung, Y.H., Lau, S.K., Woo, P.C., Suen, K.Y., Yuen, K.Y., *et al.* (2005). Identification of a novel coronavirus in bats. J. Virol. *79*, 2001–2009.

Qin, E., Shi, H., Tang, L., Wang, C., Chang, G., Ding, Z., Zhao, K., Wang, J., Chen, Z., Yu, M., *et al.* (2006). Immunogenicity and protective efficacy in monkeys of purified inactivated Vero-cell SARS vaccine. Vaccine *24*, 1028–1034.

Qu, D., Zheng, B., Yao, X., Guan, Y., Yuan, Z.H., Zhong, N.S., Lu, L.W., Xie, J.P., and Wen, Y.M. (2005). Intranasal immunization with inactivated SARS-CoV (SARS-associated coronavirus) induced local and serum antibodies in mice. Vaccine *23*, 924–931.

Roberts, A., Paddock, C., Vogel, L., Butler, E., Zaki, S., and Subbarao, K. (2005a). Aged BALB/c mice as a model for increased severity of severe acute respiratory syndrome in elderly humans. J. Virol. *79*, 5833–5838.

Roberts, A., Vogel, L., Guarner, J., Hayes, N., Murphy, B., Zaki, S., and Subbarao, K. (2005b). Severe acute respiratory syndrome coronavirus infection of golden Syrian hamsters. J. Virol. *79*, 503–511.

Rockx, B., Donaldson, E., Frieman, M., Sheahan, T., Corti, D., Lanzavecchia, A., and Baric, R.S. (2010). Escape from human monoclonal antibody neutralization affects *in vitro* and *in vivo* fitness of severe acute respiratory syndrome coronavirus. J. Infect. Dis. *201*, 946–955.

Rota, P.A., Oberste, M.S., Monroe, S.S., Nix, W.A., Campganoli, R., Icenogle, J.P., Peñaranda, S., Bankamp, B., Maher, K., Chen, M.-H., *et al.* (2003). Characterization of a novel coronavirus associated with severe acute respiratory syndrome. Science *300*, 1394–1399.

Rubenfeld, G.D., Caldwell, E., Peabody, E., Weaver, J., Martin, D.P., Neff, M., Stern, E.J., and Hudson, L.D. (2005). Incidence and outcomes of acute lung injury. N. Engl. J. Med. *353*, 1685–1693.

See, R.H., Zakhartchouk, A.N., Petric, M., Lawrence, D.J., Mok, C.P., Hogan, R.J., Rowe, T., Zitzow, L.A., Karunakaran, K.P., Hitt, M.M., *et al.* (2006). Comparative evaluation of two severe acute respiratory syndrome

(SARS) vaccine candidates in mice challenged with SARS coronavirus. J. Gen. Virol. *87*, 641–650.

Smits, S.L., de Lang, A., van den Brand, J.M., Leijten, L.M., van Ijcken, W.F., Eijkemans, M.J., van Amerongen, G., Kuiken, T., Andeweg, A.C., Osterhaus, A.D., *et al.* (2010). Exacerbated innate host response to SARS-CoV in aged non-human primates. PLoS Pathog. *6*, e1000756.

Snijder, E.J., Bredenbeek, P.J., Dobbe, J.C., Thiel, V., Ziebuhr, J., Poon, L.L., Guan, Y., Rozanov, M., Spaan, W.J., and Gorbalenya, A.E. (2003). Unique and conserved features of genome and proteome of SARS-coronavirus, an early split-off from the coronavirus group 2 lineage. J. Mol. Biol. *331*, 991–1004.

Song, H.C., Seo, M.Y., Stadler, K., Yoo, B.J., Choo, Q.L., Coates, S.R., Uematsu, Y., Harada, T., Greer, C.E., Polo, J.M., *et al.* (2004). Synthesis and characterization of a native, oligomeric form of recombinant severe acute respiratory syndrome coronavirus spike glycoprotein. J. Virol. *78*, 10328–10335.

Spruth, M., Kistner, O., Savidis-Dacho, H., Hitter, E., Crowe, B., Gerencer, M., Bruhl, P., Grillberger, L., Reiter, M., Tauer, C., *et al.* (2006). A double-inactivated whole virus candidate SARS coronavirus vaccine stimulates neutralising and protective antibody responses. Vaccine *24*, 652–661.

Stadler, K., Roberts, A., Becker, S., Vogel, L., Eickmann, M., Kolesnikova, L., Klenk, H.D., Murphy, B., Rappuoli, R., Abrignani, S., *et al.* (2005). SARS vaccine protective in mice. Emerg. Infect. Dis. *11*, 1312–1314.

Subbarao, K., McAuliffe, J., Vogel, L., Fahle, G., Fischer, S., Tatti, K., Packard, M., Shieh, W.J., Zaki, S., and Murphy, B. (2004). Prior infection and passive transfer of neutralizing antibody prevent replication of severe acute respiratory syndrome coronavirus in the respiratory tract of mice. J. Virol. *78*, 3572–3577.

Sui, J., Li, W., Murakami, A., Tamin, A., Matthews, L.J., Wong, S.K., Moore, M.J., Tallarico, A.S., Olurinde, M., Choe, H., *et al.* (2004). Potent neutralization of severe acute respiratory syndrome (SARS) coronavirus by a human mAb to S1 protein that blocks receptor association. Proc. Natl. Acad. Sci. U.S.A. *101*, 2536–2541.

Sui, J., Li, W., Roberts, A., Matthews, L.J., Murakami, A., Vogel, L., Wong, S.K., Subbarao, K., Farzan, M., and Marasco, W.A. (2005). Evaluation of human monoclonal antibody 80R for immunoprophylaxis of severe acute respiratory syndrome by an animal study, epitope mapping, and analysis of spike variants. J. Virol. *79*, 5900–5906.

Tan, Y.J., Teng, E., Shen, S., Tan, T.H.P., Goh, P.Y., Fielding, B.C., Ooi, E.E., Tan, H.C., Lim, S.G., and Hong, W. (2004). A novel severe acute respiratory syndrome coronavirus protein, U274, is transported to the cell surface and undergoes endocytosis. J. Virol. *78*, 6723–6734.

Tang, X.C., Zhang, J.X., Zhang, S.Y., Wang, P., Fan, X.H., Li, L.F., Li, G., Dong, B.Q., Liu, W., Cheung, C.L., *et al.* (2006). Prevalence and genetic diversity of coronaviruses in bats from China. J. Virol. *80*, 7481–7490.

Ter Meulen, J., Bakker, A.B., van den Brink, E.N., Weverling, G.J., Martina, B.E., Haagmans, B.L., Kuiken,

T., de Kruif, J., Preiser, W., Spaan, W., *et al.* (2004). Human monoclonal antibody as prophylaxis for SARS coronavirus infection in ferrets. Lancet *363*, 2139–2141.

Traggiai, E., Becker, S., Subbarao, K., Kolesnikova, L., Uematsu, Y., Gismondo, M.R., Murphy, B.R., Rappuoli, R., and Lanzavecchia, A. (2004). An efficient method to make human monoclonal antibodies from memory B cells: potent neutralization of SARS coronavirus. Nat. Med. *10*, 871–875.

Vennema, H., de Groot, R.J., Harbour, D.A., Dalderup, M., Gruffydd-Jones, T., Horzinek, M.C., and Spaan, W.J. (1990). Early death after feline infectious peritonitis virus challenge due to recombinant vaccinia virus immunization. J. Virol. *64*, 1407–1409.

Vogel, L.N., Roberts, A., Paddock, C.D., Genrich, G.L., Lamirande, E.W., Kapadia, S.U., Rose, J.K., Zaki, S.R., and Subbarao, K. (2007). Utility of the aged BALB/c mouse model to demonstrate prevention and control strategies for severe acute respiratory syndrome coronavirus (SARS-CoV). Vaccine *25*, 2173–2179. Epub 11 Dec 2006.

Wang, Z., Yuan, Z., Matsumoto, M., Hengge, U.R., and Chang, Y.F. (2005). Immune responses with DNA vaccines encoded different gene fragments of severe acute respiratory syndrome coronavirus inBALB/c mice. Biochem. Biophys. Res. Commun. *327*, 130–135.

Weingartl, H., Czub, M., Czub, S., Neufeld, J., Marszal, P., Gren, J., Smith, G., Jones, S., Proulx, R., Deschambault, Y., *et al.* (2004). Immunization with modified vaccinia virus Ankara-based recombinant vaccine against severe acute respiratory syndrome is associated with enhanced hepatitis in ferrets. J. Virol. *78*, 12672–12676.

Weiss, R.C., and Scott, F.W. (1981). Antibody-mediated enhancement of disease in feline infectious peritonitis: comparisons with Dengue hemorrhagic fever. Comp. Immunol. Microbiol. Infect. Dis. *4*, 175–189.

Weiss, S.R., and Navas-Martin, S. (2005). Coronavirus pathogenesis and the emerging pathogen severe acute respiratory syndrome coronavirus. Microbiol. Mol. Biol. Rev. *69*, 635–664.

Wong, C.K., Lam, C.W., Wu, A.K., Ip, W.K., Lee, N.L., Chan, I.H., Lit, L.C., Hui, D.S., Chan, M.H., Chung, S.S., *et al.* (2004). Plasma inflammatory cytokines and chemokines in severe acute respiratory syndrome. Clin. Exp. Immunol. *136*, 95–103.

Woo, P.C., Lau, S.K., Tsoi, H.W., Chen, Z.W., Wong, B.H., Zhang, L., Chan, J.K., Wong, L.P., He, W., Ma, C., *et al.* (2005). SARS coronavirus spike polypeptide DNA vaccine priming with recombinant spike polypeptide from *Escherichia coli* as booster induces high titer of neutralizing antibody against SARS coronavirus. Vaccine *23*, 4959–4968.

Woo, P.C., Lau, S.K., Li, K.S., Poon, R.W., Wong, B.H., Tsoi, H.W., Yip, B.C., Huang, Y., Chan, K.H., and Yuen, K.Y. (2006). Molecular diversity of coronaviruses in bats. Virology *351*, 180–187.

Yang, Z.Y., Kong, W.P., Huang, Y., Roberts, A., Murphy, B.R., Subbarao, K., and Nabel, G.J. (2004).A

DNAvaccine induces SARS coronavirus neutralization and protective immunity in mice. Nature *428*, 561–564.

Yang, Z.Y., Werner, H.C., Kong, W.P., Leung, K., Traggiai, E., Lanzavecchia, A., and Nabel, G.J. (2005). Evasion of antibody neutralization in emerging severe acute respiratory syndrome coronaviruses. Proc. Natl. Acad. Sci. U.S.A. *102*, 797–801.

Yang, L.T., Peng, H., Zhu, Z.L., Li, G., Huang, Z.T., Zhao, Z.X., Koup, R.A., Bailer, R.T., and Wu, C.Y. (2006). Long-lived effector/central memory T-cell responses to severe acute respiratory syndrome coronavirus (SARS-CoV) S antigen in recovered SARS patients. Clin. Immunol. *120*, 171–178.

Yasui, F., Kai, C., Kitabatake, M., Inoue, S., Yoneda, M., Yokochi, S., Kase, R., Sekiguchi, S., Morita, K., Hishima, T., *et al.* (2008). Prior immunization with severe acute respiratory syndrome (SARS)-associated coronavirus (SARS-CoV) nucleocapsid protein causes severe pneumonia in mice infected with SARS-CoV. J. Immunol. *181*, 6337–6348.

Yount, B., Roberts, R.S., Sims, A.C., Deming, D., Frieman, M.B., Sparks, J., Denison, M.R., Davis, N., and Baric, R.S. (2005). Severe acute respiratory syndrome coronavirus group-specific open reading frames encode nonessential functions for replication in cell cultures and mice. J. Virol. *79*, 14909–14922.

Yount, B., Roberts, R.S., Lindesmith, L., and Baric, R.S. (2006). Rewiring the severe acute respiratory syndrome coronavirus (SARS-CoV) transcription circuit: engineering a recombination-resistant genome. Proc. Natl. Acad. Sci. U.S.A. *103*, 12546–12551.

Zakhartchouk, A.N., Liu, Q., Petric, M., and Babiuk, L.A. (2005). Augmentation of immune responses to SARS coronavirus by a combination of DNA and whole killed virus vaccines. Vaccine *23*, 4385–4391.

Zeng, R., Chow, K.Y., Hon, C.C., Law, K.M., Yip, C.W., Chan, K.H., Peiris, J.S., and Leung, F.C. (2004). Characterization of humoral responses in mice immunized with plasmidDNAs encoding SARS-CoV spike gene fragments. Biochem. Biophys. Res. Commun. *315*, 1134–1139.

Zhang, H., Wang, G., Li, J., Nie, Y., Shi, X., Lian, G., Wang, W., Yin, X., Zhao, Y., Qu, X., *et al.* (2004). Identification of an antigenic determinant on the S2 domain of the severe acute respiratory syndrome coronavirus spike glycoprotein capable of inducing neutralizing antibodies. J. Virol. *78*, 6938–6945.

Zhao, P., Cao, J., Zhao, L.J., Qin, Z.L., Ke, J.S., Pan, W., Ren, H., Yu, J.G., and Qi, Z.T. (2005). Immune responses against SARS-coronavirus nucleocapsid protein induced by DNA vaccine. Virology *331*, 128–135.

Zhong, X., Yang, H., Guo, Z.F., Sin, W.Y., Chen, W., Xu, J., Fu, L., Wu, J., Mak, C.K., Cheng, C.S., *et al.* (2005). B-cell responses in patients who have recovered from severe acute respiratory syndrome target a dominant site in the S2 domain of the surface spike glycoprotein. J. Virol. *79*, 3401–3408.

Zhong, X., Guo, Z., Yang, H., Peng, L., Xie, Y., Wong, T.Y., and Lai, S.T. (2006). Amino terminus of the SARS coronavirus protein 3a elicits strong, potentially protective humoral responses in infected patients. J. Gen. Virol. *87*, 369–373.

Zhou, J., Wang, W., Zhong, Q., Hou, W., Yang, Z., Xiao, S.Y., Zhu, R., Tang, Z., Wang, Y., Xian, Q., *et al.* (2005). Immunogenicity, safety, and protective efficacy of an inactivated SARS-associated coronavirus vaccine in rhesus monkeys. Vaccine *23*, 3202–3209.

Zhu, M.S., Pan, Y., Chen, H.Q., Shen, Y., Wang, X.C., Sun, Y.J., and Tao, K.H. (2004). Induction of

SARS-nucleoprotein-specific immune response by use of DNA vaccine. Immunol. Lett. *92*, 237–243.

Züst, R., Cervantes-Barragán, L., Kuri, T., Blakqori, G., Weber, F., Ludewig, B., and Thiel, V. (2007). Coronavirus non-structural protein 1 is a major pathogenicity factor: implications for the rational design of coronavirus vaccines. PLoS Pathog. *3*, e109.

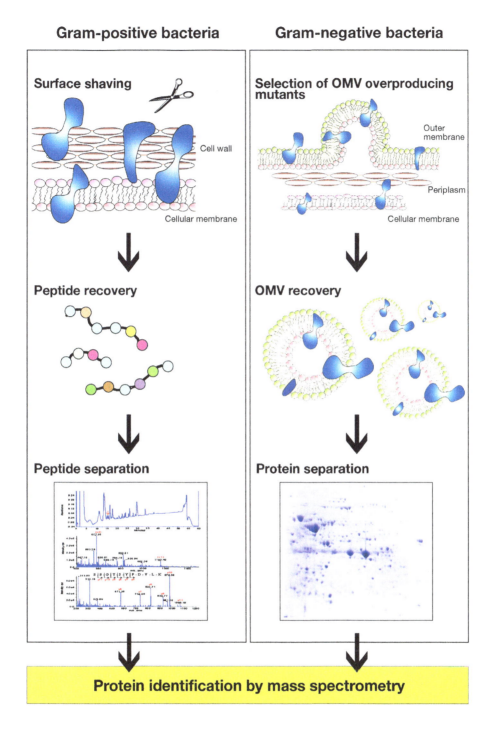

Plate 2.3 Representation of the proteomics strategy used to identify surface-exposed proteins. The methodology developed for Gram-positive bacteria consists of the use of proteases to 'shave' surface-exposed proteins (left panel). Peptides released by the digestion are recovered and then separated and analysed by LC-MS/MS. The methodology developed for Gram-negative bacteria is based on the use of OMV hyperproductive mutants (right panel). The OMVs are recovered, OMV proteins are separated by 2DE and protein spots are identified by MS. An alternative is the digestion of the OMV proteins followed by the separation and analysis of the generated peptides by LC-MS/MS.

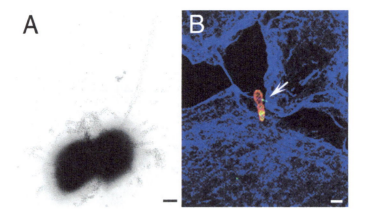

Plate 2.5 Immunoelectron-microscopy and confocal microscopy of streptococcal pili. (A) Pili expressed on the surface of *Streptococcus pneumoniae* strain TIGR4. Whole bacterial cells are incubated with polyclonal antibodies conjugated to 5-nm gold particles. The image shows the pilus backbone stained with gold-labelled antibodies raised against the main *Streptococcus pneumoniae* pilus component (RrgB). The scale bar represents 100 nm. (B) A549 cells were co-cultivated with pneumococcal strain for 1 h and imaged by confocal microscopy. Immunofluorescence shows the pilus during the epithelial cell infection experiment. 3D reconstruction of the confocal images has shown that the pilus extends from the bacterial surface contacting the host cells (arrow). Bacteria were visualized with Alexa Fluor 568 conjugated secondaries (red), A549 cells with phalloidin conjugated to Alexa Fluor 647 secondaries (blue) and the pilus with Alexa Fluor 488 conjugated secondaries (green). Scale bars, 10 µm.

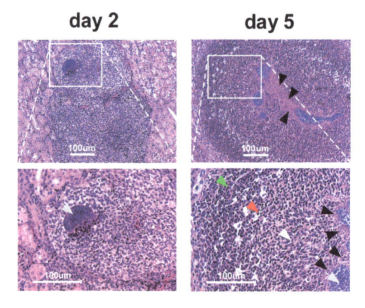

Plate 12.3 Histopathology of staphylococcal abscess communities, SACs. BALB/c mice were infected with *Staphylococcus aureus* Newman via retro-orbital injection. Thin-sectioned, H&E-stained tissues of infected kidneys on day 2 and day 5 following infection were analysed by light microscopy. On day 2, an unorganized infiltrate of PMNs are characteristic of early infectious lesions, these lesions may be accompanied by bacterial communities (grey arrow). By day 5, the abscess structure matures where the central nidus of bacteria (grey arrow) is enclosed by an amorphous, eosinophilic pseudocapsule (black arrows) and surrounded by a zone of dead PMNs (white arrow), a zone of apparently healthy PMNs (red arrow), and a rim of necrotic PMNs (green arrow), separated by a layer of eosinophilic fibrin from healthy kidney tissue.

Lightning Source UK Ltd.
Milton Keynes UK
UKOW07n0002090915

258272UK00004B/89/P

9 781904 455745